Challenges and Advai
Chemistry and Physic

Volume 17

Series Editor

Jerzy Leszczynski
Jackson State University, Jackson, Mississippi, USA

This book series provides reviews on the most recent developments in computational chemistry and physics. It covers both the method developments and their applications. Each volume consists of chapters devoted to the one research area. The series highlights the most notable advances in applications of the computational methods. The volumes include nanotechnology, material sciences, molecular biology, structures and bonding in molecular complexes, and atmospheric chemistry. The authors are recruited from among the most prominent researchers in their research areas. As computational chemistry and physics is one of the most rapidly advancing scientific areas such timely overviews are desired by chemists, physicists, molecular biologists and material scientists. The books are intended for graduate students and researchers.

More information about this series at http://www.springer.com/series/6918

Leonid Gorb • Victor Kuz'min • Eugene Muratov
Editors

Application of Computational Techniques in Pharmacy and Medicine

Editors
Leonid Gorb
Department of Molecular Biophysics
Laboratory of Computational
Structural Biology
Kiev
Ukraine

Eugene Muratov
Department of Chemical Biology
and Medicinal Chemistry
University of North Carolina, Laboratory
for Molecular Modeling
Chapel Hill
North Carolina
USA

Victor Kuz'min
Department of Molecular Structure
and Chemoinformatics
A.V. Bogatsky Physical-Chemical Institute,
National Academy of Sciences
of Ukraine
Odessa
Ukraine

ISBN 978-94-024-0696-2 ISBN 978-94-017-9257-8 (eBook)
DOI 10.1007/978-94-017-9257-8
Springer Dordrecht Heidelberg New York London

Printed on acid-free paper

Springer is part of Springer Science+Business Media (www.springer.com)

Preface

Advances in computer hardware parallel by recent enormous progress in developing computer algorithms that utilize hundreds and even thousand of computer nodes made applications of computational techniques to be indispensable in scientific research and fundamental science applications. Such research areas as computational biology, molecular pharmacy and molecular medicine are certainly among those where these computational applications are actively introduced. As the result, modern multiprocessor computers are able to treat real-life biological systems consisting of millions of atoms (ribosoms, nucleosoms, or even viruses) in a time frame of hundred nanoseconds. This tendency manifests itself even more clearly in the area of bioinformatics. Nowadays, combinatorial and high–throughput screening (HTS) technologies are widely used in both academia and industry. The pharmaceutical companies run the HTS platforms, incorporating libraries of several millions of compounds. Also, there are more and more academic centers that conduct HTS and integrate their platform with industrial drug discovery centers. Therefore, one can safely say that the computational chemist has become a respectable member of a drug design community, playing the same role as the synthetic or pharmacologists chemists. More and more often such projects have interdisciplinary character presenting an interplay between the theory and the experiment. They are intended to provide basic information as well as the data which could be used in practical pharmacological and medical applications.

The proposed volume provides basic information as well as the details of computational and computational-experimental studies improving our knowledge on functioning of alive, different properties of drugs, and predictions of new medicines. Whenever it is possible the interplay between the theory and the experiment is provided. The unique feature of the book is the fact that such different in principles computational techniques as quantum-chemical and molecular dynamic approaches on one hand and quantitative structure–activity relationships on another hand are considered inside one volume. The reviews presented in the volume cover main tendencies and priorities in application of computational methods of quantum chemistry, molecular dynamics and chemoinformatics to solve the tasks of pharmacy and medicine.

The present book is aimed at a relatively broad readership that includes advanced undergraduate and graduate students of chemistry, physics and engineering, post-doctoral associates and specialists from both academia and industry who carry out research in the fields that require molecular and QSA(P)R modeling. This book could also be useful to students in biochemistry, structural biology, bioengineering, bioinformatics, pharmaceutical chemistry, as well as other related areas, who have an interest in molecular-level computational techniques.

The book starts from the reviews that describe the studies of biological systems which are performed by the methods of quantum chemistry and classical molecular dynamics. The two initial chapters describe the theory and application of such methods as hybrid quantum-mechanical/molecular mechanical approximation, Monte-Carlo, molecular docking and molecular dynamics, in conjunction with the application of experimental techniques as Infra-Red, Raman, UV-VIS spectroscopies, and microcalorimetry. Next four chapters continue to describe the current status of the investigations in such vital area as functioning of DNA. It covers formation of DNA lesions, computational rational design of DNA polymerase inhibitors, modeling the structure of DNA quadruplexes, and the study on the structure, relative stability, and proton affinities of such building blocks as nucleotides. The seventh, eighth, ninth and tenth chapters are devoted to the application of computational and experimental techniques in such areas of medical and pharmaceutical chemistry as enzyme-inhibitor interactions, interaction of enzymes with biological membranes and a probe of polyphenol glycosides as potential remedies in kidney stones therapy and transformations of epoxided *in vivo* and *in vitro*. The following six reviews describe the advantages in the area of chemoinformatics. Most of them are devoted to developing and applications of the QSAR methodology to predict an activity and design of novel biologically active compounds. In particular, the criteria for correct QSAR models, their opprtunities and limitations are studied in the eleventh chapter. Very original QSAR methodoly named MICROCOSM is presented in chapter twelfth. The chapters 11th, 13th and 14th devoted to analysis such important properties of biologivally active compounds (possible drugs) as toxicity and farmokinetics. Very interesting methodology which combines molecular dynamics and docking approaches is described in the 15th chapter. The book is closing up by the 16th chapter which describes modern state of chemoinformatics, new problems and perspectives of treatment of avalanche-like amount of experimental chemical and biological information.

The editors of this book gratefully thank all the authors for their time and contribution. We hope that this volume may give the reader (both in academia and in an industrial pharmaceutical community) a useful overview of the computational and experimental techniques that are currently in use in the areas of computational pharmacy and medicine.

May, 2014

Leonid Gorb
Victor Kuz'min
Eugene Muratov

Contents

Contributors

Neha Agnihotri Department of Chemistry, University of Saskatchewan, Saskatoon, Canada

Vera A. Anisimova Institute of Physical and Organic Chemistry at Southern Federal University (IPOC SFU), Rostov-on-Don, Russian Federation

Xavier Assfeld Université de Lorraine, Nancy, France

Igor I. Baskin Faculty of Physics, M. V. Lomonosov Moscow State University, Moscow, Russia

I. Yu. Borisyuk A.V. Bogatsky Physico-Chemical Institute NAS of Ukraine, Odessa, Ukraine

Igor Dubey Institute of Molecular Biology and Genetics, National Academy of Sciences of Ukraine, Kyiv, Ukraine

Maxim P. Evstigneev Sevastopol National Technical University (SevNTU), Sevastopol, Ukraine

Belgorod State University (BGU), Belgorod, Russia

Denis Fourches Laboratory for Molecular Modeling, University of North Carolina, Chapel Hill, NC, USA

Julian E. Fuchs Institute of General, Inorganic and Theoretical Chemistry, University of Innsbruck, Innsbruck, Austria

R. Gancarz Organic and Pharmaceutical Technology Group, Chemistry Department, Wrocław University of Technology (WUT), Wrocław, Poland

Susanne von Grafenstein Institute of General, Inorganic and Theoretical Chemistry, University of Innsbruck, Innsbruck, Austria

N. Ya. Golovenko A.V. Bogatsky Physico-Chemical Institute NAS of Ukraine, Odessa, Ukraine

Leonid Gorb Laboratory of Computational Structural Biology, Department of Molecular Biophysics, Institute of Molecular Biology and Genetics, National Academy of Sciences of Ukraine, Kyiv, Ukraine

Nataliya A. Gurova Volgograd State Medical University (VSMU), Volgograd, Russian Federation

Tibor Hianik Department of Nuclear Physics and Biophysics, Faculty of Mathematics, Physics and Informatics, Comenius University, Bratislava, Slovak Republic, Europe

Mykola Ilchenko Institute of Molecular Biology and Genetics, National Academy of Sciences of Ukraine, Kyiv, Ukraine

Nihar R. Jena Discipline of Natural Sciences, Indian Institute of Information Technology, Design and Manufacturing, Jabalpur, India

School of Chemistry and Molecular Biosciences, University of Queensland, Brisbane, Australia

E. Klepacz Organic and Pharmaceutical Technology Group, Chemistry Department, Wrocław University of Technology (WUT), Wrocław, Poland

Vadim A. Kosolapov Volgograd State Medical University (VSMU), Volgograd, Russian Federation

Aida F. Kucheryavenko Volgograd State Medical University (VSMU), Volgograd, Russian Federation

M. A. Kulinskiy A.V. Bogatsky Physico-Chemical Institute NAS of Ukraine, Odessa, Ukraine

V. E. Kuz'min A.V. Bogatsky Physico-Chemical Institute NAS of Ukraine, Odessa, Ukraine

Alexey Lagunin Orechovich Institute of Biomedical Chemistry of Russian Academy of Medical Sciences, Laboratory of Structure-Function Based Drug Design, Moscow, Russia

Jerzy Leszczynski Interdisciplinary Center for Nanotoxicity, Department of Chemistry and Biochemistry, Jackson State University, Jackson, MS, USA

Klaus R. Liedl Institute of General, Inorganic and Theoretical Chemistry, University of Innsbruck, Innsbruck, Austria

Milan Melicherčík Department of Nuclear Physics and Biophysics, Faculty of Mathematics, Physics and Informatics, Comenius University, Bratislava, Slovak Republic, Europe

Institute of Nanobiology and Structural Biology of GCRC, Academy of Sciences of the Czech Republic, Nové Hrady, Zámek 136, Czech Republic

Phool C. Mishra Department of Physics, Banaras Hindu University, Varanasi, India

Antonio Monari Université de Lorraine, Nancy, France

E. N. Muratov A.V. Bogatsky Physico-Chemical Institute NAS of Ukraine, Odessa, Ukraine

University of North Carolina, Chapel Hill, NC, USA

Alexey Yu. Nyporko High Technology Institute, Taras Shevchenko National University of Kyiv, Kyiv, Ukraine

Sergiy Okovytyy Department of Organic Chemistry, Oles Honchar Dnipropetrovsk National University, Dnipropetrovsk, Ukrain

Gennady V. Palamarchuk SSI "Institute for Single Crystals" of National Academy of Science of Ukraine, Kharkiv, Ukraine

P. G. Polishchuk A.V. Bogatsky Physico-Chemical Institute NAS of Ukraine, Odessa, Ukraine

S. Roszak Institute of Physical and Theoretical Chemistry, Wrocław University of Technology (WUT), Wrocław, Poland

Alexander B. Rozhenko Institute of Organic Chemistry, National Academy of Sciences of Ukraine, Kyiv, Ukraine

Anna V. Shestopalova A.Usikov Institute for Radiophysics and Electronics National Academy of Sciences of Ukraine (IRE NASU), Kharkov, Ukraine

Oleg V. Shishkin SSI "Institute for Single Crystals" of National Academy of Science of Ukraine, Kharkiv, Ukraine

Alexander A. Spasov Volgograd State Medical University (VSMU), Volgograd, Russian Federation

D. Toczek Organic and Pharmaceutical Technology Group, Chemistry Department, Wrocław University of Technology (WUT), Wrocław, Poland

Ján Urban Department of Nuclear Physics and Biophysics, Faculty of Mathematics, Physics and Informatics, Comenius University, Bratislava, Slovak Republic, Europe

Pavel M. Vassiliev Volgograd State Medical University (VSMU), Volgograd, Russian Federation

Alexey Zakharov National Institutes of Health, National Cancer Institute, Chemical Biology Laboratory, Frederick, MD, USA

Nelly I. Zhokhova Faculty of Physics, M. V. Lomonosov Moscow State University, Moscow, Russia

Tetiana A. Zubatiuk SSI "Institute for Single Crystals" of National Academy of Science of Ukraine, Kharkiv, Ukraine

About the Editors

Leonid Gorb Laboratory of Computational Structural Biology, Department of Molecular Biophysics, Institute of Molecular Biology and Genetics, National Academy of Sciences of Ukraine, Zabolotnogo Str., 150, Kyiv, 03143, Ukraine

Victor Kuz'min Department of Molecular Structure and Chemoinformatics, A.V. Bogatsky Physical-Chemical Institute, National Academy of Sciences of Ukraine, 86 Lustdorfskaya Doroga, Odessa, 65080, Ukraine

Eugene Muratov Laboratory for Molecular Modeling, Department of Chemical Biology and Medicinal Chemistry, Eshelman School of Pharmacy, University of North Carolina, Chapel Hill, NC, 27599, USA

Chapter 1
Hybrid QM/MM Methods: Treating Electronic Phenomena in Very Large Molecular Systems

Antonio Monari and Xavier Assfeld

Abstract Hybrid methods, combining the accuracy of Quantum Mechanics and the potency of Molecular Mechanics, the so-called QM/MM methods, arise from the desire of theoretician chemists to study electronic phenomena in large molecular systems. In this contribution, a focus, on the Physics and Chemistry on which theses methods are based on, is given. The advantages, flaws, and limitations of each type of methods are exposed. A special emphasis is put on the Local Self-Consistent Field method, developed in our group. The latest developments are detailed and illustrated by chosen examples.

1.1 Introduction

Except some very specific experiments dealing with gas phase with very low pressure, or some particular media (interstellar space, high atmosphere, …), chemists encounter molecules in interaction with their surroundings. In fact, most of chemical or biochemical reactions take place in solution or involve macromolecules. The role of the surroundings is crucial. For example, some chemical reactions, like ethylene bromination, are quasi unfeasible in gas phase, very slow in apolar solvents, but instantaneous in water [1]. In the same vein, most biochemical reactions wouldn't be possible if not catalyzed by enzymes [2]. In addition to the role played in enzymatic catalysis, the environment is also crucial to modify, to precisely tune or to induce the response to light in photo-active systems. A paradigmatic example being for instance the role played by opsin protein in assuring an ultra-fast highly efficient photo isomerization of retinal chromophore in vision process [3, 4]. The precise understanding and tuning of light-induced responses in complex biosystems

X. Assfeld (✉) · A. Monari
Université de Lorraine, Théorie-Modélisation-Simulation, SRSMC UMR 7565,
Vandœuvre-lès-Nancy, 54506 France
e-mail: xavier.assfeld@univ-lorraine.fr

A. Monari
e-mail: antonio.monari@univ-lorraine.fr

L. Gorb et al. (eds.), *Application of Computational Techniques in Pharmacy and Medicine,*
Challenges and Advances in Computational Chemistry and Physics 17,
DOI 10.1007/978-94-017-9257-8_1

is also of seminal importance in the growing field of phototherapy [5]. For instance small organic or organometallic drugs can interact with DNA and induce permanent lesions of the nucleic acid once activated by exposition to light [6–8], so as to constitute very efficient photo-chemo-therapeutic agents. Taking care of the environment in theoretical calculations, both for ground and excited states is thus mandatory if one wants to obtain realistic results, i.e. directly comparable to experimental data, or to make reliable predictions [9–13].

The surrounding, i.e. the solvent or the bio-macromolecule, is a very large system, containing at least thousands of atoms to obtain an acceptable representation. Even if the progresses of quantum chemistry over the past decades are tremendous (algorithms, computer, new method, …), describing such large systems with quantum methods is still far beyond our computational capacities, especially when dealing with electronically excited states. We have however to acknowledge the development of linear scaling methods that allow obtaining the electronic energy of quite large molecular systems [14–17]. Nevertheless, one cannot, for the time being, carry out millions of such calculations which are nonetheless required to sample all the possible conformations, necessaries to properly describe highly flexible and dynamic systems like biomolecules. This sampling is generally realized by means of Molecular Dynamics or Monte Carlo techniques using classical force fields [18]. The main drawback of such simulations is that they cannot describe electronic phenomena (chemical reactions, electronic transitions) since electrons are only implicitly taken into account via empirical parameters. We then face an ambiguous situation where Quantum Chemistry is required but cannot be applied.

The first solution was proposed by Warshel in 1976 [19] at the semi-empirical level and by Rivail and Assfeld [20] 20 years later at the ab initio level of theory. The seminal idea is to divide the large molecular system into two communicating parts, one considered as the active part (where the electronic phenomenon takes place) is small and is described with Quantum Mechanics (QM) methods, the other, considered as the surroundings, contains the remaining thousands of atoms and is then treated with Molecular Mechanics (MM) Force Fields. This is the principle of the so-called QM/MM methods, which rely on the locality of the electronic phenomenon under investigation [9].

In regard of the large panel of available QM methods (HF, PM3, PBE0, MPn, CCSD, CI, …) [21–23] and of current MM force fields (AMBER, CHARMM, UFF, DREIDING, …) [24–30], it exist an impressive bestiary of QM/MM couplings [31–47]. Although these trivial differences can have a non-negligible effect on the potential applications one can theoretically decipher, they won't be discussed in this chapter. A contrario we will focus our discussion on the fundamental differences to treat the physical and or chemical interactions between the two sub-systems, QM and MM. Once the general review will be set in the next section, the Local Self-Consistent Field (LSCF) method developed in our group [9, 20] will be detailed in Sect. 1.3. Finally several illustrative examples will be given in the fourth section in order to show the applicability and potency of the method.

1.2 QM/MM Methods

1.2.1 QM/MM, QM:MM, QM–MM?

Before going through all the interactions between the QM and the MM parts, we will first settle a nomenclature that will be used throughout the chapter, to specify which criteria are used to specify which atoms are treated by QM methods and which are described by MM force fields. Two cases need to be considered:

- The two subsystems are not chemically bonded. For example, a solute molecule in a solution. The interactions between the fragments are then weak interaction (Keesom, Debye, London, H-bond, ...), and we will call them physical interactions. In such situations we will use the acronym QM:MM, where the colon ":" symbolizes these non-bonded interactions (in the chemical sense).
- The two subsystems are connected through covalent bonds. For example, the amino-acid residues constituting a protein. To define the QM region one has then to formally cut these strong chemical bonds. For such cases, we will use the QM–MM acronym, where the dash "–" represents the cut bond(s).
- The QM/MM acronym will be used to describe any situation, following the traditional use of the slash "/" character in Quantum Chemistry.

1.2.2 Partition of the Hamiltonian

For any system modeled with any QM/MM method, the total Hamiltonian ($\hat{H}_{tot}$) can be written as

$$\hat{H}_{tot} = \hat{H}_{QM} + \hat{H}_{MM} + \hat{H}_{QM/MM} \tag{1.1}$$

Where $\hat{H}_{QM}$ is the Hamiltonian of the QM region, $\hat{H}_{MM}$ the one of the MM region and $\hat{H}_{QM/MM}$ the Hamiltonian containing the interactions between the two parts. Most of the QM/MM methods use this additive partition of the Hamiltonian. However, a very famous method (ONIOM) developed by Morokuma [48] use a subtractive partition:

$$\hat{H}_{tot} = \hat{H}_{QM} + \left(\hat{H}_{MM}^{QM+MM} - \hat{H}_{MM}^{QM}\right) \tag{1.2}$$

Where $\hat{H}_{MM}^{QM+MM}$ is the MM Hamiltonian of the whole system, i.e. the union of the QM and the MM parts, and $\hat{H}_{MM}^{QM}$ is the MM Hamiltonian of the region described with Quantum Mechanics. The parenthesis in Eq. (1.2) is then equal to the sum of the last two terms of Eq. (1.1). One has to note that with this subtractive partitioning, the QM/MM interactions are treated at the MM level of theory.

1.3 QM/MM Embeddings

The way the QM part feels the presence of the MM surroundings is called embedding. Prior to detail the various possible embeddings, it is important to recall the elementary particles of each part and how they interact.

The MM force fields consider a set of connected atoms (point masses). The atomic interactions are divided in two families: the bonded ones (stretching, bending, torsion) and the non-bonded ones (mainly electrostatic and van der Waals interactions). Some elaborated force fields allow the atomic point charges to vary according to their environment (polarization). Albeit the lack of explicit description of the electrons and nuclei is a strong limitation of MM methods, the main restriction of the MM representation is the predefined and fixed connectivity between the atoms.

The QM part is composed by a set of nuclei surrounded by electrons, and in absence of external field, solely the coulombic interaction is considered.

The QM/MM interactions are classified in three categories listed below.

1.3.1 Mechanical Embedding (ME)

The surroundings create geometrical constrains on the QM part. For QM:MM methods, only non-bonded interactions are responsible for these constrains (mainly van der Waals, but electrostatic repulsion or attraction can also have a non-negligible effect). They define regions of space that exclude the QM atoms and consequently modify its geometry. For QM–MM methods, the terms corresponding to bonded interactions between the QM part and the close-by MM atoms also play a major role. One has to note that van der Waals parameters need to be attributed to QM atoms (most of the time these parameters are those of the used force field, but one can optimize them [51]). If the QM part doesn't feel the electrostatic field of the MM surroundings, i.e. only when bonded and van der Waals QM/MM interactions are considered, one speaks about Mechanical Embedding. In this approximation the additive and the subtractive partitioning would give the same answer, since the QM electronic wave function is only affected through geometrical polarization. This approximation is suitable only for very weak polar, or isotropic, environment.

1.3.2 Electrostatic Embedding (EE)

Most of the time, the charge distribution of the MM region is anisotropic and this results in a non uniform external electrostatic field felt by the QM fragment. As a consequence, the electronic cloud of the QM region is polarized and the physical properties are then greatly modified. This electrostatic QM/MM interaction is then of primary importance and has to be taken into account for, of course, the calculation of the total energy of the whole system, but also in the Hamiltonian which will give the electronic wave function. If this interaction is considered one speaks about Electronic (or Electrostatic) Embedding, depending if it is included (or not) in the QM Hamiltonian. Taking into account this embedding with the additive scheme is straightforward. For the subtractive portioning, special care has to be taken. One has to note that, independently of the partitioning, the charge distribution of the MM system is not perturbed by the polarization of the QM part.

1.3.3 Polarizable Embedding (PE)

Rigorously speaking, if the MM part polarized the QM region, in turn the QM region should polarize the MM part and so on until convergence is reached. Some specific force fields, which allow such procedure, are said to be polarizable (AMOEBA, TCPEp, SIBFA, …) [28–30, 49, 50, 52, 53]. Another approach, called Electronic Response of the Surrounding (ERS), that uses the dynamical part of the dielectric constant of a polarizable continuum, is developed in our group [9, 54–61]. One speaks of Polarizable Embedding when such level of sophistication is used. Although, PE is certainly the most realistic possible QM/MM simulation, it is barely used and EE is the standard. The reason behind is not the laziness of theoretical chemists but the way force fields parameters are defined. In fact, most of the (non-polarizable) force fields define the atomic point charges in such a way to reproduce condensed phase properties. Thus the MM point charges are implicitly polarized and the EE level is sufficient. It exist however some situations for which the PE approximation is mandatory: when the force field parameters are based on gas phase data, or when the electronic variations of the QM part are drastic (for example for some chemical reactions like Menshutkin's one or for electronic transitions between states of different nature). PE calculations with the additive partitioning are straightforward from a theoretical point of view. Within the subtractive decomposition of the total Hamiltonian one has to modify further the initial methodology.

Up to now, the partitioning of the Hamiltonian and the Embedding of the QM part are enough to discriminate between QM:MM methods (in addition to the trivial differentiation induced by the methods used for the QM and the MM parts). For QM–MM methods, one needs to go one step further and to consider the way they connect the MM part to the QM one, since formally covalent bonds of the total molecule are cut and have to be modeled.

1.4 QM–MM Junctions

Whatever the connection scheme, the frontier atom on the QM side must possess classical parameters for the bonded terms linking the two fragments. The most common habit is to include all MM bonded interactions when at least one MM atom is involved. However, this implies to define connectivity inside the QM region which could be incompatible with the investigated chemical process. This induces an intrinsic limitation to the smallest size the QM part can have.

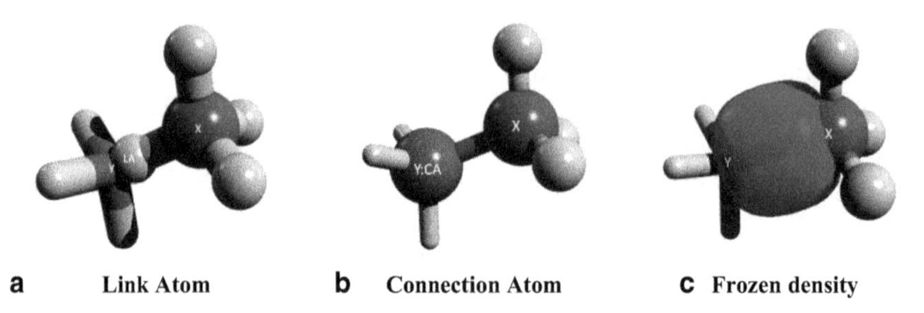

Fig. 1.1 Schematic representations of the three types of QM–MM junctions for the frontier bond X–Y where the X atom is in the QM part and the Y atom is in the MM part. The quantum part is depicted in Balls and Sticks representation and the MM one in Sticks only. **a** Link Atom (LA) approach. **b** Connection Atom (CA) method. **c** Frozen density approach, here a Strictly Localized Bond Orbital is depicted in *blue*

1.4.1 Link Atom (And Related Schemes)

The simplest way to saturate the dangling bonds is to add monovalent atoms, most of the time Hydrogen atoms, called link atoms [31–37]. These atoms are artificial in the sense that they do not exist in the initial molecule (see Fig. 1.1a). Although apparently quite simple to implement, this scheme requires some care to treat the exceeding degrees of freedom when computing energy gradient for geometry optimization or molecular dynamics. One must be aware that, when using large diffuse basis functions, the electronic density of the QM system can spill-out to the neighbor classical atoms. Although this can be true for any QM–MM junction, it is particularly evident for link atom methods for which the extra atom is very close to the classical part. In addition, if a single C–C bond polarity can be adequately modeled with a C–H bond, C–O or C–N bond polarity can hardly be reached. Some attempts involving (pseudo-)halogen atoms or atom group (CH_3 for example) have been proposed, but no universal method was given so far. Finally, multiple bonds are quite challenging to cut with this scheme.

1.4.2 Connection Atom (And Related Schemes)

The second family of methods suppresses the exceeding degrees of liberty introduced by the supplementary atoms. The second atom of the cut covalent bond is included in the QM region, and is then a quanto-classical atom having all the QM parameters (basis set, nuclear charges, semi-empirical parameters if needed) and all the MM parameters (depending of the force field used). It is called a connection atom (see Fig. 1.1b). This scheme needs an intense parameterization but is easily applicable at the ab initio level thanks to pseudopotentials [38–40]. The pseudobond approach belongs to this family. Up to now, only few atom types have been parameterized owing to the strong dependence to the MM force field used and to the QM method considered. Hence, this scheme cannot be considered as universal neither.

1.4.3 Frozen Density (And Related Schemes)

Finally, the third class of approaches encompasses all methods dealing with frozen electronic density (see Fig. 1.1c). Generally, the electronic density is obtained from orbitals (hybrid orbitals or localized molecular orbitals) determined on small molecules which contain the bond of interest [9, 19, 20]. It is then possible to cut bonds of any polarity (P–O in DNA for example), or multiplicity. It is even possible to cut peptide bond, which represent a serious advantage for the study of proteins. The universality of these methods is however accompanied by an inherent coding complexity. Among these methods, the Local Self-Consistent Field approach (LSCF) developed in our group since more than fifteen years is detailed in the next section.

1.5 The LSCF Method

The first published QM/MM method using an ab initio Hamiltonian was based on the LSCF method [19]. The basic ideas of the Local Self Consistent Field, i.e. using frozen strictly localized bond orbitals (SLBO) to describe the bonds separating the quantum to the classical subsystems, already developed for the semi-empirical level, have been applied to the ab initio or density functional levels of computation [20, 62–71]. In the latter cases a difficulty appears due to the fact that overlap between atomic orbital is no longer neglected. Therefore, the molecular orbitals of the quantum subsystem have to be kept orthogonal to each SLBO. This can be achieved by an orthogonalization of the basis set to the SLBOs, but owing to the fact that some functions of the set enter the SLBOs, a linear dependency appears between the orthogonalized functions. This inconvenience can be overcome by means of a canonical orthogonalization which yields a set of orthogonal, linearly independent basis functions which can be used to develop the molecular orbitals of the quantum subsystem. In order to recall to the reader the general equations used in the LSCF method, we present below the very basic theory [20].

Solving the LSCF problem implies optimizing a monodeterminantal wavefunction in the orbital approximation knowing that some predefined orbitals are given, and that these "external" orbitals should remain constant during the optimization procedure, *i.e.* frozen. The type of frozen orbitals is completely free. They can be monatomic, diatomic or polyatomic. In addition they can be occupied or empty. Of course to link the QM and MM together, they are doubly occupied SLBOs. The coefficients defining these frozen orbitals, are the only data needed to start the computation and are generally obtained on a simple molecule containing the bond to be mimicked. Let say that the user gives L frozen orbitals $\{\psi_i\}_{i=1,L}$ expanded on the initial basis set $\{\varphi_\mu\}_{\mu=1,K}$ composed of K real functions.

$$\left|\psi_i\right\rangle = \sum_{\mu=1}^{K} a_{\mu i} \left|\varphi_\mu\right\rangle \tag{1.3}$$

If the given frozen orbitals are non orthogonal, they can be orthogonalized by the standard Löwdin or Gramm–Schmidt procedures. Let's suppose here for simplicity and without loss of generality that these frozen orbitals are orthogonal. Each function of the initial set is projected out of the subspace spanned by the frozen orbitals.

$$\left|\phi_\mu\right\rangle = N_\mu \left[\hat{1} - \sum_{i=1}^{L} \left|\psi_i\right\rangle\left\langle\psi_i\right|\right]\left|\varphi_\mu\right\rangle \tag{1.4}$$

where N_μ is a normalization factor. This transformation can be represented by a square matrix M, of dimensions K×K, acting on the initial set to provide the new set and whose elements are given by:

$$M_{\nu\mu} = \frac{\sum_{\nu=1}^{K}\left(\delta_{\nu\mu} - \sum_{i=1}^{L} a_{\nu i} \sum_{\lambda=1}^{K} a_{\lambda i} S_{\lambda\mu}\right)}{\sqrt{1 - \sum_{i=1}^{L}\left(\sum_{\lambda=1}^{K} a_{\lambda i} S_{\lambda\mu}\right)^2}} \tag{1.5}$$

Where $S_{\lambda\mu} = \left\langle\varphi_\lambda \middle| \varphi_\mu\right\rangle$ is an overlap matrix element for the ϕ base. The new set of K functions $\left\{\phi_\mu\right\}_{\mu=1,K}$ contains however L linear dependencies as said before (at least if some linear dependencies where already present in the initial set). The linear dependencies, which would give a wavefunction exactly equal to zero everywhere, are removed thank to the canonical orthogonalization procedure. The overlap matrix R of the φ functions $\left(R_{\lambda\mu} = \left\langle\varphi_\lambda \middle| \varphi_\mu\right\rangle\right)$ is diagonalized.

$$r = A^{\dagger} R A \tag{1.6}$$

The K eigenvalues are ordered in decreasing order together with the corresponding eigenvectors, and the L eigenvalues equal to zero (or close to) are removed together with the corresponding eigenvectors from matrix A, this leaves a rectangular matrix of size $K\times(K-L)$. The orthogonalization matrix X again of dimensions $K\times(K-L)$ is then obtained by the product of the rectangular matrix of the $(K-L)$ eigenvectors (A) and the square diagonal matrix of the $(K-L)$ non zero eigenvalues at the power of minus one-half (r).

$$X_{\nu\mu} = \sum_{\eta=1}^{K-L} A_{\nu\eta} r_{\eta\mu}^{-1/2} \tag{1.7}$$

Finally, the two matrices X and M can be contracted in one performing the transformation of the initial set of K functions φ into a set of $K-L$ functions mutually orthogonal and orthogonal to the L frozen orbitals.

$$B_{\nu\mu} = \sum_{\eta=1}^{K} M_{\nu\eta} X_{\eta\mu} \tag{1.8}$$

The rectangular B matrix is used in the SCF procedure instead of the usual Löwdin matrix. Hence this method can be employed either for Hartree-Fock or Kohn-Sham equations resolution in the Roothaan formalism. The extension to the unrestricted spin case is trivial.

The only other modification one has to take care of is the construction of the total density matrix ($P^T_{\mu\nu} = P^Q_{\mu\nu} + P^F_{\mu\nu}$) to build the Fock (or Kohn-Sham) matrix elements. To the usual density matrix built over the variational orbitals ($P^Q_{\mu\nu} = \sum_i^{occ} n_i c_{\mu i} c_{\nu i}$) one has to add the contribution arising from the frozen orbitals ($P^F_{\mu\nu} = \sum_{i=1}^{L} n_i a_{\mu i} a_{\nu i}$).

The derivatives needed to allow a full optimization of geometry, or to perform molecular dynamics trajectories, have been given elsewhere, and can be obtained analytically. It appeared that during this optimization, the length of the frontier bond, i.e. the bond linking the quantum to the classical system is systematically found too short and the shape of the potential energy surface (PES) around the minimum is different from the one obtained by a full QM calculation, whatever the method used to localize the orbital. This defect is analysed as the consequence of the fact that the nucleus of the quanto-classical atom of charge Z is replaced by a charge +1 since this atom contributes for one electron to the SLBO. Therefore, the interaction between nuclei is underestimated and, in addition, the variation of the overlap between the basis function with respect to the bond length is not taken into account. This defect has been corrected by introducing a 5 parameters empirical interaction potential for the frontier bond of the form:

$$E_{X-Y} = (A + Br + Cr^2)e^{Dr} + \frac{E}{r} \tag{1.9}$$

where r is the distance between the two atoms forming the bond. The parameters have been adjusted for any pair of C, O, N atoms, either at the quanto-classical or quantum position and for various hybridization states of the carbon atom.

In order to set up a non-empirical method and then to avoid the use of an empirical potential, the analysis of the factors affecting the energy variations of the system when the length of the frontier bond is varied proved that the discrepancy comes from the fact that the quanto-classical atom is treated as a pseudo one electron atom [9, 67, 68]. Taking into account the inner shell electrons of the quanto-classical atom by means of frozen or variational core orbitals gives an elegant solution to this problem. The nuclear charge is then switched to +3 and two electrons are added in the QM system. The acronym used to specify this modification is LSCF+3, by contrast to the initial approach called LSCF+1. Many results have shown that the LSCF+3 scheme reproduces satisfactorily the position of the minimum as well as the curvature obtained with the full quantum results, the error on the equilibrium distance being less than 0.1 Å [9]. Of course, no agreement can be obtained for long interatomic distances because the SLBO is only valid in the vicinity of the distance at which it has been obtained. This scheme has been extended successfully to the peptide bond where the C atom is at the MM frontier. It has also been extended to the same bonds in which the quanto-classical atom is the nitrogen atom. In this case, adding two extra valence electrons, those contained in the orbital conjugated with the C=O π orbital, is mandatory. The acronym is then trivially LSCF+5. This method is free of fitted parameters and allows a symmetric description of the amino acid residues without the arbitrariness of adjusting the classical point charges to obtain an integer value. This procedure is particularly attractive for QM/MM calculations on proteins since it permits to directly cutting through a peptide bond, keeping the electron delocalization occurring at the amide bond. Indeed by using LSCF+3 and LSCF+5 cutting scheme we have shown that the QM/MM equilibrium geometry of a tripeptide in which only the central monomer is treated with QM reproduces well the equilibrium geometry of the full QM systems, both for bond lengths and angles. In particular the planarity of the amide groups is always perfectly respected confirming the fact that the electron delocalization is taken into account whatever atom between C and N is treated as quanto-classical. The link atom approaches which use hydrogen atom to saturate the dangling bond are of course not able to reproduce this feature.

Several localization procedures exist and many have been tested in the LSCF framework [73–80]. It has been shown that, when one is interested in relative energies, the results do not depend on the localization scheme.

One can thus consider that the LSCF method is universal in the sense that it can be applied to any MM and QM methods, to bonds of any polarity and multiplicity.

1.6 Applications

We will present here some applications of QM/MM methods to the treatment of problems related to biological systems. Although these methods have been initially developed to deal with enzymatic catalysis or biochemical reactivity in general [81, 82], they are nowadays also applied to study the photophysics or photochemistry of complex biological systems. In this chapter, we will focus on the calculations of electronic excited states and on the different effects of the environment induced on the different chromophores [83–85]. We will consider systems in which the chromophore being covalently bounded to the macromolecule the use of QM–MM methods will be necessary, together with different cases in which QM:MM allows to treat non-covalently bounded systems. Finally the role and the necessity of a proper sampling of different conformations with using MD techniques using MD techniques will be tackled and discussed.

1.6.1 Absorption of Human Serum Albumin (HSA)

HSA is a liver produced protein present in the blood where it exercises regulatory and transport functions. Despite its huge mass, it only possesses one tryptophan residue. Tryptophan is fluorescent and its optical properties are strongly dependent on the environment. The latter can be used to probe their conformation. Indeed in the protein (Fig. 1.2) the tryptophan is embedded in a pocket between different α-helices, in an environment strongly different from the one of the denatured protein where it will be mainly surrounded by water molecules.

In Fig. 1.2 (right panel) one can also see the QM/MM absorption spectrum computed at Time Dependent Density Functional (TD-DFT) level using B3LYP exchange correlation functionals and a 6–311+G(d, p) basis set [9]. The protein environment where treated using amber99 force field. Note that the transitions have been obtained as Franck-Condon vertical transitions from the ground state equilibrium geometry. The QM-MM frontier has been treated placing an SLBO between the C_α–C_β bond of the tryptophan lateral chain.

The spectrum obtained with the three different embedding schemes [9] is provided and one can see that the computed spectrum presents two well defined absorption maxima at about 250 and 275 nm, respectively. This represents a significant red-shift compared to the absorption of water solvated tryptophan, accounting for the environment effects. Notice also that the inclusion of polarization effects, treated using the ERS technique, induces a non negligible shift over the ME and EE absorption maxima, confirming the fact that differently from the case of ground state studies, in the case of electronic transitions all the three embedding effects should be taken into account.

1.6.2 Absorption of Copper Proteins (From Red to Blue Protein)

Plastocyanin (Fig. 1.3) are metallo-protein present in superior plants were they assure electron-transfer during the photosynthetic process [57, 58]. The active site of the protein is constituted by a copper ion complexed by four aminoacids residues: one deprotonated cysteine, one methionine and two hystidines. Notice that the electron-transfer is assured by the copper atoms that can reversibly convert between the +II and +I form. QM–MM calculations have allowed us to show that the protein environment regulates the necessary high rate of electron-transfer by constraining the copper environment in a geometry that is somehow mid-way between the ones of the +II and +I complex in gas phase [57] (i.e. copper coordinated by the lateral chains of the previous cited aminoacids, only), moreover in the protein geometrical differences between the oxidized and reduced form are extremely small, thus minimizing the reorganization energy of the redox process.

But oxidized plastocyanin are also known to exhibit a very peculiar absorption spectrum, characterized by a very intense absorption at about 600 nm responsible for the intense blue color, and quite different from the one of isolated copper com-

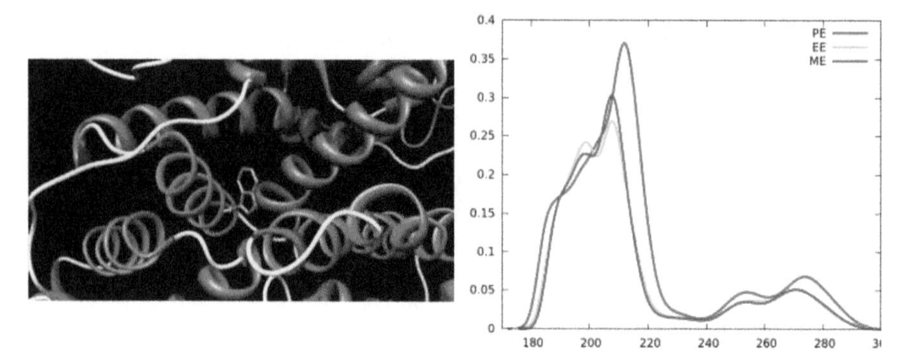

Fig. 1.2 Structure (*left*) and computed absorption spectrum (*right*) of tryptophan in HSA. For the Spectrum wavelengths in nm and intensities in arbitrary units

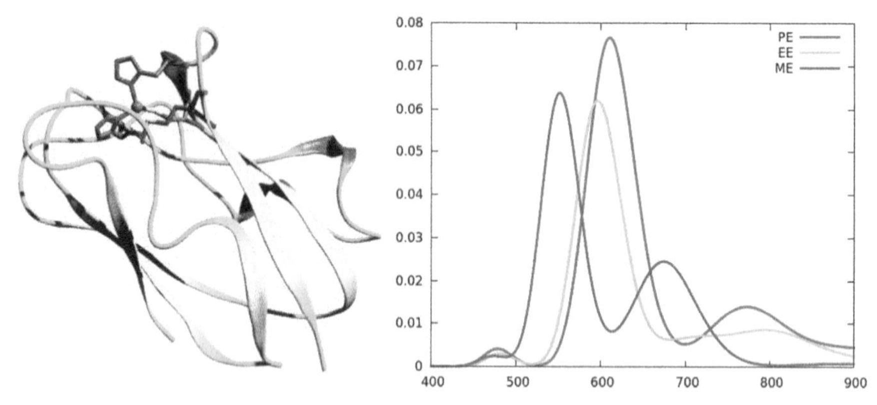

Fig. 1.3 Structure (*left*) and computed absorption spectrum (*right*) of plastocyanin. For the spectrum wavelengths in nm and intensities in arbitrary units

plexes. The QM–MM absorption spectrum obtained at TDDFT level putting only the copper ion and the ligating aminoacids in the QM part is also reported in Fig. 1.3 and one can notice the very good agreement with the experimental values, providing that the PE effect is taken into account, on the contrary ME gives totally unreliable results indicating an important effects of electrostatic and polarization effects [57]. Note also that more in detail the PE spectrum is composed of a large tail in the near infrared region while the visible part is constituted by the huge absorption band peaking at 600 nm and of a much less intense band appearing close to 490 nm. Indeed we have shown that by a selective mutation of the methionine residues complexing copper one can induce an important shift of the position of the two bands, coupled with an important change on the relative intensity ratio (Table 1.1). These two phenomena together induce an important change on the color of the protein and indeed by selectively mutating only one aminoacid, one is able to continuously pass from a blue to a red protein when methionine is substituted with the non proteinogenic aminoacid homocysteine (Hcy) [58]. Note also that these results also nicely reproduce experimental observations [86].

Table 1.1 Variation of absorption wavelengths and of the intensity ratio between first and second band in plastocyanin upon mutation of methionine

	First band λ_{max} (nm)	Second band λ_{max} (nm)	Intensity ratio
Cys	589	452	2.00
Cys-	526	447	0.44
Glu-	550	440	0.66
Hcy	452	403	0.84

1.6.3 Interactions with DNA: Light-Switch Effect and Phototherapy

The interaction of organometallic and organic metal with nucleic acid and especially DNA is nowadays a very well recognized phenomenon that is strongly exploited in chemotherapy to induce apoptosis of cancer cells or simply as DNA probe. Moreover a growing interest is devoted to the non-covalent interactions between DNA and xenobiotics [59–61, 87]. Indeed this interaction can take place by electrostatic binding to the DNA, with the interactor laying close to the minor or major groove or by intercalation and/or insertion. In these latter situations usually the xenobiotic has one large planar and conjugated moiety that can intercalate between two base pairs (intercalation) or eject one of the bases and substitute it in the double helix (insertion).

It is noteworthy to recognize that the interaction with DNA can strongly alter the photophysical and photochemical properties of many chromophores. One paradigmatic example is the so called light-switching effect in which a non luminescent Ruthenium complex becomes strongly luminescent when DNA is added to the solution. Most strikingly by just a small modification of the ligands of the complex the behavior is completely reversed, and the complex now become luminescent in water, while the emission is totally quenched by DNA. This general behavior can be interpreted in term of a competition between luminescence and a photo-induced charge transfer from DNA (most often guanine) to the organometallic complexes. In other words when in its excited state Ruthenium complexes is able to oxidize DNA and therefore to induce an irreversible lesion that can ultimately provoke the cellular death, the possible application of this feature in phototherapy is of course straightforward, and indeed some Ruthenium complexes have already entered the clinical trial phase. It is evident that to be able to produce efficient and selective phototherapeutic agents the nature of the Ruthenium complex excited states should be carefully elucidated as well as the effects induce by the DNA environment.

To this end we studied by using QM:MM methods the behavior of Ru di-bipyridyl, dipyridophenazine (Ru(bipy2,dppz)) [59, 60], whose structure is reported in Fig. 1.4, interacting with a double helix DNA pentadecamer (Fig. 1.4). After having optimized the intercalated complex at DFT level we performed TD-DFT calculations of the absorption spectrum. In all cases only the chromophore was treated at QM level, using B3LYP as exchange correlation functionals and the LANL2DZ basis, DNA as well as the water solvation box was instead treated using CHARMM force field.

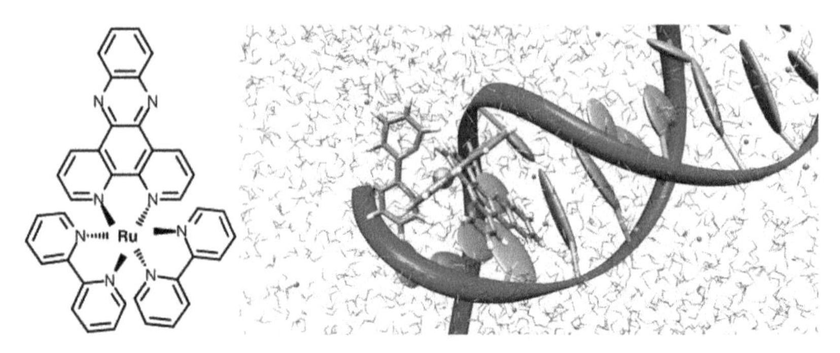

Fig. 1.4 Molecular Structure (*left*) of Ru(bipy2,dppz) cation and its interaction with DNA (*right*)

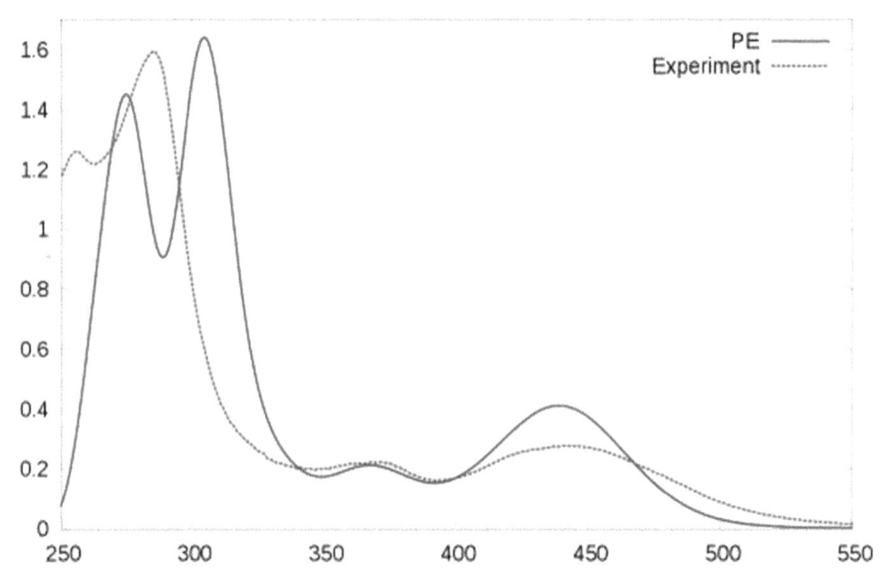

Fig. 1.5 Computed and experimental absorption spectrum of Ru(bipy2, dppz) interacting with DNA. Wavelengths in nm, intensities in arbitrary units

In Fig. 1.5 we report the QM:MM spectrum, including polarization effects, of Ru(bipy2,dppz) interacting with DNA compared to the experimental one. It is straightforward to notice the very good agreement with experimental results, in particular for the bands in the visible region. The large band situated at about 450 nm as well as the weak, but important, band appearing at about 350 nm.

By analyzing the single excitation in terms of orbital contribution, and in particular in terms of natural transition orbitals (NTO) [57, 58] we have been able to correctly interpret the spectral features. In particular the band at 450 nm is dominated by metal-to-ligand charge-transfer (MLCT) transition, while the band at 350 nm is much more complex and is composed of MLCT as well as of intra- and inter-ligand charge transfer transitions. In particular the latter are extremely important since they can leave a hole in the intercalated ligand that can favor the charge-injection from the DNA (Fig. 1.6).

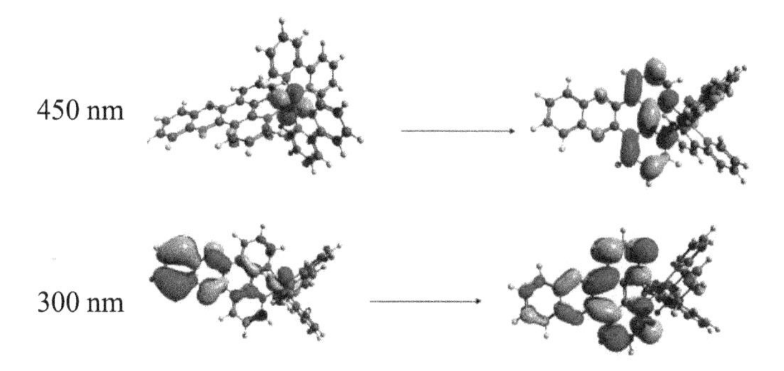

Fig. 1.6 NTOs for two selective transitions of Ru(bipy2,dppz)

Up to now the examples we have presented were obtained considering vertical transitions from the equilibrium geometry in the framework of the Franck–Condon principle. Indeed biological macromolecules such as nucleic acids are quite flexible and can explore important regions of the configuration space. The inclusion of dynamic and vibrational effects, as well as its coupling with the solvent and with environment motion can be straightforwardly tackled by using classical molecular dynamics (MD) techniques [61]. Indeed one can run a sufficiently long MD trajectory (usually some nanoseconds) and extract statistically independent snapshots. The excitation spectrum can be calculated at QM/MM level for each snapshot and the resulting final absorption spectrum will be composed by the convolution of all the individual snapshots. This technique has also the advantage to take into account the vibrational structure of the spectrum and can therefore recover the asymmetry of the absorption band or the shifts induced by vibronic coupling.

An example of a system for which such a treatment is compulsory is the one of a β-carboline, harmane, whose cation interacts with DNA via minor groove-binding or intercalation (Fig. 1.7) [61]. Due to the important out of plane vibration of the fused ring, and to its coupling with the electronic transition energy, the static approach from equilibrium geometry, gave a shift of more than 50 nm on the absorption maximum. On the other hand when considering the convolution from a MD trajectory one gets the exact experimental value, as can be seen in Fig. 1.7. Note also that MD proved that the two interaction modes where stable and almost degenerate in terms of interaction energy, they also gave rise to a practically undistinguishable absorption spectrum.

1.7 Conclusions

In this contribution we have presented the advantages and flaws of general QM/MM philosophies. Whatever the chosen method, if well parametrized, one can get reliable results and insights on very large molecular systems still unreachable by standard QM methods. However, we think that the LSCF approach is the only one that doesn't need specific parameters to be applied on any kind of (covalent) systems. In that sense, it can be qualified to be "universal".

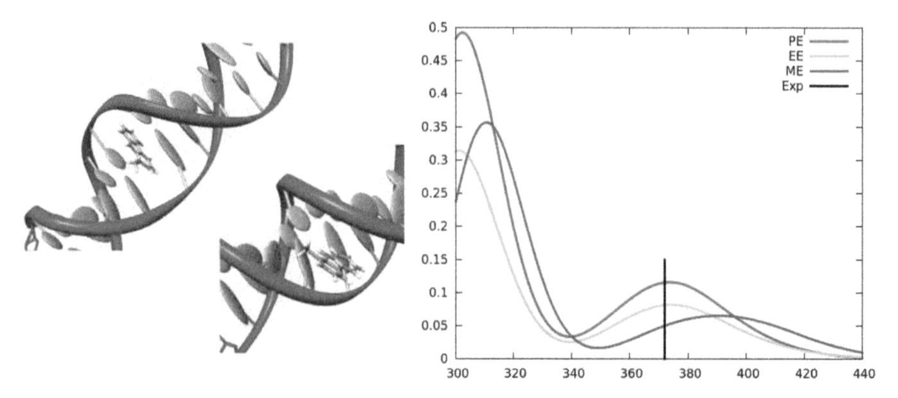

Fig. 1.7 Interaction with DNA by intercalation and groove-binding (*right*) and absorption spectrum in DNA for the harmane cation (*left*). Wavelengths in nm intensities in arbitrary units

Many factors can be of tremendous importance for large anisotropic biomolecular systems, like the sampling of the conformational space, the mutual polarization of both subsystems and so on. One must be aware that these effects are system dependent and for every study special care has to be taken. Hybrid methods are not yet black-boxes. Let's hope it will stay so for a long time.

We are really convinced that the most important feature of QM/MM methods is to give insights, and more importantly to give correct trends, on systems on which obtaining absolute number has no meaning.

The applications we have chosen to present here are just few examples of what these methods are able to tackle. However, one has to bear in mind that one can only study systems in which the electronic phenomenon is localized in a small portion of the total space. This is perhaps the area where linear scaling methods will find their most exciting applications.

References

1. Koerner T, Brown RS, Gainsforth JL, Klobulowski M (1998) Electrophilic bromination of ethylene and ethylene-d4: a combined experimental and theoretical study. J Am Chem Soc 120:5628–5636
2. Benkovic SJ, Hammes-Schiffer S (2003) A perspective on enzyme catalysis. Nature 5637:1196–1202
3. Polli D, Altoé P, Weingart O, Spillane KM, Manzoni C, Brida D, Tomasello G, Orlandi G, Kukura P, Mathies RA, Garavelli M, Cerullo G (2010) Rhodopsin isomerization probed in the visible spectral range. Nature 467:440–443
4. Gozem S, Schapiro I, Ferré N, Olivucci M (2012) The molecular mechanism of thermal noise in rod photoreceptors. Science 337:1225–1228
5. Dolmans DE, Fukumura D, Jain RK (2003) Photodynamic therapy for cancer Nature Rev. Cancer 3:380–387
6. Metcalfe C, Thomas JA (2003) Kinetically inert transition metal complexes that reversibly bind to DNA. Chem. Soc Rev 32(2):15–224

7. Zeglis BM, Pierre VC, Barton JK (2007) Metallointercalators and metalloinsertors. Chem Commun 44:4565–4579
8. Ortmans I, Elias B, Kelly JM, Moucheron C, Kirsch-De Mesmaeker A (2004) $[RU(TAP)_2(dppz)]^{2+}$: a DNA intercalating complex, which luminesces strongly in water and undergoes photo-induced proton-coupled electron transfer with guanosine-5'-monophosphate. Dalton Trans 33:668–676
9. Monari A, Rivail J-L, Assfeld X (2013) Theoretical modelling of large molecular systems. Advances in the local self consistent field method for mixed quantum mechanics/molecular mechanics calculations. Acc Chem Res 46:596–603
10. Monard G, Merz KM Jr (1999) Combined quantum mechanical/molecular mechanical methodologies applied to biomolecular systems. Acc Chem Rev 32:904–911
11. Lin H, Truhlar DG (2007) QM/MM: what we have learned, where are we, and where do we go from here? Theor Chem Acc 117:185–199
12. Senn HM, Thiel W (2009) QM/MM methods for biomolecular systems. Ang Chem Int 48:1198–1229
13. Gordon MS, Fedorov DG, Pruitt SR, Slipchenko IV (2012) Fragmentation methods: a route to accurate calculations on large molecules Chem Rev 112:632–613
14. Dixon SL, Merz KM Jr (1996) Semiempirical molecular-orbital calculations with linear system size scaling. J Chem Phys 104:6643–6649
15. Daniels AD, Millam JM, Scuseria GE (1997) Semiempirical methods with conjugate gradient density matrix search to replace diagonalization for molecular systems contianing thousands of atoms. J Chem Phys 107:425–431
16. Yang W, Lee TS (1995) A density matrix divide-and-conquer approach for electronic structure calculations of large molecules. J Chem Phys 103:5674–5678
17. Monard G, Bernal-Uruchurtu MI, van der Vaart A, Merz KM Jr, Ruiz-Lopèz MF (2005) Simulation of liquid water using semiempirical hamiltonina and the divide and conquer approach. J Phys Chem A 109:3425–3432
18. Tarek M, Delemotte L (2013) Omega currents in voltage-gated ion channels: what can we learn from uncovering the voltage-sensing mechanism using MD simulations? Acc Chem Res. doi:10.1021/ar300290u
19. Warshel A, Levitt M (1976) Theoretical studies of enzymatic reactions: dielectric, electrostatic and steric stabilization of the carbonium ion in the reaction of lysozime. J Mol Biol 103:227–249
20. Assfeld X, Rivail J-L (1996) Quantum chemical computations on parts of large molecules: the ab initio local self consistent field method. Chem Phys Lett 263:100–106
21. Jensen, F (2007) Introduction to quantum chemistry, 2nd edn. Wiley, New York
22. Piela L (2007) Ideas of quantum chemistry. Elsevier, Amsterdam
23. Cramer CJ (2004) Essentials of computational chemistry, 2nd edn. Wiley, New York
24. van Gunsteren WF, Berendsen HJC (1987) Groningen Molecular Simulation (GROMOS) library manual. Biomos, Groningen
25. Allinger NL, Yuh YH, Lii JH (1989) Molecular mechanics. The MM3 force field for hydrocarbons. J Am Chem Soc 111:8551–8566
26. Pearlman DA, Case DA, Caldwell JW, Ross WS, Cheatham TE. III, DeBolt S, Ferguson D, Seibel G, Kollman P (1995) AMBER, a package of computer programs for applying molecular mechanics, normal mode analysis, molecular dynamics and free energy calculations to simulate structural and energetic properties of molecules. Comput Phys Commun 91:1–41
27. MacKerell AD Jr, Bashford D, Bellott M, Dunbrack RL Jr, Evanseck JD, Field MJ, Fischer S, Gao J, Guo H, Ha S, Joseph-McCarthy D, Kuchnir L, Kuczera K, Lau F. TK, Mattos C, Michnick S, Ngo T, Nguyen DT, Prodhom B, Reiher WE III, Roux B, Schlenkrich M, Smith JC, Stote R, Straub J, Watanabe W, Wiórkiewicz-Kuczera J, Yin D, Karplus M (1998) All-atom empirical potential for molecular modeling and dynamics studies of proteins. J Phys Chem B 102:3586–3616
28. Dang LX, Chang T-M (1997) Molecular dynamics study of water clusters, liquid and liquid–vapor interface of water with many-body potentials. J Chem Phys 106:8149–8159

29. Dupuis M, Aida M, Kawashima Y, Hirao K (2002) A polarizable mixed hamiltonian model of electronic structure for micro-solvated excited states. I energy and gradients formulation and application to formaldehyde ($^1A_2$) J Chem Phys 117:1242–1255
30. Gresh N, Andrés Cisneros G, Darden TA, Piquemal J-P (2007) Anisotropic, polarizable molecular mechanics studies of inter- and intramolecular interactions and ligand–macromolecule-complexes. A bottom-up strategy. J Chem Theory Comput 3:1960–1986
31. Field MJ, Bash PA, Kearplus M (1990) A combined quantum mechanical and molecular mechanical potential for molecular dynamic simulations J Comput Chem 11:700–733
32. Kaminski GA, Jorgensen WL (1998) A quantum mechanical and molecular mechanical method based on cm1a charges: applications to solvent effects on organic equilibria and reactions. J Phys Chem B 102:1787–1796
33. Ferré N, Olivucci M (2003) The amide bond: pitfalls and drawback of the link atom scheme. J Mol Struct (Theochem) 632:71–84
34. Antes I, Thiel W (1999) Adjusted connection atoms for combined quantum mechanical and molecular mechanical methods. J Phys Chem A 103:9290–9295
35. Ranganathan S, Gready JE (1997) Hybrid quantum and molecular mechanical (QM/MM) studies on the pyruvate to l-lactate interconversion in l-lactate dehydrogenase. J Phys Chem B 101:5614–5618
36. Swart M (2003) AddRemove: a new link model for use in QM/MM studies. Int J Quantum Chem 91:177–183
37. Das D, Eurenius KP, Billings EM, Sherwood P, Chatfield DC, Hodoscek M, Brook BR (2002) Optimization of quantum mechanical molecular mechanical partitioning schemes: Gaussian delocalization of molecular mechanical charges and the double link atom method. J Chem Phys 117:10534–10547
38. Di Labio GA, Hurley MM, Christiansen PA (2002) Simple one-electron quantum capping potentials for use in hybrid QM/MM studies of biological molecules. J Chem Phys 116:9578–9584
39. Zhang Y, Lee T-S, Yang W (1999) A pseudobond approach to combining quantum mechanical and molecular mechanical methods. J Chem Phys 110:46–54
40. Xiao C, Zhang Y (2007) Design-atom approach for quantum mechanical/molecular mechanical covalent boundary: a design-carbon atom with five valence electrons. J Chem Phys 127:124102
41. Ferenczy GG, Rivail J-L, Surjan PR, Naray-Szabo G (1992) NDDO fragment self-consistent-field approximation for large electronic systems. J Comput Chem 13:830–837
42. Théry V, Rinaldi D, Rivail J-L, Maigret B, Ferenczy GG (1994) Quantum mechanical computations on very large molecular systems: the local self-consistent field method. J Comput Chem 15:269–282
43. Stewart JJP (1989) Optimization of parameters for semiempirical methods I. Method J Comput Chem 10:209–220
44. Gao G, Amara P, Alhambra C, Field MJ (1998) A generalized hybrid orbital (GHO) method for the treatment of boundary atoms in combined QM/MM calculations. J Phys Chem A 102:4714–4721
45. Pu J, Gao J, Truhlar DG (2004) Generalized hybrid orbital (GHO) method for combining ab initio Hartree–Fock wave functions with molecular mechanics. J Phys Chem A 108:632–650
46. Pu J, Gao J, Truhlar DG (2004) Combining self-consistent-charge density-functional tight-binding (SCC-DFTB) with molecular mechanics by the generalized hybrid orbital (GHO) method. J Phys Chem A 108:5454–5463
47. Pu J, Truhlar DG (2005) Redristributed charge and dipole schemes for combined quantum mechanical and molecular mechanical calculations J Phys Chem A 109:3991–4004
48. Maseras F, Morokuma K (1995) IMOMM, a new integrated ab initio+molecular mechanics geometry optimization scheme of equilibrium structure and transition states. J Comput Chem 16:1170–1179
49. Celebi N, Ángyán JG, Dehez F, Millot C, Chipot C (2000) Distributed polarizabilities derived from induction energies: a finite perturbation approach. J Chem Phys 112:2709–2717

50. Dehez F, Soetens JC, Chipot C, Ángyán JG, Millot C (2000) Determination of distributed polarizabilities from a statistical analysis of induction energies. J Phys Chem A 104:1293–1303
51. Martin M, Aguilar M, Chalmet S, Ruiz-Lopez MF (2001) An iterative procedure to determinate Lennard–Jone parameters for their use in quantum mechanics/molecular mechanics liquid state simulations. 284:607–614
52. Dehez F, Chipot C, Millot C, Ángyán JG (2001) Fast and accurate determination of induction energies: reduction of topologically distributed polarizability models. Chem Phys Lett 338:180–188
53. Thompson MA (1996) QM/MMpol: a consistent model for solute/solvent polarization. Application to the aqueous solvation and spectroscopy of formaldehyde, acetaldehyde and acetone. J Phys Chem 100:14492–14507
54. Jacquemin D, Perpète EA, Laurent AD, Assfeld X, Adamo C (2009) Spectral properties of self-assembled squaraine–tetralactam: a theoretical assessment. Phys Chem Chem Phys 11:1258–1262
55. Laurent AD, Assfeld X (2010) Effect of the enhanced cyan fluorescent proteinic framework on the UV/visible absorption spectra of some chromophores. Interdisc Sci Comput Life Sci 2:38–47
56. Monari A, Very T, Rivail J-L, Assfeld X (2012) A QM/MM study on the spinach plastocyanin: redox properties and absorption spectra. Comp Theoret Chem 990:119–125
57. Monari A, Very T, Rivail J-L, Assfeld X (2012) Effects of mutations on the absorption spectrum of copper proteins: a QM/MM study Theor Chem Acc 131:1221
58. Very T, Despax S, Hébraud P, Monari A, Assfeld X (2012) Spectral properties of polypyridyl ruthenium complex intercalated in DNA: theoretical insights on the surrounding effects for $[Ru(dppz)(bpy)_2]^{2+}$. Phys Chem Chem Phys 14:12496–12504
59. Chantzis A, Very T, Monari A, Assfeld X (2012) Improved treatment of surrounding effects: UV/vis absorption properties of a solvated Ru(II) complex. J Chem Theory Comp 8:1536
60. Etienne T, Very T, Perpète EA, Monari A, Assfeld X (2013) A QM/MM study of the absorption spectrum of harmane in water solution and interacting with DNA: the crucial role of dynamic effects J Phys Chem B 117:4973–4980
61. Moreau Y, Loss P-F, Assfeld X (2004) Solvent effects on the asymmetric Diels–Alder reaction between cyclopentadiene and (–)-menthyl acrylate revisited with the three-layer hybrid local self-consistent field/molecular mechanics/self-consistent reaction field method. Theor Chem Acc 112: 228–239
62. Ferré N, Assfeld X (2002) Application of the local self-consistent-field method to core-ionized and core-excited molecules, polymers, and proteins: true orthogonality between ground and excited states. J Chem Phys 117:4119–4125
63. Loos PF, Assfeld X (2007) Core-ionized and core-excited states of macromolecules. Int J Quantum Chem 107:2243–2252
64. Ferré N, Assfeld X, Rivail J-L (2002) Specific force field determination for the hybrid ab initio QM/MM LSCF method. J Comput Chem 23:610–624
65. Fornili A, Loos P-F, Sironi M, Assfeld X (2006) Frozen core orbitals as an alternative to specific frontier bond potential in hybrid quantum mechanics/molecular mechanics methods. Chem Phys Lett 427:236–240
66. Loos P-F, Fornili A, Sironi M, Assfeld X (2007) Removing extra frontier parameters in QM/MM methods: a tentative with the local self-consistent field approach. Comput Lett 4:473–486
67. Loos P-F, Assfeld X (2007) On the frontier bond location in the QM/MM description of peptides and proteins. AIP Conf Proc 963:308–315
68. Ferré N, Assfeld X (2003) A new three-layer hybrid method (LSCF/MM/Madelung) devoted to the study of chemical reactivity in zeolites. Preliminary results. J Mol Struct (Theochem) 632:83–90
69. Loos PF, Assfeld X (2007) Self-consistent strictly localized orbitals. J Chem Theory Comput 3:1047–1053

70. Boys SF (1960) Construction of some molecular orbitals to be approximately invariant for changes from one molecule to another. Rev Mod Phys 32:296–299
71. Monard G, Loos M, Théry V, Baka K, Rivail JL (1996) Hybrid classical quantum force field for modeling very large molecules. Int J Quantum Chem 58:153–159
72. Foster JM, Boys SF (1960) Canonical configurational interaction procedure. Rev Mod Phys 32:300–302
73. Magnasco V, Perico A (1967) Uniform localization of atomic and molecular orbitals. I. J Chem Phys 47:971–981
74. Weinstein H, Pauncz R (1968) Molecular orbital set determined by a localization procedure. Symp Faraday Soc 2:23–31
75. Pipek J, Mezey PG (1989) A fast intrinsic localization procedure applicable for ab initio and semiempirical linear combination of atomic orbital wave functions. J Chem Phys 90:4916–4926
76. Fornili A, Moreau Y, Sironi M, Assfeld X (2006) On the suitability of strictly localized orbitals for hybrid QM/MM calculations. J Comput Chem 27:515–523
77. Sironi M, Famulari A (2000) An orthogonal approach to determine extremely localised molecular orbitals. Theor Chem Acc 103:417–422
78. Antonczak S, Monard G, Ruiz-López MF, Rivail J-L (1998) Modeling of peptide hydrolysis by thermolysin. A semiempirical and QM/MM study. J Am Chem Soc. 120:8825–8833
79. Fornili A, Sironi M, Raimondi M (2003) Determination of extremely localized molecular orbitals and their application to quantum mechanics/molecular mechanics methods and to the study of intramolecular hydrogen bonding. J Mol Struct (Theochem) 632:157–172
80. Genoni A, Ghitti M, Pieraccini S, Sironi M (2005) A novel extremely localized molecular orbitals based technique for the one-electron density matrix computation. Chem Phys Lett 415:256–260
81. Ruiz-Pernia J, Luk L, Garcia-Meseguer R, Marti S, Loveridge J, Tuñón I, Moliner V, Allemann R (2013) Increased dynamic effects in a catalytically compromised variant of Escherichia coli dihydrofolate reductase. J Am Chem Soc. doi:10.1021/ja410519h
82. Luk L, Ruiz-Pernia J, Dawson W, Roca M, Loveridge J, Glowacki D, Harvey J, Mulholland A, Tuñón I, Moliner V, Allemann R (2013) Unravelling the role of protein dynamics in dihydrofolate reductase catalysis. Proc Nat Acad Sci U S A 110:16344–16349
83. Jacquemin D, Mennucci B, Adamo C (2011) Excited-state calculations with TD-DFT: from benchmarcks to simulations in complex environments. Phys Chem Chem Phys 13:16987–16998
84. Gonzaléz L, Escudero D, Serrano-Andrés (2012) Progress and challenges in the calculation of electronic excited states. Chem Phys Chem 13:28–51
85. Laurent AD, Jacquemin D (2013) TD-DFT benchmarks: a review. Int J Quant Chem. doi:10.1002/qua.24438
86. Clark KM, Yu Y, Marshall NM, Sieracki NA, Nilges MJ, Blackburn NJ., van de Donk WA., Lu Y (2010) Transforming a blue copper into a red copper protein: engineering cystein and homocystein into the axial position of azurin using site-directed mutagenesis and expressed protein ligation. J Am Chem Soc 132:10093–10101
87. Abtouche S, Very T, Monari A, Brahimi M, Assfeld X (2013) Insights on the interactions of polychlorobiphenyl with nucleic acid base. J Mol Model 19:581–588

Chapter 2
Structure, Thermodynamics and Energetics of Drug-DNA Interactions: Computer Modeling and Experiment

Maxim P. Evstigneev and Anna V. Shestopalova

Abstract In this chapter we demonstrate the large usefulness of using complex approach for understanding the mechanism of binding of biologically active compounds (antitumour antibiotics, mutagens etc.) with nucleic acids (NA). The applications of various biophysical methods and computer modeling to determination of structural (Infra-red and Raman vibrational spectroscopies, computer modeling by means of Monte-Carlo, molecular docking and molecular dynamics methods) and thermodynamic (UV-VIS spectrophotometry, microcalorimetry, molecular dynamics simulation) parameters of NA-ligand complexation with estimation of the role of water environment in this process, are discussed. The strategy of energy analysis of the NA-ligand binding reactions in solution is described, which is based on decomposition of experimentally measured net Gibbs free energy of binding in terms of separate energetic contributions from particular physical factors. The main outcome of such analysis is to answer the questions "What physical factors and to what extent stabilize/destabilize NA-ligand complexes?" and "What physical factors most strongly affect the bioreceptor binding affinity?"

M. P. Evstigneev (✉)
Sevastopol National Technical University (SevNTU), Universitetskaya Str., 33, Sevastopol 99053, Ukraine
e-mail: max_evstigneev@mail.ru

A. V. Shestopalova
A.Usikov Institute for Radiophysics and Electronics National Academy of Sciences of Ukraine (IRE NASU), Ak. Proskura str., 12, Kharkov 61085, Ukraine
e-mail: shestop@ire.kharkov.ua

L. Gorb et al. (eds.), *Application of Computational Techniques in Pharmacy and Medicine,* Challenges and Advances in Computational Chemistry and Physics 17,
DOI 10.1007/978-94-017-9257-8_2

2.1 Introduction

Rational design of new compounds for therapeutics requires knowledge about their structural stability and interactions with various cellular macromolecules—their molecular receptors or targets. In order to optimize the efficacy of drugs, as well as discover new ones, it is important to fully characterize the drug—bioreceptor (biopolymer) interaction [1].

Nucleic acids (NA) are common targets for antiviral, antibiotic and anticancer drugs that are used in cancer therapeutics [2] and also are viewed as a non-specific target for cytotoxic agents [3]. Many antitumour drugs are considered to exert their cytotoxic effect through DNA-specific interactions, resulting in genotoxic stress and consequent induction of programmed cell death (apoptosis) [4]. Presently, when patients can be provided with a full genome sequence as a part of their medical records, the field of drug design must be adapted and improved in order to meet this challenge [5]. Rational drug design thus requires detailed knowledge of both the structural consequences of ligation and the binding characteristics of the drug. Ideally, such information is required for DNA targets of genomic size and complexity [6].

In this regard it is important to know how small biologically active molecules—drugs or other ligands—will interact with nucleic acids [7]. One can use biophysical techniques to characterize the binding of the drugs with DNA and, based on experimental data, to expand further understanding of the binding process with an aid of molecular modelling or computer simulations. Such approach allows to get different physical parameters of the interaction in the system "molecular target (DNA)—drug" and to use them for the establishment of correlation between these parameters and drug activity *in vitro* or *in vivo* [8, 9].

2.2 Biophysical Methods for Studying DNA-Drug Complexation

One of extensively developing trends in molecular biophysics is prediction of pharmacological action of drugs at the molecular level that requires: (1) determination of the structural features of the complexes "target-drug" containing the biologically active ligands and exerting their maximal biological effectiveness; (2) determination of correlations of the physical parameters of interaction in the system "target-drug" and the biological activity of the drugs; (3) obtaining the most probable molecular models of the "target-drug" complexes based on various experimental physical methods and molecular modelling studies. As a practical outcome, one can formulate recommendations for the synthesis of new biologically active ligands with improved pharmacological properties based on information about the biomolecular target and the calculated physical parameters of ligand interaction with the target.

2.2.1 UV-VIS Spectroscopy

Absorption UV-VIS spectroscopy is one of the most widespread experimental methods used in molecular biology and biophysics for qualitative and quantitative studies of the interaction of biologically active ligands with DNA. This experimental method enables to identify the formation of complexes, to evaluate their complexation constants, to determine different types of ligand states in solution (e.g. free and bound with polymer matrix by various modes) on the basis of the shape and positions of the maxima in the corresponding spectra, to calculate the size of the binding site and the sequence specificity [10–17]. The changes in absorption spectra on addition of drugs to DNA solutions may be used for identification of different types of ligand complexation (e.g. intercalation or major/minor groove binding).

The method of spectrophotometric titration is usually used for detailed analysis of the binding modes and the structures of the molecular complexes [18]. Concentration dependencies obtained during the titration of the DNA—ligand complexes may be used for calculation of the binding parameters. The most important factor in correct determination of these parameters is the choice of the model of complexation, in which all physically possible binding modes (intercalation, binding with one of the DNA grooves, weak binding or electrostatic interactions of the ligand cation with negatively charged phosphate groups of the polynucleotide) should be taken into account. When a suitable model (or the most probable models) is selected, the equations, determining the relation between the equilibrium concentrations of the molecular components and the interaction parameters, may be used for quantitative description of the titration curve [19–21].

Thermodynamic parameters and the process of thermal denaturation of the drug—DNA complexes can be systematically studied by spectrophotometric method. Such approach yields a thermodynamic profile, i.e. standard free energy, enthalpy and entropy changes in ligand-DNA reaction of binding, using the van't Hoff plot based on determining the value of equilibrium binding constant at various temperatures. These thermodynamic parameters allow to further evaluate the enthalpic and entropic contributions to the free energy change in the DNA complexation process [22].

However, the most fruitful outcome can be achieved by a combination of the UV-VIS spectroscopy with other experimental methods.

2.2.2 Infrared and Raman Vibrational Spectroscopy

Among different physical methods of investigation of specific structural features and intermolecular interactions of nucleic acids with drugs and water, the Infrared (IR) and Raman vibrational spectroscopies occupy very important position. Both of these methods can effectively probe structural details of solution complexes between the drugs and DNA molecules of genomic size, and are used to determine the ligand binding mode, binding affinity, sequence selectivity, DNA secondary structure, and structural variations of the DNA—ligand complexes in aqueous solution [23–28].

IR and Raman spectroscopies are able to provide information on the formation of hydrogen bonds in solution. It has long been known that the formation of hydrogen bonds between the proton donor (OH, NH, NH_2, CH) and acceptor (C = O, C–O, C–N, C = N) groups is accompanied by a low-frequency shift, rising in intensity and increase in the half-width of the absorption bands of stretching vibrations in the IR spectrum [29, 30]. At the same time the absorption bands of deformation vibrations (e.g. NH_2- and OH-groups) experience high-frequency shifts [31]. These spectral features are considered as a direct evidence of H-bonds formation between the interacting molecules in solution [32]. Thus, the groups of atoms involved in stabilization of different types of DNA—ligand complexes can be identified by the vibrational spectroscopy [33–35]. In particular, analysis of Raman spectra allows to identify the atomic groups of the drugs forming hydrogen bonds with donor or acceptor atomic groups of DNA in all possible types of DNA-drug complexes [36, 37], and to determine, for example, the unwinding of double-stranded B-DNA induced by drug intercalation [38] or structural transition of DNA from B- to A-like conformation accompanying the DNA-ligand complexation [39].

2.2.3 Hydration

Since the formation of DNA—ligand complex occurs in water environment and brings one of the most significant contributions to stabilization of the DNA-ligand complexes, it is important to carry out analysis of the role, which water plays in the ligand binding processes [40–42]. From a practical point of view, understanding of hydration is valuable for rational design of novel DNA—binding drugs with predictable affinity and specificity to selected sequences of nucleic acid structures [43].

Investigations of the interaction between water molecules and DNA—ligand complexes in diluted solutions face serious difficulties due to the fact that the bound water is present only in insignificant amounts in solution. This fact implies the need to utilize highly sensitive physical methods for the study of water involvement in the complexation process. This is the reason why there is still a lack of reliable information on the distribution of water molecules in the hydration shells of various complexes, although the investigations of water surrounding of nucleic acid—ligand complexes have so far been carried out using numerous methods including X-ray crystallography [44], osmotic stress [43, 45], volumetry [46, 47] and molecular modeling methods [48–50]. In the IR spectroscopy the main difficulties are associated with the strong OH vibration of water molecules. This problem may be resolved by applying this method with respect to DNA—water systems prepared in wet films with changing water content [51–54]. Analyzing the changes in IR spectra which occur in several frequency modes (e.g. the stretching vibrations of OH- or OD-groups, the absorption band of bases—double and multiple bands, the absorption band of sugar-phosphate backbone) with an increase of the relative humidity, it is possible to observe the atomic groups, which represent hydration centers, to estimate the order and the degree of their filling with water molecules, to determine distinctive features of the formation of DNA secondary structure in the complexes with ligands and the structure of the hydration environment. This procedure enables to control the state of water and the state of individual structural groups of the biopolymer and the ligand as a function of film moistening. Such an approach also gives an opportunity to estimate thermodynamic parameters of hydration and to construct a model of hydration shell of the complexes [55].

In order to reveal the energy contribution of water to stabilization of nucleic acid structures and their complexes, it is necessary to know the thermodynamic parameters characterizing hydration of DNA, the ligands and the DNA-ligand complexes. Various physico-chemical methods may be used to solve this problem experimentally. In particular, a sufficiently sensitive piezomicrobalance or piezogravimetric method based on the use of quartz resonator, allows to obtain hydration isotherms or dependencies of sorption on relative humidity (in moles of water per mole of sorbent) [51, 52]. The isotherms measured for biopolymers or their complexes with ligands give insight into heterogeneity in the energies of interaction between the hydration sites and the sorbed water molecules.

2.2.4 Calorimetry

Structural studies are crucial for identifying the specific molecular interactions between the host DNA and the ligand, such that the overall three-dimensional shape of the complex and exact position or binding mode can be determined. But structural analysis alone can provide little knowledge on the nature of molecular forces that drive the complex formation in solution, and on the relative energetic contributions of specific molecular interactions. It is therefore essential to complement structural studies with detailed and rigorous thermodynamic analysis to fully characterize bimolecular complex formation. Differential scanning calorimetry (DSC) is one of the most convenient and informative methods for determining the energy parameters of interaction of the ligands with DNA. Direct measurement of heat effects caused by melting of DNA and its complexes enables to determine the full set of thermodynamic binding parameters and the energetic parameters of structural transitions: enthalpy, entropy and free energy changes, melting temperature and melting interval [56–61]. In order to quantify the energetic parameters of the interaction from DSC heat capacity curves, specific theoretical models must be used. The most well-elaborated approaches for the analysis of heat capacity curves have so far been developed only for protein interactions with ligands, because protein unfolding can often be described by simple two-state model [62, 63]. When the DNA-ligand system is being analysed, certain specificities of the complexation and melting of linear polymeric molecules should be taken into consideration. Recently a novel analytical approach for detailed analysis of the DNA-ligand interactions from DSC data was proposed [64]. The DNA macromolecule in this study is represented as an assembly of cooperative units, which melt according to the two-state model. Explicit account of ligand distribution on polymeric DNA and the temperature dependencies of melting and binding constants, as well as enthalpies, were considered. Such approach enables to extract the binding constant, stoichiometry, enthalpy, entropy, and heat capacity changes from multiple excess heat capacity profiles obtained at varying concentrations of the ligand (i.e. the two-dimensional DSC curves). Comparison of the binding parameters calculated by fitting of two-dimensional DSC curves with the literature data and with that obtained by alternative experimental techniques, had demonstrated that the approach presented in [64] gives satisfactory results.

2.2.5 Computer Modeling

Binding affinity of the ligands with DNA may be estimated at the molecular level based on shape complementarity of the interacting parts of the ligands and DNA, and by explicit consideration of physical interactions (electrostatic, van der Waals, hydrophobic, specific hydrogen bonding etc.). It allows to determine the extent to which the formation of the complex under investigation is energetically favourable. However, all microscopic details of the interaction cannot be identified in experiment. In such case computer simulations are commonly used as an appropriate complementary tool for modeling atomic-level interactions that produces the data about the structure of the most probable DNA—ligand complexes and on the contributions of different interactions to their stabilization with explicit account of water environment [65–71].

Molecular docking method is one of the most effective computer simulation methods, making possible a fast re-construction of all possible configurations of complexes between biological macromolecule and the ligand of interest. The molecular docking method is commonly used for estimation of specifity of protein–ligand interactions [72–74]. The docking of ligands to DNA molecules is a less frequently used approach. In this approach anticancer drugs are usually taken as the ligands [75]. In order to investigate their complexes by computer simulation, the initial coordinates must be known. If the structure of the complex under study is absent in structural databases, the investigator often faces a difficulty on how to create the binding site. The results of docking of the ligands with different DNA-targets indicate [76] that upon formation of the intercalation site it is usually enough to take into account only the most significant unwinding in one particular helical step or in the adjacent helical step of DNA double helix. The magnitude of the total unwinding of the DNA in the intercalation complex was found to be dependent on the sequence and length of the target DNA.

The application of Monte Carlo method for the study of hydration of nucleic acids, their components [77–80], and hydration of the DNA-ligand complexes (for example, dCpG with proflavine [81] and DNA with azinomycin B [82] intercalated complexes) was described in literature in detail. Monte Carlo simulations enable to evaluate the low energy conformations of various complexes of DNA fragments, including their complexes with ligands, and to determine the hydration properties of the complexes being formed [54, 55, 83, 84].

Another very valuable tool in arsenal of theoretical investigation of biological molecules is the method of molecular dynamics simulations. This computational method describes the time dependent behaviour of the given molecular system. To date an extensive use of molecular dynamics simulations has resulted in generation of a wealth of detailed information on the fluctuations and conformational changes of proteins and nucleic acids. Such methods are now routinely used to investigate the structure, dynamics and thermodynamics of biological molecules and their complexes [48–50, 65–71, 85].

2.3 Results of Experimental Investigation and Computer Simulation of DNA Complexation with New Synthetic Analogues of Anticancer Antibiotics

2.3.1 General Description of New Synthetic Analogues of Anticancer Antibiotics

As outlined above, in order to provide a scientific basis for rational design of DNA-targeted drugs, it is necessary to understand how the molecules form complexes with DNA. Another important factor is the ability to quantify such complexation in order to make meaningful comparisons of the behaviour of different drugs. This is the focus of the biophysical studies reviewed below.

Here we present the results of investigations of the physical mechanisms of the interaction with DNA of a new series of biologically-active ligands, analogues of anticancer antibiotic Actinomycin D (AMD), obtained using complex approach involving various experimental biophysical methods and molecular computer modeling.

AMD, the synthetic phenoxazone antibiotic, consists of a phenoxazone chromophore substituted with two equivalent cyclic pentapeptide lactone rings. AMD is a DNA-binding drug. Its biological activity is thought to be due to preferential intercalation of the planar phenoxazone chromophore into GC sequence of DNA with the two cyclic pentapeptide rings lying in the minor groove [86].

AMD is an anticancer drug used in treatment of tumours, but its use suffers from induction of negative side effects [87]. With a general aim to reduce the side toxicity of AMD, a new set of drugs with phenoxazone chromophore and dimethylaminoalkyl side chains (actinocin derivatives with side chains of different lengths, ActII—ActV, Fig. 2.1) have been synthesized [8].

The cytotoxic effects of the synthetic actinocin derivatives were investigated by examination of the drug-induced apoptosis and cell cycle perturbations in a human leukemia MOLT-3 cell line [88].

Examination of cytotoxic effects in leukemia cells showed that the variation in length of dimethylaminoalkyl side chains of actinocin derivatives leads to signifi-

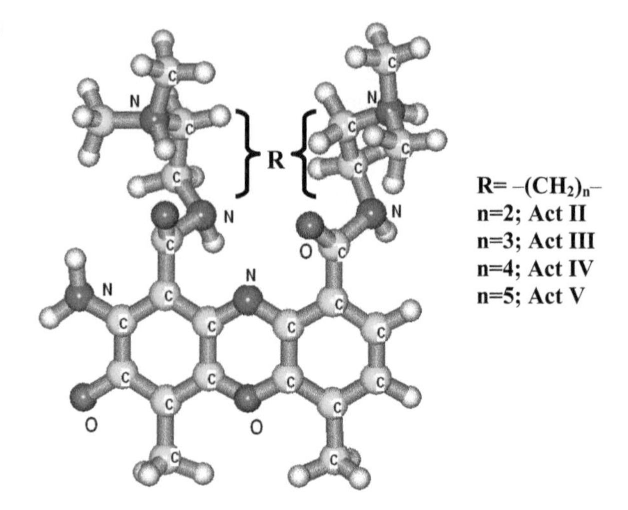

Fig. 2.1 Chemical structures of the actinocin derivatives ActII—ActV

cant variation in cytotoxic activity as a function of the number of CH_2 groups in their side chains, with pronounced maximum in cytotoxic activity for the ligand with two CH_2 groups, i.e. ActII. Hence, the antitumour activity in the series of ActII—ActV ligands was found to be very sensitive to minor modifications in the side chains of the AMD derivatives, indicating a direct correlation between structure and activity of the drugs [8].

2.3.2 Free Ligand: Investigation by Experimental and Computer Simulation Methods

The main goal of the biophysical part of these studies is to understand the nature of specificity interaction between the drugs under investigation and nucleic acids, taking into account the interaction of individual components with water molecules. The following experimental physical methods were used to solve this problem: UV-visible spectrophotometry for the study of different modes of ligand binding with DNA and the corresponding binding parameters, infrared spectroscopy and piezogravimetry, giving information on the influence of water on the formation of DNA-drug complexes, and differential scanning calorimetry for obtaining direct data on the thermostability of such complexes. In order to determine the most probable molecular models of the DNA-ligand interactions, the methods of computational analysis (molecular docking, Monte Carlo simulations and molecular dynamics) were used. It is assumed that the results obtained by these methods may be useful for directed synthesis of new drugs with improved medico-biological properties.

The first step of the study was the investigation of the solution behaviour of the synthetic drug molecules alone prior to their complexation with DNA. Investigations of the self- and hetero-association of biologically active compounds with heterocyclic planar rings (aromatic ligands) in water are interesting from the physico-chemical point of view, resulting in determination of the influence of the structures of the chromophores and side chains on association ability, and estimation of contributions of different interactions to the formation of stable aggregates. Another important issue is the pharmacological aspect, because the self- and hetero-complexations, as well as competitive binding of the drugs with bioreceptor, may influence their activity.

Distinctive features of the self-association of the actinocin derivatives were determined experimentally (by UV-VIS spectrophotometry, piezogravimetry and IR-spectroscopy) and using computer simulation (by Monte Carlo method and molecular dynamics modeling).

Analysis of both spectral and thermodynamic parameters obtained from UV-VIS spectrophotometric data for the set of synthetic actinocin derivatives enabled us to conclude that the drugs experience strong tendency to aggregate in solution and the aggregation is appreciably higher in solutions of high ionic strength. The dimerization parameters depend slightly on the number of methylene groups in the side chains of phenoxazone antibiotics. Dimerization of the investigated ligands in aqueous solution leads to significant changes in the spectral characteristics of the antibiotics ActII-ActV, which needs to be taken into account in any studies of drug complexation with DNA [89].

Formation of DNA complexes with actinocin derivatives is accompanied by hydration changes for both the DNA molecule and the intercalated ligands. In order to evaluate the energy contribution of water molecules to stabilization of these complexes, it is necessary to obtain experimental data on the energies of interaction between water molecules and the free ligands. An investigation of the adsorption of water in the films of actinocin derivatives was performed using quartz crystal microbalance (piezogravimetry). In order to identify the hydration-active centers, the IR absorption spectra of wet and dry films of the actinocin derivatives were recorded in the spectral range 900–1700 cm^{-1}, in which the DNA molecules can be characterized by the nitrogen base absorption region (1500–1700 cm^{-1}) and by the region of sugar phosphate absorption (900–1300 cm^{-1}). The main conclusion of this stage of investigation was that in contrast to the DNA molecule, the investigated actinocin derivatives demonstrate very weak absorption in the spectral region 950–1250 cm^{-1} and, thereby, analysis of the IR spectra of the DNA-drug complexes can be carried out without taking into account the drug absorption in this IR region [55].

Computer simulation of the hydrated environment of actinocin derivatives in aqueous clusters by Monte Carlo method allows to determine the most energetically favourable "ligand-water" configurations, the number and the positions of water molecules forming hydrogen bonds with actinocin derivatives or their hydrated active sites. Comparative analysis of the simulation data and the results of IR-spectroscopic and piezogravimetric studies of the actinocin derivatives' hydration had demonstrated their complementarity and general agreement.

With an aim to investigate the molecular mechanisms of actinocin derivatives complexation in water solution, the molecular dynamics simulation of both monomer and dimer forms of the ligands was carried out [90]. The hydration properties

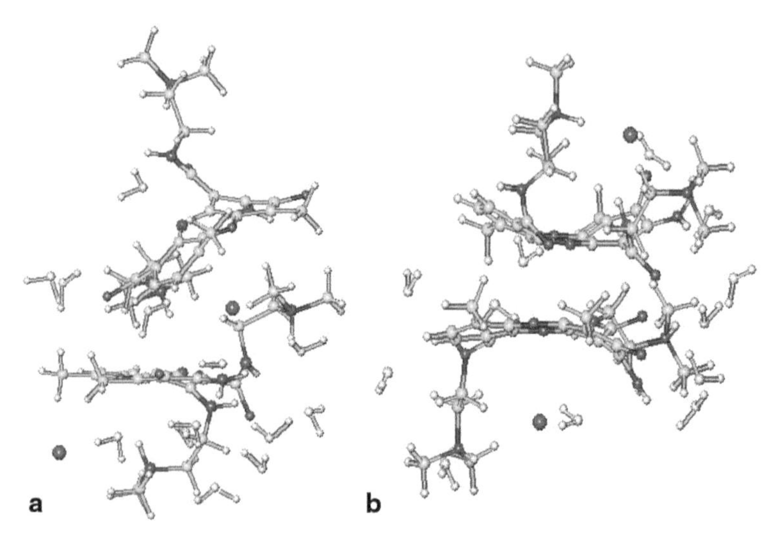

Fig. 2.2 Two stable forms of ActII dimers with the nearest water molecules (*white*) and Na^+ ions (*violet balls*). **a** Stable form I: phenoxazone chromophores are titled. **b** Stable form II: phenoxazone chromophores are parallel

of the monomer and the aggregated forms of the actinocin derivatives were determined (Fig. 2.2). The calculated values of interaction energies of the monomers in dimers show that the aggregation of these compounds in aqueous solutions is an energetically favourable process. The aggregates were stabilized by the van der Waals, electrostatic and hydrophobic interactions, and also due to formation of intermolecular hydrogen bonds [91].

In summary, the results of the first step of the investigation gave insight into the nature of the state of actinocin derivatives in aqueous solution and their interaction with solvent molecules. These results are needed for building molecular models of binding of these ligands with DNA. In particular, positions of the hydration centers indicate the sites of possible interactions of the ligands with atomic groups of DNA double helix. The information on the dimerization of the actinocin derivatives in water solution is necessary for estimation of the amount of drug molecules available for binding with DNA.

2.3.3 Investigation of the Ligand-DNA Complexation

The second step in the investigation of the activity of new synthetic anticancer antibiotics (actinocin derivatives) at the molecular level is a detailed study of their complexation with DNA.

UV-VIS spectrophotometry was used to investigate the parameters of DNA-actinocin complexation. It was shown that two types of complexes are being formed in DNA-drug solutions, viz. binding of the drug with DNA phosphate groups and intercalation or/and groove binding of the drug with DNA. The binding constant, the sizes of the drug binding sites with DNA and the cooperativity parameters for both types of complexes were calculated [36, 37, 89].

The thermodynamic parameters of the DNA-drug complexes were obtained by using UV-VIS optical melting [18] and DSC methods [89]. From the UV melting curves of DNA alone and its mixture with ligands (ActII-ActV) the melting (melting temperature, melting interval) and the binding parameters (changes of binding free energy, binding enthalpy and entropy) for all the samples studied were determined. From these data it was found that the largest value of the binding free energy change is associated with DNA-ActII complex. This result indicates that within the set of the actinocin derivatives having different number of methylene groups in dimethylaminoalkyl side chains studied in this work, specifically ActII containing two CH_2 groups features the strongest interaction with DNA.

Quantitative estimation of the binding parameters accomplished using the DSC data showed that the stability of the DNA-ligand complexes is higher than that in the case of free DNA. On decrease in the number of methylene groups in the ligands' side chains both the binding enthalpy and the free energy changes increase non-linearly reaching the maximal value for the number of CH_2 groups equal to 2 (i.e. DNA-ActII). Hence, there is a satisfactory agreement between the values of thermodynamic parameters and their dependence on the number of CH_2 groups in the ligands' side chains, obtained from DSC and UV-VIS optical melting. The magnitude of the binding enthalpy can be explained by the intercalative type of interaction, which may additionally be stabilized by hydrogen bonds and water bridges. The melting entropy of the complexes is higher by absolute value than that of free DNA. It is due to more ordered structure of the hydration environment around the complexes in comparison with free DNA.

Some peculiarities of the heat absorption curves caused by melting of DNAs having different nucleotide compositions were observed for the solutions of free DNA and its complexes with actinocin derivatives ActII-ActV. Notably, for the DNA-ActII complex the heat absorption curve is significantly distorted in high-temperature area when the GC-rich blocks of DNA are melted.

The role of water in the DNA-drug complexation was investigated by piezogravimetry and IR spectroscopy. Hydration isotherms and IR-spectra of the free DNA and the DNA-drug complexes were obtained in films. Analysis of the spectra was carried out using reliably-assigned DNA absorption bands sensitive to hydration and conformational states of nucleic acids, as well as by the absorption bands of the drugs. Investigation of the properties of water absorption to DNA and DNA-drug complexes had led to the conclusion that the energy of interaction between the water molecules and the complexes depends on the length of side chains of synthetic phenoxazone drugs ActII-ActV. The hydrated environment makes significant contribution to stabilization of double-helical structure of either free DNA and of its complexes with the drugs. Increase of relative humidity of the films in the range of 0 to 90 % leads to increase in the intensity of the IR-absorption bands for the sugar-phosphate backbone vibration, in- and out-ring groups of DNA base pairs in the drug-DNA complexes, and also of the absorption bands of C = O, C = N and $NH_2(ND_2)$ groups of the drugs. These changes in spectral parameters confirm simultaneous hydration of the drugs and DNA, which most likely originates from the water molecules acting as bridges between the drugs and DNA. In the case of actinocin derivative ActII there is also an interaction between cationic groups of the drug and DNA sugar-phosphate backbone resulting in additional stabilization of the DNA-ActII complex.

In order to determine the most probable molecular models of the hydrated actinocin drug-DNA complexes, computer simulations of the interaction of the drugs and DNA fragments (referred to as "the target") were carried out.

First, molecular docking methods were applied to the systems containing nucleic acids fragments as the targets and actinocin derivatives with different lengths of the dimethylaminoalkyl side chains as the ligands. It was found that the actinocin derivatives could form energetically favourable complexes with DNA both as intercalators and minor groove binders. The complexes of actinocin derivatives and DNA fragments were stabilized by hydrogen bonding on either, intercalation and minor groove binding. It was found that the change in solvent-accessible surface area on binding of the actinocin derivatives with DNA linearly increases with the number of CH_2 groups in the ligands' side chains. The solvation energy change on binding, calculated by the weighted solvent-accessible surface area method, was reported to be unfavourable for positively charged ligands [92].

Second, the Monte Carlo method was employed taking the solvent (water molecules) into account [89]. The following assumptions were introduced as a result of the experimental investigations of actinocin-DNA complexation reviewed above and of the molecular docking simulations: (1) the possibility of intercalation of the planar phenoxazone chromophore of actinocin drug into GC-sites of DNA, and (2) the possibility of binding of the actinocin derivatives in the minor groove of the double helix. Hence, the starting configurations of the complexes were built in agreement with these assumptions. The energy parameters of the molecular systems containing free DNA fragments and their complexes with various intercalated actinocin derivatives, and for the actinocin derivatives bound in the minor groove of DNA fragments, were obtained. Using these data, some conclusions concerning the stability of the molecular complexes could be made by comparison of the values of the average total potential energies of the systems studied and the drug-target interaction energy. For the series of actinocin drugs it was found that the highest by absolute value target-drug interaction energy was associated specifically with the ActII-target system for the both types of complexes.

Analysis of the instantaneous configurations of all the complexes had enabled us to describe the obtained structures in more detail. A remarkable feature of the structure of the free ActII in solution is the presence of intramolecular hydrogen bond between C = O and N-H groups of one of the dimethylaminoalkyl chains [90], which may hinder the conformational fitting of the side chain of the drug when interacting with DNA. This intramolecular hydrogen bond is preserved on complexation of ActII with the DNA fragment.

The complex of ActII with the DNA fragment is additionally stabilized by formation of two hydrogen bonds between O4′ atoms of the deoxyribose rings of both chains and NH_2-group of the drug chromophore and NH-group of one of the side

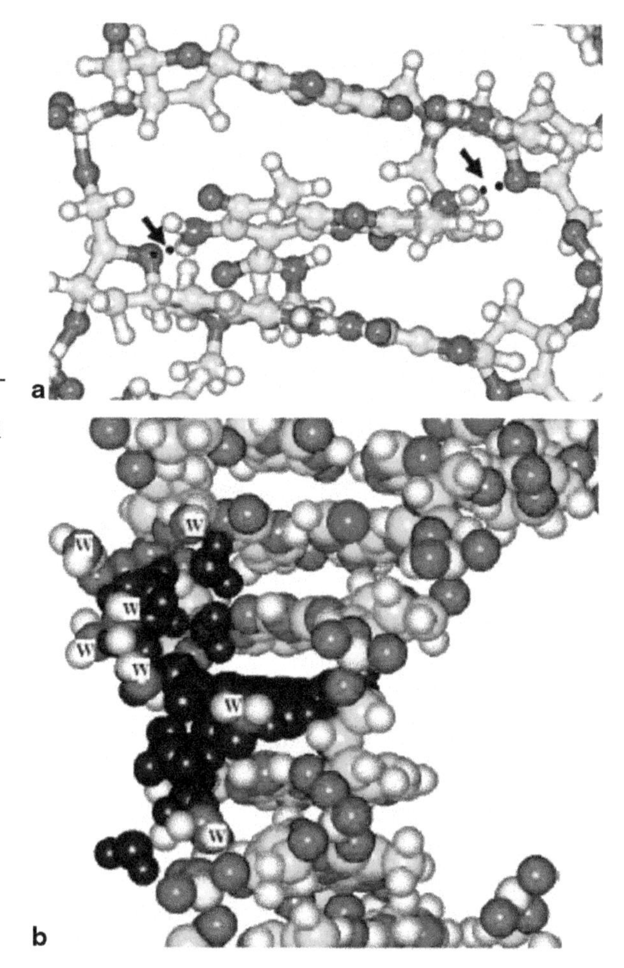

Fig. 2.3 The structure of DNA [d(GAAGCTTC)$_2$]—drug (ActII) intercalated complex (Monte Carlo computer simulation data): **a** hydrogen bonds (*black points* and *arrows*) in intercalation GC-site of the DNA fragment between the sugar-phosphate backbone atoms (O4′ of deoxyribose) and NH_2- and NH-group of the ligand, **b** water molecule (*W*) occupying bridging positions between the donor and acceptor groups of the ligand (*black balls*) and the DNA fragments in the intercalation site

chains of ActII (Fig. 2.3a). In addition, seven water molecules occupying bridging positions between the hydration centres of the drug and the DNA fragment were found to give additional stabilization to this complex (Fig. 2.3b).

Hydrogen bonds between the drugs and the sugar-phosphate backbone of the DNA fragment were not observed for the intercalated complexes of DNA fragments with ActIII—ActV drug molecules, however, two water molecules occupying bridging positions between the hydration-active centres of the drugs and the DNA fragment were found for each of these complexes.

In the ActIII—ActV complexes the hydration of both the target and the drugs is nearly the same as compared to the case when they are in the free state, thus confirming the assumption that the drug molecules are partially intercalated in the DNA duplex. It follows that the overlap of the planar phenoxazone chromophore with the planes of DNA base pairs is decreased in such complexes, resulting in a decrease of the absolute value of the interaction energy of the drugs and the target. These

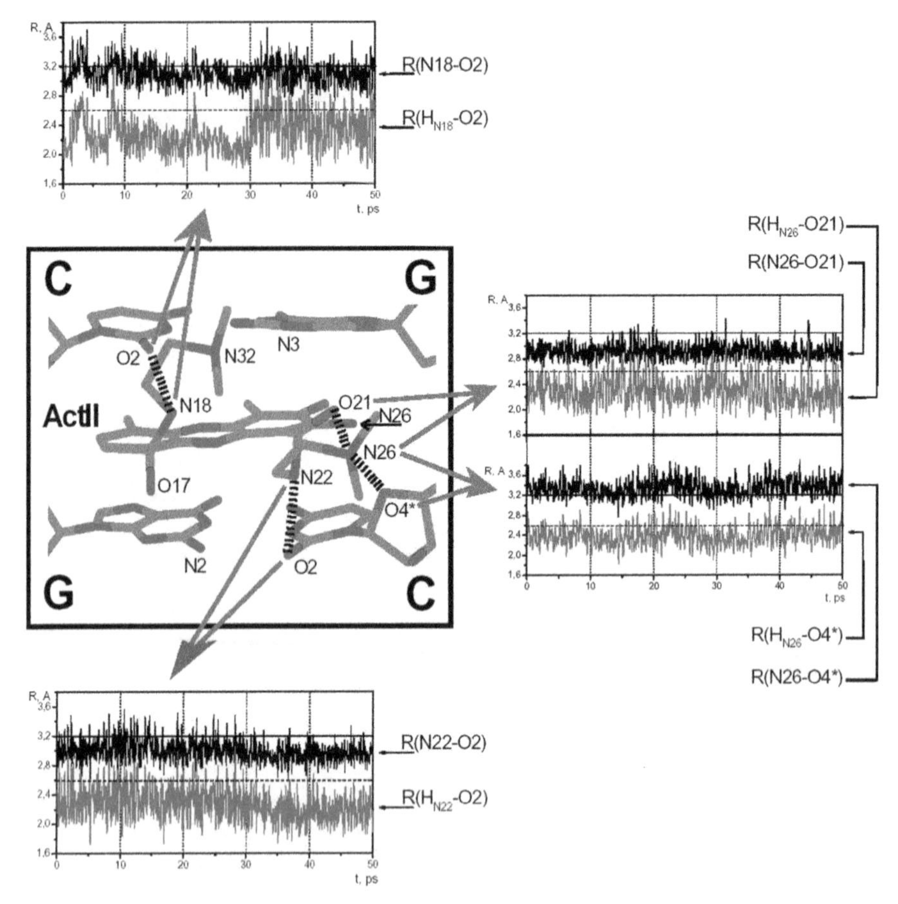

Fig. 2.4 Structure and dynamics of GC–site of DNA [5′-d(AGCT)$_2$]—drug (ActII) intercalated complex. Hydrogen bonds are shown as dotted line. Data were obtained by molecular dynamics simulation

results correlate with the values of melting temperatures obtained for the drug-DNA complexes [89].

In order to obtain more detailed information on the DNA-ActII complexation, molecular dynamics simulation was carried out. Analysis of the molecular dynamics trajectories allows one to describe in detail the structures of the investigated complexes (Fig. 2.4).

In the intercalated d(AGCT)$_2$-ActII complex the planar phenoxazone chromophore is inserted into the GC-site and the dimethylaminoalkyl side chains are located along the sugar-phosphate backbone. The complex with ActII is additionally stabilized by formation of few hydrogen bonds. Two of them formed between N-H group of ActII and C = O group of cytosine from the first strand of the DNA fragment, and N2-H group of ActII and C = O group of cytosine from the second strand of the DNA fragment, were stable during the equilibrium phase of the trajectory (Fig. 2.4). One more hydrogen bond found was of a bifurcational type. This hydrogen bond connects N-H group of ActII with C = O group of ActII (intramolecular hydrogen bond) or with the atom O4′ of the sugar-phosphate backbone of the DNA fragment (intermolecular hydrogen bond). The formation of these hydrogen bonds can explains the specificity of ActII interaction with GC-site of the DNA fragment.

Of special interest are the results of analysis of re-construction of the hydration environment in the process of complexes formation. The complexation of the ligands with DNA is accompanied by partial or full dehydration of the ligand molecules and re-construction of DNA hydration shells. In the isolated state ActII forms hydrogen bonds with 7 water molecules [90]. Ligand molecule was found to be partially dehydrated in the intercalated state in GC-site. ActII preserves hydrogen bonds with 3 water molecules. Two of these molecules occupy bridging positions between HN and C = O groups of ActII and N3 atom of guanine, and C = O group of ActII and NH group of guanine of the opposite strand.

2.3.4 Summary of the Results

It was found that at least two types of complexes may be formed between DNA and actinocin derivatives, viz. the intercalation of planar phenoxazone chromophore of the drugs into GC-sites of DNA double helix, and the binding of the drugs with the minor groove of DNA duplex. The complex in minor groove is energetically less favourable than the intercalated one. A preference is observed for both the intercalated and groove-bound types of complexes for the complexation of ActII with the DNA target. Additional stabilization of the intercalated complex may be due to formation of hydrogen bonds between the NH-group of dimethylaminoalkyl side chains of ActII and the sugar-phosphate backbone of DNA, as well as formation of specific water structure around this complex. It is likely that water molecules occupy bridging positions between the hydration-active centers of the drugs and DNA providing additional stabilization to the intercalated type of complex.

In conclusion, it is worth noting that the biophysical investigation sketched out above, including the set of experimental and computer simulations methods, has enabled us to shed new light on the molecular mechanism of biological action of a new series of biologically active ligands—analogues of anticancer antibiotic Actinomycin D. These data are in general agreement with the results of examination of cytotoxic effects of the same set of drugs on cellular level [8].

2.4 Energetics of Drug Complexation with Nucleic Acids

2.4.1 The Problem Behind the Thermodynamic Analysis

Investigation of the structure and thermodynamics of drug binding with nucleic acids performed above, demonstrates the power of using various biophysical methods in understanding the mechanism of binding, as well as provides quantitative information on thermodynamics of binding in terms of Gibbs free energy (ΔG), enthalpy (ΔH), entropy (ΔS) and heat capacity (ΔC_p) changes. In general it is considered that such approach provides a scientific basis for rational drug design, but what might be the link to designing of new drugs? In very first approximation the thermodynamical parameters of binding may be correlated with biological activity of the drug (for reviews see [93, 94]). The typical examples are the semisynthetic antibiotic Novantrone (an anthracycline derivative), which is widely used in the treatment of leukemia [95], and bis-doxorubicin (a doxorubicin derivative), which exhibits activity against multidrug-resistant tumour cells [96]. It follows that a manipulation by the parameters of drug-NA binding by means of directed chemical synthesis of the drug molecules may potentially lead to creation of new drugs. The problem behind this is that experimentally-measured ΔG, ΔH and ΔS are made up of the sum of contributions from various types of physical interactions (see Ref. [97] and references therein), viz. van der Waals, electrostatic, hydrophobic *etc*:

$$\Delta G(\text{or } \Delta H, \Delta S) = \sum_i \Delta G_i(\text{or } \Delta H_i, \Delta S_i) \tag{2.1}$$

where ΔG_i (or ΔH_i, ΔS_i) stands for the contribution of the i-th physical factor to ΔG (or ΔH, ΔS).

Any modification in the structure of a ligand in general case will likely lead to unpredictable change in magnitudes of the energy components in Eq. (2.1) and the effect of their summation in Eq. (2.1) may change the magnitudes of $\Delta G/\Delta H/\Delta S$ or even leave them unchanged. It follows that direct comparison of experimentally-measured thermodynamic parameters for different ligands is unlikely to be very meaningful and may even lead to erroneous conclusions. A common manifestation of that problem is encountered in the enthalpy-entropy compensation for binding processes in aqueous media [98, 99]. Additionally, the long-existing discussion exists in the literature on what forces (van der Waals, electrostatic or hydrophobic) or types of interactions (solute-solvent or solute-solute) dominate the stacking interactions in solution [100, 101], which makes a thermodynamic analysis intrinsically ambiguous. Nevertheless, greater understanding of the thermodynamics of drug-NA binding processes can be achieved if the problem of energy partitioning (also known as energy parsing or energy decomposition) is solved [97, 102]. This needs an independent calculation of the energy components in Eq. (2.1) and comparison of the results to the experimentally-measured total Gibbs free energy. Knowledge of these contributions is crucial in managing the properties of ligand binding with NA by manipulating the distribution of energy over various physical factors governing the reaction of complexation. However, there is a fundamental problem behind any attempt to parse experimentally-measured thermodynamic quantities ΔG, ΔH, ΔS, viz. it is generally not possible to measure independently the contribution of specific energy term to the total binding energies. Nevertheless, as we shall demonstrate below, partial overcome of this problem may be achieved using the methods of computational chemistry.

In this section we shall briefly review the solution of the energy decomposition problem for various types of NA binding drugs, viz. DNA aromatic intercalators, DNA minor groove binders (or MGB-ligands), RNA aptamer binders.

2.4.2 General Computational Approach to Study the Energetics of Binding

When studying the ligand-NA binding processes in aqueous media, two important factors must be taken into consideration:

1. binding of the ligand must be accompanied by formation of the binding site on NA, which is commonly referred to as DNA unwinding (*i.e.* transition of the DNA helix from a regular B-form into an unwound DNA) for the intercalation process [103], and DNA/RNA adaptation—for the DNA minor groove and RNA aptamer binding processes [104, 105]. Hence, the total energy of binding, ΔG_{total}, should be decomposed into two parts: the energy of NA conformational change, ΔG_{conf}, and the energy of ligand insertion, ΔG_{ins}

$$\Delta G_{total} = \Delta G_{conf} + \Delta G_{ins}. \tag{2.2}$$

2. the NA-binding process occurs in solution, which means that the total Gibbs energy should be partitioned into inter- or intra-molecular interactions of NA and ligand in vacuum, ΔG^{im}, and their interaction with solvent, ΔG^{solv} [106]:

$$\Delta G_{total} = \Delta G^{im} + \Delta G^{solv}. \tag{2.3}$$

The dissection of the total energy on solvation/intermolecular (Eq. (2.3)) and on conformation/insertion (Eq. (2.2)) terms can be incorporated into a thermodynamic cycle (Fig. 2.5).

The thermodynamic cycle suggests that at least two different ways for energy decomposition may exist [107,108]:

1. decomposition in terms of physical interactions—Eq. (2.4)

$$\Delta G_{total} = \Delta G_{conf} + \Delta G_{vdW} + \Delta G_{el} + \Delta G_{pe} + \Delta G_{hyd} + \Delta G_{HB} + \Delta G_{entr}, \tag{2.4}$$

2. further decomposition of the *"vdW"*, *"el"* and *"HB"* components in Eq. (2.4) in terms of the types of interaction (intermolecular interactions in vacuum and with solvent)—Eq. (2.5)

$$\begin{aligned} \Delta G_{total} &= \Delta G_{conf} + \Delta G^{im}_{vdW} + \Delta G^{solv}_{vdW} + \Delta G^{im}_{el} + \Delta G^{solv}_{el} + \Delta G_{pe} +, \\ &+ \Delta G_{hyd} + \Delta G^{im}_{HB} + \Delta G^{solv}_{HB} + \Delta G_{entr} \end{aligned} \tag{2.5}$$

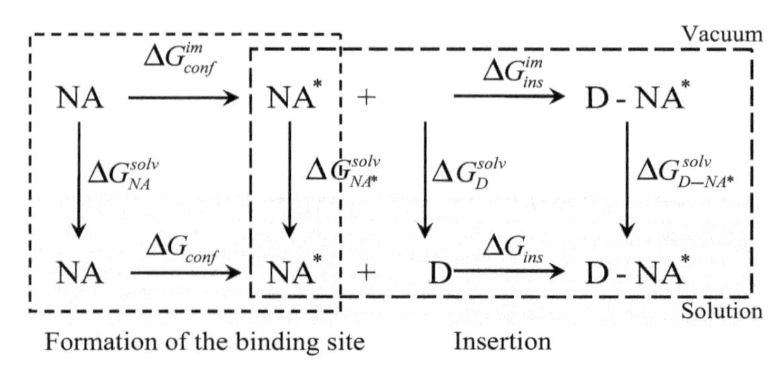

Fig. 2.5 Thermodynamic cycle for the ligand (D)—nucleic acid (NA) binding process

where ΔG_{conf} is the energetic contribution due to conformational changes of the molecules during the complexation process, ΔG_{vdW} and ΔG_{el} are the contributions from van der Waals (VDW) and electrostatic (EL) interactions, respectively, ΔG_{pe} is the polyelectrolyte contribution, ΔG_{hyd} is the hydrophobic (HYD) contribution, ΔG_{HB} is the contribution from hydrogen bonds (HB).

The entropic term, ΔG_{entr}, originates from the loss of translational (ΔG_t), rotational (ΔG_r) degrees of freedom, change in the mode of vibrations of chemical bonds (the high frequency term or type I vibrations, ΔG_v^I) and appearance of new mechanical oscillations of the ligand in the binding site (the low frequency term or type II vibrations ΔG_v^{II}), i.e.

$$\Delta G_{entr} = \Delta G_t + \Delta G_r + \Delta G_v^I + \Delta G_v^{II} \tag{2.6}$$

Recently, it has been shown that the ΔG_{entr} term also contains hidden systematic contribution from the entropy dependence on the number of bound ligands [109] and the change in rigidity of NA on sequential ligand binding [110]. However, both factors have been reported to give negligible contribution to ΔG_{entr} for the case of small ligands having much smaller dimensions than the DNA receptor, and may be excluded from the analysis of energetics.

Briefly, the computation of each of the terms in Eq. (2.4), Eq. (2.5) was performed according to the following protocols. The calculation of the VDW interactions was performed by averaging the VDW part of the interaction energy during the course of MD [107, 108]. The energies of electrostatic interactions were calculated by means of solution of non-linear Poisson-Boltzmann equation [111], and the overall approach used was shown to depend relatively weakly on the underlying method of atomic charges computation [112]. The energy of hydrophobic interactions was computed from the change in solvent accessible surface area, which had been proved to give more consistent results as compared to the alternative approaches [107, 113]. Calculation of the vibrations of chemical bonds was performed by normal mode analysis [107, 108, 114]. Calculation of mechanical vibrations of the molecules in complex was performed by means of estimation of the rigidity factor against small translational shifts [107, 108, 114].

A special note should be given to the method of explicit account of the energy of H-bonds, ΔG_{HB}. The energy contribution from hydrogen bonds to water molecules on complexation was estimated from the change in hydration index of the system (N^{im} or ΔN^{solv} representing the average number of intermolecular H-bonds and the change in the number of H-bonds to water molecules on complexation, respectively) and further calculation of ΔG_{HB} by means of formula [107, 115]

$$\Delta G_{HB} = -0.25 \cdot 9 \cdot (N^{im} + \Delta N^{solv}), \text{ kcal/mol} \tag{2.7}$$

Within the framework of such approach, it is considered that a part of the H-bond energy is already accounted in the ΔG_{vdW} and ΔG_{el} terms, hence, the ΔG_{HB} quantity bears meaning of an additional amount to the sum of VDW and electrostatic energies in order to account correctly for the total contribution due to H-bonding.

Equations (2.4), Eq. (2.5) provide the background of the methodology reviewed in this chapter. In order to make the calculated from Eq. (2.4), Eq. (2.5) energy terms meaningful the protocol for their computation must satisfy the following conditions [116]:

1. summation of the independently calculated energy terms reproduces the experimentally measured total energy of interaction within reasonable error limits. In that case the magnitudes of the calculated energies for various physical factors are meaningful and so these energies may be used in comparative analysis. The calculations must use all available experimental information on binding obtained from various biophysical methods, described above;
2. the calculations should be applied to a set of molecular systems that differ in structure and charge state. If the protocol only demonstrates satisfactory coincidence with experiment for a single system (as is often the case), the transferability to other systems will always be questionable, hence, there is no guarantee that the calculated energies are generally meaningful;
3. the calculations should be made using a similar protocol and set of parameters/restraints for each system studied. Otherwise, it appears that there may be an artificial adjustment to the results, making the calculated energies less reliable.

If the computations match these conditions, then deeper analysis of each particular energy term in (4), (5) provides an answer to the basic questions "What forces stabilize/destabilize the ligand-NA complexes in solution and what are their relative importance?" The consequence of this analysis would be an answer to a follow-up question "What physical factor exerts the highest correlation with experimental binding energy?" When answered, it may give an idea of which factor should be targeted in first instance when optimizing drug affinity to NA in rational drug design.

Below we shall review the solution of the energy decomposition problem taking as an example classical DNA intercalating reactions, and then discuss the main outcomes of solving the same task with respect to MGB-ligands and RNA binders.

2.4.3 Energy Analysis of Ligand-DNA Intercalation Reactions

To date the successful energy decomposition has been accomplished for wide variety of aromatic DNA intercalators [107, 117, 118]. We shall briefly review these results taking as an example four typical aromatic drugs [107], viz. antibiotics, daunomycin (DAU), mitoxantrone (NOV) and mutagens, ethidium bromide (EB), proflavine (PF). The conclusions to be drawn for these ligands also remain essentially the same for other aromatic ligands investigated in the cited literature.

2.4.3.1 Structure of the DNA Receptor

Although the specificity of the aromatic molecules (EB, NOV, PF, DAU) to particular DNA sequences is not great, analysis of the literature suggests that the intercalator molecules not containing heavily-branched side chains commonly have some specificity towards 5′-CG and 5′-GC sites [119, 120]. Also, taking into account that DAU exerts greater specificity to CGA triplet sites on DNA rather than to CG or GC dinucleotide sequences [121], it is reasonable to take the self-complementary fragment, $d(TCGA)_2$, flanked at both ends by CG pairs, as the minimal site for ligand binding. Hence, the 10-mer oligonucleotide duplex, $d(CGCTCGAGCG)_2$, was used as the model DNA receptor. It was shown that such length of DNA is long enough for correct reproducing of electrostatic interaction for the group of aromatic intercalators [122].

2.4.3.2 Van der Waals Energy, ΔG_{vdW}

The MD averaged van der Waals energies in the selected ligand-DNA complexes are presented in Table 2.1.

The intramolecular energies of DNA base pairs interaction, ΔG^{im}_{conf}, at the stage of unwinding are all positive, which is a result of separation of base pairs upon formation of the intercalation cavity. The energies of the solvation of the intercalation site, ΔG^{solv}_{conf}, are all negative and result from hydration of the intercalation cavity upon DNA unwinding.

At the stage of ligand insertion the intermolecular energy of ligand-DNA interaction, ΔG^{im}_{ins}, has a negative sign, which results from the attractive nature of VDW forces acting between the ligand and DNA base pairs within the intercalation site. The positive VDW energy of the interaction with solvent, ΔG^{solv}_{ins}, is due to dehydration of the ligand after its insertion into DNA interior.

The total VDW energy of insertion, ΔG_{ins}, is a relatively small value and is a result of mutual compensation from favourable intermolecular interaction between DNA and the ligand, and unfavourable interaction with the solvent. The compensation may lead to positive (PF) and negative (DAU, EB, NOV) ΔG^{ins}_{vdw} (see Table 2.1) which means that VDW interactions at the stage of insertion may either favour or

Table 2.1 Inter(intra)molecular in vacuum and with solvent van der Waals energies (kcal/mol) for ligand binding with DNA

Ligand	Unwinding			Insertion			Intercalation		
	ΔG^{im}_{conf}	ΔG^{solv}_{conf}	ΔG_{conf}	ΔG^{im}_{ins}	ΔG^{solv}_{ins}	ΔG_{ins}	ΔG^{im}_{vdW}	ΔG^{solv}_{vdW}	ΔG_{vdw}
DAU	22.3	−7.7	14.6	−84.2	67.1	−16.5	−61.9	59.4	−2.5
EB	26.9	−15.5	11.4	−52.1	41.1	−11.0	-25.2	25.6	0.4
NOV	21.9	−7.7	14.2	−57.9	41.9	−16.0	−36.0	34.2	−1.8
PF	30.2	−30.8	−0.6	−41.7	43.1	1.4	−11.5	12.3	0.8

disfavour complex formation as it depends on the interplay between the intermolecular interactions and the interactions with solvent.

The total VDW energy of binding, ΔG_{vdw}, is the sum of two large numbers, $\Delta G^{im}_{vdW} + \Delta G^{solv}_{vdW}$, of opposite sign, which results in a small net energy effect and leads to the conclusion that VDW interactions do not play a significant role in ligand-DNA binding. This is correct in terms of overall binding but not in terms of stabilization of the complexes. The portion of the total VDW energy of ligand binding, ΔG_{ins}, discussed above, which really contributes to stabilization of the complex, may have the values (see Table 2.1) higher by modulus than the experimentally-measured energies of binding, ΔG_{exp}. It means that it is necessary to take into account the contributions of VDW interactions at different stages of binding (unwinding and insertion) and for different types of interaction (in vacuum and with solvent) for energy decomposition in ligand-DNA complexation.

2.4.3.3 Electrostatic Energy, ΔG_{el}

The results of calculations of electrostatic energies are summarized in Table 2.2. It is seen from the table that the change in the electrostatic component of the energy of interaction with the surrounding water, ΔG^{solv}_{conf}, upon DNA unwinding is positive for all the ligands studied. These observations may be explained in terms of the decrease in charge density on the DNA surface as a result of unwinding, which inevitably causes the weakening of interaction with water surrounding. The contribution of coulombic interactions to the free energy of unwinding, ΔG^{im}_{conf}, for all ligands is negative (Table 2.2), i.e. this type of interaction promotes the unwinding of DNA molecule. This behaviour results from the increase in distance between the negatively charged phosphates on the formation of intercalation cavity, which as a whole is an energetically favourable process. A good correlation between the values of the untwist angle, $\Delta\Omega$, of the DNA duplex upon intercalation of the ligands and the calculated energy, ΔG^{im}_{conf}, was noted in [122]: the greater the $\Delta\Omega$ angle, the greater are the negative changes of ΔG^{im}_{conf}. The net conformational electrostatic energy, ΔG_{conf}, is a relatively small number with a sign depending on the type of ligand.

Upon ligand insertion to unwound DNA the magnitudes of the solvation component, ΔG^{solv}_{ins}, are positive for DAU, EB, PF, and negative for NOV (Table 2.2). By contrast, the change in the energy of atom-atom coulombic interaction, ΔG^{im}_{ins}, is

Table 2.2 Inter(intra)molecular in vacuum and with solvent electrostatic energies (kcal/mol) for ligand binding with DNA

Ligand	Unwinding			Insertion			Intercalation		
	ΔG_{conf}^{im}	ΔG_{conf}^{solv}	ΔG_{conf}	ΔG_{ins}^{im}	ΔG_{ins}^{solv}	ΔG_{ins}	ΔG_{el}^{im}	ΔG_{el}^{solv}	ΔG_{el}
DAU	−20.0	29.0	9.0	−127.7	130.6	2.9	−147.7	159.6	11.9
EB	−29.1	26.4	−2.7	−144.5	149.9	5.4	−173.5	176.3	2.8
NOV	−39.2	35.2	−4.0	20.7	−8.1	12.6	−18.5	27.2	8.7
PF	−27.1	24.4	−2.7	−125.3	127.3	2.0	−152.4	151.7	−0.7

negative for DAU, EB, PF and positive for NOV. Such a pattern can be explained by the fact that NOV is neutral, whereas the other ligands carry a single positive charge. When the positively-charged molecules intercalate into the DNA double helix, their charges are compensated by the negative charge of the neighbouring DNA phosphates, which leads to an overall weakening of electrostatic interaction with solution. It also provides energetically favourable electrostatic interaction between positively-charged molecules and negatively-charged phosphates which, in turn, leads to large-by-absolute-value negative magnitude of ΔG_{ins}^{im}. For the neutral molecule, NOV, the quantities ΔG_{ins}^{im} and ΔG_{ins}^{solv} are opposite in sign (Table 2.2), probably because on intercalation a part of the charge distributed over the ligand atoms falls inside the intercalation cavity and is shielded from solution. At the same time the portion of the charge remaining on atoms of the ligand and protruding into solution effectively interacts with it by ion-dipole-like interactions, which makes ΔG_{ins}^{solv} quantity negative.

The total electrostatic energy, ΔG_{ins}, is the sum of two numbers, in which the quantities ΔG_{ins}^{im} and ΔG_{ins}^{solv} are large in value and have opposite signs (Table 2.2) resulting in small values. In total ΔG_{ins} is positive and relatively small for aromatic intercalators. Hence, the total electrostatic interactions at the stage of ligand insertion appear to be energetically unfavourable and hinder the formation of the complexes between DNA and aromatic molecules.

The magnitude of the total change in electrostatic energy ΔG_{el} is relatively small by absolute value but, on average, is comparable to the experimental energy of binding ΔG_{exp}, and is the sum of components (ΔG_{el}^{solv} and ΔG_{el}^{im}) with large values but opposite in sign, similar to the situation found for the van der Waals energies (see above). Analysis of the results for wide variety of aromatic ligands [107, 123] enabled us to conclude that there is no significant correlation between the type of ligand and the total electrostatic energy (ΔG_{el}), although some correlation was observed above at the level of DNA unwinding (ΔG_{conf}) and ligand insertion (ΔG_{ins}). It is likely that any link between ΔG_{el} and the structure/charge of the ligand becomes masked on summation of ΔG_{conf} and ΔG_{ins} in Eq. (2.4), Eq. (2.5). This conclusion drawn with respect to electrostatic energy resembles a problem of enthalpy/entropy compensation in biomolecular interactions [98] which makes analysis of total Gibbs energy to certain extent ambiguous. Hence, it is concluded that any search for a correlation between the structure of a ligand and its energy of complexation should only be made at the level of separate steps of the complexation process and appropriate components of the electrostatic energy rather than in terms of total electrostatic energy [107, 122, 123]. The same conclusion was drawn above with respect to the van der Waals energy.

2.4.3.4 Polyelectrolyte Energy, ΔG_{pe}

The polyelectrolyte contribution contains both an enthalpic term, originating from coulombic interaction of solute molecule with counterions present in solution, and an entropic term, coming from disordering of ion atmosphere upon ligand intercalation. Within the framework of the approach being reviewed in this chapter, the ΔG_{pe} component is separated from the total electrostatic energy because it can be measured experimentally and the experimental values for some drugs are available in the literature (for review see [124]). It is also assumed that the main contribution to the polyelectrolyte energy comes from the ligand insertion stage.

Analysis of literature suggests that the magnitudes of ΔG_{pe} for aromatic intercalators show very similar values even though the ligands may have very different structures. By that reason it is reasonable to take the average value, $\Delta G_{pe} \approx -1.1$ kcal/mol, for the set of aromatic molecules studied [107]. In general, the polyelectrolyte contribution favours formation of complexes. The effect is predominantly entropic in origin and is due to entropically favourable ion release upon ligand insertion into DNA.

2.4.3.5 Hydrogen Bonding Energy, ΔG_{HB}

Hydrogen bonding in the complexation of ligands with DNA comes from [107, 115]:

1. formation of intermolecular H-bonds between the ligand and DNA within the intercalation cavity, characterized by the number of intermolecular H-bonds, N^{im}, and
2. the loss of hydrogen bonds to water due to dehydration of the ligand upon insertion into the intercalation site, characterized by change in hydration index, ΔN^{solv}.

As noted in section 2.4.2, within the framework of the methodology used for energy analysis the magnitude of ΔG_{HB} does not bear the meaning of real energy of H-bonding, and an analysis of the number of hydrogen bonds, N^{im} and ΔN^{solv}, is considered to be more appropriate.

It is seen from Table 2.3 that the sign of ΔN^{solv}_{conf} value is positive indicating the predominant solvation of the intercalation cavity on DNA unwinding, whereas the ΔN^{solv}_{ins} value is negative pointing out on the net removal of water molecules on ligand insertion. The resulting effect of the whole complexation process is negative, $\Delta N^{solv} < 0$, indicating that there is a dehydration of the ligand and DNA molecules during the intercalation.

The dehydration on insertion (ΔN^{solv}_{ins}) is not overbalanced by the sum of hydration on unwinding (ΔN^{solv}_{conf}) and formation of intermolecular H-bonds (N^{im}_{ins}), resulting in net positive contribution of H-bonding to the total energy of the intercalation

Table 2.3 Energetic contribution of Hydrogen bonds (kcal/mol)

Ligand	Unwinding	Insertion		Intercalation	
	ΔN^{solv}_{conf}	N^{im}_{ins}	ΔN^{solv}_{ins}	ΔN^{solv}	ΔG_{HB}
DAU	2.0	3	−7.5	−5.5	5.8
EB	3.1	1	−5.9	−2.8	4.1
NOV	2.0	1	−4.6	−2.6	3.6
PF	3.0	1	−6.3	−3.3	5.3

Table 2.4 Calculated values of the hydrophobic contribution (kcal/mol)

Ligand	Unwinding	Insertion	Intercalation
DAU	14.2	−43.9	−29.7
EB	10.7	−34.5	−23.8
NOV	11.8	−37.0	−25.2
PF	10.7	−26.7	−16.0

reaction, $\Delta G_{HB} > 0$. It follows that the net effect of H-bonding is destabilizing with respect to the intercalation.

It is worth noting that there is much controversy in literature regarding the question whether the ligand-DNA binding process is accompanied by uptake or release of water molecules (reviewed in [41, 115]). The results presented here clearly show that there is a net removal of water molecules during the intercalation.

2.4.3.6 Hydrophobic Energy, ΔG_{hyd}

It is seen from Table 2.4 that the hydrophobic contribution is favourable and much higher by absolute value than the experimental total energy (see also Table 2.6). Inspection of the data in Table 2.4 suggests that the hydrophobic energy decreases in the sequence: DAU > NOV > EB > PF.

Such a sequence qualitatively correlates with the degree of branching of the side chains of the investigated ligands, i.e. from large side chains for DAU down to two relatively small hydrophilic aminogroups in PF. The bulkyness of the side chains influences the effectiveness of removal of water molecules from the hydration layers of DNA and the ligand on complexation and, hence, the hydrophobic contribution is usually higher for bulky side chains of the intercalators buried in the DNA grooves. The effect of removal of water molecules is also confirmed by the fact that the total change in solvent accessible surface area for the stages of DNA unwinding and ligand insertion is negative (data not shown). This result agrees with the general view on water involvement in complexation reactions but is unable to shed any light on the results of osmotic stress measurements [41], which had led to the completely opposite conclusion, viz. binding of DAU/PF ligands is accompanied by uptake instead of release of water molecules.

Table 2.5 Energetic contribution of entropic factors (kcal/mol)

Ligand	Translational	Rotational	Vibrational of type I	Vibrational of type II	ΔG_{entr}
DAU	10.5	10.3	−4.3	−8.0	8.4
EB	10.0	9.4		−7.6	7.5
NOV	10.3	10.3		−9.0	7.3
PF	9.7	8.6		−7.6	6.3

Table 2.6 Partition of the total energy of ligand-DNA binding (kcal/mol)

Ligand	Unwinding	ΔG for ligand insertion					ΔG_{total}	ΔG_{exp}
		VDW	el+HB	pe	hyd	entr		
DAU	33.2	−16.5	13.3	−1.0	−43.9	8.4	−6.6	−9.0
EB	12.3	−11.0	16.7	−1.2	−34.5	7.5	−10.3	−9.5
NOV	17.4	−16.0	20.8	−1.1	−37.0	7.3	−8.6	−9.5
PF	0.5	1.4	14.2	−1.1	−26.7	6.3	−5.4	−6.0

2.4.3.7 Entropic Contribution, ΔG_{entr}

The results of calculations of the entropic contributions are summarised in Table 2.5. The major contribution to the quantities ΔG_t and ΔG_r is entropic in nature ($|T\Delta S|>|\Delta H|$) [107, 114] which is unfavourable and therefore yields a positive sign of ΔG. This result is quite expected and is due to the entropically unfavourable loss of three translational and rotational degrees of freedom upon complexation.

The mean values of ΔG_t and ΔG_r averaged over various aromatic ligands equal to $\langle \Delta G_t \rangle = (10.3 \pm 0.3)$ kcal/mol and $\langle \Delta G_r \rangle = (10.0 \pm 0.8)$ kcal/mol [107, 114]. The differences in these energies for the different types of ligand are relatively small and so the mean energies $\langle \Delta G_t \rangle$ and $\langle \Delta G_r \rangle$ can effectively be used in analysis of the contributions for different aromatic ligands. The mean sum, $\langle \Delta G_t + \Delta G_r \rangle = (20.2 \pm 1.1)$ kcal/mol [107, 114], is close to but slightly higher than the empirical value $\Delta G_{t+r} = 15$ kcal/mol, used previously [124] for energy partitioning of ligand-DNA interactions.

Analysis of the results of calculations for the change in type I vibrations (ΔG_v^I) suggests that this factor is enthalpically unfavourable but entropically favourable [107, 114], which can be interpreted in terms of formation of new vibrational degrees of freedom. In total the entropic factor overwhelms and type I vibrations appear to favour complex formation for the ligands studied.

The values of type II vibrations exhibit only small deviations from the mean value $\langle \Delta G_v^{II} \rangle = -(8.1 \pm 0.5)$ kcal/mol for different ligands and so one value may be used for different aromatic ligands as found above for the contributions of translational and rotational energies [107, 114]. The type II vibrations are mainly entropic in origin and favour formation of the complexes, which is the result of creating new vibrational degree of freedom due to the appearance of mechanical oscillation of a ligand on intercalation. It is important to note that the magnitudes of ΔG_v^I and ΔG_v^{II} are commensurable to the experimental energy of binding, which means that consideration of these components is important in the energy analysis of ligand-DNA interactions.

The sum of all entropic terms $\langle \Delta G_{entr} \rangle = \langle \Delta G_t \rangle + \langle \Delta G_r \rangle + \langle \Delta G_v^I \rangle + \langle \Delta G_v^{II} \rangle$ ≈ 7.8 kcal/mol is, on the whole, unfavourable and destabilizes ligand-DNA binding [107, 114]. As discussed above, each entropic component differs little from the mean value, which suggests that the total sum $\langle \Delta G_{entr} \rangle = 7.8$ kcal/mol can be used in the energy analysis of ligand-DNA complexation for different aromatic ligand with non-heavily branched side chains.

2.4.3.8 The Total Energy of the Ligand-DNA Binding Process

The total energy of ligand-DNA binding according to Eq. (2.4) is summarised in Table 2.6 using the values of the various contributions to the energy in Tables 2.1, 2.2, 2.3, 2.4, 2.5 and data in Ref. [107]. The H-bonding term, ΔG_{HB} (see Table 2.3), was included in the electrostatic energy for the stages of unwinding and insertion. The unwinding energy, ΔG_{conf}, was taken as a sum of all contributing factors from Tables 2.1, 2.2, 2.3, 2.4.

As seen from Table 2.6, the sum of 6 different energy components for various ligands has ended up with values, which differ from the experimental energies, on average, for 1.2 kcal/mol. This result is considered to be successful and makes possible further analysis of individual energy components. Note that the selected ligands (as a small subset of the ligands studied in literature) have very different structures and charge states and very different approaches were used to calculate each energy component.

As seen from Table 2.6 the DNA unwinding stage is always unfavourable for ligand binding and is the main contributor to the activation energy for the reaction of ligand-DNA complexation. Other unfavourable contributions are the net effect of electrostatic interactions, entropic factors and hydrogen bonding. The main stabilization comes from hydrophobic, van der Waals (except that for PF) and polyelectrolyte terms, which is in general agreement with what is known about stacking of molecules with aromatic surfaces in solution [125, 126]. The van der Waals and hydrophobic forces are the most important and the latter one is dominant for all the ligands studied.

Another important issue is the fact that the small value of the total Gibbs free energy of ligand-DNA complexation (*ca.* −9 kcal/mol) is the result of summation of components with large magnitude but of opposite sign (see the components in Tables 2.1 and 2.2 having the magnitude of dozens and hundreds of kcal/mol). It can be seen that the ligand binding to DNA is governed by the effect of compensation of energy contributions at the levels of physical forces, different stages of ligand binding and inter(intra)molecular/to-solvent interactions in vacuum. This fact was shown to be the reason why the net energies in Eq. (2.4) does not generally correlate with the physico-chemical properties of the ligand and such correlation can be observed only on the level of the energy components in Eq. (2.5) [107, 123]. In fact, similar conclusions have been drawn with respect to the energy of π-stacking interactions in solution for variety of aromatic molecules [116], largely resembling the process of DNA intercalation.

Search of a correlation (r) of the energy terms in Eq. (2.5) with the experimental energy, ΔG_{exp}, has shown that the highest impact on it is provided by the VDW energy of ligand insertion, ΔG_{vdW}^{ins} ($r = 0.66$) and VDW energy of DNA unwinding, ΔG_{vdW}^{uw} ($r = -0.67$). The rest terms give lower correlation not exceeding $|r| = 0.5$. The correlation between ΔG_{exp} and ΔG_{hyd} was equal to $r = 0.42$. This result is in accord with the above-made conclusion on the importance of VDW interactions in the net energetics of binding. Noteworthy, the EL energy, formally featuring the largest magnitude of the energy components (see Table 2.2), appears to be relatively unimportant in the modulating the binding affinity in the intercalation reactions. This result highlights the key role of the intermolecular VDW forces in managing the affinity of aromatic drugs to DNA and points out the way to modify the ligand structure with an aim to increase the binding strength with DNA.

2.4.4 Energy Analysis of DNA Minor Groove Binding Reactions

Ligand molecules which exert predominant affinity to DNA minor groove commonly contain a set of hetero-cycles linked by single bonds and closely matching the shape of DNA minor groove. Typical examples of MGB-ligands are Hoechst33258, Netropsin, Berenil, Distamycin. The DNA-binding and medico-biological properties of the MGB-ligands have been extensively reviewed and these molecules are currently considered as promising agents in chemotherapy of cancer [127, 128]. The MGB-ligands exert major specificity to AT sites of DNA, covering approximately 4 base pairs when binding within the minor groove [127], hence, the non-selfcomplementary dodecamer d(CGCA$_4$GCG)/(CGCT$_4$GCG) may be selected as a receptor in energy analysis [108].

The full energy analysis of MGB-ligands binding with DNA was accomplished in Ref. [108] using the methodology generally similar to that reviewed above for DNA intercalation. The general patterns of the sign and magnitude of various energy terms, already discussed above for DNA intercalators, were reported to be preserved in the case of MGB-binding as well. In particular, the compensatory effect, the absence of apparent correlation of the net energies in Eq. (2.4) with the properties of ligand, and the coincidence of ΔG_{total} and ΔG_{exp}, remain valid. It was found that there are at least three major stabilizing factors, appearing in Eq. (2.5), which govern the binding process of the MGB-ligands with DNA, placed in descending order according to the absolute value of the energy change: intermolecular electrostatic interactions (ΔG_{el}^{im}), intermolecular van der Waals interactions (ΔG_{vdW}^{im}) and hydrophobic interactions (ΔG_{hyd}). The stabilization of the complexes is also provided by the formation of intermolecular H-bonds (N^{im}), formation of residual mechanical vibrations in the binding site (ΔG_{v}^{II}) and the polyelectrolyte factor (ΔG_{pe})—the latter two giving minor contribution as compared to other factors. The major factors which destabilize complexes of the MGB-ligands with DNA are the electrostatic

(ΔG_{el}^{solv}) and van der Waals (ΔG_{vdW}^{solv}) desolvation, loss of H-bonds to-water (ΔN^{solv}), change in the number of translational (ΔG_t), rotational (ΔG_r), vibrational (ΔG_v^I) degrees of freedom, and restriction of internal rotations in MGB molecules (ΔG_{conf}). The hydrogen bonding factor among the rest energy terms was shown to be more important specifically for the MGB-ligands than for the intercalators [129].

The net energies in Eq. (2.4), which stabilize complexes, can be placed in descending order by the absolute value: $\Delta G_{hyd} > \Delta G_{vdW} > \Delta G_{pe}$, whereas the order of destabilizing factors is $\Delta G_{entr} \geq \Delta G_{HB} > \Delta G_{el}$ [108].

With the aim of searching the physical factor the most strongly affecting the binding affinity, the correlation coefficients of experimental energy with the solvation and intermolecular components for all the factors in Eq. (2.5) were calculated [108]. It was found that the highest correlation is observed for the electrostatic energy, which suggests that the major effect on variation of DNA binding affinity with the type of ligand is provided by the electrostatic component. This is in agreement with the qualitative estimations of other authors [130, 131].

2.4.5 Energy Analysis of RNA Binding Reactions

Ligand binding with RNA is probably the most difficult object to study as compared with DNA binding reactions due to large variability of RNA binding sites. In Ref. [132] the energy analysis of binding of 11 small molecules to RNA aptamers was accomplished using the methodology reviewed in section 2.4.2. Although the details of specific adaptation of the ligand to the binding site, currently considered to be important in case of RNA binding ligands [133, 134], were not unveiled in this work, the general patterns of distribution of energy over various energy terms were reported to be similar to DNA intercalation and minor-groove binding, presumably reflecting the general pattern of binding reactions in aqueous media [123, 129].

The most important contribution to the binding energetics in terms of the net absolute energies in Eq. (2.4) is given by the ΔG_{hyd} and ΔG_{vdW} ($\Delta G_{hyd} > \Delta G_{vdW}$) factors, and the destabilization originates from ΔG_{entr} and ΔG_{HB}, which is qualitatively similar to what was found above for the DNA intercalation and minor-groove binding. The electrostatic factor ΔG_{el} is relatively unimportant for the ligands with no charge or bearing single charge, whereas the doubly- or more-charged ligands elevate ΔG_{el} to the level commensurable with ΔG_{vdW}.

The stabilizing energy terms in Eq. (2.5) can be placed in the sequence by extent of their contribution: $\Delta G_{vdW}^{im} > \Delta G_{hyd} > \Delta G_v^{II}$. The sequence for the destabilizing energies is: $\Delta G_{vdW}^{solv} \geq \Delta G_v^I \geq \Delta G_{t,r}$. The VDW energies were found to depend strongly on the type and dimensions of side chains of the ligand and the efficacy of π-stacking with RNA bases.

Intermolecular (ΔG_{el}^{im}) and to-water (ΔG_{el}^{solv}) electrostatic energies by the magnitude and sign strongly depend on the charge of the ligand, viz. ΔG_{el}^{im} is favourable and ΔG_{el}^{solv} is unfavourable for positively charged ligands, whereas the signs of these terms get reversed for negatively charged ligands.

The correlation coefficients of the terms in Eq. (2.5) with ΔG_{exp} do not exceed 0.5 and do not allow selecting the terms which exert the maximal impact on the binding affinity. It means that the principal physical factors, such as VDW, EL, HYD, HB, give approximately equal contribution to the variability of ΔG_{exp} with the type of ligand.

2.4.6 *Summary of the Results, and Implications on the Use of Energy Analysis in Rational Drug Design*

Energy analysis of ligand-NA interactions, sketched out above for typical DNA intercalators, DNA minor groove binders and RNA binders, enables us to answer the key question, viz. "What physical factors stabilize/destabilize the ligand-NA complexes in solution and what are their relative importance?" The follow-up question now is "How one can use the results of the energy analysis, say, in rational design of new drugs?"

The set of stabilizing and destabilizing energies aligned in descending order, as the main outcome of the energy analysis, provides a fundamental knowledge on energetics of binding reactions in solution but, in fact, gives little idea on the way how one can manipulate the magnitude of ΔG_{exp} and, eventually, the medico-biological effect of the NA-binding drugs [123, 129]. It is considered that the search of the factor which is most strongly correlated with the equilibrium binding constant $K = \exp\left(-\frac{\Delta G_{\exp}}{RT}\right)$ may give this idea. If it is known what factor modulates the ligand affinity to DNA (VDW, hydrophobic, electrostatics or else), it becomes more clear what type of atomic group must be chemically added/substituted in the ligand structure in order to amplify the contribution of this particular physical factor to the net energy of binding, resulting in increase of ΔG_{exp}. In particular, it was shown above that in the case of DNA intercalators the "managing" of the binding affinity may be achieved via the VDW factor, whereas for the group of DNA minor groove binders the EL energy appears to be the key factor. However, such approach may be of value if the binding affinity is the target property to be manipulated, or if no sufficient data on biological activity of the studied group of ligands is available, and the amplification of the binding affinity to bioreceptor remains the only possible strategy. The case if relevant biological data are available, search of correlations between the biological activity and specific energy terms may have real practical outcome. Let us consider such possibility taking as an example the results of analysis performed for the group of MGB binders in Ref. [129].

Table 2.7 contains the calculated values of the energy terms for the set of MGB-ligands and the ID_{50} factor for the same ligands (which is a micromolar concentration of the drug, needed for 50 % suppression of L1210 leukemia cell growth).

It must be noted that rather limited dataset presented in Table 2.7 does not allow reporting on statistically reliable correlation, nevertheless, the qualitative level of correlation may be considered. It is seen that the highest correlation of the ID_{50}

Table 2.7 Correlation (r) of the energy terms (kcal/mol) and equilibrium binding constant, K (M^{-1}), with the measure of biological activity of the drug, ID_{50}

MGB-ligand	ID_{50}	ΔG^{solv}_{vdW}	ΔG^{im}_{vdW}	ΔG^{solv}_{el}	ΔG^{im}_{el}	ΔG_{hyd}	ΔN^{solv}	N^{im}	$\Delta N^{solv}+N^{im}$	K
SN6999	0.02	54.0	−64.9	258	−255	−45.7	−11.5	1	−10.5	$2.0 \cdot 10^6$
Hoechst33258	1.5	52.7	−64.1	141	−140	−46.3	−14.1	4	−10.1	$3.2 \cdot 10^6$
Distamycin	9	61.4	−65.8	138	−136	−53.0	−16.4	10	−6.4	$2.0 \cdot 10^5$
Netropsin	10	75.4	−63.0	266	−260	−44.6	−11.2	11	−0.2	$1.0 \cdot 10^5$
Berenil	10.4	43.3	−45.9	267	−264	−34.7	−7.7	2	−5.7	$1.3 \cdot 10^7$
r		0.27	0.48	0.20	−0.20	0.25	0.24	0.58	0.83	0.27

factor is the case with the change in the net number of hydrogen bonds on binding: $N^{im}+\Delta N^{solv}$, hence, it is suggested that this factor might be modified in first instance in rational drug design. Interestingly, the ID_{50} factor does not show apparent correlation with the MGB−DNA binding constant (as an integral measure of the net energetics of binding, see Table 2.7). Although this result may be considered as preliminary, it clearly demonstrates the potential importance of the energy analysis in designing new drugs.

Taking as a whole, the strategies based on energy analysis of drug-NA binding reactions and reviewed above, may extend the existing approaches in rational drug design based on computer modeling.

2.5 Concluding Remarks

The results of application of various biophysical methods in understanding the mechanism of drug binding with DNA, reviewed above, demonstrate the power and mutual complementarity of experiment and computer modeling in solving particular scientific problem. Taking the derivatives of antitumour antibiotic Actinomycin D as an example, we have shown that initial experimental evidence, provided by the Nature on *in vitro* level, can be fully investigated in detail on molecular level yielding complete structural/thermodynamic/energetic picture of the drug's binding with bioreceptor, and eventually resulting in understanding the key factors governing this process. Further manipulation of the governing factors by means of chemical modification of the drug provides scientific background of the strategy of rational drug design. Its main outcome is the possibility to create new drugs with improved pharmacological properties, which currently remains one of most important challenges in biomolecular sciences.

Acknowledgments The authors express their special thanks to the following people which, in part, created the background, contributed and stimulated further the results reviewed in this chapter: Professor Vladimir Ya. Maleev (IRE NASU), Professor Mikhail A. Semenov (IRE NASU), Dr. Elena B. Kruglova (IRE NASU), Dr. Ekaterina G. Bereznyak (IRE NASU), Dr. Viktor V. Kostjukov (SevNTU). Support from the Ministry of Education and National Academy of Sciences of Ukraine via the grants 0103U002268 (2002–2006), 0107U001331 (2007–2009), 0107U001079 (2007–2011), 0110U001683 (2010–2012), F27/60-2010 is greatly acknowledged.

References

1. Ramos MJ, Fernandes PA (2006) Atomic-level rational drug design Curr Comp-Aided Drug Des 2:57–81
2. Nelson SM, Ferguson LR, Denny WA (2004) DNA and the chromosome—varied targets for chemotherapy. Cell Chromosome 3(1):1–26
3. Hurley LH (2002) DNA and its associated processes as targets for cancer therapy. Nat Rev Cancer 2(3):188–199
4. Au JL, Panchal N, Li D, Gan Y (1997) Apoptosis: a new pharmacodynamic endpoint. Pharm Res 14(12):1659–1671
5. Selwood DL (2013) Beyond the hundred dollar genome—drug discovery futures. Chem Biol Drug Des. 81(1):1–4
6. Ren J, Chaires JB (1999) Sequence and structural selectivity of nucleic acid binding ligands. Biochemistry 38(49):16067–16075
7. Dervan PB (2001) Molecular recognition of DNA by small molecules. Bioorg Med Chem 9(9):2215–2235
8. Veselkov AN, Maleev VYa, Glibin EN, Karawajew L, Davies DB (2003) Structure–activity relation for synthetic phenoxazone drugs. Evidence for a direct correlation between DNA binding and pro-apoptotic activity. Eur J Biochem 270(20):4200–4207
9. Murthy VR, Raghuram DV, Murthy PN (2007) Drug, dosage, activity, studies of antimalarials by physical methods—II. Bioinformation 2(1):12–16.
10. Sobell HM, Jain SC (1972) Stereochemistry of actinomycin binding to DNA. II. Detailed molecular model of actinomycin-DNA complex and its implications. J Mol Biol 68(1):21–34.
11. Porumb H (1978) The solution spectroscopy of drugs and the drug-nucleic acid interactions. Prog Biophys Mol Biol 34(3):175–195
12. Yielding LW, Yielding KL (1984) Ethidium binding to deoxyribonucleic acid: spectrophotometric analysis of analogs with amino, azido, and hydrogen substituents. Biopolymers 23(1):83–110
13. Barcelo F, Ortiz-Lombardia M, Portugal J (2001) Heterogenous DNA binding modes of berenil. Biochim Biophys Acta 1519(3):175–184
14. Barcelo F, Capo D, Portugal J (2002) Thermodynamic characterization of the multiplay binding of chartreusin to DNA. Nucleic Acids Res 30(20):4567–4573
15. Sovenyhazy K, Bolderon J, Petty J (2003) Spectroscopic studies of the multiple binding modes of trimetine-bridget cyanine dye with DNA. Nucleic Acids Res 31(10):2561–2569
16. Ghosh R, Bhowmik S, Bagchi A, Das D, Ghosh S (2010) Chemotherapeutic potential of 9-phenyl acridine: biophysical studies on its binding to DNA. Eur Biophys J 39(8):1243–1249
17. Kumar S, Pandya P, Pandav K, Gupta SP, Chopra AN (2012) Structural studies on ligand–DNA systems: a robust approach in drug design. J Biosci 37(3): 553–561
18. Kruglova EB, Gladkovskaya NA, Maleev VY (2005) The use of the spectrophotometry analysis for the calculation of the thermodynamic parameters in actinocin derivative-DNA systems. Biophysics 50(2):253–264
19. McGhee JD, von Hippel PH (1974) Theoretical aspects of DNA-protein interactions: co-operative and non-co-operative binding of large ligands to a onedimensional homogeneous lattice. J Mol Biol 86(2):469–489
20. Nechipurenko YuD (1984) Cooperative effects on binding of large ligands to DNA. II. Contact cooperative interactions between bound ligand molecules. Mol Biol 18(6):1066–1079
21. Kruglova EB, Gladkovskaya NA (2002) Comparison of the binding of the therapeutically active nucleotides to DNA molecules with different level of lesions. Proceedings of SPIE 4938:241–245 and Iermak Ie (2011). Light-absorption spectroscopy of mutagen—DNA complex: binding model selection and binding parameters calculation J Appl Electromagn 13(1):15–22

22. Hajan R, Guan HT (2013) Spectrophotometric studies on the thermodynamics of the ds-DNA interaction with irinotecan for a better understanding of anticancer drug-DNA interactions. J Spectrosc. ID 380352. http://dx.doi.org/10.1155/2013/380352
23. Neault JF, Tajmir-Rihi HA. (1996) Diethylstilbestrol-DNA interaction studied by Fourier transform infrared and Raman spectroscopy. J Biol Chem 271(14):8140–8143
24. Neault, J.-F. & Tajmir-Riahi, H. A. (1998). DNA-chlorophyllin interaction. J Phys Chem B 102(4):1610–1614
25. Deng H, Bloomfield VA, Benevides JM, Thomas GJ (1999) Dependence of the Raman signature of genomic B-DNA on nucleotide base sequence. Biopolymers 50(6):656–666
26. Quameur AA, Tajmir-Riahi H-A (2004) Structural analysis of DNA interactions with biogenic polyamines and cobalt(III)hexamine studied by Fourier transform infrared and capillary electrophoresis. J Biol Chem 279(40):42041–42054
27. Deng H, Bloomfield VA, Benevides JM (2000) Structural basis of polyamine-DNA recognition: spermidine and spermine interactions with genomic B-DNAs of different GC content probed by Raman spectroscopy. Nucleic Acids Res 28(17):3379–3385
28. Benevides JM, Thomas GJ (2005) Local conformational changes induced in B-DNA by ethidium intercalation. Biochemistry 44(8):2993–2999
29. Kyogoku Y, Lord RC, Rich A (1967) The effect of substituents on the hydrogen bonding of adenine and uracil derivatives. J Am Chem Soc 89(3):496–504
30. Starikov EB, Semenov MA, Maleev VYa, Gasan AI (1991) Evidental study of correlated events in biochemistry: physico-chemical mechanisms of nucleic acids hydration as revealed by factor analysis. Biopolymers 31(2):255–273
31. Hartman KA, Lord RC, Thomas GJ (1973) Structural studies of nucleic acids and polynucleotides by infrared and Raman Spectroscopy In: J. Duchesne (ed) Physio–chemical properties of nucleic acids. Academic, New York, pp. 1–89
32. Semenov MA, Blyzniuk IuN, Bolbukh TV, Shestopalova AV, Evstigneev MP, Maleev VY (2012) Intermolecular hydrogen bonds in hetero-complexes of biologically active aromatic molecules probed by the methods of vibrational spectroscopy. Spectrochimica Acta Part A: Mol Biomol Spectrosc 95(2):224–229
33. Martin JC, Wartell RM, O'Shea I (1978) Conformational features of distamycin-DNA and netropsin-DNA complexes by Raman spectroscopy. Proc Natl Acad Sci USA 75(12):5483–5487
34. Smulevich G, Angeloni L, Marzocchi MP (1980) Raman exitation profiles of actinomycin D. Biochim Biophys Acta 610(2):384–391
35. Ruiz-Chica J, Medina MA, Sanchez F (2001) Fourier transform Raman study of the structural specificities on the interaction between DNA and biogenic polyamines. Biophys J 80(2):449–454
36. Kruglova EB, Bolbukh TV, Gladkovskaya NA, Bliznyuk JuN (2005) The binding of actinocin antibiotics to polyphosphate matrix. Biopolym Cell 21(2):358–364
37. Bliznyuk YuN, Kruglova EB, Bolbukh TV, Ovchinnikov DV (2009) Influence of solution acidity on structure of actinocin derivatives and their affinity to DNA studies as a function of pH by Raman spectroscopy. Spectrosc Lett 42(3):498–505
38. Tsuboi M, Benevides JM, Thomas GJ (2009) Raman tensors and their application in structural studies of biological systems. Proc Jpn Acad Ser B Phys Biol Sci 85(1):83–97
39. Blyzniuk IuN, Bolbukh TV, Kruglova OB, Semenov MA, Maleev VYa (2009) Investigation of complexation of ethidium bromide with DNA by the method of Raman spectroscopy. Biopolym Cell 25(1):126–132
40. Lane AN, Jenkins TC (2000) Thermodynamics of nucleic acids and their interactions with ligands. Q Rev Biophysics 33(3):255–306
41. Qu X, Chaires JB (2001) Hydration changes for DNA intercalation reactions. J Am Chem Soc 123(1):1–7
42. Pal SK, Zhao L, Zewail AH (2003) Water at DNA surfaces: ultrafast dynamics in minor groove recognition. Proc Natl Acad Sci USA 100(14):8113–8118

43. Parsegian VA, Rand RP, Rau DC (2000) Osmotic stress, crowding, preferential hydration, and binding: a comparison of perspectives. Proc Natl Acad Sci USA 97(8):3987–3992
44. Schneider B, Ginell SL, Berman HM (1992) Low temperature structures of dCpG-proflavine conformational and hydration effects. Biophys J 63(6):1572–1578
45. Shimizu S (2004) Estimating hydration changes upon biomolecular reactions from osmotic stress, high pressure, and preferential hydration experiments. Proc Nat Acad Sci USA 101(5):1195–1199
46. Marky LA, Kupke DW, Kankia BI (2001) Volume changes accompanying interaction of ligands with nucleic acids. Methods Enzymol 340:149–165
47. Han F, Chalikian TV (2003) Hydration changes accompanying nucleic acid intercalation reactions: volumetric characterizations. J Am Chem Soc 125(24):7219–7229
48. Auffinger P, Westhof R (1999) Role of hydration on the structure and dynamics of nucleic acids In: Ross YH (ed) Water management in the design and distribution of quality foods. Technomic Publishing Co, Basel, pp 165–198
49. Korolev N, Lyubartsev AP, Laaksonen A (2002) On the competition between water, sodium ions, and spermine in binding to DNA: a molecular dynamics computer simulation study. Biophys J 82(6):2860–2875
50. Korolev N., Lyubartsev AP, Laaksonen A (2003) A molecular dynamics simulation study of oriented DNA with polyamine and sodium counterions: diffusion and averaged binding of water and cations. Nucleic Acids Res 3(20):5971–5981
51. Maleev VYa, Semenov MA, Gasan AI, Kashpur VA (1993) Physical properties of the system DNA-water. Biophysics 38(3):768–790
52. Semenov MA, Bolbukh TV, Maleev VYa (1997) Infrared study of the influence of water on DNA stability in the dependence on AT/GC composition. J Mol Struct 408/409(2):213–217
53. Semenov MA, Bereznyak EG (2000) Hydration and stability of nucleic acids in the condensed state. Comments Mol Cell Biophys 10(1):1–23
54. Maleev V, Semenov M, Kashpur V, Bolbukh T, Shestopalova A, Anishchenko D (2002) Structure and hydration of polycytidylic acid from the data of infrared spectroscopy, EHF dielectrometry and computer modeling. J Mol Struct 605(1):51–61
55. Bereznyak EG, Semenov MA, Bol'bukh TV, Dukhopel'nikov EV, Shestopalova AV, Maleev VYa (2002) A study of the effect of water on the interaction of DNA with actinoxcin derivatives having different lengths of aminoalkyl chains by the methods of IR spectroscopy and computer simulation. Biophysics 47(6):1019–1026
56. Marky LA, Blumenfeld KS, Breslauer KJ (1983) Calorimetric and spectroscopic investigation of drug-DNA interactions. I. The binding of netropsin to poly d(AT). Nucleic Acids Res 11(9):2857–2870
57. Marky LA, Snyder JG, Breslauer KJ (1983) Calorimetric and spectroscopic investigation of drug-DNA interactions: II. Dipyrandium binding to poly d(AT). Nucleic Acids Res 11(16):5701–5715
58. Jelesarov I, Bosshard HR (1999) Isothermal titration calorimetry and differential scanning calorimetry as complementary tools to investigate the energetics of biomolecular recognition. J Mol Recognit 12(1):3–18
59. Cooper A (1999) Thermodynamic analysis of biomolecular interactions. Curr Opin Chem Biol 3(5):557–563
60. O'Brien R, Haq I (2004) Applications of biocalorimetry: binding, stability and enzyme kinetics. In: Ladbury JE, Michael Doyle M (eds) Biocalorimetry 2. Wiley. Budapest, Hungary
61. Bruylants G, Wouters J, Michaux C (2005) Diï¬erential scanning calorimetry in life science: thermodynamics, stability, molecular recognition and application in drug design. Curr Med Chem 12(17):2011–2020
62. Celej S, Fidelio G, Dassie S (2005) Protein unfolding coupled to ligand binding: differential scanning calorimetry simulation approach. J Chem Educ 82(1):85–92
63. Celej S, Dassie S, Gonzalez M, Bianconi M, Fidelio G (2006) Differential scanning calorimetry as a tool to estimate binding parameters in multiligand binding proteins. Anal Biochem 350(2):277–284

64. Dukhopelnikov EV, Bereznyak EG, Khrebtova AS, Lantushenko AO, Zinchenko AV (2012) Determination of ligand to DNA binding parameters from two-dimensional DSC curves. J Therm Anal Calorim. doi:10.1007/s10973–012-2561–6
65. Orozco M, Luque FJ (2000) Theoretical methods for the description of the solvent effect in biomolecular systems. Chem Rev 100(11):4187–4225
66. Lazaridis T (2002) Binding affinity and specificity from computational studies. Cur Organ Chem 6(14):1319–1332
67. Schlick T (2010) Molecular modelling and simulation: an interdisciplinary guide, 2nd edn. Springer, New York
68. Cheatham TE III (2004) Simulation and modeling of nucleic acid structure, dynamics and interactions. Curr Opin Struct Biol 14(3):360–367
69. Dolenc J, Oostenbrink Ch, Koller J, van Gunsteren WF (2005) Molecular dynamics simulation and free energy calculations of netropsin and distamycin binding to AAAAA DNA binding site. Nucleic Acids Res 33(2):725–733
70. Ruiz R, García B, Ruisi G, Silvestri A, Barone G (2009) Computational study of the interaction of proflavine with d(ATATATATAT)2 and d(GCGCGCGCGC)2. J Mol Struct: THEOCHEM 915(1):86–92
71. Sasikala WD, Mukherjee A (2012) Molecular mechanism of direct proflavine–DNA Intercalation: evidence for drug-induced minimum base-stacking penalty pathway. J Phys Chem B 116(40):12208–12212
72. Schneider G, Bohm H-J (2002) Virtual screening and fast automated docking methods. Drug Discov Today 7(1):64–70
73. Halperin I, Ma B, Wolfson H, Nussinov R (2002) Principles of docking: an overview of search algorithms and a guide to scoring functions. Proteins 47 (4):409–415
74. Smith GR, Sternberg MJE (2002) Prediction of protein–protein interactions by docking methods. Curr Opin Struct Biol 12(1):28–35
75. Lauria A, Diana P, Barraja P, Montalbano A, Dattolo G, Cirrincione G (2004) Docking of indolo- and pyrrolo-pyrimidines to DNA. New DNA-interactive polycycles from amino-indoles/pyrroles and BMMA. ARKIVOC 5(2):263–271
76. Miroshnychenko KV, Shestopalova AV (2010) The effect of drug-DNA interactions on the intercalation site formation. Int J Quant Chem 110(1):161–176
77. Danilov VI, Tolokh IS (1990) Hydration of uracil and thymine methylderivatives: a Monte Carlo simulation. J Biomol Struct Dyn 7(5):1167–1183
78. Danilov VI, Zheltovsky NV, Slyusarchuk ON, Poltev VI, Alderfer JL (1997) The study of the stability of Watson-Crick nucleic acid base pairs in water and dimethyl sulfoxide: computer simulation by the Monte Carlo method. J Biomol Struct Dyn 15(1):69–80
79. Teplukhin AV, Malenkov GG, Poltev VI (1998) Monte Carlo simulation of DNA fragment hydration in the presence of alkaline cations using novel atom-atom potential functions. J Biomol Struct Dyn 16(2):289–300
80. Alderfer JL, Danilov VI, Poltev VI, Slyusarchuk ON (1999) A study of the hydration of deoxydinucleoside monophosphates containing thymine, uracil and its 5-halogen derivatives: Monte Carlo simulation. J Biomol Struct Dyn 16(5):1107–1117
81. Resat H, Mezei M (1996) Grand canonical ensemble Monte Carlo simulation of the dCpG/proflavine crystal hydrate. Biophysical J 71(3):1179–1190
82. Alcaro S, Coleman RS (2000) A molecular model for DNA cross-linking by the antitumor agent azinomycin B. J Med Chem 43(15):2783–2788
83. Shestopalova AV (2002) Hydration of nucleic acids components in dependence of nucleotide composition and relative humidity: a Monte Carlo simulation. Europ Phys J D 20(1):331–337
84. Shestopalova AV (2007) The binding of actinocin derivative with DNA fragments (Monte Carlo simulation). Biopolym Cell 23(1):35–44
85. Auffinger P, Westhof E (1997) Molecular dynamics: simulations of nucleic acids. Rev Comp Chem 11(2):317–328

86. Chen H, Liu X, Patel DJ (1996) DNA binding and unwinding associated with Actinomycin D antibiotics bound to partially overlapping sites in DNA. J Mol Biol 258(3):457–479
87. Takusagawa F, Carlson RG, Weaver RF (2001) Anti-Leukemia selectivity in Actinomycin Analogues. Bioorg Med Chem 9(3):719–725
88. Karawajew L, Ruppert V, Wutcher C, Kosser A, Schappe M, Dorken B, Ludwing WD (2000) Inhibition in vitro spontaneous apoptosis by IL-7 correlates with upregulation of Bcl-2, cortical/mature immunophenotype, and bettercytoreduction in childhood T-ALL. Blood 98(1):297–306
89. Maleev VYa, Semenov MA, Kruglova EB, Bolbukh TV, Gasan AI, Bereznyak EG, Shestopalova AV (2003) Spectroscopic and calorimetric study of DNA interaction with a new series of actinocin derivatives. J Mol Struct 645(1):145–158
90. Shestopalova AV (2006) The investigation of the association of caffeine and actinocin derivatives in aqueous solution: a molecular dynamics simulation. J Mol Liquids 127 (1):113–117
91. Shestopalova AV (2006) Computer simulation of the association of caffeine and actinocin derivatives in aqueous solution. Biophysics 51(3):389–401
92. Miroshnychenko KV, Shestopalova AV (2005) Flexible docking of DNA fragments and actinocin derivatives. Mol Simulation 31(8):567–574
93. Demeunynck M, Bailly C, Wilson WD (eds) (2003) Small molecule DNA and RNA binders: from synthesis to nucleic acid complexes, vol 2. Wiley-VCH, Weinheim, p 483
94. Ihmels H, Otto D (2005) Intercalation of organic dye molecules into double-stranded DNA—general principles and recent developments. Top Curr Chem 258:161–204
95. Armitage OJ (2002) The role of mitoxantrone in non-Hodgkin's lymphoma. Oncology 16(4):490–512
96. Portugal J, Cashman DJ, Trent JO, Ferrer-Miralles N, Przewloka T, Fokt I, Priebe W, Chaires JB (2005) A new bisintercalating anthracycline with picomolar DNA binding affinity. J Med Chem 48(26):8209–8219
97. Haq I (2002) Thermodynamics of drug–DNA interactions Arch. Biochem Biophys 403(1):1–15
98. Gilli P, Ferretti V, Gilli G, Borea PA (1994) Enthalpy-entropy compensation in drug-receptor binding. J Phys Chem 98(5):1515–1518
99. Dill KA (1997) Additivity principles in biochemistry. J Biol Chem 272(2):701–704
100. McKay SL, Haptonstall B, Gellman SH (2001) Beyond the hydrophobic effect: attractions involving heteroaromatic rings in aqueous solution. J Am Chem Soc 123(6):1244–1245
101. Luo R, Gilson HSR., Potter MJ, Gilson MK (2001) The physical basis of nucleic acid base stacking in water. Biophys J 80(1):140–148
102. Ren J, Jenkins TC, Chaires JB (2000) Energetics of DNA intercalation reactions. Biochemistry 39(29):8439–8447
103. Mukherjee A, Lavery R, Bagchi B, Hynes JT (2008) On the molecular mechanism of drug intercalation into DNA: a simulation study of the intercalation pathway, free energy, and DNA structural changes. J Am Chem Soc 130(30):9747–9755
104. Treesuwan W, Wittayanaraku K, Anthony NG, Huchet G, Alniss G, Hannongbua S, Khalaf AI, Suckling CJ, Parkinson JA, Mackay SP (2009) A detailed binding free energy study of 2:1 ligand–DNA complex formation by experiment and simulation. Phys Chem Chem Phys 11(45):10682–10693
105. Chow CS, Bogdan FM (1997) A structural basis for RNA-ligand interactions. Chem Rev 97(5):1489–1513
106. Gilson MK, Given JA, Bush BL, McCammon JA (1997) The statistical-thermodynamical basis for computation of binding affinities: a critical review. Biophys J 72(3):1047–1069
107. Kostjukov VV, Khomytova NM, Evstigneev MP (2009) Partition of thermodynamic energies of drug–DNA complexation. Biopolymers 91(9):773–790
108. Kostjukov VV, Hernandez Santiago AA, Rodriguez FR, Castilla SR, Parkinson JA, Evstigneev MP (2012) Energetics of ligand binding to the DNA minor groove. Phys Chem Chem Phys 14(16):5588–5600

109. Beshnova DA, Lantushenko AO, Evstigneev MP (2010) Does the ligand-biopolymer equilibrium binding constant depend on the number of bound ligands? Biopolymers 93(11):932–935
110. Kostjukov VV, Evstigneev MP (2012) Relation between the change in DNA elasticity on ligand binding and the binding energetics. Phys Rev E 86(3 Pt 1):031919
111. Rocchia W, Alexov E, Honig B (2001) Extending the applicability of the nonlinear Poisson-Boltzmann equation: multiple dielectric constants and multivalent ions. J Phys Chem B 105(28):6507–6514
112. Kostjukov VV, Khomytova NM, Hernandez Santiago AA, Licona Ibarra R, Davies DB, Evstigneev MP (2011) Calculation of the electrostatic charges and energies for intercalation of aromatic drug molecules with DNA. Int J Quantum Chem 111(3):711–721
113. Kostjukov VV, Khomutova NM, Lantushenko AO, Evstigneev MP (2009) Hydrophobic contribution to the free energy of complexation of aromatic ligands with DNA. Biopolym Cell 25(2):133–141
114. Kostyukov VV, Khomutova NM, Evstigneev MP (2009) Contribution of changes in translational, rotational, and vibrational degrees of freedom to the energy of complex formation of aromatic ligands with DNA. Biophysics 54(4):606–615
115. Kostjukov VV, Khomytova NM, Evstigneev MP (2010) Hydration change on complexation of aromatic ligands with DNA: molecular dynamics simulations. Biopolym Cell 26(1):36–44
116. Kostjukov VV, Khomytova NM, Hernandez Santiago AA, Tavera A-M C, Alvarado JS, Evstigneev MP (2011) Parsing of the free energy of aromatic–aromatic stacking interactions in solution. J Chem Thermodyn 43(10):1424–1434
117. Kostyukov VV (2011) Energy of intercalation of aromatic heterocyclic ligands into DNA and its partition into additive components. Biopolym Cell 27(4):264–272
118. Kostyukov VV (2011) Energetics of complex formation of the DNA hairpin structure d(GCGAAGC) with aromatic ligands. Biophysics 56(1):28–39
119. Neidle S, Pearl LH, Herzyk P, Berman HM (1988) A molecular model for proflavine-DNA intercalation. Nucleic Acids Res 16(18):8999–9016
120. Brana MF, Cacho M, Gradillas A, de Pascual-Teresa B, Ramos A (2001) Intercalators as anticancer drugs. Curr Pharm Des 7(17):1745–1780
121. Pullman B (1989) Molecular mechanism of specificity in DNA-antitumor drug interactions. Adv Drug Res 18(1):2–112
122. Kostjukov VV, Khomytova NM, Davies DB, Evstigneev MP (2008) Electrostatic contribution to the energy of binding of aromatic ligands with DNA. Biopolymers 89(8):680–690
123. Kostjukov VV, Evstigneev MP (2014) Energy analysis of non-covalent ligand binding to nucleic acids: present and future. Biophysics 59(4):673–677
124. Chaires JB (1997) Energetics of Drug-DNA interactions. Biopolymers 44(3):201–215
125. Kubar T, Hanus M, Ryjacek F, Hobza P (2005) Binding of cationic and neutral phenanthridine intercalators to a DNA oligomer is controlled by dispersion energy: quantum chemical calculations and molecular mechanics simulations. Chem Eur J 12(1):280–290
126. Buisine E, de Villiers K, Egan TG, Biot C (2006) Solvent-induced effects: self-association of positively charged π systems. J Am Chem Soc 128(37):12122–12128
127. Nelson SM, Ferguson LR, Denny WA (2007) Non-covalent ligand/DNA interactions: minor groove binding agents. Mutation Res 623(1):24–40
128. Cai X, Gray PJ, Von Hoff DD (2009) DNA minor groove binders: back in the groove. Cancer Treatment Rev 35(5):437–450
129. Kostjukov VV, Rogova OV, Evstigneev MP (2014) General features of the energetics of complex formation between ligand and nucleic acids. Biophysics 59(4):666–672
130. Shaikh SA, Ahmed SR, Jayaram B (2004) A molecular thermodynamic view of DNA–drug interactions: a case study of 25 minor-groove binders. Arch Biochem Biophys 429(1):81–99
131. Dolenc J, Borstnik U, Hodoscek M, Koller J, Janezic D (2005) An ab initio QM/MM study of the conformational stability of complexes formed by netropsin and DNA. The importance of van der Waals interactions and hydrogen bonding. J Mol Struct 718(1):77–85

132. Kostjukov VV, Evstigneev MP (2012) Energy of ligand-RNA complex formation. Biophysics 57(4):450–463
133. Latham MP, Zimmermann GR, Pardi A (2009) NMR chemical exchange as a probe for ligand-binding kinetics in a theophylline-binding RNA aptamer. J Am Chem Soc 131(14):5052–5053
134. Lee SW, Zhao L, Pardi A, Xia T (2010) Ultrafast dynamics show that the theophylline and 3-methylxanthine aptamers employ a conformational capture mechanism for binding their ligands. Biochemistry 49(13):2943–2951

Chapter 3
Formation of DNA Lesions, its Prevention and Repair

Nihar R. Jena, Neha Agnihotri and Phool C. Mishra

Abstract The present review discusses three important aspects which are intimately related to human health at the molecular level. The first aspect is the formation of DNA lesions caused due to the reactions of DNA with certain free radicals known as reactive oxygen species (ROS) and reactive nitrogen oxide species (RNOS). Some of these free radicals are constantly formed in biological systems during the metabolic activities while others can be ascribed to the exposure to radiation or pollution. These species react with DNA bases, particularly guanine, leading to base misparing, mutation and several diseases including cancer. The mechanisms of reactions of certain ROS and RNOS with the DNA bases, particularly guanine or its modifications are discussed. The second aspect discussed here is the role of antioxidants some of which are present inside biological systems while others can be taken from external sources as food supplements. Certain endogenous anti-oxidants present in biological systems inhibit the formation of reactive free radicals while others, particularly those taken from outside, scavenge the same through appropriate chemical reactions. The molecular mechanisms of action of several anti-oxidants are discussed. The third aspect discussed here is the working of the complex and intelligent molecular machinery which is constantly active in biological systems and removes or repairs the damaged bases. Molecular mechanisms involved in some of these activities are reviewed. Details of recent theoretical studies on all the three aspects mentioned above are discussed.

P. C. Mishra (✉)
Department of Physics, Banaras Hindu University, Varanasi 221 005, India
e-mail: pcmishra_in@yahoo.com

N. R. Jena
Discipline of Natural Sciences, Indian Institute of Information Technology, Design and Manufacturing, Khamaria, Jabalpur 482005, India
e-mail: nrjena@iiitdmj.ac.in

N. Agnihotri
Department of Chemistry, University of Saskatchewan, Saskatoon S7N 5C9, Canada
e-mail: n.agnihotri6@gmail.com

L. Gorb et al. (eds.), *Application of Computational Techniques in Pharmacy and Medicine*, Challenges and Advances in Computational Chemistry and Physics 17,
DOI 10.1007/978-94-017-9257-8_3

3.1 Introduction

Several reactive species are continuously produced inside living cells during the normal metabolic activities. Formation of these species in low concentration is beneficial for living cells as they are involved in mitogenic response, cellular signal transduction, neurotransmission, blood pressure regulation etc. [1–3]. However, *in vivo* overproduction of these reactive species can be quite harmful for living systems as they perturb structures and functions of important biomolecules such as DNA, proteins and lipids severely. These reactive species can be subdivided into three main categories, i.e. (1) reactive oxygen species (ROS), (2) reactive nitrogen oxide species (RNOS), and (3) reactive halogen species (RHS). Oxygen in ROS, nitrogen in RNOS, and the halogen atom in RHS are the main constituents that readily react with different biomolecules in living systems. Although these reactive species can damage almost all biomolecules in living cells, damages to DNA are most abundant and harmful. As a consequence, several genotoxic and mutagenic lesions are created which ultimately induce life degrading diseases such as aging, cancer, neurodegenerative disorders including Alzheimer's disease, Parkinson's disease, and acute central nervous system injuries [2, 4]. In addition to these *in vivo* reactants, pre-solvated electrons [5, 6], other chemical agents [7], particularly pollutants [8], and high energy radiation [9–13] can also damage DNA by participating in complex biochemical reactions involving the various components of DNA. Several drugs [14, 15] used for cancer therapy can also potentially interfere with the normal functioning of DNA, thereby inducing other complicated diseases. It should be noted that in comparison to exogenous factors, endogenous factors are more responsible for the formation of several DNA damaged products which cause different pathological consequences.

Damage to DNA can occur in various possible ways such as base modification [16–18], conformational changes, formation of abasic sites [19], DNA strand breaks [20, 21], DNA strand cross-links [22, 23] and DNA-protein cross-links [24–26]. Among all forms of DNA damage, base modifications by reactive species are the simplest and very significant, following which a plethora of base lesions are generated in cells. Occurrence of these base lesions is frequently observed in disease-prone cells and tissues, particularly in tumours [27, 28]. It is believed that the various base lesions that are formed mainly due to oxidation of DNA accumulate in both mitochondrial and nuclear DNA with increasing age. This accumulation followed by failure of cellular defence mechanisms to repair or excise different DNA base lesions ultimately induces diseases. Not only these base lesions are capable of directly generating stable mutations, cancer and other pathological conditions, they can also be involved in the creation of several complicated processes indirectly by participating in the formation of complex tandem lesions. For example, generation of base damages in DNA can interfere with other bases on the same or the other strand, thereby creating intra-strand or inter-strand crosslinks respectively. Similarly the interference of these base lesions with proteins can also create different DNA-protein crosslink products [24–26].

Although a plethora of different lesions are formed in living cells every day, under normal conditions, majority of cells remain disease free. This is due to the fact that living cells have complex defense mechanisms operating against cell alterations mediated by reactive species. The main components of these defence mechanisms include several anti-oxidants which can be classified into enzymatic and non-enzymatic categories [29, 30]. Enzymatic anti-oxidants such as superoxide dismutase (SOD), catalase, glutathione, glutathione peroxidase and reductase in general inhibit the formation of reactive species in cells [31] while non-enzymatic anti-oxidants like vitamin E (α-tocopherol), vitamin C (ascorbic acid), carotenoids, flavonoids etc. scavenge the reactive species. These anti-oxidants ensure that there is minimal damage to DNA and other cellular components. The other and most important mechanism of cellular defense involves direct repair of the damaged bases [32–34] or their excision out of DNA followed by insertion of a new appropriate base at the corresponding location [35, 36]. This function is basically performed by different proteins that have inherent catalytic abilities. Biological systems have evolved with many such enzymes that have specific repair functions. For example, oxidative DNA base lesions are repaired by DNA glycosylases while alkylated base lesions are repaired by DNA alkyl transferases. The enzymatic DNA repair pathways have been shown to consist of two important processes i.e. damage recognition and catalysis.

Due to a spectrum of DNA damage lesions produced in cells, multiple cellular defense mechanisms are sometimes unable to prevent and repair these lesions. As a consequence, cells get affected by diseases. Therefore, understanding of formation of different biochemical reaction intermediates and products and multiple functions of various cellular defense mechanisms will certainly enrich our knowledge which can enable us to devise techniques to protect cells from diseases, e.g. by designing appropriate drugs. Application of theoretical methods can be immensely valuable towards understanding molecular mechanisms involved in the functioning of DNA, anti-oxidants and proteins. It is established that use of density functional theory (DFT) and molecular dynamics simulation can greatly help in the pursuit of explaining structures and functions of different biomolecules. For relatively smaller molecules, DFT calculations can predict structures and even reaction energetics fairly accurately. Therefore, we will mainly discuss results of DFT studies regarding mechanisms of formation of different DNA base lesions, action of anti-oxidants and repair of different base lesions in DNA by enzymes.

3.2 Endogenous Formation of Reactive Species

It is established that during the normal metabolic activities, some electrons (1–3 % of the total number) leak away from the mitochondrial electron transport chain and get bound to normal molecular oxygen, producing superoxide radical anion ($O_2^{\bullet-}$) [37, 38]. Electrons also leak from enzymatic sources such as NAD(P)H and xanthine oxidases and produce $O_2^{\bullet-}$. Superoxide radical anion ($O_2^{\bullet-}$) is quite reactive, having the half-life of 10^{-6} s, and has been observed in many pathological conditions including cancer [37, 38]. It is rapidly converted to nonreactive H_2O_2 by superoxide dismutatse (SOD) [39]. H_2O_2 is quite unreactive and does not participate in biochemical reactions directly. Further, enzymatic anti-oxidants like catalase and glutathione peroxidase transform H_2O_2 to H_2O.

It is believed that under stress conditions, in living cells, larger numbers of $O_2^{\bullet-}$ are formed which act as oxidants for certain enzymes which release Fe^{2+}. Subsequently, Fe^{2+} catalyzes the formation of $OH^{\bullet}$ from H_2O_2 following the Fenton reaction (Eq. 3.1). Further, horseradish peroxidase catalyses the formation of $OH^{\bullet}$ from H_2O_2 and $O_2^{\bullet-}$ involving Fe^{+2}/Fe^{+3} by the Haber-Weiss reaction (Eq. 3.2) [40]. $OH^{\bullet}$ is the most reactive among all *in vivo* reactants and can perturb structures and functions of all components of DNA. It has been found that $OH^{\bullet}$ has a very short half-life i.e. 10^{-9} s due to which it cannot diffuse to large distances [40]. Therefore, it reacts with DNA and other biomolecules only when formed in their close proximity.

$$Fe^{2+} + H_2O_2 \rightarrow Fe^{3+} + OH^{\bullet} + OH^{-} \tag{3.1}$$

$$O_2^{\bullet-} + H_2O_2 \rightarrow O_2 + OH^{\bullet} + OH^{-} \tag{3.2}$$

Peroxyl radicals ($ROO^{\bullet}$) are also formed in cells via lipid peroxidation. These radicals are quite stable and can diffuse to remote cellular locations. It is estimated that the half-lives of peroxyl radicals are upto a few seconds [41]. Among several $ROO^{\bullet}$, $HOO^{\bullet}$ is the simplest peroxyl radical formed due to the protonation of $O_2^{\bullet-}$ in living cells. It can modify fatty acids, proteins and DNA. $HOO^{\bullet}$ mediated damage to DNA mainly occurs through its reactions with the bases and sugar moieties. Similarly, metal induced catalysis of organic peroxyl radicals can generate alkoxyl radicals ($RO^{\bullet}$), which are even more reactive than $ROO^{\bullet}$. However, the half-life of an alkoxyl radical is much shorter than that of a typical $ROO^{\bullet}$.

Nitric oxide ($NO^{\bullet}$) is generated in living cells during nitric oxide synthase (NOS) mediated conversion of arginine to citruline [42]. It is quite beneficial for cells and is involved in insulin secretion, neural development, immune regulation, muscle relaxation, blood pressure regulation, neurotransmission etc. [43]. $NO^{\bullet}$ is quite stable and has a half-life of a few seconds. However, during oxidative stress, immune cells produce $NO^{\bullet}$ and $O_2^{\bullet-}$ in excess. As a result, these two species react rapidly to form peroxynitrite ($ONOO^{-}$) [44, 45]. $ONOO^{-}$ is an RNOS which is more reactive than $NO^{\bullet}$. Although reactivity of $ONOO^{-}$ is much less than that of $OH^{\bullet}$, due to its stability and large diffusion constant, it can react with DNA, proteins and lipids very effectively. Further, $ONOO^{-}$ itself is capable of generating other RNOS that are very reactive. For example, on protonation, it can generate ONOOH [46], which upon homolytic dissociation generates the reactive species nitrogen dioxide ($NO_2^{\bullet}$) and $OH^{\bullet}$ [47].

During the respiratory burst, H_2O_2 and $O_2^{\bullet-}$ are rapidly released by immune cells to kill pathogens. During this process, reaction of H_2O_2 with chloride anion (Cl^-), catalysed by heme myeloperoxidases produces hypochlorous acid (HClO) [48]. HClO is a powerful RHS and is quite cytotoxic. Due to this property, neutrophil cells use HClO to kill bacteria and different pathogens. However, it has been found that HOCl interferes with gene function by modifying different components of DNA. Another RHS, HBrO is also generated in living cells by human eosinophils that readily brominate DNA bases [49]. NO_2Cl is believed to be produced in cells by reaction of HOCl with nitrite (NO_2^-) [50]. NO_2Cl is quite cytotoxic and is capable of causing oxidation, nitration and halogenation of almost all biomolecules including DNA and RNA [51]. It has also been observed that when HOCl is added to a buffer containing Cl^-, it generates $^1O_2$(Eq. 3.3) [52], and in the presence of metal cations (M^{n+}), it generates $OH^{\bullet}$ (Eq. 3.4) [53]. Thus it is clear that although metals such as iron, manganese, copper etc. are essential for the maintenance of cell homeostasis, their excess presence in cells enhances production of different *in vivo* reactive species [54].

$$2HOCl \rightarrow {}^1O_2 + 2Cl^- + 2H^+ \tag{3.3}$$

$$HOCl + M^{n+} \rightarrow OH^{\bullet} + Cl^- + M^{n+1} \tag{3.4}$$

3.3 DNA Base Modifications

As discussed earlier, guanine (G), adenine (A), cytosine (C), and thymine (T) in DNA can be modified by *in vivo* reactive species or different exogenous factors giving rise to a plethora of DNA base lesions. *In vitro* studies with DNA or model compounds have shown that almost seventy different types of base lesions can be produced in DNA [55]. However, quantification of all these lesions in cellular DNA is difficult. As a result, only about 20 different base and sugar lesions have been identified so far in cells [56]. These include both simple and complex base lesions. Simple base lesions mainly consist of different oxidatively damaged products of DNA bases whereas complex base lesions mainly include bulky lesions like DNA-DNA and DNA-protein crosslinks. It should be noted that as the oxidation potential of G is the lowest among all the bases [57, 58], it is more frequently attacked by different reactive species than others. Therefore, we will here mainly consider the formation of different simple base lesions related to the modifications of guanine in DNA by different *in vivo* reactive species.

According to several model *in vitro* studies, oxidation of guanine can generate different lesions. These lesions include 8-oxoguanine (8-oxoG) [59–76], 2,6-diamino-4-oxo-5-formamidopyrimidine (FapyG) [77–82], 2,5-diamino-4H-imidazolone

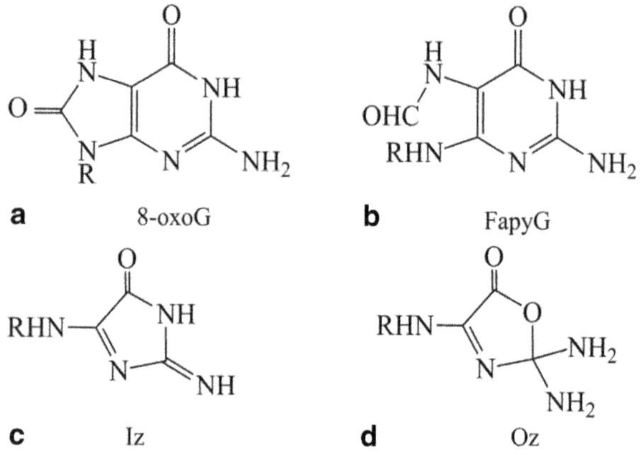

Fig. 3.1 Structures of different oxidative lesions of guanine observed in cellular DNA [55]. Here R stands for the sugar group

a 8-oxoG
b FapyG
c Iz
d Oz

(Iz) [83–90], 2,2,4-triamino-5(2H)-oxazolone (Oz) [83–91] etc. as the primary oxidation products and guanidinohydantoin (Gh) [92–96], spiroiminodihydantoin (Sp) [96, 97], cyanuric acid (Ca) [98], oxaluric acid (Oa) [98–103], etc. as secondary oxidation products. The secondary oxidation products of guanine arise due to the degradation of 8-oxoG in DNA [95–105]. Among these lesions, the formation of only 8-oxoG, FapyG, and Iz or Oz [78, 106] has been accurately quantified in cellular DNA. Structures of these oxidative lesions of guanine are shown in Fig. 3. 1(a-d).

Biological implications of different guanine oxidation products are not yet fully understood since studies regarding their structures and properties in cellular DNA are limited. However, it is expected that these guanine lesions would play important roles in different disorders [76, 107] and diseases [7]. Among the nitration lesions in cellular DNA, only 8-nitroguanine (8-NO_2G) and 5-guanidino-4-nitroimidazole (NI) have been quantified [107–114]. An overview of mechanisms of the formation of different mutagenic species due to *in vivo* reactions of certain common reactive species with guanine is presented below.

3.3.1 By Hydroxyl Radical ($OH^{\cdot}$)

The hydroxyl radical reacts preferentially with four carbon centres of guanine, thereby generating C2-OH, C4-OH, C5-OH, and C8-OH radical adduct intermediates [59, 74, 78]. The formation of the C8-OH radical adduct is also possible from the C4-OH and C5-OH radical adducts by subsequent dehydration and hydration. It has been shown that dehydration (elimination of OH^-) of C4-OH and C5-OH radical adducts generates guanine radical cation (G^+), which upon hydration (addition of OH^-) at the C8 position ultimately yields C8-OH neutral radical intermediate (Fig. 3.2) [115].

It has been shown that once the C8-OH radical adduct is formed, it undergoes subsequent oxidation to yield 8-oxoG which is far more stable than the C8-OH

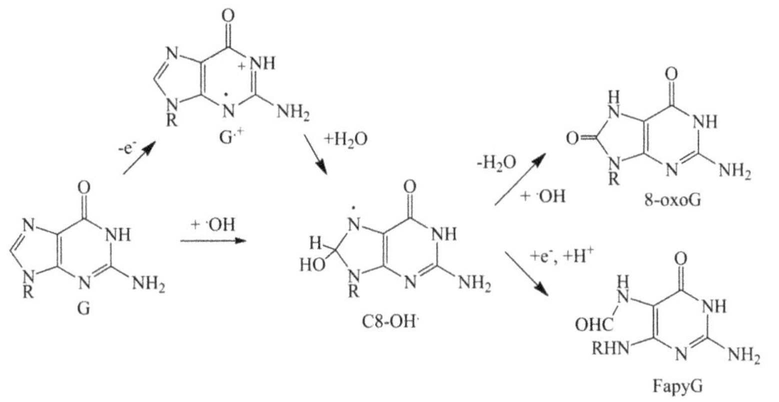

Fig. 3.2 Mechanisms of the formation of 8-oxoG and FapyG [76, 78]

radical intermediate [59]. On the basis of DFT studies, it is shown that under high concentration of OH radicals, reaction of the C8-OH with another $OH^{\bullet}$ can yield either 8-hydro,8-hydroxyguanine (8-OHG) or an imidazole ring-opened intermediate, which subsequently rearranges to 8-oxoG [59]. It is further found that the formation of 8-oxoG via this ring-opened intermediate is the most preferred pathway of 8-oxoG formation [59, 76, 78]. In addition to 8-oxoG, FapyG can also be formed from the C8-OH radical adduct and the ring-opened intermediate following reduction [77, 80, 81, 83] (Fig. 3.2). On the basis of a DFT study, it is proposed that the C8-OH radical adduct may undergo imidazole ring-opening followed by simultaneous protonation and reduction, or vice versa, to yield FapyG [77]. Existence of both 8-oxoG and FapyG in DNA may lead to guanine to thymine transversion mutation which is observed in a number of tumours [116].

3.3.2 *By Superoxide Radical Anion ($O_2^{\bullet-}$)*

It should be noted that guanine radical cation ($G^{\bullet+}$) formed from C4-OH and C5-OH radical adducts or direct oxidation of guanine by loss of an electron is prone to quick deprotonation (rate constant $= 1.8 \times 10^7\ s^{-1}$) to produce the guanine radical $G(-H)^{\bullet}$ [57, 58, 117, 118]. This reaction channel operates in competition with the one that leads to the formation of 8-oxoG in DNA [119]. Electron paramagnetic resonance and laser flash photolysis studies in aqueous media at the normal temperature have shown that the lifetime of the $G(-H)^{\bullet}$ is a few seconds. Thus under normal conditions, $G(-H)^{\bullet}$ is quite stable and does not degrade to other reaction products. However, its oxidation by $O_2^{\bullet-}$ has been observed to yield Iz, which on subsequent hydration yields Oz [103]. Oz is suggested to be the most stable oxidation product of guanine. Although $O_2^{\bullet-}$ mediated oxidation of guanine can result in the formation of 8-oxoG, due to its low oxidation potential, it further reacts with $O_2^{\bullet-}$ to form the most stable guanine oxidation product Oz. It is found that degradation of 8-oxoG to Oz readily occurs at neutral pH, while formation of Iz is relatively more favoured at basic pH [120].

3.3.3 By Peroxynitrite ($ONOO^-$)

It has been suggested that $ONOO^-$ is capable of initiating both oxidation and nitration of guanine. $ONOO^-$ mediated oxidation of guanine mainly generates 8-oxoG and Oz, while its nitration produces 8-NO_2G [61, 65, 106, 108, 109] and NI [110–114]. However, in cellular DNA, NI has been observed to be the dominant nitration product of guanine. A DFT study has revealed that $ONOO^-$ mediated oxidation of guanine generates 8-oxoG and NO_2^-, while nitration of guanine can yield either 8-NO_2G^-+H_2O or 8-NO_2G+OH^- [61]. Further, DFT studies have revealed that the reactivity of $ONOO^-$ gets enhanced in the presence of carbon dioxide (CO_2) that catalyses its reaction with guanine to produce 8-oxoG as discussed below [65]. A theoretical study of the reaction of $ONOO^-$ with guanine in the presence of CO_2 was carried out [65] at the B3LYP/6-31G(d, p) and B3LYP/AUG-cc-pVDZ levels of density functional theory [121, 122]. Geometry optimization calculations were carried out in gas phase while bulk solvent effect in aqueous media was treated by single-point energy calculations at the B3LYP/AUG-cc-pVDZ level of theory employing the polarizable continuum model (PCM) [123, 124]. An important catalytic role was found to be played by CO_2 in the reactions of $ONOO^-$ with guanine as discussed below.

Initially, $ONOO^-$ and CO_2 react together to form the nitrosoperoxycarbonate anion ($ONOOCO_2^-$) complex and subsequently this complex reacts with guanine [65]. Certain details of this reaction are as follows (Fig. 3.3). (a) The cis-conformer of nitrosoperoxycarbonate anion is more stable than its trans-conformer by about 1 kcal/mol. Further, the cis-conformer of nitrosoperoxycarbonate anion makes a stronger hydrogen bonded complex with the H9 atom of guanine than its trans-conformer. (b) The reactions between $ONOOCO_2^-$ and guanine occurring through different schemes mainly produce 8-oxoG or 8-NO_2G^-. (c) An analysis of the structures of products and barrier energies reveals that CO_2 acts as a catalyst in these reactions. It is found that $ONOOCO_2^-$ is broken into the CO_3 radical anion and $NO_2^{\bullet}$ due to dissociation of the OO bond while reacting with guanine. Intermediacy of $CO_3^{\bullet-}$ and $NO_2^{\bullet}$ appears to be the main cause of the catalytic action of CO_2. (d) As revealed by the calculated total energies of the products, 8-oxoG would be produced in much more abundance than 8-NO_2G^-. Therefore, $ONOO^-$ in complexation with CO_2 would cause mutation mainly through the formation of 8-oxoG. (e) The bulk solvent effect of water plays an important role in reducing the reaction barrier energies.

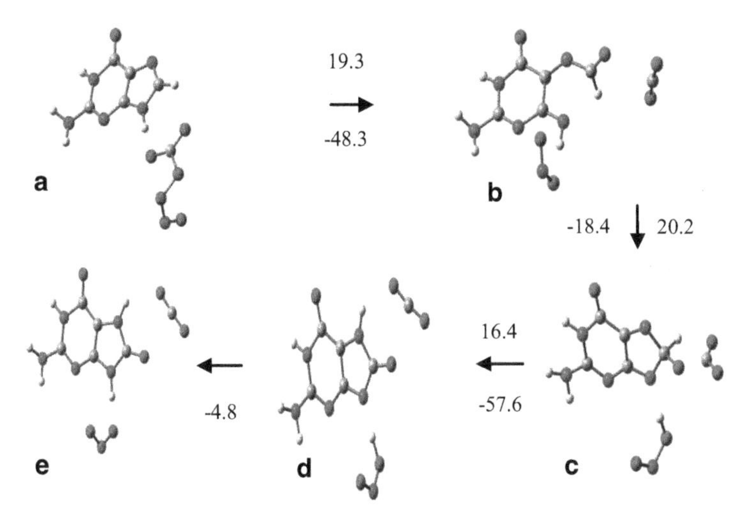

Fig. 3.3 Reaction of nitrosoperoxycarbonate anion with guanine. **a** Initial reactant complex, **b**, **c**, **d** Intermediates and **e** Product complex (8-oxoG+NO_2^-+CO_2). Gibbs barrier (positive) and released (negative) energies (kcal/mol) at each step obtained at the B3LYP/AUG-cc-pVDZ level in aqueous media are given near the arrows. The last step is barrierless [65]

3.3.4 By Nitrogen Dioxide ($NO_2^•$)

Formation of the guanine radical G(-H)$^•$ has been observed experimentally [118]. Reaction of G(-H)$^•$ with $NO_2^•$ generates 8-NO_2G as shown by DFT calculations [67]. Details in terms of structures and energies for a particular scheme of reaction between G(-H9$^•$) and $NO_2^•$ leading to the formation of 8-NO_2G obtained by geometry optimization at the BHandHLYP/6-31G(d, p) level in gas phase followed by single point energy calculations at the MP2/Aug-cc-pVDZ level of theory in aqueous media are shown in Fig. 3.4. The polarizable continuum model (PCM) [123, 124] was used to study solvation in bulk aqueous media. Water molecules facilitate proton transfer between different sites. In this case, proton transfer takes place from C8 to N9 in two steps. It should be noted that addition of $NO_2^•$ to G(-H)$^•$ where H9 is removed would not be relevant to the actual DNA since in DNA N9 is bonded to deoxyribose sugar. Reaction of G(-H)$^•$ with $NO_2^•$ has also been suggested to yield NI [78]. This in general occurs due to the addition of $NO_2^•$ at the C5 position of G(-H)$^•$ followed by subsequent hydration and decarboxylation. Formation of NI from 2′,3′,5′-Tri-O-acetyl-guanosine has been observed in UV/vis spectroscopy, ESI-MS, and NMR studies [113]. Further, it has been found that NI does not degrade to other products and is quite stable in aqueous solutions at both acidic and basic pH.

Formation of guanine radical cation ($G^{•+}$) has been observed experimentally when the specimen is exposed to high energy radiation [118]. The reaction of $G^{•+}$ with $NO_2^•$ has been shown to yield 8-NO_2G^+ [63]. Details of a particular scheme

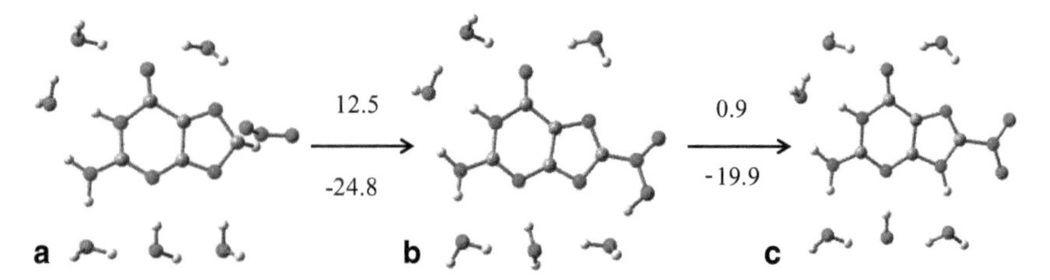

Fig. 3.4 **a** Reactant complex of G(-H9)$^{\cdot}$, $NO_2^{\cdot}$ and six water molecules, **b** Intermediate complex, and **c** Product complex (8-NO_2G+6H_2O). Water molecules facilitate proton transfer. Barrier (positive) and released (negative) energies (kcal/mol) at each step are given near the arrows [67]

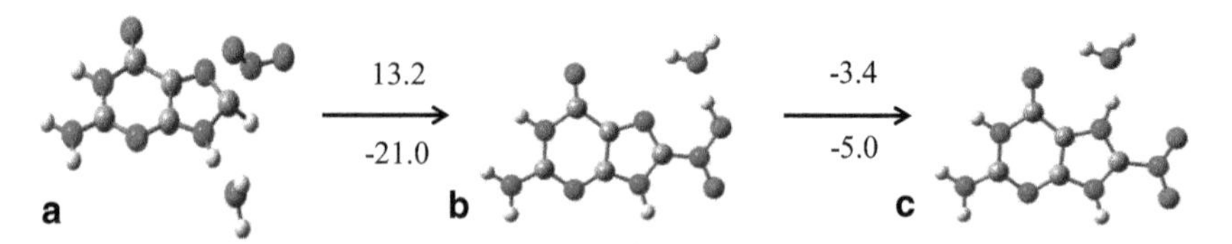

Fig. 3.5 **a** Reactant complex, **b** Intermediate complex, and **c** Product complex consisting of $G^{\cdot +}$ with $NO_2^{\cdot}$ in presence of a water molecule. The water molecule facilitates proton transfer. Gibbs barrier (*above* the arrows) and released (*below* the arrows) energies (kcal/mol) at each step are given. A negative barrier energy implies a barrierless reaction [63]

of this reaction obtained by geometry optimization at the B3LYP/AUG-cc-pVDZ level in gas phase followed by single point energy calculations at the MP2/Aug-cc-pVDZ level in aqueous media are shown in Fig. 3.5. Several barrier energies were appreciably lowered down due to the bulk solvent effect of water. A water molecule facilitates proton transfer from C8 to N7 in two steps. A comparison of binding energies of seven base pairs between one of the normal DNA bases and any one of 8-oxoG, 8-NO_2G or 8-NO_2G^+ showed that the initially formed 8-NO_2G^+-adenine base pair stabilized as 8-NO_2G-adenine$^+$ base pair and it was the most stable among all the base pairs considered [63]. It shows that 8-NO_2G^+ is highly mutagenic. In human respiratory tract epithelial cells, direct exposure of guanine to $NO_2^{\cdot}$ has also been observed to yield xanthine due to the deamination of guanine instead of its oxidation leading to 8-oxoG [125].

3.3.5 By Carbonate Radical Anion ($CO_3^{\cdot -}$)

Formation of carbonate radical anion ($CO_3^{\cdot -}$) as a site-selective oxidizing agent of guanine in double-stranded oligonucleotides leading to the formation of 8-oxoG has been observed experimentally [126–128]. In view of this observation, the reaction between guanine radical cation ($G^{\cdot +}$) and $CO_3^{\cdot -}$ was studied theoretically [73]. The relevant geometries were fully optimized in gas phase at the B3LYP/6-31G(d, p), BHandHLYP/AUG-cc-pVDZ, and B3LYP/AUG-cc-pVDZ levels of density functional theory. It was followed by single point energy calculations at the MP2/AUG-cc-pVDZ level in chlorobenzene using the gas phase geometries optimized at the

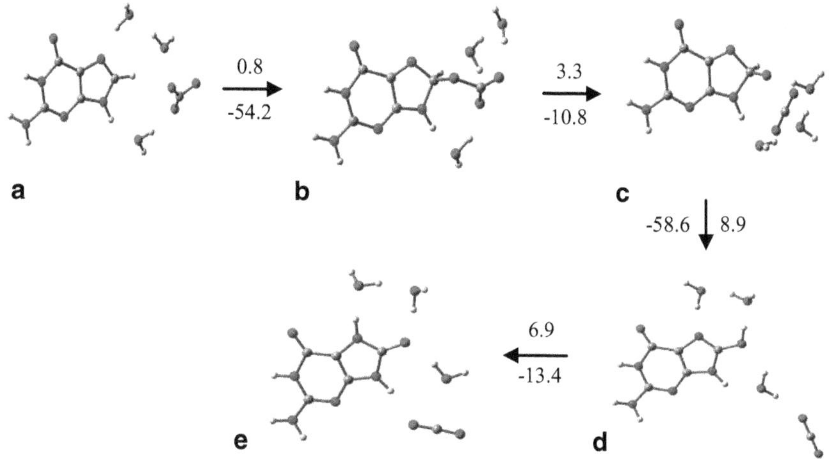

Fig. 3.6 **a** Reactant complex of $G^{\bullet+}$, $CO_3^{\bullet-}$ and three water molecules in the medium of chlorobenzene, **b**, **c**, **d** Intermediate complexes, and **e** Product complex (8-oxoG + CO_2 + $3H_2O$). Gibbs barrier (positive) and released (negative) energies (kcal/mol) at each step are given near the arrows [73]

BHandHLYP/AUG-cc-pVDZ level. The Gibbs barrier and released energies thus obtained are shown in Fig. 3.6. The calculated binding energy of the reactant complex (RC) between $G^{\bullet+}$ and $CO_3^{\bullet-}$ and three specific water molecules in gas phase was found to be large negative while that in bulk water using the PCM was found to be large positive. It was a clear indication that the reaction between $G^{\bullet+}$and $CO_3^{\bullet-}$ would not occur in pure water. The RC was not stabilized in aqueous media since the individual charged reactants would polarize the aqueous medium to much larger extents than their electrically neutral RC. It is to be noted that the experimental medium is very complex having several additional chemical species, and it is also photoirradiated [128]. In order to find an equivalent medium to the experimental one satisfying the condition that the RC was satisfactorily stabilized, bulk solvent effect on its stability in various solvent media corresponding to different dielectric constants was investigated at the level of single point energy calculations in toluene, acetone, dimethylsulfoxide, chlorobenzene, dichloroethane and water employing the PCM. It was thus found that the experimental medium would be represented by chlorobenzene fairly closely. The calculated results in chlorobenzene and certain other media showed that the reaction between $G^{\bullet+}$ and $CO_3^{\bullet-}$ leading to the formation of 8-oxoG would occur efficiently (Fig. 3.6). Thus $CO_3^{\bullet-}$ is found to be an efficient ROS.

3.3.6 By HOCl

It is established that HOCl is a potent oxidant that readily oxidises and chlorinates DNA, in particular guanine. The major reaction products of guanine thus formed are 8-oxoG and 8-chloroG [62, 129]. As suggested earlier, HOCl reacts mainly at the C8 position of guanine to generate the above mentioned species. In addition to this, HOCl also reacts with the C8 position of adenine (A) and C5 position of cytosine (C), thereby generating 8-chloroA and 5-chloroC respectively. It is further found that HOCl facilitated the formation of 8-chloroG which is more efficient than that of 8-oxoG. As activated neutrophils secrete myeloperoxidase produced HOCl, 8-chloroG can be a potential biomarker of inflammation [130, 131].

3.3.7 By NO_2Cl

It has been shown that the reaction of NO_2Cl with guanine can lead to its oxidation, chlorination and nitration, thereby forming 8-oxoG, 8-chloroG and 8-NO_2G respectively. Among these reactions, NO_2Cl has been proposed to efficiently induce nitration of guanine producing 8-NO_2G [68, 132]. However, epithelial cells at the site of inflection near the human stomach and respiratory tract can form NO_2Cl-mediated oxidized guanine lesions efficiently, as speculated earlier [132]. Both 8-oxoG and 8-NO_2G are mutagenic and can induce guanine to thymine transversion mutation in mammalian cells. In addition to this, 8-NO_2G can cause depurination to produce different cytotoxic products in DNA [132].

3.4 Prevention of Formation of Lesions by Anti-oxidants

As mentioned earlier, anti-oxidants can readily scavenge reactive species from biological media. Fruits, vegetables, whole grains and certain Indian spices are in general good sources of anti-oxidants [133]. Depending on their ability to scavenge reactive species, they can be classified in different categories [133]. There are three different important mechanisms by which anti-oxidants scavenge free radicals. If we consider R as a free radical and A as an anti-oxidant, we can express these three mechanisms as follows.

Single electron transfer (SET):

$$A + R^{\bullet} \rightarrow A^{\bullet +} + R^{-}$$

Hydrogen atom transfer (HAT) (or hydrogen abstraction):

$$A + R^{\bullet} \rightarrow [TS] \rightarrow A(\text{-H})^{\bullet} + RH$$

Radical adduct formation (RAF):

$$A + R^{\bullet} \rightarrow [TS'] \rightarrow AR^{\bullet}$$

Here TS and TS′ stand for transition states for hydrogen abstraction and addition reactions respectively. This section describes roles of SET, HAT and RAF mechanisms with regard to the reactions of various nutritional anti-oxidants including different vitamins and certain other important anti-oxidants which scavenge the different free radicals.

3.4.1 Superoxide Radical Anion Scavengers

As superoxide radical anion ($O_2^{\bullet-}$) is a negatively charged radical, its behaviour in charge transfer processes differs significantly from those of the uncharged ROS and RNOS. In addition to superoxide dismutase (SOD), dietary polyphenols including flavonoids and non-flavonoids can potentially scavenge $O_2^{\bullet-}$ from cells. Among other chemical agents, quercetins, and chlorogenic acids are shown to be potential superoxide scavengers [134, 135]. In an early study, it has been shown that ascorbic acid has a better $O_2^{\bullet-}$ scavenging ability than SOD [136]. However, in a subsequent study, on the basis of calculated rate constants involving reactions between different superoxide scavengers and $O_2^{\bullet-}$, it has been shown that the superoxide inhibitory ability follows the order SOD > L-ascorbic acid > eugenol > guaiacol > phenol [137].

Carotenoids (Figs. 3.7a–3.7c) are the naturally occurring organic pigments abundant in human diet mainly in carrots, pumpkins and sweet potatoes. Among this general class of carotenoids, β-carotene is distinguished by its beta rings at both the ends of the molecule. In nature, β-carotene is a precursor (inactive form) to vitamin A and the corresponding reaction occurs via the action of β-carotene 15,15'-monooxygenase. Carotenoids are very strong anti-oxidants having ability to scavenge almost all ROS and RNOS [138]. They have been found to have an excellent ability to directly scavenge $O_2^{\bullet-}$ [139, 140]. Carotenoids scavenge $O_2^{\bullet-}$ through the SET mechanism. In a recent theoretical study [141], it was proposed that $O_2^{\bullet-}$ inverts the direction of electron transfer in comparison to other ROS and RNOS. The SET mechanism can be expressed as follows.

$$\text{Carotenoids} + O_2^{\bullet-} \rightarrow \text{Carotenoids}^{\bullet-} + O_2$$

Here carotenoids act as electron acceptors while $O_2^{\bullet-}$ acts as an electron donor. Thus, the anti-oxidant property of carotenoids lies in their ability to convert $O_2^{\bullet-}$ into O_2. Reactions of $O_2^{\bullet-}$ with different carotenoid molecules including β-carotene (BC), adonirubin (ADO), astaxanthin (ASTA), canthaxanthin (CAN), β-doradexanthin (BDOR), 4-oxo-rubixanthin (OXO), torulene (TOR), lycopene (LYC) etc. were studied theoretically [141] using the B3LYP functional of DFT along with the 6-311G(d) basis set. Solvent effects in polar media e.g. water and non-polar media e.g. benzene were treated employing the integral equation formalism of the PCM (IEF-PCM) [141]. Since carotenoids are hydrophobic molecules, these are expected to be located mainly in the lipid portions of membranes. Therefore, their high reactivity towards $O_2^{\bullet-}$ in non-polar media e.g. benzene would play an important role with regard to their ability to prevent lipid peroxidation. The carotenoids which were found to have high scavenging ability towards $O_2^{\bullet-}$ in non-polar media are ADO, ASTA, CAN, BDOR, OXO, etc.

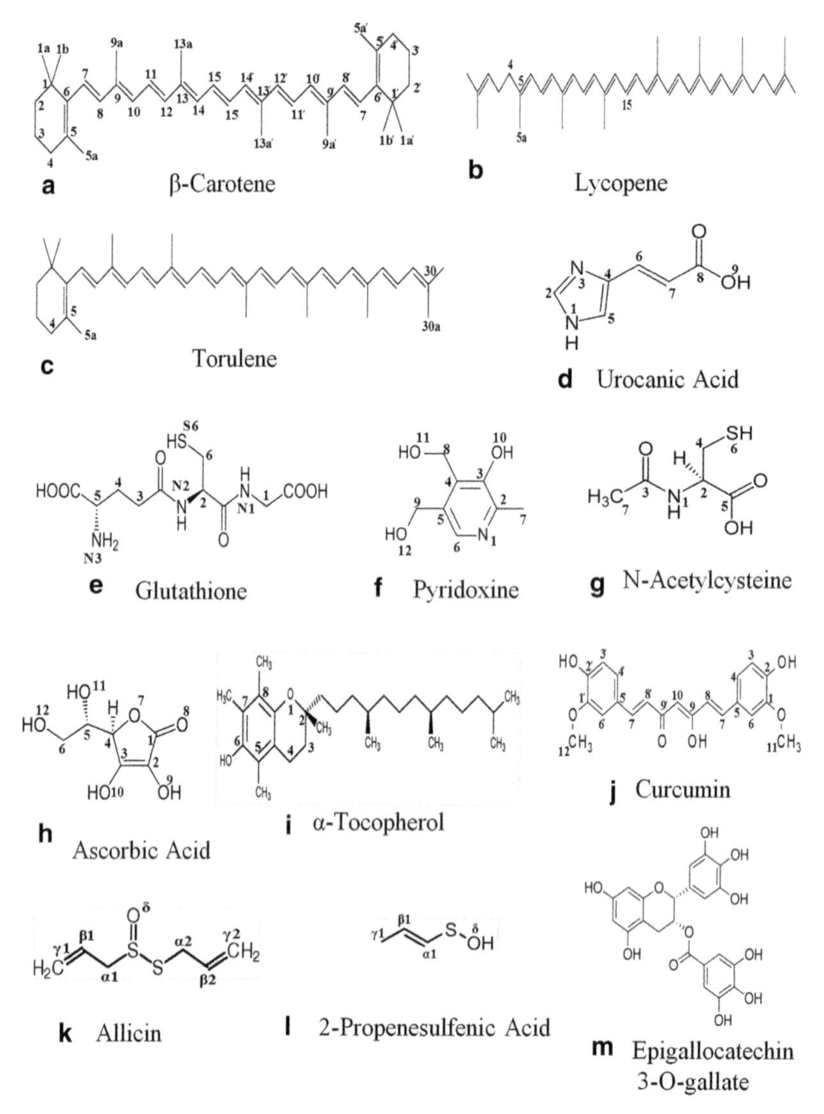

Fig. 3.7 Structures of some potential ROS/RNOS scavangers [133]

3.4.2 OH Radical Scavengers

The hydroxyl radical is the most reactive ROS present in biological systems. Urocanic acid (UCA) (Fig. 3.7d), a metabolite of histidine, is located in the upper layers of the human skin epidermis and acts as an absorber of UV-radiation. It has been recently shown that UCA can effectively scavenge OH radicals [142]. As high energy radiation produces OH• by water splitting and dissociation of H_2O_2, UCA can play a vital role in protecting genome from OH• mediated mutagenesis.

It was found that when an OH• attacks UCA, the RAF mechanism takes place and the lowest Gibbs barrier energy for this reaction corresponds to the C5 site (Fig. 3.7d). Barrier energies for the corresponding reactions at the C6 and C7 sites were found to be similar. The second OH• attack at all these three sites led to adduct formation barrierlessly. Binding energy analysis of the thus formed adducts revealed that the most stable adduct was formed by the addition of an OH• at the C6 site of the C7-OH• adduct. The third OH• abstracts the hydrogen atom of OH group bonded to the C6 or C7 site leading to two different mechanims which are initiated barrierlessly. Attack of the fourth OH• leads to the formation of imidazole-4-carboxaldehyde, glyoxylic acid and two water molecules as the primary products [142]. Formation of imidazole-4-carboxaldehyde and glyoxylic acid as the primary products has been observed experimentally [143, 144]. It was found that a positive barrier energy was involved only in the addition reaction of the first OH• to urocanic acid, while the reactions of the other three OH• were barrierless each [142]. Thus UCA has been shown to be an effective anti-oxidant. It is believed that UCA can serve as a better OH• scavenger than the traditional anti-oxidants like vitamins C and E.

Recently, it has been shown that compounds with phenolic rings in general can act as good OH• scavengers. For example, resveratrol and salicylates have been shown to have protective roles in this respect [145–147]. Curcumin, an extract of turmeric, has been shown to act as a potential OH• scavenger (Fig. 3.7j). It is known to have protective roles in cancer and the Alzheimer's disease [148]. On the basis of density functional theoretical studies, it has been shown that curcumin exhibits its anti-oxidant property by donating an electron to OH•, by undergoing hydrogen abstraction or by addition reactions with OH radical [149]. Ferulic acid which can be derived from curcumin [150] also scavenges OH• in a similar manner as curcumin. It is present in leaves and seeds of brown rice, whole wheat, apple, orange etc. In addition to the above, phenolic compounds can also scavenge OH radicals following addition of OH• to their double bonds in aromatic rings. Natural polyene and polyphenol classes of substances, such as flavonoids, present mostly in fruits and vegetables can also protect biological systems against OH• mediated oxidative damage.

A detailed theoretical study of the anti-oxidant activity of curcumin has been performed [149]. All geometry optimization calculations were carried out at the BHandHLYP/6-31G(d, p) level of density functional theory in the gas phase. It was followed by single-point energy calculations in the gas phase at the B3LYP/aug-cc-pVDZ and BHandHLYP/aug-cc-pVDZ levels of theory. Solvent effect in aqueous media was treated at the level of single point energy calculations at all the levels of theory mentioned above and employing the PCM along with the gas phase optimized geometries at the BHandHLYP/6-31G(d, p) level. Geometry optimization in aqueous media was also performed for certain reaction steps, and the Gibbs barrier energies thus obtained were quite similar to those obtained by single-point energy calculations. The SET mechanism was found to be more likely in polar media than in the gas phase. For hydrogen abstraction and OH• addition, all the possible sites were considered (Fig. 3.7j). It was observed that the most favorable site for hydrogen abstraction is the OH group attached to C2 while OH• addition was found to be the most favored at the C10 site of the heptadiene chain (Figs. 3.7j, 3.8). In

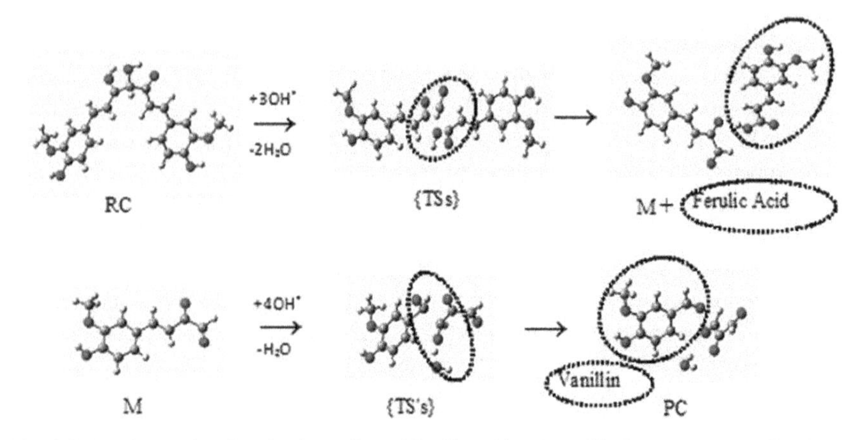

Fig. 3.8 A scheme showing the formation of ferulic acid and vanillin by a sequence of hydrogen abstraction and OH$^{\bullet}$ addition reactions between curcumin or its degradation products and eight OH$^{\bullet}$ [149]

Fig. 3.8, {TSs} and {TS′s} symbolically represent sets of three and four transition states respectively. The structures shown near {TSs} and {TS′s} are those of the last transition states in each set. RC and PC stand for reactant complex and product complex respectively. M stands for the molecule lying near ferulic acid in Fig. 3.8. In the RC of Fig. 3.8, an OH radical is already added at the C10 site. A sequence of addition and hydrogen abstraction reactions (Fig. 3.8) lead to the formation of ferulic acid and vanillin. Thus curcumin and its degradation products in total scavenge eight OH radicals.

Glutathione (Fig. 3.7e) is a major detoxifying agent inside our body particularly in the brain. Reduced glutathione (GSH) plays a central physiological role in protecting cells against exogeneous and endogenous oxidants, toxicants, DNA damaging agents e.g. OH$^{\bullet}$ and carcinogens [151–155]. It is produced in biological systems from the amino acids cysteine, glutamic acid and glycine. Glutathione is produced inside cells and cannot be ingested as a supplement as it is too large a molecule to pass through the intestinal walls. Further, glutathione levels cannot be increased by ingesting cysteine orally because oral cysteine is potentially toxic and is spontaneously destroyed in the gastronomical tract [156]. However, N-acetylcysteine (NAC), can be ingested orally, is the bioavailable form of cysteine and acts as a precursor for glutathione synthesis [157, 158].

A theoretical study on the OH$^{\bullet}$ scavenging ability of glutathione in its neutral non-zwitterionic form through the HAT mechanism has recently been carried out [155]. Hydrogen abstraction was considered from all the possible sites of glutathione (Fig. 3.7e). All the geometry optimization calculations were performed at the B3LYP/6-31G(d, p) level followed by M06/Aug-ccpVDZ and M06-2X/Aug-ccpVDZ level single point energy calculations in water [155]. In this work, abstractions of twelve hydrogen atoms attached to different carbon or nitrogen atoms of glutathione were found to be associated with small positive or negative Gibbs barrier energies. Further, the Gibbs barrier energies for abstractions of hydrogen atoms attached to four different sites including that attached to the sulphur atom were found to be moderate while the Gibbs barrier energy for abstraction of a hydrogen atom attached a nitrogen atom was found to be high. Thus glutathione has been shown to be an excellent scavenger of OH• [155]. Glutathione exists in an anionic form at physioplogical pH where the glycine moiety is deprotonated while the glutamic acid moiety is in the zwitterionic form [159]. A theoretical study was carried out at the M05-2X/6-311+G(d, p) level of theory on the OH• scavenging ability of glutathione where its anionic form (GS^-) was considered [153]. To treat solvent effect in water, the SMD continuum model was employed [160]. The SET mechanism was found to be endergonic and hence was ruled out. It was concluded that glutathione acts as an anti-oxidant exclusively by the HAT mechanism. Further, the most reactive site was found to be that of sulphur where the hydrogen abstraction reaction occurred in a barrierless manner and with a high rate constant (1.16×10^9 $M^{-1}s^{-1}$) [153]. It is clear that the barrier energy for hydrogen abstraction from the SH group of glutathione is small. However, whether it is negligibly small or not is still not established [153, 155].

Vitamin B_6 (Fig. 3.7f), also named as pyridoxine, is one of the eight water soluble vitamins of class B [161–163]. A high OH• quenching ability was reported for vitamin $B_6$and it was also found to be as effective as Vitamin E (Fig. 3.7i) [161]. Pyridoxine was found to be the most reactive among the vitamin B_6 sub-class of molecules i.e. pyridoxine, pyridoxal, pyridoxamine and pyridoxal-5-phosphate. In a recent theoretical study performed on this system at the B3LYP/6-31G(d, p) level of theory, it was found that it can scavenge up to eight OH• [163]. Thus vitamin B_6would be very beneficial as an OH• scavenger. Hydrogen abstraction reactions between OH• and pyridoxine were found to occur preferentially either from the C8 or the C9 site at the first step (Fig. 3.7f) [163]. In subsequent reactions, addition and cyclization were also considered. Thus vitamin B_6 is also shown to be an efficient OH• scavenger.

Interestingly, recently, high concentration of molecular hydrogen (H_2) has been shown to scavenge OH•. Drinking of water that contains higher level of H_2 has been shown to decrease urinary 8-oxoG significantly by mainly scavenging OH•. It is further suggested that the use of H_2 can be beneficial in the prevention of rheumatoid arthritis that mainly arises due to OH• mediated oxidative stress [164].

A density functional theoretical study was performed on the OH• scavenging property of N-acetylcysteine (NAC) which is a precursor of glutathione (Fig. 3.7g) [165]. Solvent effect in water was treated employing the PCM. N-acetylcysteine was found to effectively scavenge OH• through the HAT mechanism. This reaction at the sulphur site was found at the BHandHLYP/Aug-cc-pVDZ level of theory in polar media to take place barrierlessly and with a high rate constant. However, hydogen abstraction from two carbon sites was also found to contribute significantly to the OH• scavenging ability of N-acetylcysteine.

The mechanism of action of vitamin C (ascorbic acid) as an anti-oxidant towards OH• is well known. At physiological pH, it exists in a monoanionic form (AA^-). A theoretical study of the reactions between AA^- and OH• was performed employing density functional theory [166]. Solevnt effect in water was treated employing the

PCM. The SET mechanism being extremely endothermic has been ruled out. It was found that scavenging of OH• was possible by the HAT and RAF mechanisms. Hyrogen abstraction reactions from the O9, O10, O11 and C4 sites (Fig. 3.7h) were found to be barrierless. Thus vitamin C is shown to be an excellent scavenger of OH•. Its most preferred site for OH• addition was shown to be C2 (Fig. 3.7h) for which the Gibbs barrier energy at the B3LYP/6-311++G(3df,2pd) level was found to be ~ 1.25 kcal/mol in solution.

A theoretical study of the OH• scavenging ability of vitamin E (α-Tocopherol) (Fig. 3.7i) has revealed the following information [167]. The reaction was found to proceed through the HAT and RAF mechanisms. Hydrogen abstraction reactions from the phenolic OH group and methyl groups and OH• addition reactions at several positions of the aromatic ring were studied at the BHandHLYP/6-311++G(2d,2p) level of theory. The most probable site for hydrogen abstraction with a low barrier energy (0.8 kcal/mol) was found to be C2 while OH• addition was found to take place at the C6 site in a barrierless manner. Reaction rate constants for the HAT and RAF mechanisms were found to be 2.2×10^8 $M^{-1}s^{-1}$ and 5.6×10^7 $M^{-1}s^{-1}$ respectively [167]. Thus vitamin E is also shown to be an effective OH• scavenger like vitamin C.

3.4.3 OOH Radical Scavengers

Garlic is a well known for its medicinal properties. It has been shown to be an excellent anti-oxidant. It effectively scavenges free radicals due to its active ingredients allicin (2-propenyl 2-propenethiosulfinate) and 2-propenesulfenic acid (Fig. 3.7k, l) [168–170]. Both of these molecules can effectively scavenge OOH•. In addition, carotenoids are also well known OOH• scavengers. In recent theoretical studies [171, 172], reactions of OOH• with several carotenoids have been studied where these molecules have shown to be potent scavengers of OOH•. The OOH• scavenging ability of allicin has been found to be 1000 times less than that of 2-propenesulfenic acid [170]. A theoretical study of both the ingredients of garlic was carried out at the BHandHLYP/6-311++G(d, p) level of theory followed by CBS-QB3 calculations to overcome limitations arising due to the truncation of basis set and spin contamination. Solvent effect in water was treated by the IEF formalism of the PCM [141]. All the three reaction mechanisms mentioned earlier were taken into account. It was found that the SET mechanims would not contribute significantly to the reactions of OOH• with allicin and 2-propenesulfenic acid. Therefore, the HAT and RAF mechanisms are important in these reactions. Hydrogen abstraction was considered from the α1 and α2 sites of allicin and α1 and δ sites of 2-propenesulfenic acid (Fig. 3.7k, l). It was found that 2-propenesulfenic acid is much more reactive towards OOH• by the HAT mechanism than allicin. However, allicin was found to be somewhat more reactive towards OOH• by the RAF mechanism than 2-propenesulfenic acid. On the whole, the 2-propenesulfenic acid ingredient of garlic was shown to be a better scavenger of OOH• than the other ingredient allicin.

Carotenoids have also been found to scavenge $OOH^{\bullet}$. Their reactions with $OOH^{\bullet}$ can take place along more than one pathways. Certain theoretical studies have suggested that the SET mechanism is highly unfavourable for the reaction between carotenoids and $OOH^{\bullet}$ [173]. A theoretical study of reactions between carotenoids and $OOH^{\bullet}$ was performed at the BPW91/6-31G(d, p) level of theory [171]. Solvent effect of water and benzene on the reactions was studied employing the IEF formalism of the PCM [141]. A comprehensive study of HAT and RAF mechanisms was carried out considering three different carotenoids, namely, β-carotene (BC), lycopene (LYC) and torulene (TOR). For the HAT mechanism, in the cases of BC and LYC, lowest Gibbs barrier energies were found for hydrogen abstraction from the sites 5a and 4 while in the case of TOR, the lowest Gibbs barrier energy sites were 30 and 4 (Fig. 3.7a–c) in both water and benzene solvents. Among the three carotenoids, TOR was found to be the most efficient as an $OOH^{\bullet}$ scavenger through the HAT mechanism [171]. For RAF mechanism [171] in non-polar media, the C5 site corresponds to lowest barrier energy for both BC and LYC, while in the case of TOR, the lowest barrier energy corresponds to addition at the C30 site followed by that at the C5 site. However, polar media alters the order of reactivities of the different sites. For BC, Gibbs barrier energies for additions at C5, C7 and C9 sites were found to be comparable, and in the case of LYC, the barrier energy for addition at C15 was also found to be similar to that for addition at C5. In the case of TOR, the lowest barrier energy corresponds to the C5 site, and it was followed by the C30 site. Therefore, there would be a wider product distribution in polar media than in non-polar ones. The calculated reaction rate constants suggested that TOR is appreciably more reactive than BC through the RAF and HAT mechanisms in non-polar media. The RAF mechanism seems to be much less important in the context of reactions of carotenoids with $OOH^{\bullet}$ than HAT. On the whole, reactivities of carotenoids towards $OOH^{\bullet}$ are predicted to follow the order [171]: LYC > TOR > BC in non-polar media and TOR > LYC > BC in polar media. Adducts of $OOH^{\bullet}$ are predicted to be formed mainly at the terminal C5 site of the conjugated polyene chains.

3.4.4 NO_2 Radical Scavengers

Carotenoids are also highly reactive towards $NO_2^{\bullet}$. Among all carotenoids, β-carotene scavenges $NO_2^{\bullet}$ most effectively, preventing cardiovascular diseases [174]. Several experimental studies had been carried out for the reaction between β-carotene and $NO_2^{\bullet}$ in different environments but the favourable reaction mechanism and solvent effects could not be conclusively established [175–178]. Certain pulse radiolysis experiments had suggested that the reaction would proceed through electron transfer [177] while in other experiments, it was suggested to take place through the RAF mechanism [178]. All the three mechanisms (SET, HAT, RAF) for the reaction between β-carotene and $NO_2^{\bullet}$ were studied theoretically at the B3LYP/6-31G(d) level [174]. Solvent effect was treated employing the PCM in the solvents heptane, methanol and water having low, medium and high polarities respectively.

This theoretical study [174] revealed that the SET mechanism would operate only in polar media while the HAT mechanism would operate in all the solvents though it would also be more favoured in polar media. The most favoured site of β-carotene for hydrogen abstraction was found to be C4 of the β-ionone ring (Fig. 3.7a).

Structural changes occurred in β-carotene due to addition of $NO_2^{\bullet}$ at any of the carbon atoms of the polyene chain [174]. It resulted in breaking of the conjugated system and partial loss of planarity of the chain. Addition of $NO_2^{\bullet}$ at any of the C5 and C6 positions of the β-ionone ring caused ring twisting which had a noticeable effect since the double bond of the ring remained only partially conjugated with the polyene chain. The adduct BC(C5)-$NO_2^{\bullet}$ was found to be the most stable among all radical adducts confirming this site to be the most favoured one for the RAF mechanism [174]. A greater stability of the BC(C5)-$NO_2^{\bullet}$ adduct than that of BC(C6)-$NO_2^{\bullet}$ was due to a combination of two effects i.e. stabilization by resonance of the unpaired electron which does not occur in the latter, and partial extension of planarity of the conjugated polyene chain to the end groups by proper twisting of the β-ionone rings.

3.4.5 *$ONOO^-$ Scavengers*

Xanthine and hypoxanthine are formed from guanine following its oxidation. Uric acid (UA) is formed as a reaction product of xanthine and hypoxantine during the metabolic activities. UA has been shown to have $ONOO^-$ scavenging ability that significantly reduces nitration of tyrosine. Thus UA prevents inflammatory cell invasion into the central nervous system [179, 180]. Similarly, marine extracts like 2,3,6-tribromo-4,5-dihydroxybenzyl methyl ether can also act as potential $ONOO^-$ scavengers [181]. It has been shown that compounds that contain galloyl group exhibit high $ONOO^-$ scavenging activity. For example, several components of the green tea tannin such as epigallocatechin 3-o-gallate (Fig. 3.7m) and gallocatechin 3-o-gallate have been shown to scavenge $ONOO^-$ actively [182].

The above discussion shows that there is a strong support in favour of ability of the anti-oxidant molecules to scavenge ROS and RNOS which cause DNA damage and produce mutagenic products. Application of density functional theory has been particularly very useful in this context.

3.5 Enzymatic DNA Repair

Enzymatic DNA repair mainly depends on three vital processes, i.e. (1) protein translocation, (2) identification of the target, and (3) lesion repair by catalysis. These processes are discussed below.

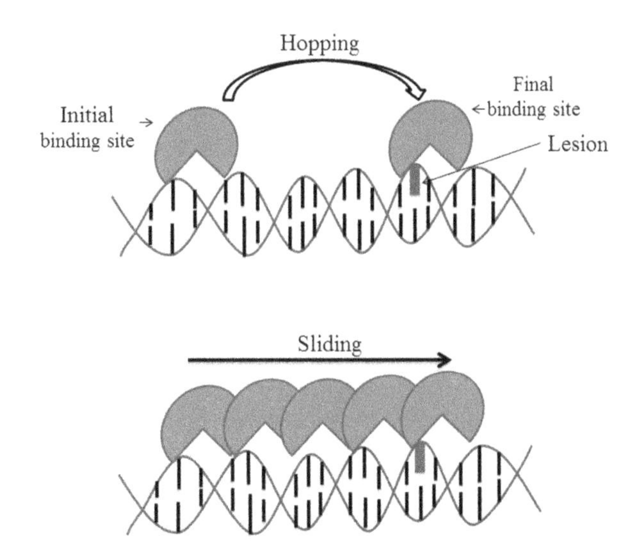

Fig. 3.9 Protein translocation by hopping and sliding on DNA [194]

3.5.1 Protein Translocation

It has been proposed that proteins find their target by diffusing along DNA in several possible ways such as hopping or jumping, sliding, intersegment transfer etc. [183]. Protein translocation by hopping involves scanning of DNA by the protein by making various microscopic associations and dissociations [183]. During hopping, proteins need to be associated with DNA initially by a non-specific binding. During this process, a protein jumps by dissociating from one segment of DNA and associating at another segment of the same (Fig. 3.9). It is believed that due to a small diffusion constant, after dissociation, a protein molecule spends some time near the initial site of binding. As a consequence, the next binding occurs by short range hops involving a few base pairs. However, during this process, a protein may also jump farther distances involving many base pairs away from its initial site of binding [184, 185].

During sliding, proteins remain bound at a site of DNA without being dissociated for a fairly long time so that translocation by one dimensional diffusion can occur accurately [186, 187] (Fig. 3.9). However, due to random thermal diffusion, the protein may move forward or backward on DNA from its initial site of binding. It has been suggested that if the length of DNA which is being scanned by a protein is relatively small, the rate of protein translocation by sliding gets accelerated. In intersegment transfer, proteins move from one segment of DNA to another via loops [188]. In this case, proteins bind to DNA at two different sites simultaneously and then dissociate from one end to move to the other. Further, intersegmental transfer requires a mean step size of about 400 base pairs and two parts of DNA binding surface to complete translocation [189, 190]. For this reason, it is not the most preferred mode of protein translocation. Recent studies involving both experimental and theoretical analyses performed on model systems in vitro suggest that proteins diffuse by employing the sliding mechanism when the length of DNA is less than 100 base pairs and by hopping if it is longer than 100 base pairs [191, 192]. It is also possible that proteins may undergo rotation along the helix axis of DNA during its translocation [193].

3.5.2 Target Recognition

Once a protein reaches the target site on DNA, it needs to recognize the lesion completely, as it can be a simple base lesion or a bulky cross-link product or an abasic lesion. Depending on the nature of the lesion, repair proteins adopt different strategies to accurately identify them. In the case of a simple base lesion, lesion recognition and repair occur by the nucleotide flipping mechanism where the base and the sugar get flipped out of the DNA double helix into the active site of the protein for further processing [194–199]. Two different mechanisms of nucleotide flipping have been proposed in some recent experimental and theoretical studies where either a protein is directly involved in base flipping or base flipping occurs without the involvement of a protein. According to the first mechanism, specific binding of a protein with DNA at the lesion site followed by subsequent squeezing can make the lesion extrahelical in DNA [200]. Alternatively, after binding to DNA, proteins may recruit one or more amino acid residues to intercalate into the DNA to push the damaged nucleotide out of the helix into its active site [201–203]. The space thus generated due to the affected base extrusion is ultimately filled by an amino acid that provides the necessary interaction required to stabilize the complementary base on one of the DNA strands [201–213]. It is proposed that due to a base modification, DNA gets locally distorted. This distortion is sensed by the protein during its translocation. As a result, the protein specifically binds at the lesion site and the protein translocation ends after recognizing the correct nucleotide. Several enzymes such as different DNA glycosylases like human (hOGG1) or bacterial (FPG) 8-oxoguanine-DNA glycosylase, human (AAG) and bacterial (AlkA/AlkB) alkyl adenine-DNA glycosylase, uracil-DNA glycosylase (UDG) [204–216], etc. and different DNA transferases like O6-alkylguanine-DNA alkyltransferase (AGT) [217], cytosine-5-methyltransferase [218, 219], etc. are proposed to identify and process the base lesion following the above mentioned mechanism. In addition to the above mentioned enzymes, endonuclease V (EndoV) [220, 221], which repairs bulky DNA lesions like thymine dimers has also been proposed to identify and process the lesion by the above protein facilitated nucleotide flipping mechanism.

According to the second mechanism of nucleotide flipping, intrinsic dynamics of DNA pushes the modified base significantly out of the DNA double helix, which then gets captured by the protein during its translocation on the DNA surface [222, 223]. This is argued to be possible due to differences in stabilities of base pairs and stacking interactions between the modified and complementary bases in DNA. In this situation, the repair protein may undergo conformational changes in such a way that the damaged base enters exactly into its active site where its repair can be completed accurately. It is also argued that repair proteins possess an inherent intriguing gate keeping strategy by which they deny unmodified bases access to its active site [224].

Molecular dynamical studies have recently emerged as a valuable tool to understand DNA repair by nucleotide flipping. Analysis of distributions of DNA bending and angle opening can indicate how nucleotide flipping occurs in DNA. Thermodynamical and kinetic factors associated with nucleotide flipping can be calculated which may give valuable insights into the mechanism of this process. Mechanisms of nucleotide flipping of thymine dimers by EndoV and uracil by UDG have been studied recently by molecular dynamics simulation [224]. This study revealed that due to base damage, DNA becomes quite flexible and the energy difference between the closed and open states decreases. This flexibility in DNA is sensed by the protein which binds to DNA at the damaged site. Due to this binding, DNA gets distorted which diminishes the barrier energy required for nucleotide flipping. In another similar study, it has been found that DNA bending due to base modification and internal DNA dynamics is the initial stage of base flipping where DNA becomes heavily distorted at the site of a mutated base pair. As a result, the mutated base may become slightly extrahelical which will then be captured by the protein that would subsequently push it into its active site after making it completely extrahelical [195]. As during nucleotide flipping, only a few bases close to the lesion site are involved, DFT can be employed to understand the detailed mechanism of nucleotide flipping. For example, using DFT, a two-step mechanism of nucleotide flipping has recently been proposed for the repair of O6-methylgunine (O6MG) by AGT [196]. According to this mechanism, at the first-step, AGT recruits one of its amino acids (Arg128) to intercalate into DNA at the lesion site that perturbs base pairing interactions of O6MG with C. In the second-step, Arg128 pushes O6MG out of the DNA double helix into the enzyme active site for catalysis and takes the vacant position of O6MG by making necessary hydrogen bonds with C to stabilize the DNA [196].

3.5.3 Repair by Catalysis

After placing the nucleotide into the protein's active site correctly, the enzyme initiates the catalytic reaction to repair the damage. Different enzymes use different catalytic processes depending on the nature of the lesion [76]. For example, alkylated DNA base damages are repaired by removing the alkyl group attached to the base by DNA alkyl transferases [225, 226]. However, oxidative damages are generally removed from DNA by different DNA glycosylases [224]. Monofunctional glycosylases cleave the *N*-glycosidic bond of one of the bases in the affected base pair, thereby creating an isolated base and apurinic or apyrimidinic (AP) site [208, 210, 212, 224, 225]. AP sites are then further processed by AP endonucleases forming 3'-hydroxyl and 5'-deoxyribose phosphate termini (Fig. 3.10a). In contrast, bifunctional glycosylases not only cleave the *N*-glycosidic bond but can also cleave the AP

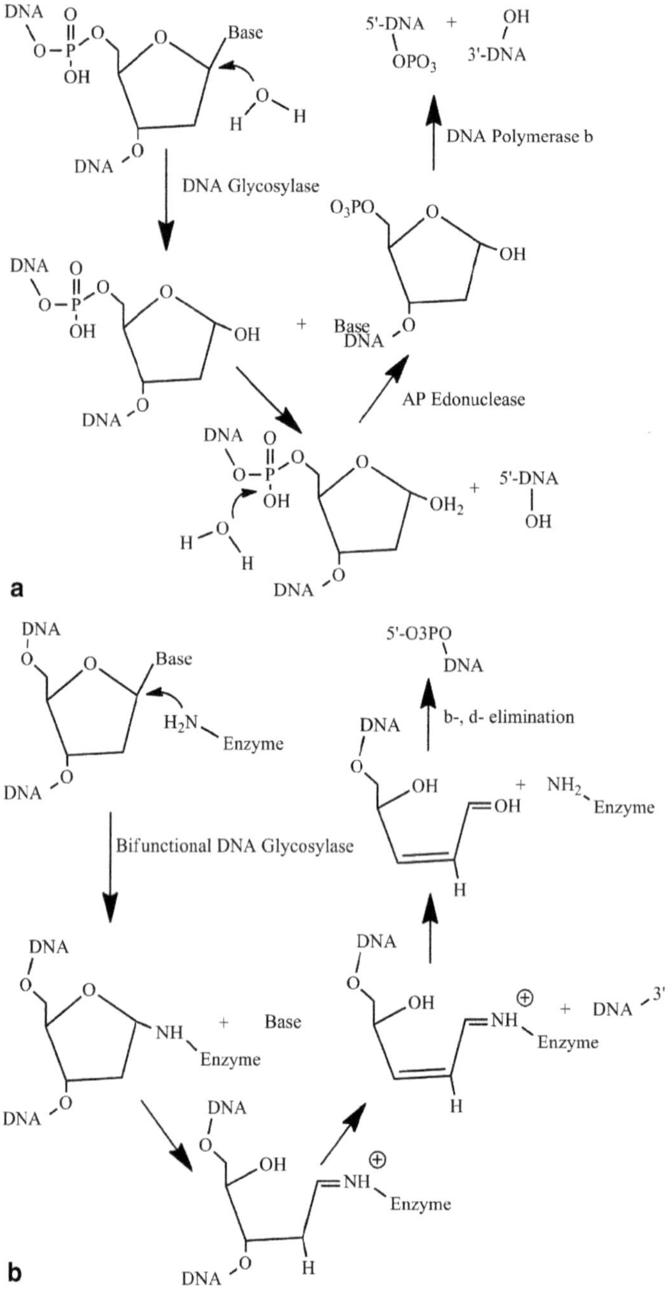

Fig. 3.10 Mechanisms of *N*-glycosidic bond dissociation by **a** monofunctional and **b** bifunctional DNA glycosylases [208]

site, leaving a 5'-phosphate and a 3'-α,β-unsaturated aldehyde [208] (Fig. 3.10b). Subsequently, a new nucleotide is synthesized by DNA polymerase β to be placed in the vacant position in DNA which then gets sealed to the nearest backbone by a DNA ligase [208].

As described above, catalysis by DNA glycosylases involves the dissociation of the *N*-glycosidic bond which can be achieved by its hydrolysis [208]. Generally a nucleophile located proximal to the glycosidic bond helps in the hydrolysis of the glycosidic bond [227]. After *N*-glycosidic bond dissociation, the corresponding base carries an extra electron which needs to be stabilized by an acid. It has been shown that in the case of monofunctional DNA glycosylases [228], a water molecule acts as a nucleophile leading to hydrolysis of the glycosidic bond. However, in the case of bifunctional glycosylases, an amino acid residue of the enzyme acts as the necessary nucleophile [208, 210, 212, 224, 225]. Other amino acids that directly or indirectly (through water mediated hydrogen bonds) stabilize the transition states of the glycosidic bond cleavage reaction are of paramount importance in reducing the barrier energy. As proposed earlier, a complete glycosidic bond cleavage may occur in multiple steps depending on the enzyme and the affected DNA base. In addition to this, processing of the phosphate and sugar moieties previously attached to the affected base also occurs in multiple steps and ultimately the DNA polymer gets sealed by completing the repair process [208, 210, 212, 224, 225].

The enzyme adenine DNA glycosylase which is also known as MutY catalyzes base excision repair by removing adenine from the abnormal base pair between 2′-deoxyadenosine and 8-oxo-2′-deoxyguanosine. In their study, McCann and Berti [228] studied the crystal structure of Escherichia coli MutY, obtained the transition state structures of MutY catalyzed DNA hydrolysis and also computed energetics of the reaction mechanism employing B3PW91/6-31þG(d, p) level of density functional theory. Gibbs free energy changes involved in the reaction mechanism proposed by McCann and Berti [228] were calculated at the MP2/AUG-cc-pVDZ level of theory in the gas phase using the B3LYP/6-31G(d, p) level optimized geometries [228]. It was found that in the model proposed by McCann and Berti [228], the second barrier energy was too high to be overcome in the biological medium. This difficulty was resolved by showing that the formation of the product having dissociated *N*-glycosidic bond of 2′-deoxyadenosine from the intermediate formed after the first step which has a moderate barrier energy would occur directly and barrierlessly without involving any other step [229]. This example shows that detailed quantum chemical studies of reactions can be immensely valuable to investigate mechanisms operating in complex biological systems.

Acknowledgment PCM is thankful to the University Grants Commission (New Delhi) for financial support. NRJ is thankful to the Indian National Science Academy (INSA) for Indo-Australia early career visiting fellowship. NA gratefully acknowledges use of facilities of the Department of Chemistry, University of Saskatchewan, Canada.

References

1. Khan AU, Wilson T (1995) Reactive oxygen species as cellular messengers. Chem Bio 2:437–445
2. Alfadda AA, Sallam RM (2012) Reactive oxygen species in health and disease. J Biomed Biotech 2012:936486
3. Yoneyama M, Kawada K, Gotoh Y, Shiba T, Ogita K (2010) Endogenous reactive oxygen species are essential for proliferation of neural stem/progenitor cells. Neuro chem Intern 56:740–746
4. Smith JA, Park S, Krause JS, Banik NL (2013) Oxidative stress, DNA damage, and the telomeric complex as therapeutic targets in acute neurodegeneration. Neurochem Int 62:764–775
5. Gu J, Leszczynski J, Schaefer III HF (2012) Interactions of electrons with bare and hydrated biomolecules: from nucleic acid bases to DNA segments. Chem Rev 112:5603–5640
6. Alizadeh E, Sanche L (2012) Precursors of solvated electrons in radiobiological physics and chemistry. Chem Rev 112:5578–5602
7. Poirier MC (2004) Chemical-induced DNA damage and human cancer risk. Nat Rev Cancer 8:630–637
8. DeMarini DM, Claxton LD (2006) Outdoor air pollution and DNA damage. Occup Environ Med 63:227–229
9. Alizadeh E, Sanz AG, Garcia G, Sanche L (2013) Radiation damage to DNA: the indirect effect of low-energy electrons. J Phys Chem Lett 4:820–825
10. Swiderek P (2006) Fundamental processes in radiation damage of DNA. Angew Chem Int Ed Engl 45:4056–4059
11. Becker D, Sevilla MD (1993) The chemical consequences of radiation damage to DNA. Adv Radiat Biol 17:121–180
12. Cadet J, Douki T (2011) Oxidatively generated damage to DNA by UVA radiation in cells and human skin. J Invest Dermatol 131:1005–1007
13. Kumar A, Sevilla MD, (2010) Proton-coupled electron transfer in DNA on formation of radiation-produced ion radicals. Chem Rev 110:7002–7023
14. Smith BL, Bauer GB, Povirk LF (1994) DNA damage induced by bleomycin, neocarzinostatin, and melphalan in a precisely positioned nucleosome. Asymmetry in protection at the periphery of nucleosome-bound DNA. J Biol Chem 269:30587–30594
15. Basu A, Krishnamurthy S (2011) Cellular responses to cisplatin-induced DNA damage. J Nucleic Acids 2011:201367
16. Dizdaroglu M (2012) Oxidatively induced DNA damage: mechanisms, repair and disease. Cancer Lett 327:26–47
17. Dizdaroglu M, Jaruga P (2012) Mechanisms of free radical-induced damage to DNA. Free Radic Res 46:382–419
18. Cooke M, Evans MD, Dizdaroglu M, Lunec J (2003) Oxidative DNA damage: mechanisms, mutation, and disease. The FASEB J 17:1195–1214
19. Takeshita M, Eisenberg W (1994) Mechanism of mutation on DNA templates containing synthetic abasic sites: study with a double strand vector. Nucleic Acids Res 22:1897–1902
20. Li X, Sevilla MD, Sanche L (2003) Density functional theory studies of electron interaction with DNA: can zero eV electrons induce strand breaks. J Am Chem Soc 125:13668–13669
21. Leenhouts HP, Chadwick KH (1974) Radiation induced DNA double strand breaks and chromosome aberrations. Theor Appl Genet 44:167–172
22. Noll DM, Mason TM, Miller PS (2006) Formation and repair of interstrand cross-links in DNA. Chem Rev 106:277–301
23. Deans AJ, West SC (2011) DNA interstrand crosslink repair and cancer. Nat Rev Cancer 11:467–480
24. Kow YW, Bao G, Minesinger B, Jinks-Robertson S, Siede W, Jiang YL, Greenberg MM (2005) Mutagenic effects of abasic and oxidized abasic lesions in Saccharomyces cerevisiae. Nucleic Acids Res 33:6196–6202

25. Shoulkamy M, Nakano T, Ohshima M, Hirayama R, Uzawa A, Furusawa Y, Ide H (2012) Detection of DNA–protein crosslinks (DPCs) by novel direct fluorescence labeling methods: distinct stabilities of aldehyde and radiation-induced DPCs. Nucleic Acids Res 40:1–13
26. Madison AL, Perez ZA, To P, Maisonet T, Rios EV, Trejo Y, Ochoa-Paniagua C, Reno A, Stemp EDA (2012) Dependence of DNA-protein cross-linking via guanine oxidation upon local DNA sequence as studied by restriction endonuclease inhibition. Biochemistry 51:362–369
27. Blanpain C, Mohrin M, Sotiropoulou PA, Passegue E (2011) DNA-damage response in tissue-specific and cancer stem cells. Cell Stem Cell 8:16–29
28. Redon CE, Dickey JS, Nakamura AJ, Kareva IG, Naf D, Nowsheen S, Kryston TB, Bonner WM, Georgakilas AG, Sedelnikova OA (2010) Tumors induce complex DNA damage in distant proliferative tissues in vivo. Proc Natl Acad Sci U S A 107:17992–17997
29. Limon-Pacheco J, Gonsebatt ME (2009) The role of anti-oxidants and anti-oxidant-related enzymes in protective responses to environmentally induced oxidative stress. Mutat Res 674:137–147
30. Blokhina O, Virolaineni E, Fagerstedt KV (2003) Anti-oxidants, oxidative damage and oxygen deprivation stress. a review. Ann Bot 91:179–194
31. Hu P, Tirelli N (2012) Scavenging ROS: superoxide dismutase/catalase mimetics by the use of an oxidation-sensitive nanocarrier/enzyme conjugate. Bioconjug Chem 23:438–449
32. Jena NR, Mishra PC, Suhai S (2009) Protection against radiation-induced DNA damage by amino acids: a DFT study. J Phys Chem B 113:5633–5644
33. Eker AP, Quayle C, Chaves I, Van der Horst GT (2009) DNA repair in mammalian cells: direct DNA damage reversal: elegant solutions for nasty problems. Cell Mol Life Sci 66:968–980
34. Mishina Y, Duguid EM, He C (2006) Direct reversal of DNA alkylation damage. Chem Rev 106:215–232
35. Yasui A, McCready SJ (1998) Alternative repair pathways for UV-induced DNA damage. Bioessays 20:291–297
36. Robertson AB, Klungland A, Aognes T, Leiros I (2009) DNA repair in mammalian cells: base excision repair: the long and short of it. Cell Mol Life Sci 66:981–993
37. De Bont R, van Larebeke N (2004) Endogenous DNA damage in humans: a review of quantitative data. Mutagenesis 19:169–185
38. Liu Y, Fiscum G, Schubert D (2002) Generation of reactive oxygen species by the mitochondrial electron transport chain. J Neurochem 80:780–787
39. Loschen G, Azzi A, Flohe L (1974) Superoxide radicals as precursors of mitochondrial hydrogen peroxide. FEBS Lett 42:68–72
40. Chen SX, Schopfer P (1999) Hydroxyl-radical production in physiological reactions. A novel function of peroxidase. Eur J Biochem 260:726–735
41. Marnett LJ (1987) Peroxyl free radicals: potential mediators of tumor initiation and promotion. Carcinogenesis 8:1365–1373
42. Tjalkens RB, Carbone DL, Wu G (2011) Detection of nitric oxide formation in primary neural cells and tissues. Methods Mol Biol 758:267–277
43. Moncada S, Palmer RMJ, Higgs EA (1991) Nitric oxide: physiology, pathophysiology and pharmacology. Pharmacol Rev 43:109–142
44. Beckman JS, Beckman TW, Chen J, Marshall PA, Freeman BA (1990) Apparent hydroxyl radical production by peroxynitrite: implications for endothelial injury from nitric oxide and superoxide. Proc Natl Acad Sci U S A 87:1620–1624
45. Beckman JS (1990) Ischaemic injury mediator. Nature 345:27–28
46. Merenyi G, Lind J (1998) Free radical formation in the peroxynitrous acid (ONOOH)/peroxynitrite ($ONOO^-$) system. Chem Res Toxicol 4:243–246
47. Beckman JS, Beckman TW, Chen J, Marshall PA, Freeman BA (1990) Apparent hydroxyl radical production by peroxynitrite: implications for endothelial injury from nitric oxide and superoxide. Proc Natl Acad Sci U S A 87:1620–1624

48. Gungor N, Knaapen AM, Munnia A, Peluso M, Haenen GR, Chiu RK, Godschalk RWL, van Schooten FJ (2010) Genotoxic effects of neutrophils and hypochlorous acid. Mutagenesis 25:149–154
49. Weiss SJ, Test ST, Eckmann CM, Roos D, Regiani S (1986) Brominating oxidants generated by human eosinophils. Science 234:200–203
50. Eiserich JP, Hristova M, Cross CE, Jones AD, Freeman BA, Halliwell B, van der Vliet A (1998) Formation of nitric oxide-derived inflammatory oxidants by myeloperoxidase in neutrophils. Nature 391:393–397
51. Eiserich JP, Cross CE, Jones AD, Halliwell B, van der Vliet A (1996) Formation of nitrating and chlorinating species by reaction of nitrite with hypochlorous acid. A novel mechanism for nitric oxide-mediated protein modification. J Biol Chem 271:19199–19208
52. Khan AU, Kasha M (1994) Singlet molecular oxygen evolution upon simple acidification of aqueous hypochlorite: application to studies on the deleterious health effects of chlorinated drinking water. Proc Natl Acad Sci U S A 91:12362–12364
53. Candeias LP, Stratford MR, Wardman P (1994) Formation of hydroxyl radicals on reaction of hypochlorous acid with ferrocyanide, a model iron(II) complex. Free Radic Res 20:241–249
54. Farina M, Avila DS, da Rocha JBT, Aschner M (2013) Metals, oxidative stress and neurodegeneration: a focus on iron, manganese and mercury. Neurochem Int 62:575–594
55. Cadet J, Douki T, Badouard C, Favier A, Ravanat J (2007) Oxidatively generated damage to cellular DNA. In: Evans MD, Cooke MS (eds) Oxidative damage to nucleic acids. Landes Bioscience & Springer, Austin, pp 1–13
56. Loft S, Poulsen HE (1996) Cancer risk and oxidative DNA damage in man. J Mol Mod 74:297–312
57. Steenken S, Jovanovic S (1997) How easily oxidizable is DNA? One-electron reduction potentials of adenosine and guanosine in aqueous solution. J Am Chem Soc 119:617–618
58. Fukuzumi S, Miyao H, Ohkubo K, Suenobu T (2005) Electron-transfer oxidation properties of DNA bases and DNA oligomers. J Phys Chem A 109:3285–3294
59. Jena NR, Mishra PC (2005) Mechanisms of formation of 8-oxoguanine due to reactions of one and two OH radicals and the H_2O_2 molecule with guanine: a quantum computational study. J Phys Chem B 109:14205–14218
60. White B, Smyth MR, Stuart JD, Rusling JF (2003) Oscillating formation of 8-oxoguanine during DNA oxidation. J Am Chem Soc 125:6604–6605
61. Jena NR, Mishra PC (2007) Formation of 8-nitroguanine and 8-oxoguanine due to reactions of peroxynitrite with guanine. J Comput Chem 8:1321–1335
62. Jena NR, Kushwaha PS, Mishra PC (2008) Reaction of hypochlorous acid with imidazole: formation of 2-chloro- and 2-oxoimidazoles. J Comput Chem 29:98–107
63. Agnihotri N, Mishra PC (2009) Mutagenic product formation due to reaction of guanine radical cation with nitrogen dioxide. J Phys Chem B 113:3129–3138
64. Shukla PK, Mishra PC (2007) H_2O_3 as a reactive oxygen species: formation of 8-oxoguanine from its reaction with guanine. J Phys Chem B 111: 4603–4615
65. Shukla PK, Mishra PC (2008) Catalytic involvement of CO_2 in the mutagenesis caused by reactions of $ONOO^-$ with guanine. J Phys Chem B 112:4779–4789
66. Kumar N, Shukla PK, Mishra PC (2010) Reaction of the OOH radical with guanine: mechanisms of formation of 8-oxoguanine and other products. Chem Phys 375:118–129
67. Agnihotri N, Mishra PC (2010) Formation of 8-nitroguanine due to reaction between guanyl radical and nitrogen dioxide: catalytic role of hydration. J Phys Chem B 114:7391–7404
68. Shukla PK, Mishra PC (2008) Reaction of NO_2Cl with imidazole: a model study for the corresponding reactions of guanine. J Phys Chem B 112:7925–7936
69. Mishra PC, Singh AK, Suhai S (2005) Interaction of singlet oxygen and superoxide radical anion with guanine and formation of its mutagenic modification 8-oxoguanine. Int J Quant Chem 102:282–301
70. Candeias LP, Steenken S (2000) Reaction of $OH^{\bullet}$ with guanine derivatives in aqueous solution: formation of two different redox-active OH-adduct radicals and their unimolecular transformation reactions. Properties of G(-H). Chem A Eur J 6:475–484

71. Steenken S (1989) Purine bases, nucleosides, nucleotides: aqueous solution redox chemistry and transformation reactions of their radical cations and e^- and OH adducts. Chem Rev 89:503–520
72. Reynisson J, Steenken S (2002) DFT calculations on the electrophilic reaction with water of the guanine and adenine radical cations: a model for the situation in DNA. Phys Chem Chem Phys 4:527–532
73. Yadav A, Mishra PC (2011) Quantum theoretical study of mechanism of the reaction between guanine radical cation and carbonate radical anion: formation of 8-oxoguanine. Int J Quantum Chem 112:2000–2008
74. Agnihotri N, Mishra PC (2011) Reactivities of radicals of adenine and guanine towards reactive oxygen species and reactive nitrogen oxide species: $OH^\bullet$ and $NO_2^\bullet$. Chem Phys Lett 503:305–309
75. Kasai H, Yamaizumi Z, Berger M, Cadet J (1992) Photosensitized formation of 7,8-dihydro-8-oxo-2'-deoxyguanosine (8-hydroxy-2'-deoxyguanosine) in DNA by riboflavin: a non-singlet oxygen mediated reaction. J Am Chem Soc 114:9692–9694
76. Jena NR (2012) DNA damage by reactive species: mechanisms, mutation and repair. J Biosci 3:503–517
77. Munk BH, Burrows CJ, Schlegel HB (2007) Exploration of mechanisms for the transformation of 8-hydroxy guanine radical to FAPyG by density functional theory. Chem Res Toxicol 20:432–444
78. Jena NR, Mishra PC (2012) Formation of ring-opened and rearranged products of guanine: mechanisms and biological significance. Free Radic Biol Med 53:81–94
79. Chatgilialoglu C, Neill PO (2001) Free radicals associated with DNA damage. Exp Gerontol 36:1459–1471
80. Tudek B (2003) Imidazole ring-opened DNA purines and their biological significance. J Biochem Mol Biol 36:12–19
81. Dizdaroglu M, Kirkali G, Jaruga P (2008) Formamidopyrimidines in DNA: mechanisms of formation, repair and biological effects. Free Radic Biol Med 45:1610–1620
82. Gates KS (2009) An overview of chemical processes that damage cellular DNA: spontaneous hydrolysis, alkylation and reactions with radicals. Chem Res Toxicol 22:1747–1760
83. Douki T, Spinelli S, Ravanat JL, Cadet J (1999) Hydroxyl radical-induced degradation of 2'-deoxyguanosine under reducing conditions. J Chem Soc Perkin Trans 2:1875–1880
84. Cui L, Ye WJ, Prestwich EG, Wishnok JS, Taghizadeh K, Dedon PC, Tannenbaum SR (2013) Comparative analysis of four oxidized guanine lesions from reactions of DNA with peroxynitrite, singlet oxygen, and gamma-radiation. Chem Res Toxicol 26:195–202
85. Raoul S, Berger M, Buchko GW, Joshi PC, Morin B, Weinfeld M, Cadet J (1996) H-1, C-13 and N-15 nuclear magnetic resonance analysis and chemical features of two main radical oxidation products of 2'-deoxyguanosine: oxazolone and imidazolone nucleosides. J Chem Soc Perkin Trans 2:371–381
86. Epe B (1996) DNA damage profiles induced by oxidizing agents. Rev Physiol Biochem Pharmacol 127:223–249
87. Breen AP, Murphy JA (1995) Reactions of oxyl radicals with DNA. Free Radic Biol Med 18:1033–1077
88. Matter B, Malejka-Giganti D, Csallany AS, Tretyakova N (2006) Quantitative analysis of the oxidative DNA lesion, 2,2-Diamino-43,5-di-O-acetyl-2-deoxy-b-D-erthro-sepentofuranosyl) amino]-5-(2H)-oxazolone (oxazolone) in vitro and in vivo by isotope dilution-capillary HPLC-ESI-MS/MS. Nucleic Acids Res 34:5499–5460
89. Luo WC, Muller JG, Rachlin EM, Burrows CJ (2001) Characterization of hydantoin products from one-electron oxidation of 8-oxo-7,8-dihydroguanosine in a nucleoside model. Chem Res Toxicol 14:927–938
90. Neeley WL, Essigmann JM (2006) Mechanisms of formation, genotoxicity and mutation of guanine oxidation products. Chem Res Toxicol 19:491–505
91. Matter B, Malejka-Giganti D, Csallany AS, Treyakova N (2006) Quantitative analysis of the oxidative DNA lesion, 2,2-diamino-4-(2-deoxy-β-d-erythro-pentofuranosyl)amino]-5(*2H*)-oxazolone (oxazolone), in vitro and in vivo by isotope dilution-capillary HPLC-ESI-MS/MS. Nucleic Acids Res 34:5449–5460

92. Hah SS, Kim HM, Sumbad RA, Henderson PT (2005) Hydantoin derivative formation from oxidation of 7,8-dihydro-8-oxo-2'-deoxyguanosine (8-oxodG) and incorporation of 14C-labeled 8-oxodG into the DNA of human breast cancer cells. Bioorg Med Chem Lett 15:3627–3631
93. Vialas C, Pratviel G, Claporols C, Meunier B (1998) Efficient oxidation of 2'-deoxyguanosine by Mn-TMPyP/KHSO5 to imidazolone without formation of 8-oxodG. J Am Chem Soc 120:11548–11553
94. Crean C, Geacintov NE, Shafirovich V (2008) Pathways of arachidonic acid peroxyl radical reactions and product formation with guanine radicals. Chem Res Toxicol 21:358–373
95. Vialas C, Claparols C, Pratviel G () Meunier B (200) Guanine oxidation in double-stranded DNAby Mn-TMPyP/KHSO5: 5,8-dihydroxy-7,8-dihydroguanine residue as a key precursor of imidazolone and parabanic acid derivatives. J Am Chem Soc 122:2157–2167
96. Niles JC, Wishnok JS, Tannenbaum SR (2001) Spiroiminohydantoin is the major product of 8-oxo-7,8-dihydroguanosine reaction with peroxynitrite in the presence of thiols, and guanosine photooxidation by methylene blue. Org Lett 3:763–766
97. Luo WC, Muller JC, Rachlin E, Burrows CJ (2000) Characterization of spiroiminohydantoin as a product of one-electron oxidation of 8-oxo-7,8-dihydroguanosine. Org Lett 2:613–617
98. Raoul S, Cadet J (1996) Photosensitized reaction of 8-oxo-7,8-dihydro-2'-deoxyguanosine. identification of 1-(2-deoxy-beta-D-erythro-penofuransoyl)cyanuric acid as the major singlet oxygen oxidation product. J Am Chem Soc 118:1892–1898
99. Sheu C, Foote CS (1995) Photosensitized oxygenation of a 7,8-dihydro-8-oxoguanosine derivative. Formation of dioxetane and hydroperoxide intermediates. J Am Chem Soc 117:474–477
100. Duarte V, Gasparutto D, Yamaguchi LF, Ravanat JL, Martinez GR, Medeiros MHG, Di Mascio P, Cadet J (2000) Oxaluric acid as the major product of singlet oxygen mediated oxidation of 8-oxo-7,8-dihydroguanine in DNA. J Am Chem Soc 122:12622–12628
101. Niles JC, Wishnok JS, Tannenbaum SR (2000) A novel nitration product formed during the reaction of peroxynitrite with 2',3',5'-tri-O-acetyl-7,8-dihydro-8-oxoguanosine. Chem Res Toxicol 13:390–396
102. Niles JC, Burney S, Singh SP, Wishnok JS, Tannenbaum SR (1999) Peroxynitrite reaction products of 3',5'-di-O-acetyl-8-oxo-7,8-dihydro-2'-deoxyuanosine. Proc Natl Acad Sci U S A 96:11729–11734
103. Tretyakova NY, Niles JC, Burney S, Wishnok JS, Tannenbaum SR (1999) Peroxynitrite-induced reactions of synthetic oligonucleotides containing 8-oxoguanine. Chem Res Toxicol 12:459–466
104. Ravanat JL, Saint-Pierre C, Cadet J (2003) One-electron oxidation of the guanine moiety of 2'-deoxyguanosine: influence of 8-oxo-7,8-dihydro-2'-deoxyguanosine. J Am Chem Soc 125:2030–2031
105. Jena NR, Mishra PC (2006) Addition and hydrogen abstraction reactions of an OH radical with 8-oxoguanine. Chem Phys Lett 422:417–423
106. Bashir S, Harris G, Denman MA, Blake DR, Winyard PG (1993) Oxidative DNA damage and cellular sensitivity to oxidative stress in human autoimmune diseases. Ann Rheum Dis 52:659–666
107. Jena, NR, Mishra PC (2013) Is FapyG mutagenic: evidence from the DFT study. Chemphyschem 14:3263–3270.
108. Akaike T, Okamoto S, Sawa T, Yoshitake J, Tamura F, Ichimori K, Miyazaki K, Sasamoto K, Maeda H (2003) 8-Nitroguanosine formation in viral pneumonia and its implication for pathogenesis. Proc Natl Acad Sci U S A 100:685–690
109. Tuo J, Liu L, Poulsen HE, Weimann A, Svendsen O, Loft S (2000) Importance of guanine nitration and hydroxylation in DNA in vitro and in vivo. Free Radic Biol Med 29:147–155
110. Halliwell B (2007) Oxidative stress and cancer: have we moved forward? Biochem J 401:1–11

111. Gu F, Stillwell WG, Wishnok JS, Shallop AJ, Jones RA, Tannenbaum SR (2002) Peroxynitrite-induced reactions of synthetic oligo 2'-deoxynucleotides and DNA containing guanine: formation and stability of a 5-guanidino-4-nitroimidazole lesion. Biochemistry 41:7508–7518
112. Niles JC, Wishnok JS, Tannenbaum SR (2006) Peoxynitrite-induced oxidation and nitration products of guanine and 8-oxoguanine: structures and mechanisms of product formation. Nitric Oxide 14:109–121
113. Niles JC, Wishnok JS, Tannenbaum SR (2001) A novel nitroimidazole compound formed during the reaction of peroxynitrite with 2',3',5'-tri-O-acetyl-guanosine. J Am Chem Soc 49:12147–12151
114. Neeley WL, Henderson PT, Delaney JC, Essigmann JM (2004) In vivo bypass efficiencies and mutational signatures of the guanine oxidation products 2-aminoimidazolone and 5-guanidino-4-nitroimidazole. J Biol Chem 279:43568–43573
115. Cullis PM, Malone ME, MersonDavies LA (1996) Guanine radical cations are precursors of 7,8-dihydro-8-oxo-2'-deoxyguanosine but are not precursors of immediate strand breaks in DNA. J Am Chem Soc 118:2775–2781
116. Sarasin A, Bounacer A, Lepage F, Schlumberger M, Suarez HG (1999) Mechanisms of mutagenesis in mammalian cells. Application to human thyroid tumours. C R Acad Sci III 322:143–149
117. Kumar A, Venkata P, Sevilla MD (2011) Hydroxyl Radical Reaction with Guanine in an Aqueous Environment: a DFT Study. J Phys Chem B 115:15129–15137
118. Kobayashi K, Tagawa S (2003) Direct observation of guanyl radical cation deprotonation in duplex DNA using pulse radiolysis. J Am Chem Soc 125:10213–10218
119. Rokhlenko Y, Geacintov NE, Shafirovich V (2012) Lifetimes and reaction pathways of guanine radical cations and neutral guanine radicals in an oligonucleotide in aqueous solutions. J Am Chem Soc 134:4955–4962
120. Luo W, Muller, JG, Burrows CJ (2001) The pH-dependent role of superoxide in riboflavin-catalyzed photooxidation of 8-oxo-7,8-dihydroguanosine. Org Lett 3:2801–2804
121. Hohenberg P, Kohn W (1964) Inhomogeneous electron gas. Phys Rev 136B:864–871
122. Kohn W, Sham LJ (1965) Self-consistent equations including exchange and correlation effects. Phys Rev A 140:1133–1138
123. Barone V, Cossi M, Tomasi J (1997) A new definition of cavities for the computation of solvation free energies by the polarizable continuum model. J Chem Phys 107:3210–3221
124. Cossi M, Barone V, Mennucci B, Tomasi J (1998) Ab-initio Study of Ionic Solutions by a Polarizable Continuum Dielectric Model. Chem Phys Lett 286:253–260
125. Spencer JPE, Whiteman M, Jenner A, Halliwell B (2000) Nitrite-induced deamination and hypochlorite-induced oxidation of DNA in intact human respiratory tract epithelial cells. Free Radic Biol Med 28:1039–1050
126. Shafirovich V, Dourandin A, Huang W, Geacintov NE (2001) The carbonate radical is a site-selective oxidizing agent of guanine in double-stranded oligonucleotides. J Biol Chem 276:24621–24626
127. Rokhlenko Y, Geacintov NE, Shafirovich V (2012) Lifetimes and reaction pathways of guanine radical cations and neutral guanine radicals in an oligonucleotide in aqueous solutions. J Am Chem Soc 134:4955–4962
128. Joffe A, Geacintov NE, Shafirovich V (2003) DNA lesions derived from the site selective oxidation of guanine by carbonate radical anions. Chem Res Toxicol 16:1528–1538
129. Masuda M, Suzuki T, Friensen MD, Ravanat JL, Cadet J, Pignattelli B, Nishino H, Ohshima H (2001) Chlorination of guanosine and other nucleosides by hypochlorous acid and myeloperoxidase of activated human neutrophils catalysis by nicotine and trimethylamine. J Biol Chem 276:40486–40496
130. Badouard C, Masuda M, Nishino H, Cadet J, Favier A, Ravanat JL (2005) Detection of chlorinated DNA and RNA nucleosides by HPLC coupled to tandem mass spectrometry as potential biomarkers of inflammation. J Chromatogr B Analyt Technol Biomed Life Sci 827:26–31

131. Whiteman M, Spencer JPE, Jenner A, Halliwell B (1999) Hypochlorous acid induced DNA base modification: potentiation by nitrite: biomarkers of DNA damage by reactive oxygen species. Biochem Biophy Res Commun 257:572–576
132. Ohshima R, Sawa T, Akaike T (2006) 8-Nitroguanine, a product of nitrative DNA damage caused by reactive nitrogen species: formation, occurrence, and implications in inflammation and carcinogenesis. Antioxid Redox Signal 8:1033–1045
133. Dauchet L, Amouyel P, Dallongeville J (2009) Fruits, vegetables and coronary heart disease. Nature Rev Cardiol 6:599–608
134. Slemmer JE, Shacka JJ, Sweeney MI, Weber JT (2008) Anti-oxidants and free radical scavengers for the treatment of stroke, traumatic brain injury and aging. Curr Med Chem 15:404–414
135. Chun OK, Kim DO, Lee CY (2003) Superoxide radical scavenging activity of the major polyphenols in fresh plums. J Agric Food Chem 51:8067–8072
136. Nandi A, Chatterjee IB (1987) Scavenging of superoxide radical by ascorbic acid. J Biosci 11:435–441
137. Yasuhisa T, Hideki H, Muneyoshi Y (1993) Superoxide radical scavenging activity of phenolic compounds. Int J Biochem 25:491–494
138. Edge R, McGarvey DJ, Truscott TG (1997) The carotenoids as anti-oxidants-a review. J Photochem Photobiol B: Biol 41:189–200
139. Cardounel AJ, Dumitrescu C, Zweier JL, Lockwood SF (2003) Direct superoxide anion scavenging by a disodium disuccinate astaxanthin derivative: relative efficacy of individual stereoisomers versus the statistical mixture of stereoisomers by electron paramagnetic resonance imaging. Biochem Biophys Res Commun 307:704–712
140. Foss BJ, Sliwka HR, Partali V, Cardounel AJ, Zweierb JL, Lockwood SF (2004) Direct superoxide anion scavenging by a highly water-dispersible carotenoid phospholipid evaluated by electron paramagnetic resonance (EPR) spectroscopy. Bioorg Med Chem Lett 14:2807–2812
141. Galano A, Vargas R, Martínez A (2010) Carotenoids can act as anti-oxidants by oxidizing the superoxide radical anion. Phys Chem Chem Phys 12:193–200
142. Tiwari S, Mishra PC (2011) Urocanic acid as an efficient hydroxyl radical scavenger: a quantum theoretical study. J Mol Model 17:59–72
143. Kammeyer A, Eggelte TA, Overmars H, Bootsma JD, Bos JD, Teunissen MBM (2001) Oxidative breakdown and conversion of urocanic acid isomers by hydroxyl radical generating systems. Biochim Biophys Acta 1526:277–285
144. Gibbs NK, Tye J, Norval M (2008) Recent advances in urocanic acid photochemistry, photobiology and photoimmunology. Photochem Photobiol Sci 7:655–667
145. Lu KT, Chiou RY, Chen LG, Chen MH, Tseng WT (2006) Neuroprotective effects of resveratrol on cerebral ischemia-induced neuron loss mediated by free radical scavenging and cerebral blood flow elevation. J Agric Food Chem 54:3126–3131
146. Lipinski B (2011) Hydroxyl radical and its scavengers in health and disease. Oxid Med Cell Longev 2011:809696
147. Ghiselli A, Laurenti O, DeMattia G, Maiani G, Ferro-Luzzi A (1992) Salicylate hydroxylation as an early marker of in vivo oxidative stress in diabetic patients. Free Rad Biol Med 13:621–626
148. Hatcher H, Planalp R, Cho J, Torti FM, Torti SV (2008) Curcumin: from ancient medicine to current clinical trials. Cell Mol Life Sci 65:1631–1652
149. Agnihotri N, Mishra PC (2011) Scavenging mechanism of curcumin toward the hydroxyl radical: a theoretical study of reactions producing ferulic acid and vanillin. J Phys Chem A 115:14221–14232
150. Srinivasan M, Sudheer AR, Menon VP (2007) Ferulic acid: therapeutic potential through its anti-oxidant property. J Clin Biochem Nutr 40:92–100
151. Balendiran GK, Dabur R, Fraser D (2004) The role of glutathione in cancer. Cell Biochem Funct 22:343–352

152. Kerksick C, Willoughby DJ (2005) The anti-oxidant role of glutathione and N-Acetylcysteine supplements and exercise-induced oxidative stress. Int Soc Sport Nutr 2:38–44
153. Galano A, Alvarez-Idaboy JR (2011) Glutathione:mechanism and kinetics of its non-enzymatic defense action against free radicals. RSC Adv 1:1763–1771
154. Alvarez-Idaboy JR, Galano A (2012) On the chemical repair of DNA radicals by glutathione: hydrogen vs electron transfer. J Phys Chem B 116:9316–9325
155. Yadav A, Mishra PC (2013) Modeling the activity of glutathione as a hydroxyl radical scavenger considering its neutral non-zwitterionic from. J Mol Model 19:767–777
156. Hemat RAS (2003). Principles of Modern Urology, Publisher Urotext (urotext@urotext.com)
157. Morley N, Curnow A, Salter L, Campbell S, Gould DJ (2003) N-acetyl-l-cysteine prevents DNA damage induced by UVA, UVB and visible radiation in human fibroblasts. Photochem Photobiol B Biol 72:55–60
158. Aruoma OI, Halliwell B, Hoey BM, Butler J (1989) The anti-oxidant action of N-acetylcysteine: its reaction with hydrogen peroxide, hydroxyl radical, superoxide, and hypochlorous acid. Free Rad Biol Med 6:593–597
159. Han Y, HaoMiao Z, Shen J (2007) Molecular dynamics simulation study on zwitterionic structure to maintain the normal conformations of glutathione. Sci China Ser Chem B 50:660–664
160. Marenich AV, Cramer CJ, Truhlar DG (2009) Universal solvation model based on solute electron density and on a continuum model of the solvent defined by the bulk dielectric constant and atomic surface tensions. J Phys Chem B 113:6378–6396
161. Ehrenshaft M, Bilski P, Li M, Chignell CF, Daub ME (1999) A highly conserved sequence is a novel gene involved in de novo vitamin B6 biosynthesis. Proc Natl Acad Sci 96:9374–9378
162. Matxain JM, Ristilä M, Strid Å, Eriksson LA (2006) Theoretical study of the anti-oxidant properties of pyridoxine. J Phys Chem A 110:13068–13072
163. Matxain JM, Padro D, Ristilä M, Strid Å, Eriksson LA (2009) Evidence of high OH• radical quenching efficiency by vitamin B_6. J Phys Chem B 113:9629–9632
164. Ishibashi T, Sato B, Rikitake M, Seo T, Kurokawa R, Hara Y, Naritomi Y, Hara H, Nagao T (2012) Consumption of water containing a high concentration of molecular hydrogen reduces oxidative stress and disease activity in patients with rheumatoid arthritis: an open-label pilot study. Med Gas Res 2:27
165. Agnihotri N, Mishra PC (2009) Mechanism of scavenging action of N-acetylcysteine for the OH radical: a quantum computational study. J Phys Chem B 113:12096–12104
166. Li P, Shen Z, Wang W, Ma Z, Bi S, Sun H, Bu Y (2010) The capture of H• and OH• radicals by vitamin C and implications for the new source for the formation of the anion free radical. Phys Chem Chem Phys 12:5256–5267
167. Navarrete M, Rangel C, Corchado JC, Espinosa-García J (2005) Trapping of the OH radical by α-Tocopherol: a theoretical study. J Phys Chem A 109:4777–4784
168. Okada Y, Tanaka K, Fujita I, Sato E, Okajima H (2005) Anti-oxidant activity of thiosulfinates derived from garlic. Redox Rep 10:96–102
169. Vaidya V, Ingold KU, Pratt DA (2009) Garlic: source of the ultimate anti-oxidants-sulfenic acids. Angew Chem Int Ed 48:157–160
170. Galano A, Francisco-Marquez M (2009) Peroxyl radical scavenging activity of garlic: 2-propenesulfenic acid vs allicin. J Phys Chem B 113:16077–16081
171. Galano A, Francisco-Marquez M (2009) Reactions of OOH radical with β-carotene, lycopene, and torulene: hydrogen atom transfer and adduct formation mechanisms. J Phys Chem B 113:11338–11345
172. Martínez A, Vargas R, Galano A (2010) Theoretical study on the chemical fate of adducts formed through free radical addition reactions to carotenoids. Theor Chem Acc 127:595–603
173. Galano A (2007) Relative antioxidant efficiency of a large series of carotenoids in terms of one electron transfer reaction. J Phys Chem B 111:12898–12908

174. Cerón-Carrasco JP, Bastida A, Requena A, Zúñiga J (2010) A theoretical study of the reaction of β-carotene with the nitrogen dioxide radical in solution. J Phys Chem B 114:4366–4372
175. Böhm F, Tinkler J, Truscott T (1995) Carotenoids protect against cell membrane damage by the nitrogen dioxide radical. Nat Med 1:98–99
176. Kikugawa K, Hiramoto K, Tomiyama S, Asano Y (1997) β-Carotene effectively scavenges toxic nitrogen oxides: nitrogen dioxide and peroxynitrous acid. FEBS Lett 404:175–178
177. Everett SA, Dennis MF, Patel KB, Maddix S, Kundu SC, Willson RL (1996) Scavenging of nitrogen dioxide, thiyl, and sulfonyl free radicals by the nutritional antioxidant β-carotene. J Biol Chem 271:3988–3994
178. Khopde SM, Priyadarsini KI, Bhide MK, Sastry MD, Mukherjee T (2003) Spin-trapping studies on the reaction of NO_2 with β-carotene. Res Chem Intermediat 29:495–502
179. Kean RB, Spitsin SV, Mikheeva T, Scott GS, Hooper DC (2000) The peroxynitrite scavenger uric acid prevents inflammatory cell invasion into the central nervous system in experimental allergic encephalomyelitis through maintenance of blood-central nervous system barrier integrity. J Immunol 165:6511–6518
180. Hooper DC, Spitsin S, Kean RB, Champion JM, Dickson GM, Chaudhry I, Koprowski H (1998) Uric acid, a natural scavenger of peroxynitrite, in experimental allergic encephalomyelitis and multiple sclerosis. Proc Natl Acad Sci U S A 95:675–680
181. Chung HY, Choi HR, Park HJ, Choi JS, Choi WC (2001) Peroxynitrite scavenging and cytoprotective activity of 2,3,6-tribromo-4,5-dihydroxybenzyl methyl ether from the marine alga Symphyocladia latiuscula. J Agric Food Chem 49:3614–3621
182. Chung HY, Yokozawa T, Soung DY, Kye IS, No JK, Baek BS (1998) Peroxynitrite-scavenging activity of green tea tannin. J Agric Food Chem 46:4484–4486
183. Hedglin M, O'Brien PJ (2010) Hopping enables a DNA repair glycosylase to search both strands and bypass a bound Protein. ACS Chem Biol 5:427–436
184. Berg OG (1978) On diffusion-controlled dissociation. Chem Phys 31:47–57
185. Misteli T (2001) Protein dynamics: implications for nuclear architecture and gene expression. Science 291:843–847
186. Gowers DM, Halford SE (2003) Protein motion from non-specific to specific DNA by three-dimensional routes aided by supercoiling. EMBO J 22:1410–1418
187. Berg OG, Blomberg C (1976) Association kinetics with coupled diffusional fows. special application to the lac repressor-operator system. Biophys Chem 4:367–381
188. Berg OG, Blomberg C (1977) Association kinetics with coupled diffusion. An extension to coiled-chain macromolecules applied to the lac repressor-operator system. Biophys Chem 7:33–39
189. Bellomy GR, Record MT (1990) Stable DNA loops in vivo and in vitro: roles in gene regulation at a distance and in biophysical characterization of DNA. Prog Nucl Acid Res Mol Biol 30:81–128
190. Vologodskii A, Cozzarelli NR (1996) Effect of supercoiling on the juxtaposition and relative orientation of DNA sites. Biophys J 70:2548–2556
191. Gowers DM, Wilson GG, Halford SE (2005) Measurement of the contributions of 1D and 3D pathways to the translocation of a protein along DNA. Proc Natl Acad Sci U S A 102:15883–15888
192. Hagerman PJ (1988) Flexibility of DNA. Annu Rev Biophys Biophys Chem 17:265–286
193. Gowers DM, Wilson GG, Halford SE (2005) Measurement of the contributions of 1D and 3D pathways to the translocation of a protein along DNA. Proc Natl Acad Sci U S A 102:15883–15888
194. Blainey PC, Luo G, Kou SC, Mangel WF, Verdine GL, Bagchi B, Xie XS (2009) Nonspecifically bound proteins spin while diffusing along DNA. Nat Struct Mol Biol 16:1224–1229
195. Halford SE, Marko JF (2004) How do site-specific DNA-binding proteins find their targets. Nucl Acid Res 32:3040–3052

196. Stivers JT (2004) Site-specific DNA damage recognition by enzyme-induced base flipping. Prog Nucleic Acid Res Mol Biol 77:37–65
197. Jena NR, Bansal M (2011) Mutagenicity associated with O6-methylguanine-DNA damage and mechanism of nucleotide flipping by AGT during repair. Phys Biol 8:046007
198. Friedman JI, Stivers JT (2010) Detection of damaged DNA bases by DNA glycosylase enzymes. Biochemistry 49:4957–4967
199. Lyons DM, O'Brien PJ (2009) Efficient recognition of an unpaired lesion by a DNA repair glycosylase. J Am Chem Soc 131:17742–17745
200. Yang CG, Garcia K, He C (2009) Damage detection and base flipping in direct DNA alkylation repair. Chembiochem 10:417–423
201. Yu B, Edstrom WC, Benach J, Hamuro Y, Weber PC, Gibney BR, Hunt JF (2006) Crystal structures of catalytic complexes of the oxidative DNA/RNA repair enzyme AlkB. Nature 439:879–884
202. David SS, O'Shea VL, Kundu S (2007) Base-excision repair of oxidative DNA damage. Nature 447:941–950
203. Kunkel TA, Wilson SH (1996) DNA repair. Push and pull of base flipping. Nature 384:25–26
204. Scharer OD, Campbell AJ (2009) Wedging out DNA damage. Nat Struct Mol Biol 16:102–104
205. Qi Y, Spong MC, Nam K, Banerjee A, Jiralerspong S, Karplus M, Verdine GL (2009) Encounter and extrusion of an intrahelical lesion by a DNA repair enzyme. Nature 462:762–766
206. Wolfe AE, O'Brien PJ (2009) Kinetic mechanism for the flipping and excision of 1,N(6)-ethenoadenine by human alkyladenine DNA glycosylase. Biochemistry 48:11357–11369
207. Rubinson EH, Eichman BF (2012) Nucleic acid recognition by tandem helical repeats. Curr Opin Struct Biol 22:101–109
208. Scharer OD, Jiricny J (2001) Recent progress in the biology, chemistry and structural biology of DNA glycosylases. Bioessays 23:270–281
209. Stivers JT, Jiang YL (2003) A mechanistic perspective on the chemistry of DNA repair glycosylases. Chem Rev 103:2729–2759
210. Huffman JL, Sundheim O, Tainer JA (2005) DNA base damage recognition and removal: new twists and grooves. Mutat Res 577:55–76
211. Fromme JC, Banerjee A, Verdine GL (2004) DNA glycosylase recognition and catalysis. Curr Opin Struct Biol 14:43–49
212. Friedman JI, Stivers JT (2010) Detection of damaged DNA bases by DNA glycosylase enzymes. Biochemistry 49:4957–4967
213. Dalhus B, Laerdahl JK, Backe PH, Bjørås M (2009) DNA base repair–recognition and initiation of catalysis. FEMS Microbiol Rev 33:1044–1078
214. Li GM (2010) Novel molecular insights into the mechanism of GO removal by MutM. Cell Res 20:116–118
215. Hollis T, Lau A, Ellenberger T (2000) Structural studies of human alkyladenine glycosylase and E. Coli 3-methyladenine glycosylase. Mut Res 460:201–210
216. Slupphaug G, Mol CD, Kavil B, Arvai AS, Krokan HE, Tainer JA (1996) A nucleotide-flipping mechanism from the structure of human uracil-DNA glycosylase bound to DNA. Nature 384:87–92
217. Daniels DS, Woo TT, Luu KX, Noll DM, Clarke ND, Pegg AE, Tainer JA (2004) DNA binding and nucleotide flipping by the human DNA repair protein AGT. Nat Struct Mol Biol 11:714–720
218. Huang N, Banavali NK, MacKerell AD Jr (2003) Protein-facilitated base flipping in DNA by cytosine-5-methyltransferase. Proc Natl Acad Sci U S A 100:68–73
219. Shieh FK, Youngblood B, Reich NO (2006) The role of Arg165 towards base flipping, base stabilization and catalysis in M.HhaI. J Mol Biol 362:516–527

220. Morikawa K, Matsumoto O, Tsujimoto M, Katayanagi K, Ariyoshi M, Doi T, Ikehara M, Inaoka T, Ohtsuka E (1992) X-ray structure of T4 endonuclease V: an excision repair enzyme specific for a pyrimidine dimer. Science 256:523–526
221. Vassylyev DG, Kashiwagi T, Mikami Y, Ariyoshi M, Iwai S, Ohtsuka E, Morikawa K (1995) Atomic model of a pyrimidine dimer excision repair enzyme complexed with a DNA substrate: structural basis for damaged DNA recognition. Cell 83:773–782
222. Cao C, Jiang YL, Stivers JT, Song F (2007) Dynamic opening of DNA during the enzymatic search for a damaged base. Nat Struct Mol Biol 11:1230–1236
223. Parker JB, Bianchet MA, Krosky DJ, Friedman JI, Amzel LM, Stivers JT (2007) Enzymatic capture of an extrahelical thymine in the search for uracil in DNA. Nature 449:433–437
224. Qi Y, Spong MC, Nam K, Karplus M, Verdine GL (2010) Entrapment and structure of an extrahelical guanine attempting to enter the active site of a bacterial DNA glycosylase MutM. J Biol Chem 285:1468–1478
225. Osman R, Fuxreiter M, Luo N (2000) Specificity of damage recognition and catalysis of DNA repair. Computer Chem 24:331–339
226. Brooks SC, Adhikary S, Rubinson EH, Eichman BF (2013) Recent advances in the structural mechanisms of DNA glycosylases. Prot Proteom 1834:247–271
227. Jena NR, Shukla PK, Jena HS, Mishra PC, Suhai S (2009) O6-Methylguanine repair by O6-alkylguanine-DNA alkyltransferase. J Phys Chem B 113:16285–16290
228. McCann JAB, Berti PJJ (2008) Transition state analysis of the DNA repair enzyme MutY. J Am Chem Soc 130:5789–5797
229. Tiwari S, Agnihotri N, Mishra PC (2011) Quantum theoretical study of cleavage of the glycosidic bond of 2'-deoxyadenosine: base excision-repair mechanism of DNA by MutY. J Phys Chem B 115:3200–3207

Chapter 4
DNA Dependent DNA Polymerases as Targets for Low-Weight Molecular Inhibitors: State of Art and Prospects of Rational Design

Alexey Yu. Nyporko

Abstract DNA dependent DNA polymerases (DNA pols) are key enzymes providing the processes of DNA replication and reparation in living systems. Exceptional importance of DNA pols makes them to be attractive targets for specific low-molecular weight inhibitors, which can be used (and are actually used) as molecular tuning tools in molecular biology investigations, and as antineoplastic and antiviral drugs as well. Detailed comprehension of structural insights of pol–inhibitor interaction would not only give a possibility to design new drugs with highly selective activity with respect to the targeted polymerases, but would essentially extend our understanding of the structural basis of replicative/reparative processes as a whole. Several computational approaches including sophisticated modeling of protein structure, blind and site-oriented docking of inhibitor molecules, molecular dynamics simulation of pol–inhibitor complexes and free energy decomposition analysis are useful tools to improve the quality of structural analysis of pol–inhibitor interactions as well as selectivity of pols' inhibitors developed *de novo*. Extended application of these methods is principle tendency in modern rational design, including search and/or design of new inhibitors of DNA polymerases.

4.1 Introduction

Life on Earth would be impossible without the processes of reproduction, storage, repairing and transmission of hereditary information in a series of generations. Key factors providing these processes are DNA dependent DNA polymerases (DNA pols). These enzymes catalyze the synthesis of a new DNA chain based on pre-existing DNA molecule, according to the principle of complementarity of nitrogenous bases in the "mother" and "daughter" chains. Such process of matrix DNA synthesis is known as **DNA replication**. The elementary chemical act of catalysis

---

A. Yu. Nyporko (✉)
High Technology Institute, Taras Shevchenko National University of Kyiv, 60, Volodymyrska St, 01601 Kyiv, Ukraine
e-mail: anyporko@univ.net.ua; dfnalex@gmail.com

L. Gorb et al. (eds.), *Application of Computational Techniques in Pharmacy and Medicine*, Challenges and Advances in Computational Chemistry and Physics 17,
DOI 10.1007/978-94-017-9257-8_4

by DNA polymerases during replication is the addition of nucleoside monophosphate deriving from appropriate nucleoside triphosphate to OH-group on 3′-end of growing DNA chain [1].

Second important DNA polymerases' function in living systems is participation in the **DNA repair**, aimed at correcting errors of the DNA synthesis during replication as well as numerous injuries that occur in DNA as a result of chemical and physical factors [2]. Most of the reparation process involves removing the damaged fragment with subsequent single-stranded DNA synthesis that is performed by DNA polymerases.

Besides their essential tasks *in vivo*, DNA polymerases are now the key tool in numerous important molecular biological and medical core technologies, such as the widely applied polymerase chain reaction (PCR), cDNA cloning, genome sequencing, nucleic acids based diagnostics, and in techniques to analyze the ancient and otherwise damaged DNA [3].

Like the other cellular and molecular "bottlenecks", the DNA dependent DNA polymerases are attractive targets for low-molecular weight inhibitors. These compounds can be used (and are actually used) as molecular tuning tools in molecular biology investigations, and as antineoplastic and antiviral drugs as well. Despite numerous investigations, devoted to the search for DNA pol inhibitors and their development, some fundamental problems in this field are still unresolved. The most important of them is the problem of structural insights of inhibitor–pol interactions and the inhibitor selectivity to different DNA polymerases. Detailed comprehension of these insights would not only give a possibility to design new drugs with highly selective activity with respect to the targeted polymerases, but would essentially extend our understanding of the structural basis of replicative/reparative processes as a whole.

4.2 Diversity of DNA Dependent DNA Polymerases

According to present views, all of the known DNA polymerases are divided into seven families based on their sequence homology (especially, sequences of the catalytic domain) and the structure of catalytic domain. Six of them—A, B, C, D, X and Y—are DNA-dependent DNA polymerases, one is a DNA polymerase of a different nature, namely it is RNA-dependent (more commonly known as 'reverse transcriptases' or RT family) [3]. The spectra of DNA polymerases' families are individual for different organic kingdoms. For example, bacteria usually contain DNA polymerases of A, B, C, X and Y families, while archaea have members of families B, D, X and Y. Among eukaryotic DNA polymerases, we can find various members of A, B, X, Y families (and at least one RT-member—telomerase supplying synthesis of the terminal fragment of chromosome which cannot be synthesized in the matrix way). Viral DNA polymerases are presented by families A, B and X (as well as RT—reverse transcriptases of retroviruses) (see Table 4.1).

Table 4.1 Classification of DNA dependent DNA polymerases

Family	Kingdom	Members	Main functions
A	Virus	T3, T5 and T7 DNA pol(s)	Replication
	Bacteria	DNA pol(s) I (*E. coli*, *T. aquaticus*, etc.)	Replication, repair
	Eukaryotes	DNA pol γ	Mitochondrial DNA replication
		DNA pol θ	Replication of cross links, base excision repair
B	Virus	T4, T6, RB69, Adeno, HSV-1, Vaccinia, Phi29 DNA pol(s)	Replication
	Archaea	DNA pol(s) BI, BII	Replication, repair(?)
		DNA pol BIII	Repair, replication(?)
	Bacteria	DNA pol II	Repair, replication(?)
	Eukaryotes	DNA pol α	Replication (priming)
		DNA pol δ	Replication (lagging strand), repair
		DNA pol ε	Replication (leading strand), repair
		DNA pol ζ/Rev3	Translesion synthesis (extension)
C	Bacteria	DNA pol(s) III (*E. coli*, *T. aquaticus*, *B. subtilis*, etc.), DNA pol E (*B. subtilis*)	Replication
D	Archaea	DNA pol D	Replication
X	Virus	ASFV DNA pol	Repair
	Archaea	DNA pol(s) × (various species)	Repair
	Bacteria	DNA pol(s) × (various species)	Repair
	Eukaryotes	DNA pol β (pol IV in *S. cerevisiae*)	Base excision repair
		DNA pol λ (pol LSP in *S. pombe*)	Base excision repair, double-strand break repair, immunoglobulin recombinational repair, translesion synthesis
		DNA pol μ	Immunoglobulin recombinational repair
		DNA pol σ[a]	Sister chromatid cohesion
		TdT	Antigen receptor diversity
Y	Archaea	Dpo4 DNA pol, Dbh DNA pol	Translesion synthesis
	Bacteria	DNA pol IV, DNA pol V (*E. coli*)	Translesion synthesis
	Eukaryotes	DNA pol η, DNA pol κ, DNA pol ι, Rev1	Translesion synthesis (incorporation)

[a] DNA pol σwas later shown to be a poly (A) RNA polymerase [4]

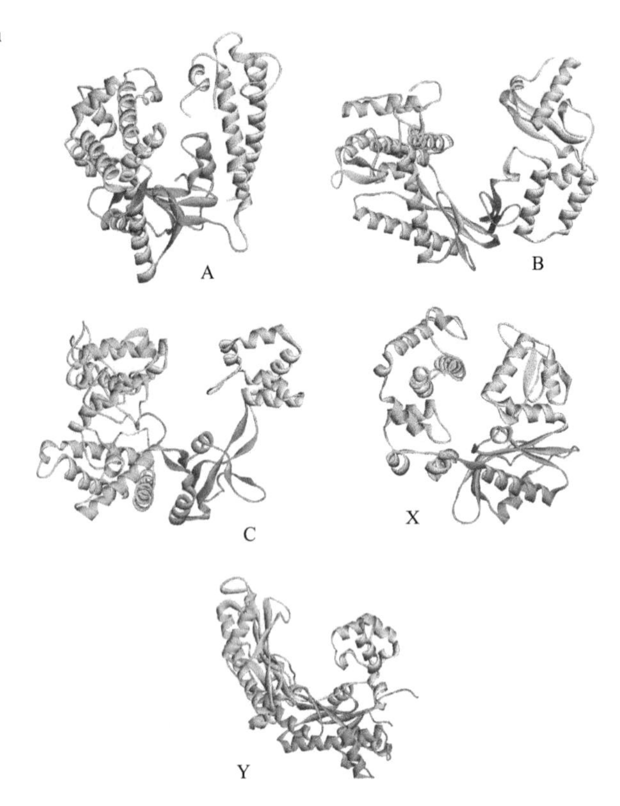

Fig. 4.1 Spatial organization of catalytic domains of DNA dependent DNA polymerases from different families. Fingers, palms and thumbs colored by *green*, *blue* and *yellow*, respectively

4.2.1 Common Peculiarities of DNA-Dependent DNA Polymerases

The available data on the spatial organization of DNA polymerases from different families testifies that, independently of their detailed domain structures, all polymerases, whose structures are known presently, appear to share a common overall architectural feature. Their polymerase domains have a shape that can be compared with something like a right hand and has been described as consisting of "thumb," "palm," and "fingers" subdomains[1] [5] (Fig. 4.1). The function of the palm subdomain appears to be the catalysis of the phosphoryl transfer reaction, whereas that of the fingers domain includes important interactions with the incoming nucleoside triphosphate as well as the template base to which it is paired. The thumb, on the other hand, may play a role in positioning the duplex DNA and in processivity and translocation. Although the palm domain appears to be homologous among the pol I, pol α, and RT families, the fingers and thumb domains are different in all of these families for which structures are known to date [3, 5].

[1] In the literature, these components of polymerase domain are also often named "domains", which makes sense due to their different spatial folds. However, to avoid terminological confusion, we propose to consider them as subdomains of a unified polymerase domain.

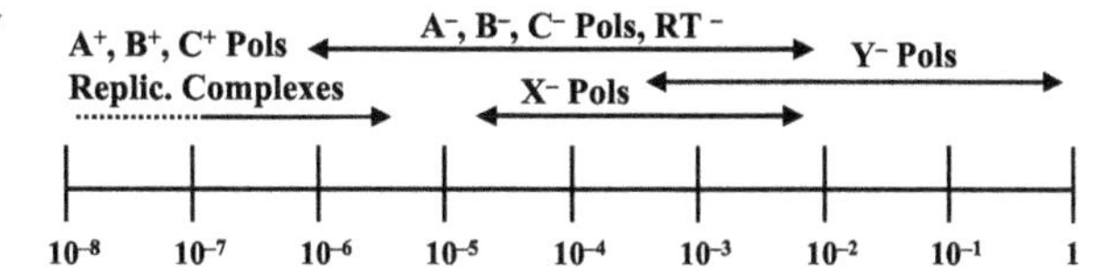

Fig. 4.2 Comparative fidelity of different DNA polymerase families. (Adapted with permission from [7])

Each DNA polymerase can be characterized by at least two significant parameters—fidelity and processivity. Fidelity is a common property of polymerase enzymes (DNA and RNA polymerases) to reproduce a polynucleotide chain with certain accuracy. Distinct DNA pols families have different levels of fidelity (Fig. 4.2).

Structural mechanisms providing the distinct fidelity of DNA polymerases from different families are investigated well enough. First, it is known that DNA polymerases having proofreading exonuclease activity are significantly more precise because of the ability to recognize and remove the wrongly embedded nucleotides. The proofreading activity is realized with 3′–5′ exonuclease domains in the composition of appropriate DNA polymerases and improves the synthesis fidelity by three to four orders of magnitude. Second, active sites of DNA polymerases from different families have individual features of amino acid composition that result in distinct substrate specificity. It is indicative that the amino acid substitutions in active site can reduce fidelity of exact DNA polymerases [6]. However, the structural basis of the DNA polymerase fidelity phenomenon itself are not completely clear, despite numerous investigations in this field [1, 7–10].

DNA polymerase processivity is a measure of the average number of nucleotides added by the enzyme per one association/disassociation with the template. DNA polymerases associated with the DNA replication tend to be highly processive, while those associated with the DNA repair tend to have low processivity. Because the binding of the polymerase to the template is the rate-limiting step in the DNA synthesis, the overall rate of the DNA replication during the S phase of the cell cycle is dependent on the processivity of the DNA polymerases performing the replication [11].

4.2.2 *Features of Different DNA Polymerase Families*

4.2.2.1 Family A

This family includes viral, bacterial and eukaryotic DNA polymerases. The main functions of these enzymes are replication and, to less extent, repair of the genetic material. For instance, representatives of the A family provide replication of mitochondrial DNA in eukaryotic cells [3].

The most known bacterial member of this protein family is the DNA polymerase I (pol I). This enzyme is encoded by the *polA* gene and is ubiquitous among prokaryotes. This repair polymerase is involved in excision repair with 3′–5′ and 5′–3′ exonuclease activity and processing of Okazaki fragments generated during the lagging strand synthesis [3]. Pol I is the most abundant polymerase accounting for >95 % of polymerase activity in *E. coli*, yet cells lacking pol I have been found, suggesting that pol I activity can be replaced by the other four polymerases. Pol I adds ~15–20 nucleotides per second, thus showing poor processivity. Instead, pol I starts adding nucleotides at the RNA primer–template junction known as the origin of replication (ori). Approximately ~400 bp downstream from the origin, the pol III holoenzyme is assembled and takes over replication at a highly processive speed and nature [12].

Pol γ, encoded by the *polG* gene, is the only **mitochondrial** DNA polymerase and therefore replicates and repairs **mitochondrial** DNA. In addition, pol γ has proofreading 3′–5′ exonuclease (like the other A-member polymerases) and 5′ dRP lyase activities. Any mutation that leads to a limited or non-functioning pol γ has significant effect on mtDNA and is the most common cause of autosomal inherited mithochondrial disorders (pol γ contains a C-terminus polymerase domain and a N-terminus 3′–5′ exonuclease domain that are connected via the linker region binding the accessory subunit. The accessory subunit binds DNA and is required for the proccessivity of pol γ. Point mutation A467T in the linker region is responsible for more than one third of all pol γ-associated mitochondrial disorders [13].

While a lot of pol θ homologs, encoded by the *polQ* genes, are found in eukaryotes, their functions are not clearly understood. The sequence of amino acids in the C-terminus is what classifies pol θ as family A polymerase, although the error rate for pol θ is more closely related to family Y polymerases. pol θ may extend the mismatched primer termini and can bypass abasic sites by adding a nucleotide opposite the lesion.

Among viral members of A family, one should mention the T3 and T5 DNA polymerases that provide the replication of DNA of those viruses.

4.2.2.2 Family B

This polymerase family includes enzymes with the highest level of replication fidelity. It is clearly understood that the members of this family are actually the main catalysts of DNA replication in viruses, archea and eukaryotes. Functions of bacterial B DNA polymerases are more complicated.

The DNA polymerase II (pol II) is the most known bacterial B-family polymerase. It is coded by *polB* gene and has 3′–5′ exonuclease activity [14]. Pol II participates in the DNA repair and replication restart to bypass lesions. Its cell presence can vary from 30 to 50 copies per cell to 200–300 during SOS induction. Pol II is also thought to be a backup to pol III as it can interact with holoenzyme proteins and assumes a high level of processivity. The special role of pol II is thought to be the ability to direct polymerase activity at the replication fork and its help to the stalled pol III in bypassing terminal mismatches [14].

Eukaryotic family B DNA pols (α, δ, ε and ζ) are multisubunit enzymes which, with the exception of the DNA pol ζ, are responsible for the replication of nuclear DNA in all eukaryotic organisms.

Pol α (primase) consists of four subunits, two α and two-subunit primase which are encoded by the *POLA1* and *POLA2* genes. The primary role of this enzyme is the initiation of the leading strand DNA replication and in the repeated priming of Okazaki fragments during lagging-strand DNA replication. The DNA pol α holoenzyme possesses two distinct yet functionally interacting active sites: one in the large subunit responsible for the DNA synthesis, and one in the dimeric primase responsible for the RNA synthesis [15]. Once primase has created the RNA primer, pol α starts replication elongating the primer with ~20 nucleotides.

Due to their high processivity, pol ε and pol δ take over from pol α the leading and lagging strand synthesis, respectively [16]. Pol δ is expressed by genes *polD1*, creating the catalytic subunit, *polD2*, *polD3*, and *polD4* creating the other subunits that interact with the Proliferating Cell Nuclear Antigen (PCNA) which is a DNA clamp that allows pol δ to possess processivity [3, 16]. Pol ε is encoded by the *polE*, the catalytic subunit, *polE2*, and *polE3* genes. While pol ε's main function is to extend the leading strand during replication, pol ε's C-terminus region is thought to be essential to cell vitality as well. The C-terminus region is thought to provide a checkpoint before entering anaphase, to provide stability to the holoenzyme, and to add proteins to the holoenzyme, necessary for the initiation of replication [3, 17].

Pol ζ, another B family polymerase, is made of two subunits Rev3, the catalytic subunit, and Rev7, which increases the catalytic function of the polymerase, and is involved in the translesion synthesis [18]. Pol ζ lacks 3′–5′ exonuclease activity, and is unique in that it can extend primers with terminal mismatches. Rev1 has three regions of interest in the BRCT domain, ubiquitin-binding domain, and C-terminal domain, and has dCMP transferase ability, which adds deoxycytidine opposite lesions that would stall replicative polymerases pol δ and pol ε. These stalled polymerases activate ubiquitin complexes, which in turn disassociate replication polymerases and recruit pol ζ and Rev1. Together, pol ζ and Rev1 add deoxycytidine and pol ζ extends past the lesion. Through a yet undetermined process, pol ζ disassociates and replication polymerases reassociate and continue replication. pol ζ and Rev1 are not required for replication, but loss of REV3 gene in budding yeast can cause increased sensitivity to DNA-damaging agents due to collapse of replication forks where replication polymerases have stalled [18].

4.2.2.3 Family C

Family C of DNA polymerases is presented exceptionally by bacterial enzymes involved in replicative processes. So, the DNA polymerase III holoenzyme is the main enzyme realizing the DNA replication in *Escherichia coli*, *Bacillus subtilis,* and belongs to family C polymerases. It consists of three assemblies: the pol III core, the beta sliding clamp processivity factor and the clamp-loading complex. The core consists of three subunits—α, the polymerase activity hub, ε, exonucleolytic proofreader, and θ, which may act as a stabilizer for ε. The holoenzyme contains two cores, one for each strand, the lagging and leading [19]. The beta sliding clamp processivity factor is also present in duplicate, one for each core, to create a clamp that encloses DNA allowing for high processivity [20]. The third assembly is a seven-subunit ($\tau 2\gamma\delta\delta'\chi\psi$) clamp loader complex.

4.2.2.4 Family D

D family consists exclusively of archaea polymerases. The DNA polD enzyme is a heterodimer composed of large DP2 and small DP1 subunits [21]. DP2 is the catalytic subunit, while DP1 serves as an accessory factor. The interaction of the two subunits has been reported to be necessary for the optimal DNA pol and $3' \rightarrow 5'$ exonuclease activity [21]. Accordingly, DP1 has been shown to possess an intrinsic proofreading activity [22]. The DP1 subunit shows homology to small, non-catalytic subunits of eukaryotic DNA pols α (p70 subunit), DNA pol δ (Cdc27p) and DNA pol ε (p55 subunit). Similarly to archaeal DNA pol B, DNA pol D is also stimulated by PCNA and RF-C. However, the DP1 subunit also directly interacts with RadB, a homolog of the eukaryotic proteins Dmc1 and Rad51 in *Pyrococcus furiosus,* suggesting that DNA pol D may participate in recombination and/or repair in addition to its role in replication [23].

4.2.2.5 Family X

Family X polymerases contain the well-known eukaryotic polymerase pol β—the main reparative polymerase of eukaryotes, as well as other eukaryotic polymerases such as pol σ, pol λ, pol μ, and terminal deoxynucleotidyl transferase (TdT) [24]. Family X polymerases are mainly found in vertebrates and a few are found in plants and fungi. These polymerases have highly conserved regions that include two helix–hairpin–helix motifs that are imperative in the DNA–polymerase interactions. One motif is located in the N-terminal lyase domain (8 kDa domain) that interacts with the downstream DNA and one motif is located in the thumb domain that interacts with the primer strand. Pol β, encoded by *polB* gene, is required for short-patch base excision repair, a DNA repair pathway that is essential for repairing alkylated or oxidized bases as well as abasic sites. Pol λ and pol μ, encoded by the *polL* and *polM* genes respectively, are involved in non-homologous end-joining, a mechanism for rejoining DNA double-strand breaks due to the influence of the hydrogen peroxide and ionizing radiation, respectively. TdT is expressed only in lymphoid tissue, and adds “n nucleotides” to double-strand breaks formed during V(D)J recombination to promote immunological diversity [24].

4.2.2.6 Family Y

In *E. coli*, DNA polymerase IV (pol 4) is an error-prone DNA polymerase involved in non-targeted mutagenesis [25]. Pol IV is a family Y polymerase expressed by the *dinB* gene that is switched on via SOS induction caused by stalled polymerases at the replication fork. During the SOS induction, pol IV production is increased 10-fold and one of its functions during this time is to interfere with the pol III holoenzyme processivity. This creates a checkpoint, stops replication, and allows time to repair DNA lesions via the appropriate repair pathway [26]. Another function of pol IV is to perform translesion synthesis at the stalled replication fork, for example, bypassing N2-deoxyguanine adducts at a faster rate than transversing undamaged DNA. Cells lacking *dinB* gene have a higher rate of mutagenesis caused by DNA damaging agents [27].

DNA polymerase V (pol V) is a Y-family DNA polymerase that is involved in SOS response and translesion synthesis DNA repair mechanisms [28]. Transcription of pol V via the *umuDC* genes is highly regulated to only produce pol V when damaged DNA is present in the cell, generating the SOS repsonse. Stalled polymerases cause RecA to bind to the ssDNA, which causes the LexA protein to autodigest. LexA then loses is ability to repress the transcription of the umuDC operon. The same RecA-ssDNA nucleoprotein post-translationally modifies the umuD protein into the umuD′ protein. UmuD and umuD′ form a heterodimer that interacts with umuC, which in turn activates the umuC's polymerase catalytic activity on the damaged DNA [29].

Pol η, pol ι, and pol κ are family Y DNA polymerases involved in the DNA repair by translesion synthesis and encoded by genes *polH*, *polI* and *polK*, respectively. Members of family Y have 5 common motifs to aid in binding the substrate and primer terminus and they all include the typical right-hand thumb, palm and finger domains, with additional domains like little finger (LF), polymerase-associated domain (PAD), or wrist. The active site, however, differs between family members due to the different lesions being repaired [30]. Polymerases in family Y have low fidelity, but have been proven to do more good than harm as mutations can cause various diseases, such as skin cancer and xeroderma pigmentosum variant (XPS). The importance of these polymerases is evidenced by the fact that one refers to the gene encoding DNA polymerase η as XPV, because the loss of this gene results in the disease xeroderma pigmentosum variant [30]. Pol η is particularly important for allowing accurate translesion synthesis of the DNA damage resulting from ultraviolet radiation or UV. The functionality of pol κ is not completely understood, but researchers have found two probable functions. Pol κ is thought to act as an extender or an inserter of a specific base at certain DNA lesions. All three translesion synthesis polymerases, along with Rev1, are recruited to damaged lesions via stalled replicative DNA polymerases. There are two pathways of damage repair leading researchers to conclude that the chosen pathway depends on which strand contains the damage, the leading or the lagging strand [30].

4.3 Pol Inhibitors and Possible Mechanisms of Action

All known inhibitors can be subdivided into several groups according to certain analytical criteria. Mostly, inhibitors are sorted by their chemical nature or by specificity regarding to certain polymerases and polymerase families.

According to the chemical nature, the pol inhibitors can be divided into two large groups—substrate analogs (nucleoside/nucleotide/dNTP compounds) and all other compounds[2]. This division reflects the special role played by nucleotides in replicative/repair processes. Due to the chemical similarity to natural DNA polymerase substrate, practically any nucleoside/nucleotide analogs are able to bind with active site, thus competing with the natural substrates of DNA polymerases. At first glance, selectivity of these inhibitors is not out of the question. However, one should mention that amino acid compositions of dNTP-binding sites are distinct in different DNA pol families and, thus, affinity of various nucleotide inhibitors to them also can be substantially different.

There are at least three types of nucleoside inhibitors' behavior at the DNA polymerase active site:

- the nucleoside analog incorporates in the growing DNA chain and terminates the next DNA synthesis due to the lack or modification of the 3′-OH group in its composition. Strictly speaking, such compound does not influence the DNA polymerase activity itself, but prevents the addition of the next nucleotide only;
- the nucleoside analog binds at the active center of DNA polymerase, but does not incorporate in the DNA chain;
- the nucleoside analog incorporates in the growing DNA chain and (possibly) deforms the spatial geometry of the active center, which results in a slowdown of the further inclusion of natural nucleotides.

The inhibitors of DNA polymerases have been found among dNTP derivates modified in the base (2-substituted dATP and N^2-substituted dGTP analogs, derivatives with altered base-pairing specificity), derivates modified in the sugar (arabinonucleotides, 2′,3′-dideoxynucleotides, acyclonucleotides) and derivates modified in the triphosphate group (for example, phosphorothioates of dATP) [31–33].

Among nucleotide inhibitors one should specially mention acyclic nucleoside phosphonates, in particular, (S)-[3-hydroxy-2-(phosphonomethoxy)propyl] nucleosides (HPMP) [34, 35]. These compounds exhibit activity with respect to a wide spectrum of viral and eukaryotic DNA polymerases—herpes simplex virus 1 (HSV-1) DNA polymerase [36], human cytomegalovirus HCMV DNA polymerase [37, 38], vaccinia virus DNA polymerase [39, 40], human DNA polymerases α, δ, and ε [41] as well as reverse transcriptases [42]. Several antiviral drugs with a wide spectrum of targets are developed on the basis of acyclic nucleoside phosphonates. Among them one can mention **cidofovir** (commercial name *vistide*®) (Fig. 4.3) that is used against polyoma-, papilloma-, adeno-, herpes-, irido- and poxviruses [34, 43, 44], **tenofovir** (*viread*®) acting on hepata- and retoviruses [34, 43], **adefovir** (*hepsera*®) (Fig. 4.4), which is effective against all above mentioned groups of viruses [34, 43]. A feature of the HPMP interaction with DNA replicative machine is that they are not simple terminators of the DNA chain growth. These compounds can be incorporated into DNA and this, in principle, permits the continuation of DNA synthesis, but it becomes significantly slower [35].

[2] Some researchers [31] consider the pyrophosphate analogs as a separate inhibitor group (product analogs' group).

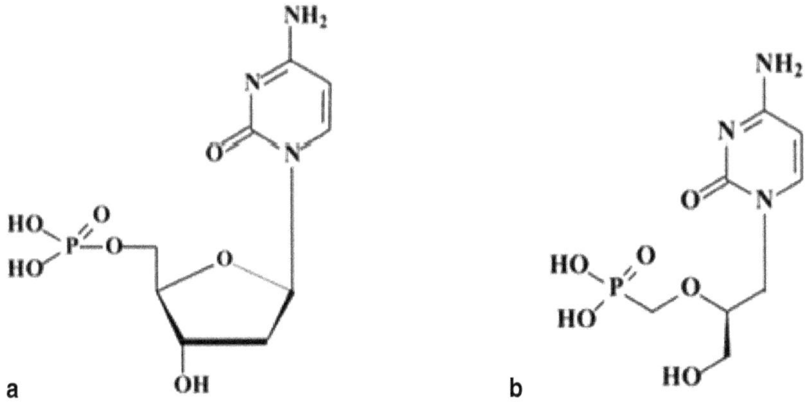

Fig. 4.3 Comparative structure of natural (cyclic) and acyclic cytosine nucleotides: **a** natural nucleotide 2′-deoxycytidine-5′-monophospate, **b** acyclic nucleotide cidofovir

One should remark that the most discovered nucleoside/nucleotide inhibitors are compounds decreasing the activity of **viral** DNA-dependent DNA polymerases. Consequently, a lot of them are used as antiviral drugs [33]. In addition to the substances mentioned above, among them one finds **acyclovir** which is a synthetic guanosine analogue used for treating herpes simplex virus (HSV) and varicella zoster virus (VZV) infections [45, 46], **brivudin** which is a 5′-halogenated thymidine nucleoside analogue highly active against HSV-1 and VZV [47, 48], **ganciclovir**—an acyclic 2′-deoxyguanosine analogue for the management of CMV [49], **penciclovir**—an acyclic guanine analogue chemically similar to acyclovir that is efficient against HSV-1, HSV-2, and VZV, and, to less extent, against EBV [50], **farmcyclovir**—a diacetyl 6-deoxy analogue of penciclovir with the same antiviral activity, **valacyclovir**, the L-valyl ester prodrug of acyclovir [51] approved for VZV treatment [33]. Structures of these compounds are shown on Fig. 4.4.

In contrast to DNA pol substrate analogs, non-nucleotide inhibitors (NNI) of DNA-dependent DNA polymerases are potentially able to specifically interact with different regions on DNA pol surface. They can be competitive in relation to the dNTP binding site (in this case we can talk about nucleotide mimics), bind to DNA template binding area and directly prevent initial DNA interaction, have allosteric binding site and, thus, exhibit non-competitive type of inhibition.

Non-nucleotide pol inhibitors have been discovered among different classes of natural and synthetic chemical compounds. The main of them are long-chain fatty acids, fatty acid derivatives, bile acid derivatives, steroid derivatives, triterpenoids, cerebrosides, alkaloids, flavonoids, anthocyanins, glycolipids, catechins, coenzyme Qs, isosteviols, dipeptide alcohols, vitamins, etc. [52, 53]. The most well-studied non-nucleotide inhibitors are presented in Table 4.2.

As opposed to nucleotide ones, the lion's share of known non-nucleotides inhibitors are inhibitors of eukaryotic DNA polymerases. In turn, most of them are inhibitors of animal pol β. Discovered non-nucleotide inhibitors are characterized by different selectivity with respect to different polymerases. Selectivity can reveal

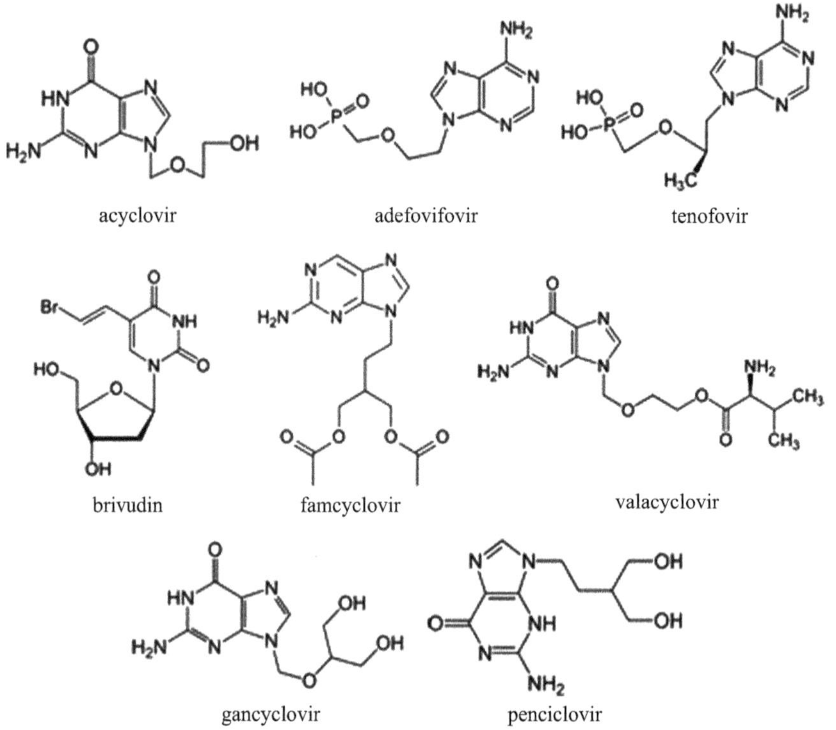

acyclovir adefovifovir tenofovir

brivudin famcyclovir valacyclovir

gancyclovir penciclovir

Fig. 4.4 Structures of DNA pol nucleoside inhibitors approved as antiviral drugs

itself at the level of pol families as well as at the level of individual polymerases. So, classical antibiotic aphidicolin is able to inhibit only animal (human in particular) pol α [52, 56], curcumin derivates are able to oppress only pol λ [85, 86], but glycyrrhetinic acid inhibits effectively the mammalian pols α (B family), k (Y family), β and λ (X family) [70], and kohamaic acid A derivative 11 decreases activity of the all eukaryotic pol families [76].

Besides of eukaryotic pols, non-nucleotide inhibitors are known for DNA pols from viruses. Non-nucleotide inhibitors of viral pols are presented by few classes of compounds. So, the 4-oxo-dihydroquinoline-3-carboxamides (4-oxo-DHQ) demonstrated inhibition of HCMV, HSV, and VZV polymerases in subnanomolar concentrations [91, 92]. High specificity for viral DNA polymerases compared to human pols α, γ and δ is observed. 4-Oxo-DHQs are inactive against unrelated DNA or RNA viruses, indicating specificity for herpesviruses. A strong correlation between the inhibition of viral DNA polymerases and the antiviral activity for this class of compounds supports inhibition of the viral DNA polymerase as the mechanism of antiviral activity. The 4-oxo-DHQs were found to be competitive inhibitors of nucleoside binding [107].

Further SAR studies led to the discovery of 4-oxo-4,7-dihydrothienopyridines (DHTPs) [108, 109] and 7-oxo-4,7-dihydrothieno [3, 2-b]pyridine-6-carboxamides

Table 4.2 The most well-studied non-nucleotide inhibitors of DNA dependent DNA polymerases

ID	Compound	Structure	Target	Reference
	Anacardic acid	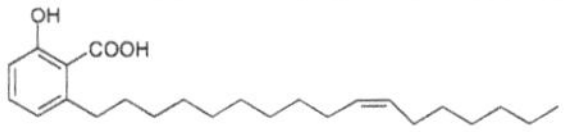	pol β	[54]
	Aphidicolin	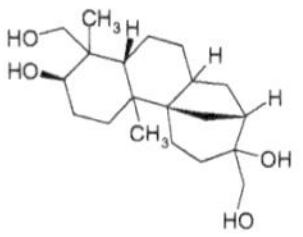	pol α	[52, 55, 56]
	Betulinic acid	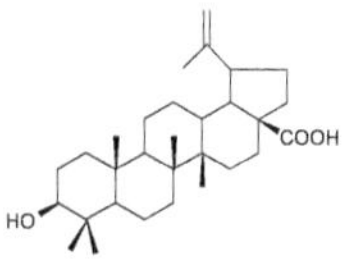	pol β	[57]
	Bis-5-alkylresorcinols derivative	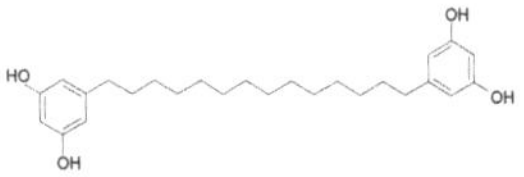	pol β	[58]
	Bredinin-5′-monophosphate (BreMP)	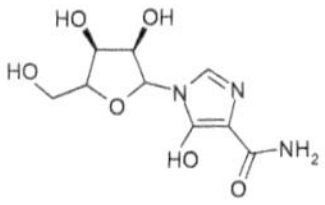	pol α, pol β	[59]
	Carbonyldiphosphonate	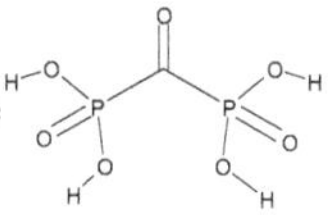	Pol δ	[60]
	Cephalomannine	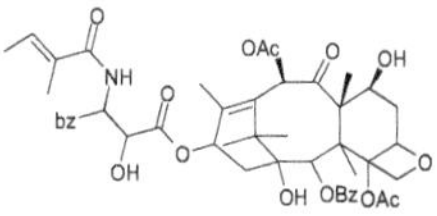	pol α*	[61]
	Cilexetil	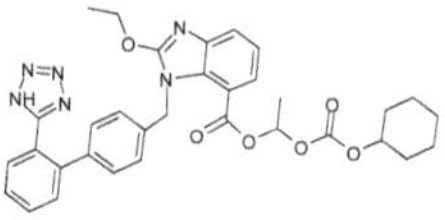	pol κ	[62]
	3-cis-p-Coumaroyl maslinic acid	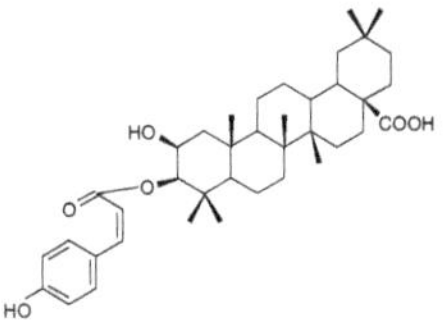	pol β	[57]
	3-trans-p-Coumaroyl maslinic acid	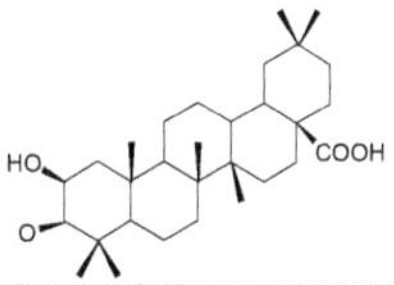	pol β	[57]

Table 4.2 (continued)

Dehydroaltenusin	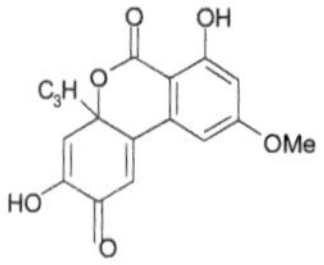 O OH C_3H O OMe HO O	pol α*	[63]
DRB	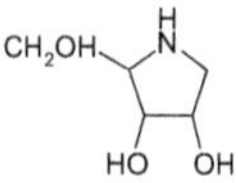 H CH_2OH N HO OH	pol α, pol β, pol δ, pol ε polymerases and glycosidases	[64]
Edgeworin	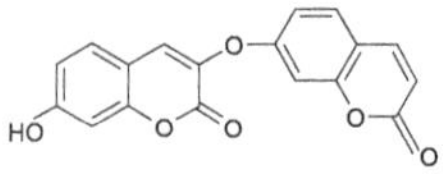 O HO O O O O	pol β lyase activity	[65]
Epicatechin	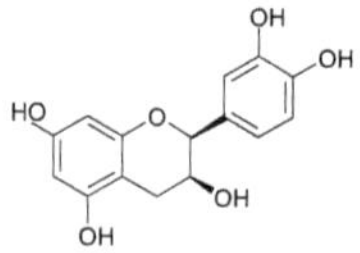 OH OH HO O OH OH	pol β lyase activity	[66]
EPO	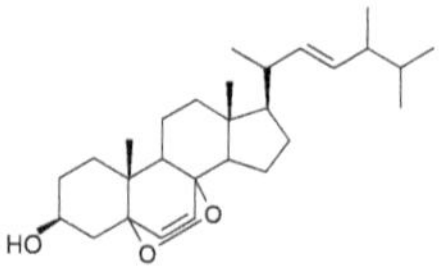 O HO O	Pol β (weakly) intense LA action	[67]
Fomitellic acid A	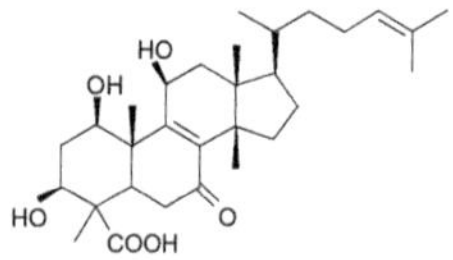 HO OH HO COOH O	pol α, pol β	[68, 69]
Glycyrrhetinic acid	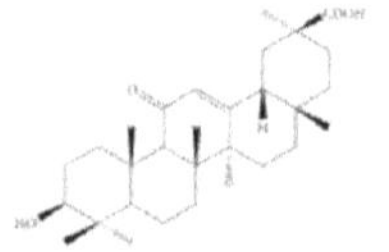	pol α,pol β, pol λ, pol k	[70]
Harbinatic acid	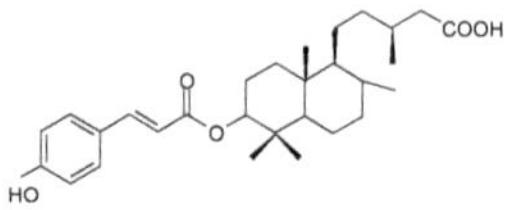 COOH O O HO	pol β	[71]
HMI	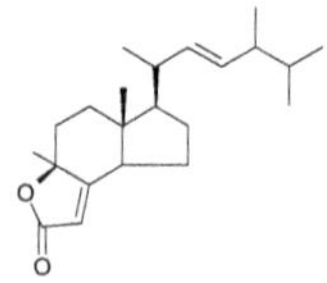 O O	pol α	[72]
2-α-Hydroxyursolic acid	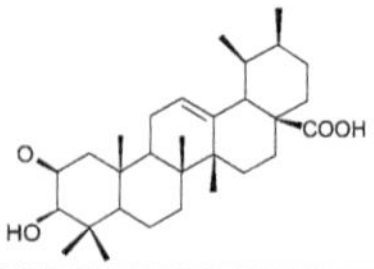 COOH O HO	pol β	[73]

Table 4.2 (continued)

KAG	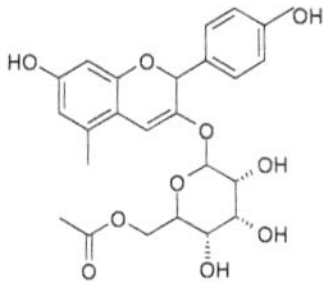	Pol α, pol δ, pol e	[74]
KN-208	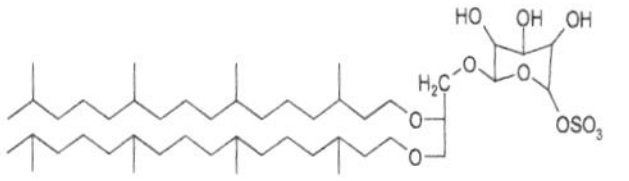	pol α, pol β, pol I E. coli	[75]
Kohamaic acid A derivative 11	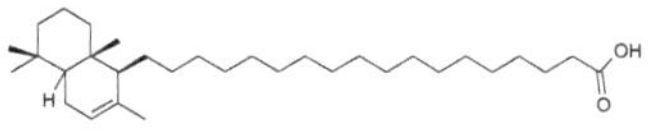	pol α,polβ, polγ, polδ, polε, polη, polι, polκ, polλ	[76]
Koetjapic acid	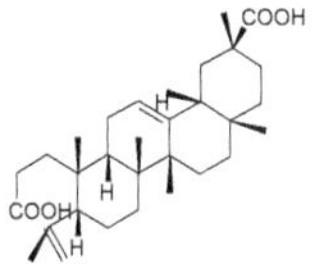	pol β	[77, 78]
Linoleic acid (LA)	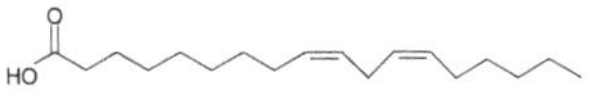	pol α, pol β	[79,80, 81]
Lithocholic acid (LCA)	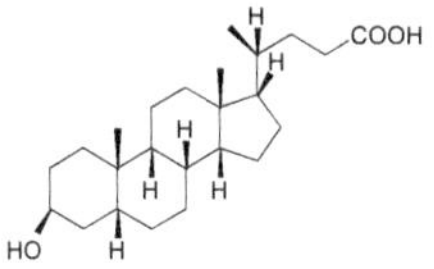	pol α, pol β	[82, 83]
Lupane triterpenoids derivative	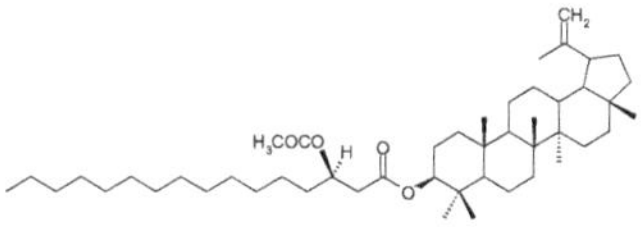	pol β lyase activity	[84]
Manoalide	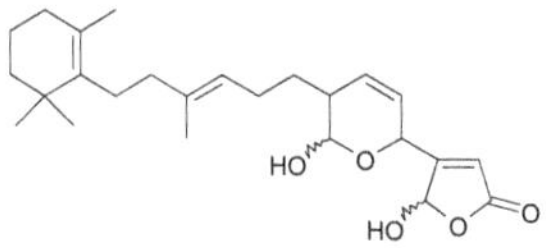	pol κ	[62]
MK-886	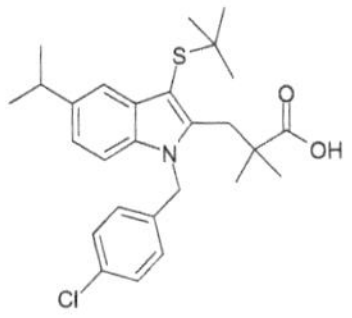	pol κ, pol ι,	[62]
Monoacetyl curcumin	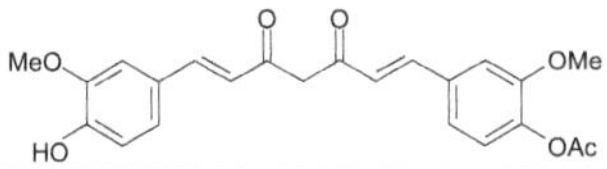	Pol λ	[85, 86]

Table 4.2 (continued)

Myristinin A	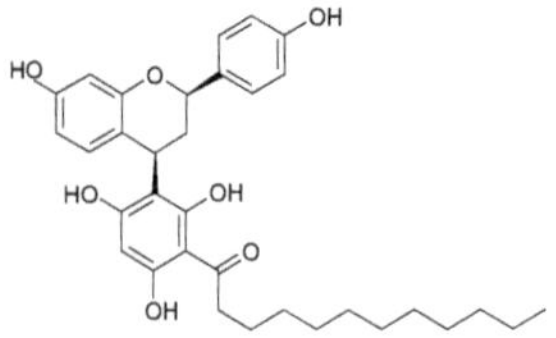	pol β	[87]
Myricetin	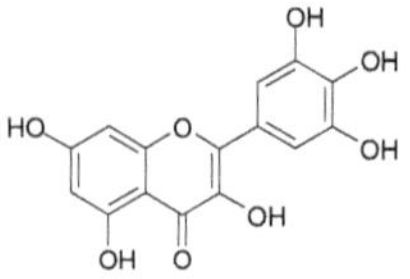	all animal pols	[88]
Nervonic acid (NA)	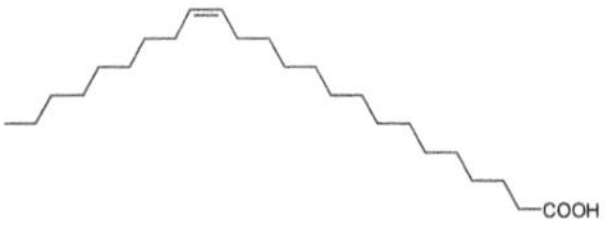	pol α, pol β	[79,80, 81]
Neolignan-1	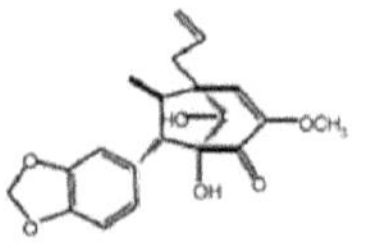	pol β lyase activity	[89]
Neolignan-3	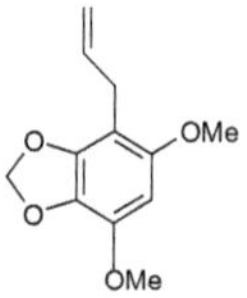	pol β lyase activity	[89]
Oleanolic acid	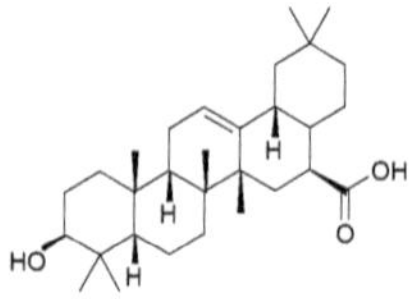	pol β	[90]
Oleic acid	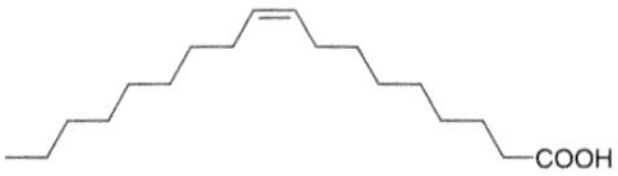	pol β	[54]
4-Oxo-dihydroquinoline-3-carboxamides	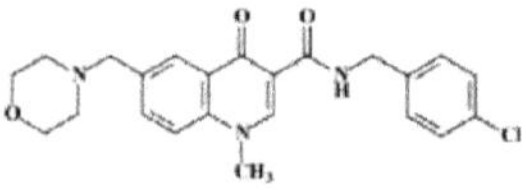	HCMV, HSV, and VZV pols	[91, 92]
Pamoic acid (PA)	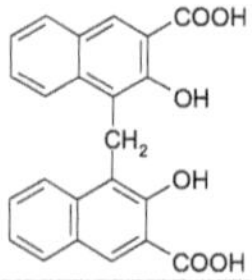	pol β	[78, 93]

Table 4.2 (continued)

PGG	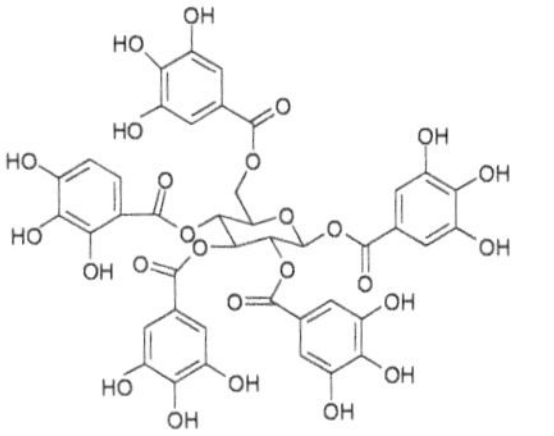	pol α, pol β, pol k	[94]
Petasiphenol	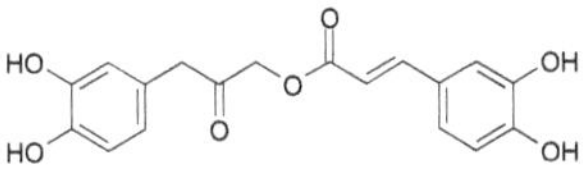	pol λ	[95]
Pyridoxal 5'-phosphate	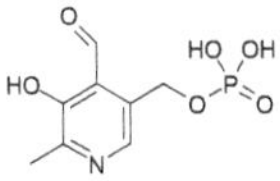	pol α, pol ε	[96]
QAG	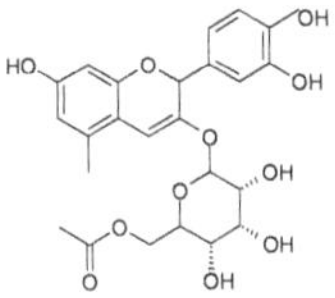	pol α, pol δ, pol ε	[74]
Resveratrol derivates	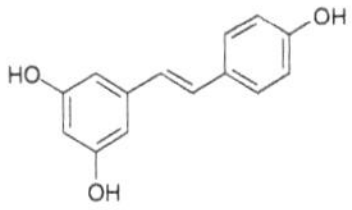	pol α, pol λ, SV40 pol	[97]
Rhodanines		pol λ	[98]
Sodium cholesterol sulfate	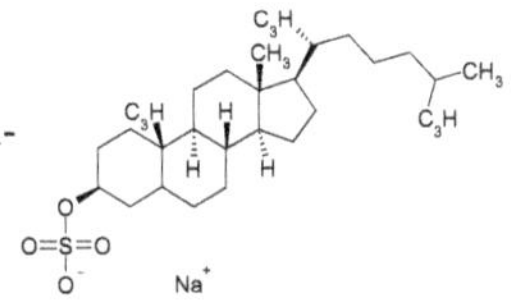	pol α, pol β, pol γ, pol ε, pol δ, pol λ	[99]
Sculezonone-A	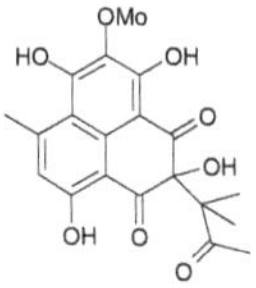	pol α, pol γ, pol β, pol ε	[100]
β-Sitosterol-3-O-β-d-glucopyranoside	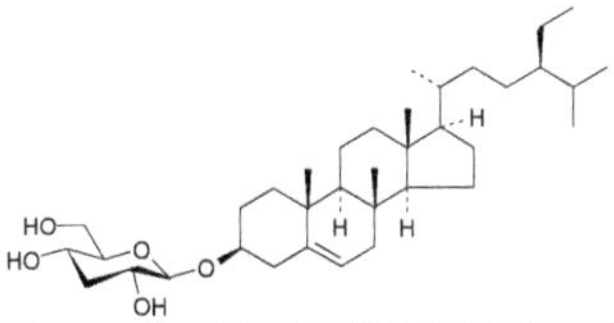	pol λ	[101]

Table 4.2 (continued)

Compound	Structure	Target	Ref.
SQAG derivates	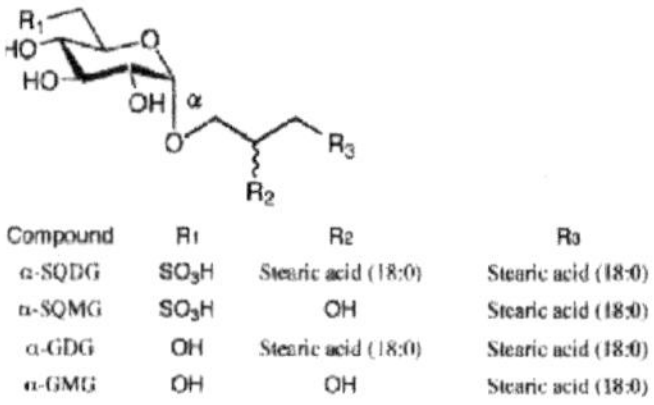 Compound / R_1 / R_2 / R_3 α-SQDG / SO_3H / Stearic acid (18:0) / Stearic acid (18:0) α-SQMG / SO_3H / OH / Stearic acid (18:0) α-GDG / OH / Stearic acid (18:0) / Stearic acid (18:0) α-GMG / OH / OH / Stearic acid (18:0)	pol α, pol β	[102]
Solanapyrone A	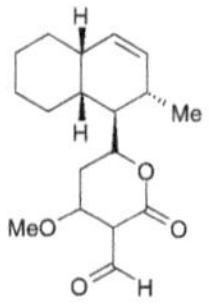	pol β, polλ	[103]
SQMG	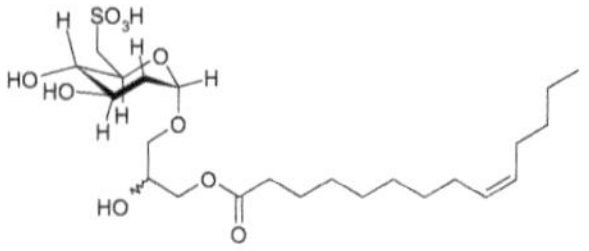	pol β	[104]
Sulfolipid derivative 1	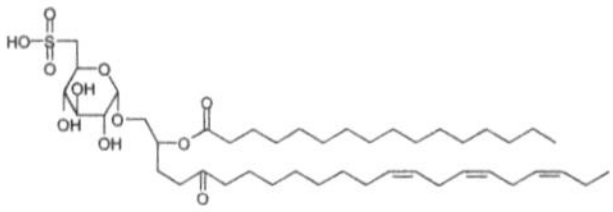	pol α pol β	[81]
Taxinine	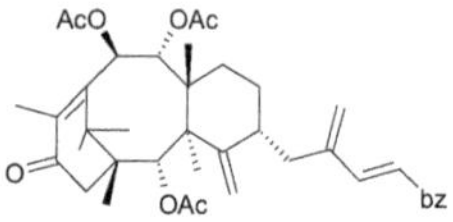	pol α pol β	[61]
Taxole derivate	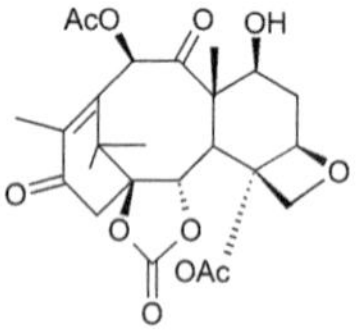	pol α*	[61]
Tocotrienol	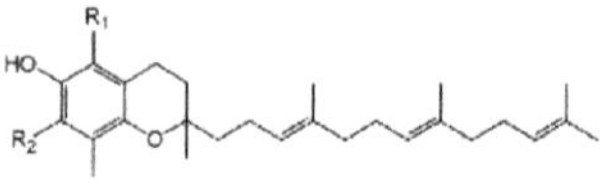	pol λ	[105]
Triterpenoid-derivative	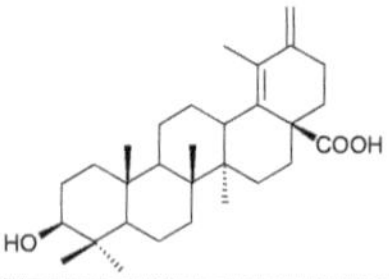	pol β	[106]

Abbreviations: DRB 1,4-dideoxy-1,4-imino-D-ribitol, *HCMV* human cytomegalovirus, *HSV* herpes simplex virus, *HMI* 4-Hydroxy-17-methylincisterol, *KA-A 11* (1S*,4aS*,8aS*)-17-(1,4,4a,5,6,7,8,8a-octahydro-2,5,5,8atetramethylnaphthalen-1-yl)heptadecanoic acid, *KAG* kaempferol 3-O-(600-acetyl)-b-glucopyranoside, *KN-208* 1-O-(6′-sulfo-alpha-D-glucopyranosyl)-2,3-di-O-phytanyl-sn-glycerol, *MK-866* 1H-Indole-2-propanoic acid, *QAG* quercetin 3-O-(600-acetyl)-b-glucopyranoside, *PGG* penta-O-galloyl-beta-D-glucose, *SQAG* sulfoquinovosyl-acylglycerol, *SQMG* sulfoquinovosylmonoacylglycerol, *VZV* varicella zoster virus

[110]. Some of these compounds demonstrated broad-spectrum inhibition of the herpesvirus polymerases HCMV, HSV-1, EBV, and VZV with high specificity compared to human DNA polymerases. DHTPs, in contrast to the kinetics determined for the 4-oxo-DHQs, proved to be competitive inhibitors of dTTP incorporation into primer template by HCMV DNA polymerase [108].

However, as we mentioned above, the majority of non-nucleotide inhibitors influence the activity of eukaryotic DNA-dependent pols. Despite the significant number of known ones, search and development of new selective NNI is continuing up to now. For instance, inhibitors of several DNA polymerases were found among flavonoide derivates. Shiomi et al. [88] investigated the inhibitory activities of 16 major bioflavonoids against mammalian DNA polymerases. Myricetin (3,3′,4′,5,5′,7-hexahydroxyflavone) was the most potent inhibitor of pols among the compounds tested, with IC_{50} values of 21.3–40.9 μM. This compound did not affect the activities of plant (cauliflower) pol α or prokaryotic pols. Myricetin also inhibited human DNA topoisomerase II (topo II) activity with an IC_{50} value of 27.5 μM, but did not inhibit the activities of other DNA metabolic enzymes tested [88]. Myricetin also did not influence the direct binding to double stranded DNA as determined by the thermal transition analysis. It was found to prevent the proliferation of human colon HCT116 carcinoma cells with an LD_{50} of 28.2 μM, halt the cell cycle in G2/M phase, and induce apoptosis. These results suggest that the decrease of proliferation may be a result of the inhibition of cellular topoisomerase (topo) II rather than pols [88].

Significant inhibitory effectiveness against representatives of B and Y pol families is demonstrated for natural plant gallotannin penta-O-galloyl-beta-D-glucose (PGG) that has been shown to inhibit the *in vivo* growth of several types of tumors without evident adverse side effects [111–113].

PGG exhibits a selective inhibition against the activities of pol α and pol κ in nanomolar concentrations. The inhibitory effect of PGG on pol α is the strongest among known low-weight inhibitors, with IC_{50} value of 13 nM. PGG activity against pol κ is slightly less—IC_{50} in this case is 30 nM [94]. PGG is also able to inhibit pol β, but its potency is an order of magnitude less than that against pol α—the corresponding value of IC_{50} is in the range of 108–160 nM. PGG inhibition of pol α and κ activity is non-competitive with respect to the DNA template-primer and the dNTP substrate; in contrast to the inhibition of pol β activity which is competitive [94]. The structural model of 'pol β–PGG' interaction is also proposed (see the next chapter for details). PGG seems to be a very promising compound for fundamental investigations (distinct inhibitory mechanisms for different pol families) as well as for practical application (potential anticancer drug).

4.4 Structural Analysis of Pol–Inhibitor Interactions

There is essential difference between analytical approaches suitable for the analysis of structural mechanisms/features of DNA polymerase interactions with nucleoside and non-nucleoside inhibitors. Being the analogs of incoming dNTP, the nucleoside inhibitors *a priori* bind into appropriate active site in the pol structure and compete with natural enzyme substrates. Consequently, the structure of polymerase complex with nucleoside inhibitor can be easy reconstructed *in silico* with high accuracy and without any additional experimental data. On the contrary, the structural analysis of DNA polymerase interaction with non-nucleotide compounds requires the preliminary identification of appropriate binding site(-s), which is a very sophisticated task without additional input. Direct *in silico* identification of the interactive site, based on structures of protein and ligand only (blind docking) in many cases is not completely exact. So, it is not surprising that one usually uses a combination of computational and instrumental approaches to identify non-nucleoside compounds' sites on the DNA polymerase surface.

Mizushina et al. [82, 114] identified for the first time the mode of interaction between the human DNA pol β and the lithocholic acid (LCA) (see Table 4.2). The 39-kDa pol β was separated proteolytically into two fragments corresponding to the template-primer binding domain (8 kDa) and the catalytic domain (31 kDa). It was shown that LCA bound tightly to the 8-kDa fragment but not to the 31-kDa fragment. In ^{1}H–^{15}N HMQC NMR analysis of pol β with LCA, the 8-kDa domain bound to LCA as a 1:1 complex with a dissociation constant (K_D) of 1.56 mM. The chemical shifts were observed only in residues mainly in helix-3, helix-4, and the 79–87 turn of the same face. No significant shifts were observed for helix-1, helix-2, and other loops of the 8-kDa domain [82]. The maximal shift was observed for three amino acid residues—Lys60, Leu77, and Thr79 of pol β on the LCA. Obtained data were used for the further docking of LCA molecule into the appropriate region on the pol β surface and interaction interface reconstruction [114].

Later, the same group reported about reconstruction pol β complexes with LCA derivates—3-alpha-methoxy-5-beta-cholan-24-oic acid (compound **2**) and 3-alpha-O-lauroyl-5-beta-cholan-24-oic acid (compound **9**) [115]. The docking was carried out using a fixed docking procedure in the *Affinity* module within Insight II[3] modeling software (Accerlys Inc., San Diego, CA). The calculations used a CVFF force field in the *Discovery* module and the Monte Carlo strategy in the *Affinity* module of Insight II. Each energy-minimized final-docking position of LCA derivatives was evaluated using interactive score function in the *Ludi* module. The *Ludi* score includes the contribution of the loss of translational and rotational entropy of the fragment, number and quality of hydrogen bonds, and contributions from ionic and lipophilic interactions to the binding energy. According to the data obtained, compound 2 shares the same binding site with LCA. Compound 9, containing long fatty acid moiety, is one of the strongest pol β inhibitors ($K_D = 1.7$ nM) and binds into a different site in the surface 8-kDa pol β domain. A critical role in the compound 9 binding belongs to amino acid residues Leu11, Lys35, His51, and Thr79 [115]. Second site is also able to bind another pol β inhibitor—nervonic acid (NA) [115].

Inhibitor of different eukaryotic DNA polymerase (pol α, pol δ, pol ε, pol γ, pol ι, pol κ and TdT) sulfoquinovosylmonoacylglycerol (SQMG) also was found to identify its binding mode with pol β [104]. According to the data of the NMR

[3] Now Accerlys Inc. has discounted a development and support of Insight II. Functionality of this software is transferred to Discovery Studio and Pipeline Pilot program suites.

chemical shift mapping and subsequent molecular docking (*Affinity* and *Discovery* modules of Insight II software, ESFF force field [116]), the two potential binding were proposed. Site I includes amino acid residues Leu22, Phe25, Glu26, Asn28, Ile33, Lys35, Asn37, site II—residues Lys60, Leu62, Gly64, Gly66 and Ala70, consequently. Both discovered sites are located on template DNA binding interface of the 8-kDa domain, and SQMG incorporation into any of them would prevent interaction with the DNA template [104]. The above-described binding sites for LCA and its derivates, as well as for NA, are situated on the template DNA binding interface, too [82, 114, 115]. Thus, one can conclude that all pol β inhibitors identified by industrious japanese researchers as interacting with 8-kDA domain have a very similar structural mechanism of action. These compounds interact with the DNA binding area and compete with the DNA template in a similar fashion.

Similar strategy combining spectroscopic and computational approaches was applied by Hazan et al. [93] to identify the pamoic acid binding site on the surface of DNA polymerase β. The ability of pamoic acid to interact with the 8-kDa template-primer binding domain was previously demonstrated by Hu et al. [78]. The molecular docking of pamoic acid to the 8 kDa domain of pol β was carried out using AutoDock 3.0.5 software [117, 118].

Structures generated by AutoDock have been ranked according to their binding energy and 100 lowest energy structures were selected for the further analysis. With the force field used by AutoDock, the energy values for the best ligands varied from −9.58 to −8.96 kcal/mol. Systematic analysis of the 100 best docked structures revealed that all of them were located at a single site, although pamoic acid could move freely around the 8 kDa domain during docking. Close atomic contacts between pairs of protein–ligand atoms (with a distance cutoff of 2 Å) were computed. Nine residues—His34, Lys35, Asn37, Ala38, Lys41 on helix 2 and Gly64, Gly66, Lys68, Lys69 on helix 4—were frequently found to be close to the pamoic acid. In fact, in more than 50 % of the resulting conformations, at least one proton of Ala38, Lys68 and Ile69 was located within 2 Å from the pamoic acid. For residues His34, Lys35, Asn37, Lys41, Gly64 and Gly66, over 20 % of the 100 best docked structures contained a pair of protein–ligand atoms with a separation below 2 Å. Mapping these residues onto the 8 kDa domain structure indicated that they form a single positively charged groove at the protein surface (Fig. 4.5a). Interestingly, Lys35, Lys60 and Lys68, which have been shown to be responsible for single-stranded DNA binding by site-directed mutagenesis [119] are located in the groove where pamoic acid binds to. As this groove is the one where DNA binds, pamoic acid is likely to interfere with single-stranded DNA recognition.

Clustering the 100 best ligand structures has been performed using the RMSD (root mean square deviation) cutoff value of 2 Å. The five resulting clusters indicated that the ligands adopt five different ensembles of conformations in the binding site described above [93]. Calculated conformation ensembles were verified using NMR chemical shift mapping [120], NOE (Nuclear Overhauser Enhancement) spectroscopy and STD (Saturation Transfer Difference) experiments [121, 122]. Only one reconstructed complex between the 8 kDa domain and pamoic acid was consistent with the entire NMR data (Fig. 4.5b).

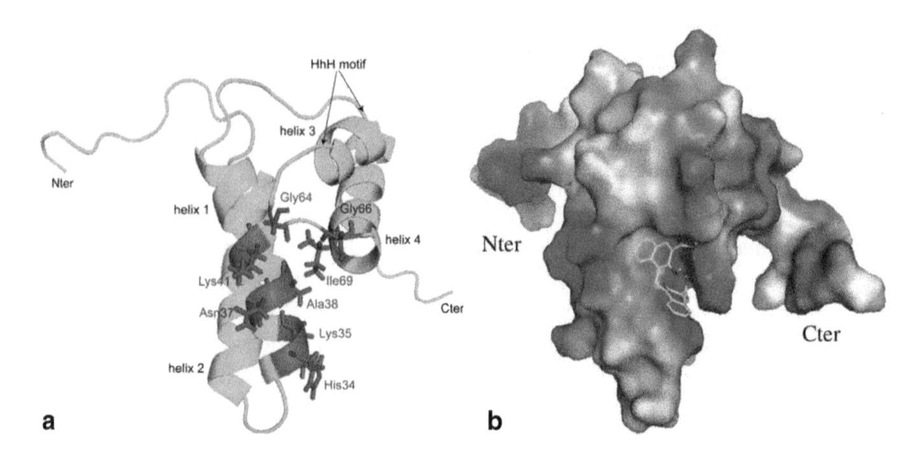

Fig. 4.5 **a** Amino acid microenvironment of pamoic acid bound to 8 kDa domain of pol β, **b** pamoic acid docked into pol β. (Adapted from [93])

Hazan et al. [93] have analyzed in detail the contacts between the validated ligand structure and amino acid microenvironment. The aromatic groups of pamoic acid participate in hydrophobic interactions with the main amino acids of the binding site, such as Tyr39, Ala42, Gly64 and Gly66. Furthermore, numerous lysine residues present in the site can form electrostatic interactions with both carboxyl groups. One of the carboxyl groups is oriented towards His34 and Lys35. It makes close contacts with Ile69 amide proton and electrostatic interaction with the terminal NH_3^+ group of the Lys68 side chain. The other carboxyl group forms hydrogen bonds with the amide proton of Lys68 (distance of 1.67 Å) and with the hydroxyl group of Thr67 (distance of 1.94 Å). Obviously, the two carboxyl groups contribute to pamoic acid affinity for the 8 kDa domain.

One more pol β inhibitor with 8 kDa domain affinity—solanapyrone A—was studied in molecular docking experiments [103]. Docking procedure was performed in the same way as for the above-described LCA, NA and SQMGA [82, 104, 114, 115] with the subsequent decomposition analysis of the binding energy using *Ludi* module of Insight II modeling software. Solanapyrone A binding site is located on the protein-DNA template contact interface, similar to the interactive sites described above, and consists of amino acids Ile53, Gly56, Ala59, Lys60, Ala70 and Ile73. The main contribution to the total binding energy is made by the binding energy between NH_3^+ of Lys60 and the ketone groups in solanapyrone A—−28.230 kcal/mol by hydrogen bond, and the binding force consists of the Coulomb force (−27.212 kcal/mol) and van der Waals forces (–1.018 kcal/mol). The distances between the two ketone groups of solanapyrone A and the NH_3^+ residue of Lys60 were 2.01 and 2.41 Å (Fig. 4.6). The sum of binding energy between the benzene backbone of solanapyrone A and the hydrophobic amino acids (i.e. Ile53, Gly56, Ala59, Ala70, and Ile73) is only −7.682 kcal/mol.

Thus, all known DNA pol β inhibitors with a specific affinity to the template–primer binding domain have similar structural mechanisms of the inhibitory action.

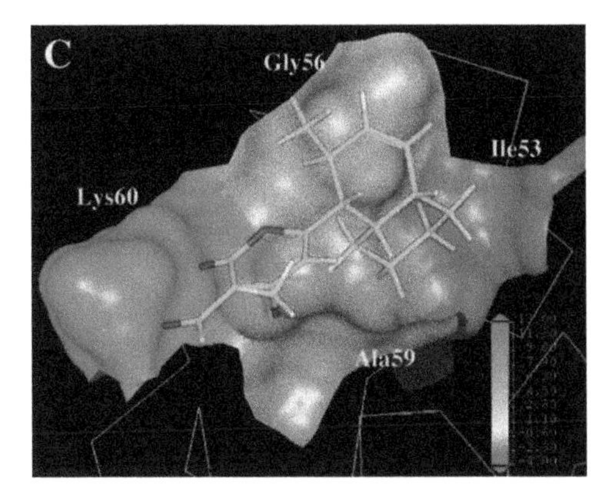

Fig. 4.6 Hydrophobicity map on molecular surface of solanapyrone A binding site. (Adapted with permission from Mizushina et al. [103])

Appropriate binding sites are localized on the DNA template binding interface sufficiently close to each other and can contain shared amino acids. The binding of any of the above-considered low-weight molecular compounds directly prevents interaction with ssDNA and defines the competitive character of polymerase inhibition.

The computational modeling of protein–ligand interactions was performed also for compounds inhibiting other DNA-dependent DNA polymerases. So, structural insights of actions were investigated *in silico* for high specific pol λ inhibitors—petasiphenol [95] as well as curcumin and its derivate monoacetylcurcumin [86].

The petasiphenol is a natural phenolic compound produced from a higher plant, a Japanese vegetable (*Petasites japonicus*) that was originally found to be a bioantimutagen in UV-induced mutagenic *Escherichia coli* WP2 B/r Trp-isolated from the same plant [123]. It was established experimentally that petasiphenol binds to N-terminal BRCT domain of pol λ with IC_{50} of 7.6 μM and does not bind to the C-terminal catalytic domain including the pol β-like core of pol λ [95].

Spatial structure of pol λ BRCT domain was reconstructed via homology modeling using molecular modeling software Insight II (module *Homology*) (Accelrys Inc., San Diego, CA). The spatial structure of human XRCC1 (PDB accession code is 1CDZ [124]) was used as a template for the modeling. Molecular docking of petasiphenol molecule into pol λ BRCT domain and the further binding evaluation were performed according to the procedures described above for solanapyrone A. It was revealed that the N-terminal BRCT domain of pol λ (residues 36–132) consists of three α-helices and four β-sheets. The petasiphenol-binding region in the BRCT domain of pol λ is assumed to consist of the two loops (residues 74–81 and 84–107) between the β-sheet (residues 82–83) and includes amino acids Gln76, Ile83, Asp90, Glu92, Arg93 Ala94 Leu95, Arg96, Leu98, Arg99, Leu100, Gln102, Leu103 and Pro104 [95].

The hydroxyl and ketone groups of petasiphenol may show a preference for binding to the hydrophilic residue of Gln76, Arg93, and Arg99, and, on the other hand, the benzene groups may be absorbed to the hydrophobic amino acids in the loops. The binding energies between NH_2 of Gln76, NH_2^+ of Arg93, or NH_2^+ of Arg99 and the hydrophilic groups in petasiphenol were −9.400, −3.652 and −4.642 kcal/mol respectively, and the binding force consisted of the Coulomb force (− 7.904, −2.220

and −3.186 kcal/mol, respectively) and van der Waals forces (−1.496, −1.432, and −1.656 kcal/mol, respectively). The distances between the three hydroxyl groups of petasiphenol and the hydrophilic residues of Gln76, Arg93, and Arg99 were 1.69, 1.79 and 2.00–2.04 Å, respectively. The binding energy between the other hydrophilic amino acids (Asp90, Glu92, Ala94, Arg96, Gln102 and Pro104) and petasiphenol is −11.284 kcal/mol, and the binding energy between the benzene backbone of petasiphenol and the hydrophobic amino acids (Ile83, Leu95, Leu98, Leu100 and Leu103) was − 15.342 kcal/mol. On the BRCT domain of pol λ, petasiphenol was smoothly intercalated into the pocket of the loops, and the residues around the amino acid site consisting of hydrogen bonds (i.e., Gln76, Arg93, and Arg99) appear to be most important for petasiphenol binding [95].

Takeuchi et al. [86] have investigated the structural insights of the interaction between the pol λ BRCT domain and curcumin derivates. The curcumin is known as an antichronic inflammatory agent and an anti-oxidative compound. The procedures of BRCT homology modeling, monoacetylcurcumin molecular docking and binding energy decomposition analysis were carried out in the same way as for the petasiphenol docking experiment [95]. It was shown that monoacetylcurcumin binding site on the pol λ surface does not coincide with the petasiphenol binding site and consists of residues Thr51, Gly52, Gly54, Ala58, Glu59, Glu62, Lys63, Val66, Val85, Glu87 and Ala113 belonging to β-sheet1 (Thr51, theα-helix-1 (residues 57–69) and two loops (residues 51–56 and 70–75)). The main contribution to the total binding energy is made by the interaction between curcumin and Lys63—the energy of this interaction is −37.93930 kcal/mol (the Coulomb energy is −29.59488 and the van der Waals energy is −8.34442 kcal/mol). The energies of interaction with dicarboxylic amino acids Glu59 and Glu62 are −7.55664 and −9.11299 kcal/mol, respectively [86].

The distances between the three hydroxyl groups of monoacetylcurcumin and the hydrophilic residues of Glu59, Glu62, and Lys63 are 2.65, 2.74–2.87 and 2.43 Å respectively. The binding energy between the other hydrophilic and neutral amino acids (Thr51 and Glu87) and monoacetylcurcumin is −7.48746 kcal/mol, the binding energy between the benzene backbone of monoacetylcurcumin and the hydrophobic amino acids (Gly52, Gly54, Ala58, Val66, Val85 and Ala113) is −23.12777 kcal/mol. On the BRCT domain of pol λ, monoacetylcurcumin is smoothly intercalated into the pocket of the loops, and the side of the nonacetoxy group on it is just fitted into pocket of the BRCT domain. The residues around the amino acid site consisting of a covalent bond (i.e., Cys73) and five hydrogen bonds (i.e. Glu59, Glu62 and, Lys63) appear to be important for binding to monoacetylcurcumin [86].

Non-nucleotide inhibitors are also able to get bound to the DNA polymerase active site. So, the above-mentioned gallotannin PGG inhibiting polymerases α, β and k were docked into the active site on the pol β surface [94]. The docking results show that PGG could form several favorable interactions with the polymerase catalytic pocket/binding site for the incoming dNTP. The free energy of the binding is predicted to be −10.26 kcal/mol and the docking runs gave only one possible spatial geometry (Fig. 4.7). In addition, the compound seems to bind in a way that sterically obstructs two amino acids (Asp192 and Asp196) that are part of the

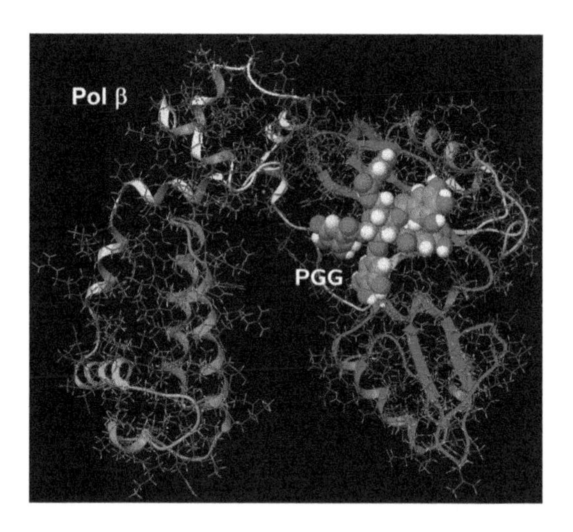

Fig. 4.7 Localization of PGG binding site in space of pol β. (Reprinted from Ref. [94], © (2010), with permission from Elsevier [94])

catalytic core of the protein, potentially inhibiting the enzyme activity. The docking simulation provides important molecular insights into how PGG inhibits pol β in a competitive manner with respect to the dNTP substrate and DNA template–primer [94]. It is extremely exciting that PGG binding by polymerases from different families proceeds by distinct mechanisms (it is non-competitive for pol α and pol k but competitive for pol β), and future investigations of these processes can essentially extend our knowledge about structural insights of specificity and selectivity of low-weight molecular compound in relation to various biomolecular targets.

Recently, the structural mechanisms of Y-family pol k inhibition by1H-Indole-2-propanoic acid (MK-866) were investigated with *in silico* docking (Ketkar et al. [125]). Docking runs were performed using the PDB files 4EBC (pol ι), 3MR2 (pol η) and 2OH2 (pol κ) downloaded from the Protein Data Bank [126], for target Y-family DNA polymerases, either with the DNA coordinates (binary) in place or after removing the DNA atoms (apoenzyme). The protein PDB files were preliminarily prepared for docking using the Dock Prep tool [127] available in the free software package UCSF Chimera [128]. This involved the addition of hydrogens, removal of water and other extra molecules, and assigning partial charges (using the AMBER99 force field).

The spatial coordinates for the MK886 molecule were generated using the Marvin Sketch free software tool in the ChemAxon package (http://www.chemaxon.com/products/marvin/marvinsketch). Automated *in silico* docking was performed using the web-based docking server SwissDock (http://www.swissdock.ch/) that is based on the docking algorithm EADock DSS [129]. The processed coordinates file (as described above) for each of the proteins and for the ligand MK886 were uploaded, and docking runs were performed using the "Accurate" parameters option, which is the most exhaustive in terms of the number of binding modes sampled. Docking runs were performed as blind, covering the entire protein surface, and not defining any specific region of the protein as the binding pocket in order to avoid sampling bias. Output clusters were obtained after each docking run and were

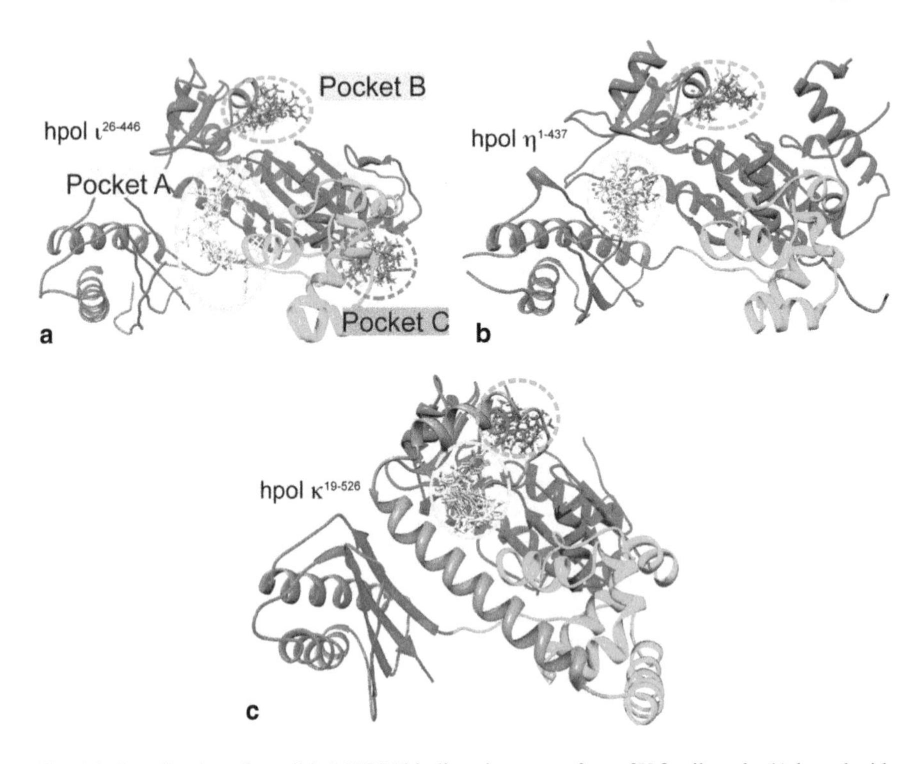

Fig. 4.8 Localization of possible MK886 binding sites on surface of Y family pols. (Adapted with permission from Ref. [125]. Copyright (2013) American Chemical Society [125])

ranked according to the FullFitness (FF) scoring function specified by the Swiss-Dock algorithm (cluster 0 being the cluster with the best FullFitness score). Within each cluster, the individual binding poses were further arranged and ranked based on their FF score.

According to docking results, three MK886 possible binding pockets for pol ι (Fig. 4.8A) and only two MK886 binding pockets for pol η (Fig. 4.8B) and pol κ (Fig. 4.8C) were consistently identified. For all three polymerases, the highest number of clusters was found to localize at the interface between the DNA-binding cleft and the active site of each Y-family member, which we refer to as pocket A (Fig. 4.8). In addition to this pocket, two more distinct binding pockets were observed for pol ι. The first of these, which we call pocket B, lies at the junction between the finger and palm subdomains of pol ι (Fig. 4.8A). Pocket B was also identified in docking analyses with pol η and pol κ. The final binding pocket for MK886 on pol ι (pocket C) lies at the junction between the palm and thumb subdomains (Fig. 4.8A). Consistently, it was observed that at least one cluster from the top 5 (top 10 in the case of pocket C) localized at these three pockets on pol ι. It is interesting to note that nearly all of the binding modes identified in docking analyses for the polymerases localized at one of these three pockets, with only an occasional outlier cluster, which localized at a completely different region of the polymerase. The consistent identification of MK886 binding pockets on the DNA polymerases is in stark contrast to what was observed with BSA where the top 10 binding modes changed positions dramatically between docking. Docking runs were performed with two versions of the target polymerase PDB files, one in which the coordinates for the DNA atoms were left in place (binary form) and the other in which these were removed to give an "apo" form. The only difference observed in the results obtained with these two versions was in the number of clusters observed in pocket A.

Unfortunately, information about structural details of Y family pols interaction with MK886 presented in paper of Ketkar et al. [125] is very poor. The negative impact upon catalysis if MK886 binds to pocket A is obvious for all three polymerases. In the case of pol ι, the interactions between MK886 and the side chains of Arg103 and Arg331 would disrupt key electrostatic interactions between the polymerase and the template strand. Similar effects upon DNA binding could be predicted for pol η and pol κ. Consideration of inhibitor binding to pocket B in pol ι shows that electrostatic interaction between the docked MK886 molecule and Asn216 stations the inhibitor near the "gate" to the dNTP binding cleft, which could also reasonably be assumed to interfere with the productive binding of the incoming nucleotide triphosphate. Likewise, binding of MK886 to pocket B of either pol η or pol κ could conceivably interfere with the productive dNTP binding. Finally, pocket C is only observed with pol ι, where MK886 is found to interact with residues in the thumb and palm domains. The potential inhibitory effect of pocket C is less obvious than that of either pocket A or pocket B. In pocket C, MK886 interacts with residues that are near the base of the αH helix (Gln227) in the thumb domain, located not far from the binding site of the third metal ion. Transient coordination of the third metal ion was recently shown to play a role in the catalysis by pol η, and metal ion coordination has been suggested to be the rate-limiting step in the catalysis by pol η [130, 131]. Other studies have revealed that conformational changes in the thumb domain play a role in nucleotide selection by Y family polymerases [132, 133]. Thus, it would appear that binding of MK886 to pocket C near the pol ι thumb region may contribute to the more potent inhibitory effect by interfering with either conformational changes or metal ion dynamics and that this effect is not observed for either pol η and pol κ since they do not possess a well-formed pocket C.

In silico approaches are also used for the analysis of structural insights of bacterial polymerases inhibition. Martin et al. [134] have studied possible structural mechanisms of Taq polymerase I inhibition by 6,10,2′,6′-tetraacetyl-O-catalpol (Table 4.2) with the wide spectrum of computational methods. Classical and semi-empirical methods were used to characterize the conformational preferences of this organic compound in solution. The Gabedit software package [135] was used to generate a catalpol starting geometry which was initially optimized using the classic quasi-Newton method, followed by semiempirical optimization using the software MOPAC [136]. Minimized conformation was used to obtain the parameters and topology files for the GROMACS software [137, 138] using the ProDGR server [139]. The analysis of catalpol conformational space was performed using two different approaches—simulated annealing (SA) and molecular dynamics (MD). SA was performed with GROMACS software using appropriate protocol. The force field used was ffG53a6, the solvent was explicitly simulated using the SPC model, periodic boundary conditions were implemented using the PME (particle mesh Ewald) algorithm [140] and the cut-off values were 1.4 and 0.9 nm for the van der Waals and electrostatic interactions, respectively. MD was carried out with Yasara Dynamics 10 software [141]. The time step of 2 fs was used. The Amber03 force field [142] was selected; the simulation box allowed at least 10 Å around all of the atoms in 6,10,2′,6′-tetraacetyl-O-catalpol, and was filled with the TIP3 water model. Periodic boundary conditions were used as implemented in the PME algorithm. Na^+ and Cl^- ions were added in order to properly simulate the ion strength in physiological solution. Two simulations of 300 productive nanoseconds each were performed starting from the best energy conformation identified during the SA procedure and minimized with the Amber03 force field, and the same conformation with the torsion angle α rotated 180° and minimized with the Amber03 force field.

Using docking simulations, the most probable binding mode was found, and the stabilities of the docked solutions were tested in a series of molecular dynamics experiments. The docking was performed with Autodock 4.0 software [118] using two different modes—blind docking search and binding site restricted search. The coordinates of the target molecule (including the Klentaq fragment and a short DNA portion) were taken from the Protein Data Bank [126] deposited under the accession code 2KTQ [143].

The ligand position with the lowest energy (the best energy solution of −4.45 kcal/mol) is located at the active site of the enzyme [134]. This result is in line with previous experimental observations regarding the inhibitory mechanism of catalpol [144], and supports the hypothesis of a competitive inhibitory mechanism for 6,10,2′,6′-tetraacetyl-O-catalpol. Spatial microenvironment of catalpol in binding site consists of amino acids Asp610, Tyr611, Ser612, Gln613, Ile614, Glu615, Leu616, Lys663, Phe667, Leu670, Tyr671, Asp785 and Glu786. All of them except Leu616 participate in the incoming dNTP binding as well. The time stability of calculated 'Taq polymerase–catalpol' complex was confirmed by a set of molecular dynamics simulations [134]. Trajectory analysis indicated four time-stable hydrogen bonds between 6,10,2′,6′-tetraacetyl-O-catalpol and the enzyme that were present for $>20\,\%$ of the simulation time. Spatial structure of reconstructed complex can be used as a starting point for directed optimization of DNA pol inhibitors based on catalpol derivates.

Computational analysis is also applied to the investigation of structural mechanisms of actions of viral DNA polymerase inhibitors. So, Li et al. [145] have used *in silico* approaches, including homology modeling, docking, MD simulation and MM/PBSA free energy analysis, to study structural insights underlying the influence of DNA polymerase from different genotypes of hepatitis B virus (HBV) on the binding affinity of acyclic nucleotide adefovir (ADV). An important feature of HBV pol is its ability to act as a matrix for the complimentary DNA synthesis of both DNA and RNA molecules. Spatial structure of HBV pols of B and C genotypes was reconstructed by the homology modeling using the automated modeling module, MODELLER [146] in the Discovery Studio (DS) software 3.0 (Accelrys, San Diego, CA, USA). After the energy minimization, the geometric quality of the modeled structures was evaluated by PROCHECK [147].

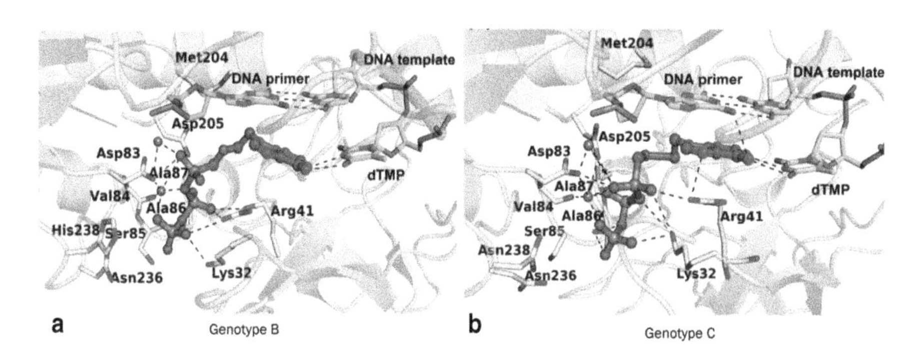

Fig. 4.9 The best possible binding modes of ADV-DP in the HBV pol active sites of genotypes B and C are. (Adapted by permission from Macmillan Publishers Ltd.: [145]. © 2013 [145])

The molecular docking study was performed with the Glide program. One should mentioned that ADV is a prodrug, and due to its bioactive form, ADV diphosphate (ADV-DP), which is generated through stepwise phosphorylation after the oral administration of ADV, was modeled,. The 3D conformation of ADV-DP was constructed according to the coordinates of the ligand, TDV-DP, which was deposited in the template structure, by deleting a methyl group.

The molecular docking was performed using standard precision protocols with default parameters. The docking poses were ranked by their glide scores, and the best predicted conformation in each system was used for the subsequent molecular dynamics simulation and binding energy analysis.

Classical molecular dynamics simulations of docked conformation-HBV polymerase complexes were performed using the AMBER 9.0 suite [148]. The electrostatic potential of ADV-DP was computed with Gaussian03 at the HF/6-31G* level [149], then the charges were assigned using the RESP (restrained electrostatic potential) methodology [150]. The charges and force field parameters for ADV-DP were generated by an Antechamber [151]. The polyphosphate parameters of ADV-DP in studied system were identical to those developed by Meagher et al. [152]. The AMBER03 force field [142] and the general AMBER force field (GAFF) [153] were chosen to create the potential of the proteins and ADV-DP, respectively. The systems were then solvated in a truncated octahedral box of TIP3P water molecules 10 Å away from the protein. Counter-ions of Cl^- were then added to obtain the electrostatic neutrality of the systems. The ADV-DP binding free energies (ΔG_{bind}) HBV pols of B and C genotypes were calculated using the MM-PBSA method [154]. The best possible binding modes of ADV-DP in the HBV pol active sites of genotypes B and C are illustrated in Fig. 4.9.

Both complexes are stabilized by extensive hydrogen bonding networks. The nitrogen base of ADV-DP displayed two hydrogen bonds with the complementary base pair dTMP in the template chain. Moreover, it formed a π–π stacking interaction with the DNA base in the primer chain. The β-phosphate and γ-phosphate of the ligand formed a hydrogen bond with the backbone amide NH of Ala86 and Ala87. The side chains of Arg41 and Lys32 were involved in ionic interactions with the phosphate group of ADV-DP. Arg41 in the systems of genotypes B and C formed three and two hydrogen bonds with ADV-DP, respectively. The carbonyl groups of Asp83, Asp205 and Val84 together with the three phosphate groups of ADV-DP formed a metal chelating interaction with the two Mg^{2+} ions that were present in the active site [145].

Sequence analyses revealed that residue 238 near the binding pocket was not only a polymorphic site but also a genotype-specific site (His238 in genotype B, Asn238 in genotype C). The calculated binding free-energy for the HBV pol from genotypes C and B is − 147.81 and − 126.85 kcal/mol, respectively. It confirms the hypothesis that the HBV pol from genotype C is more sensitive to the ADV treatment than one from genotype B. By using the MD simulation trajectory analysis and binding free energy decomposition, some energy variation in the residues around the binding pocket was observed. According to the energy decomposition data, residues Lys32, Arg41, Asp83, Ser85, Ala86, Ala87 and dTMP make a main contribution into the ADV-DP binding, due to the hydrogen bonding. The purine ring of ADV-DP formed a strong π–π stacking interaction with the primer DNA base, which also showed an obvious hydrophobic interaction with Phe88. Phe88 is located within the hydrophobic pocket comprised of Ala87, Phe88, Ile180, and Met204. This result is consistent with the report by Daga et al. [155]. In addition, for most of the key residues, a slightly stronger binding energy contribution was found in the genotype C system [145].

Thus, investigations of structural insights of pol–inhibitor interaction have received a substantial boost during the last years, which provides the opportunity to use their results in rational design of new compounds with directed antipolymerases activity.

4.5 Computational Approaches in Rational Design of DNA Polymerase Inhibitors

Modern strategies of rational design of specific/selective effectors for biomolecular targets naturally combine the computational approaches, used for the detailed analysis of the structural mechanisms of ligand–target interaction, and predictions of the ligand affinity with instrumental methods of activity and selectivity assessment for the developed compounds [156]. Actually, the rational design procedure consists of the following steps:

- the choice of a biomolecular target (protein in the most common case);
- the analysis of individual spatial structure features of the functional and allosteric sites of the target protein;
- high throughput receptor-based virtual screening of libraries of low-weight molecular organic compounds, and identification of classes of compounds characterized by the highest affinity to the target protein *in silico*;
- experimental verification of the inhibitory activity and selectivity of the most promising compounds according to the previous stage on set close to the target protein *in vitro*;

- correlation analysis of the "structure-activity" and selecting the "compound hits" for directed (purposeful) chemical optimization based on the results of biochemical assays;
- chemical synthesis and optimization of new inhibitors based on the analysis of "structure–activity", selectivity and computer simulation data.

From this viewpoint the clear understanding of structural aspects of protein–ligand interaction is a key factor for the correct prediction of inhibitor affinity. Unfortunately, structural information about binding sites for non-nucleoside inhibitors on the surface of DNA-dependent DNA polymerases is rather limited (see the previous section).

At the same time, researchers start to use various computational approaches in the rational design procedure of new (nucleoside analogs or nucleoside containing) inhibitors of DNA-dependent DNA polymerases (especially, human pols). Spectrum of applicable approaches is sufficiently broad and includes all the above-mentioned methods of the structural analysis *in silico*.

Richartz and co-workers [157] have successfully applied the methods of molecular docking and molecular dynamics for the analysis of structural action mechanisms of several nucleotide analogs—potential inhibitors of human DNA polymerase α (pol α), and of the well-known non-nucleotide pol α inhibitor aphidicolin [56, 157].

In molecular dynamics simulations, aphidicolin occupied the catalytic centre, but acted in a not truly competitive manner with respect to nucleotides. It destabilized the replicating "closed" form of the pol alpha and transferred the enzyme into the inactive "open" conformation [157]. This result is consistent with recent experiments on the binding mode of aphidicolin. Unfortunately, aphidicolin could not be introduced into therapy because of its toxicity and rapid metabolism after systemic application [158]. Among studied 'nucleotides', the highest potential for selective pol α inhibition was established for 2-butylanilino-dATP (BuAdATP). The butylphenyl moiety of BuAdATP occupies a lipophilic pocket, formed by the residues Leu960, Leu972, Val976, Ile869, Tyr865 and Tyr957. These lipophilic interactions, coupled with hydrogen bonds between BuAdATP, template nucleotide and side chains of residues Tyr865 and Lys950, are likely to be responsible for the good inhibition efficiency of the butylanilino derivatives. The lowest abilities to inhibit human pol α were demonstrated for lamivudine-TP and zidovudine-TP (both compounds in three phosphate form) [157].

Later Höltje and co-workers (from the same research team) reported the development of several new human polymerase α inhibitors applicable for skin tumor treatment (in order to design new drugs for actinic keratosis and squamous cell carcinoma) [159] (Fig. 4.10). To study the binding modes of these compounds, the same computational approaches as described in Richartz et al. [157] were used.

It was shown that the compound HM1-TP forms two hydrogen bonds with the DNA template nucleotide, two hydrogen bonds with the side chain of Lys950 and one hydrogen bond each with the backbone of Tyr865 and the side chain of Arg922 in the active center of human pol α. At the same time BuP-OH-TP forms two hydrogen bonds with the DNA template nucleotide, two hydrogen bonds with the backbone and side chain of Tyr865, and one hydrogen bond with Lys950 side chain.

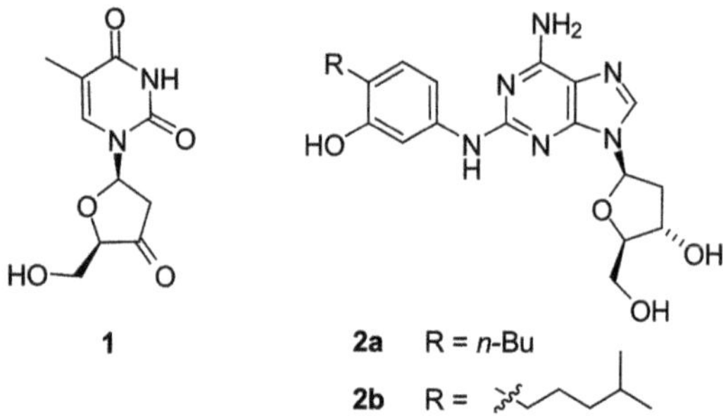

Fig. 4.10 Structure of human pol α inhibitors HM-1 (*1*), BuP-OH (*2a*) and iso-Hex-OH (*2b*)

Assays in the squamous cancer cell line SCC25 have shown that the developed compounds exhibit cytotoxicity and antiproliferative activity in the nanomolar range of concentrations [160]. Thus, they can be considered as promising antitumor drugs.

In further investigations, the BuP-OH was used as an initial compound for the design of new pol α inhibitors [161]. Activity of new BuP-OH derivates was evaluated *in silico* (by docking into the enzyme active site and molecular dynamics of obtained complexes), after that the predicted hits were assayed *in vitro* on the culture of NHK and SCC-25 cells. It was demonstrated that 2 new derivates OxBu and OxHex are able to efficiently inhibit the grow of neoplastic cells without registered effects on normal keratocytes. Thus, results of these investigations confirm the productivity of drug design approach including the computational procedure of drug affinity evaluation.

Computational approaches also have started to apply for design low-weight molecular inhibitor of bacterial and viral polymerases. So, Karampuri et al. [162] reported about design of new inhibitors of HSV pol on the base of α-pyrone (4-oxo-dihydroquinoline-3-carboxamide). Drug prototypes were constructed on the base of Lipinski rules and undergone in conformation search procedure, after that their low energy conformation were docked into spatial structure of HSV DNA polymerase [162]. It was found that designed compounds 5h (Fig. 4.13a) is more active against HSV than well-known acyclovir and is more selective in relation to HSV-1 compared to HSV-2. (Fig. 4.11).

4.6 Conclusions

Initial stage of development of DNA polymerases' inhibitors can be characterized as a pre-structural. The absolute majority of known pol inhibitors were either discovered among natural metabolites from plants and fungi or among synthesized analogs of dNTP. However, currently we can observe the principal paradigm shift in this field. Inclusion of computational approaches into procedures of structural analysis and rational design let to perform the target-directed development of new pol inhibitors. Clear comprehension of structural insights of pol–inhibitor interactions

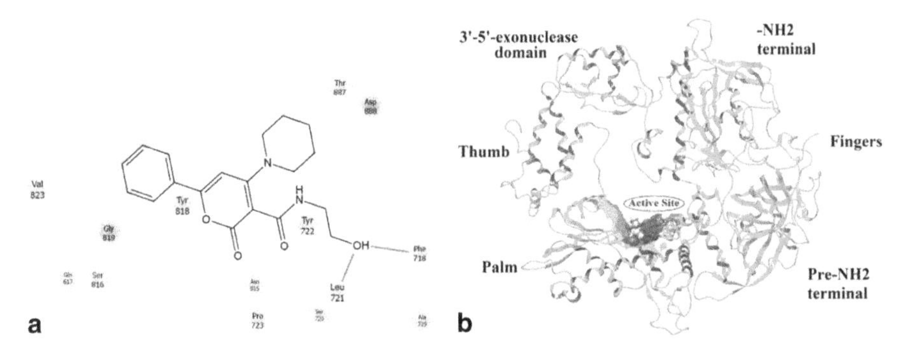

Fig. 4.11 **a** Amino acid microenvironment of 5h in HSV pol binding state, **b** localization of α-pyrones binding site in HSV pol space. (Adapted from [162], © 2012, with permission from Elsevier)

gives a possibility to optimize the known inhibitory compounds towards enhanced affinity and selectivity in relation to desirable polymerases. One should note that this paradigm shift reflects the common tendency in development of new drugs, herbicides, fungicides and other biologically active compounds.

The next development of pol inhibitors can be associated with search of allosteric binding sites on the DNA polymerases. These compounds have to be a priori more selective than ones binding into active site of enzyme. Unfortunately, among the all known pol inhibitors only MK886—inhibitor of Y family pols—can pretend to be really allosteric. This state of facts is a result of absence of appropriate information about allosteric sites' localization. However, it can be corrected via analysis of correlated motions in proteins calculated from molecular dynamics trajectories. This approach lets to reveal behavior coordination between spatially distant parts of macromolecule. The ligand bound to one from such "coordinated" parts appropriately causes the structural changes in another. Using this approach it's possible to reveal the all sites on protein surface, their changes can significantly influence on spatial organization of enzyme active site.

Acknowledgements Author would like to sincere gratitude to Prof. Leonid Gorb, Prof. Alexei Kolezhuk, Oleg Lytuga, Tamara Limanska and Fedor Lavrik for their invaluable aid in preparing of this article.

References

1. Berdis AJ (2009) Mechanisms of DNA polymerases. Chem Rev 109(7):2862–2879
2. Bebenek K, Kunkel TA (2004) Functions of DNA polymerases. Adv Protein Chem 69:137–65
3. Hübscher U, Spadari S, Villani G, Maga G (2010) DNA polymerases: discovery, characterization and functions in cellular DNA transactions. World Scientific, New Jersey
4. Haracska L, Johnson RE, Prakash L, Prakash S (2005) Trf4 and Trf5 proteins of *Saccharomyces cerevisiae* exhibit poly(A) RNA polymerase activity but no DNA polymerase activity. Mol Cell Biol 25(22):10183–10189

5. Steitz TA (1999) DNA polymerases: structural diversity and common mechanisms. J Biol Chem 274(25):17395–17398
6. Prindle MJ, Schmitt MW, Parmeggiani F, Loeb LA (2013) A substitution in the fingers domain of DNA polymerase δ reduces fidelity by altering nucleotide discrimination in the catalytic site. J Biol Chem 288(8):5572–5580
7. Kunkel TA (2004) DNA replication fidelity. J Biol Chem 279(17):16895–16898
8. Tsai YC, Johnson KA (2006) A new paradigm for DNA polymerase specificity. Biochemistry 45(32):9675–9687
9. Johnson KA (2010) The kinetic and chemical mechanism of high-fidelity DNA polymerases. Biochim Biophys Acta 1804(5):1041–1048
10. Gouge J, Ralec C, Henneke G, Delarue M (2012) Molecular recognition of canonical and deaminated bases by *P. abyssi* family B DNA polymerase. J Mol Biol 423:315–336
11. Breyer WA, Matthews BW (2001) A structural basis for processivity. Protein Sci 10(9): 1699–1711
12. Camps M, Loeb LA (2004) When pol I goes into high gear: processive DNA synthesis by pol I in the cell. Cell Cycle 3(2):116–118s
13. Stumpf JD, Copeland WC (2011) Mitochondrial DNA replication and disease: insights from DNA polymerase γ mutations. Cell Mol Life Sci 68 (2): 219–233
14. Banach-Orlowska M, Fijalkowska IJ, Schaaper RM, Jonczyk P (2005) DNA polymerase II as a fidelity factor in chromosomal DNA synthesis in *Escherichia coli*. Mol Microbiol 58(1):61–70
15. Muzi-Falconi M, Giannattasio M, Foiani M, Plevani P (2003) The DNA polymerase α-primase complex: multiple functions and interactions. ScientificWorldJournal 17(3):21–33
16. Hübscher U, Maga G, Spadari S (2002) Eukaryotic DNA polymerases. Annu Rev Biochem 71:133–163
17. Edwards S, Li CM, Levy DL, Brown J, Snow PM, Campbell JL (2003) *Saccharomyces cerevisiae* DNA polymerase epsilon and polymerase sigma interact physically and functionally, suggesting a role for polymerase epsilon in sister chromatid cohesion. Mol Cell Biol 23(8): 2733–2748
18. Gan GN, Wittschieben JP, Wittschieben BØ, Wood RD (2008) DNA polymerase zeta (pol zeta) in higher eukaryotes. Cell Res 18(1):174–183
19. Kelman Z, O'Donnell M (1995) DNA polymerase III holoenzyme: structure and function of a chromosomal replicating machine. Annu Rev Biochem 64:171–200
20. O'Donnell M, Jeruzalmi D, Kuriyan J (2001) Clamp loader structure predicts the architecture of DNA polymerase III holoenzyme and RFC. Curr Biol 11(22):R935–946
21. Cann IK, Komori K, Toh H, Kanai S, Ishino Y (1998) A heterodimeric DNA polymerase: evidence that members of Euryarchaeota possess a distinct DNA polymerase. Proc Natl Acad Sci U S A 95(24):14250–14255
22. Jokela M, Eskelinen A, Pospiech H, Rouvinen J, Syväoja JE (2004) Characterization of the 3' exonuclease subunit DP1 of *Methanococcus jannaschi*i replicative DNA polymerase D. Nucleic Acids Res 32(8):2430–2440
23. Hayashi I, Morikawa K, Ishino Y (1999) Specific interaction between DNA polymerase II (PolD) and RadB, a Rad51/Dmc1 homolog, in *Pyrococcus furiosus*. Nucleic Acids Res 27(24):4695–4702
24. Yamtich J, Sweasy JB (2010) DNA polymerase family X: function, structure, and cellular roles. Biochim Biophys Acta 1804(5):1136–1150
25. Goodman MF (2002) Error-prone repair DNA polymerases in prokaryotes and eukaryotes. Annu Rev Biochem 71:17–50
26. Mori T, Nakamura T, Okazaki N, Furukohri A, Maki H, Akiyama MT (2012) *Escherichia coli* DinB inhibits replication fork progression without significantly inducing the SOS response. Genes Genet Syst 87(2):75–87
27. Jarosz DF, Godoy VG, Walker GC (2007) Proficient and accurate bypass of persistent DNA lesions by DinB DNA polymerases. Cell Cycle 6(7):817–822
28. Patel M, Jiang Q, Woodgate R, Cox MM, Goodman MF (2010) A new model for SOS-induced mutagenesis: how RecA protein activates DNA polymerase V. Crit Rev Biochem Mol Biol 45(3):171–184

29. Sutton MD, Walker GC (2001) Managing DNA polymerases: coordinating DNA replication, DNA repair, and DNA recombination. Proc Natl Acad Sci U S A 98(15):8342–8349
30. Ohmori H, Hanafusa T, Ohashi E, Vaziri C (2009) Separate roles of structured and unstructured regions of Y-family DNA polymerases. Adv Protein Chem Struct Biol 78:99–146
31. Öberg B (2006) Rational design of polymerase inhibitors as antiviral drugs. Antiviral Res. 71(2–3):90–95
32. Wright GE, Brown NC (1990) Deoxyribonucleotide analogs as inhibitors and substrates of DNA polymerases. Pharmacol Ther 47(3):447–497
33. Razonable RR (2011) Antiviral drugs for viruses other than human immunodeficiency virus. Mayo Clin Proc 86(10):1009–1026
34. De Clercq E (2007) The acyclic nucleoside phosphonates from inception to clinical use: historical perspective. Antiviral Res 75:1–13
35. Magee WC, Evans DH (2012) The antiviral activity and mechanism of action of (S)-[3-hydroxy-2-(phosphonomethoxy)propyl] (HPMP) nucleosides. Antiviral Res 96(2):169–180
36. Merta A, Votruba I, Rosenberg I, Otmar M, Hrebabecký H, Bernaerts R, Holý A (1990) Inhibition of herpes simplex virus DNA polymerase by diphosphates of acyclic phosphonylmethoxyalkyl nucleotide analogues. Antiviral Res 13(5):209–218
37. Xiong X, Smith JL, Kim C, Huang ES, Chen MS (1996) Kinetic analysis of theinteraction of cidofovir diphosphate with human cytomegalovirus DNApolymerase. Biochem Pharmacol 51:1563–1567
38. Xiong X, Smith JL, Chen MS (1997) Effect of incorporation of cidofovir into DNA by human cytomegalovirus DNA polymerase on DNA elongation. Antimicrob Agents Chemother 41:594–599
39. Magee WC, Hostetler KY, Evans DH (2005) Mechanism of inhibition of vaccinia virus DNA polymerase by cidofovir diphosphate. Antimicrob Agents Chemother 49, 3153–3162
40. Magee WC, Aldern KA, Hostetler KY, Evans DH (2008) Cidofovir and (S)-9-[3-hydroxy-(2-phosphonomethoxy)propyl]adenine are highly effective inhibitors of vaccinia virus DNA polymerase when incorporated into the template strand. Antimicrob Agents Chemother 52, 586–597
41. Birkus G, Rejman D, Otmar M, Votruba I, Rosenberg I, Holy A (2004) The substrate activity of (S)-9-[3-hydroxy-(2-phosphonomethoxy)propyl]adenine diphosphate toward DNA polymerases alpha, delta and epsilon. Antivir Chem Chemother 15:23–33
42. Magee WC, Valiaeva N, Beadle JR, Richman DD, Hostetler KY, Evans DH (2011) Antimicrob Agents Chemother 55(11):5063–5072
43. De Clercq E (2011) The clinical potential of the acyclic (and cyclic) nucleoside phosphonates. The magic of the phosphonate bond. Biochem Pharmacol 82:99–109
44. Andrei G, Snoeck R (2010) Cidofovir activity against poxvirus infections. Viruses 2(12): 2803–2830
45. Corey L, Benedetti J, Critchlow C, Mertz G, Douglas J, Fife K, Fahnlander A, Remington ML, Winter C, Dragavon J (1983) Treatment of primary first-episode genital herpes simplex virus infections with acyclovir: results of topical, intravenous and oral therapy. J Antimicrob Chemother 12(Suppl B):79–88
46. Serota FT, Starr SE, Bryan CK, Koch PA, Plotkin SA, August CS (1982) Acyclovir treatment of herpes zoster infections: use in children undergoing bone marrow transplantation. JAMA 247:2132–2135
47. De Clercq E (2004) Discovery and development of BVDU (brivudin) as a therapeutic for the treatment of herpes zoster. Biochem Pharmacol 68:2301–2315
48. Superti F, Ammendolia MG, Marchetti M (2008) New advances in anti-HSV chemotherapy. Curr Med Chem 15:900–911
49. Nichols WG, Boeckh M (2000) Recent advances in the therapy and prevention of CMV infections. J Clin Virol 16:25–40
50. Sarisky RT, Bacon TH, Boon RJ, Duffy KE, Esser KM, Leary J, Locke LA, Nguyen TT, Quail MR, Saltzman R (2003) Profiling penciclovir susceptibility and prevalence of resistance of herpes simplex virus isolates across eleven clinical trials. Arch Virol 148(9):1757–1769

51. Acosta EP, Fletcher CV (1997) Valacyclovir. Ann Pharmacother 31:185–191
52. Mizushina Y (2009) Specific inhibitors of mammalian DNA polymerase species. Biotechnol Biochem 73(6):1239–1251
53. Barakat KH, Gajewski MM, Tuszynski JA (2012) DNA polymerase beta (pol β) inhibitors: a comprehensive overview. Drug Discov Today 17(15–16):913–920
54. Chen J, Zhang YH, Wang LK, Sucheck SJ, Snow AM, Hecht SM (1998) Inhibitors of DNA polymerase β from *Schoepfia californica.* J Chem Soc Chem Commun 24:2769–2770
55. Spadari S, Pedrali-Noy G, Falaschi MC, Ciarrocchi G (1984) Control of DNA replication and cell proliferation in eukaryotes by aphidicolin. Toxicol Pathol 12(2):143–148
56. Arabshahi L, Brown N, Khan N, Wright G. (1988) Inhibition of DNApolymerase alpha by aphidicolin derivatives. Nucleic Acids Res 16:5107–5113
57. Ma J, Starck SR, Hecht SM (1999) DNA polymerase β inhibitors from *Tetracera boiviniana.* J Nat Prod 62:1660–1663
58. Deng JZ, Starck SR, Hecht SM (1999) Bis-5-alkylresorcinols from *Panopsis rubescens* that inhibit DNA polymerase β. J Nat Prod 62:477–480
59. Mizushina Y, Matsukage A, Sakaguchi K (1998) The biochemical inhibition mode of bredinin-5'-monophosphate on DNA polymerase β. Biochim Biophys Acta 1403(1):5–11
60. Talanian RV, Brown NC, McKenna CE, Ye TG, Levy JN, Wright GE (1989) Carbonyldiphosphonate, a selective inhibitor of mammalian DNA polymerase δ. Biochemistry 28(21): 8270–8274
61. Oshige M, Takenouchi M, Kato Y, Kamisuki S, Takeuchi T, Kuramochi K, Shiina I, Suenaga Y, Kawakita Y, Kuroda K, Sato N, Kobayashi S, Sugawara F, Sakaguchi K (2004) Taxol derivatives are selective inhibitors of DNA polymerase α. Bioorg Med Chem 12(10):2597–2601
62. Yamanaka K, Dorjsuren D, Eoff RL, Egli M, Maloney DJ, Jadhav A, Simeonov A, Lloyd RS (2012) A comprehensive strategy to discover inhibitors of the translesion synthesis DNA polymerase κ. PLoS One 7(10): e45032
63. Maeda N, Kokai Y, Ohtani S, Sahara H, Kuriyama I, Kamisuki S, Takahashi S, Sakaguchi K, Sugawara F, Yoshida H, Sato N, Mizushina Y (2007) Anti-tumor effects of dehydroaltenusin, a specific inhibitor of mammalian DNA polymerase alpha. Biochem Biophys Res Commun 352(2):390–396
64. Mizushina Y, Xu X, Asano N, Kasai N, Kato A, Takemura M, Asahara H, Linn S, Sugawara F, Yoshida H, Sakaguchi K (2003) The inhibitory action of pyrrolidine alkaloid, 1,4-dideoxy-1,4-imino-D-ribitol, on eukaryotic DNA polymerases. Biochem Biophys Res Commun 304:78–85
65. Li SS, Gao Z, Feng X, Hecht SM (2004) Biscoumarin derivatives from *Edgeworthia gardneri* that inhibit the lyase activity of DNA polymerase β. J Nat Prod 67(9):1608–1610
66. Feng X, Gao Z, Li S, Jones SH, Hecht SM (2004) DNA polymerase β lyase inhibitors from *Maytenus putterlickoides*. J Nat Prod 67:1744–1747
67. Mizushina Y, Watanabe I, Togashi H, Hanashima L, Takemura M, Ohta K, Sugawara F, Koshino H, Esumi Y, Uzawa J, Matsukage A, Yoshida S, Sakaguchi K (1998) An ergosterol peroxide, a natural product that selectively enhances the inhibitory effect of linoleic acid on DNA polymerase β. Biol Pharm Bull 21(5):444–448
68. Mizushina Y, Tanaka N, Kitamura A, Tamai K, Ikeda M, Takemura M, Sugawara F, Arai T, Matsukage A, Yoshida S, Sakaguchi K (1998) The inhibitory effect of novel triterpenoid compounds, fomitellic acids, on DNA polymerase beta. Biochem J 330(Pt 3):1325–1332
69. Tanaka N, Kitamura A, Mizushina Y, Sugawara F, Sakaguchi K (1998) Fomitellic acids, triterpenoid inhibitors of eukaryotic DNA polymerases from a basidiomycete, *Fomitella fraxinea.* J Nat Prod 61(2):193–197
70. Ishida T, Mizushina Y, Yagi S, Irino Y, Nishiumi S, Miki I, Kondo Y, Mizuno S, Yoshida H, Azuma T, Yoshida M (2012) Inhibitory effects of glycyrrhetinic acid on DNA polymerase and inflammatory activities. Evid Based Complement Alternat Med 2012:650514
71. Deng JZ, Starck SR, Hecht SM, Ijames CF, Hemling ME (1999) Harbinatic acid, a novel and potent DNA polymerase β inhibitor from *Hardwickia binata.* J Nat Prod 62:1000–1002

72. Togashi H, Mizushina Y, Takemura M, Sugawara F, Koshino H, Esumi Y, Uzawa J, Kumagai H, Matsukage A, Yoshida S, Sakaguchi K (1998) 4-Hydroxy-17-methylincisterol, an inhibitor of DNA polymerase-α activity and the growth of human cancer cells in vitro. Biochem Pharmacol 56(5):583–590
73. Cao S, Gao Z, Thomas SJ, Hecht SM, Lazo JS, Kingston DG (2004) Marine sesquiterpenoids that inhibit the lyase activity of DNA polymerase β. J Nat Prod 67:1716–1718
74. Mizushina Y, Ishidoh T, Kamisuki S, Nakazawa S, Takemura M, Sugawara F, Yoshida H, Sakaguchi K (2003) Flavonoid glycoside: a new inhibitor of eukaryotic DNA polymerase alpha and a new carrier for inhibitor-affinity chromatography. Biochem Biophys Res Commun 301(2):480–487
75. Ogawa A, Murate T, Izuta S, Takemura M, Furuta K, Kobayashi J, Kamikawa T, Nimura Y, Yoshida S (1998) Sulfated glycoglycerolipid from archaebacterium inhibits eukaryotic DNA polymerase α, β and retroviral reverse transcriptase and affects methylmethanesulfonate cytotoxicity
76. Mizushina Y, Manita D, Takeuchi T, Sugawara F, Kumamoto-Yonezawa Y, Matsui Y, Takemura M, Sasaki M, Yoshida H, Takikawa H (2009) The inhibitory action of kohamaic acid A derivatives on mammalian DNA polymerase β. Molecules 14(1):102–121
77. Sun DA, Starck SR, Locke EP, Hecht SM (1999) DNA polymerase beta inhibitors from *Sandoricum koetjape*. J Nat Prod 62: 1110–1113
78. Hu HY, Horton JK, Gryk MR, Prasad R, Naron JM, Sun DA, Hecht SM, Wilson SH, Mullen GP (2004) Identification of small molecule synthetic inhibitors of DNA polymerase β by NMR chemical shift mapping. J Biol Chem 279:39736–39744
79. Mizushina Y, Tanaka N, Yagi H, Kurosawa T, Onoue M, Seto H, Horie T, Aoyagi N, Yamaoka M, Matsukage A, Yoshida S, Sakaguchi K (1996) Fatty acids selectively inhibit eukaryotic DNA polymerase activities *in vitro*. Biochim Biophys Acta 1308(3):256–262
80. Mizushina Y, Yoshida S, Matsukage A, Sakaguchi K (1997) The inhibitory action of fatty acids on DNA polymerase β. Biochim Biophys Acta 1336(3):509–521
81. Mizushina Y, Watanabe I, Ohta K, Takemura M, Sahara H, Takahashi N, Gasa S, Sugawara F, Matsukage A, Yoshida S, Sakaguchi K (1998) Studies on inhibitors of mammalian DNA polymerase α and β: sulfolipids from a pteridophyte, *Athyrium niponicum*. Biochem Pharmacol 55(4):537–541
82. Mizushina Y, Ohkubo T, Sugawara F, Sakaguchi K. (2000) Structure of lithocholic acid binding to the N-terminal 8-kDa domain of DNA polymerase β. Biochemistry 39(41): 12606–12613
83. Ogawa A, Murate T, Suzuki M, Nimura Y, Yoshida S (1998) Lithocholic acid, a putative tumor promoter, inhibits mammalian DNA polymerase β. Jpn J Cancer Res 89(11):1154–1159
84. Chaturvedula VS, Gao Z, Jones SH, Feng X, Hecht SM, Kingston DG (2004) A new ursane tri-terpene from *Monochaetum vulcanicum* that inhibits DNA polymerase β lyase. J Nat Prod 67:899–901
85. Mizushina Y, Ishidoh T, Takeuchi T, Shimazaki N, Koiwai O, Kuramochi K, Kobayashi S, Sugawara F, Sakaguchi K, Yoshida H (2005) Monoacetylcurcumin: a new inhibitor of eukaryotic DNA polymerase λ and a new ligand for inhibitor-affinity chromatography. Biochem Biophys Res Commun 337(4):1288–1295
86. Takeuchi T, Ishidoh T, Iijima H, Kuriyama I, Shimazaki N, Koiwai O, Kuramochi K, Kobayashi S, Sugawara F, Sakaguchi K, Yoshida H, Mizushina Y (2006) Structural relationship of curcumin derivatives binding to the BRCT domain of human DNA polymerase λ. Genes Cells 11(3):223–235
87. Maloney DJ, Deng JZ, Starck SR, Gao Z, Hecht SM (2005) (ı)-Myristinin A, a naturally occurring DNA polymerase β inhibitor and potent DNA-damaging agent. J Am Chem Soc 127(12): 4140–4141
88. Shiomi K, Kuriyama I, Yoshida H, Mizushina Y (2013) Inhibitory effects of myricetin on mammalian DNA polymerase, topoisomerase and human cancer cell proliferation. Food Chem 139(1–4):910–918

89. Prakash Chaturvedula VS, Hecht SM, Gao Z, Jones SH, Feng X, Kingston DG (2004) New neolignans that inhibit DNA polymerase beta lyase. J Nat Prod 67(6): 964–967
90. Deng JZ, Starck SR, Hecht SM (2000) Pentacyclic triterpenoids from *Freziera* sp. that inhibit DNA polymerase β. Bioorg Med Chem 8:247–250
91. Brideau RJ, Knechtel ML, Huang A, Vaillancourt VA, Vera EE, Oien NL, Hopkins TA, Wieber JL, Wilkinson KF, Rush BD, Schwende FJ, Wathen MW (2002) Broad-spectrum antiviral activity of PNU-183792, a 4-oxo-dihydroquinoline, against human and animal herpesviruses. Antiviral Res 54(1):19–28
92. Oien NL, Brideau RJ, Hopkins TA, Wieber JL, Knechtel ML, Shelly JA, Anstadt RA, Wells PA, Poorman RA, Huang A, Vaillancourt VA, Clayton TL, Tucker JA, Wathen MW (2002) Broad-spectrum antiherpes activities of 4-hydroxyquinoline carboxamides, a novel class of herpesvirus polymerase inhibitors. Antimicrob Agents Chemother 46(3):724–730
93. Hazan C, Boudsocq F, Gervais V, Saurel O, Ciais M, Cazaux C, Czaplicki J, Milon A (2008) Structural insights on the pamoic acid and the 8 kDa domain of DNA polymerase β complex: towards the design of higher-affinity inhibitors. BMC Struct Biol 8:22
94. Mizushina Y, Zhang J, Pugliese A, Kim SH, Lü J (2010) Anti-cancer gallotannin penta-O-galloyl-beta-D-glucose is a nanomolar inhibitor of select mammalian DNA polymerases. Biochem Pharmacol 80(8):1125–1132
95. Mizushina Y, Kamisuki S, Kasai N, Ishidoh T, Shimazaki N, Takemura M, Asahara H, Linn S, Yoshida S, Koiwai O, Sugawara F, Yoshida H, Sakaguchi K (2002) Petasiphenol: a DNA polymerase λ inhibitor. Biochemistry 41(49):14463–14471
96. Mizushina Y, Xu X, Matsubara K, Murakami C, Kuriyama I, Oshige M, Takemura M, Kato N, Yoshida H, Sakaguchi K (2003) Pyridoxal 5'-phosphate is a selective inhibitor *in vivo* of DNA polymerase α and e. Biochem Biophys Res Commun 312(4):1025–1032
97. Locatelli GA, Savio M, Forti L, Shevelev I, Ramadan K, Stivala LA, Vannini V, Hübscher U, Spadari S, Maga G (2005) Inhibition of mammalian DNA polymerases by resveratrol: mechanism and structural determinants. Biochem J 389(Pt 2): 259–268
98. Strittmatter T, Bareth B, Immel TA, Huhn T, Mayer TU, Marx A (2011) Small molecule inhibitors of human DNA polymerase λ. ACS Chem Biol 6(4):314–319
99. Ishimaru C, Yonezawa Y, Kuriyama I, Nishida M, Yoshida H, Mizushina Y (2008) Inhibitory effects of cholesterol derivatives on DNA polymerase and topoisomerase activities, and human cancer cell growth. Lipids 43(4):373–382
100. Perpelescu M, Kobayashi J, Furuta M, Ito Y, Izuta S, Takemura M, Suzuki M, Yoshida S (2002) Novel phenalenone derivatives from a marine-derived fungus exhibit distinct inhibition spectra against eukaryotic DNA polymerases. Biochemistry 41:7610–7616
101. Mizushina Y, Nakanishi R, Kuriyama I, Kamiya K, Satake T, Shimazaki N, Koiwai O, Uchiya-ma Y, Yonezawa Y, Takemura M, Sakaguchi K, Yoshida H (2006) Beta-sitosterol-3-O-beta-D-glucopyranoside: a eukaryotic DNA polymerase λ inhibitor. J Steroid Biochem Mol Biol 99(2–3):100–107
102. Mizushina Y, Kasai N, Iijima H, Sugawara F, Yoshida H, Sakaguchi K (2005) Sulfo-quinovosyl-acyl-glycerol (SQAG), a eukaryotic DNA polymerase inhibitor and anti-cancer agent. Curr Med Chem Anticancer Agents 5(6):613–625
103. Mizushina, Y. et al. (2002) A plant phytotoxin, solanapyrone A, is an inhibitor of DNA polymerase beta and lambda. J. Biol. Chem 277:630–638
104. Kasai N, Mizushina Y, Murata H, Yamazaki T, Ohkubo T, Sakaguchi K, Sugawara F (2005) Sulfoquinovosylmonoacylglycerol inhibitory mode analysis of rat DNA polymerase β. FEBS J 272(17):4349–4361
105. Mizushina Y, Nakagawa K, Shibata A, Awata Y, Kuriyama I, Shimazaki N, Koiwai O, Uchiyama Y, Sakaguchi K, Miyazawa T, Yoshida H (2006) Inhibitory effect of tocotrienol on eukaryotic DNA polymerase λ and angiogenesis. Biochem Biophys Res Commun 339(3): 949–955
106. Deng JZ, Starck SR, Hecht SM (1999) DNA polymerase β inhibitors from *Baeckea gunniana*. J Nat Prod 62:1624–1626

107. Andrei G, De Clercq E, Snoeck R (2009) Viral DNA polymerase inhibitors. In: Raney KD, Götte M, Cameron CE (eds) Viral genome replication. Springer, New York, p 481–526
108. Schnute ME, Cudahy MM, Brideau RJ, Homa FL, Hopkins TA, Knechtel ML, Oien NL, Pitts TW, Poorman RA, Wathen MW, Wieber JL (2005) 4-Oxo-4,7-dihydrothieno[2,3-b] pyridines as non-nucleoside inhibitors of human cytomegalovirus and related herpesvirus polymerases. J Med Chem 48(18):5794–5804
109. Schnute ME, Anderson DJ, Brideau RJ, Ciske FL, Collier SA, Cudahy MM, Eggen M, Genin MJ, Hopkins TA, Judge TM, Kim EJ, Knechtel ML, Nair SK, Nieman JA, Oien NL, Scott A, Tanis SP, Vaillancourt VA, Wathen MW, Wieber JL (2007) 2-Aryl-2-hydroxyethylamine substituted 4-oxo-4,7-dihydrothieno[2,3-b]pyridines as broad-spectrum inhibitors of human herpesvirus polymerases. Bioorg Med Chem Lett 17(12):3349–3353
110. Larsen SD, Zhang Z, DiPaolo BA, Manninen PR, Rohrer DC, Hageman MJ, Hopkins TA, Knechtel ML, Oien NL, Rush BD, Schwende FJ, Stefanski KJ, Wieber JL, Wilkinson KF, Zamora KM, Wathen MW, Brideau RJ (2007) 7-Oxo-4,7-dihydrothieno[3,2-b]pyridine-6-carboxamides: synthesis and biological activity of a new class of highly potent inhibitors of human cytomegalovirus DNA polymerase. Bioorg Med Chem Lett 17(14):3840–3844
111. Hu H, Zhang J, Lee HJ, Kim SH, Lü J (2009) Penta-O-galloyl-beta-D-glucose induces S- and G(1)-cell cycle arrests in prostate cancer cells targeting DNA replication and cyclin D1. Carcinogenesis 30(5):818–823
112. Zhang J, Li L, Kim SH, Hagerman AE, Lu J (2009) Anti-cancer, anti-diabetic and other pharmacologicand biological activities of penta-galloyl-glucose. Pharm res 26:2066–2080
113. Chai Y, Lee HJ, Shaik AA, Nkhata K, Xing C, Zhang J, Jeong SJ, Kim SH, Lu J (2010) Penta-O-galloyl-beta-D-glucose induces G1 arrest and DNA replicative S-phase arrest independently of cyclin-dependent kinase inhibitor 1A, cyclin-dependent kinase inhibitor 1B and P53 in human breast cancer cells and is orally active against triple negative xenograft growth. Breast Cancer Res 12(5):R67
114. Mizushina Y, Kasai N, Sugawara F, Iida A, Yoshida H, Sakaguchi K (2001) Three-dimensional structural model analysis of the binding site of lithocholic acid, an inhibitor of DNA polymerase β and DNA topoisomerase II. J Biochem 130(5):657–664
115. Mizushina Y, Kasai N, Miura K, Hanashima S, Takemura M, Yoshida H, Sugawara F, Sakaguchi K (2004) Structural relationship of lithocholic acid derivatives binding to the N-terminal 8-kDa domain of DNA polymerase β. Biochemistry 43(33):10669–10677
116. Shi S, Yan L, Yang Y, Fisher-Shaulsky J, Thacher T (2003) An extensible and systematic force field, ESFF, for molecular modeling of organic, inorganic, and organometallic systems. J Comput Chem 24(9):1059–1076
117. Goodsell DS, Morris GM, Olson AJ (1996) Automated docking of flexible ligands: applications of AutoDock. J Mol Recognit 9:1–5
118. Morris GM, Goodsell DS, Halliday RS, Huey R, Hart WE, Belew RK, Olson AJ (1998) Auto-mated docking using a lamarckian genetic algorithm and and empirical binding free energy function J Comput Chem 19: 1639–1662
119. Murakami S, Kamisuki S, Takata K, Kasai N, Kimura S, Mizushina Y, Ohta K, Sugawara F, Sakaguchi K (2006) Site-directed mutational analysis of structural interactions of low molecule compounds binding to the N-terminal 8 kDa domain of DNA polymerase β. Biochem Biophys Res Commun 350(1):7–16
120. Clarkson J, Campbell ID (2003) Studies of protein-ligand interactions by NMR. Biochem Soc Trans 31(Pt 5):1006–1009
121. Mayer M, Meyer B (2001, Jun 27) Group epitope mapping by saturation transfer difference NMR to identify segments of a ligand in direct contact with a protein receptor. J Am Chem Soc 123(25):6108–6117
122. Meyer B, Klein J, Mayer M, Meinecke R, Möller H, Neffe A, Schuster O, Wülfken J, Ding Y, Knaie O, Labbe J, Palcic MM, Hindsgaul O, Wagner B, Ernst B (2004) Saturation transfer difference NMR spectroscopy for identifying ligand epitopes and binding specificities. Ernst Schering Res Found Workshop 44:149–167

123. Iriye R, Furukawa K, Nishida R, Kim C, Fukami H (1992) Isolation and synthesis of a new bio-antimutagen, petasiphenol, from scapes of *Petasites japonicum*. Biosci Biotechnol Biochem. 56(11):1773–1775
124. Zhang X, Moréra S, Bates PA, Whitehead PC, Coffer AI, Hainbucher K, Nash RA, Sternberg MJ, Lindahl T, Freemont PS (1998) Structure of an XRCC1 BRCT domain: a new protein–protein interaction module. EMBO J 17(21):6404–6411
125. Ketkar A, Zafar MK, Maddukuri L, Yamanaka K, Banerjee S, Egli M, Choi JY, Lloyd RS, Eoff RL (2013) Leukotriene biosynthesis inhibitor MK886 impedes DNA polymerase activity. Chem Res Toxicol 26(2):221–232
126. Berman HM (2000) The protein data bank. Nucleic Acids Res 28:235–242
127. Lang PT, Brozell SR, Mukherjee S, Pettersen EF, Meng EC, Thomas V, Rizzo RC, Case DA, James TL, Kuntz ID (2009) DOCK 6: combining techniques to model RNA-small molecule complexes. RNA 15(6):1219–1230
128. Pettersen EF, Goddard TD, Huang CC, Couch GS, Greenblatt DM, Meng EC, Ferrin TE (2004) UCSF Chimera—a visualization system for exploratory research and analysis. J Comput Chem 25(13):1605–1612
129. Grosdidier A, Zoete V, Michielin O (2011) Fast docking using the CHARMM force field with EADock DSS. J Comput Chem 32(10): 2149–2159
130. Ummat A, Silverstein TD, Jain R, Buku A, Johnson RE, Prakash L, Prakash S, Aggarwal AK (2011) Human DNA polymerase eta is pre-aligned for dNTP binding and catalysis. J Mol Biol 415: 627–634
131. Nakamura T, Zhao Y, Yamagata Y, Hua YJ, Yang W (2012) Watching DNA polymerase η make a phosphodiester bond. Nature 487(7406):196–201
132. Beckman JW, Wang Q, Guengerich FP (2008) Kinetic analysis of correct nucleotide insertion by a Y-family DNA polymerase reveals conformational changes both prior to and following phosphodiester bond formation as detected by tryptophan fluorescence. J Biol Chem 283(52):36711–36723
133. Eoff RL, Sanchez-Ponce R, Guengerich FP (2009) Conformational changes during nucleotide selection by *Sulfolobus solfataricus* DNA polymerase Dpo4. J Biol Chem 284(31): 21090–21099
134. Martin OA, Garro HA, Kurina Sanz MB, Pungitore CR, Tonn CE (2011) *In silico* study of the inhibition of DNA polymerase by a novel catalpol derivative. J Mol Model 17(10): 2717–2723
135. Allouche AR (2011) Gabedit—a graphical user interface for computational chemistry softwares. J Comput Chem 32(1):174–182
136. Stewart JJ (2007) Optimization of parameters for semiempirical methods V: modification of NDDO approximations and application to 70 elements. J Mol Model 13:1173–1213
137. Van der Spoel D, Lindahl E, Hess B, Groenhof G, Mark AE, Berendsen HJ (2005) GROMACS: fast, flexible, and free. J Comput Chem 26:1701–1718
138. Hess B, Kutzner C, van der Spoel D, Lindahl E (2008) GROMACS 4: algorithms for highly efficient, load-balanced, and scalable molecular simulation. J Chem Theor Comput 4: 435–447
139. Schüttelkopf AW, van Aalten DM (2004) PRODRG: a tool for high-throughput crystallography of protein–ligand complexes. Acta Crystallogr D 60:1355–1363
140. Essmann U, Perera L, Berkowitz ML, Darden T, Lee H, Pedersen LG (1995) A smooth particle mesh Ewald method. J Chem Phys 103:8577–8593
141. Krieger E, Darden T, Nabuurs SB, Finkelstein A, Vriend G (2004) Making optimal use of empirical energy functions: force-field parameterization in crystal space. Proteins 57: 678–683
142. Duan Y, Wu C, Chowdhury S, Lee MC, Xiong G, Zhang W (2003) A point-charge force field for molecular mechanics simulations of proteins based on condensed-phase quantum mechanical calculations. J Comput Chem 24:1999–2012
143. Li Y, Korolev S, Waksman G (1998) Crystal structures of open and closed forms of binary and ternary complexes of the large fragment of *Thermus aquaticus* DNA polymerase I: structural basis for nucleotide incorporation. EMBO J 17:7514–7525

144. Pungitore CR, Ayub MJ, García M, Borkowski EJ, Sosa ME, Ciuffo G, Tonn CE (2004) Iridoids as allelochemicals and DNA polymerase inhibitors. J Nat Prod 67:357–361
145. Li J, Du Y, Liu X, Shen QC, Huang AL, Zheng MY, Luo XM, Jiang HL (2013) Binding sensi-tivity of adefovir to the polymerase from different genotypes of HBV: molecular modeling, docking and dynamics simulation studies. Acta Pharmacol Sin 34(2):319–328
146. Sali A, Potterton L, Yuan F, van Vlijmen H, Karplus M (1995) Evaluation of comparative protein modeling by MODELLER. Proteins 23: 318–326
147. Laskowski RA, Macarthur MW, Moss DS, Thornton JM (1993) Procheck—a program to check the stereochemical quality of protein structures. J Appl Crystallogr 26:283–291
148. Case DA, Cheatham TE 3rd, Darden T, Gohlke H, Luo R, Merz KM Jr, Onufriev A, Simmerling C, Wang B, Woods RJ (2005) The Amber biomolecular simulation programs. J Comput Chem 26: 1668–1688
149. Frisch MJ, Trucks GW, Schlegel HB, Scuseria GE, Robb MA, Cheeseman JR et al (2004) Gaussian 03, Revision B.05. Gaussian, Inc, Wallingford, CT
150. Bayly CI, Cieplak P, Cornell W, Kollman PA (1993) A well-behaved electrostatic potential based method susing charge restraints for deriving atomic charges: the RESP model. J Phys Chem 97(40):10269–10280
151. Wang JM, Wang W, Kollman PA (2001) Automatic atom type and bond type perception in molecular mechanical calculations. J Mol Graph Model 25(2):247–260
152. Meagher KL, Redman LT, Carlson HA (2003) Development of polyphosphate parameters for use with the AMBER force field. J Comput Chem 24:1016–1025
153. Wang J, Wolf RM, Caldwell JW, Kollman PA, Case DA (2004) Development and testing of a general amber force field. J Comput Chem 25:1157–1174
154. Kollman PA, Massova I, Reyes C, Kuhn B, Huo S, Chong L, Lee M, Lee T, Duan Y, Wang W, Donini O, Cieplak P, Srinivasan J, Case DA, Cheatham TE 3rd (2000) Calculating structures and free energies of complex molecules: combining molecular mechanics and continuum models. Acc Chem Res 33:889–897
155. Daga PR, Duan J, Doerksen RJ (2010) Computational model of hepatitis B virus DNA polymerase: molecular dynamics and docking to understand resistant mutations. Protein Sci 19:796–807
156. Yarmolyuk SM, Nyporko AYu, Bdzhola VG (2013) Rational design of protein kinase inhibitors. Biopolym Cell 29(4):339–347
157. Richartz A, Höltje M, Brandt B, Schäfer-Korting M, Höltje HD (2008) Targeting human DNA polymerase α for the inhibition of keratinocyte proliferation. Part 1. Homology model, active site architecture and ligand binding. J Enzyme Inhib Med Chem 23(1):94–100
158. Edelson RE, Gorycki PD, MacDonald TL (1990) The mechanism of aphidicolin bioinactivation by rat liver *in vitro* systems. Xenobiotica 20:273–287
159. Höltje M, Richartz A, Zdrazil B, Schwanke A, Dugovic B, Murruzzu C, Reissig HU, Korting HC, Kleuser B, Höltje HD, Schäfer-Korting M. (2010) Human polymerase α inhibitors for skin tumors. Part 2. Modeling, synthesis and influence on normal and transformed keratinocytes of new thymidine and purine derivatives. J Enzyme Inhib Med Chem 25(2):250–265
160. Schwanke A, Murruzzu C, Zdrazil B, Zuhse R, Natek M, Höltje M, Korting HC, Reissig HU, Höltje HD, Schäfer-Korting M (2010) Antitumor effects of guanosine-analog phosphonates identified by molecular modelling. Int J Pharm 397(1–2):9–18
161. Zdrazil B, Schwanke A, Schmitz B, Schäfer-Korting M, Höltje HD (2011) Molecular modelling studies of new potential human DNA polymerase α inhibitors. J Enzyme Inhib Med Chem 26(2):270–279
162. Karampuri S, Bag P, Yasmin S, Chouhan DK, Bal C, Mitra D, Chattopadhyay D, Sharon A (2012) Structure based molecular design, synthesis and biological evaluation of α-pyrone analogs as anti-HSV agent. Bioorg Med Chem Lett 22(19):6261–6266

Chapter 5
Molecular Structures, Relative Stability, and Proton Affinities of Nucleotides: Broad View and Novel Findings

Tetiana A. Zubatiuk, Gennady V. Palamarchuk, Oleg V. Shishkin, Leonid Gorb and Jerzy Leszczynski

To the memory of Dr. Oleg Shishkin, our friend and colleague, for all inspiration he had continuously provided.

Abstract In this chapter we analyze and systematize the data related to intramolecular hydrogen bonds and their impact on molecular geometry of nucleotides. The application of various non-empirical methods of quantum chemistry to determination of conformational characteristics of anions of the canonical 2′-deoxyribonucleotides and their methyl esters, as well as their energetics, is discussed. We revealed an existence of novel intramolecular interactions of the canonical 2′-deoxyribonucleotide anions. They are caused by incorporation of 2′-deoxyribonucleotide anions into DNA as well as by the impact of the nucleobases on the conformational features of the nucleotides and intramolecular interactions of these molecules. The efficient strategy of the evaluation of proton affinity for the different types of nucleotides is described.

T. A. Zubatiuk (✉) · G. V. Palamarchuk · O. V. Shishkin
SSI "Institute for Single Crystals" of National Academy of Science of Ukraine,
Lenina Ave., 60, Kharkiv 61001, Ukraine
e-mail: tklimenko@xray.isc.kharkov.com

G. V. Palamarchuk
e-mail: dandygp@list.ru

O. V. Shishkin
e-mail: shishkin@xray.isc.kharkov.com

L. Gorb
Laboratory of Computational Structural Biology, Department of Molecular Biophysics, Institute of Molecular Biology and Genetics, National Academy of Sciences of Ukraine, Key State Laboratory in Molecular and Cell Biology, Zabolotnogo Str., 150, Kyiv 03143, Ukraine
e-mail: lgorb@icnanotox.org

J. Leszczynski
Interdisciplinary Center for Nanotoxicity, Department of Chemistry and Biochemistry, Jackson State University, P.O. Box 17910, Lynch Str., 1325, Jackson, MS 39217, USA
e-mail: jerzy@icnanotox.org

L. Gorb et al. (eds.), *Application of Computational Techniques in Pharmacy and Medicine*,
Challenges and Advances in Computational Chemistry and Physics 17,
DOI 10.1007/978-94-017-9257-8_5

It is based on the analysis of consequences of nucleobases protonation along with the details of intramolecular interactions in 2'-deoxyribonucleotide anions. The results of our molecular simulations cast light on relationship between the conformational dynamics of a molecule and the tautomeric transitions in the components of nucleotides.

5.1 Introduction

The canonical 2′-deoxyribonucleotides (DNTs) represent the monomeric unit of DNA macromolecules [1–3]. Substituted or modified nucleotides are widely used as antibiotics, hormones, coenzymes, etc. [1–5]. In addition, nucleotides have other, independent functions of being cofactors, allosteric effectors, they are incorporated into coenzymes and directly involved into metabolic and accumulation processes, as well as into energy transfer. Nucleosides and nucleotides interact with proteins in all stages of their metabolism. Interestingly, the derivatives of adenosine in living cells perform a variety of biological functions, e.g. they are the inhibitors of protein synthesis. In the form of di- and triphosphate, adenosine is the energy source for a set of enzymatic reactions and muscle contractions. Therefore, knowledge of the structure and nature of intermolecular interactions in nucleotides is essential for understanding of the molecular mechanisms that take place in living cells.

The structures of monomeric DNTs and their derivatives were extensively studied by experimental methods—mainly by X-ray diffraction [6–8] and NMR spectroscopy [9]. Information concerning the structure of DNTs as building blocks of DNA was derived from experimental data for various oligonucleotides. As a result of such studies it is generally accepted that nucleotides are not rigid molecules [10–12]. The nonrigidity of nucleotides is described by the rotation of the nucleobase and furanose moiety relative to each other around the corresponding σ-bonds, and the phenomenon of the pseudorotation of the furanose ring. It was also demonstrated that DNTs adopt two preferable conformations which have close orientation of the nucleobase and phosphate with respect to sugar moiety, but have different conformations of the furanose ring. This finding was supported by *ab initio* quantum chemical investigations of the molecular structure of DNTs [12–18].

Typically, the experimental studies of DNTs were carried out in the condensed states where their conformations are significantly affected by intermolecular interactions (hydrogen bonds, interactions with counter ions). This makes uncertain what exactly have been studied and taken into account: intramolecular properties of DNTs or the influence of the environment on molecular structure of DNTs. Therefore, experimental data may not reflect the intrinsic conformational properties of DNTs. Such information may be obtained using gas phase experiments. However, this information is not available for 2′-deoxyribonucleotides. Therefore, in this case a missing data could be obtained from investigation of the intrinsic conformational characteristics of DNTs using computational methods.

Ab initio quantum chemical methods allow not only to calculate the equilibrium geometry, but also to examine the nature of the electron density distribution in

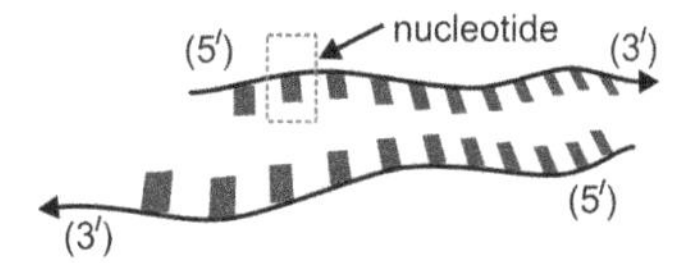

Fig. 5.1 The two strands of DNA separated, showing a nucleotide. Each nucleotide is about 6 Å wide

the molecule. This facilitates investigation of intramolecular interactions [19–21]. One needs to acknowledge that modern computer resources are adequate for conformational studies of molecules at a high theory level, revealing data of the experimental accuracy. This opens up new possibilities for the study of conformational characteristics and intramolecular interactions of fundamental biologically active molecules, such as nucleotides.

In the past conformational characteristics of DNTs were studied using force field [10–12] and semi-empirical quantum-chemical, e.g. [13] methods. Although these methods deliver valuable insight into the conformational features of DNTs they are not able to provide accurate quantitative data related to the conformational characteristics of these molecules. The most reliable structural data of DNTs may be obtained from static and dynamic *ab initio* and DFT quantum-mechanical (QM) calculations. However, such methods are much more time and resources demanding than the classical MD simulations. By now, published results of DFT molecular dynamic simulations and QM studies on simple DNA constituents revealed huge amount of data. Such studies have been reported for nucleobases and base pairs, e.g. [22–30], nucleosides and nucleotides, e.g. [31–36]. They provide vital information about molecular and electronic structure, conformational flexibility, tautomerism, and interactions with metals, water, and other molecules.

In this review, we present the recent results of the comprehensive studies of the conformational and energy characteristics of the anions of the canonical DNTs, their methyl ethers, and protonated methyl ethers anions. The special attention is paid to the analysis and classification of the ample set of intramolecular hydrogen bonds which are an essential part of the structures of the nucleotide molecules. We discuss special criteria which allow delineating the hydrogen bonds with some stable electrostatic interactions in a nucleotide. Noticeable consideration is given to data that explain the effect of hydrogen bonding on structural geometry changes in nucleotides. The specific biological relevant composition of DNA is described from the point of view of the non-standard "orthogonal" syn-conformers of 2′-deoxycytidine-5′-phosphate and 2′-deoxyadenosine-5′-phosphate, which are stabilized due to unusual strong intramolecular hydrogen bonds N–H…O between the amino group of nucleobase and the oxygen atom of the phosphate group.

5.2 Structure of 2′-Deoxyribonucleotides in DNA Macromolecules and Oligonucleotides

The main interest in canonical DNTs is caused by the fact that they are building blocks of DNA. If we uncoil the two strands, as shown in Fig. 5.1, then each strand may be seen to consist of a series of nucleotides units. These are linked to one

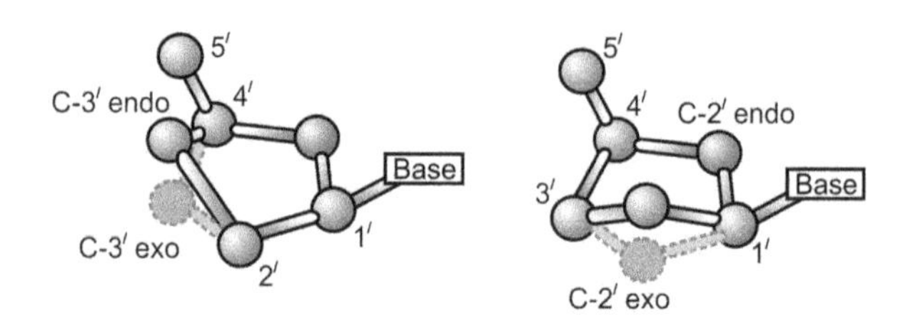

Fig. 5.2 The ribose ring conformations

another with a certain "directionality", known technically as "5-prime to 3-prime", in a head-to-tail sense. The two strands run in opposite directions, as shown in Fig. 5.1 by the labels 5′ and 3′, and by the arrows.

Each nucleotide is made of about 20 atoms, such as carbon, nitrogen, and oxygen. These atoms can again be grouped into smaller parts which are connected in a particular way. The three parts of a nucleotide are its sugar, phosphate, and base. Numerous experimental and theoretical studies of 2′-deoxyribonucleotides, e.g. [37–39] indicate that these molecules are very flexible and they can adopt many different conformations. These conformations may be classified based on geometrical parameters of deoxyribose ring (SU), sugar–phosphate backbone (BB), and orientation of the base (BU) with respect to deoxyribose ring. These main fragments of DNT can adopt different stable conformations leading to numerous conformers of DNTs with different combination of configurations of its fragments.

From the whole set of possible conformations the ribose ring in DNTs adopts two conformations with C2′ (C2′-endo, south) or C3′ (C3′-endo, north) atoms lying on one side of average plane of a ring with the C5′ atom (Fig. 5.2). The DNA base can display two orientations with respect to SU due to rotation around glycosidic C–N bond (syn and anti). These orientations are described by a value of torsion angle χ. The χ value is within $-115° \div -180°$ for the anti conformers and $60° \div 80°$ for syn conformers.

In DNA macromolecules two oxygen atoms of phosphate residue are involved in the formation of the phosphodiester bridge to neighboring nucleotides. Therefore, the charge of the phosphate group is -1. Experimental studies of various DNA and oligonucleotides showed that the nucleotides exist within these macromolecules as a monoanions [1]. In this case, the negative charges of the phosphate groups are compensated by the counterions: K^+, Na^+, or Mg^{2+}.

The BB includes atoms of a phosphate group and the C3′–C5′ carbon atoms of ribose. Conformation of the BB for each form of DNA is described by unique set of torsion angles α, β, γ, δ, and ζ (Fig. 5.3). Previous numerous experimental and theoretical investigations, e.g. [1, 2, 39] shown that each of these torsion angles have the most populated range of values associated with different types of DNA conformation. The recent systematization includes five main forms of DNA: A, BI, BII, ZI and ZII [39–42]. The values of the torsion angles corresponding to each of the types of DNA are shown in Table 5.1. Additionally, comprehensive

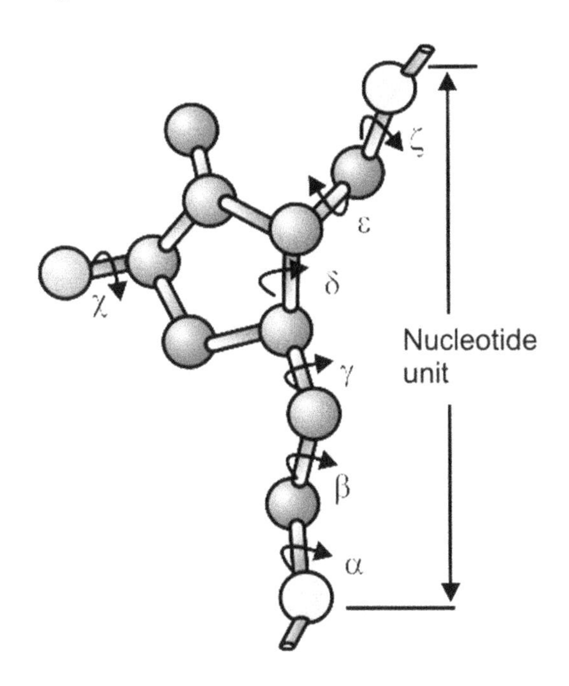

Fig. 5.3 The sugar–phosphate backbone

Table 5.1 The average values of torsion angles in different forms of DNA (The data were obtained through the analysis of 118 naked (noncomplexed) DNA structures, see Ref. [39])

DNA form	Torsion angles of backbone (deg.)						
	α	β	γ	δ	ε	ζ	χ
Canonic A-DNA (AI)	295	173	54	82	206	285	201
AII-form	146	192	183	85	197	289	203
Canonic B-DNA (BI)	299	179	48	133	182	263	250
BII-form	293	143	46	143	251	168	278
ZI-form, Y-R step	66	186	54	147	264	76	205
ZI-form, R-Y step	210	233	177	96	242	292	63
ZII-form, R-Y step	169	162	179	95	187	187	58

Y pyrimidine, *R* purine

study by Schneider et al. [39] clearly demonstrated differences in backbone torsion angles for A, BI, BII, ZI, and ZII types of DNA. In A-DNA north conformation of SU is observed while BI-, and BII-DNA contain south conformation of SU. In Z-DNA both south and north conformations of sugar are observed. The A, BI, and BII forms of DNA include conformers of nucleotides only with the anti-orientation of a base with respect to the SU. In the case of Z-DNA purine nucleotides have a syn-orientation of a base [2, 39].

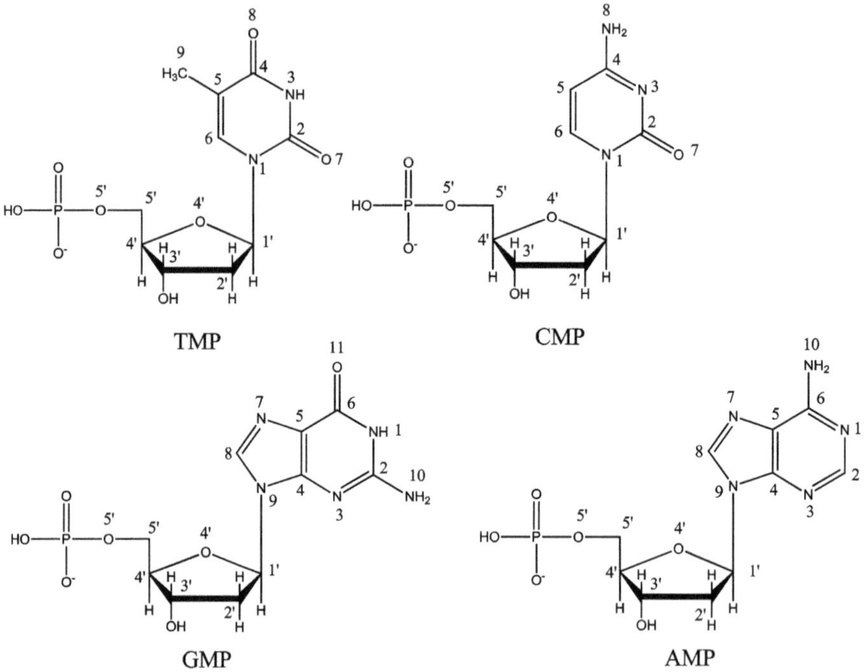

TMP CMP GMP AMP

Fig. 5.4 Structure and nomenclature of the canonical anionic 2′-deoxyribonucleotides

5.3 The Structure and the Relative Stability of 2′-Deoxorybonucleotides Conformers

Theoretical studies [19, 20] of the molecular structure of canonical DNTs, namely, 2′-deoxythymidine-5′-phosphate (TMP), 2′-deoxycytidine-5′-phosphate (CMP), 2′-deoxy-adenosine-5′-phosphate (AMP), and 2′-deoxyguanosine-5′-phosphate (GMP), have revealed the presence of numerous intramolecular N–H…O and C–H…X (X=O, N) hydrogen bonds. The structures of discussed monoanions are present in Fig. 5.4. It was found that every conformation of each DNT contains up to four intramolecular hydrogen bonds. Formation of these hydrogen bonds significantly influences the equilibrium conformation of the nucleotides. In particular, this concerns the orientation of the nucleobase and phosphate group with respect to a sugar fragment. The revealed examples include south/anti and north/anti conformers of all molecules (Fig. 5.5), south/syn and north/syn conformers of GMP (Fig. 5.6), and north/syn conformers of CMP and AMP with orthogonal orientation of base with respect to SU (Fig. 5.6). An especially strong influence of intramolecular hydrogen bonds is observed for dianions of DNTs, where the presence of very strong C–H…O bonds results in significant deformation of the SU conformation. Formation of N–H…O intramolecular bonds is also responsible for the stabilization of conformers with a syn orientation of nucleobases [12, 20] in the case of GMP.

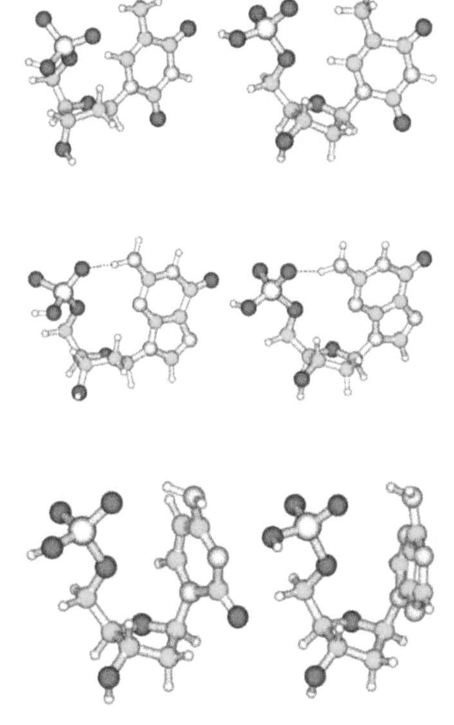

Fig. 5.5 The structure of S/anti and N/anti conformers of anionic 2′-deoxyribonucleotides. TMP molecule is shown as representative example

Fig. 5.6 The structure of S/syn and N/syn conformers of anion of GMP molecule

Fig. 5.7 The structure of syn-conformers of CMP and AMP with an orthogonal orientation of the nucleobase relatively deoxyribose ring

The most striking differences in the molecular structure of DNTs, revealed by the results of calculations performed using an extended basis set 6-31++G(d,p) (compared to 6-31 (d)), are found for conformers with a south/syn orientation of the nucleobase. The obtained data concluded that pure south/syn conformers correspond to local minima on the potential energy surface only in the case of GMP (Fig. 5.6). Earlier it was demonstrated [12, 20] that only south/syn conformers of GMP are stabilized by the formation of a strong intramolecular N–H...O hydrogen bond between the amino and the phosphate groups. All other south/syn conformers were stabilized by the C–H…O hydrogen bond or by other intramolecular interactions. Therefore, a more rigorous computational treatment of the anionic states of DNTs, due to the application of diffuse functions results in the disappearance of these conformers also for TMP and AMP. Previously, similar results were obtained for dianions of DNTs [19, 20]. In the case of monoanions the absence of syn-conformers was observed only for CMP [19].

In addition, the CMP and AMP conformers with an almost orthogonal orientation of the nucleobase with respect to the C1′–H bonds are found instead of the north/syn conformers (Fig. 5.7). Such an orientation of the cytosine and adenine moiety is stabilized by the intramolecular N–H...O hydrogen bond between the amino and the phosphate groups following significant out-of-plane deformation of the pyrimidine ring (Tables 5.2 and 5.3). The formation of this type of hydrogen bond is impossible

Table 5.2 Selected geometrical parameters of the pyrimidine 2′-deoxyribonucleotides calculated at the B3LYP/6-31++G(d,p) level

Parameters	Nucleotide, conformation				
	TMP	TMP	CMP	CMP	CMP
	S/anti	N/anti	S/anti	N/anti	N/syn
C2(C4)–N10(N8)H2 (Å)			1.377	1.376	1.373
N9(N1)–C1′ (Å)	1.468	1.488	1.473	1.487	1.493
C1′–O4′ (Å)	1.429	1.412	1.423	1.416	1.418
C4′–O4′ (Å)	1.449	1.443	1.450	1.444	1.457
C3′–O3′ (Å)	1.438	1.425	1.438	1.426	1.437
O5′–P (Å)	1.687	1.689	1.680	1.694	1.676
Σ(NH2) (deg.)	–	–	349.4	350.8	335.8
OH–P–O5′–C5′ (deg.)	83.5	76.6	61.2	93.2	69.0
β (deg.)	180.0	178.4	142.6	155.3	159.3
γ (deg.)	10.6	55.3	40.1	53.0	42.2
δ (deg.)	143.5	86.6	142.3	85.9	98.2
χ (deg.)	−115.7	−148.5	−144.1	−155.8	162.1
P	181.4	12.9	184.9	5.1	330.7
vmax	33.3	28.9	34.3	30.9	34.5
Sugar conformation	3T2	3T2	3T2	3T2	2T1

Table 5.3 Selected geometrical parameters of the purine 2′-deoxyribonucleotides calculated at the B3LYP/6-31++G(d,p) level

Parameters	Nucleotide, conformation						
	AMP	AMP	AMP	GMP	GMP	GMP	GMP
	S/anti	N/anti	N/syn	S/anti	N/anti	S/syn	N/syn
C2(C4)–N10(N8)H2 (Å)	1.367	1.365	1.367	1.393	1.389	1.353	1.355
N9(N1)–C1′ (Å)	1.456	1.470	1.473	1.451	1.477	1.451	1.454
C1′–O4′ (Å)	1.423	1.414	1.420	1.425	1.414	1.425	1.421
C4′-O4′ (Å)	1.452	1.446	1.454	1.445	1.457	1.447	1.437
C3′-O3′ (Å)	1.439	1.425	1.429	1.446	1.415	1.445	1.427
O5′-P (Å)	1.684	1.692	1.691	1.689	1.679	1.661	1.657
Σ(NH2) (deg.)	354.0	353.8	335.1	338.7	340.7	349.4	346.1
OH–P–O5′–C5′ (deg.)	68.4	−72.7	73.4	150.3	−106.5	66.4	65.1
β (deg.)	132.9	−163.2	174.2	101.0	96.7	174.0	−166.8
γ (deg.)	30.6	53.2	51.2	30.6	−66.9	42.7	47.4
δ (deg.)	142.9	90.1	100.2	134.3	99.7	139.1	86.5
χ (deg.)	−130.2	−139.6	122.4	−110.7	−117.3	70.9	62.0
P	171.9	3.3	320.9	146.5	358.7	158.2	44.0
vmax	36.5	27.3	35.1	40.2	28.6	36.4	26.8
Sugar conformation	2T3	3T2	1T2	2T1	2T3	2T1	4T3

in south conformers of CMP and AMP because of the equatorial orientation of the phosphate group. The absence of an amino group makes the presence of any syn-conformers of TMP impossible, unlike the previously reported case [19].

Table 5.4 The relative energy of conformers of the 2′-deoxyribonucleotides anions, kcal/mol

		B3LYP/6-31++G(d,p)			MP2/aug-cc-pvdz//B3LYP/6-31++G(d,p)		
Molecule	Conformer	ΔE_{DFT}	ΔG^{0K}	ΔG^{298}	ΔE_{MP2}	$\Delta G^{0K a}$	$\Delta G^{298 a}$
TMP	S/anti	0	0	0	0	0	0
TMP	N/anti	3.35	3.15	3.19	4.49	4.28	4.32
CMP	S/anti	0	0	0	1.33	0.80	1.09
CMP	N/anti	0.44	0.48	0.44	2.62	2.15	2.39
CMP	N/syn	3.14	3.67	3.38	0	0	0
AMP	S/anti	0	0	0	0	0	0
AMP	N/anti	4.15	4.00	4.14	5.56	5.40	5.54
AMP	N/syn	7.56	8.08	7.76	1.66	2.18	1.86
GMP	S/anti	9.64	9.82	10.02	9.32	9.68	9.77
GMP	N/anti	4.74	5.66	5.60	5.62	6.54	6.48
GMP	S/syn	0	0	0	0	0	0
GMP	N/syn	4.58	4.40	4.51	5.57	5.40	5.51

[a] Values of ZPE and ΔG^t were obtained from the B3LYP/6-31++G(d,p) calculations

The application of an extended basis set allows revealing the clear dependence of lengths of the C3′–O3′ bonds on the conformation of the SU. This bond in the south conformer is systematically longer compared to the north conformer. The biggest differences are found for GMP ($\Delta\ell$ is 0.031 Å for anti- and 0.018 Å for syn-conformers), while this value for other DNTs is considerably smaller ($\Delta\ell$ is 0.012–0.014 Å). The existence of such dependence clearly indicates that the C3′–O3′ bond is involved in some stereo electronic interactions, such as the anomeric effect, despite the absence of a second heteroatom bound to the C3′ atom.

Among south and north conformers of DNTs the south/anti conformer is significantly more stable for TMP and AMP (Table 5.4). In the case of CMP the difference in energy between south/anti and north/anti conformers is considerably smaller. Nevertheless, the south/anti conformer remains the most stable one. North/syn conformers of CMP and AMP possess significantly higher relative energy. An unexpected inversion of stability of conformers is observed for GMP. Conformers with a syn orientation of guanine are more stable compared to anti conformers, in agreement with the previous study [21], with the south/syn conformer being the most favorable one. However, among anti-conformers, the north/anti conformer possesses significantly lower energy compared to the south/anti conformer. Similar results were obtained earlier only for dianions of GMP [20, 21]. Probably, a more rigorous treatment of the anionic state and the intramolecular hydrogen bond due to an application of extended 6-31++G(d,p) basis set is responsible for the inversion of the relative stability of conformers of GMP.

A different explanation of the unusual stability of the north/anti conformer of GMP may be derived from an analysis of intramolecular hydrogen bonds. This conformer contains the O–H…O bond between the phosphate and the O3′–H hydroxyl groups, leading to a significant decrease in the energy of the molecule. However, such a hydrogen bond is impossible in DNA, where the O3′ atom is involved in a phosphodiester linkage with a neighboring nucleotide. Thus, high stability of the north/anti conformer of GMP may be observed only in isolated DNT.

To verify the obtained results, a more rigorous account of electron correlation using the MP2/aug-cc-pvdz method has been performed. As follows from the data presented in Table 5.4, the application of the MP2 level does not change the order of stability of anti conformers of DNTs. Only some increase of differences in energy between south/anti and north/anti conformers is observed. However, the MP2 method significantly decreases the relative energy of orthogonal (north/syn) conformers of CMP and AMP (Table 5.4). In the case of CMP this conformer becomes the most stable, and the energy of this conformer in AMP is lower, compared to the north/anti conformer. Such stabilization may be caused by differences in energy of deformation of the fragments of these DNTs, calculated at the DFT and MP2 levels of theory. In particular, it was recently demonstrated [43] that the MP2 method slightly underestimates the conformational flexibility of the pyrimidine ring in uracil, compared to the more accurate, CCSD(T) data. The value of ring deformation energy calculated within the density functional theory is slightly higher, compared to the MP2 data. Similar effects may be expected for other fragments of DNTs. Therefore, the contribution of these differences in deformation energy may be considerable.

An analysis of the geometrical parameters and relative stability of conformers discussed above indicates that the application of diffuse functions is required for a correct description of the molecular structure and energetics of DNTs. An absence of diffuse functions may lead to the appearance of artificial local minima on the potential energy surface. A comparison of the molecular structure of DNTs calculated using the 6-31(d) basis set as reported in [19, 20] and the 6-31++G(d,p) basis (Tables 5.2, 5.3) reveals that an increase in the size of the basis set results in slight changes in the geometrical parameters of the considered molecules. A decrease of differences between the O4′–C bond lengths and the degree of pyramidality of the amino groups are among molecular parameters which changes are observed when going from the 6-31G(d,p) to the 6-31++ G(d,p) basis set.

5.4 Intramolecular Hydrogen Bonds in the 2′-deoxyribonucleotide

According to the AIM theory [44], the presence of a hydrogen bond that appears in topological analysis of the electron density distribution, like any chemical bond, must correspond to the existence of a bond path between the hydrogen atom and the acceptor containing bond critical points (BCP). This is the requirement and first criteria for the existence of any chemical bond. In the case of hydrogen bonds several additional criteria were developed [45, 46]. Two of them concern properties of the BCP, namely, the value of electron density (ρ) at the BCP should be between 0.002 and 0.035 a.u. and the value of the Laplacian of electron density $\nabla^2(\rho)$ should be within 0.024–0.139 a.u. Besides that, some useful information about the stability of hydrogen bonds may be retrieved from the values of bond ellipticity at the BCP [47] and the distance between the BCP and the ring critical points (RCP) [48]. RCP is defined as a (3, +1) critical point and exists whenever a succession of bond paths closes into a ring. An abnormally high value of ellipticity and the short distance between BCP and RCP usually indicate locally unstable topology of

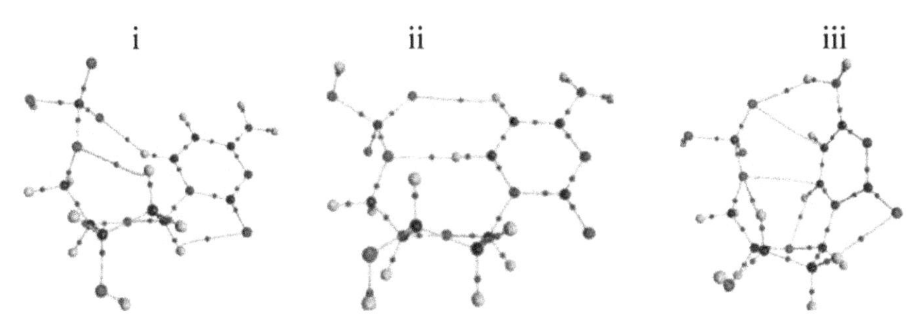

Fig. 5.8 Molecular graph of CMP`s conformers according to AIM theory: (**i**) south/anti, (**ii**) north/anti, (**iii**) north/syn (orthogonal)

electron density, a so-called bifurcation point [44], which is not observed for normal hydrogen bonds. Taking into account that all criteria mentioned above deal with the properties of the BCP, they can be referred to as BCP criteria.

The most immediate evidence of bonding within the AIM theory is the existence of a bond path containing BCP between two atoms. The collection of bond paths within a molecule represents a molecular graph showing all intramolecular bonding interactions, including also hydrogen bonds. An example of a molecular graph for all located by computational study conformers of CMP is visualized in Fig. 5.8. The molecular graph from AIM analysis demonstrates the presence of a network of bond paths corresponding to all chemical bonds in agreement with the Lewis model of molecules.

The results of the calculations reveal also the existence of bond paths corresponding to potential intramolecular hydrogen bonds in DNTs. The main part of these hydrogen bonds represents the interaction of nucleobases with phosphate and sugar. Only a few examples of interactions between phosphate and sugar, or between different atoms of sugar are observed. Geometrical parameters, characteristics of BCP, and the distance between BCP and RCP of these potential hydrogen bonds are listed in Table 5.5. Thus, all these interactions meet the first criteria for hydrogen bonds.

All revealed interactions could be divided into three groups, based on geometrical parameters and values of electron density and Laplacian of the electron density. The *first group*, which may be called "well-defined" hydrogen bonds, includes first of all classical N–H…O and O–H…O hydrogen bonds in GMP, characterized by the shortest H…O distances and the highest values of ρ and $\nabla^2(\rho)$. On the basis of the sum of van der Waals radii of the hydrogen and oxygen (2.45 Å, for reliability we used the shortest radii by Zefirov and Zorkii [49]) and geometrical criteria [50] of classic hydrogen bonds (H…A<2.3 Å, D–H…A>130°), the N–H…O bond in the north/syn conformer of CMP and some C–H…O bonds in all DNTs should be also considered as representatives of this group.

Among "well-defined" C–H…O bonds the main part includes interactions with participation of the C6–H atom of pyrimidine and the C8–H atom of purine fragments, for conformers with anti orientation of the base. This agrees well with the conclusion by Hocquet about the ability of these hydrogen atoms to form stable intramolecular hydrogen bonds in 2′-deoxyribonucleosides [48, 51]. However, in

Table 5.5 Geometrical parameters, characteristics of (3, −1) BCPs, and distances between BCP and RCP for all potential intramolecular hydrogen bonds. True hydrogen bonds are underlined

Molecule	Conformer	D–H…A	H…A (Å)	D–H…A (deg.)	ρ (Å/a.u.^{3})	$\nabla\rho$ (ε/a.u.^{5})	ε	L_{RCP} (a.u.)
AMP	S/anti	C8–H…O–P	2.028	177.6	0.0228	0.0610	0.0366	2.57
		C2′–H…O5′	2.519	110.0	0.0110	0.0398	0.4462	0.82
AMP	N/anti	C8–H…O5′	2.200	154.8	0.0160	0.0472	0.0800	1.78
AMP	N/syn	N6–H…O–P	2.215	129.3	0.0148	0.0510	0.1760	3.92
		C2′–H…N3	2.322	120.9	0.0157	0.0518	0.1158	1.46
GMP	S/anti	C2′–H…O–P	2.262	153.7	0.0154	0.0441	0.0574	1.93
		C8–H…O–P	2.030	162.1	0.0223	0.0640	0.0396	2.14
GMP	N/anti	C3′–H…O–P	2.490	135.1	0.0108	0.0085	0.0795	1.40
		O3′–H…O(H)–P	1.792	166.2	0.0357	0.0263	0.0620	2.37
GMP	S/syn	N10–H…O–P	1.681	174.6	0.0457	0.1335	0.0262	2.82
		C2′–H…N3	2.359	127.2	0.0138	0.0456	0.0489	1.32
		C2′–H…O5′	2.638	108.7	0.0081	0.0325	2.0929	0.31
GMP	N/syn	N10–H…O–P	1.714	161.5	0.0426	0.1248	0.0427	2.61
		C3′–H…N3	2.893	111.7	0.0055	0.0192	0.4419	0.92
TMP	S/anti	C9–H…O–P	2.331	166.3	0.0125	0.0376	0.0437	1.89
		C1′–H…O7	2.211	110.7	0.0198	0.0785	0.8871	0.46
		C2′–H…O5	2.495	109.4	0.0112	0.0413	0.2387	1.03
		C6–H…O5′	2.161	160.5	0.0182	0.0512	0.0670	1.73
TMP	N/anti	C9–H…O–P	2.253	178.6	0.0143	0.0424	0.0317	2.26
		C6–H…O5′	2.128	171.2	0.0188	0.0529	0.0529	1.52
CMP	S/anti	C6–H…O–P	2.195	149.4	0.0174	0.0477	0.294	2.69
		C1′–H…O7	2.238	104.0	0.0197	0.0804	0.9646	0.44
		C2′–H…O5′	2.513	111.0	0.0109	0.0393	0.3163	0.93
CMP	N/anti	C5–H…O–P	2.785	121.2	0.0052	0.0201	0.1969	1.06
		C6–H…O5′	2.069	161.5	0.0220	0.0611	0.0650	1.56
CMP	N/syn	N8–H…O–P	2.150	141.2	0.0169	0.0523	0.0861	1.75
		C6–H…O4′	2.255	99.7	0.0196	0.0855	1.6105	0.39
		C3′–H…O5′	2.332	107.6	0.0144	0.0609	1.6211	0.30
		C2′-H…O7	2.90	118.1	0.0159	0.0561	0.2114	1.17

DNTs, a clear difference is revealed between purine and pyrimidine nucleotides. In the case of TMP and CMP, the C6–H…O5′ hydrogen bond is observed in all conformers except the south/anti conformer of CMP where the C6–H atom interacts with the oxygen of the phosphate group. An opposite situation is found for purine nucleotides. The C8–H…O-P hydrogen bond is located in south/anti conformers of AMP and GMP, while the C8–H…O5′ bond is formed in the north/anti conformer of AMP. Unlike these conformers, interactions with the participation of the C8′–H atom were not found in the north/anti conformer of GMP (Table 5.5).

The values of electron density and Laplacian of electron density at the BCP for all of these hydrogen bonds meet Popelier's criteria. The range of variation of the properties of the BCP for well-defined hydrogen bonds ($\rho > 0.015$ a.u., $\nabla^2(\rho) > 0.044$ a.u.) may be used for further classification of other intramolecular interactions in DNTs. An analysis of the ellipticity of these hydrogen bonds ($\varepsilon < 0.09$) and the distance between BCP and RCP ($L_{RCP} > 1.5$ a.u.) does not indicate any instability of these interactions, and these values also may be used as a reference for further consideration.

The *second group* of hydrogen bonds includes interactions that do not meet BCP criteria. The value of the Laplacian of the electron density at the BCP is smaller than 0.024 a.u. for the C5–H…O–P bond in the north/anti conformer of CMP and the C3′–H…N3 bond in the north/syn conformer of GMP (Table 5.5). Therefore these interactions should be considered to be electrostatic interactions rather than real hydrogen bonds. They are characterized by the longest H…O distances (more than 2.7 Å), small values of the C–H…O angles (less than 122°), and the electron density at BCP ($\rho < 0.006$). In addition, these interactions have relatively high ellipticity at the BCP ($\varepsilon > 0.19$) and a shorter distance to the RCP ($L_{RCP} < 1.1$ a.u.).

The remaining interactions belong to the *third group*, and they require additional careful consideration. Using characteristics of the two groups of interactions mentioned above, we can classify these potential intramolecular hydrogen bonds in DNTs. First of all we should consider the N–H…O interaction in the north/syn conformer of AMP (Table 5.5). This interaction demonstrates all properties of the inherent hydrogen bond, namely, clear directionality and influence on the conformation of the molecule. The formation of this hydrogen bond results in out-of-plane deformation of the pyrimidine ring of adenine and stabilization of an unusual conformer with orthogonal orientation of the base with respect to sugar. All BCP properties of this hydrogen bond (Table 5.5) are within the ranges typical for well-defined hydrogen bonds mentioned above, except the N–H…O angle that is only 129.3° and the value of ellipticity at BCP that is too high for normal hydrogen bonds (Table 5.5). Therefore, this interaction undoubtedly should be classified as a real hydrogen bond. The high value of ellipticity probably reflects the weak character of this bond, which is in agreement with the relatively long H…O distance and the small value of the N–H…O angle. Consideration of the N–H…O interaction as a true hydrogen bond allows the criteria for the existence of hydrogen bonds to be refined. In particular, this concerns the value of ellipticity at the BCP that should be less than 0.18.

An analysis of the BCP properties of the remaining potential intramolecular hydrogen bonds demonstrates that all values of ρ and $\nabla 2(\rho)$ meet Popelier's criteria (Table 5.5). Therefore, these parameters cannot be used for the classification of the interactions under consideration. The values of bond ellipticity and the distance between the BCP and the RCP are much more informative for this purpose. The C1′–H…O7 interaction in the south/anti conformer of TMP and CMP, the C6–H…O4′ and the C3′–H…O5′ interactions in the north/syn conformer of CMP, and the C2′–H…O5′ interaction in the south/syn conformer of GMP are characterized by extremely large values of ellipticity ($\varepsilon > 0.8$) and very short distances between the BCP and the RCP ($L_{RCP} < 0.5$ a.u.). This indicates that the topology of the electron density distribution at these BCPs is very unusual and lies out of range for normal hydrogen bonds. Therefore, these interactions cannot be considered as hydrogen bonds and should be treated as strong electrostatic interactions. It should be noted that the H…O distances for some of these interactions are relatively short (2.21–2.33 Å), but the values of the C–H…O angle are considerably smaller than 120° (99.7–110.7°). This demonstrates that the H…O distance cannot be a good indicator for the existence of hydrogen bonds. It should be combined with the value of the C–H…O angle that is normally larger than 120°.

At least the characteristics of two interactions (C3′–H…O–P in the north/anti conformer and C2′–H…N3 in the south/syn conformer of GMP) are very similar to the characteristics of well-defined hydrogen bonds. Only the H…O distance in the C3′–H…O–P bond (2.490 Å) is considerably longer, compared to other hydrogen bonds (Table 5.5). However, this probably is compensated by a higher value of the C–H…O angle (135.1°). Therefore, these interactions undoubtedly should be considered as true hydrogen bonds.

The most interesting situation is observed for the C2′–H…O7 interaction in the north/syn conformer of CMP. The geometrical parameters of this potential hydrogen bond are almost appropriate for hydrogen bonding. The C–H…O angle is slightly smaller than 120° (Table 5.5). The values of ellipticity and L_{RCP} are also close to suitable ranges, but are slightly outside the normal values. Therefore, it is possible to consider this interaction as a hydrogen bond, but it represents a borderline case for the classification of such interactions.

The remaining interactions (C2′–H…O5′ in the south/anti conformer of TMP, CMP, and AMP) are characterized by unsuitable geometrical parameters, a shorter distance between the BCP and RCP, and considerably higher ellipticity at the BCP (Table 5.5). Therefore these interactions also should be considered as electrostatic interactions, rather than the hydrogen bonds.

In summary, a topological analysis of the electron density distribution in DNTs reveals numerous intramolecular D–H…A interactions that may be considered as potential hydrogen bonds. However, a more detailed analysis of the properties of BCP for these interactions allows a clear enough distinction between true hydrogen bonds and strong electrostatic interactions of atoms with opposite charges to be made. On the basis of data presented in [52] it is demonstrated that geometrical parameters and values of the electron density and the Laplacian of electron density cannot be used for classification of such interactions. Only the values of the bond ellipticity (ε) and distance between BCP and ring critical points (RCP) allow a distinction between true hydrogen bonds and strong electrostatic interactions to be made. According to the analysis true hydrogen bonds should be characterized by the following values of BCP properties: $\varepsilon < 0.1$ and $L_{RCP} > 1.4$ a.u. for well-defined hydrogen bonds and $\varepsilon < 0.22$ and $L_{RCP} > 1.1$ a.u. for all hydrogen bonds, including very weak hydrogen bonds.

5.5 Deformation of 2′-Deoxyribonucleotides Inside DNA Macromolecule

As was previously shown, investigation of structure of canonical 2′-deoxyribonucleosides and 2′-deoxyribonucleotides [12, 15, 18–20, 48, 52] revealed existence of numerous N–H…O and C–H…X (where X=O, N) intramolecular hydrogen bonds. Purine and pyrimidine nucleotides are characterized by the unique set of intramolecular interactions defined by the nature of the base and conformation of SU and BU. Formation of intramolecular hydrogen bonds significantly influences conformations of molecules [15, 19, 48, 52]. In particular, they are responsible for

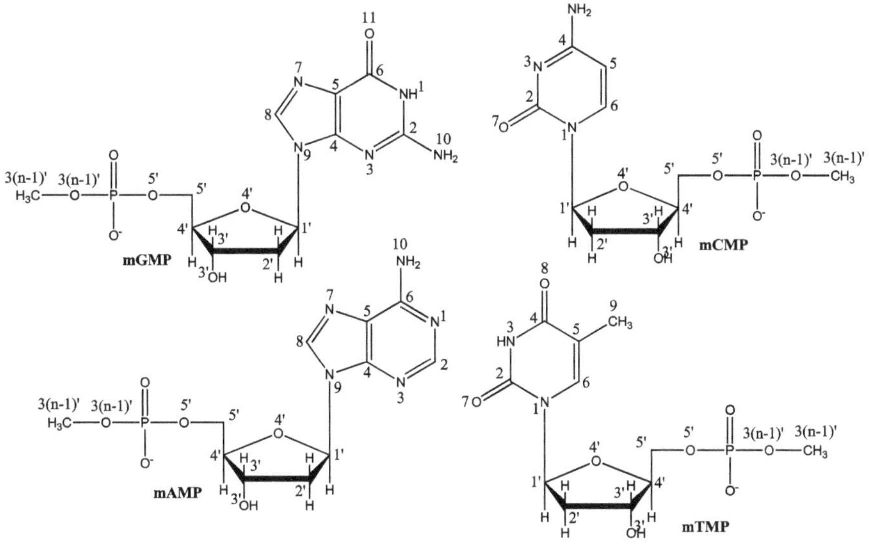

Fig. 5.9 Numbering of atoms in methyl esters of 2′-deoxyribonucleotides

orientation of base and a pronounced deformation of ribose in dianionic nucleotides [19, 20].

Comparison of values of torsion angles describing conformation of isolated nucleotides with average values of these angles for different forms of DNA indicates that incorporation of DNTs into DNA macromolecule results in some deformation of geometry of nucleotides. This also should be accompanied by changes in intramolecular hydrogen bonds. However, the energy values related to variety of such deformations are still unknown.

Our study [53] reveals that the change of nucleotides geometry also leads to change of intramolecular hydrogen bonds pattern which become unique for every form of DNA. Thus, some corrections should be made for modeled structures of nucleotides, as far as previous investigations of DNTs concluded that compensation of negative charge of phosphate group by hydrogen atom lead to appearance of "artificial" intramolecular hydrogen bonds with participation of the P-O-H fragment. That also resulted in changes of equilibrium conformation and relative stability of different conformers of DNTs. In order to prevent formation of such hydrogen bonds we use model of monomethyl esters of DNTs (Fig. 5.9) where the carbon atom of methyl group corresponds to the C3′ atom of neighboring SU.

The investigations were carried out using the density functional theory approach. The molecular structures of methyl ethers of DNTs namely thymidine-5′-phosphate (mTMP), 2-deoxycytidine-5′-phosphate (mCMP), 2-deoxyadenosine-5′-phosphate (mAMP), and 2′-deoxyguanosine-5′-phosphate (mGMP) were optimized applying the Becke's three-parameter exchange functional, the gradient-corrected functional of Lee et al. [54–56] and the standard aug-cc-pvdz basis set. Local minima were verified by establishing that the matrix of energy second derivatives (Hessian) has

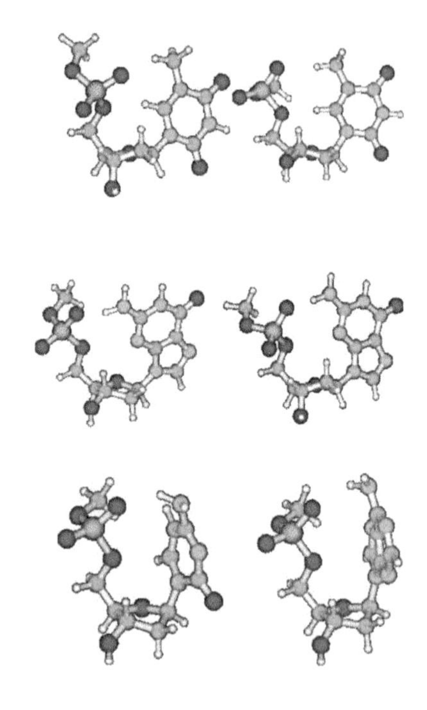

Fig. 5.10 The structure of S/anti and N/anti conformers of anions of 2′-deoxyribonucleotides. mTMP molecule is shown as representative example

Fig. 5.11 The structure of S/syn and N/syn conformers of mGMP

Fig. 5.12 The structure of north/syn conformers of mCMP and mAMP with orthogonal orientation of base with respect to ribose

only positive eigenvalues. The energy of zero point vibrations was calculated at the same level of theory (B3LYP/aug-cc-pvdz) within the harmonic approximation. Influence of environment on relative stability of conformers was estimated by single point calculations using Polarized Continuum Model (PCM) [57–59] and water as a solvent at the same level of theory.

The results of calculations demonstrate that the set of stable conformers of methyl esters of DNTs corresponds well to the structures considered in the previous studies [19, 20, 52]. The revealed species include south/anti and north/anti conformers of all molecules (Fig. 5.10), south/syn and north/syn conformers of mGMP (Fig. 5.11), and north/syn conformers of mCMP and mAMP with orthogonal orientation of base with respect to ribose (Fig. 5.12). But in addition to our previous investigations, south/syn conformer of mAMP was also found. Existence of this rare conformer is possible because presence of intramolecular C–H…N hydrogen bond. This conformer corresponds to minimum on energy surface which is nearest to syn-conformer of mAMP in ZI–ZII–DNA form. Such a distribution of minima on the PES of the molecule of methyl nucleotide repeats and confirms the results of the conformational analysis performed for DNTs anions with P–OH fragment.

An absence of "artificial" intramolecular hydrogen bonds results in a change of relative stability of anti-conformers of mGMP. Unlike previous studies, here the south/anti conformer of mGMP is more stable than the north/anti form. However, the syn conformers of this nucleotide are considerably more stable (Table 5.6) due to the formation of intramolecular N–H…O hydrogen bond, in agreement with the previous findings [19, 20, 52].

Table 5.6 B3LYP/aug-cc-pvdz relative energies (kcal/mol) of conformers of methyl ethers of 2′-deoxyribonucleotides in gas phase and PCM

Nucleotide	Conformer	Vacuum		PCM
		ΔE_{DFT}	ΔG^{0K}	ΔE_{DFT}
mAMP	S/anti	0	0	0.46
mAMP	N/anti	7.16	6.72	0.11
mAMP	S/syn	13.93	13.28	0
mAMP	N/syn	12.14	12.19	10.96
mGMP	S/anti	7.59	7.54	0.45
mGMP	N/anti	16.01	15.84	0.45
mGMP	S/syn	0	0	0
mGMP	N/syn	4.41	4.15	0.46
mTMP	S/anti	0	0	0.66
mTMP	N/anti	5.07	4.83	0
mCMP	S/anti	0	0	0.85
mCMP	N/anti	3.42	3.15	0
mCMP	N/syn	8.33	8.53	11.76

The results of calculations in a solution (PCM model) reveal drastic changes in relative stability of DNTs conformers compared to the gas phase data (Table 5.6). In the case of mTMP and mCMP the north/anti conformer becomes the most stable. In the case of mGMP and mAMP the south/syn conformer remains the most stable. However, relative energy of other conformers is only slightly higher (less than by 1 kcal/mol). The exceptions are south/syn conformers of mAMP and mCMP. Relative energy of these conformers is significantly higher (close to 11 kcal/mol). These data demonstrate that relative stability of DNTs conformers may be easily regulated by environment effects.

The comparison of the geometrical parameters of DNTs containing P–OH and P–OCH_3 fragments demonstrates that the replacement of a hydroxyl group by methoxy group leads to some changes of the torsion angles defining the conformation of nucleotide (Table 5.7). A significant reduction of range of variation of torsion angles within BB, especially values of β and γ angles, is predicted. It is possible to ascertain, that with elongation of backbone, the deformability of DNTs molecules slightly decreases and their equilibrium conformations becomes closer to each other.

The comparison of values of the torsion angles in equilibrium conformations of DNTs with the average values of these angles for each type of DNA (Table 5.1) demonstrates that the smallest difference is accounted for the A-DNA. However, the results of calculations demonstrate some decrease of energy of nucleotides with backbone constrained to values corresponding to A-DNA (Table 5.8). Detailed inspection of molecular structure of nucleotides indicates that such geometry of backbone is favorable for attractive interactions between O3′ hydroxyl group and phosphate residue. This leads to change of orientation of the hydrogen atom of OH group toward the oxygen atom of the PU, as compared to equilibrium conformation of nucleotides. The O…H distance in constrained DNTs is out of range of usual values for the O–H…O hydrogen bond. However, such attractive interactions results in additional stabilization of molecule. Thus, deformation energy in

Table 5.7 B3LYP/aug-cc-pvdz geometrical parameters (deg.) of conformers of methyl ethers of 2′-deoxyribonucleotides

Nucleotide	Conformer	Backbone torsion angles							
		α	β	γ	δ	ζ	χ	P	υ_{max}
mTMP	S/anti	−87.7	−105.6	63.1	141.6	−71.6	−113.5	171.8	30.1
	N/anti	−70.7	−166.6	57.6	87.6	−73.5	−144.7	11.6	28.1
mCMP	S/anti	−85.1	−107.9	60.7	141.5	−72.0	−124.3	171.1	30.5
	N/anti	−65.8	−177.0	59.0	85.9	−73.8	−153.3	6.2	30.8
	N/syn	−65.3	−160.9	58.8	95.5	−75.4	161.0	330.6	33.1
mAMP	S/anti	−102.3	−101.0	61.3	144.7	−70.9	−99.0	183.2	29.8
	N/anti	−75.8	−161.9	54.1	89.5	−73.8	−138.5	3.8	27.7
	S/syn	74.2	157.4	−172.0	115.4	73.0	79.8	120.0	40.3
	N/syn	−66.9	−169.5	58.9	99.2	−77.1	121.0	322.5	35.2
mGMP	S/anti	−105.3	−99.6	60.0	145.3	−71.1	−95.1	183.4	30.4
	N/anti	−80.3	−155.4	52.6	92.7	−73.8	−133.7	6.3	23.7
	S/syn	−68.2	−110.7	58.2	134.2	−70.0	72.7	155.0	31.8
	N/syn	−90.0	−139.9	53.4	85.7	−67.8	65.8	40.6	27.3

Table 5.8 B3LYP/aug-cc-pvdz deformation energies (kcal/mol) of isolated and hydrated (within parentheses) methyl ethers of 2′-deoxyribonucleotides resulted from incorporation into DNA

DNA form	Nucleotide			
	mTMP	mCMP	mAMP	mGMP
A[a]	−1.8(−0.1)	−1.6(−0.1)	−1.8(−0.6)	−2.0(−0.2)
BI[b]	5.1(−0.1)	4.4(0.2)	6.6(−0.4)	6.3(0.2)
BII[b]	12.0(0.5)	4.2(0.5)	6.7(0.4)	5.9(0.4)
ZI[b]	5.8(1.1)	5.0(1.3)	10.8(0.6)	10.8(−0.3)
ZII[b]	6.4(0.8)	5.1(0.6)	10.8(0.6)	10.8(−0.3)
ZI[a], ZII[a]			9.7(0.0)	9.8(0.9)
ZI[c], ZII[c]			0.5(2.8)	10.1(−10.6)
ZI[d], ZII[d]			1.0(1.1)	13.0(0.2)

[a] N/anti-conformers
[b] S/anti-conformers
[c] N/syn-conformers
[d] S/syn-conformers

the case of A-DNA does not correspond to real deformation of nucleotides due to incorporation into macromolecule. Nevertheless, taking into account rather small differences in values of torsion angles for equilibrium conformation of DNTs and average values for A-DNA it is possible to suggest that deformation energy in this case should be the smallest. The largest increase of energy is found for the Z-forms of DNA (Table 5.8).

The obtained data demonstrate that the polarization effect of environment results in considerable decrease of deformation energy of nucleotides during incorporation into all types of DNA (Table 5.8). Only for A-DNA notable decrease of deformation energy is not found.

In the case of A-DNA amount of energy for the deformation of the geometry of nucleotides is practically equal for all molecules. However, for BI, BII, ZI, and ZII-forms of DNA (south/anti conformation of nucleotides) a difference between purine

Table 5.9 Selected geometrical parameters of methyl ethers of 2′-deoxyribonucleotides in equilibrium and DNA-like conformations, calculated at the B3LYP/ aug-cc-pvdz level

Nucleotide	Parameter	Equilibrium		DNA-like conformation				
		S/anti	N/anti	A[a]	BI[b]	BII[b]	ZI[b]	ZII[b]
mTMP	C1′–O4′ (Å)	1.426	1.415	1.415	1.417	1.426	1.419	1.414
	C4′–O4′ (Å)	1.447	1.448	1.447	1.451	1.461	1.449	1.442
	χ (deg.)	−113.5	−144.7	−150.5	−121.4	−106.7	−132.7	−143.1
	P	171.8	11.6	11.4	159.9	155.3	171	177.2
mCMP	C1′–O4′ (Å)	1.428	1.418	1.419	1.420	1.422	1.421	1.414
	C4′–O4′ (Å)	1.447	1.448	1.447	1.447	1.449	1.449	1.443
	χ (deg.)	−124.3	−153.3	−154.6	−129.9	−146.7	−137.8	−152.1
	P	171.1	6.2	8.6	159.8	165.3	170.7	180.6

[a] N/anti conformers
[b] S/anti conformers

and pyrimidine nucleotides is observed. In all cases purine nucleotides require more energy for deformation of backbone (largest difference is observed for ZI- and ZII-DNA—about 5 kcal/mol). In the case of north conformers of purine nucleotides in Z-DNA a similar to south/anti conformations values of energy are found. For the syn conformers of purine nucleotides in Z-form of DNA drastic changes between mAMP and mGMP are observed. In the case of mAMP, energy of deformation is less than 1 kcal/mol, but for mGMP the analogous energy amounts more than 10 kcal/mol.

Interestingly, the energy of deformation of isolated base pairs of: A–T and G–C in each form of DNA is almost identical. They amount to 3.6 kcal/mol for A-DNA, 10.9–11.4 kcal/mol for BI- and BII-DNA, and 15.8–17.2 kcal/mol for Z-types of DNA (Table 5.8). Thus, a formation of isolated Watson–Crick A–T and G–C base pairs of nucleotides is practically equivalent from viewpoint of energy required for deformation of nucleotides geometry. However, a presence of polar environment makes a distinction between deformation energy of nucleotides in A–T and G–C pairs (Table 5.8).

The comparison of geometrical parameters of nucleotides in equilibrium and DNA-like conformations (Tables 5.9–5.11) indicates that the changes of conformations of the considered species do not lead to their appreciable variations. The only significant difference revealed is related to the C–O bond lengths within the C4′–O4′–C1′ fragment. In the case of Z-forms of DNA considerable increase of these bond lengths ($\Delta\ell = 0.028–0.057$ Å) is predicted. This is usually explained by strengthening of anomeric interactions.

An incorporation of nucleotides into different types of DNA results in considerable changes of ribose conformation (Tables 5.9–5.11). Comparison of the values of pseudo rotation angles in equilibrium and DNA-like conformations demonstrates that these values in B-forms of DNA are systematically lower ($\Delta P = 36.9°$) as compared to the equilibrium conformation. At the same time we observe noticeable changes of the values of pseudo rotation angles among different nucleotides in similar form of DNA, and for one nucleotide in different types of DNA. These data

Table 5.10 Selected geometrical parameters of methyl ethers of 2′-deoxyribonucleotides in equilibrium, A-DNA and B-DNA like conformations calculated at the B3LYP/ aug-cc-pvdz level

Nucleotide	Parameter	Equilibrium		DNA-like conformation		
		S/anti	N/anti	A[a]	BI[b]	BII[b]
mGMP	C1′–O4′ (Å)	1.426	1.413	1.415	1.416	1.419
	C4′–O4′ (Å)	1.447	1.449	1.445	1.447	1.446
	χ (deg.)	−95.1	−133.7	−146.6	−120.9	−133.2
	P	183.4	6.3	16.7	146.3	165.4
mAMP	C1′–O4′ (Å)	1.426	1.415	1.416	1.416	1.419
	C4′–O4′ (Å)	1.446	1.449	1.446	1.447	1.447
	χ (deg.)	−99.0	−138.5	−147.5	−119.5	−137.4
	P	183.2	3.8	11.3	162.3	166.0

[a] N/anti conformers
[b] S/anti conformers

Table 5.11 Selected geometrical parameters of methyl ethers of 2′-deoxyribonucleotides in equilibrium and DNA-Z like conformations, calculated at the B3LYP/ aug-cc-pvdz level

Nucleotide	Parameter	Equilibrium		DNA-like conformation		
		S/anti	N/anti	ZI[a]	ZII[a]	ZI[b], ZII[b]	ZI[c], ZII[c]
mGMP	C1′–O4′ (Å)	1.426	1.413	1.412	1.404	1.406	1.414
	C4′–O4′ (Å)	1.447	1.449	1.460	1.457	1.457	1.464
	χ (deg).	−95.1	−133.7	−140.6	−161.0	44.9	69.1
	P	183.4	6.3	167.6	17.6	6.8	177.8
mAMP	C1′–O4′ (Å)	1.426	1.415	1.411	1.403	1.410	1.414
	C4′–O4′ (Å) (Å)	1.446	1.449	1.460	1.460	1.453	1.459
	X (deg.)	−99.0	−138.5	−150.0	−169.4	79.1	81.4
	P	183.2	3.8	171.0	2.1	36.0	169.3

[a] N/anti conformers
[b] N/syn conformers
[c] S/syn conformers

confirm the earlier suggested assumptions [19, 20, 52] that the SU plays a role of a soft buffer, changing the conformation according to the interactions between the BB and the BU.

Based on the analysis of the electron density distribution it was demonstrated that conformers with anti orientation of the base are characterized by the C6/C8–H…O5′, or the C6/C8–H…O-P hydrogen bond (Table 5.12). However, in the case of some conformations of backbone such reference hydrogen bond converts to another form. Thus, in BII-form of DNA in case of all nucleotides, the C6/C8–H…O3′ (n−1) hydrogen bond is observed. In the case of mGMP in BI-form of DNA this hydrogen bond is weaker and it can be classified as rather strong electrostatic interaction rather than an intramolecular hydrogen bond. An absence of such reference H-bonds for anti-conformers of purine nucleotides in case of ZI–ZII-form of DNA (Table 5.13) is observed. Presence of all these reference hydrogen bonds (C6/C8–H…O5′, C6/C8–H…O–P, and C6/C8–H…O3′ (n−1)) can be summarized

Table 5.12 B3LYP/aug-cc-pvdz intramolecular interaction in methyl ethers of 2′-deoxyribonucleotides in equilibrium and DNA like conformations. True hydrogen bonds are underlined

Interaction	Nucleotide					
	A-DNA		BI-DNA		BII-DNA	
	H...A (Å)	D–H...A (deg.)	H...A (Å)	D–H...A (deg.)	H...A (Å)	D–H...A (deg.)
mTMP						
C6–H...O3′ (n−1)	3.84	150.0	3.54	132.4	2.48	127.8
C6–H...O5′	2.14	173.0	2.23	163.4	2.50	159.2
C9–H...O–P	2.38	176.2	2.48	173.4	2.53	167.6
mCMP						
C6–H...O3′ (n−1)	3.75	149.8	3.39	139.6	2.16	155.2
C6–H...O5′	2.09	168.3	2.23	165.7	2.78	141.5
mAMP						
C8–H... O3′(n−1)	3.97	149.6	3.51	141.3	2.20	168.4
C8–H...O5′	2.36	155.8	2.55	144.2	3.02	132.8
C8–H...O–P	3.35	154.8	3.14	163.6	3.32	138.9
mGMP						
C8–H...O3′ (n−1)	4.13	148.6	3.98	146.2	2.26	164.2
C8–H...O5′	2.53	153.5	3.07	138.8	3.04	133.7
C8–H...O–P	3.53	160.2	3.76	174.2	3.31	143.3

as C–H…O hydrogen bond between C6/C8–H group of base and oxygen atom of backbone. In anti-conformers of purine nucleotides in ZI–ZII-forms of DNA an absence of such H-bonds may be explained by some screening of the oxygen atoms of backbone by hydrogen atoms of the C5′–H2 methyl groups, which in our investigations, corresponds to the C3′ atom of previous nucleotide.

Two other reference hydrogen bonds are observed in syn-conformers of nucleotides in equilibrium and Z-forms of DNA (Table 5.14). First of them is the N10–H…O–P hydrogen bond. This H-bond occurs in all syn-conformers of mGMP. This hydrogen bond is also found in case of north/syn conformer of mAMP. This reference H-bonds are also responsible for the stabilization of the syn-conformers of mCMP, mAMP, and mGMP. Other reference hydrogen bond is the C2′–H…N3 hydrogen bond. In case of north/syn conformer of mGMP in equilibrium, and north/syn conformers in Z-forms of DNA such bond is weaken. Therefore, it is possible to conclude that it corresponds to rather strong electrostatic interaction rather than to intramolecular hydrogen bond. These H-bonds stabilize syn-conformers when formation of the N–H…O–P hydrogen bond between the amino and phosphate groups is impossible. Interestingly, in some cases an existence of two such reference hydrogen bonds (C2′–H…N3 and N–H…O–P) at the same time is observed.

There are three specific hydrogen bonds in the structure of methyl ethers of DNTs. These are the C–H…O–P bonds between the methyl and phosphate groups

Table 5.13 B3LYP/aug-cc-pvdz intramolecular interaction in anti-conformers of methyl ethers of 2′-deoxyribonucleotides in Z-DNA like conformations. True hydrogen bonds are underlined

Interaction	Nucleotide			
	ZI-DNA		ZII-DNA	
	H…A (Å)	D–H…A (deg.)	H…A (Å)	D–H…A (deg.)
mTMP				
C6–H…O3′ (n−1)	3.67	137.6	3.72	125.5
C6-H…O5′	2.26	159.7	2.43	144.9
C9–H…O–P	–	–	2.46	164.7
mCMP				
C6–H…O3′ (n−1)	3.69	135.7	3.78	117.9
C6–H…O5′	2.28	156.0	2.48	135.4
mAMP				
C8–H…O3′ (n−1)	3.31	165.4	3.31	165.4
C8–H…O5′	3.47	143.9	3.47	143.9
C8–H…O–P	------	------	------	------
mGMP				
C8–H…O3′(n−1)	3.39	178.2	3.39	178.2
C8–H…O5′	3.67	137.4	3.67	137.4
C8–H…O–P	------	------	------	------

Table 5.14 B3LYP/aug-cc-pvdz intramolecular interaction in syn-conformers of methyl ethers of 2′-deoxyribonucleotides in Z-DNA like conformations. True hydrogen bonds are underlined

Interaction	Nucleotide			
	ZI-DNA		ZII-DNA	
	H…A (Å)	D–H…A (deg.)	H…A (Å)	D–H…A (deg.)
mAMP				
C2′–H…N3	2.67	107.0	2.36	129.3
C5′–H…N3	3.74	125.1	3.32	139.6
N10–H…O–P	–	–	–	–
mGMP				
C2′–H…N3	3.06	108.9	2.44	125.0
C5′–H…N3	2.35	140.8	2.56	141.6
N10–H…O–P	1.93	154.0	2.01	144.3
C3′–H…O–P	4.93	100.3	4.96	102.9

in north/anti conformer of mTMP. However, it should be noted that these H-bonds are rather weak and probably do not influence significantly the properties of DNTs. In syn conformer of mGMP in Z-forms of DNA the C–H…N hydrogen bond between the H5′ hydrogen atom of sugar and the N3 nitrogen atom of base is revealed. On the other hand, in south/syn conformer of mGMP the C3′–H…O–P hydrogen bond (Table 5.14) is found.

The results of calculations demonstrate that a general pattern of reference intramolecular hydrogen bonds in methyl esters of DNTs is almost the same as in species containing P–OH fragment. Only in the case of C3′-endo/anti conformer of mGMP replacement of the hydrogen atom by methyl group results in disappearance of "artificial" O3′–H...O(H)–P hydrogen bond, which cannot be formed in DNA.

The comparison of intramolecular hydrogen bonds in nucleotides for equilibrium and DNA-like conformations (Tables 5.12–5.14) demonstrates conformational dependence of hydrogen bond characteristics that is typical for weak H-bonds. A change of conformation sometimes leads to appreciable variation of reference hydrogen bonds. In particular, H-bonds between the oxygen atoms of backbone and the hydrogen atoms of sugar in DNA-like conformations are not revealed. This confirms our earlier conclusion that the effective (or specific) hydrogen bonds, because of their weakness, should be viewed as a kind of electrostatic interactions, rather than the real hydrogen bonds, as they do not affect the structure and conformational characteristics of nucleotides. Recognition of such interactions can be made on the basis of the Bader`s analysis of the electron density distribution.

5.6 Structure of Protonated 2′-Deoxorybonucleotides and Relative Stability of Conformers

Protonation, in some sense, is one of the simplest acid-base chemical reactions that are observed in both living systems and inorganic species. In case of DNA, the protonation of nucleobases significantly influences on it structure and function. In particular, the protonated cytosine makes significant contribution to the stabilization of DNA triplexes [3, 60, 61]. The protonation can also cause mutations in the DNA via mispairing of complementary bases [62–65]. It was suggested [66] that the structures of so-called rare tautomers stabilized by transition metals could also appear in complexes between protonated bases and a metal. Protonation is considered as a catalytic factor for the hydrolytic cleavage of the N-glycosidic bond [67–70], high reactivity of the C8 atom in purine bases [71–73], and it is closely related to the conformational dynamics of nucleotides [74]. Being so important, acid-base equilibrium involving nucleic acid bases has been widely studied by experimental and theoretical methods both in gas and condensed phases. There are several fundamental questions for these studies to address, namely, comparative proton affinity of different nucleobases, preferable sites of protonation within each nucleic acid base, and changes of the molecular structure and conformational characteristics of DNA constituents induced by a protonation of nucleobases.

More than 10 years ago the structure of protonated DNTs containing a neutral phosphate group was investigated only by the semiempirical AM1 method [75, 76]. These studies were focused on calculations of values of the PAs of nucleobases, without analysis of the conformation of protonated molecules. It was concluded that appearance of a phosphate group in DNTs results in a change of preferable protonation sites. In the case of neutral CMP, GMP, and AMP molecules, the highest PAs were found for the N3 atom, while in TMP the oxygen atom of the C4=O carbonyl group remains the most preferable site for protonation. In the case of anionic DNTs containing a deprotonated phosphate group, it was concluded on the basis of calculations by the AM1 method that the N7 atom has the highest PA value for GMP [77]. Other anionic nucleotides have the same preferable sites of protonation as molecules with a neutral phosphate group. In the case of anionic AMP, the highest stability of tautomer with the proton located at the N3 atom was also confirmed by calculations using DFT methods [78]. However, contrary to AM1 data, it was found that the N7 atom of adenine is the most preferable site for protonation of a molecule with a neutral phosphate group. It should be noted that analysis of the molecular structure of protonated DNTs represents a considerably more complex task as compared to nucleobases. As it was discussed above, the nucleotides can adopt several stable conformations differing in geometrical parameters and energy [19, 20, 52]. Moreover, the presence of negative charge on the phosphate group significantly influences the relative stability and geometry of molecules. Protonation of such molecules may lead to significant changes in their conformations and energetics. For example, investigation of protonated AMP indicated [78] that attachment of a proton to the N3 atom results in switching of the base orientation from anti to syn because of the formation of strong intramolecular hydrogen bond. A significant increase of strength of usually weak C-H...O hydrogen bonds was found in protonated AMP due to electrostatic attraction between the negatively charged phosphate group and protonated adenine. Taking into account these data, it is possible to assume that each tautomer of protonated nucleotide can exist in several stable conformations characterized by different energy. Therefore, comprehensive evaluation of PAs requires careful consideration of the population of conformers of non-protonated anionic DNTs and each tautomer of protonated molecules.

Before starting an analysis of the influence of a protonation on the molecular structure and relative stability of conformers of protonated molecules, it is necessary to summarize conformational characteristics of non-protonated nucleotides, discussed above. Despite the high conformational flexibility of nucleotides, they adopt only four conformational states, being incorporated into DNA macromolecules. These states are characterized by the conformation of a SU belonging to the south or north region of a pseudo rotation cycle and by the syn or anti orientation of BU. Therefore, only these conformations are considered usually for DNTs as related to DNA. Earlier, [52, 53] it was found that conformers with a syn orientation of base are absent in pyrimidine nucleotides as well as in AMP. However, in the case of CMP and AMP, it was found that minima on the potential energy surface correspond to conformers with an almost orthogonal orientation of base with respect to the C1′–H bond and geometry of the furanose ring belonging to the north region of the pseudo rotation cycle. Thus, non-protonated DNTs contain two (mTMP), three (mCMP, mAMP), or four (mGMP) stable conformers. On the basis of the relative Gibbs energy of these conformers, it is possible to conclude that for every nucleotide only one of conformers dominates in the gas phase state (Table 5.15). There is an S/anti conformer in mTMP, mCMP, and mAMP and an S/syn conformer in mGMP. The latter conformer is stabilized by a strong intramolecular N–H...O hydrogen bond between the amino and phosphate groups.

The relative stability of tautomers was calculated as the difference in average Gibbs free energies as compared to the most stable tautomer. Average Gibbs free energies were calculated using population of conformers of tautomers with Eq. (5.1):

Table 5.15 B3LYP/aug-cc-pvdz relative Gibbs free energy (kcal/mol), 298 K and population (%)

Nucleotide	Tautomer	Conformer	ΔG	P
Conformers of non-protonated 2′-deoxyribonucleotides				
mTMP		S/anti	0	100
		N/anti	4.85	0
mCMP		S/anti	0	99.96
		N/anti	3.22	0.04
		N/ort	8.22	0
mAMP		S/anti	0	100
		N/anti	6.95	0
		N/ort	12.00	0
mGMP		S/anti	8.11	0
		N/anti	16.28	0
		S/syn	0	100
		N/syn	4.18	0
Conformers of protonated tautomers of 2′-deoxyribonucleotides[a]				
mTMP-H	H7cis	S/ort	11.67	0
	H7cis	N/anti	19.61	0
	H7trans	C4′-endo/syn	0	100
	H8cis	S/ort	5.65	0
	H8cis	S/anti	6.11	0
	H8trans	S/anti	0	86.2
	H8trans	N/anti	1.09	13.8
mCMP-H	H3	S/anti	4.06	0.1
	H3	N/anti	5.35	0
	H3	C2′-exo/ort	0	99.9
	H7cis	S/anti	1.04	14.3
	H7cis	N/anti	3.55	0.2
	H7cis	C2′-exo/ort	2.06	2.6
	H7trans	N/anti	10.36	0
	H7trans	C4′-endo/syn	0	82.9
	H7trans	C2′-exo/ort	9.37	0
mGMP-H	H3	S/syn	0	100
	H7	C3′-exo/ort	0	97.5
	H7	S/syn	2.50	2.5
	H10	C3′-exo/ort	0	100
	H11	C3′-exo/ort	16.13	0
	H11	S/syn	0	100
mAMP-H	H1	S/syn	9.24	0
	H1	S/ort	0	100
	H3	C3′-exo/ort	0	100
	H7	C3′-exo/ort	0	100
	H10	C3′-exo/ort	0	100

[a] Energies for non-protonated molecules are taken from ref [53]

$$G_{av} = \sum_{i=1}^{M} G_i P_i \tag{5.1}$$

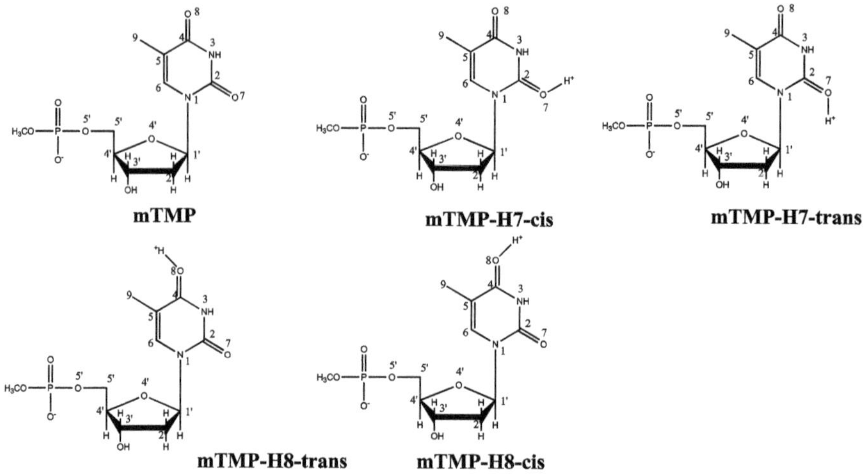

mTMP mTMP-H7-cis mTMP-H7-trans

mTMP-H8-trans mTMP-H8-cis

Fig. 5.13 Tautomers of protonated mTMP

where G_i is the calculated Gibbs free energy of the i-th conformer and P_i is the population of this conformer ($0 \leq P\ i \leq 1$). The relative stability of conformers and tautomers were calculated at 298 K.

Starting geometries of protonated nucleotides were generated from each stable conformer of non-protonated molecule by addition of a proton to the heteroatom. Therefore, it is possible to expect that every tautomer of protonated nucleotides will have two to four stable conformers. However, results of calculations demonstrated (Table 5.15) that protonation results in significant changes of the number of stable conformers. As follows from obtained results, up to six conformers are observed for tautomers of protonated pyrimidine nucleotides and only one to two conformers for tautomers of mGMP and mAMP.

5.6.1 Protonated mTMP

The mTMP anion has only two sites for protonation, namely, the oxygen atoms of carbonyl groups. However, each of these protonated tautomers possesses an additional degree of freedom caused by rotation around the C–O bond, leading to existence of conformers with cis and trans orientation of the hydrogen atom of the protonated carbonyl group with respect to the N3 atom (Fig. 5.13).

Results of calculations reveal that the most stable conformers of both tautomers of mTMP (mTMP–H7 and mTMP–H8) have the hydrogen atom of the protonated carbonyl group being oriented away from the N3 atom of the pyrimidine ring (Table 5.15). The furanose ring adopts only slightly different conformations: south for mTMP–H8trans and C4′-endo for mTMP–H7trans. The C4′-endo conformation is close to the south region of the pseudo rotation cycle. The thymine moiety has an opposite orientation: syn in mTMP–H7 trans and anti-orientation in mTMP–H8 trans. In both cases, the orientation of the base is stabilized by intramolecular hydrogen bonds, namely, strong O–H...O in the C4′-endo/syn conformer of mTMP–H7 trans and considerably weaker C–H...O bonds in the S/anti conformer of mTMP–H8 trans. It should be noted that only the last type of the hydrogen bonds is observed in all other conformers of mTMP forms. Some of such hydrogen bonds are quite strong because of opposite charge assistance, as was mentioned earlier [78]. Nevertheless, the influence of relatively strong C–H...O interactions on the relative energy of conformers is considerably smaller, as compared to the strong conventional O–H...O hydrogen bond. It is possible to conclude that the mTMP–H7 tautomer exists exclusively as the C4′-endo/syn conformer, while the conformational state of the mTMP–H8 tautomer may be described as S/anti with a minor supplement of N/anti conformer (Table 5.15).

5.6.2 Protonated mCMP

In the case of mCMP, it is possible to suggest existence of three tautomers with a protonated ring nitrogen atom, carbonyl and amino group (Fig. 5.14). However, results of calculations demonstrated that the mCMP-H8 tautomer does not correspond to a minimum on the potential energy surface. A proton transfer from the protonated amino group to the phosphate group was revealed during optimization of its molecular geometry. Therefore, only tautomers with a protonated ring nitrogen and carbonyl group should be considered. Taking into account two possible orientations of the OH bond of a protonated carbonyl group, one can conclude that tautomers of protonated mCMP have three stable conformers similar to a non-protonated molecule (Table 5.15). Only in the case of the mCMP-H7 trans tautomer, protonation leads to disappearance of the S/anti conformer because of the transition of the base from anti to syn orientation accompanied by deformation of the furanose ring.

Results of calculations show that protonation of the mCMP yields significant stabilization of conformers with orthogonal orientation of cytosine with respect to the C1′-H bond. This conformer possesses the lowest energy among the mCMP-H3 tautomers, and it has only slightly higher energy for the mCMP-H7 tautomer. For the last tautomer, the most stable conformer is the C4′-endo/syn with trans orientation of the hydrogen atom of a protonated carbonyl group with respect to the N3 atom of the pyrimidine ring (Table 5.15, Fig. 5.14). It should be noted that the conformers with the lowest energy in both tautomers are stabilized by the N–H…O (mCMP-H3) and O-H…O (mCMP-H7) hydrogen bonds. The N-H…O bonds also are found in the C2′-exo/ort conformers of the mCMP-H7 tautomer. However, its energy is considerably smaller, as compared to the C4′-endo/syn conformer of mCMP-H7trans. This allows suggesting that the strength of the N-H…O or O-H…O hydrogen bonds plays a very important role in stabilization of conformers of protonated mCMP.

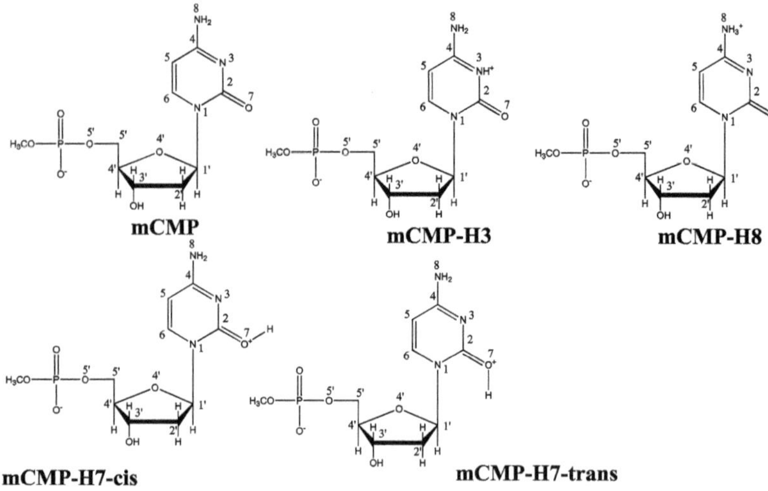

mCMP mCMP-H3 mCMP-H8

mCMP-H7-cis mCMP-H7-trans

Fig. 5.14 Tautomers of protonated mCMP

5.6.3 Protonated mGMP

In the molecule of 2′-deoxyguanosine number of possible sites of protonation is 4. This includes two nitrogen atoms in the purine ring (N3, N7), the nitrogen atom of the amino group (N10) and the carbonyl oxygen (O11). According to the stability of mGMP conformers with syn orientation of the base respectively to the furanose ring, one can expect a significant conformational variety of tautomers for protonated mGMP (Fig. 5.15). Interestingly, in the case of mGMP, protonation leads to complete disappearance of conformers with anti orientation of the guanine moiety (Table 5.15). Minima on the potential energy surface are found only for conformers with syn and orthogonal orientation of the base.

Besides that, all stable conformers of protonated mdGMP possess S or C3′-exo conformation of the furanose ring in contrast to a non-protonated nucleotide (Table 5.15). The number of stable conformers is limited to one (H3 and H10 tautomers) or two (H7 and H11 tautomers).

One may suppose that the significant decrease of conformational space of protonated mGMP is caused by the formation of strong intramolecular hydrogen bonds. Especially strong hydrogen bonding is observed between the amino and phosphate group in mGMP–H3 and mGMP–H11 tautomers. Geometrical parameters and energy of the N–H...O bonds in these conformers allows suggesting almost free transition of the hydrogen atom between interacting heteroatoms. However, only one minimum on the potential energy surface corresponding to the location of the hydrogen atom at the nitrogen of the amino group was found.

The C3′-exo/ort conformers of the mGMP–H7 and mGMP–H10 tautomers are stabilized only by the C–H...O hydrogen bonds. However, their strength is increased significantly by electrostatic interactions between the nucleobase and phosphate group possessing opposite charges. Earlier, it was demonstrated [74, 78] that

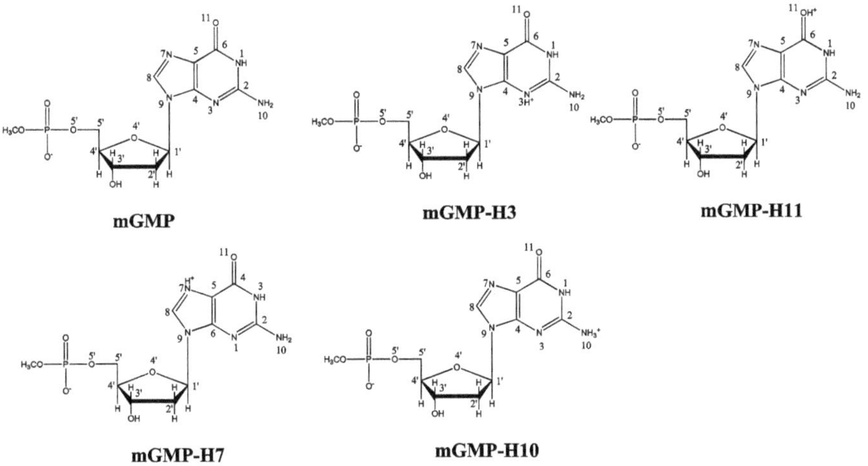

mGMP

mGMP-H3

mGMP-H11

mGMP-H7

mGMP-H10

Fig. 5.15 Tautomers of protonated mGMP

an opposite charge assistance in hydrogen bonding may lead to transformation of usually weak C–H…X hydrogen bonds into very strong bonds. Such situation is observed for the C8–H…O–P hydrogen bond in the C3′-exo/ort conformer of the mGMP–H7 tautomer. Geometrical parameters and estimated energy of bonding indicate that C—H…O bond is stronger than the quite strong conventional N1–H… O5′ hydrogen bond in the S/syn conformer of the mGMP–H3.

5.6.4 *Protonated mAMP*

In the molecule of mAMP only protonation of nitrogen atoms of purine ring is possible (N1, N3, N7), and of nitrogen atom of amino group (N10). Based on the similarity of the nitrogen atoms of purine one can expect a set of base protonated tautomers (Fig. 5.16) in different conformations. However, the results of computational study showed that mAMP protonation leads to a decrease of conformational space.

Only one stable conformer was located for all tautomers, except the mAMP–H1. All the most stable conformers have the S or C3'-exo conformation of the furanose ring with orthogonal orientation of base with respect to the C1'-H bond (Table 5.15). They are stabilized by strong intramolecular N–H…O or C–H…O hydrogen bonds, as was described earlier [78]. The second conformer of the mAMP–H1 tautomer has a syn orientation of base with respect to sugar and considerably higher energy, as compared with the S/ort conformer (Table 5.15).

More detailed analysis of geometrical parameters of protonated nucleotides demonstrates that, besides the conformation of the molecules, the protonation also results in significant changes of bond lengths within the C4′-O4′-C1′-N fragment. It is well-known that protonation is a first stage of hydrolytic cleavage of the N-glycosidic bond [69, 70, 79]. The transition state of this process is highly dissocia-

mAMP mAMP-H1 mAMP-H10

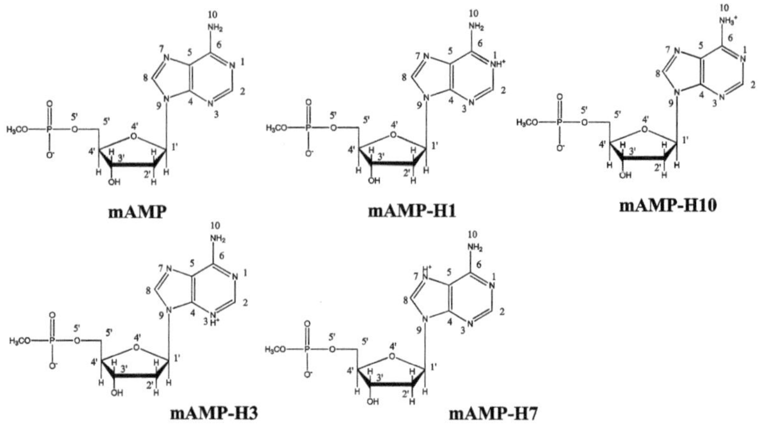

mAMP-H3 mAMP-H7

Fig. 5.16 Tautomers of protonated mAMP

tive with substantial elongation of the C1′–N bond [79–81]. Therefore, increase of the N-glycosidic bond length due to protonation creates favorable conditions for disruption of this bond. It should be noted that this process has been investigated mainly for the purine nucleotides because it represents the first step in the base excision repair pathway [82, 83]. In agreement with previous findings, [19, 20, 52, 53] the length of the N-glycosidic bond C1′–N in non-protonated purine nucleotides is smaller than that for pyrimidine ones (average values are 1.464 and 1.482 Å, respectively). This is caused by the nature of the base [18]. The same situation is observed in protonated species. The average length of the C1′-N bond in the purine nucleotides (1.481 Å) remains shorter than that in the pyrimidine nucleotides (1.519 Å). However, in all cases, the protonation results in significant elongation of this bond, causing its weakening. This effect is more pronounced in the pyrimidine nucleotides than in the mAMP and mGMP (differences between average values of the C1′–N bond lengths are $\Delta\ell = 0.037$ Å for pyrimidine and $\Delta\ell = 0.017$ Å for purine nucleotides). Results of our calculations [84] demonstrate that weakening of the N-glycosidic bond is more pronounced in the pyrimidine nucleotides than in the mAMP and mGMP. This means that cleavage of the N-glycosidic bond in the pyrimidine nucleotides should be even easier than in the purine ones, especially taking into account C1′–N bond`s length.

5.7 Proton Affinity of Nucleobases in 2′-Deoxyribonucleotides

Experimental [85–92] and theoretical [92–96] studies of PAs of nucleobases in the gas phase demonstrated clear differences in PA values of DNA bases. For nucleobases the following inequality holds PA (G) > PA (C) > PA (A) >> PA (T) [94]. In particular, PAs of thymine were found to be considerably lower (8.9–9.2 eV) as compared to guanine, cytosine, and adenine (9.6–9.9 eV). It should also be noted that there is a good agreement between experimental and theoretical data. Appearance of a sugar fragment in molecules of 2′-deoxyribonucleosides results in an increase of PAs of all nucleobases [75, 88–99]. Nevertheless, the PA value for 2′-deoxythymidine (9.8 eV) remains lower than that for other nucleosides (10.1–10.3 eV). The values of PA for nucleoside follow the trend: PA (dG) > PA (dC) ≥ PA (dA) >> PA (dT) [88]. According to experimental data [75, 76], the presence of a neutral phosphate group in DNTs does not influence the PAs of the nucleobases. For instance, the value of PA for TMP is 9.7–9.8 eV and it amount to 10.1–10.3 eV for CMP, GMP, and AMP. Whereas for nucleotides the analogous trend is as follows: PA (GMP) ≈ PA (CMP) ≈ PA (AMP) >> PA (TMP) [96]. In general, these data agree well with results of semiempirical quantum-chemical calculations by the AM1 method [75].

Further increase of the PA of the nucleobases was found in anions of DNTs [77]. As it was expected, appearance of negative charge due to deprotonation of the phosphate group results in an increase of the PA values by 2.7–2.9 eV. However, similar to results for isolated bases, the base in the TMP anion has the lowest PA. Besides that, the PA value for CMP also becomes slightly lower than that for AMP and GMP (ΔPA ≈ 0.3 eV). In the discussed here investigations, P–O group of phosphate is not considered as a protonation sites. The PA of P–O is always much higher than PA of any nucleobase atom.

It should be noted some experimental problems in study of protonation of nucleotides. At the first stage the protonation of phosphate (PU) always happens. This is explained by the significant difference in the values of PA of nucleobases protonation sites and PU. Therefore, the experimental data reflect the process of protonation of neutral forms of deoxyribonucleotides. But it is well known, that *in vivo*, nucleotides have a negative charge, which is balanced by the metal cations (Na^+or K^+). Thus, the issue of nucleobases protonation in monoanionic DNTs remains unanswered in experimental investigations.

According to semiempirical AM1 study [75, 76] of PA of protonated DNTs containing a neutral phosphate group it was concluded that appearance of a phosphate group in DNTs results in a change of preferable protonation sites. In the case of neutral CMP, GMP, and AMP molecules, the highest PA were found for the N3 atom, while in TMP the oxygen atom of the C4=O carbonyl group remains the most preferable site for protonation. In the case of anionic DNTs containing a deprotonated phosphate group, it was concluded on the basis of calculations by the AM1 method that the N7 atom has the highest PA value for GMP [77]. Other anionic nucleotides have the same preferable sites of protonation as molecules with a neutral phosphate group. In the case of anionic AMP, the highest stability of tautomer with the proton located at the N3 atom was also confirmed by calculations using DFT methods [78]. However, contrary to AM1 data, it was found in the DFT study that the N7 atom of adenine is the most preferable site for protonation of a molecule with a neutral phosphate group.

Taking into account that fact one concludes that protonation of anionic nucleotides may lead to significant changes in their conformations and energetic. It is possible to assume that each tautomer of protonated nucleotide can exist in several stable conformations characterized by different energy. Therefore, comprehensive evaluation of PA requires careful consideration of the population of conformers of non-protonated anionic DNTs and each tautomer of protonated molecules and should be calculated using the analysis of the population of the stable conformers.

Taking into account that each tautomer exists in several conformations, the observed PA represents a average value. Therefore, a PA for every protonation site was calculated using Eq. (5.2):

$$PA = -\left[\sum_{i=1}^{M}\left(E_{tot}\left(mXMP-H\right)_i + E_{corr}\left(mXMP-H\right)_i\right)P_i - \sum_{j=1}^{N}\left(E_{tot}\left(mXMP\right)_j + E_{corr}\left(mXMP\right)_j\right)P_j\right] + 5/2RT \tag{5.2}$$

where E_{tot} is the total energy of protonated (mXMP-H) and non-protonated (mXMP) nucleotide obtained from DFT calculations, E_{corr} is the thermal correction to enthalpy, and the term of 5/2RT includes ΔnRT for acid–base reaction and translational energy of proton. M represents the number of conformers of a protonated tautomer of nucleotide, N is the number of conformers of non-protonated nucleotide, and P is the population of each conformer calculated using a Boltzmann distribution function ($0 \leq P\,i \leq 1$).

The PA of nucleobases in DNTs was calculated in the same way, taking into account the population of all tautomers of protonated nucleotide using the following Eq. (5.3):

$$PA = -\left[\sum_{n=1}^{K}P_n\left(\sum_{i=1}^{M}(E_{tot}\left(mXMP-H\right)_i + E_{corr}\left(mXMP-H\right)_i\right)P_i - \sum_{j=1}^{N}\left(E_{tot}\left(mXMP\right)_j + E_{corr}\left(mXMP\right)_j\right)P_j)\right] + 5/2RT \tag{5.3}$$

where K is the number of tautomers for protonated nucleotide and P_n is the population of each tautomer. PA values for all conformers and tautomers were calculated at 298 K.

The analysis of population of different conformers of monoanionic DNTs showed that in the gas phase, they exist in form of only one conformer. This is S/anti conformer for mTMP, mCMP, mAMP and S/syn-conformer for mGMP. The appearance of the polar environment, simulated with PCM model, leads to larger conformational space. The greatest population in this space has N/anti conformer for mTMP, mCMP and mAMP. S/syn-conformer retains a dominant position for

Table 5.16 B3LYP/aug-cc-pvdz level relative energies (ΔE, kcal/mol), populations (P, %), and proton affinities (PA, kcal/mol) for 2′-dexyrobonucleotides in gas phase and hydrated state (PCM)

Molecule	Tautomer	ΔE	P	PA	P (PCM)	PA (PCM)
mTMP	H7	0	65.92	12.07	0	11.26
	H8	0.39	34.07	12.06	100	11.47
mCMP	H3	0	99.98	13.33	100	12.05
	H7	5.0	0.02	12.77	0	11.77
mGMP	H3	0	100.0	13.59	99.4	12.40
	H7	12.93	0.0	13.19	0.6	12.29
	H10	74.83	0.0	10.51	0	11.58
	H11	19.03	0.0	12.58	0	11.68
mAMP	H1	14.84	0.0	13.18	8.3	12.31
	H3	0	100.0	13.30	91.65	12.37
	H7	11.14	0.0	12.82	0.03	12.17
	H10	31.96	0.0	11.29	0	11.72

mGMP. However, its population is reduced by more than half. The protonation of the nitrogen atoms in the gas phase always yields one dominant conformer. Protonation of the oxygen atom of the carbonyl group leads to a mixture of two conformers. The exception is only in the case of tautomers H7 and H11 for mTMP mGMP. The appearance of the polar environment leads to a mixture of conformers for all possible tautomers.

Results of calculations of the relative stability of tautomers in protonated DNTs demonstrate an existence of the tautomeric equilibrium only in protonated mTMP molecule (Table 5.16). Other protonated nucleotides have only one stable tautomer. This especially concerns the purine nucleotides where differences in energy between the most stable H3 tautomer and other tautomers are higher than 10 kcal/mol. The analogous tautomer is also the most stable form for the mCMP. However, in the case of this nucleotide, the difference in energy between the H3 and H7 tautomers is considerably smaller. For mCMP, mGMP and mAMP the greatest value of PA is associated with a nitrogen atom N3, and protonated nucleotides exist in form of tautomers H3. For mTMP, in the gas phase protonated nucleotide exists as a mixture of tautomers H7 and H8. Thus, the most preferable protonation sites of anionic nucleotides are the same as for neutral nucleotides [75, 76]. In contrast to previous conclusions based on AM1 data, [77] the more accurate DFT level calculations reveals that the N7 atom of the mGMP has a significantly smaller PA value as compared to the N3 site.

In general, the highest PA values in the gas phase and in a polar environment have a nucleobase of mGMP. In the gas phase values of PA for mCMP and mAMP tautomers are almost identical, but in the polar environment the PA of mAMP is slightly higher. The smallest PA is predicted for mTMP tautomers. It should be noted that the PA values for mCMP, mGMP and mAMP tautomers are very close and significantly higher than those for mTMP. This is in agreement with experimental data. Thus one concludes that purine bases are more attractive to protonation than pyrimidine.

Table 5.17 Proton affinities (eV) of nucleobases, nucleosides, neutral, and anionic nucleotides obtained from theoretical and experimental studies

Molecules	Studies	Thy	Cyt	Gua	Ade
Nucleobases	Exp. Ref. [88]	9.09	9.79	9.86	9.72
	Exp. Ref. [89]	9.05	9.70	9.67	9.69
	MP4(SDTQ)/6-31+G(d,p)// MP2/6-31+G(d,p), Ref. [96]	8.94	9.91	9.86	9.75
Nucleosides	Exp. Ref. [88]	9.75	10.11	10.16	10.13
Neutral nucleotides	Exp. Ref. [75]	9.72	10.27	10.28	10.29
Anionic Nucleotides	B3LYP/aug-cc-pvdz	12.07	13.33	13.59	13.30

On the basis of PAs and populations of the individual tautomers of each protonated molecule, it is possible to calculate the PAs for anionic nucleotides as whole. According to these values, the mGMP possesses the highest PA and the mTMP has the lowest PA value (Table 5.17). PAs of the mCMP and mGMP are almost the same. Comparison of the PA values of nucleic acid bases in different molecules (Table 5.17) indicates that the PA depends on the presence of the substituent and its charge. Appearance of the sugar moiety in nucleosides or the sugar −phosphate substituent in neutral nucleotides results in an increase of the PA values for all bases. Further increase of the PA is observed for anionic nucleotides (Table 5.17). Therefore, one may suggest that the variation of the degree of neutralization of negative charge of the phosphate group of DNA nucleotides may be used as a tool for tuning of the proton affinity of base.

5.8 An Unusually Strong Intramolecular Hydrogen Bonds In The Protonated 2′-Deoxorybonucleotides

As mentioned above, the protonation of a DNTs base in some cases leads to abnormal hydrogen bonds C-H...O. Usually, such bonds represent weak hydrogen bonds and are characterized by the O...H distance of about 2.1–2.5 Å. However, in some tautomers of protonated nucleotides this distance is less than 2 Å. The shortest bonds are observed in purine rings, where the H...O distance is reduced to 1.6 Å, which is typical bond length for strong classical hydrogen bond. This is also in agreement with energy of a hydrogen bond (more than 10 kcal/mol), calculated from the characteristics of the BCP. Such abnormally strong intramolecular hydrogen C–H...O bonds have been discussed in [78] from the point of view of the possible tautomeric transitions and their impact on the conformational dynamics of nucleotides. The results of calculations showed that protonation of N7 atom in the mAMP molecule leads to the strengthening of the C8–H...O hydrogen bond. The obvious reason for this unusual behavior of the C8–H...O bond is electrostatic attraction between oppositely charged nucleobases and phosphate groups. Thus, this bond becomes the strongest, as measured by the geometric parameters and the characteristics of the electron density distribution.

Table 5.18 The geometrical parameters of the intramolecular hydrogen bonds in the monoanionic nucleotide mAMP and its protonated tautomers mAMP-H7 and mAMP-H7* calculated at the B3LYP/aug-cc-pVDZ level

Hydrogen bond	Parameter	mAMP	mAMP-H7	mAMP-H7*
C8–H…O–P	H…O (Å)	2.281	1.640	1.670
	C–H…O (deg.)	164.6	168.9	175.5
C3′–H…O–P	H…O (Å)	2.365	2.398	2.398
	C–H…O (deg.)	140.0	130.0	130.0

The geometrical parameters (Table 5.18) of the C8–H...O–P hydrogen bond in mAMP-H7 tautomer make it possible to characterize this interaction as a strong hydrogen bond with characteristics close to the classical O–H...O and N–H...O hydrogen bonds. In the case of C–H...O hydrogen bonds, these features suggest the unique event of the hydrogen bond strengthening. Due to the presence in the considered species of the strong electrostatic interactions these hydrogen bonds can be classified as opposite charges assisted hydrogen bonds.

To verify that feature, we have manually transferred the proton to create the C8–H…O–P→C8…HOP type of proton transfer. The obtained in this way structure was optimized and verified for the presence of imaginary frequencies. No imaginary frequencies have been found. Therefore, another local minimum having C8…HOP type of hydrogen bond has been established. An analysis of energy profile for this process as a function of the H…O distance (Fig. 5.17) also reveals existence of the second minimum on the potential energy surface corresponding to C8…H–OP tautomer (mAMP-H7*) lying 3.3 kcal/mol higher as compared to mAMP-H7 tautomer. We have also found that the total energy difference between a local minimum and the transition state for proton transfer process is 3.6 kcal/mol in forward direction, and just 0.33 kcal/mol in reverse direction.

As we already mentioned, the value of barrier of reverse proton transfer C8…H–OP→C–H…OP is very small. Therefore, in order to verify a population of this minimum on the potential energy surface we have applied the approximation which we have already used earlier [100]. It based on the well known quantum-chemical conclusion that minimum is really populated, if at least one vibrational level could be placed inside of it. Otherwise, such an area on potential surface should be considered as an area of large amplitude vibration. The results of calculations reveal that energy of ground vibrational level for the O–H stretching vibration of mAMP-H7* tautomer is 3.2 kcal/mol. This indicated that the considered level lies above barrier of proton transfer from the oxygen to the carbon atoms. Moreover, an analysis of energy of vibrational levels for stretching vibration of the C8–H bond in mAMP-H7 tautomer reveals that energy of ground vibrational level (3.7 kcal/mol) is also higher than the barrier of tautomeric transition [78]. Thus, one can expect that mAMP-H7 species might exist as the structure when the hydrogen atom possesses barrierless motion between the C8 and oxygen atom of the phosphate group in the wide range of the distances. Accordingly, protonation of N7 atom significantly increases the reactivity of adenine, especially at the C8 atom.

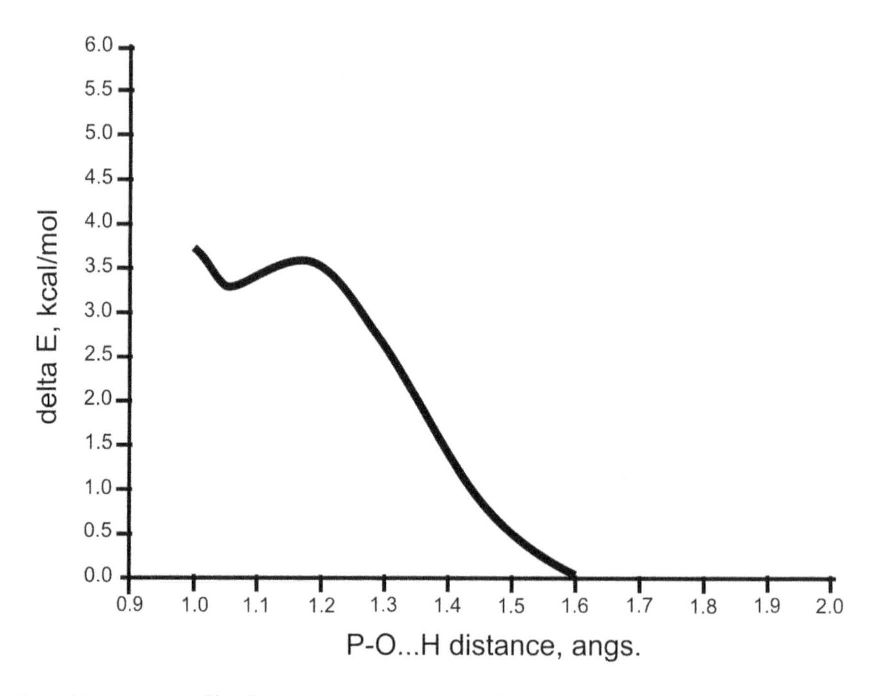

Fig. 5.17 Energy profile of proton transfer in mAMP-H7 tautomer (B3LYP/aug-cc-pvdz)

Interestingly, analysis of published data showed that the carbon atom C8 is the most preferred for bonding with various chemical agents, particularly in the oxidation reactions [101, 102]. The process of oxidative damage of DNA chain is one of the most likely causes of mutagenesis and carcinogenesis [103]. That is why the activation of C8 carbon molecules in adenine is a negative factor for the functioning of the DNA chain. Taking into account the high-performance system of DNA repairing, one can assume that the protonation of the nitrogen atom N7 at nucleotide mAMP should lead to some deformation of the geometry of the molecule to avoid the formation of a strong hydrogen bond C8–H…O, in order to deactivate the carbon atom C8.

Rotation of nucleobase around the glycosidic bond is one of the easiest ways to deactivation. *Ab initio* molecular dynamics Car-Parrinello (CPMD) studies show that the tautomer mAMP-H7 is stable during the first 3.5 ps of simulation [74]. This indicates that this tautomer corresponds to the actual minimum of the potential energy surface. After 3.5 ps mAMP-H7 carries out significant conformational changes, which are expressed in a rotation of nucleobase around the glycosidic bond. This leads to the disappearance of the hydrogen bond C8–H…O. Thus, the interaction between the oxygen atom of phosphate and the hydrogen atom of protonated nitrogen (N7) significantly increase. Further rotation around the glycosidic bond leads to notable increase in the interaction PO…H–N7, which makes possible proton transfer to the oxygen atom. This process corresponds to the transformation of mAMP-H7 into

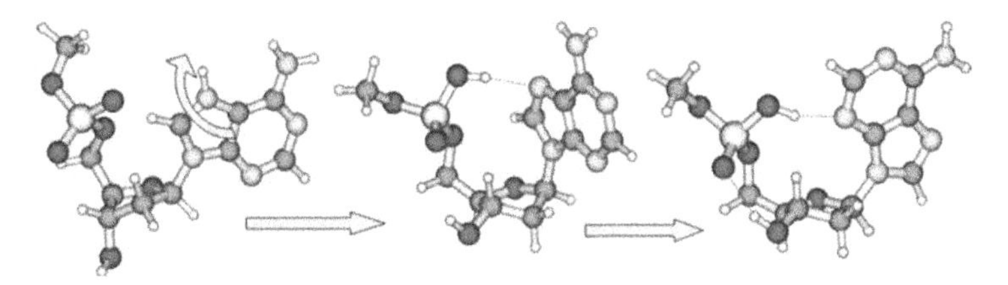

Fig. 5.18 B3LYP/aug-cc-pVDZ level predicted structures of tautomers of protonated mAMP, from *left* to *right*: mAMP-H7, mAMP-OH1 and mAMP-OH2

another tautomer—mAMP-OH1 (Fig. 5.18). This process is accompanied by significant conformational changes. B3LYP/aug-cc-pVDZ geometry optimization of this new tautomers showed that this structure corresponds to a minimum on the potential energy surface [74]. CPMD simulation at 300K revealed that the energy barrier of this transformation on the potential energy surface is about 1 kcal/mol. This value is less than kT calculated with static quantum-chemical method (about 2.1 kcal/mol). The obtained results clearly show that syn to anti conformational change of adenine causes the proton transfer and further modifications of the nucleotide.

5.9 Summary

During the last few decades computational studies become an important source of information related to DNA fragments. Based on the data obtained from *ab initio* and DFT investigations vital data related to biomolecules have emerged.

Among the frequently studied DNA components are nucleotides and nucleosides. A number of the DNA-like conformations of the isolated nucleotides have been revealed. The S/anti conformation is the most favorable for 2′-deoxythymidine monophosphate (mTMP), 2′-deoxycytidine (mCMP), and 2′-deoxyadenosine (mAMP); S/syn conformation is favorable for 2′-monophosphate deoxyguanosine (mGMP). Conformers with the syn orientation of nucleobase, relative to the sugar moiety, exist only in the case of mGMP. However, conformers with an orthogonal orientation of the nucleobase relative to the sugar moiety are found in mCMP mAMP molecules.

All conformations of deoxyribonucleotides are stabilized by weak C-H...O hydrogen bonds. Based on the obtained results, the existing hydrogen bonds may be divided into two groups: reference (stable) hydrogen bonds encountered in many conformations of all DNTs and specific (effective) hydrogen bonds that exist only for given (instantaneous) nucleotide conformation and are easily destroyed by the small changes in the conformation of the molecule. The differences between these two types of hydrogen bonds could be derived from characteristics of the bond critical point (3, − 1) of electron density distribution. Reference hydrogen bonds are also responsible for stabilization of the syn-conformers of CMP, AMP, and GMP.

The presence of polar environment around the deoxyribonucleotides significantly affects the relative stability of their conformers. In the case of mTMP and mCMP molecules the N/anti-conformation is most advantageous in terms of the relative energy, whereas in the case of mAMP and mGMP the S/syn conformation is the most preferable. The difference in conformer's energies changes noticeable in polar solvent.

An incorporation of nucleotides into A-DNA macromolecules requires the smallest deformation energy. Nucleotides mAMP and mGMP undergo the greatest deformations, during their incorporation into Z-DNA. Change of DNTs conformation causes switch between different types of intramolecular H-bonds and results in different energetic effects for purine and pyrimidine during their incorporation into DNA structure. Every type of DNA possesses unique set of intramolecular hydrogen bonds in nucleotides.

Protonation of nucleobases in anions of canonical 2′-deoxyribonucleotides demonstrated that this process leads to significant decrease of conformational space of purine nucleotides. Interestingly, almost all conformers found for non-protonated molecules correspond to minima of the potential energy surface for protonated mTMP and mCMP. However, in all nucleotides, only one conformer is populated. It concerns all tautomers of protonated molecules except of mTMP and mCMP with the proton attached to the carbonyl group. In these two cases also a minor population of second conformer is observed. Protonation of nucleobase leads to significant elongation of the N-glycosidic bond. These findings agree well with suggestions that protonation of nucleobase is a first step in cleavage of the glycosidic bond. The oxygen atoms of both carbonyl groups of thymine and the N3 atom of the pyrimidine ring of cytosine, guanine, and adenine represent the most preferable sites for protonation of anions of 2′-deoxyrobonucleotides. The highest proton affinity is observed for the base in mGMP and the lowest for the thymine moiety in mTMP. It should be noted that calculated values of the proton affinities in anionic nucleotides are significantly higher (by 2–3 eV) than for nucleosides and neutral nucleotides. This emphasizes that the proton affinity of the base in DNA macromolecule may be tuned by changing the extent of shielding or neutralization of negative charge of the phosphate group.

Relationship between the conformational dynamics of nucleotides and their tautomeric transitions demonstrated that deprotonization of nucleobase is carried out due to rotation of nucleobase around the glycosidic bond. This leads to the proton transfer from the nitrogen atom of nucleobase to the oxygen atom of the phosphate group. Such deformation of the geometry of the molecule prevent the formation of a strong C8–H...O hydrogen bond, in order to deactivate the nucleobase`s carbon atom C8. This carbon is the most preferred for bonding in oxidation reaction, particularly in the process of oxidative damage of DNA chain.

Acknowledgments The authors thank the National Science Foundation for financial support through NSF/CREST Award (HRD-0833178). This research was supported in part by the Extreme Science and Engineering Discovery Environment (XSEDE) by National Science Foundation grant number OCI-1053575 and XSEDE award allocation number DMR110088. Authors thank to the Mississippi Center for Supercomputer Research (Oxford, MS) for the generous allotment of computer time. The support from computational facilities of joint computational cluster of SSI "Institute for Single Crystals" and Institute for Scintillation Materials of National Academy of Science of Ukraine incorporated into Ukrainian National Grid is gratefully acknowledged.

References

1. Saenger W (1988) Principles of nucleic acid structures. Springer, New York
2. Neidle S (1994) DNA Structure and recognition. Oxford University Press, Oxford
3. Sinden RSR (1994) DNA structure and function. Academic Press, San Diego
4. Hecht SM (1996) Bioorganic chemistry: nucleic acids. Oxford University Press, Oxford
5. Chu CK, Baker DC (1993) Nucleosides and nucleotides as antitumor and antiviral agents. Plenum Press, New York
6. Sato T (1984) Structure of calcium thymidine 5′-phosphate dihydrate, Ca2+.C10H13N2O8P2–.2H2O. Acta Crystallogr Sect C Cryst Struct Commun 40:736–738. doi:10.1107/S0108270184005539
7. Trueblood KN, Horn P, Luzzati V (1961) The crystal structure of calcium thymidylate. Acta Crystallogr 14:965–982. doi:10.1107/S0365110X61002801
8. Lalitha HN, Ramakumar S, Viswamitra MA (1989) Structure of 5-methyl-2′-deoxycytidine 5′-monophosphate dihydrate. Acta Crystallogr Sect C Cryst Struct Commun 45:1652–1655. doi:10.1107/S0108270189005445
9. Jardetsky O, Roberts GCK (1981) NMR in molecular biology. Academic Press, New York
10. Sundaralingam M (1973) Conformation of biological molecules and polymers. Jerus Symp Quant Chem Biochem 5:417.
11. Sundaralingam M (1975) Structure and conformation of nucleic acid and protein–nucleic acid interactions. University of Baltimore, Baltimore
12. Foloppe N, Hartmann B, Nilsson L, MacKerell AD (2002) Intrinsic conformational energetics associated with the glycosyl torsion in DNA: a quantum mechanical study. Biophysical J 82:1554–1569. doi:10.1016/S0006-3495(02)75507-0
13. Leulliot N, Ghomi M, Scalmani G, Berthier G (1999) Ground state properties of the nucleic acid constituents studied by density functional calculations. I. Conformational features of ribose, dimethyl phosphate, uridine, cytidine, 5′-methyl phosphate-uridine, and 3′-methyl phosphate-uridine. J Phys Chem A 103:8716–8724. doi: 10.1021/jp9915634
14. Leulliot N, Ghomi M, Jobic H et al (1999) Ground state properties of the nucleic acid constituents studied by density functional calculations. 2. Comparison between calculated and experimental vibrational spectra of uridine and cytidine. J Phys Chem B 103:10934–10944. doi:10.1021/jp9921147
15. Hocquet A, Leulliot N, Ghomi M (2000) Ground-state properties of nucleic acid constituents studied by density functional calculations. 3. Role of sugar puckering and base orientation on the energetics and geometry of 2′-deoxyribonucleosides and ribonucleosides. J Phys Chem B 104:4560–4568. doi:10.1021/jp994077p
16. Foloppe N, MacKerell AD (1999) Intrinsic conformational properties of deoxyribonucleosides: implicated role for cytosine in the equilibrium among the A, B, and Z forms of DNA. Biophysical J 76:3206–3218. doi:10.1016/S0006-3495(99)77472-2
17. Gaigeot M-P, Leulliot N, Ghomi M et al (2000) Analysis of the structural and vibrational properties of RNA building blocks by means of neutron inelastic scattering and density functional theory calculations. Chem Phys 261:217–237. doi:10.1016/S0301-0104(00)00224-X
18. Shishkin OV, Pelmenschikov A, Hovorun DM, Leszczynski J (2000) Molecular structure of free canonical 2′-deoxyribonucleosides: a density functional study. J Mol Struct 526:329–341. doi:10.1016/S0022-2860(00)00497-X
19. Shishkin O V, Gorb L, Zhikol OA, Leszczynski J (2004) Conformational analysis of canonical 2-deoxyribonucleotides. 1. Pyrimidine nucleotides. J Biomol Struct Dyn 21:537–554. doi:10.1080/07391102.2004.10506947

20. Shishkin OV, Gorb L, Zhikol OA, Leszczynski J (2004) Conformational analysis of canonical 2-deoxyribonucleotides. 2. Purine nucleotides. J Biomol Struct Dyn 22:227–244. doi:10.1080/07391102.2004.10506998
21. Gorb L, Shishkin O, Leszczynski J (2005) Charges of phosphate groups. A role in stabilization of 2′-deoxyribonucleotides. A DFT investigation. J Biomol Struct Dyn 22:441–454. doi: 10.1080/07391102.2005.10507015
22. Šponer J, Leszczynski J, Hobza P (1996) Structures and energies of hydrogen-bonded DNA base pairs. A nonempirical study with Inclusion of electron correlation. J Phys Chem 100:1965–1974. doi:10.1021/jp952760f
23. Šponer J, Leszczyński J, Hobza P (1996) Nature of nucleic acid–base stacking: nonempirical ab initio and empirical potential Characterization of 10 stacked base dimers. comparison of stacked and h-bonded base pairs. J Phys Chem 100:5590–5596. doi:10.1021/jp953306e
24. Šponer J, Burda J V., Sabat M et al (1998) Interaction between the guanine–cytosine Watson–Crick DNA base pair and hydrated group IIa (Mg 2+, Ca 2+, Sr 2+, Ba 2+) and group IIb (Zn 2+, Cd 2+, Hg 2+) metal cations. J Phys Chem A 102:5951–5957. doi:10.1021/jp980769m
25. Řeha D, Kabeláčı M, Ryjáčĺek F et al (2002) Intercalators. 1. nature of stacking interactions between intercalators (ethidium, daunomycin, ellipticine, and 4′,6-diaminide-2-phenylindole) and DNA base pairs. Ab initio quantum chemical, Density functional theory, and empirical potential study. J American Chem Soc 124:3366–3376. doi:10.1021/ja011490d
26. Spacková N, Cheatham TE, Ryjácek F et al (2003) Molecular dynamics simulations and thermodynamics analysis of DNA–drug complexes. Minor groove binding between 4′,6-diamidino-2-phenylindole and DNA duplexes in solution. J Am Chem Soc 125:1759–1769. doi:10.1021/ja025660d
27. Shishkin O V, Gorb L, Luzanov A V et al (2003) Structure and conformational flexibility of uracil:a comprehensive study of performance of the MP2, B3LYP and SCC-DFTB methods. J Mol Struct: Theochem 625:295–303. doi:10.1016/S0166-1280(03)00032-0
28. Gorb L, Kaczmarek A, Gorb A et al (2005) Thermodynamics and kinetics of intramolecular proton transfer in guanine. Post Hartree-Fock study. J Phys Chem B 109:13770–13776. doi:10.1021/jp050394m
29. Isayev O, Furmanchuk A, Shishkin O V et al (2007) Are isolated nucleic acid bases really planar? A Car-Parrinello molecular dynamics study. J phys chem B 111:3476–3480. doi:10.1021/jp070857j
30. Samijlenko SP, Yurenko YP, Stepanyugin A V, Hovorun DM (2010) Tautomeric equilibrium of uracil and thymine in model protein–nucleic acid contacts. Spectroscopic and quantum chemical approach. J phys chem B 114:1454–1461. doi:10.1021/jp909099a
31. Shishkin O V, Palamarchuk G V, Gorb L, Leszczynski J (2006) Intramolecular hydrogen bonds in canonical 2′-deoxyribonucleotides: an atoms in molecules study. J phys chem B 110:4413–4422. doi:10.1021/jp056902+
32. Slavícek P, Winter B, Faubel M et al (2009) Ionization energies of aqueous nucleic acids: photoelectron spectroscopy of pyrimidine nucleosides and ab initio calculations. J Am Chem Soc 131:6460–6467. doi:10.1021/ja8091246
33. Sapse DS, Champeil É, Maddaluno J et al (2008) An ab initio study of the interaction of DNA fragments with methyllithium. C R Chim 11:1262–1270. doi:10.1016/j.crci.2008.04.009
34. Yurenko YP, Zhurakivsky RO, Ghomi M et al (2007) How many conformers determine the thymidine low-temperature matrix infrared spectrum? DFT and MP2 quantum chemical study. J Phys Chem B 111:9655–9663. doi:10.1021/jp073203j
35. Yurenko YP, Zhurakivsky RO, Ghomi M et al (2007) Comprehensive conformational analysis of the nucleoside analogue 2′-beta-deoxy-6-azacytidine by DFT and MP2 calculations. J Phys Chem B 111:6263–6271. doi:10.1021/jp066742h
36. Shishkin OV, Pelmenschikov A, Hovorun DM, Leszczynski J (2000) Molecular structure of free canonical 2′-deoxyribonucleosides: a density functional study. J Mol Struct 526:329–341. doi:10.1016/S0022-2860(00)00497-X
37. Hobza P, Šponer J (1999) Structure, energetics, and dynamics of the nucleic acid base pairs: nonempirical ab initio calculations. Chem Rev 99:3247–3276. doi:10.1021/cr9800255

38. Sponer J, Leszczynski J, Hobza P (2002) Electronic properties, hydrogen bonding, stacking, and cation binding of DNA and RNA bases. Biopolymers 61:3–31. doi:10.1002/1097-0282 (2001) 61:1<3::AID-BIP10048>3.0.CO;2–4
39. Svozil D, Kalina J, Omelka M, Schneider B (2008) DNA conformations and their sequence preferences. Nucleic acids res 36:3690–3706. doi:10.1093/nar/gkn260
40. Schneider B, Neidle S, Berman HM (1997) Conformations of the sugar-phosphate backbone in helical DNA crystal structures. Biopolymers 42:113–124. doi:10.1002/(SICI)1097-0282(199707)42:1<113::AID-BIP10>3.0.CO;2–O
41. Drew HR, Wing RM, Takano T et al (1981) Structure of a B-DNA dodecamer: conformation and dynamics. PNAS 78:2179.
42. Grzeskowiak K, Yanagi K, Privé GG, Dickerson and RE (1991) The structure of B-helical C-G-A-T-C-G-A-T-C-G and comparison with C-C-A-A-C-G-T-T-G-G. The effect of base pair reversals. J Biol Chem 266:8861.
43. Shishkin OV, Gorb L, Luzanov AV et al (2003) Structure and conformational flexibility of uracil: a comprehensive study of performance of the MP2, B3LYP and SCC-DFTB methods. J Mol Struct: Theochem 625:295–303. doi:10.1016/S0166-1280(03)00032-0
44. Bader RFW (1990) Atoms in molecules. A quantum theory. Clarendon, Oxford
45. Popelier PLA (1998) Characterization of a dihydrogen bond on the basis of the electron density. J Phys Chem A 102:1873–1878. doi: 10.1021/jp9805048
46. Koch U, Popelier PLA (1995) Characterization of C–H–O hydrogen bonds on the basis of the charge density. J Phys Chem 99:9747–9754. doi:10.1021/j100024a016
47. Cremer D, Kraka E, Slee TS et al (1983) Description of homoaromaticity in terms of electron distributions. J Am Chem Soc 105:5069–5075. doi:10.1021/ja00353a036
48. Hocquet A (2001) Intramolecular hydrogen bonding in 2′-deoxyribonucleosides: an AIM topological study of the electronic density. Phys Chem Chem Phys 3:3192–3199. doi:10.1039/b101781k
49. Zefirov YV., Zorky PM (1989) No title. Uspekhi Khimii (Russian Chemical Rev) 58:713.
50. Thomas Steiner (2002) The hydrogen bond in the solid state. Angew Chem Int Ed 41:48. doi:10.1002/1521-3773(20020104)41:1<48::AID-ANIE48>3.0.CO;2–U
51. Hocquet A, Ghomi M (2000) The peculiar role of cytosine in nucleoside conformational behaviour: hydrogen bond donor capacity of nucleic bases. Phys Chem Chem Phys 2:5351–5353. doi:10.1039/b007246j
52. Shishkin OV, Palamarchuk GV, Gorb L, Leszczynski J (2006) Intramolecular hydrogen bonds in canonical 2′-deoxyribonucleotides: an atoms in molecules study. J Phys Chem B 110:4413–4422. doi:10.1021/jp056902+
53. Palamarchuk GV, Shishkin OV, Gorb L, Leszczynski J (2009) Dependence of deformability of geometries and characteristics of intramolecular hydrogen bonds in canonical 2′-deoxyribonucleotides on DNA conformations. J Biomol Struct Dyn 26:653–662. doi:10.1080/07391102.2009.10507279
54. Lee C, Yang W, Parr RG (1988) Development of the Colle-Salvetti correlation-energy formula into a functional of the electron density. Phys Rev B 37:785–789. doi:10.1103/PhysRevB.37.785
55. Becke AD (1988) Density-functional exchange-energy approximation with correct asymptotic behavior. Phys Rev A 38:3098–3100. doi: 10.1103/PhysRevA.38.3098
56. Parr RG, Yang W (1989) Density functional theory of atoms and molecules. Oxford University Press, New York
57. Cammi R, Mennucci B, Tomasi J (2000) Fast evaluation of geometries and properties of excited molecules in solution: a Tamm-Dancoff model with application to 4-dimethylaminobenzonitrile. J Phys Chem A 104:5631–5637. doi:10.1021/jp000156l
58. Cossi M, Scalmani G, Rega N, Barone V (2002) New developments in the polarizable continuum model for quantum mechanical and classical calculations on molecules in solution. J Chem Phys 117:43. doi: 10.1063/1.1480445
59. Miertuš S, Scrocco E, Tomasi J (1981) Electrostatic interaction of a solute with a continuum. A direct utilizaion of AB initio molecular potentials for the prevision of solvent effects. Chem Phys 55:117–129. doi:10.1016/0301-0104(81)85090-2

60. Lee H-T, Khutsishvili I, Marky LA (2010) DNA complexes containing joined triplex and duplex motifs: melting behavior of intramolecular and bimolecular complexes with similar sequences. J Phys Chem B 114:541–548. doi:10.1021/jp9084074
61. Frank-Kamenetskii MD, Mirkin SM (1995) Triplex DNA structures. Annu Rev Biochem 64:65–95. doi:10.1146/annurev.bi.64.070195.000433
62. Jissy AK, Datta A (2010) Designing molecular switches based on DNA–base mispairing. J Phys Chem B 114:15311–15318. doi:10.1021/jp106732u
63. Hunter WN, Brown T, Anand NN, Kennard O Structure of an adenine–cytosine base pair in DNA and its implications for mismatch repair. Nature 320:552–555. doi:10.1038/320552a0
64. Lowdin PO (1965) Quantum genetics and the aperiodic solid: some aspects on the biological problems of heredity, mutations, aging, and tumors in view of the quantum theory of the dna molecule. Adv Quant Chem 2:213–354. doi:10.1016/S0065-3276(08)60076-3
65. Florián J, Leszczyński J (1996) Spontaneous DNA mutations induced by proton transfer in the guanine·cytosine base pairs: an energetic perspective. J Am Chem Soc 118:3010–3017. doi:10.1021/ja951983g
66. Šponer J, Šponer JE, Gorb L et al (1999) Metal-stabilized rare tautomers and mispairs of DNA bases: N6-metalated adenine and N4-metalated cytosine, theoretical and experimental views. J Phys Chem A 103:11406–11413. doi:10.1021/jp992337x
67. O'Brien PJ, Ellenberger T (2003) Human alkyladenine DNA glycosylase uses acid-base catalysis for selective excision of damaged purines. Biochemistry 42:12418–12429. doi:10.1021/bi035177v
68. Rios-Font R, Rodríguez-Santiago L, Bertran J, Sodupe M (2007) Influence of N7 protonation on the mechanism of the N-glycosidic bond hydrolysis in 2′-deoxyguanosine. A theoretical study. J phys chem B 111:6071–6077. doi: 10.1021/jp070822j
69. Chen X-Y, Berti PJ, Schramm VL (2000) Ricin a-chain: kinetic isotope effects and transition state structure with stem-loop RNA †. J Am Chem Soc 122:1609–1617. doi:10.1021/ja992750i
70. Francis AW, Helquist SA, Kool ET, David SS (2003) Probing the requirements for recognition and catalysis in Fpg and MutY with nonpolar adenine isosteres. J Am Chem Soc 125:16235–16242. doi:10.1021/ja0374426
71. Kennedy SA, Novak M, Kolb BA (1997) Reactions of ester derivatives of carcinogenic N-(4-Biphenylyl)hydroxylamine and the corresponding hydroxamic acid with purine nucleosides. J Am Chem Soc 119:7654–7664. doi:10.1021/ja970698p
72. McClelland RA, Ahmad A, Dicks AP, Licence VE (1999) Spectroscopic characterization of the initial c8 intermediate in the reaction of the 2-fluorenylnitrenium Ion with 2′deoxyguanosine. J Am Chem Soc 121:3303–3310. doi:10.1021/ja9836702
73. Qi S-F, Wang X-N, Yang Z-Z, Xu X-H (2009) Effect of N7-protonated purine nucleosides on formation of C8 adducts in carcinogenic reactions of arylnitrenium ions with purine nucleosides: a quantum chemistry study. J Phys Chem B 113:5645–5652. doi:10.1021/jp811262x
74. Shishkin OV, Dopieralski P, Palamarchuk GV, Latajka Z (2010) Rotation around the glycosidic bond as driving force of proton transfer in protonated 2′-deoxyriboadenosine monophosphate (dAMP). Chem Phys Lett 490:221–225. doi:10.1016/j.cplett.2010.03.044
75. Green-Church KB, Limbach PA (2000) Mononucleotide gas-phase proton affinities as determined by the kinetic method. J Am Soc Mass Spectrom 11:24–32. doi:10.1016/S1044-0305(99)00116-6
76. Green-Church KB, Limbach PA, Freitas MA, Marshall AG (2001) Gas-phase hydrogen/deuterium exchange of positively charged mononucleotides by use of Fourier-transform ion cyclotron resonance mass spectrometry. J Am Soc Mass Spectrom 12:268–277. doi:10.1016/S1044-0305(00)00222-1
77. Pan S, Verhoeven K, Lee JK (2005) Investigation of the initial fragmentation of oligodeoxynucleotides in a quadrupole ion trap: charge level-related base loss. J Am Soc Mass Spectrom 16:1853–1865. doi:10.1016/j.jasms.2005.07.009
78. Shishkin OV, Palamarchuk GV, Gorb L, Leszczynski J (2008) Opposite charges assisted extra strong C–H…O hydrogen bond in protonated 2′-deoxyadenosine monophosphate. Chem Phys Lett 452:198–205. doi:10.1016/j.cplett.2007.12.052

79. Ebrahimi A, Habibi-Khorassani M, Bazzi S (2011) The impact of protonation and deprotonation of 3-methyl-2′-deoxyadenosine on N-glycosidic bond cleavage. Phys Chem Chem Phys: PCCP 13:3334–3343. doi:10.1039/c0cp01279c
80. Berti PJ, Tanaka KSE (2002) No title. Adv Phys Org Chem 37:239–314.
81. Loverix S, Geerlings P, McNaughton M et al (2005) Substrate-assisted leaving group activation in enzyme-catalyzed N-glycosidic bond cleavage. J Biol Chem 280:14799–14802. doi:10.1074/jbc.M413231200
82. Mol CD, Parikh SS, Putnam CD et al (1999) DNA repair mechanisms for the recognition and removal of damaged DNA bases. Annu Rev Biophys and Biomol Struct 28:101–128. doi:10.1146/annurev.biophys.28.1.101
83. Cao C, Kwon K, Jiang YL et al (2003) Solution structure and base perturbation studies reveal a novel mode of alkylated base recognition by 3-methyladenine DNA glycosylase I. J Biol Chem 278:48012–48020. doi: 10.1074/jbc.M307500200
84. Palamarchuk GV, Shishkin OV, Gorb L, Leszczynski J (2013) Nucleic acid bases in anionic 2′-deoxyribonucleotides: a DFT/B3LYP study of structures, relative stability, and proton affinities. J Phys Chem B 117:2841–2849. doi:10.1021/jp311363c
85. Gonnella NC, Nakanishi H, Holtwick JB et al (1983) Studies of tautomers and protonation of adenine and its derivatives by nitrogen-15 nuclear magnetic resonance spectroscopy. J Am Chem Soc 105:2050–2055. doi:10.1021/ja00345a063
86. Brown RD, Godfrey PD, McNaughton D, Pierlot AP (1989) A study of the major gas-phase tautomer of adenine by microwave spectroscopy. Chem Phys Lett 156:61–63. doi:10.1016/0009-2614(89)87081-2
87. Lias SG, Liebman JF, Levin RD (1984) Evaluated gas phase basicities and proton affinities of molecules; heats of formation of protonated molecules. J Phys Chem Ref Data 13:695. doi:10.1063/1.555719
88. Greco F, Liguori A, Sindona G, Uccella N (1990) Gas-phase proton affinity of deoxyribonucleosides and related nucleobases by fast atom bombardment tandem mass spectrometry. J the American Chem Soc 112:9092–9096. doi:10.1021/ja00181a009
89. Meot-Ner M (1979) Ion thermochemistry of low-volatility compounds in the gas phase. 2. Intrinsic basicities and hydrogen-bonded dimers of nitrogen heterocyclics and nucleic bases. J Am Chem Soc 101:2396–2403. doi:10.1021/ja00503a027
90. Kurinovich MA, Lee JK (2000) The acidity of uracil from the gas phase to solution: the coalescence of the N1 and N3 sites and implications for biological glycosylation. J Am Chem Soc 122:6258–6262. doi:10.1021/ja000549y
91. Kurinovich MA, Lee JK (2002) The acidity of uracil and uracil analogs in the gas phase: four surprisingly acidic sites and biological implications. J Am Soc Mass Spectrom 13:985–995. doi:10.1016/S1044-0305(02)00410-5
92. Liu M, Li T, Amegayibor FS et al (2008) Gas-phase thermochemical properties of pyrimidine nucleobases. J Org Chem 73:9283–9291. doi:10.1021/jo801822s
93. Bonaccorsi R, Pullman A, Scrocco E, Tomasi J (1972) The molecular electrostatic potentials for the nucleic acid bases: adenine, thymine, and cytosine. Theor Chim Acta 24:51–60. doi:10.1007/BF00528310
94. Russo N, Toscano M, Grand A, Jolibois F (1998) Protonation of thymine, cytosine, adenine, and guanine DNA nucleic acid bases: theoretical investigation into the framework of density functional theory. J Comput Chem 19:989–1000. doi:10.1002/(SICI)1096-987X(19980715)19:9<989::AID-JCC1>3.0.CO;2–F
95. Colominas C, Luque FJ, Orozco M (1996) Tautomerism and protonation of guanine and cytosine. implications in the formation of hydrogen-bonded complexes. J Am Chem Soc 118:6811–6821. doi:10.1021/ja954293l
96. Podolyan Y, Gorb L, Leszczynski J (2000) Protonation of nucleic acid bases. A comprehensive post-Hartree–Fock study of the energetics and proton affinities. J Phys Chem A 104:7346–7352. doi:10.1021/jp000740u
97. Liguori A, Napoli A, Sindona G (2000) Survey of the proton affinities of adenine, cytosine, thymine and uracil dideoxyribonucleosides, deoxyribonucleosides and ribonucleosides. J Mass Spectrom: JMS 35:139–144. doi:10.1002/(SICI)1096-9888(200002)35:2<139::AID-JMS921>3.0.CO;2–A

98. Di Donna L, Napoli A, Sindona G, Athanassopoulos C (2004) A comprehensive evaluation of the kinetic method applied in the determination of the proton affinity of the nucleic acid molecules. J Am Soc Mass Spectrom 15:1080–1086. doi:10.1016/j.jasms.2004.04.027
99. Wilson MS, McCloskey JA (1975) Chemical ionization mass spectrometry of nucleosides. Mechanisms of ion formation and estimations of proton affinity. J Am Chem Soc 97:3436–3444. doi:10.1021/ja00845a026
100. Gorb L, Podolyan Y, Dziekonski P et al (2004) Double-proton transfer in adenine-thymine and guanine–cytosine base pairs. A post-Hartree-Fock ab initio study. J Am Chem Soc 126:10119–10129. doi:10.1021/ja049155n
101. Lukin M, De Los Santos C (2006) NMR structures of damaged DNA. Chem Rev 106:607–686. doi:10.1021/cr0404646
102. Boussicault F, Robert M (2008) Electron transfer in DNA and in DNA-related biological processes. Electrochem insights. Chem Rev 108:2622–2645. doi:10.1021/cr0680787
103. Giovangelle C, Sun JS, Helene C (1996) In comprehensive supramolecular chemistry. Pergamon Press, Oxford, p 177

Chapter 6
Quantum Chemical Approaches in Modeling the Structure of DNA Quadruplexes and Their Interaction with Metal Ions and Small Molecules

Mykola Ilchenko and Igor Dubey

Abstract Certain guanine-rich DNA and RNA sequences can fold into unique biologically significant high-order structures called G-quadruplexes (G4) formed by stacked arrays of guanine quartets connected by non-canonical hydrogen bonds. Novel anticancer strategy is based on the use of organic molecules that specifically target quadruplex structures present in telomeres and some other regions of the genome. We provide a brief overview of the structural features of quadruplex nucleic acids and main mechanisms of G4-ligand interaction. Current methods for the molecular modeling of quadruplex DNA structures and their ligand binding are discussed in the review. We mainly focus on quantum chemical computational approaches to model the interaction of G4 DNA and its structural elements with metal cations and small molecules, including hybrid QM/MM approaches.

6.1 Introduction

The molecular basis of the formation of biologically functional structures of biomacromolecules (proteins, nucleic acids, etc.) and their specific interactions with low-molecular ligands remains one of the most exciting problems of biomedical science and a foundation of modern drug design.

Quite recently emerged the antitumor strategy based on the use of small molecules that specifically target telomeres and telomerase [1–4]. Telomeres are guanine-rich DNA sequences localized at the ends of the chromosomes. They protect chromosomal DNA from degradation, prevent end-to-end fusion and other forms of aberrant recombination, and allow it to be completely replicated without loss of genetic material. The length of the telomeres correlates with the ability of a cell to

I. Dubey (✉) · M. Ilchenko
Institute of Molecular Biology and Genetics, National Academy of Sciences of Ukraine, 150 Zabolotnogo str., 03680 Kyiv, Ukraine
e-mail: dubey@imbg.org.ua

L. Gorb et al. (eds.), *Application of Computational Techniques in Pharmacy and Medicine,* Challenges and Advances in Computational Chemistry and Physics 17,
DOI 10.1007/978-94-017-9257-8_6

undergo a large number of cell divisions. Normally telomeric DNA is shortened by 50–200 nucleotides upon each cellular division that may control the proliferative capacity of normal somatic cells. However, this does not occur in tumor cells due to high activity of telomerase, an enzyme which is responsible for maintaining the telomere length and synthesizes the lost telomeric sequences by adding telomeric repeats (5'-TTAGGG-3' sequence in humans), that leads to uncontrolled proliferation. Inhibition of telomerase activity induces senescence in cancer cells followed by their death. In contrast, normal somatic cells are devoid of telomerase activity, so high level of enzyme expression is directly associated with cancer. Indeed, increased telomerase activity was detected in 85–90 % of human tumors [1, 3]. As a result, telomerase system is now considered a promising biological target for novel anticancer drugs.

Telomerase is a multicomponent highly specialized enzyme responsible for the synthesis of telomeres. Its catalytic subunit (TERT, telomerase reverse transcriptase) utilizes the RNA component of the enzyme (TR) as a template to synthesizes telomeric DNA repeats. A number of strategies for telomerase inhibition by low-molecular drugs have been proposed. They include the application of nucleoside and non-nucleoside reverse transcriptase inhibitors, antisense oligonucleotides and their analogues against TR RNA, ribozymes and siRNA directed against TR and TERT components of the enzyme, etc. [5–9]. These approaches are rather traditional for the inhibition of enzymes of nucleic acids biosynthesis. Totally different approach is based on the presence of unique structural motifs in telomeric DNA called G-quadruplexes (G4).

Certain guanine-rich DNA sequences readily fold into the four-stranded structures formed by stacked arrays of guanine quartets (or tetrads) – square planar arrangements of four guanine bases connected by Hoogsteen-type hydrogen bonds (Fig. 6.1).

These stable higher-order DNA arrangements were shown to play a crucial biological role in a living cell. DNA sequences able to adopt quadruplex structures are prevalent in telomeres as telomeric repeats, although they have been also found in a number of gene promoter regions, first of all in proto-oncogenes, like *c-myc, c-kit* or *k-ras*, that can be also targeted by drugs [10–15]. RNA sequences can also form quadruplex structures as recent finding demonstrated that telomere DNA is transcribed into telomeric repeat-containing RNA [11, 16–18].

Perhaps the formation and dissociation of quadruplex structures in nucleic acids is one of the universal ways of the regulation of gene expression *in vivo*. So the development of specific quadruplex ligands, besides the development of anticancer compounds, would allow controlling many fundamental biological processes. Therefore, quadruplex nucleic acids are an important new target for drug design, and there is growing interest in the development of small molecules targeting these structures with high affinity and selectivity.

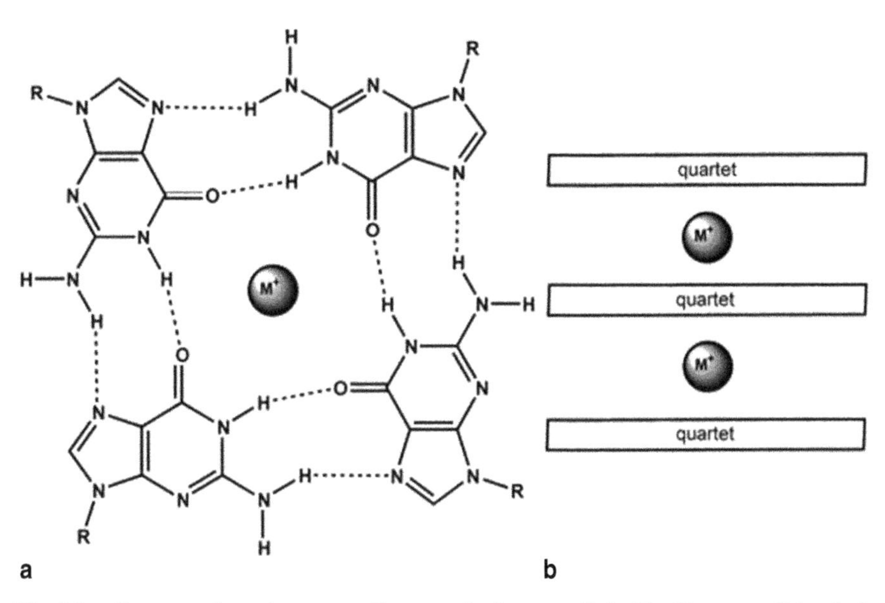

Fig. 6.1 **a** Structure of guanine quartet. Four guanine bases are linked by Hoogsteen H-bonds. **b** simple quadruplex model: side view of the stack of three G-quartets containing two monovalent metal cations. Ions are located in the channel formed by guanine residues

6.2 G-Quadruplex DNA and its Ligands

G-quadruplexes are specific structures that include planar G-quartet stacks and four grooves providing different geometries and spatial distribution of functional group as compared to duplex DNA. This difference allows specific recognition of quadruplexes by low-molecular ligands and binding selectivity over duplex DNA, i.e. the ability to interact only with quadruplex but not duplex nucleic acids. Selective G4 ligands stabilizing G4 structures may disturb the binding of enzyme to telomeric DNA and thus block its elongation that results in anticancer activity [4, 5, 7, 10]. In other words, single-stranded DNA is a substrate of telomerase, whereas G-quadruplex DNA is not.

It is interesting to note that in this approach enzyme inhibition is achieved due to the interaction of a ligand with telomerase substrate, i. e. telomeric DNA, rather than with the enzyme itself.

Thermodynamic and kinetic data suggest that quadruplex stability depends on a number of factors, including the type of structure adopted by the DNA strand (or strands), strand sequence, the size of intervening loops, base and phosphate modifications, pH and the presence of cations [19]. Small molecules may stabilize quadruplex DNA (or facilitate DNA folding into quadruplex structures) due to shifting the competitive equilibrium between the single-stranded or Watson-Crick duplex and quadruplex DNA towards the latter form [19, 20]. Inhibition activity of G4 ligands depends mainly on the stability of their complexes with telomeric DNA quadruplexes.

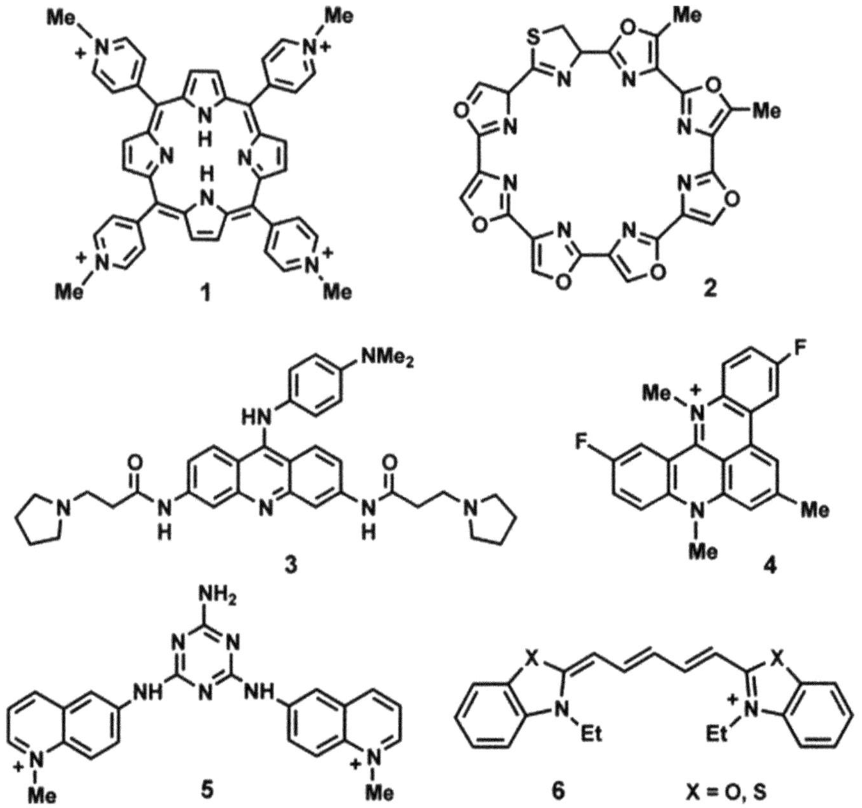

Fig. 6.2 Chemical structures of classic G4 ligands: TMPyP4 (1), telomestatin (2), BRACO-19 (3), RHPS4 (4), triazine (5) and cyanine (6) derivatives

A number of efficient quadruplex binding/stabilizing ligands with anticancer properties were reported, e.g. TMPyP4, BRACO-19, telomestatin, BMVC, quarfloxin. Typical G4 ligands (Fig. 6.2) are usually based on heteroaromatic polycyclic structures like acridines, anthraquinones, carbazoles, macrocyclic polyoxazoles, etc. [5, 7, 21–23].

Generally, stabilization of quadruplex DNA conformations by small organic molecules can occur via the π-π interaction of cationic or neutral aromatic fragments with G-quartets (usually external stacking of the ligand at the terminal G-quartet) and by the electrostatic interaction of positively charged ligands, either aromatic cores or cationic or easily protonated basic substituents, with G4 DNA polyphosphate backbone. The design of quadruplex ligands is mainly based on planar polycyclic aromatic scaffolds able to interact with G-quartets via the stacking mechanism.

Since G-quartet (or G-tetrad) consists of four guanine bases, its square is at least twice as large as the square of usual DNA purine-pyrimidine base pair. So a specific quadruplex binder should contain a large aromatic/heteroaromatic core, larger than that required for common duplex DNA binding compounds. Only in this case G4 ligand would be able to ensure an efficient overlap with guanine tetrad and thus provide good quadruplex selectivity of the drug over the duplex DNA [7, 21, 22]. From this point of view, large ligands like porphyrins and telomestatin may be preferable as their molecules perfectly overlap with G-tetrads. At the same time, relatively small size of the central aromatic core can be efficiently compensated by the substituents of cationic nature.

There are numerous experimental methods to investigate quadruplex DNA and monitor and quantify its interactions with low-molecular ligands [24–26]. Besides biochemical and electrophoretic methods [27], various biophysical approaches are available, including e.g. absorption spectroscopy [28, 29], circular dichroism spectroscopy [29, 30], fluorescence resonance energy transfer (FRET) [31, 32], fluorescence melting assays [33] and other fluorescence-based techniques, and mass spectrometry [25]. Another popular approach for studying biomolecular interactions in G-quadruplexes is surface plasmon resonance (SPR) [34–36]. These methods provide valuable and diverse structural, kinetic and thermodynamic data.

At the same time, the main sources of precise structural information on G-quadruplexes and their complexes with small molecules are X-ray crystallography [37–41] and NMR spectrometry [38, 42, 43]. Dozens of 3D structures of various forms of quadruplexes are currently available from Protein Data Bank and other sources. These crystallographic or NMR structures are of great importance as they are the basis for molecular modelling and modern drug design.

6.3 Theoretical Studies on G-Quadruplex Structures and Their Interactions With Low-Molecular Ligands

The accurate modelling of the structures of biomacromolecules and their complexes with small molecules and determining their thermodynamic parameters is still a complicated and challenging problem due to the large size and complexity of molecular systems. Nevertheless, computer modelling of the structures of nucleic acids and proteins and their interaction with low-molecular ligands is now an integral part of drug design. Molecular modelling based on docking or molecular dynamics is very common in this field. At the same time, quantum chemical (quantum mechanical) approaches are not so common; however, they are able to provide information that cannot be obtained by the other methods.

6.3.1 Computer Modelling Methods in the Study of Biomacromolecules

In general, computational approaches to modelling a molecular system may be divided into two broad categories: quantum mechanics (QM) [44] and molecular mechanics (MM) [45]. QM methods provide more accurate modelling results than MM-based approaches, although they are generally much more demanding in terms of a computation power. It is now widely recognized that both methods reinforce one another in an attempt to understand chemical and biochemical behaviour of biomolecules at the molecular level. From a practical point of view, the complexity of the system, time limits, available computation resources and other limiting factors determine which method is feasible [46].

MM may be used to model biomacromolecular systems to which even semi-empirical QM calculations can be applied effectively. In MM, molecular motions are determined by the masses of atoms and the forces acting on them, whereas wave functions or electron densities are not computed. MM is widely used in chemistry and biochemistry to obtain molecular models since this approach is much faster and requires less computation power than QM methods. It allows the modelling of large molecular systems. However, MM energies have little meaning as absolute values and can be used rather to compare relative energies obtained for several molecular structures [46]. Moreover, MM approaches often cannot succeed with molecular systems where electronic interactions are dominant, including π-π-stacking interactions which are the basis of most ligands binding to G-quadruplex DNA structures. In this case, QM calculations should be used to obtain accurate results. At the same time, despite some severe intrinsic limitations in MD approaches, base stacking can be reasonably approximated based on well-calibrated force fields [47].

It should be noted that the molecular mechanics provide only a static view of the flexible molecular system. The most common approach used for the simulation of biomolecules motion on the atomic level is the molecular dynamics (MD) [48]. The forces acting on atoms are usually calculated here using MM methods. MD can provide information on the possible conformations and dynamics of the system, as well as its thermodynamic parameters.

MD simulations have some intrinsic limitations, e.g. force-field imperfections and often insufficient simulation times. Nevertheless, MM-based methods, including MD, have become very popular research tools in biochemistry and drug design, including the studies on G-quadruplexes and G4-ligand complexes. MD simulation of G-quadruplex structure and dynamics [49–52] and the interactions of quadruplexes of various topologies with cations [49, 50, 53–56] and low-molecular organic compounds [50–52, 57–63] has been widely used to understand the basic properties of quadruplex DNA and to improve the recognition of quadruplexes by small molecules in the design of efficient G4 ligands, as well as to complement available experimental data (see e.g. [25] and references therein). Such common approaches to modern drug design as molecular docking and virtual screening have also been successfully applied to the development of potential G4 ligands as antitumor agents [57, 62–67]. However, in the present review we will mainly concentrate on purely quantum chemical (i. e. quantum mechanical) methods being currently used to model G-quadruplex DNA and quadruplex-drug complexes with high accuracy. Moreover, non-QM methods of molecular modelling of G4 structures and G4-ligand complexes, in particular MD approaches, have been recently reviewed in a number of works [50, 52, 60, 63–66], including a detailed methodology review by Haider and Needle [68].

6.3.2 *Quantum Chemical Calculations on G-Quadruplexes*

Recent advances in computational processing power and modelling algorithms have resulted in the development of new efficient computer-aided methods for the discovery of novel drugs interacting with biomacromolecules. The availability of crystallographic and NMR data for G4 structures strongly facilitates the design of potent antitumor compounds using computational methods.

6.3.2.1 G-Quadruplex Structures and Ligand Binding

Despite recent theoretical advances, the complexity of quadruplex architectures remains a great challenge for computer modelling of G4 and their complexes. One of the main problems for any molecular modelling approach is unusually broad structural polymorphism of quadruplex DNA. This polymorphism results mainly from the conformational flexibility of DNA chains and non-covalent (hydrophobic, stacking, electrostatic) interactions of quadruplex fragments, both heterocyclic nucleic bases and sugar-phosphate backbone. There are intra- and intermolecular (dimeric, tetrameric) quadruplexes, with parallel, antiparallel or mixed (hybrid) type of G4 structures. The geometric parameters of guanine quartets may differ to some extent as well. The topology of a quadruplex is determined by a number of factors, including e.g. nucleotide sequence, pH, the nature (Na^+, K^+ or NH_4^+) and concentration of cations present in the medium, etc. Moreover, depending on the conditions the same nucleic acid sequence may form several quadruplex structures with different conformations, or their equilibrium mixtures [5, 7, 10–15, 37–43, 69–72]. It is widely accepted that the crystal (X-ray) and solution (determined by NMR) structures may be also different for the same oligonucleotide sequence [38, 72, 73]. Some examples of diverse quadruplex topologies are presented in Fig. 6.3.

To understand the functioning of G-quadruplexes and its recognition by small molecules, a deep analysis of structural and energetic properties of G4 fragments, first of all guanine quartets and their stacks as a key element of quadruplex structures, is required.

In general, there are several possible binding modes for quadruplex ligands: they can stack externally upon a terminal G-quartet (mainly via π-π interactions), intercalate between two G-quartets, and bind to the quadruplex grooves between two adjacent DNA chains. Additional electrostatic interaction of a ligand with phosphate groups, most often provided by basic/cationic side chains, is usually required to ensure high binding affinity. Ligand interaction with quadruplex loops may further increase the binding specificity [5, 7, 10, 12, 21–23]. Taking into account the structural diversity and polymorphism of quadruplex structures, design of efficient G4 ligands is a complicated and challenging task, especially when specific binding to a particular topologic form of a quadruplex should be achieved.

Nevertheless, all forms of quadruplexes have common structural elements as these assemblies are formed by DNA chains and contain the stacks of G-quartets.

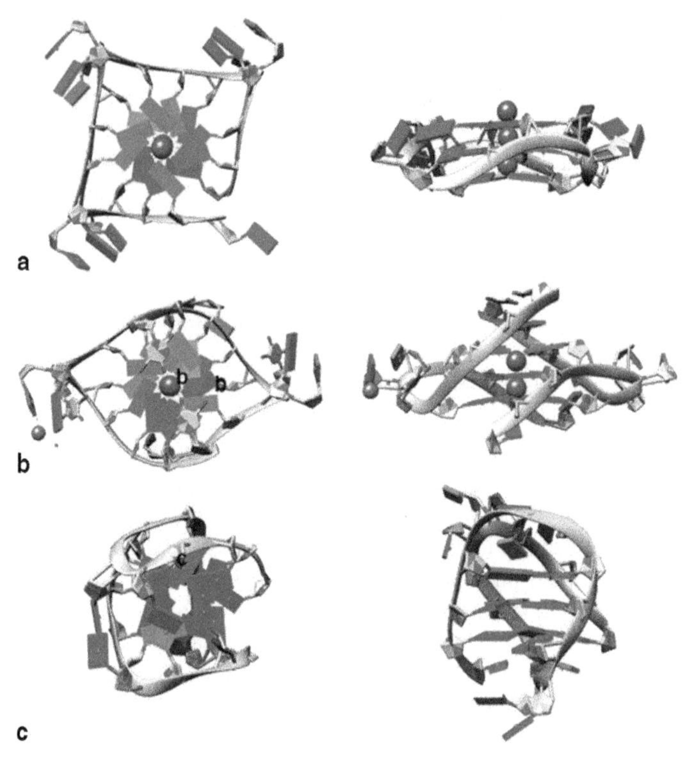

Fig. 6.3 Top (*left*) and side (*right*) views of G-quadruplex structures of various topologies: **a** – parallel (Protein Data Bank entry 1KF1). **b** – mixed (PDB 1K8P) [37]. **c** – antiparallel (PDB 143D [74]) intramolecular human telomeric DNA quadruplexes. PDB structures were visualized with a Chimera package [75].

As a result, known to date G4 ligands share some common structural features as well: they are based on large planar polycyclic heteroaromatic scaffolds able to interact with guanine tetrads, and usually contain one or several cationic substituents for strengthening the interaction with anionic DNA backbone (sometimes the aromatic core of the ligand can be cationic itself, e.g. in ethidium-based compounds or acridinium derivative RHPS4, although most known G4 ligands are based on the neutral heteroaromatic systems).

There are also G4 ligands that bind at the grooves of quadruplexes, with structures similar to those for duplex DNA minor groove binding ligands [76, 77].

6.3.2.2 DFT Calculations of the Structure of G-Quartets and Their Interaction With Metal Cations

Since G-quartets and their stacked arrays are a unique component of all topological types of G-quadruplexes, much effort were made to study guanine quartets and octets by quantum chemical methods. QM-based approaches are very appropriate for

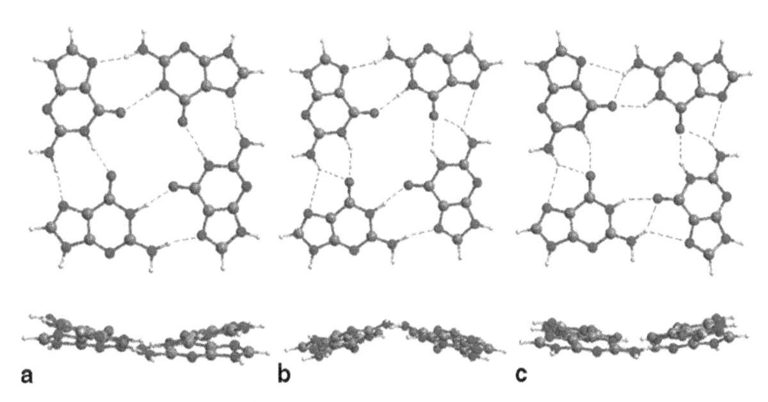

Fig. 6.4 Top and side views of G-quartets with Hoogsteen (**a**), mixed (**b**) and bifurcated (**c**) systems of hydrogen bonding. Structures were obtained by the authors using B3LYP/6–311+G(d, p) DFT calculations in vacuum

these studies since the size of G-quartet molecular system is not too large for QM methods which are at the same time much more accurate than the methods based on molecular mechanics, including MD, and provide more complete information on the structure of quadruplex building blocks – guanine quartets. Since the latter are the principal targets for virtually all G4-binding drugs, the availability of accurate molecular structures of G-tetrads is of critical importance for successful drug design.

Guanine bases in quadruplexes are held together by a number of non-covalent interactions: hydrogen bonds responsible for the organization of four guanine fragments into the planar quartet; π-π-stacking interactions between individual quartets enable the assembly of their stacks, and the presence of a monovalent cation (typically Na^+, K^+ or NH_4^+) between the quartet planes within the quadruplex channel stabilizes G4 structure by neutralizing the electrostatic repulsion between guanine O^6 oxygen atoms that form the inner rim of the quartet [78–81]. Some variations in the geometry of guanine quartets are possible that depend on the type of hydrogen bonding (classical Hoogsteen or so-called bifurcated system of H-bonds) (Fig. 6.4). All these structural and electronic factors should be thoroughly considered when performing the molecular modelling studies.

Quantum chemical theoretical studies on the structure of G-quartets and their stacks most often employ the DFT (Density Functional Theory) method [82, 83].

A systematic DFT study of nucleic acid G-quartets was performed in [84]. A number of functionals, including B3LYP [85, 86], M05-2X [87] and M06-2X [88], were used in the calculations. M05-2X and M06-2X functionals effectively incorporate the long-range dispersion forces [89] which are important for modelling the stacking-based systems. RI-DFTD [90, 91] calculations at BLYP/TZVPP [92] level were also performed to account for dispersion interactions using TURBOMOLE 6.0 program system [93, 94]. The polarizable continuum model (PCM) [95] was used for the solvent calculations which were performed at B3LYP/6-31G(d) level. In this fundamental theoretical work [84] the structures of G-quartets with Hoogsteen-type bonding (S_4 symmetry), with two bifurcated bonds (C_2 symmetry) and with all bifurcated H-bonds (S_4 symmetry). The structures of G-octet with two bifurcated bonds (C_2 symmetry) was also investigated. Analysis of binding energies demonstrated that G-quartet with two bifurcated bonds (C_2 symmetry) is the most stable structure in gas phase. It was also shown that G-quartets with no bifurcated bonds (C_2, S_4, C_{4h} symmetries) did not give a stable structure at the M05-2X/6-31+G(d, p) level in gas phase. At the M06-2X/6-31+G(d, p) level, optimization of the three different types of quartets studied resulted in the quartet with all bifurcated H-bonds with S_4 symmetry. We have independently come to analogous conclusions on the stability of G-quartets n vacuum [96]. As to the effect of aqueous medium, our data strongly differ from those obtained in [84]. We have found that full optimization of G-quartets in water results in immediate zero barrier transformation of all types of G-quartets into a single structure with Hoogsteen-type bonding. These data are supported by the results presented in [97] demonstrating that the structures of S_4 symmetry are global minima in water for G-quartets with Hoogsteen-type system of hydrogen bonds. We also suppose that Jissy et al. [84] were not correct while mentioning that there is one quartet with two bifurcated bonds in PDB crystal structure of G-quartet 1LVS. In our opinion, quartet of this type more closely resembles the deformed G-quartets with Hoogsteen bonding. We have found only one more paper in the literature [98] where the analogous mixed quartet form was briefly mentioned, although with no comparison with other possible structures. The authors studied guanosine 5'-hydrazide self-assemblies in the gel state by the combination of spectral data and B3LYP/6-31G** DFT calculations and concentrated mainly on the dependence of H-bond parameters on the orientation of ribose fragments in nucleoside.

It was recognized early that the ability to stabilize guanine gels depends on the cation nature and that the ionic radius is important for complex stability; in the alkali series K^+ promotes the most stable complexes [80]. Cations play a critical role in stabilizing G-quadruplex structures to the extent when changing e.g. potassium cation for Na^+ can completely alter the whole topology of G-quadruplex [37–43, 69–74]. It is not surprising therefore that the role of metal ions in formation and stability of G-quadruplex structures was studied thoroughly in a number of theoretical works.

In addition to the above mentioned research, the authors of [84] have also performed calculations for Li^+, Na^+, K^+, Be^{2+}, Mg^{2+} and Ca^{2+} complexes of G-tetrads. Calculations showed that for an isolated quartet, the metal ion with the smallest ionic radius in their respective groups (IA and IIA) form more stable complexes. Other properties such as the HOMO-LUMO gap and polarizability have also been analyzed. The variation in the polarizability has been studied with respect to the movement of cations along the central cavity of the quartet to show that such movement leads to a large anisotropy of polarization and hence the refractive index (η) thereby creating optical birefringence which have potential applications in biomolecular imaging.

The structures and interaction energies of guanine and uracil quartets have been determined by B3LYP hybrid density functional calculations in [99]. The total interaction energy of the C_{4h}-symmetric guanine quartet consisting of Hoogsteen-type base pairs with two hydrogen bonds between two neighbour bases was found to be −66.07 kcal/mol at the B3LYP/6-311G(d, p) level of theory. Complexes of metal ions with G-quartets can be classified into different structure types. The one with Ca^{2+} in the central cavity adopts a C_{4h}-symmetric structure with coplanar bases, whereas the energies of the planar and nonplanar Na^{+} complexes are almost identical. Metal cations with small radii (Li^{+}, Be^{2+}, Cu^{+}, and Zn^{2+}) and a high charge enforce a nonplanarity of the base quartets and may thus prevent a stacking of G-quartet, unlike Na^{+} and K^{+} cations. The electrostatic potential of G-quartets provides probably favourable binding sites for metal ions between the stacked quartet planes, whereas isolated quartets have the region of most negative electrostatic potential in the central cavity. Uracil quartets in the orientation with N3–H3…O4 H-bond are probably also capable of binding cations at the centre.

The complexes of metal cations Fe^{2+}, Co^{2+}, Ni^{2+}, Cu^{2+} and Zn^{2+} with guanine tetrads (G_4) of C_{4h}, C_4 and S_4 symmetry) were studied in [100]. The system contained two water molecules above and under the G_4-cation plane, with six-coordinated metal ion. G_4–Co^{2+} and G_4–Cu^{2+} being open shell species were treated using unrestricted method UB3LYP and 6-31G(d) basis set. BSSE (Basis Set Superposition Error) correction was evaluated according to the counterpoise method of Boys and Bernardi [101]. Bader's AIM (atoms in molecules) theory [102] was applied to determine a strong hydrogen bond [103]. The main conclusions were as follows: (a) the stability sequence is $Ni^{2+}>Cu^{2+}>Co^{2+}>Fe^{2+}>Zn^{2+}$ when including BSSE correction, and $Ni^{2+}>Fe^{2+}>Co^{2+}>Cu^{2+}>Zn^{2+}$ after hydration energy correction; (b) the sequence for G_4–M–water complexes is $Co^{2+}>Fe^{2+}>Ni^{2+}>Cu^{2+}>Zn^{2+}$ with BSSE correction; (c) electron density and its Laplacian at the bond critical points well correlate with the hydrogen bond length in the tetrads.

Structural properties and the effect of interaction of alkali (Li^{+}, Na^{+}, K^{+}) and alkaline earth (Be^{2+}, Mg^{2+}, Ca^{2+}) metal cations with guanine and thioguanine (SG) tetrads were studied [104]. Complex formation was investigated using *ab initio* and DFT methods. In some cases MP2/6-311G** single point energy calculation was performed for the geometries optimized by B3LYP/6-311G** level of theory. Single point energy calculations were carried out to study the solute–solvent interaction using the self-consistent reaction field theory (SCRF) [95] at B3LYP/6-311G** level of theory. This method is based on Tomasi's polarized continuum model (PCM), which defines the cavity as the union of a series of interlocking atomic spheres. The calculations revealed that cation-G and SG-tetrad complexes adopt normal four-stranded Hoogsteen bonded structures. The substitution of cations on guanine and SG-tetrads affects their geometries and charge distributions. The gas phase binding sequence for cation-G and SG-tetrads follows the interaction energy and metal ion affinity order $Li^{+}>Na^{+}>K^{+}$, $Be^{2+}>Mg^{2+}>Ca^{2+}$. The smaller ions are tightly bonded to the quartets suggesting the domination of electrostatic interaction in the cation–tetraplexes systems. The solvent interaction with the molecular systems has increased the stability of both guanine and thioguanine quartets and their complexes. The two and three-body interaction energies have been used to analyze the influence of a metal cation on the stability of tetrads. AIM theory was also used to study the hydrogen bonds in the metal interacting complexes.

Theoretical study of incorporating 6-thioguanine into a guanine tetrad and their influence on the metal ion–guanine tetrad was performed by Meng et al. [105]. The initial structure of the G-quartet has been generated from the coordinates of the human telomeric DNA (PDB code 1KF1) [37]. The calculation method used was B3LYP, and the basis set was the standard double-zeta with polarization functions 6-31G*. The geometries, energies and charge distributions were discussed. The effects of different cations (K^+ and Na^+) on the various tetrads were studied as well. The calculation results demonstrated that upon the increase of SG units number the quartet becomes more and more unstable. Without hydration correction, the Na^+ cation was found to bind more tightly with the tetrad than that of potassium, whereas when hydration effects were considered the stability sequence changed to $K^+ > Na^+$. More favorable binding of potassium ion comparing to Na^+ in solution is due to higher dehydration energy of the latter, although Na^+ cation has higher intrinsic propensity to bind tightly to DNA quadruplexes [106].

Effect of external electric field on H-bonding and π-stacking interactions in guanine aggregates were studied by Jissy and Datta [107]. The DFT calculations were carried out at the M05-2X level of theory with the 6-31+G(d, p) basis set. The structure and electronic properties of guanine oligomers and π-stacks of guanine quartets with circulenes were investigated under an external field through first-principles calculations. The binding energy of the circulenes with G-quartets were fond to increase on application of an electric field along the stacking direction. Besides that, the stability of G-quartet–circulene π-stacks was shown to depends on the phase of the dipole moment (in-phase or out-of-phase) induced by an external electric field. The stability of stacks of bowl-shaped circulenes with G-quartets depended on the direction of the applied field.

At the end of this paragraph we would like to mention two pioneering papers published by Leszczynski, Gu and Bansal [108, 109], where the first HF and DFT calculations on the stability and structure of G-quartets were performed and the possibility of the formation of the structures with bifurcated hydrogen bonds was demonstrated.

Thus, guanine quartets were studied in depth by quantum chemical methods. Their structural diversity based on different possible patterns of hydrogen bonding was demonstrated and the key role of metal cations in stabilizing guanine assemblies was shown.

6.3.2.3 DFT Studies on the Structure of G-Octets and Their Metal Complexes

Guanine octets are more complex structural elements of quadruplex systems than the quartets. G-octet is the system of two stacked G-quartets interacting via the π- π-stacking mechanism. This molecular assembly is closer to natural G-quadruplexes which usually contain three or four guanine quartets, thus its modelling is able to provide more realistic structural and thermodynamic data to be extrapolated to quadruplex DNA. At the same time, G-octet molecular system is twice as large as guanine quartet, and forces stabilizing them include π-π-interactions between the quartets, thus quantum chemical calculations with G-octets are much more complicated and require more computational resources.

Characterization of the monovalent ion position and hydrogen-bond network in guanine quartets and octets by DFT calculations of NMR parameters was performed by van Mourik and Dingley [110]. The structures of the guanine quartets at C_{4h} and S_4 symmetry (G4 and G4-M^+ systems) were fully optimized using the B3LYP and B97 [111] functionals employing basis sets ranging from 6-31G(d) to 6-311++G(d, p) with NWChem software [112]. At the same time, only constrained optimization was carried out for more complex G4-M^+-G4 and G4-M^+-G4-M^+ systems keeping the quartet-quartet distances fixed. Similar calculations using Gaussian 03 package failed to converge or converged to an alternative structure containing bifurcated hydrogen bonds, whereas using NWChem both the Hoogsteen H-bonded structure obtained in crystallographic and NMR studies, as well as the bifurcated structure could be optimized. It was shown that the presence of a monovalent ion in the centre of G-quartet led to the contraction of the quartet O6–O6 distance. This effect was largest for the smallest ion, thus showing that the contraction of the G-quartet facilitates the optimal coordination of the monovalent ion with the O6 atoms of the guanine bases. In addition, cation localization sites were found for G-octets with the distances between G-quartet planes ranging from 3.3 to 5.2 Å. The results for the G_4-M^+-G_4 model showed that at quartet–quartet distances observed in the DNA quadruplex crystal structures, the smaller Na^+ and Li^+ cations have two shallow minima located at 0.55 and 0.95 Å outside the plane of the quartet, respectively. At the same time, the larger K^+ ion has a minimum centred between successive G-quartets. At increasing quartet–quartet distances the Na^+ and Li^+ ions converged to a position coplanar with the G-quartet, whereas the optimal K^+ ion position converged to a location just outside the G-quartet. Apparently at shorter quartet–quartet distances the sodium and lithium cations are weakly attracted to the second G-quartet and therefore do not favour a coplanar position with the G-quartet. Increasing the quartet–quartet distance reduces this weak attraction to zero and the Na^+ and Li^+ ions shift to an energetically favoured coplanar position. The attraction of the ion to both G-quartets at quartet–quartet distances observed in DNA quadruplex structures may facilitate the transport of the ions through the DNA quadruplex central channel. The smaller Li^+ and Na^+ ions have rather low energy barriers separating the minima in the quartet-ion-quartet model that under physiological conditions are most likely overcome by vibrational and thermodynamic effects. Consequently, their movement through the channel is energetically unimpeded, in contrast to larger K^+ ion that will not move as freely.

An interesting work [98] that we have already mentioned as the paper where the mixed G-quartet structure was presented, applied the electrospray ionization mass spectrometry (ESI-MS) to investigate hydrogen-bonded G-quartets and their complexes. ESI analysis displayed magic numbers of guanine tetramer adducts with Na^+, Li^+ and K^+, not only for guanine, but also for xanthine bases. The optimized structures of guanine and xanthine quartets have been determined by B3LYP hybrid DFT calculations. The optimized structures obtained for each quartet explained the gas-phase experimental results. The gas-phase binding sequence between the monovalent cations and the xanthine quartet follows the order $Li^+ > Na^+ > K^+$, which is consistent with that obtained for the guanine quartet in the literature. The smallest stabilization energy of K^+ and its position versus the other alkali metal ions in guanine and xanthine quartets is consistent with the fact that the potassium cation can be located between two guanine or xanthine quartets, for providing a $[(G\ or\ Xan)_8 + K]^+$ octamer adduct. While an octamer adduct with K^+ for xanthine was detected by ESI-MS, it was not the case for guanine. The formation of tetrameric and octameric aggregates of guanine analog 3-methylxanthine with NH_4^+, Na^+ and K^+ ions has been also observed in the gas phase in ESI-MS spectra to confirm the results of computational studies performed at the BLYP-D/TZ2P level of theory [113].

Nucleic acid tetraplexes and lipophilic self-assembling G-quadruplexes contain stacked base tetrads with intercalated metal ions as basic building blocks. In [114] quantum-chemical methods were also used to systematically explore the geometric and energetic properties of base tetrads with and without metal ions. The structures were optimized with the B3LYP hybrid density functional method and the DZVP basis sets. Sandwiched G-, C-, U-, and T-tetrads with Na^+ and K^+ ions at different symmetries were studied. The detailed information on total energies as well as on metal ion tetrad and base– base interaction energies was obtained. The geometrical parameters of the sandwiched metal ion complexes were compared to both experimental structures and to calculated geometries of complexes of single tetrads with metal ions. A microsolvation model was successfully applied to explain the ion selectivity preference of K^+ over Na^+ in a qualitative sense.

A systematic DFT study of sandwiched isoguanine (iG) complexes with intercalating alkali metal ions was carried out in [115]. The study of sandwiched isoguanine tetrad and pentad complexes consisting of two polyads with Na^+, K^+ and Rb^+ ions was performed at the B3LYP level. In iG sandwich structures, the ion-base interaction energy is slightly larger than in the corresponding guanine sandwich complexes. Because the base–base interaction energy is even more increased in passing from guanine to isoguanine, the iG sandwiches are thus far the only examples where the base–base interaction energy is larger than that of the base–metal ion interaction. Stacking interactions have been studied in smaller models consisting of two bases, retaining the geometry from the complete complex structures. From the data obtained at the B3LYP and BH&H levels and with Møller-Plesset perturbation theory, one can conclude that the B3LYP method overestimates the repulsion in stacked base dimers. For the complexes studied in this work, this is only of minor importance because the direct inter-tetrad or inter-pentad interaction is supplemented by a strong metal ion-base interaction. Using a microsolvation model, the metal ion preference $K^+ \approx Rb^+ > Na^+$ was found for tetrad complexes. On the other hand, for pentads the corresponding ordering is $Rb^+ > K^+ > Na^+$. In the latter case experimental data are available that agree with this theoretical prediction.

DFT calculations at the M052X/6-31G(d) level and PCM/TD-PBE0/6-31G(d) level were performed in [116] to get insights into the effect of metal ions on the excited states of guanine nanostructures, short $d(TG_4T)_4$ quadruplexes and long G_4-wires. As a first step, the ground state geometry of short $d(TG_4T)_4$ quadruplexes was optimized at the PCM/M052X/6-31G(d) level. Although the absence of the sugar-phosphate backbone does not allow detailed comparison of computation results with experimental data on G-quadruplex DNA, the developed model was able to reproduce the different location of K^+ and Na^+ ions within the G-quadruplexes. A similar inter-quartet distance was obtained for Na^+(3.35 Å) and K^+(3.36 Å), in agreement with NMR measurements. As expected, independently of the type of cation, the two 9-Me-G involved in the electronic transition adopt the structure of a 9-Me-G^+ cation and a 9-Me-G^- anion. The position of all the ions changes with respect to that found in the ground state minimum. The metal ions lose the symmetric arrangement with respect to the G-quadruplex axis; they move farther from the 9-Me-G^+ cation and get closer to the 9-Me-G^- anion.

As to the cation-free octets, we have determined the structures and energies of guanine quartets and octets in water by DFT calculations using M06-2X functional and 6-31G(d, p) basic set [96]. Guanine quartets in vacuum were found to have not only the Hoogsteen or bifurcated, but also mixed system of hydrogen bonds; in water the latter two forms are transformed into the classic Hoogsteen-type structure, as it has been mentioned above. Four stable configurations of G-octets with D_4, C_4 and S_4 symmetry formed by the pairs of guanine quartets with Hoogsteen, bifurcated or mixed system of H-bonds were identified. In contrast to G-quartets, the most stable structure of G-octet in aqueous medium was shown to be S_4-symmetric assembly consisting of the pair of mixed Hoogsteen-bifurcated type G-quartets.

Protonation of guanine quartets and a two-plane guanine quartet stack was studied [117]. For G-quartets, the optimized geometries were obtained at the B3LYP/LACVP** level of theory. Relative energies were obtained by performing single point calculations at the B3LYP/6-311+G(2df, p) level for G_4 and B3LYP/6-311+G(d, p) level for G-octet. The singly protonated G_4 complex prefers protonation at the Watson–Crick face of the O6 moiety. However, all multi-protonated G_4 complexes were found to favour protonation at the Hoogsteen face of the O6 base centres. The proton affinities were also calculated for the addition of one, two, three and four protons to the central oxygens of G_4 and compared with those of monomeric guanine and other biochemically appropriate bases. These results suggest that guanine quartet unit might reasonably readily accept two protons. For the singly to quadruply protonated octets, the added protons prefer to distribute over both planes with maximally two per plane. Furthermore, unlike the (G_4-nH^+) complexes ($n>2$), protonation at the Watson–Crick faces of the O6 moieties was found to be preferred for all protonation states. In addition, (2G_4-nH^+) complexes ($n=1$–4) were also obtained in which inter-plane hydrogen bonds were formed, effectively enabling the protons to "sit between" the planes.

Thus QM methods, primarily DFT calculations, have been successfully used to study guanine quartets and octets and their interactions. It should be remembered however that while the conventional DFT is much superior to MM force fields and can accurately calculate H-bonding in G-quartets and guanine-cation interactions, most DFT functionals do not account for π-π-stacking and therefore cannot correctly describe the interactions between different G-quartets [68]. In order to accurately calculate stacking interactions, one can alternatively employ e.g. the MP2 method. Nice example is a recent work [118] examining how guanine base stacking influences the stability of G-quadruplexes. Quartet models were created by first performing QM geometry optimization of a single G-tetrad at the MP2/6-31(d, p) level. The optimized tetrad was then used to create models for single-point energy calculations and structural characterization, which contain two parallel G-tetrads with a single central K^+ ion. The geometries of stacked tetrads were varied by their separation and relative rotation. Single-point interaction energy calculations used to create energy landscapes were performed at the MP2 level using the modified split valance basis set 6-31G*(0.25). The calculations of stacked G-tetrads revealed large energy differences of up to 12 kcal/mol between different experimentally observed geometries at the interface of stacked G-quadruplexes. Energy landscapes were also computed using an AMBER molecular mechanics description of stacking energy and were shown to agree quite well with QM calculated landscapes.

Nowadays, old issue of deficient description of stacking interactions in DFT has been satisfactorily resolved, and many current DFT approaches include quite well the dispersion energy. The most popular and computationally most effective way to do so is to add well-calibrated empirical dispersion force-field-like correction to the DFT electronic structure calculations [47]. New empirical dispersion corrections such as D3 (DFT-D3) can be successfully applied for the study of G-quartet systems [119].

Gradient optimization of a two-quartet structure (i.e. G-octet) can result in BSSE error originating from the incompleteness of the basis set of atomic orbitals and causing an artefactual stabilization of complexes. As has been already mentioned, this error can be corrected for single-point calculations by employing the standard counterpoise method [101]. It should be mentioned that empirical dispersion corrections are also able to absorb small BSSE effects [120].

6.3.2.4 QM and QM/MM Modelling of Small Organic Molecules Binding to G4 DNA

Determining the accurate structures of guanine quartets and their stacks is a key step in the development of specific G4 ligands using the computer modelling approaches. However, drug design is based on studying the interactions of potential ligands with their biomolecular targets, in this case quadruplex DNA. Molecular dynamics is most often employed in these studies. G-quadruplexes for which X-ray and NMR structures are available are large molecular systems (from 500 to 1000 atoms) that cannot be easily computed with non-empirical QM methods. For this reason, quadruplex fragments can be used as G4 models to perform quantum chemical studies on ligand binding. Of course, the optimization of ligand structures is routinely performed by QM calculation methods, usually DFT (often followed by the manual docking of the optimized ligand onto G4 target and MD simulations of quadruplex-ligand interaction).

Theoretical calculations applying the "pure" QM method to the studies on quadruplex-ligand binding are quite rare. We have already mentioned a work studying the circulene stacking complexes with G-quartet by DFT method [107]. We have proposed a G-octet as a simple and convenient model to study quadruplex-ligand binding, quite adequate at least in the case of neutral ligands that do not interact with phosphate DNA backbone [96].

"Pure" QM methods have some intrinsic limitations, including e.g. insufficient sampling of conformational space and difficulties with modeling the systems evolving in time. These problems can be successfully overcome by MD approaches. In the studies of large biomacromolecules QM and MM methods are sometimes combined into a single relatively fast QM/MM approach where only a functionally important part of the system is modelled with QM, whereas the most of the molecule is modelled using MM [121]. Application of combined quantum and molecular mechanical methods focuses on predicting activation barriers and the structures of stationary points for organic and biomolecular reactions. Characterization of the factors that stabilize transition structures in solution and in active sites of biomolecules provides a basis for design and optimization of catalysts and drugs [122, 123].

Combined QM and MM methods were applied to investigate the nature of stacking interactions in triple stacks of guanine quartets which is a central part of the human telomeric DNA and a drug target [124]. In this fundamental theoretical work Clay and Gould studied in detail the differences in the human telomeric structures, including the structural changes observed upon changing the potassium to sodium cation. The QM calculations were carried out at the DFT B3LYP and HF levels of theory using the 3–21G* and 6–31G** basis sets. The molecular dynamics simulations were carried out using the AMBER8 suite of programs with the Cornell force field [125, 126]. It was concluded that the sodium filled guanine core may appear to be energetically more stable than for the potassium case from the QM calculations, but the partial QM optimization indicated that the guanine core is not stable and the MD simulations showed that even with the DNA structure present, the core does not remain stable. This could be due to the fact that the structure was based on the potassium form and not the sodium one which has a different strand pattern and the bases are a mixture of *syn* and *anti* conformations rather than just *anti*.

Another work that employed the combined QM/MM method studied the interaction of preclinical 9-aminoacridine anticancer derivatives with a human telomeric quadruplex [127]. The mixed pseudo-bond *ab initio* QM/MM approach was used along with a molecular docking and MD simulations of G4-ligand complexes. For the QM/MM calculations, the DNA-ligand system resulting from the docking study was first partitioned into a QM subsystem and an MM subsystem. The reaction system used a smaller QM subsystem consisting of the ligand and bases within 3.5 Å, whereas the rest of the system (the MM subsystem) was treated using the AMBER force field, together with a low memory convergence algorithm. The boundary problem between the QM and MM subsystems was treated using the pseudo-bond approach. With this quadruplex-substrate QM/MM system, an iterative optimization procedure was applied, using B3LYP/3-21G* QM/MM calculations, leading to an optimized structure for the reactants. The convergence criterion used was set to obtain an energy gradient below 10^{-4}, using the twin-range cut-off method for non-bonded interactions, with a long-range cut-off of 14 Å and a short-range cut-off of

8 Å. It was shown that 9-aminoacridines selectively bind to G-quadruplex sequence between A and G-tetrads, involving significant π- π-interactions and several strong hydrogen bonds. The specific interactions between different moieties of the ligands to the DNA were shown to play a key role in governing the overall stabilities of G4 complexes. The ligands were found to induce different level of structural stabilization through intercalation. This unique property of altering structural stability is likely a contributing factor for affecting telomerase function and, subsequently, the observed differences in the anticancer activities between the studied 9-aminoacridines.

The molecular modelling studies on binding of novel dimethylamino-ethyl-acridine analogues to G-quadruplex DNA were described in [128]. The comparison of force field and quantum polarized docking methods was performed. The docking study was conducted at three levels: (a) Glide XP [129] force field docking, (b) Quantum Polarized Ligand Docking (QPLD) using Jaguar software [130] with B3LYP density functional method and LACVP basis sets, and (c) QPLD docking with B3LYP density function method and LACVP* basis sets. Ultimately, the results from each of these methods were compared and contrasted for obtaining useful insights. Binding energies were calculated for a number of ligands to identify three drug-like molecules for future optimization.

As we have already mentioned, "pure" quantum chemical methods are now in common use when G4 ligand structure optimization is considered. It is of great importance for the subsequent modelling of ligand-quadruplex binding by any method, including docking or MD simulations, and at the same time can provide an interesting information helping to understand experimental data. For example, our quantum chemical study of the molecular structure of efficient telomerase inhibitors, cationic porphyrin-imidazophenazine conjugates and their metal complexes [131] supplemented and explained spectral data. Calculations were performed by DFT method using M06 and M06-2X [88, 132] functionals that are known to adequately describe stacking interactions, and 6-31G(d) and 6-31G(d, p) basic sets. Full geometry optimization was performed in vacuum and in water, employing the supermolecular approximation and CPCM model [133] to consider the solvent effects. Calculations demonstrated that conjugates could form stable intramolecular complexes due to either stacking interaction or metal coordination between the chromophores. Both in vacuum and water, two types of complexes are formed. Non-metalated conjugate was found to adopt the conformation with coplanar chromophores stabilized by π-π-stacking. At the same time, a hybrid containing Zn(II) porphyrin complex forms different structure where the metal ion coordinates a nitrogen atom of Imidazophenazine fragment. The folding of linear conjugates to form intramolecular complexes is energetically very favourable; e.g., for Zn(II) complex ΔG^{298} of the process in vacuum and water is 15.61 and 12.34 kcal/mol, respectively. The computation results fully confirmed the experimental data obtained for these conjugates and their interaction with Tel22 G-quadruplex [134, 135]. The absorption and fluorescence studies of the hybrids revealed the formation of intramolecular heterodimers based on strong electronic interaction between the cationic porphyrin and imidazophenazine heterocycle, which obviously affects the binding of these telomerase inhibitors to intramolecular G4.

We would like to mention here also the recent work by Nicoludis et al. [136], where B3LYP/6-31G(d) calculations were successfully applied to the detailed study of the structure of mesoporphyrin IX and N-methyl mesoporphyrin IX and possible ways of binding of these specific G4 binders to human telomeric DNA sequence d[AGGG(TTAGGG)$_3$] (Tel22).

6.4 Conclusions

Increasing understanding of the molecular basis of cancer has resulted in the identification of a number of novel molecular targets for anticancer drugs, including telomerase and quadruplex nucleic acids that play critical role in the development of tumors and other pathologies. The variability of DNA structures (from common single- and double-stranded to more complex triplex and quadruplex forms) and their conformational flexibility are the key factors in diverse biological functions of DNA. Among possible DNA structures, G-quadruplexes deserve a special attention. The formation of these non-canonical assemblies in telomeres and some gene promoter regions is a way of regulating the variety of basic biological processes in a living cell. Taking into account important biological functions of G-quadruplexes, the understanding of the structural, electronic and thermodynamic properties of these DNA arrangements, their topology, dynamics, stability and mechanisms of interaction with small molecules is of fundamental interest to biology, biomedical science and pharmacology, as well as supramolecular chemistry and nanotechnology (see e.g. [78, 79, 137] and references therein). In this review we have discussed biological functions and structural features of G-quadruplex DNA and quadruplex-binding compounds, and focused on molecular modelling methods being used in the studies of these specific assemblies and their interaction with low-molecular ligands, including metal cations and small organic molecules of potential interest to pharmacology. As the structures of more and more quadruplexes and G4-ligand complexes become available, modern computational approaches, along with biochemical and biophysical experimental methods, are increasingly considered indispensable tools in the study of DNA quadruplexes. These *in silico* tools allow deeper insights into quadruplex structures and energetic features and, even more important, predicting of many important properties of G4 and quadruplex-ligand complexes. A number of successful anticancer drugs have already emerged from molecular modelling studies. However, the real challenge in medicinal chemistry over the next years will be the development of novel drugs that would not only be selective to quadruplex over duplex nucleic acids to efficiently bind to G4 structures, but would be also able to discriminate between the unique quadruplex topologies. Quantum chemical computational approaches will undoubtedly be a key player in achieving this exciting goal.

Acknowledgments Molecular visualization was performed with the UCSF Chimera package developed by the Resource for Biocomputing, Visualization, and Informatics at the University of California, San Francisco (supported by NIGMS P41-GM103311).

References

1. Harley CB (2008) Telomerase and cancer therapeutics. Nature Rev Cancer 8:167–179.
2. Rudolph KL (2010) Telomeres and Telomerase in Aging, Disease, and Cancer. Springer-Verlag, Berlin-Heidelberg
3. Tian X, Chen B, Liu X (2010) Telomere and telomerase as targets for cancer therapy. Appl Biochem Biotechnol 160:1460–1472
4. Ruden M, Puri N (2012) Novel anticancer therapeutics targeting telomerase. Cancer Treat Rev doi:10.1016/j.ctrv.2012.06.007
5. De Cian A, Lacroix L, Douarre C, Temime-Smaali N, Trentesaux C, Riou J-F, Mergny J-L (2008) Targeting telomeres and telomerase. Biochimie 90:131–155
6. Tarkanyi I, Aradi J (2008) Pharmacological intervention strategies for affecting telomerase activity. Further prospects to treat cancer and degenerative diseases. Biochimie 90:156–172
7. Xu Y (2011) Chemistry in human telomere biology: structure, function and targeting of telomere DNA/RNA. Chem Soc Rev 40:2719–2740
8. Andrews L, Tollefsbol TO. (eds) (2011) Telomerase Inhibition: Strategies and Protocols. Humana Press, New York.
9. Chen H, Li Y, Tollefsbol TO. (2009) Strategies targeting telomerase inhibition. Mol Biotechnol 41:194–199
10. Neidle S (2010) Human telomeric G-quadruplex: the current status of telomeric G-quadruplexes as therapeutic targets in human cancer. FEBS J 277:1118–1125
11. Collie G.W, Parkinson GN (2011) The application of DNA and RNA G-quadruplexes to therapeutic medicines. Chem Soc Rev 40:5867–5892.
12. Kaushik M, Kaushik S, Bansal A, Saxena S, Kukreti S. (2011) Structural diversity and specific recognition of four stranded G-quadruplex DNA. Curr Mol Med 11:744–769
13. Duchler M (2012) G-quadruplexes: targets and tools in anticancer drug design. J Drug Target 20:389–400
14. Huppert JL, Balasubramanian S (2007) G-quadruplexes in promoters throughout the human genome. Nucleic Acids Res. 35:406–413
15. Balasubramanian S, Hurley LH, Neidle S (2011) Targeting G-quadruplexes in gene promoters: a novel anticancer strategy?. Nature Rev Drug Discov 10:261–275
16. Ji X, Sun H, Zhou H, Xiang J, Tang Y, Zhao C (2011) Research progress of RNA quadruplex. Nucleic Acid Ther 21:185–200
17. Millevoi S, Moine H, Vagner S (2012) G-quadruplexes in RNA biology. Wiley Interdisc. Rev RNA 3:495–507
18. Xu Y, Komiyama M (2012) Structure, function and targeting of human telomere RNA. Methods 57:100–105
19. Hardin CC, Perry AG, White K (2001) Thermodynamic and kinetic characterization of the dissociation and assembly of quadruplex nucleic acids. Biopolym 56:147–194
20. Kumar N, Maiti S (2005) The effect of osmolytes and small molecule on Quadruplex-WC duplex equilibrium: a fluorescence resonance energy transfer study. Nucleic Acids Res 33:6723–6732.
21. Monchaud D, Teulade-Fichou M-P (2008). A hitchhiker's guide to G-quadruplex ligands. Org Biomol Chem 6:627–636
22. Le TVT, Han S, Chae J, Park H-J (2012). G-quadruplex binding ligands: from naturally occurring to rationally designed molecules Curr Pharm Des 18:1948–1972
23. Paul A, Bhattacharya S (2012) Chemistry and biology of DNA-binding small molecules. Curr Sci 102:212–231
24. Gonzalez-Ruiz V, Olives AI, Martin MA, Ribelles P, Ramos MT, Menendez JC (2011) An overview of analytical techniques employed to evidence drug-DNA interactions. Applications to the design of genosensors. In: Olsztynska S. (ed.). Biomedical engineering. trends, research and technologies. InTech, Rijeka, p. 65–90

25. Murat P, Singh Y, Defrancq E (2011) Methods for investigating G-quadruplex DNA/ligand interactions. Chem Soc Rev 40:5293–307
26. Palchaudhuri R, Hergenrother PJ (2007) DNA as a target for anticancer compounds: methods to determine the mode of binding and the mechanism of action. Curr Opin Biotechnol 18:497–503
27. Sun D, Hurley H (2010) Biochemical techniques for the characterization of G-quadruplex structures: EMSA, DMS footprinting and DNA polymerase stop assay. Methods Mol Biol 608:65–79
28. Mergny J-L, Lacroix L (2009) UV melting of G-quadruplexes. Curr Protoc Nucleic Acid Chem 37:17.1.1–17.1.15. doi: 10.1002/0471142700.nc1701s37
29. Olsen CM, Marky LA (2010) Monitoring the temperature unfolding of G-quadruplexes by UV and circular dichroism spectroscopies and calorimetry techniques. Methods Mol Biol 608:147–158
30. Vorlíčková M, Kejnovská I, Sagi J, Renčiuk D, Bednářová K, Motlová J, Kypr J (2012) Circular dichroism and guanine quadruplexes. Methods 57:64–75
31. Juskowiak B, Takenaka S (2006) Fluorescence resonance energy transfer in the studies of guanine quadruplexes. Methods Mol Biol. 335:311–341
32. Okumus B, Ha T (2010) Real-time observation of G-quadruplex dynamics using single-molecule FRET microscopy. Methods Mol Biol 608: 81–96
33. De Cian A, Guittat L, Kaiser M, SaccÃ B, Amrane S, Bourdoncle A, Alberti P, Teulade-Fichou M-P, Lacroix L, Mergny J-L (2007) Fluorescence-based melting assays for studying quadruplex ligands. Methods 42:183–195
34. Zhao Y, Kan Z-y, Zeng Z-x, Hao Y-h, Chen H, Tan Z (2004) Determining the folding and unfolding rate constants of nucleic acids by biosensor. application to telomere G-quadruplex. J Am Chem Soc 126:13255–13264
35. Redman JE (2007) Surface plasmon resonance for probing quadruplex folding and interactions with proteins and small molecules. Methods 43:302–312
36. Schlachter C, Lisdat F, Frohme M, Erdmann VA, Konthur Z, Lehrach H, Glökler J (2012) Pushing the detection limits: The evanescent field in surface plasmon resonance and analyte-induced folding observation of long human telomeric repeats. Biosens Bioelectron 31:571–574
37. Parkinson GN, Lee MP, Neidle S (2002) Crystal structure of parallel quadruplexes from human telomeric DNA. Nature 417:876–880
38. Li J, Correia JJ, Wang L, Trent JO, Chaires JB (2005) Not so crystal clear: the structure of the human telomere G-quadruplex in solution differs from that present in a crystal. Nucleic Acids Res 33:4649–4659
39. Campbell NH, Parkinson GN (2007) Crystallographic studies of quadruplex nucleic acids. Methods 43:252–263
40. Neidle S, Parkinson GN (2008) Quadruplex DNA crystal structures and drug design. Biochimie 90:1184–1196
41. Campbell N, Collie GW, Neidle S (2012) Crystallography of DNA and RNA G-quadruplex nucleic acids and their ligand complexes. Curr Protoc Nucleic Acid Chem 50:17.6.1–17.6.22. doi: 10.1002/0471142700.nc1706s50
42. Webba da Silva M (2007) NMR methods for studying quadruplex nucleic acids. Methods 43:264–277
43. Adrian M, Heddi B, Phan AT (2012) NMR spectroscopy of G-quadruplexes. Methods 57:11–24
44. Hehre JW, Radom L, Schleyer P, Pople J (1986) Ab initio molecular orbital theory. John Wiley and Sons, New York
45. Berkert U, Allinger NL (1982) Molecular Mechanics. American Chem Soc Washington, DC
46. Tsai SC (2007) Biomacromolecules. Introduction to Structure, Function and Informatics. Wiley-Liss, New York p. 249–288
47. Šponer J, Šponer JE, Mládek A, Jurečka P, Banáš P, Otyepka M (2013) Nature and magnitude of aromatic base stacking in DNA and RNA: Quantum chemistry, molecular mechanics, and experiment. Biopolym, 99, 978–988.

48. McCammon JA, Harvey SC (1987) Dynamics of Proteins and Nucleic Acids. Cambridge University Press, New York
49. Špačková N, Berger I, Šponer J (2001) Structural dynamics and cation interactions of DNA quadruplex molecules containing mixed guanine/cytosine quartets revealed by large-scale MD simulations. J Am Chem Soc 123:3295–3307
50. Šponer J, Špačková N (2007) Molecular dynamics simulations and their application to four-stranded DNA. Methods 43:278–290
51. Haider S, Parkinson GN, Neidle S (2008) Molecular dynamics and principal components analysis of human telomeric quadruplex multimers. Biophys J 95:296–311
52. Šponer J, Cang X, Cheatham TE III (2012) Molecular dynamics simulations of G-DNA and perspectives on the simulation of nucleic acid structures. Methods 57:25–39
53. Chowdhury S, Bansal M (2000) Effect of coordinated ions on structure and flexibility of parallel G-quandruplexes: a molecular dynamics study. J Biomol Struct Dyn 18:11–28
54. Cavallari M, Calzolari A, Garbesi A., DiFelice R (2006) Stability and migration of metal ions in G4-wires by molecular dynamics simulations. J Phys Chem B 110:26337–26348
55. Akhshi P, Acton G, Wu G (2012) Molecular dynamics simulations to provide new insights into the asymmetrical ammonium ion movement inside of the $[d(G_3T_4G_4)]_2$ G-quadruplex DNA structure. J Phys Chem B 116:9363–9370
56. Novotny J, Kulhanek P, Marek R (2012) Biocompatible xanthine-quadruplex scaffold for ion-transporting DNA channels. J Phys Chem Lett 3:1788–1792
57. Read MA, Wood AA, Harrison JR, Gowan SM, Kelland LR, Dosanjh HS, Neidle S (1999) Molecular modelling studies on G-quadruplex complexes of telomerase inhibitors: structure-activity relationships. J Med Chem 42, 4538–4546
58. Gavathiotis E, Heald RA, Stevens MFG, Searle MS (2003) Drug recognition and stabilisation of the parallel-stranded DNA quadruplex $d(TTAGGGT)_4$ containing the human telomeric repeat. J Mol Biol 334:25–36
59. Agrawal S, Ojha RP, Maiti S (2008) Energetics of the human Tel-22 quadruplex-telomestatin interaction: a molecular dynamics study. J Phys Chem B 112:6828–6836
60. Hou J-Q, Chen S-B, Tan J-H, Ou T-M, Luo H-B, Li D, Xu J, Gu L-Q, Huaang Z-S (2010) New insights into the structures of ligand-quadruplex complexes from molecular dynamics simulations. J Phys Chem B 114:15301–15310
61. Li M-H, Luo Q, Li Z-S (2010) Molecular dynamics study on the interactions of porphyrin with two antiparallel human telomeric quadruplexes. J Phys Chem B 114:6216–6224.
62. Li J, Jin X, Hu L, Wang J, Su Z (2011) Identification of nonplanar small molecule for G-quadruplex grooves: molecular docking and molecular dynamic study. Bioorg Med Chem Lett 21:6969–6972
63. Hou J-Q, Chen S-B, Tan J-H, Luo H-B, Li D, Gu L-Q, Huaang Z-S. (2012) New insights from molecular dynamic simulation studies of the multiple binding sodes of a ligand with G-quadruplex DNA. J Comput Aided Mol Des 26:1355–1368
64. Holt PA, Chaires JB, Trent JO (2008) Molecular docking of intercalators and groove-binders to nucleic acids using Autodock and Surflex. J Chem Inf Model 48:1602–1615
65. Ma D-L, Chan DS-H, Lee P, Kwan MH-T, Leung C-H (2011) Molecular modeling of drug-DNA interactions: virtual screening to structure-based design. Biochimie 93:1252–1266
66. Ma D-L, Ma VP-Y, Chan DS-H, Leung K-H, Zhong H-J, Leung C-H. (2012) In silico screening of quadruplex-binding ligands. Methods 57:106–114
67. Alcaro S, Musetti C, Distinto S, Casatti M, Zagotto G, Artese A, Parrotta L, Moraca F, Costa G, Ortuso F, Maccioni E, Sissi C (2013).Identification and characterization of new DNA G-quadruplex binders selected by a combination of ligand and structure-based virtual screening approaches. J Med Chem 56:843–855
68. Haider S, Neidle S (2010) Molecular modeling and simulation of G-quadruplexes and quadruplex-ligand complexes. In: Baumann P. (ed.). G-Quadruplex DNA: Methods and Protocols. Methods Mol Biol, vol 608. Humana Press, New York, p. 17–37
69. Lee JY, Okumus B, Kim DS, Ha T (2005) Extreme conformational diversity in human telomeric DNA. Proc Natl Acad Sci USA 102:18938–18943

70. Burge S, Parkinson GN, Hazel P, Todd AK, Neidle S (2006) Quadruplex DNA: sequence, topology and structure. Nucleic Acids Res 34:5402–5415
71. Dai J, Carver M, Yang D (2008) Polymorphism of human telomeric quadruplex structures. Biochimie 90:1172–1183
72. Yang D, Okamoto K (2010) Structural insights into G-quadruplexes: towards new anticancer drugs. Future Med Chem 2:619–646
73. Ambrus A, Chen D, Dai J, Bialis T, Jones RA, Yang D (2006) Human telomeric sequence forms a hybrid-type intramolecular G-quadruplex structure with mixed parallel/antiparallel strands in potassium solution. Nucleic Acids Res 34:2723–2735
74. Wang Y, Patel DJ (1993) Solution structure of the human telomeric repeat d[AG3(T2AG3)3] G-tetraplex. Structure 1:263–282
75. Pettersen EF, Goddard TD, Huang CC, Couch GS, Greenblatt DM, Meng EC, Ferrin TE (2004) UCSF Chimera – a visualization system for exploratory research and analysis. J Comput Chem 25:1605–1612
76. Jain AK, Bhattacharya S (2011) Interaction of G-quadruplexes with nonintercalating duplex-DNA minor groove binding ligands. Bioconjugate Chem 22:2355–2368
77. Trotta R, De Tito S, Lauri I, La Pietra V, Marinelli L, Cosconati S, Martino L, Conte MR, Mayol L, Novellino E, Randazzo A (2011) A more detailed picture of the interactions between virtual screening-derived hits and the DNA G-quadruplex: NMR, molecular modelling and ITC studies. Biochimie 93:1280–1287
78. Williamson JR (1994) G-quartet structures in telomeric DNA. Annu Rev Biophys Biomol Struct 23:703–730
79. Davies JT (2004) G-quartets 40 years later: from 5'-GMP to molecular biology and supramolecular chemistry. Angew. Chem Int Ed 43:668–698
80. Riley KE, Hobza P (2011) Noncovalent interactions in biochemistry. Wiley Interdisc. Revs Comput Mol Sci 1:3–17
81. Bryan TM, Baumann P (2011) G-Quadruplexes: from guanine gels to chemotherapeutics. Mol. Biotechnol 49:198–208
82. Chong DP (ed) (1995) Recent advances in density functional methods. V. 1. recent advances in density functional methods (Part I). World Scientific Publishing, Singapore
83. Wesolowski TA, Wang YA (eds) (2013) Recent advances in computational chemistry. V. 6. recent progress in orbital-free density functional theory. World Scientific Publishing, Singapore
84. Jissy AK, Ashik UPM, Datta A (2011) Nucleic acid G-quartets: insights into diverse patterns and optical properties. J Phys Chem C 115:12530–12546
85. Becke AD (1993) Density-functional thermochemistry. III. The role of exact exchange. J Chem Phys 98:5648–5652
86. Lee C, Yang W, Parr RG (1988) Development of the Colle-Salvetti correlation-energy formula into a functional of the electron density. Phys Rev B 37:785–789
87. Zhao Y, Schultz NE, Truhlar DG (2006) Design of density functionals by combining the method of constraint satisfaction with parametrization for thermochemistry, thermochemical kinetics, and noncovalent interactions. J Chem Theory Comput 2:364–382
88. Zhao Y, Truhlar DG (2008) The M06 suite of density functionals for main group thermochemistry, thermochemical kinetics, noncovalent interactions, excited states, and transition elements: two new functionals and systematic testing of four M06-class functionals and 12 other functionals. Theor Chem Acc 120:215–241
89. Jissy AK, Datta A (2010) Designing molecular switches based on DNA-base mispairing. J Phys Chem B 114:15311–15318
90. Kendall RA, Fruchtl HA (1997) The impact of the resolution of the identity approximate integral method on modern ab initio algorithm development. Theor Chem Acc 97:158–163
91. Vahtras O, Almlof J, Feyereisen MW (1993) Integral approximations for LCAO-SCF calculations. Chem Phys Lett 213:514–518

92. Schlund S, Schmuck C, Engels B (2005) Knock-out" analogues as a tool to quantify supramolecular processes: a theoretical study of molecular interactions in guanidiniocarbonyl pyrrole carboxylate dimers J Am Chem Soc 127:11115–11124
93. Ahlrichs R, Bär M, Häser M, Horn H, Kölmel C (1989) Electronic structure calculations on workstation computers: the program system turbomole. Chem Phys Lett 162:165–169.
94. Von Arnim M, Ahlrichs R (1998) Performance of parallel TURBOMOLE for density functional calculations. J Comput Chem 19:1746–1757
95. Miertus S, Scrocco, E, Tomasi J (1981) Electrostatic interaction of a solute with a continuum. A direct utilizaion of ab initio molecular potentials for the prevision of solvent effects. Chem Phys 55:117–129
96. Ilchenko MM, Dubey I Ya (2011) Density functional study of the structure of guanine octets in aqueous medium. Int Rev Biophys Chem 2:82–86
97. Fonseca Guerra C, van der Wijst T, Poater J, Swart M, Bickelhaupt MF (2010) Adenine versus guanine quartets in aqueous solution: dispersion-corrected DFT study on the differences in π-π-stacking and hydrogen-bonding behavior. Theor Chem Acc 125:245–252
98. Mezzache S, Alves S, Paumard J-P, Pepe C, Tabet J-C (2007) Theoretical and gas-phase studies of specific cationized purine base quartet. Rapid Commun Mass Spectrom 21:1075–1082
99. Meyer M, Steinke T, Brandl M, Sühnel J (2001) Density functional study of guanine and uracil quartets and of guanine quartet/metal ion complexes. J Comput Chem 22:109–124
100. Meng F, Wang F, Zhao X, Jalbout AF (2008) Guanine tetrad interacting with divalent metal ions ($M = Fe^{2+}$, Co^{2+}, Ni^{2+}, $Cu^{2+}$and Zn^{2+}): a density functional study. J Mol Struct: THEOCHEM 854:26–30.
101. Boys SF, Bernardi F (1970) Calculations of small molecular interaction by the difference of separate total energies. Some procedures with reduced error. Mol Phys 19:553–566
102. Bader RFW (1990) Atoms in Molecules: A Quantum Theory. Clarendon Press, Oxford, UK
103. Rozas I, Alkorta I, Elguero J (1997) Unusual hydrogen bonds: H…π interactions. J Phys Chem A 101:9457–9463
104. Deepa P, Kolandaivel P, Senthilkumar K (2011) Structural properties and the effect of interaction of alkali (Li^{+}, Na^{+}, K^{+}) and alkaline earth (Be^{2+}, Mg^{2+}, Ca^{2+}) metal cations with G and SG-tetrads. Comput Theor Chem 974:57–65
105. Meng F, Xu W, Liu C (2004) Theoretical study of incorporating 6-thioguanine into a guanine tetrad and their influence on the metal ion–guanine tetrad. Chem Phys Lett 389:421–426
106. Yurenko YeP, Novotný J, Sklenář V, Marek R (2013) Exploring non-covalent interactions in guanine- and xanthine-based model DNA quadruplex structures: A comprehensive quantum chemical approach. Phys. Chem Chem Phys DOI: 10.1039/C3CP53875C
107. Jissy AK, Datta A (2012) Effect of external electric field on H-bonding and π-stacking interactions in guanine aggregates. Chem Phys Chem 13:4163–4172
108. Gu J, Leszczynski J, Bansal M (1999) A new insight into the structure and stability of Hoogsteen hydrogen-bonded G-tetrad: an ab initio SCF study. Chem Phys Lett 311:209–314
109. Gu J, Leszczynski J (2000) A remarkable alteration in the bonding pattern: an HF and DFT study on the interactions between the metal cations and the Hoogsteen hydrogen-bonded G-tetrad. J Phys Chem A 104:6308–6313
110. van Mourik T, Dingley AJ (2005) Characterization of the monovalent ion position and hydrogen-bond network in guanine quartets by DFT calculations of NMR parameters. Chem Eur J 11:6064–6079
111. Becke AD (1997) Density-functional thermochemistry. V. Systematic optimization of exchange-correlation functionals. J Chem Phys 107:8554–8560
112. NWChem (2003) Version 4.5, High Performance Computational Chemistry Group, Pacific Northwest National Laboratory, Richland WA

113. Szolomájer J, Paragi G, Batta G, Fonseca Guerra C, Bickelhaupt FM, Kele Z, Pádár P, Kupihára Z, Kovács L (2011) 3-Substituted xanthines as promising candidates for quadruplex formation: computational, synthetic and analytical studies. New J Chem 35:486–482
114. Meyer M, Hocquet A, Sühnel J (2005) Interaction of sodium and potassium ions with sandwiched cytosine-, guanine-, thymine-, and uracil-base tetrads. J Comput Chem 26:352–364
115. Meyer M, Steinke T, Sühnel J (2007) Density functional study of isoguanine tetrad and pentad sandwich complexes with alkali metal ions. J Mol Model 13:335–345
116. Hua Y, Changenet-Barret P, Improta R, Vaya I, Gustavsson T, Kotlyar AB, Zikich D, Šket P, Plavec J, Markovitsi D (2012) Cation effect on the electronic excited states of guanine nanostructures studied by time-resolved fluorescence spectroscopy. J Phys Chem C 116:14682–14689
117. Liu H, Gauld JW (2009) Protonation of guanine quartets and quartet stacks: insights from DFT studies. Phys Chem Chem Phys 11:278–287
118. Lech CJ, Heidi B, Phan AT (2013) Guanine base stacking in G-quadruplex nucleic acids. Nucl Acids Res 41:2034–2046
119. Šponer J, Mládek A, Špačková N, Cang X, Cheatham TE, Grimme S (2013) Relative stability of different DNA guanine quadruplex stem topologies derived using large-scale quantum-chemical computations. J Am Chem Soc 135:9785–9796
120. Fonseca Guerra C, Zijlstra H, Paragi G, Bickelhaupt FM (2011) Telomere structure and stability: Covalency in hydrogen bonds, not resonance assistance, causes cooperativity in guanine quartets. Chem Eur J 17:12612–12622
121. Warshel A. (1991). Computer Modeling of Chemical Reactions in Enzymes and Solutions. John Wiley, New York.
122. Acevedo O, Jorgensen WL (2010) Advances in quantum and molecular mechanical (QM/MM) simulations for organic and enzymatic reactions. Acc Chem Res 43:142–151
123. Banáš P, Jurečka P, Walter NG, Šponer J, Otyepka M (2009) Theoretical studies of RNA catalysis: hybrid QM/MM methods and their comparison with MD and QM. Methods 49:202–216
124. Clay EH, Gould IR (2005) A combined QM and MM investigation into guanine quadruplexes J Mol Graph Model 24:138–146
125. Pearlman DA, Case DA, Caldwell JW, Ross WS, Cheatham TE III, DeBolt S, Ferguson D, Seibel G, Kollman P (1995) AMBER, a package of computer programs for applying molecular mechanics, normal mode analysis, molecular dynamics and free energy calculations to simulate the structural and energetic properties of molecules. Comput Phys Comm 91:1–41
126. Cornell WD, Cieplak P, Bayly CI, Gould IR, Merz KM, Ferguson DM, Spellmeyer DC, Fox T, Caldwell JW, Kollman PA (1995) A second generation force field for the simulation of proteins, nucleic acids, and organic molecules. J Am Chem Soc 117:5179–5197
127. Ferreira R, Artali R, Benoit A, Gargallo R, Eritja R, Ferguson DM, Sham YY, Mazzini S (2013) Structure and stability of human telomeric G-quadruplex with preclinical 9-amino acridines. PLOS One, 8, e57701. doi: 10.1371/journal.pone.0057701
128. Subramanian AK, Cardin CJ (2012) Molecular modelling studies of binding of DACD derivatives into G-quadruplex DNA: comparison of force field and quantum polarized ligand docking methods. Int J Pharm Pharm Sci 4:509–514
129. Glide (2009) version 5.5, Schrodinger, Inc, New York
130. Jaguar (2011) version 7.8, Schrodinger, Inc, New York
131. Dubey LV, Ilchenko MM, Zozulya VN, Ryazanova OA, Pogrebnoy PV, Dubey IYa (2011) Synthesis, structure and antiproliferative activity of cationic porphyrin-imidazophenazine conjugate. Int Rev Biophys Chem 2:147–152
132. Zhao Y, Truhlar DG (2006) Comparative DFT study of van der Waals complexes: rare-gas dimers, alkaline-earth dimers, zinc dimer, and zinc-rare-gas dimers. J Phys Chem A 110:5121–5129
133. Cossi M, Scalmani G, Rega N, Barone V (2002) New developments in the polarizable continuum model for quantum mechanical and classical calculations on molecules in solution. J Chem Phys 117:43–54

134. Zozulya VN, Ryazanova OA, Voloshin IM, Dubey LV, Dubey IYa (2011) Spectroscopic studies on binding of porphyrin-phenazine conjugate to intramolecular G-quadruplex formed by 22-mer oligonucleotide. Int Rev Biophys Chem 2:112–119
135. Negrutska VV, Dubey LV, Ilchenko MM, Dubey IYa (2013) Design and study of telomerase inhibitors based on G-quadruplex ligands. Biopolym. Cell 29:169–176
136. Nicoludis JM, Miller ST, Jeffrey PD, Barrett SP, Rablen PR, Lawton TJ, Yatsunyk LA (2012) Optimized end-stacking provides specificity of N-methyl mesoporphyrin IX for human telomeric G-quadruplex DNA. J Am Chem Soc 134:20446–20456
137. König SLB, Evans AC, Huppert JL (2010) Seven essential questions on G-quadruplexes. BioMol Concepts 1:197–213

Chapter 7
Density Functional Theory Calculations of Enzyme–Inhibitor Interactions in Medicinal Chemistry and Drug Design

Alexander B. Rozhenko

Abstract The density functional theory (DFT) is currently predominating theoretical approach in quantum chemistry. It is suitable for investigating structures up to several hundreds of atoms, studying of reaction pathways and calculating precisely reaction energy values. The usage of the DFT approach for studying enzyme–substrate interactions could be a prospective way for elaborating new efficient enzyme inhibitors. This is a direct way to discovery of new drugs and modification of the existing drugs. While enzymes are still too large for the computational analysis using DFT, numerous efforts have been exerted in the last years in this field using simplified enzyme models or calculating for the substrate some valuable properties, important in the enzyme–substrate interactions. These examples have been analyzed in the current review. A rapid development of new efficient calculation routines makes it possible to increase the role of the DFT methods in medicinal chemistry in the nearest future.

Listing of Used Acronyms

ACE	angiotensin-converting enzyme
AChE	acetylcholinesterase
AD	Alzheimer's disease
AIDS	acquired immune deficiency syndrome
BACE-1	betasite of APP-cleaving enzyme-1
BChE	butyrylcholinesterase
Cat B	cathepsin B
DFT	density functional theory
DNA	deoxyribonucleic acid
EP	electrostatic potential

A. B. Rozhenko (✉)
Institute of Organic Chemistry, National Academy of Sciences of Ukraine,
Murmans'ka str. 5, Kyiv 02660, Ukraine
e-mail: a_rozhenko@ukr.net

L. Gorb et al. (eds.), *Application of Computational Techniques in Pharmacy and Medicine*,
Challenges and Advances in Computational Chemistry and Physics 17,
DOI 10.1007/978-94-017-9257-8_7

EPS	electrostatic potential surface
FAAH	fatty acid amide hydrolase
FEP	free energy perturbation
HIV	human immunodeficiency virus
HMGR	3-hydroxy-3-methylglutaryl-coenzyme A reductase
HOMO	highest occupied molecular orbital
IC_{50}	half maximal inhibitory concentration
IEF	integral equation formalism
IN	integrase
LUMO	lowest unoccupied molecular orbital
MD	molecular dynamics
MD/MM	molecular mechanics/molecular dynamics
MEP	molecular electrostatic potential
MFCC	molecular fractionation with conjugate caps approach
MMP	matrix metalloproteinase
MNDO	modified neglect of diatomic overlap
MO	molecular orbital
MP2	second-order Møller-Plesset perturbation theory
PCM	polarizable continuum model
PDE	phosphodiesterase
PES	potential energy surfaces
PLA_2	phospholipases A_2 enzymes
PM3	parameterized model number 3 (Stewart's semi-empirical approach)
PMF	potential of mean force
QM/MM	quantum mechanic/molecular mechanics hybrid approach
QSAR	quantitative structure–activity relationship
RHF	restricted Hartree-Fock method
RI	resolution of the identity
RNA	ribonucleic acid
SCC-DFTB	self-consistent charge-density functional tight binding
SCRFPCM	self-consistent reaction field polarizable continuum model
SIBFA	sum of interactions between fragments *ab initio* computed
TSS	transition state structures
XO	xanthine oxidase

7.1 Introduction

Over the last few decades, quantum chemical calculations became a powerful alternative to experimental methods in medicinal and drug chemistry in creating novel drug candidates [1–3]. One of these modern approaches is based on elaboration of small molecules, binding to the active site of an enzyme and inhibiting the reaction catalyzed by the enzyme [4]. These chemical substances should also exhibit an unique selectivity for the target enzyme and exert their biological effect at low doses. Nowadays, enzyme inhibition is one of the key approaches to the drug design in the research and industry [5]. As reported in the recent review devoted to drug predictions methods [6], the drugs using the enzymes as targets amount to 24% from the total number of small-molecule medicaments. Currently the computational techniques and software have become suitable for the theoretical analysis of the enzyme–substrate interactions [2]. The molecular docking [7], based on the molecular mechanics/molecular dynamics techniques, and quantitative structure–activity relationship (QSAR) methods [8], are widely used for this purpose.

An efficient enzyme–inhibitor interaction is usually characterized by a negative free binding energy, ΔG. This is equivalent to the reaction free energy, widely used in the quantum chemical description of chemical processes. Therefore, a maximization of the enzyme–substrate negative interaction energy under control of classical quantum-mechanical methods, *ab initio* or density functional theory (DFT), would probably be the most direct way to discover new efficient drug substances and to modify the existing drugs. The detailed outlook of using the DFT for calculations of ligand–protein complexes is given in the recent review of Utkov et al. [9]. However, despite the rapid development observed in computational techniques and routines, the quantum-mechanical methods (*ab initio* and DFT approaches) are still too slow and size-restricted or too inexact (molecular mechanics or semi-empirical methods) to provide a quantitative description for kinetics and thermodynamics of enzyme inhibition processes. One way to overcome the size problem in DFT calculations is to consider the interactions with every amino acid in the polypeptide chain separately and then to sum all the contributions. This approach is realized, for example, in the molecular fractionation with conjugate caps (MFCC) approach [10, 11]. In the other approaches, such as QM/MM calculations [12–14] or ONIOM routine implemented in the Gaussian sets of programs [12], the DFT calculations are used only for the small fragments of the active site and inhibitor structure, whereas the other part is described semi-empirically or by a classic force field [15]. The enzyme–inhibitor interactions investigated by using QM/MM methods are analyzed in the several recent reviews [16–18]. DFT calculations of the enzyme inhibition processes can be also performed for the truncated (up to several hundreds atoms) enzyme binding sites [19]. Recently some important steps have been undertaken for the rapid development of DFT-based drug chemistry. First of all, several efficient linear-scaling techniques [20] such as the *Resolution of the Identity* (*RI*) [21–27] have been proposed as an efficient solution of the size problem by quantum chemistry calculations and implemented into several popular quantum mechanic program sets (see, for example Ref. [28, 29]). Other well known indirect applications of the DFT methods in medicinal chemistry and drug design should be mentioned here: (a) evaluation of structure, conformation and properties, which can directly correlate with the inhibition activity of a molecules-candidate, such as molecular orbitals (MOs) [30, 31], electron density distribution, dipole moments [32], electrostatic potential surfaces (EPS) [33–36] etc.; (b) calculations of properties of structures, subsequently used as descriptors for QSAR analysis [8, 31–39].

The current short review is not exhaustive in the field. It covers mainly the last five years and gives examples of successful using the DFT methods for elaborating new drugs based on enzyme–inhibitor interactions.

7.2 Enzyme–Substrate Interaction Modelling Using DFT Methods

7.2.1 Hydrolase

Fatty acid amide hydrolase (FAAH) catalyzes hydrolysis of several fatty acid amides, in particular, transforms arachidonoylethanolamide to arachidonic acid and ethanolamine. The FAAH inhibition has an attractive therapeutic effect for the treatment of several central nervous system disorders. The inhibition mechanisms for two efficient FAAH inhibitors, O-aryl carbamate (**1**) and piperidinyl/piperazinyl-arylurea (**2**), have been recently studied by Lodola et al. [40] using the QM/MM approach. These inhibitors carbamoylate the active-site nucleophile Ser241. The theoretical model included self-consistent charge-density functional tight binding (SCC-DFTB), the approximate density functional theory method as the quantum-mechanical part. For the crucial steps of deacylation and decarbamoylation reactions, potential energy surfaces (PESs) were calculated and compared to that for deacylation of FAAH by the acylated substrate oleamide. A carbamic group bound to Ser241 substantially increased the activation energy for the hydrolysis reaction. Moreover, the activation energy derived theoretically for **1** was lower than that found for **2**, which is in line with the experimentally found for **1** and **2** reversible and irreversible inhibition, respectively.

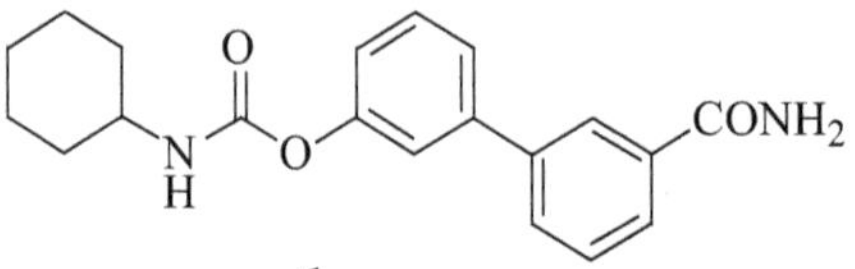

1 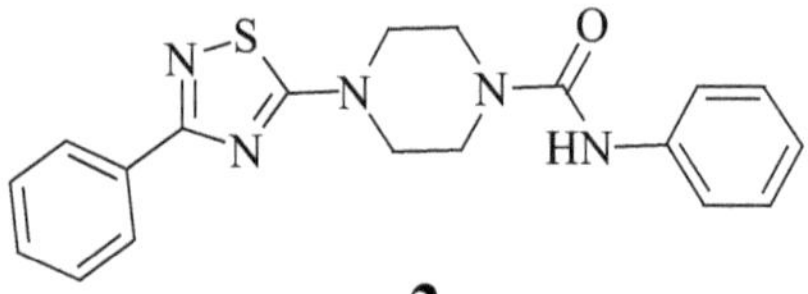 **2**

7.2.2 APP-Cleaving Enzyme-1

The betasite of APP-cleaving enzyme-1 (BACE-1) was used as target for the semi-empirical, DFT and MP2 calculations [41] for the adducts with a series of 14 hydroxyethylamines with a general formula **3** taken from Brookhaven Protein Data Bank. BACE-1 is a key enzyme in the production of Amyloid-β peptides, a major pathological feature of Alzheimer's disease [42, 43].

The interaction energy was determined for the complexes formed between the hydroxyethylamine as the BACE-1 inhibitors and 24 residues in the BACE-1 active site. After a short molecular dynamics simulation the structures were optimized at the semiempirical level of theory. The optimized structures were used for single-point energy calculations with the M062X [44] and X3LYP DFT functional [45], which account for London dispersion forces, proper hydrogen-bonding and van der Waals complexes; alternatively the MP2 level of theory was employed. The active site cavity was separated into individual fragments to isolate each residue energy contribution when interacting with each ligand. The polar interactions were predominant in the system studied. In particular, the most remarkable role played the negatively charged aspartate residues with positively charged ligand moiety, providing a main contribution (over 90 %) to the total attractive interaction energy. On the other hand, the positively charged ARG296 residue exhibited the most repulsive ion–ion interaction that should also be reduced to improve the complex stability. The interactions with non-polar residuals, such as π,π-interactions, were less important, but taking them into account at the M062X or MP2 level of theory was required for providing better agreement with the experiment in the studied series of BACE-1 inhibitors.

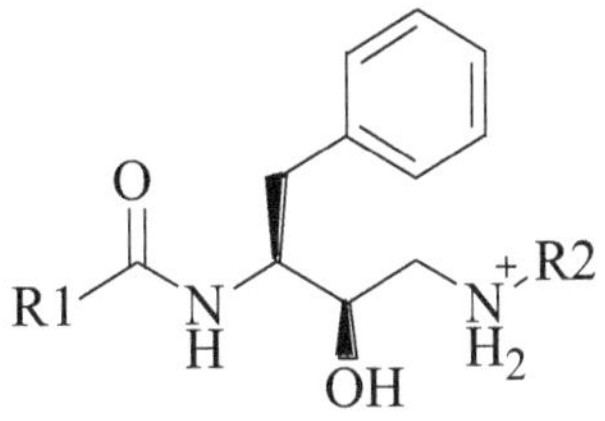

3

7.2.3 *Reductase*

A high level of cholesterol in blood (also called hypercholesterolaemia) [46] often cause the hardening and narrowing of arteries (atherosclerosis) in the major vascular systems. The cholesterol moderating statin drugs inhibit the second step in the biosynthetic pathway of producing cholesterol by binding to the active site of 3-hydroxy-3-methylglutaryl-coenzyme A reductase (HMGR), blocking the natural substrate of HMGR and disabling the synthesis of cholesterol [47]. Cafiero et al. [48] investigated theoretically (at the MP2 and DFT levels of theory) structures **4–7** and **8–10** (Fig. 7.1) – the products of modification of the existing drugs, the efficient HMGR inhibitors rosuvastatin and simvastatin, respectively. The rosuvastatin and simvastatin moieties interact with one end of the active site (Ser684, Asp690, Lys691 and Lys692), whereas the novel products bind to another end of the active site (Tyr479). The calculated interaction energies between Tyr479 and fragment **11** increased with increasing electron acceptor effects of the substituents X_1–X_3 (Table 7.1). The popular B3LYP DFT method was inadequate for this type of system and the local functional SVWN was used instead. The highest interaction

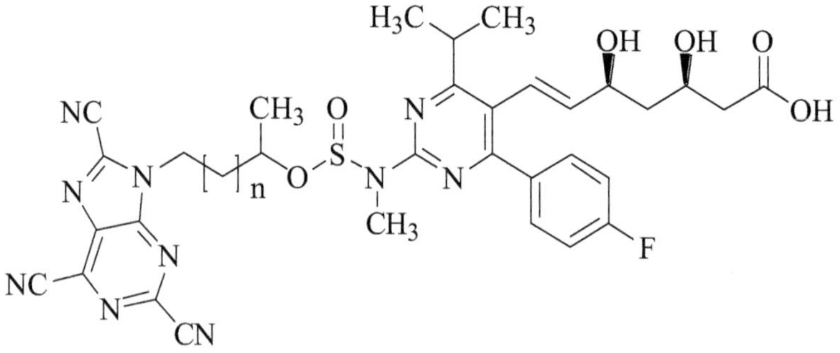

4-7

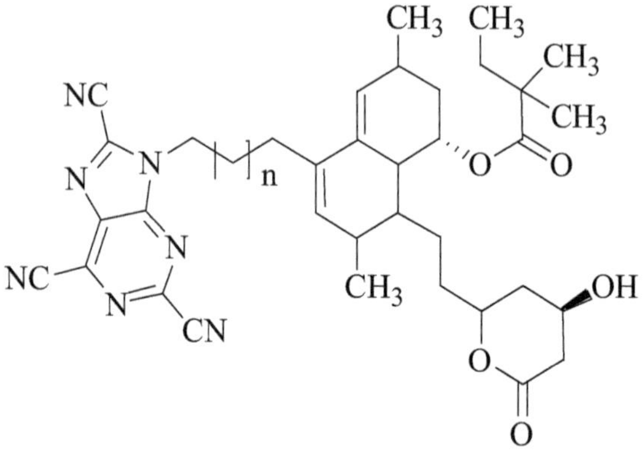

8-10

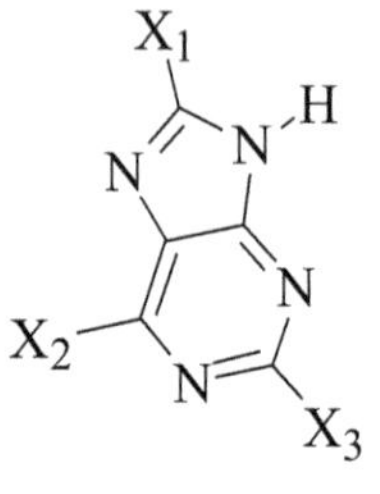

11

Fig. 7.1 Structure of drug candidates **4** (n=3), **5** (n=4), **6** (n=5), **7** (n=6), **8** (n=8), **9** (n=9) and **10** (n=10), and structure **11** as a novel drug fragment (see Table 7.1 for substituents X_1–X_3). (Reproduced with permission from Ref. [48]. Copyright © 2011 Elsevier)

Table 7.1 Counterpoise corrected interaction energies (kcal/mol) between Tyr479 and bicyclic fragment **11** by variation of the substituents X_1–X_3 (6-311++G(d,p) basis sets were used). (Reproduced with permission from Ref. [48]. Copyright © 2011 Elsevier)

Calc. method	NH_2 (X_1)	CN (X_2)	CN (X_1, X_2)	CN (X_3)	CN (X_1–X_3)	CN (X_1–X_3)	Cl (X_1–X_3)	NO_2 (X_1–X_3)
MP2	−5.45	−6.68	−8.51	−6.09	−7.94	−9.27	−8.46	−11.73
B3LYP	2.75	2.00	0.51	2.33	0.83	0.01	1.03	−3.48
SVWN	−4.31	−5.17	−6.75	−4.80	−6.34	−7.34	−6.55	−13.04

energies were predicted for X_1–X_3=NO_2, but it was not suitable for creating novel drug molecules because the bulky NO_2 group prevented the molecule's ability to penetrate the enzyme's cavity. The CN group was chosen as a good compromise for modifying the heterocyclic site.

Table 7.2 Interaction energies (kcal/mol) for novel candidate molecules with the HMG-CoA reductase active site calculated using DFT in combination with 6-311++G** basis sets and semi-empirical methods. (Reproduced with permission from Ref. [48]. Copyright © 2011 Elsevier)

	B3LYP	SVWN	HCTH407	AM1
HMG-CoA	−1.11	−39.93[3][a]	1.06	–
Rosuvastatin	−5.01	−31.51	−5.14	−106.14
4	−23.12	−57.72	−18.27	9.89
5	10.71	−26.84	12.78	5.40
6	19.18	−6.27	18.56	9.67
7	−74.22	−104.46	−73.45	−41.95
Simvastatin	−5.81	−27.64	−7.01	1.20
8	−14.43	−40.28	−14.00	−6.30
9	1.28	−32.46	1.89	−1531
10	−6.56	−29.82	−5.53	−4.70

[a] MP2 value for comparison is: −20.49 using 6-311+G* basis sets

The candidates **4–10** were docked in the active site of HMGR and then DFT and AM1 calculations were performed for the final structures of the molecules. The calculated interaction energy values (Table 7.2) indicated that the SVWN approach provided better agreement with the data obtained at the MP2 level of theory than the B3LYP and HCTH functionals.

The highest interaction energy was predicted for **7**: proton transfer occurs between the carboxyl group of **7** and NH_2 group of Lys692, leading to the very strong charge–charge interaction, whereas **1** reveals approximately twice a lower interaction energy. All three simvastatin-based candidates **8–10** seemed to interact stronger with HMGR than the original drug. Thus, the modified drugs might also possess higher efficacy.

Russo et al. [49] applied DFT for studying binding mode of flavonoids brutieridin (**12**) and melitidin (**13**). These structural analogs of statins, extracted from bergamot, inhibit HMGR, lower lipid concentration and cholesterol levels and reduce the risks of stroke [50]. Similarly to statins, brutieridin and melitidin were expected to interact effectively with the active site of the human HMGR enzyme. The active site of the enzyme was modeled starting from the X-ray structure of the adduct of simvastatin with HMGR. After the crude geometry optimization using the MD/MM simulation, the structure was truncated to the size suitable for the quantum chemical description (approx. 150 atoms) [51], considering only 17 amino acids and substituting some of them by more simple moieties. Also for modeling brutieridin and melitidin the smaller structures, **14** and **15** (Fig. 7.2) were utilized. The B3LYP/6-31+G* approach was used for geometry optimization. One H atom of each amino acid residue coming from the protein was kept frozen at its crystallographic position. The energy values were then defined more exactly at the B3LYP/6-311++G** level of approximation. The solvent effects were taken into account within the framework of Self Consistent Reaction Field Polarizable Continuum Model (SCRFPCM) using the IEF-PCM approach. The B3LYP-optimized structure of the HMGR complex with **14** is shown in Fig. 7.3a. The found binding energy (ΔE), not corrected

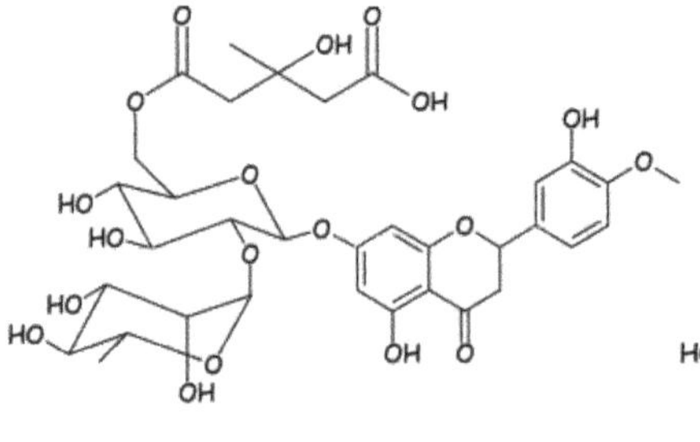

12

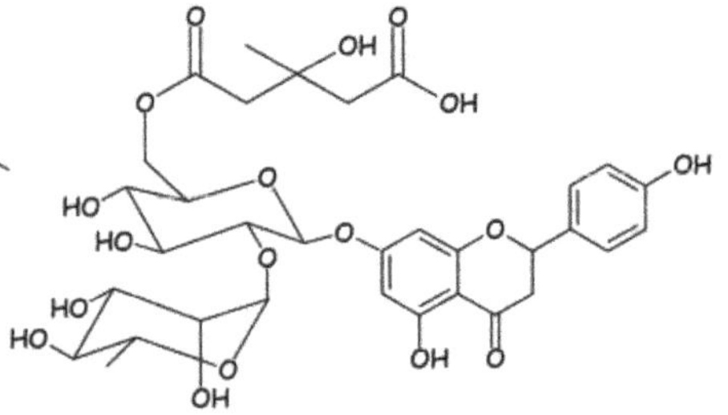

13

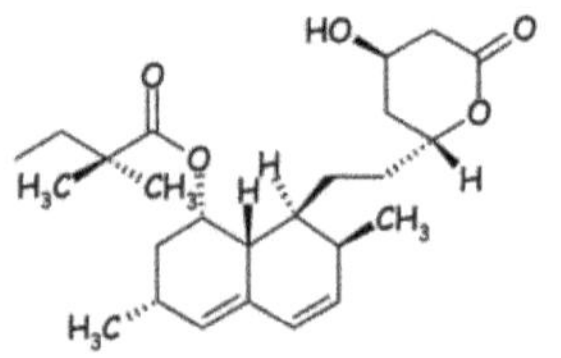

14

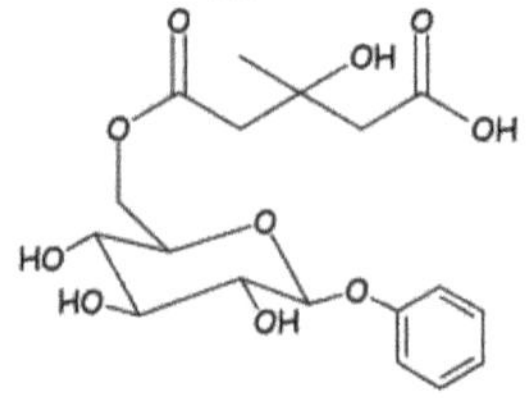

15

Fig. 7.2 Structures of brutieridin (**12**) and melitidin (**13**) and the model structures **14** and **15** used for calculations. (Reproduced with permission from Ref. [49]. Copyright © 2010 American Chemical Society)

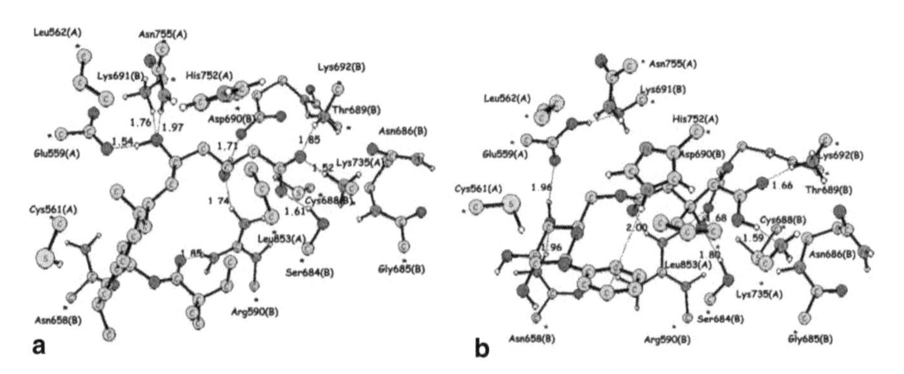

Fig. 7.3 B3LYP optimized geometry of the **3**-HMGR (*left*) and **4**-HMGR (*right*) complexes. For clarity, unimportant hydrogen atoms are omitted. (Reproduced with permission from Ref. [49]. Copyright © 2010 American Chemical Society)

for entropy effects, was rather high (−101.1 kcal/mol). The interactions in the adduct were mainly of electrostatic nature and include numerous hydrogen bonds. The equilibrium structure of the complex with **15** (Fig. 7.3b) was only slightly less favored (ΔE = −90.8 kcal/mol). Therefore, **14** and **15** are good basis structures for a development of new anticholesterolemic drugs.

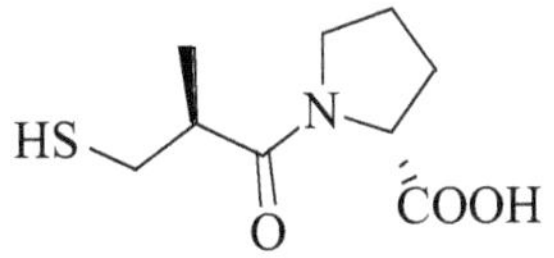

HS COOH O **16-D** HS COOH O **16-L** HS Ph COOH **17**

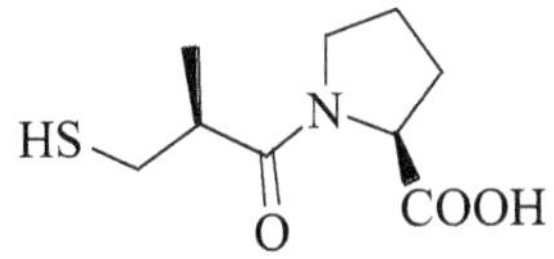

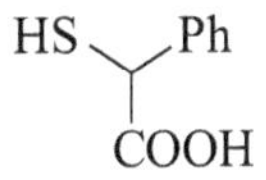

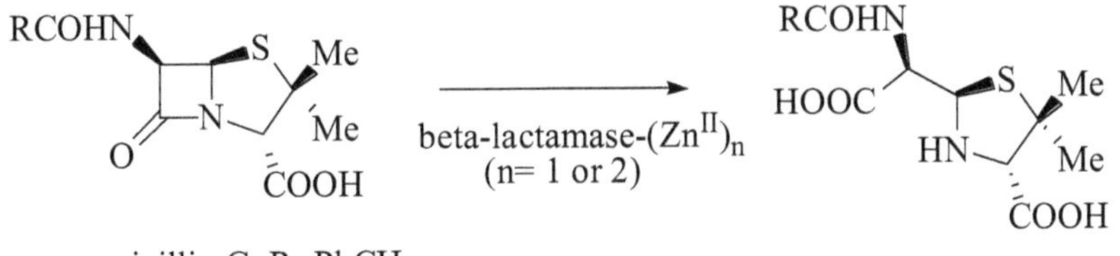

RCOHN S Me Me O COOH penicillin G, R=$PhCH_2$ → beta-lactamase-$(Zn^{II})_n$ (n= 1 or 2) RCOHN HOOC S Me HN Me COOH

Scheme 7.1 Cleavage of the lactam ring of penicillin in presence of metallo-β-lactamase

7.2.4 *Metallo-β-lactamase*

The D- and L-captopril (**16**) [52] and D- and L-thiomandelate (**17**) [53], were studied theoretically as the simplified models for potential inhibitors of bacterial Zn^{2+} metallo-β-lactamase from *B. fragilis.* This enzyme cleaves the lactam ring of penicillin (Scheme 7.1), cephalosporin, and carbapenem antibiotics, strongly reducing their efficiency, which is a serious problem in medicine.

The theoretical investigations included molecular dynamics, SIBFA (Sum of Interactions Between Fragments *Ab initio* computed), molecular mechanics, HF and DFT calculations (on models of inhibitor–enzyme complexes on small model complexes including 88 atoms, extracted from the 104-residue complexes [53]. Calculations were carried out both with uncorrelated (HF) as well as correlated (DFT, MP2) quantum chemical approaches.

7.2.5 *Topoisomerase II and T7 RNA Polymerase*

Authors [54] studied the structure–activity relationship of four new polypyridyl ruthenium(II) complexes ($[Ru(4dmb)_2(ppd)]^{2+}$ (4dmb = 4,40-dimethyl-2,20-bipyridine, ppd = pteridino[6,7-f][1,10]phenanthroline-1,13(10H,12H)-dione) (**18**), $[Ru(5dmb)_2(ppd)]^{2+}$ (5dmb = 5,50-dimethyl-2,20-bipyridine) (**19**), $[Ru(dip)_2(ppd)]^{2+}$ (dip = 4,7-diphenyl-1,10-phenanthroline) (**20**), and $[Ru(ip)_2(ppd)]^{2+}$ (ip = imidazole[4,5-f][1,10]phenanthroline) (**21**)) as topoisomerase II and T7 RNA polymerase inhibitors and potential antitumor drugs. The frontier MOs were derived using the optimized geometries of the complexes. It was suggested that the lowest unoccupied MO (LUMO) provided a more effective overlap with the highest occupied MO (HOMO) of DNA. This correlated well with the DNA affinities of the complexes, but the experimentally found trend of topoisomerase II inhibition activity was somewhat different from the DNA binding ability.

7.2.6 Cathepsin B Cysteine Protease

A series of organometallic compounds were studied using the DFT approach on the inhibitory properties against cathepsin B (cat B), a lysosomal papain-family cysteine protease [55]. Cat B is involved in cellular metabolism processes and implicated in the tumor progression and metastasis and hence it is the widely used target in medicinal chemistry. The Ru and Os complexes **22** and **23** and antimetastatic compound NAMI-A (**29**) showed similar enzyme inhibition properties *in vitro* (with IC_{50} values in the low μM range), whereas the Rh(III) and Ir(III) compounds (**24**–**28**) turned out to be inactive. As the direct coordination of the metal centre to the active site cysteine occurs, the different activities of the investigated organometallic complexes toward cat B were expected to be essentially determined by the strength of the corresponding covalent M–S bonds between the metal and cysteine residue. The authors used N-acetyl-L-cysteine-N′-methylamide (CH_3CO–NH–$CH(CH_2SH)$–CO–$NHCH_3$) as the model of cat B target, which mimics the cysteine side chain of the enzyme active site and neighboring peptide groups.

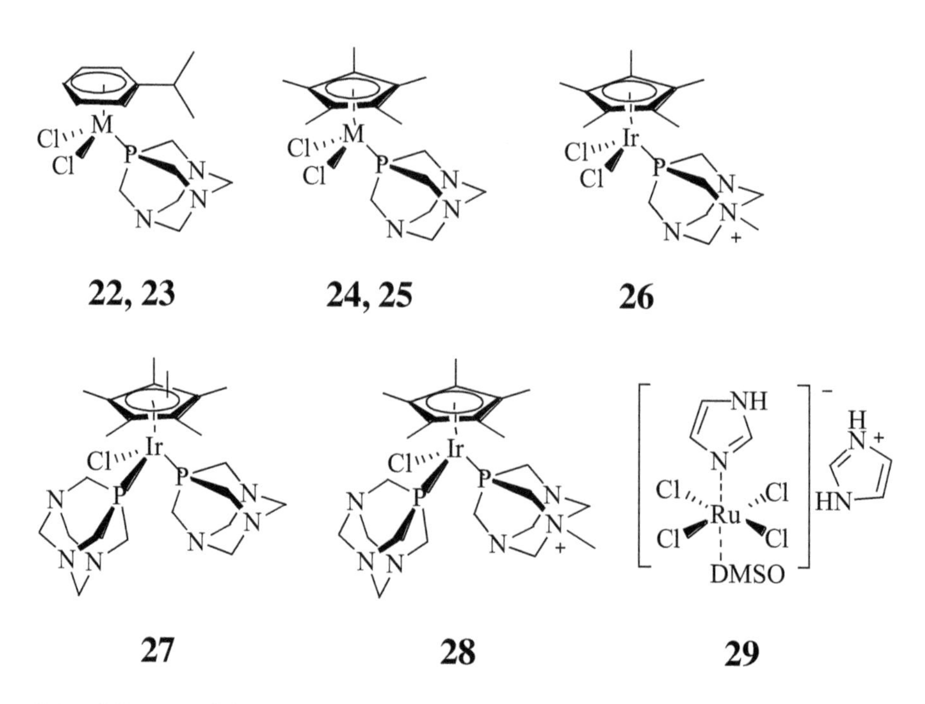

M= **22**: Ru, **23**: Os, **24**: Rh, **25**: Ir

The calculations predicted for some compounds thermodynamically favorable binding (negative ΔG values), whereas the inactive compounds were characterized by slightly positive binding free energy values. Thus, in contrast to Ru(II), Os(II) or Ru(III) complexes, the Rh(III) and Ir(III) compounds possessed a weak inhibition activity toward cat B, probably because the corresponding M–S bonds formed by these metal ions. They were characterized by *ca.* 20–30 kJ mol^{-1} lower bond energies than the more active complexes.

Shokhen et al. [56] analyzed possible mechanisms for the reversible formation of the complex between papain, a prototype enzyme of cysteine proteases, and peptidyl aldehyde inhibitors, using the quantum mechanical (DFT)/self consistent reaction field (virtual solvent) approach.

7.2.7 Acetylcholinesterase

Inhibitors of the acetylcholinesterase (AChE, E.C. 3.1.1.7) and butyrylcholinesterase (BChE, E.C. 3.1.1.8) activity demonstrate good results in the treatment of Alzheimer's disease (AD) [57, 58]. In the last years, the pathogenesis of AD has been associated with both cholinesterases, resulting in several studies that have targeted these two enzymes [59, 60]. Authors [61] investigated 88 N-aryl-substituted structures with general formulas **30** and **31** as potential inhibitors of the AChE and BChE residues using docking and density functional theory (DFT) methods. Some compounds were synthesized and their activities were tested *in vitro*. Among the candidates studied, several structures with the electron-acceptor substituents attached to the aromatic ring were predicted to be the most potent AChE inhibitors. These results demonstrated the importance of the electronic effects on ligand recognition and prompted authors to analyze HOMO and LUMO energies. They were suggested to correlate with biological activity [62]. The interaction between the amino-acids at the active site and inhibitors were considered to be determined by energies of frontier orbitals of the cholinesterase and substrate. This approach seems to be prospective for the design of new efficient AChE inhibitors.

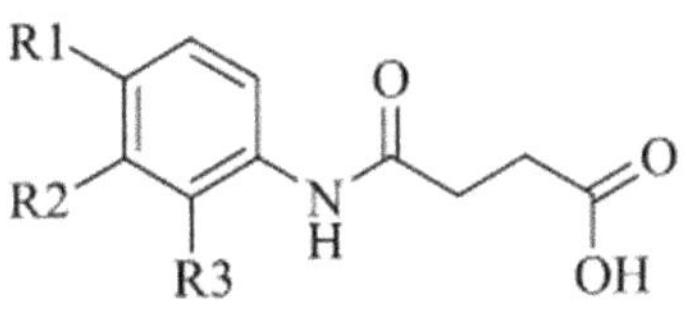

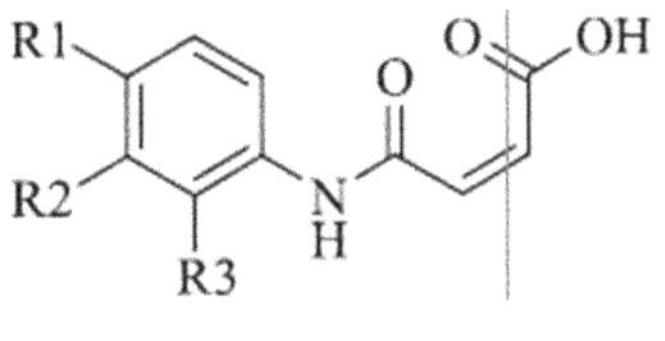

In the other work [63], semi-empirical, restricted Hartree-Fock (RHF) and DFT calculations were carried out to study the well-known acetylcholinesterase inhibitors tacrine (**32**), galantamine (**33**), donepezil (**34**), tacrine dimer (**35**), and physostigmine (**36**). Some electronic and structural properties were evaluated (Table 7.3),

Table 7.3 Electronic and geometrical parameters data for optimized AChEI structures by B3LYP/6-31+G(d,p) method. (Reproduced with permission from Ref. [63]. Copyright © 2008 Elsevier)

Property	**32**	**33**	**34**	**35**	**36**
HOMO (eV)	−5.76	−5.05	−5.95	−5.90	−5.46
Gap (eV)[a]	4.49	5.53	4.41	4.43	5.12
Volume (Å^3)	236	329	454	606	321
C–N (Å)	1.386	1.468	1.465	1.443	1.362
N–H (Å)	1.009	–	–	1.013	1.008
C–O (Å)	–	1.436	1.364	–	1.373
O–H (Å)	–	0.967	–	–	–
H–H (Å)	1.683	2.360	2.342	1.998	2.319
Charge H$^+$	0.30	0.34	0.15	0.26	0.32
Molecular size (E)	9.516	10.290	17.254	19.386	12.927

[a] Difference of energy between LUMO and HOMO orbitals

such as charge distribution, dipole moments, frontier orbital energies, acidicity of hydrogens, molecular size, molecular volume, distance between the most acid hydrogens (H–H), and the molecular electrostatic potential (MEP). The calculated properties were used to correlate an inhibitory activity of the studied compounds towards acetylcholinesterase with their molecular structure.

Ganguly et al. analyzed the reaction of the sarin- [64] and VX-inhibited AChE [65] with nucleophiles by means of DFT [B3LYP/6-311G(d,p)] calculations. The hydroxylamine anion turned out to be more efficient in the reactivation process than other nucleophiles, for instance formoximate anion, and can be used as a good antidote agent against sarin and VX.

7.2.8 *Matrix Metalloproteinases*

Matrix metalloproteinases (MMPs) is a primary target for drug design, because it is involved in many biological processes, such as embryonic development [66], tissue remodeling and repair [67], neurophathic pain processes [68], cancers [69] [70], and other diseases. (4-Phenoxyphenylsulfonyl)methylthiirane also known as SB-3CT (**37b**) is the selective inhibitor of matrix metalloproteinase 2 (MMP2). The coupled deprotonation and ring-opening mechanism of SB-3CT inhibition of MMP2 (Scheme 7.2) as well as by 4-(phenoxyphenylsulfinyl)methylthiiranes (**38b**, **39b**), the sulfoxide analogue of SB-3CT, was examined computationally using DFT and QM/MM approaches [71].

For the model structures **38a** and **39a**, the complete conformational analysis was performed at the DFT (B3LYP/6-31+G(d)) level of theory. Nine conformational minima were identified for **38a** with energy differences between 0.2 and 4.8 kcal/mol. For the concerted deprotonation/ring-opening reaction, five different transition state structures (TSS) were located for the (*R*, *R*) diastereomer **38a** with the barrier

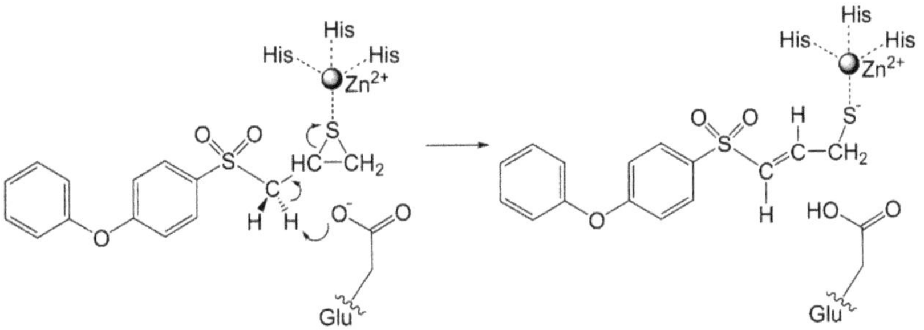

Scheme 7.2 Coupled deprotonation and ring-opening mechanism of the SB-3CT inhibition of MMP2. (Reproduced with permission from Ref. [71]. Copyright © 2010 American Chemical Society)

energies from 16.9 to 23.3 kcal/mol. For six located TSS for (*S, R*) diastereomer **39a** amplitudes of the relative energy variation was similar (from 16.9 to 22.0 kcal/mol). The lowest energies for the transition states turned out to be higher than that found for **37a**, modeling SB-3CT structure (16.9 *vs*. 11.3 kcal/mol, respectively). Therefore, the sulfoxide is less disposed to the concerted deprotonation/ring-opening reaction, probably due to the lower acidity of alkylarylsulfoxides compared with the corresponding sulfones.

Relative energies for the MMP2 complexes of **37b**, **38b** and **39b** (Fig. 7.4) were calculated using the ONIOM(B3LYP/6-311+G(d,p):AMBER) approach. In the active site of MMP2, the barriers for ring-opening for the sulfoxide analogues of SB-3CT (23.3 kcal/mol for **38b** and 28.5 kcal/mol for **39b**) were higher than that for SB-3CT (19.9 kcal/mol) [72], and overall the reactions for **38b** and **39b** (−17.2 and −17.3 kcal/mol, respectively) were less favored than that for SB-3CT (−21.0 kcal/mol), both kinetically and thermodynamically. The sulfoxide analogue of SB-3CT was found to be a linear competitive inhibitor, whereas SB-3CT itself was classified as a slow binding inhibitor.

One more important aspect of using the MMP inhibitors as drugs is their specific inhibition of one of 25 MMPs known in humans, or at least of one specific subgroup of this family of enzymes [73]. Guillaume et al. [74] reported a DFT (B3LYP/6-31G**+LANL2DZ) study for a series of potentially efficient inhibitors of matrix metalloprotease (MMP), N-acetyl-N′-sulfonylhydrazides (**40–48**). N-acetohydroxamic acid (**49**) and N-phenylsufonylglycine (**50**) were used for comparison.

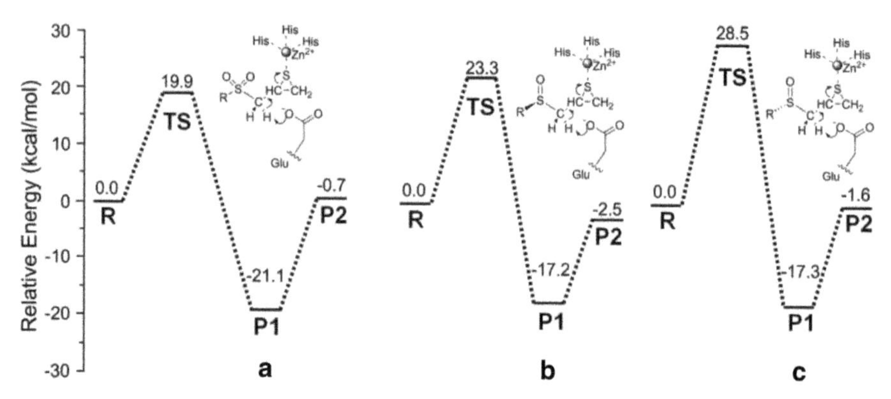

Fig. 7.4 Energy profiles for **37b** (**a**) and its sulfoxide analogues **38b** (**b**) and **39b** (**c**) in the MMP2 active site. R: reagents, TS: transition state, P1: products 1, P2: products 2. (Reproduced with permission from Ref. [71]. Copyright © 2010 American Chemical Society)

40-48 **49** **50**

R: **40** Me; **41** Ph; **42** 4-MeC_6H_4; **43** 4-PhC_6H_4; **44** 4-PhOC_6H_4; **45** 4-FC_6H_4; **46** 3-$NO_2C_6H_4$; **47** 2-$NO_2C_6H_4$; **48** 2,4,6-(i-Pr)$_3C_6H_2$

The authors [74] evaluated the zinc-binding ability of ligands in an enzyme active site model composed of a Zn^{2+} ion using the DFT level of theory. Three main types of Zn–ligand interactions were found in the series of complexes studied. Type I corresponded to a bidentate zinc coordination involving the oxygen of the carbonyl group and the sulfonamide nitrogen atom. Type II resulted from an additional interaction with one of the sulfonyl oxygen atoms. As three 4-methylimidazole ligands were attached, the sulfamide nitrogen atom did not participate anymore in the complexation (the corresponding Zn–N distances were ~3.35 Å). Similar, but even more stable complexes were formed with a deprotonated form of the ligands.

A subsequent docking study resulted in the different coordination ability of **40–50** towards enzymes MM-1, MMP-2, MMP-9, MMP-12 and MMP-14 and demonstrated that these species can be used for the drug design as the efficient and selective MMP inhibitors.

Zhang and co-authors studied theoretically pyrogallic acid (**51**) and myricetin (**52**) as the potential non-peptide inhibitors of MMP-1 and MMP-3 [75]. The corresponding docked complexes with the model active sites were optimized at the B3LYP/6-31G* level of theory. Total calculated interaction energies for MMP-1 with **51** and **52** are −77.07 and −108.39 kcal/mol, respectively). Therefore, myricetin bound to MMP-1 more tightly than pyrogallic acid, which agreed with the

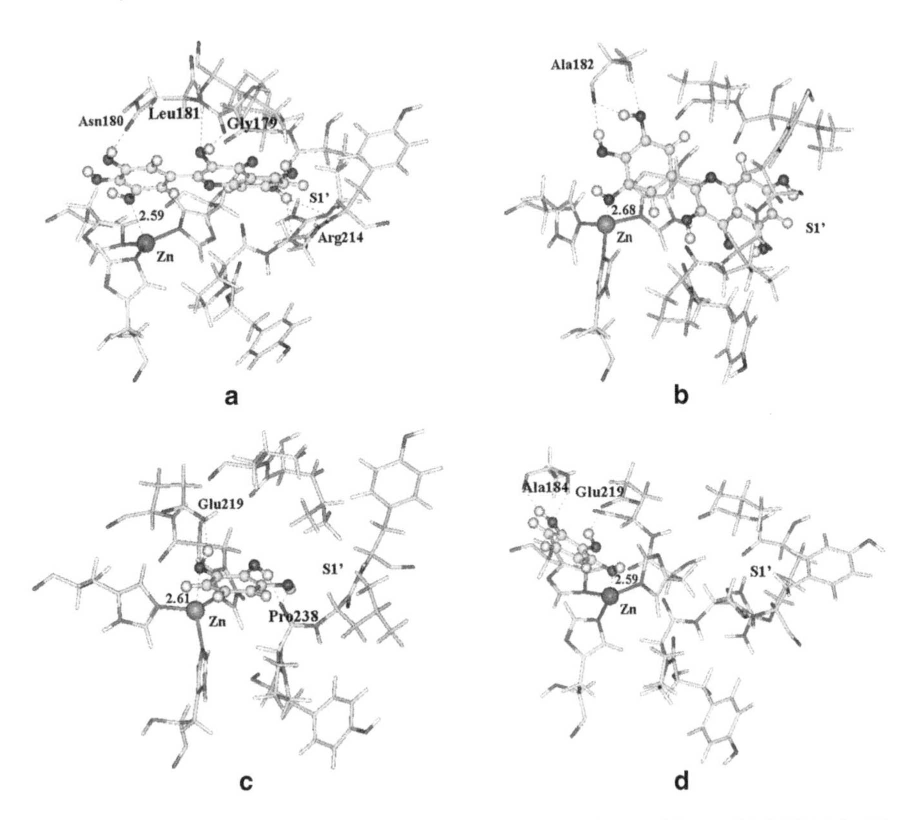

Fig. 7.5 Detailed representation of adducts of **51** (**a**, **c**) and **52** (**b**, **d**) adducts with MMP-1 (**a**, **b**) and MMP-3 (**c**, **d**) active sites. (Reproduced with permission from Ref. [73]. Copyright © 2011 Elsevier)

experimentally derived IC_{50} values for **51** and **52** (2.57 and 1.01 μM, respectively). The determined binding affinities with MMP-3 were found to be significantly lower (−52.37 and −74.56 kcal/mol, respectively), whereas the corresponding IC_{50} magnitudes were essentially higher (12.47 and 4.18 μM, respectively). Thus, **52** indicated better potency on both MMP-1 and MMP-3 than pyrogallic acid. The authors explained such result by a favorable interaction of the S′ cavity in the both MMPs with the benzopyran-4-one substituent in adducts of MMPs with **52** (Fig. 7.5), whereas by binding with **51** it remained empty.

51 **52**

7.2.9 Phospholipases

The phospholipases A_2 (PLA_2) enzymes are responsible for the hydrolysis of membrane phospholipids that release arachidonic acid. The latter serves as substrate for pro-inflammatory mediators, such as prostaglandins and leucotriens. Inhibition of the enzymatic activity and edema induction by (PLA_2), extracted from the venom of *Crotalus adamanteus,* was explored by da Silva et al. [76]. Five different polyhydroxy phenolic compounds **53–57** were studied, both theoretically and experimentally, as the potential PLA_2 inhibitors. Molecular mechanics optimization indicated that the substrate binding occurred mainly via Asp49. This destabilized the interaction with calcium playing an important role in the catalytic activity of PLA_2. The electrostatic potential surface (EPS), calculated for compounds **53–57**, explained differences in inhibition of enzymatic activity of PLA_2: compounds **53–55** possessed the positive EPSs (approx. 0.7 eV) favorable for the formation of complexes with PLA_2. Compound **56** indicated the even higher positive magnitude of EPS (0.912 eV), strengthening the formed complex. Compound **57**, showed the EPS around 0.7 eV, but the hydroxyl groups in this molecule were sterically hindered by acetyl group. This prevented the formation of the inhibitor complex with PLA_2. Both **55** and **57** formed internal hydrogen bonds between the hydroxyl from position 2 and the carbonyl group. Thus, structures **53** and **54** demonstrated the highest inhibition activity.

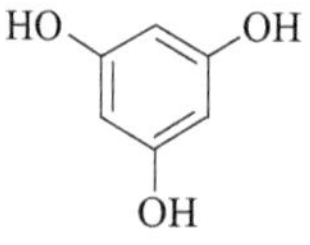

53 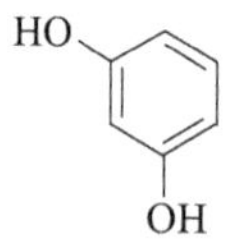 **54** 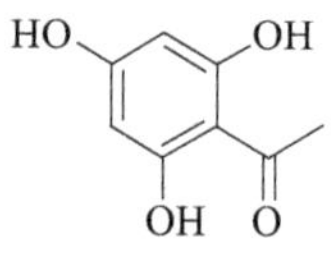 **55** 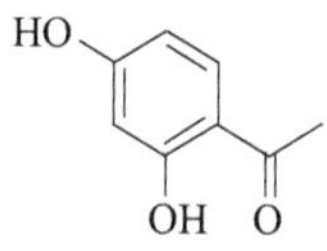 **56** 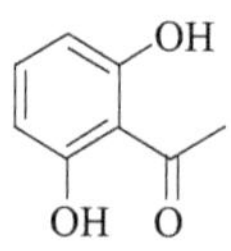 **57**

7.2.10 Angiotensin-Converting Enzyme

Šramko et al. investigated thermodynamics of the interaction of angiotensin-converting enzyme (ACE, EC 3.4.15.1) inhibitors with a truncated zinc metallopeptidase active site (Fig. 7.6) using DFT (B3LYP) and two-layered ONIOM B3LYP:MNDO approaches [77]. The authors investigated binding of various ACE inhibitors with $[Zn^{2+}(imidazole)_2CH_3COO^-]^+$ as the model binding site of ACE in neutral and anionic forms and calculate interaction and dissociation enthalpies and Gibbs energies (see Table 7.4). Several tested inhibitors were widely used drugs, effective in the treatment of hypertension, congestive heart failure, post-myocardial infarction and diabetic nephropathy [78], whereas the other ones were products of their structural modification. The 6-31+G(d,p) basis set was used for zinc, the dissociating functional groups and their closest vicinity, and the standard 6-31G(d)

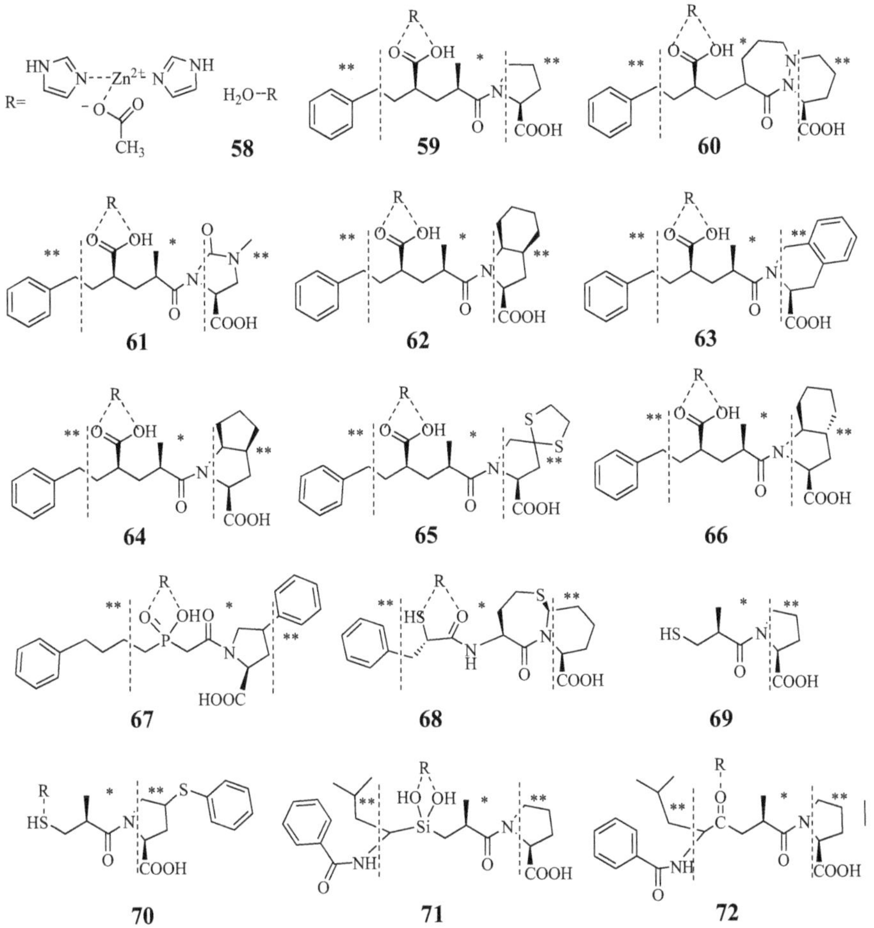

Fig. 7.6 Structural formulas of complexes of enzyme ("receptor") part (R) and neutral inhibitors with atom numbering. *Asterisks* exhibit different parts of molecules treated at the different levels of theory. (Reproduced with permission from Ref. [77]. Copyright © 2011 Elsevier)

basis sets were used for all the other atoms in the high layer. The ionized species originated by removal of proton from the acidic functional group of ACE inhibitor.

Complexes containing ionized (non-protonated) ACE inhibitors are marked with "**a**" and complexes containing neutral (protonated) ACE inhibitors are marked with "**b**". The structures of the optimized complexes **59a–66a** were very similar, with dissociated N-terminal carboxyl group of ACE inhibitors bound to zinc cation partially bidentately and acetate anion bound to zinc monodentately. Interestingly, the structural fragment of complex **59a** with inhibitor enalaprilat taken from Protein Data Bank (reference code 1UZE) was similar to the calculated structure, whereas in the optimized complex **69a** with inhibitor captopril, the inhibitor molecule was turned to the opposite side compared to the crystal structure (reference code 1UZF).

Table 7.4 Calculated gas-phase affinities of neutral ACE inhibitors to R expressed as interaction enthalpies, Gibbs Energies and entropies (at T=298 K). (Reproduced with permission from Ref. [73]. Copyright © 2011 Elsevier)

Complex	Inhibitor	ΔH^{298}	ΔG^{298}	ΔS^{298}	$\Delta\Delta G^{298}$
58b	H_2O	−13.56	−2.58	−36.82	0.0
59b	Enalaprilat	−22.07	−10.57	−38.60	−7.99
60b	Cilazaprilat	−23.21	−11.93	−37.82	−9.36
61b	Imidaprilat	−23.26	−11.78	−38.50	−9.20
62b	Perindoprilat	−21.31	−10.94	−34.78	−8.36
63b	Quinaprilat	−22.19	−11.20	−36.87	−8.62
64b	Ramiprilat	−21.80	−10.59	−37.60	−8.02
65b	Spiraprilat	−22.53	−11.68	−36.41	−9.10
66b	Trandolaprilat	−21.89	−10.95	−36.68	−8.37
67b	Fosinoprilat	−24.93	−11.92	−43.63	−9.35
68b	Omapatrilat	−17.68	−5.94	−39.38	−3.36
69b	Captopril	−14.85	−2.32	−42.03	0.25
70b	Zofenoprilat	−15.10	−3.25	−39.75	−0.67
71b	Silanediol	−13.67	−0.94	−42.71	1.64
72b	Keto-ACE	−21.07	−6.14	−50.07	−3.56

[a] Relative interaction Gibbs free energy values against water complex (**58b**)

In general, interaction enthalpies and Gibbs free energies of negatively charged ionic ACE inhibitors (Table 7.4) were significantly larger than those of neutral inhibitors (not shown here), but the acidity of the inhibitors considerably increases upon chelation: deprotonation Gibbs free energies (ΔG^{298}) of ACE inhibitors in complex with the model active site R were by the average of 85.96 kcal/mol lower than the deprotonation Gibbs energies of the free inhibitors. The results of the study [77] demonstrate that all structures within the investigated series are able to form stable tetra- or penta-coordinated complexes with R system. The highest binding affinity was observed for N-terminal anion of captopril ($\Delta G^{298} = -96.66$ kcal/mol). The model used was proven to be suitable for the analysis of the potential angiotensin-converting enzyme inhibitors and can be treated using DFT methods.

7.2.11 Phosphodiesterase

Inhibitory properties for series of 54 phosphodiesterase 7 (PDE7) inhibitors (spiroquinazolinones), previously reported by Lorthiois and coworkers [79, 80], was studied using docking and DFT methods with the aim of identifying the characteristics that distinguish between potent and weak inhibitors [81]. The conformations from docking studies were further used for DFT (B3LYP/6-31G*) geometry optimization. It is generally suggested that molecules with similar electrostatic potential (EP) surfaces may bind well to the same receptor [82–84]. The EPs calculated for the series of inhibitors (Fig. 7.7) were compared with the experimentally determined pIC_{50} magnitudes. The relative nucleophilicity of N1 with respect to N3 in

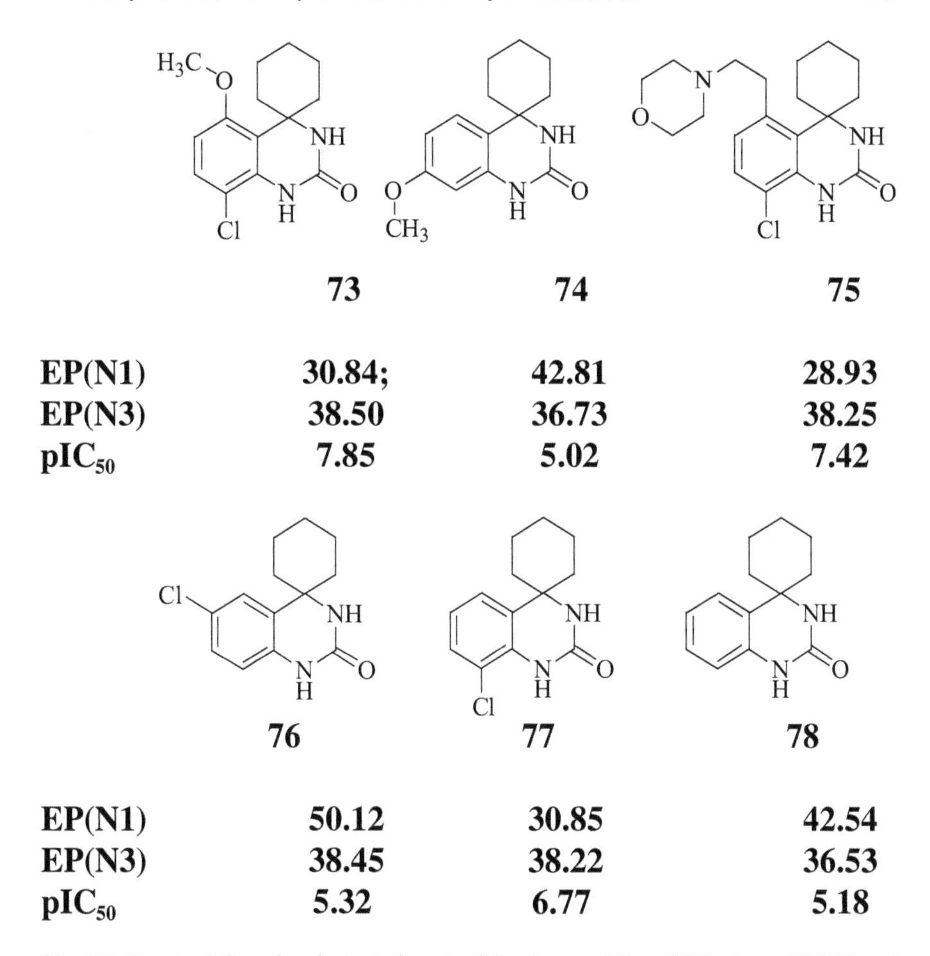

Fig. 7.7 Structural formula, electrostatic potential values on N1 and N3 atoms (EP(N1) and EP(N3), respectively) and activity in molar concentration (pIC_{50}) for structures **73–78**

the spiroquinazolines was found to be important for the activity: all the active molecules had low EP(N1) and EP(N1) < EP(N3).

Thus, the electron density distribution in the molecules and steric factors are equally important for binding the molecules to the receptor. This computational study should aid in design of new molecules in this class with improved PDE7 inhibition.

7.2.12 HIV-1 Transcriptase

Searching for new effective anti-AIDS drug remains the challenge for drug chemistry [85]. One of the prospective ways is the development of efficient inhibitors of the human immunodeficiency virus reverse transcriptase (HIV-1 RT). Authors [86] studied theoretically the binding of five different ligands **79–83** to the HIV-1 RT using molecular dynamics (MD) simulations within hybrid QM/MM potentials. Both potential of mean force (PMF) and free energy perturbation (FEP) methods presented **81** as the best candidate to inhibit the HIV-1 RT with binding energies −57.2 and −30.3 kcal/mol, respectively while for **83** the lowest negative binding energy was predicted (−16.4 and −9.0 kcal/mol, respectively). The EPS were derived from B3LYP/6-31+G(d,p) calculations. The active site displays the large positive electrostatic potential at the positions of the magnesium cations and nitrogen backbone atoms of Asp443, Glu478, Asp549 and His539, whereas the large negative regions were found at the positions of oxygen atoms of backbone belonging to His539, Asp443, Glu478, Asp498 and Asp549. The negative EP in the fragment of DNA-chain was at the oxygen atoms of the phosphate groups, hence there is a reasonable complementarity between the active site of the enzyme and its natural substrate.

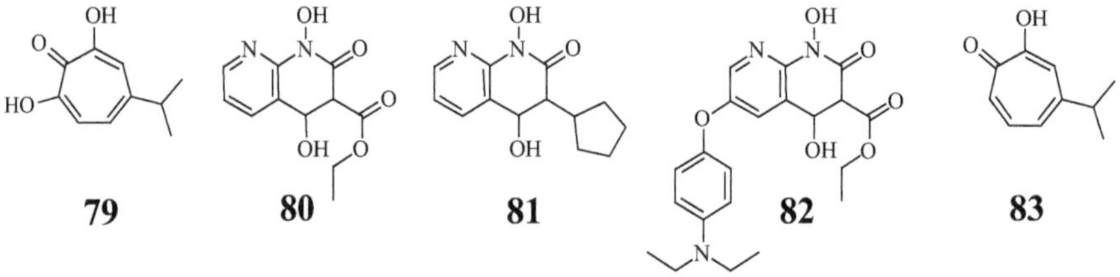

79 **80** **81** **82** **83**

Liang and Chen [87] investigated the interaction between a potential anti-AIDS drug dapivirine and and the HIV-1 RT binding site using the ONIOM2 (B3LYP/6-31G(d,p): PM3) approach and calculating the energy at the B3LYP/6-31G(d,p) level of theory. The interaction energy was divided into several contributions coming from interactions with individual residues of the active site. The calculations predicted two hydrogen bonds between 2-aminopyrimidine groups of dapivirine with the carbonyl oxygen and amino hydrogen of Lys101. Additionally, two aromatic residues, Tyr181 and Tyr188, exhibited H···π and π···π interactions with the aromatic ring of dapivirine.

7.2.13 HIV-1 Aspartic Protease

Fleurat-Lessard et al. [88] analyzed the methods suitable for modeling human immunodeficiency virus type 1 aspartic protease (HIV-1 PR) enzyme. The semiempirical methods failed to describe the geometry of the protease active site. Within DFT, the best results were obtained with hybrid GGA B3LYP or X3LYP and with hybrid meta GGA functionals with a fraction of exact exchange around 30–40%, such as in the M06, B1B95, or BMK functionals. In the more recent work, Fleurat-Lessard et al. [89] studied using QM/MM method new HIV-1 drug candidates, potential inhibitors of HIV-1 PR. Though rigid structures are usually more efficient inhibitors of HIV-1 PR, they are less amenable to adapt to shape modifications of the enzymatic binding site induced by mutations. Hasserodt et al. [90–92] proposed previously a new type of more mobile aspartic protease inhibitors, amino-aldehyde peptides, which adopted their form based on a non-covalent interaction of a tertiary amine nitrogen with a carbonyl group, the so-called N···CO bond. However, the calculations exhibited that the presence of water molecule W301 induced a systematic competition between formation/dissociation of the N···CO bond and the interaction network involving the structural water molecule. Probably, this competition determined the poor inhibition activity of amino-aldehyde peptides [90] and might be avoided by the proper design of non-peptidic cyclic hydrazino-urea derivatives.

Another way for the development of drugs using HIV-1 PR as target with minimizing the drug resistance effect of HIV-1 is an irreversible inhibition. It consists in the chemical modification of the binding site of HIV-1 PR, in particular, at the key Asp 25 and Asp25′ amino acid fragments resulted in the complete lost of catalytic activity. One possible way is to include the oxyrane ring in the potential drug structures [93–95]. Kóňa [96] analyzed two possible mechanisms of the irreversible inhibition of HIV-1 protease by epoxide inhibitors by means of *ab initio* (MP2) and DFT (B3LYP, MPW1K and M05-2X) calculations. In the first version of the reaction mechanism, the water molecule participated in the reaction, but another mechanism with a direct proton transfer from the acid catalyst to the inhibitor was shown to be more preferable. The structures (118 atoms) modeling both the local minima and transition state were located at the DFT [B3LYP/6-31+G(d,p)] level of theory. The activation energy was predicted to be *ca.* 15–21 kcal/mol. The process of irreversible inhibition exhibited significantly large negative reaction energy. The most probable mechanisms of modifying the model inhibitor structure were discussed.

7.2.14 HIV-1 Integrase

HIV-1 integrase (IN) is the relatively new and highly promising target for developing anti-AIDS drugs [97–99]. Understanding the inhibition mechanism of known inhibitors would make possible testing new perspective drug candidates using the DFT methods. However, it is still not known how the enzyme binds the inhibitors or its substrate, viral DNA [100]. The active site of HIV-1 IN is characterized by the dinuclear magnesium center, coordinated by carboxylate groups of three amino acid. Therefore, the main aim of theoretical efforts for the future development of the HIV-1 IN inhibitors with novel scaffolds is to provide a suitable ligand capable of chelating two Mg^{2+} ions [101]. Noteworthy, some efficient IN inhibitors exist in the multiple tautomeric forms [102], which were not studied in detail. Even less was known about the tautomerism of the ligands in the binding site of HIV-1 IN. The most stable tautomeric forms and rotamers for the known inhibitors of HIV-1 IN: α,γ-diketoacids (**84**), α,γ-diketotriazole (**85**), dihydroxypyrimidine carboxylate (**86**) and 4-quinolone-3-carboxylic acid (**87**) were calculated at the B3LYP/6-311++G(d,p) level of theory by Liao and Nicklaus [100]. Next, for the studied structures the chelating complexes with two magnesium ions in the moiety modeling the active site of HIV-1 IN were calculated. As objects for DFT calculations, the magnesium ions were surrounded with three formic acids and four water molecules in order to mimic the IN binding site. In the optimized structures, the most stable forms in water solution included deprotonated, enolized or phenolic hydroxy groups, with the two eight-coordinated magnesium ions separated by a distance 3.70–3.74 Å. Replacing one water in the complex with one molecule of methanol mimiced the terminal 3′-OH of viral DNA, and the chelating complex remained stable. Probably, after 3′-processing, in the binding site of IN the terminal 3′-OH of viral DNA interacts with one Mg^{2+} by chelation.

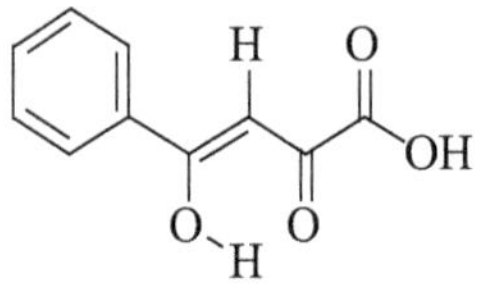

84 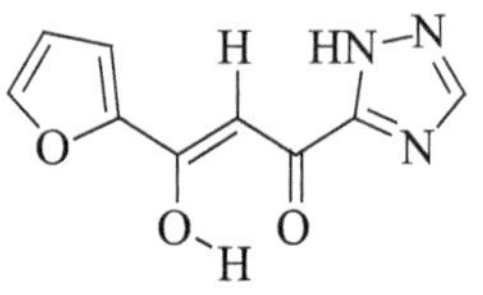 **85** 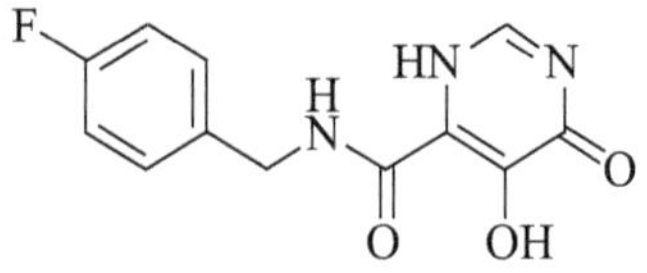 **86**

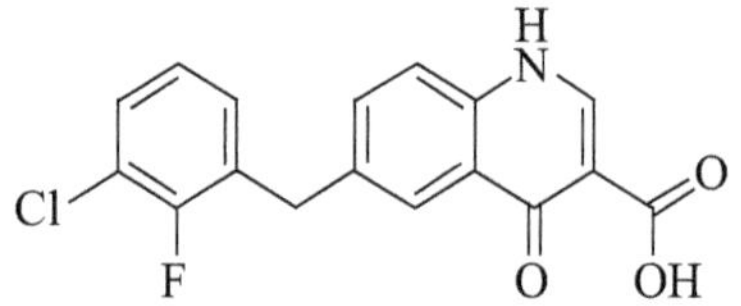

87

Wolschann et al. [103] used DFT calculations to identify the protonation state of HIV-1 IN, in particular, residues Lys156 and Lys159, which are of importance for binding 5CITEP inhibitor (**88**). The most favored conformations of 5CITEP were derived at the B3LYP/6-31G(d,p) level of theory by a variation of two torsion angles, Tor1 (C19-C11-C9-C8) and Tor2 (N3-C5-C6-C8) (Fig. 7.8, left). The potential energy surface (PES) was analyzed at the HF/3-21G level with subsequent reoptimization of the structures corresponding to local minima at the more superior B3LYP/6-31G(d,p) theoretical approach. The initial geometry of the IN/5CITEP adduct was taken from Protein Data Bank (PDB, entry code 1QS4). Interestingly, 5CITEP in the complex with IN in the X-ray determined structure, differs slightly from the equilibrium conformation of the free ligand, probably, due to additional interactions arising between 5CITEP and the surrounding amino acids.

Seven different structures of the complex were generated including both neutral and deprotonated forms of 5CITEP as well as both neutral and protonated forms of two lysines (Lys156 and Lys159) and then optimized by fixing the C_α atoms in the amino acids. The lowest energy structures for the protonated and noncharged states of the adduct are shown in Fig. 7.9. 5CITEP is in its neutral form, where the hydrogen atom is attached to N3 of the tetrazole ring while Lys156 and Lys159 are non-protonated and protonated, respectively. The structure of the

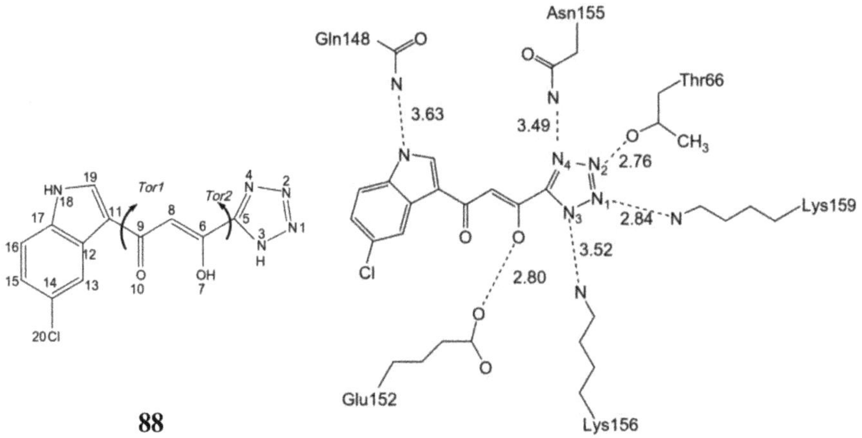

88

Fig. 7.8 Conformational analysis of 5CITEP (*left*); modeling the adduct of 5CITEP with the binding site of HIV-1 IN (*right*). (Reproduced with permission from Ref. [103]. Copyright © 2007 Elsevier)

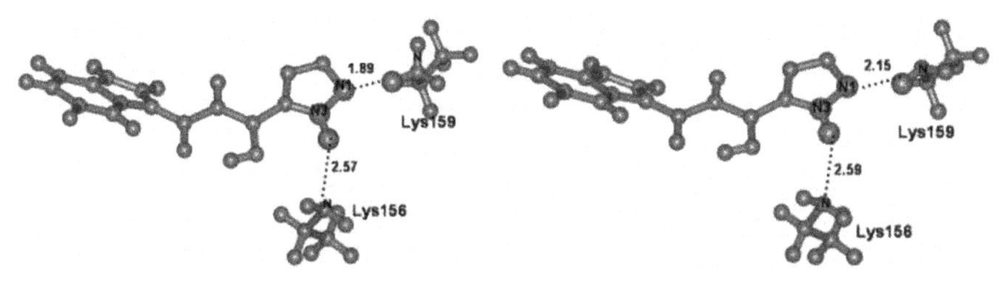

Fig. 7.9 The lowest energy configurations for the protonated (*left*) and non-charged (*right*) states of the adduct of **88** with Lys156 and Lys159 from HIV-1 IN. (Reproduced with permission from Ref. [103]. Copyright © 2007 Elsevier)

non-charged state is similar, but in contrast to the protonated form, both lysine residues are non-protonated.

For determining the binding energy, the most important six amino acids (Thr66, Gln148, Glu152, Asn155, Lys156, and Lys159) were selected (Fig. 7.8, right). Although the calculated structure of 5CITEP was slightly different from that found in the co-crystal structure, its binding energy (−41.33 kcal mol^{-1}) indicated the energetically favorable system. In contrast to the aforementioned adduct of 5CITEP with two lysine molecules (Lys156 and Lys159), in the complex including six amino acids, where the side chains of amino acids were allowed to change their position, proton at N3 of the ligand was transferred to Lys156.

Alves and co-authors [104] compared the activity of HIV-1 IN effective inhibitor S-1360 (**89**), that underwent clinical trials, with two its analogues **90** and **91**. While the QM/MM calculations (using BLYP/6-31G* level for the QM region) predicted lower interaction energies for **90** and **91** (−611.7 and −622.7 kJ/mol) than for **89** (−667.0 kJ/mol), they exhibited strong interactions with the residues of the active site. The EP surfaces were analyzed for all three structures.

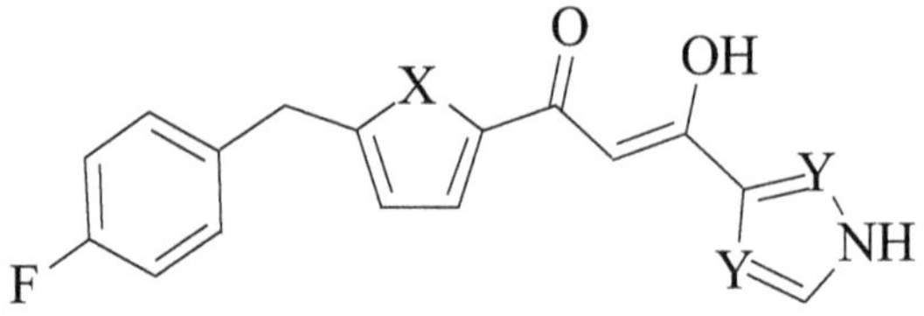

89: X=O, Y=N; **90** X=NH; Y=N; **91** X=O, Y=H

7.2.15 Xanthine Oxidase

Xanthine oxidase (XO) is a flavoprotein enzyme which catalyzes the oxidative hydroxylation of purine substrates. Because of its availability (it is abundant in cow's milk), XO has become a well-established target of drugs against gout and hyperuricemia. The reduction of molecular oxygen by XO produces free radicals which can cause damage to surrounding tissues. The activation of XO generates superoxide and hydrogen peroxide, hence it is generally seen as a potentially destructive agent in the vasculature. The paper of Lespade and Bercion [105] is devoted to the computational (DFT) study of one of the possible mechanisms of XO inhibition: the attraction and anchorage of the molecule inside the cavity. Two classes of potential inhibitors were tested as inhibitors: the series of flavonoids of natural origin [106, 107]: luteolin (**92**), apigenin (**93**), chrysin (**94**), kaempferol (**95**), galangin (**96**), myricetin (**97**), quercetin (**98**), morin (**99**); and gallic acid derivatives [108]: gallic acid (**100**), ellagic acid (**101**) and ellagic acid-4-*O*-*β*-D-xylopyranoside (**102**). For this purpose, electrostatic interactions between the molybdopterin moiety and two series of inhibitors were calculated at the DFT level of theory, in order to evaluate the interconnection between the electrostatic potential and inhibition forces. The most stable conformations were determined for the inhibitors using B3LYP/6-31+G(d,p) approach. As this functional poorly reproduces electron dispersion, the energies of the conformations were calculated at the MP2/6-31+G(d,p) level of theory. The authors of [105] concluded that the most potent inhibitors in the investigated series should be polar, possess a longitudinal dipole moment, and weakly dissociate at physiological pH.

7.2.16 Trombin

The activity of trombin is responsible for the cleavage of fibrogen to form fibrin that then polymerizes with forming a network of fibers. This determines not only the positive wound-healing process, but also such diseases as myocardial infarction, pulmonary embolism and stroke [109]. Understanding the binding mechanism of the known inhibitors to thrombin would provide a valuable information for a discovery of new more efficient and selective inhibitors.

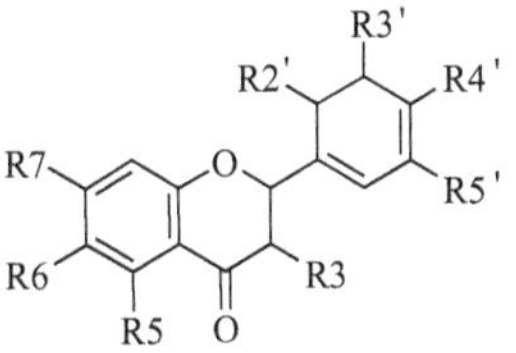

101 R=H
102 R=

92-99 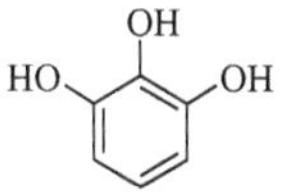 **100** 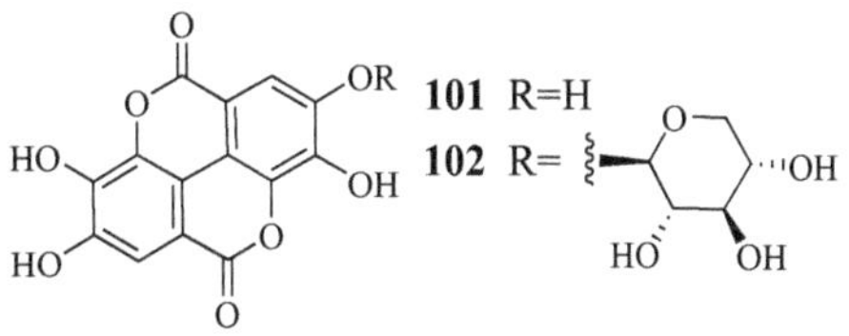 **101, 102**

92 R3, R6,R2',R5'=H, R5, R7, R3', R4'=OH; **93** R3, R6, R2', R3', R5'= H, R5, R7, R4'= OH **94** R3, R6, R2', R3', R4', R5'=H, R5, R7=OH; **95** R6, R2', R3', R5'= H, R3, R5, R7, R4'= OH; **96** R6, R2', R3', R4', R5'= H, R3, R5, R7, = OH; **97** R6, R2', = H, R3, R5, R7, R3', R4', R5' = OH; **98** R6, R2', R5' = H, R3, R5, R7, R3', R4' = OH; **99** R6, R3', R5' = H, R3, R5, R7, R2', R4', = OH.

The authors [110] studied interactions of two pyrazinone- and prolyne-based macrocyclic inhibitors (**103** and **104**, respectively) using molecular dynamics simulations, DFT and molecular mechanics calculations. An analysis of binding interactions became possible by applying molecular fractionation with conjugate caps (MFCC) approach [10, 11]. In the case of **103** main binding attractions were provided by six residues with individual gas-phase binding energies > 2 kcal/mol: Ser^{214}, Trp^{215}, Gly^{216}, Glu^{217}, Asp^{102}, and Asp^{189}. Similarly, for **104** interactions with Asp^{189}, Ser^{214}, Trp^{215}, Gly^{216}, Asp^{102}, and Glu^{146} were of importance. The fragment interaction energies calculated at the MP2/6-311G* level of theory agreed well with those derived using the DFT (B3LYP/6-31G*) method. A good agreement was observed between the calculated and experimental binding free energy values.

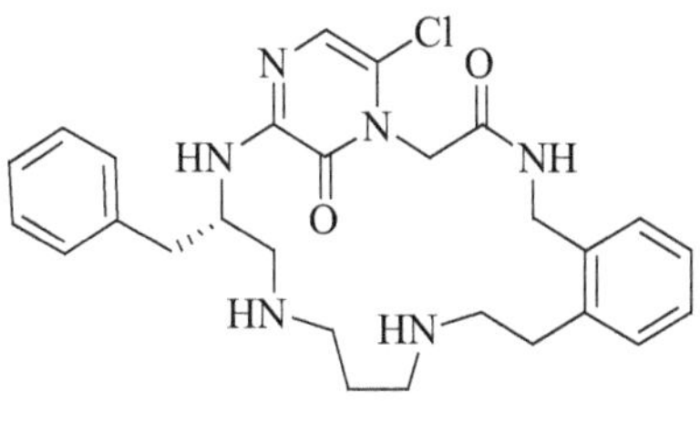

103 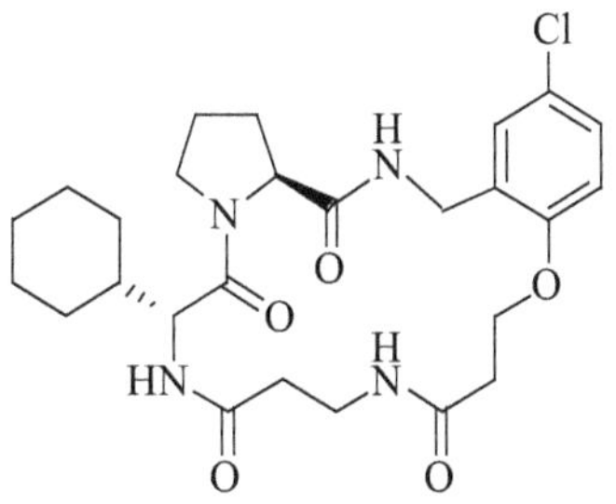 **104**

7.2.17 Lipase B

De Oliveira et al. [111] carried out quantum chemical (DFT) calculations for adducts of three flavonoids, quercetin (**98**), isoquercitrin (**105**) and rutin (**106**), docked in the mini-model that mimicked the catalytic site of *Candida antarctica* lipase B (CALB). The analysis of these results showed that an ester bond with the carbonyl C atom of the Ser105-bound acetate was expected for the rhamnose 4‴-O of rutin and for the glucose 6″-O of isoquercitrin, but no ester bond was predicted to be formed with the B-ring of 3′-O of quercetin. The mechanism of coordination was modeled calculating non-covalently bound as well as covalently bound intermediates. The theoretical results agreed well with the experiment.

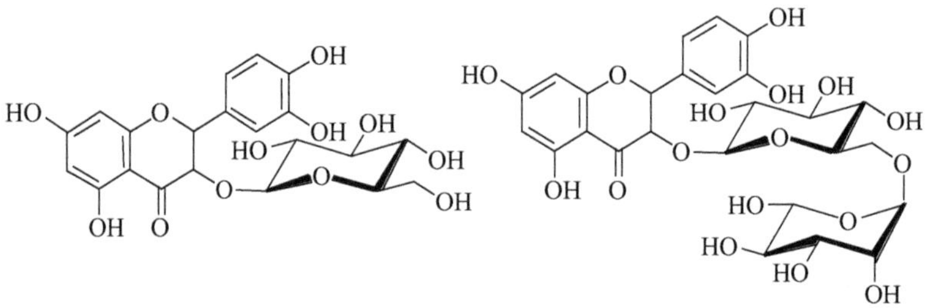

105 **106**

7.2.18 Urease

Urease (urea amidohydrolase, EC 3.5.1.5) is involved in a number of diseases, such as pyelonephritis, ammonia encephalopathy, hepatic coma, peptic ulcers and formation of kidney stones [112, 113], hence the urease inhibitors could be useful as efficient drugs. Leopoldini et al. [114] explored at the DFT (B3LYP using LANL2DZ basis set for Ni atoms and 6-311G** for all other atoms) boric acid as a rapid reversible inhibitor of urease. Two models of different size were analyzed. The smaller one included truncated amino acids from the first coordination shell of two Ni^{2+} ions: the histidine residues (His137, His139, His249, His275), the carbamylated lysine (Lys220) and the Asp363 were simulated by imidazole rings, a carboxylated methylamine (CH_3NHCOO^-) and an acetate group (CH_3COO^-) (Fig. 7.10). The B–O interactions were strong and possessed the covalent character. The bonding character did not change when the binding site model was extended by adding other amino acid residues and two water molecules involved in the inhibitor binding mode (totally 122 atoms). The boric acid molecule seemed to be firmly anchored to the enzyme and thus prevented the urease catalytic reaction.

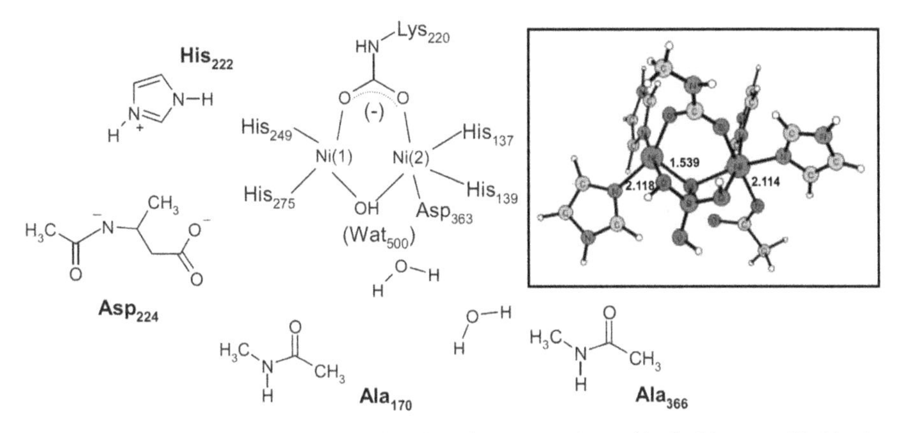

Fig. 7.10 Model used for the urease active site. The outer amino acids (*bold*) were added in the larger model. In the inset: optimized geometry of the complex of boric acid with the model of the urease binding site. The distances are in Å. (Reproduced with permission from Ref. [114]. Copyright © 2008 Wiley Periodicals, Inc.)

7.2.19 Other Results

Several 1,3-bisphospho-D-glyceric acid analogs were studied theoretically (using QM/MM molecular dynamics and DFT-based EPS) as potential inhibitors of glyceraldehyde-3-phosphate dehydrogenase, and as new drugs against Chagas disease [15]. The first detailed QM/MM study on the possible mechanisms for the reaction of proteasome with a representative peptide inhibitor, Epoxomicin was reported [115]. The obtained novel mechanistic insights should be valuable for a future rational design of more efficient proteasome inhibitors.

7.3 Conclusions and Prospects

The DFT approach is widely used nowadays in the pure form or in combination with other, less computationally demanding approaches for modeling enzyme–ligand adducts towards understanding mechanisms of enzyme catalyzed reactions and constructing novel drugs based on enzyme–inhibitor interactions. The recently developed efficient linear-scaling techniques like the *Resolution of the Identity (RI)* provide a new quantum chemical methodology for modeling ligand–protein interactions. Recently, the novel efficient linear-scaling methods have also been proposed for the hybrid functionals, such as the popular B3LYP [116, 117]. The frontiers of using DFT for modeling the enzyme–inhibitor interactions exceed now several hundreds atoms. This makes it possible to increase significantly the size of the DFT-calculated enzyme binding site fragments for the virtual (*in silico*) construction of the enzyme–inhibitor adducts. For example, the RI-DFT molecular dynamics methods, RI-DFT-based geometry optimization and RI-MP2 single-point energy calculations can be combined for the model adducts of relatively large size. This gives us believe to assert that, similarly to quantum chemistry of small molecules, such kind of calculations will sooner or later turn into a routine.

Finally, the special event should be mentioned here as the scientific and public recognition of achievements of computational chemistry over the last decades and its great prospects in the future: the Nobel Prize in Chemistry 2013 awarded jointly to Martin Karplus, Michael Levitt and Arieh Warshel "for the development of multiscale models for complex chemical systems". Inter alia, the laureates laid the foundation for the modern QM/MM approach [118] based nowadays on the *ab initio* or DFT and MM/MD approaches. There is no doubt that the extent of the DFT constituent will grow, increasing the reliability of the method by modeling enzyme–inhibitor reactions.

References

1. Putz MV, Mingos DMP (eds) (2013) Applications of density functional theory to biological and bioinorganic chemistry. Springer, Berlin. doi:10.1007/978-3-642-32750-6
2. Dahan A, Khamis M, Agbaria R, Karaman R (2012) Targeted prodrugs in oral drug delivery: the modern molecular biopharmaceutical approach. Expert Opin Drug Del 9(8):1001–1013. doi:10.1517/17425247.2012.697055
3. Kortagere S (ed) (2013) In silico models for drug discovery: methods in molecular biology, vol 993. Humana Press, Totowa. doi:10.1007/978-1-62703-342-8
4. Jorgensen WL (2010) Drug discovery: pulled from a protein's embrace. Nature 466(7302):42–43. doi:10.1038/466042a
5. Sharma R (ed) (2012) Enzyme inhibition and bioapplications. InTech, Rijeka
6. Barril X (2012) Druggability predictions: methods, limitations, and applications. WIREs Comput Mol Sci. doi:10.1002/wcms.1134. doi:10.1002/wcms.1134
7. Morris GM, Lim-Wilby M (2008) Molecular docking. In: Methods in molecular biology. Spinger, Clifton, p 365–382
8. Puzyn T, Leszczynski J, Cronin MT (eds) (2010) Recent advances in QSAR studies, in series: modern techniques and applications, vol 8. Springer, Netherlands. doi:10.1007/978-1-4020-9783-6
9. Utkov H, Livengood M, Cafiero M (2010) Using density functional theory methods for modeling induction and dispersion interactions in ligand–protein complexes. Ann Rep Comput Chem 6:96–112. doi:10.1016/S1574-1400(10)06007-X
10. Zhang DW, Zhang JZH (2003) Molecular fractionation with conjugate caps for full quantum mechanical calculation of protein–molecule interaction energy. J Chem Phys 119(7):3599–3605. doi:10.1063/1.1591727
11. He X, Mei Y, Xiang Y, Zhang DW, Zhang JZH (2005) Quantum computational analysis for drug resistance of HIV-1 reverse transcriptase to nevirapine through point mutations. Proteins 61(2):423–432. doi:10.1002/prot.20578
12. York DM, Lee T-S (eds) (2009) Multi-scale quantum models for biocatalysis. In: Modern techniques and applications, vol. 7, Springer, Dordrecht. doi:10.1007/978-1-4020-9956-4
13. Mulholland AJ (2007) Chemical accuracy in QM/MM calculations on enzyme-catalysed reactions. Chem Cent J 1(1):19–24. doi:10.1186/1752-153X-1-19
14. Söderhjelm P, Aquilante F, Ryde U (2009) Calculation of protein–ligand interaction energies by a fragmentation approach combining high-level quantum chemistry with classical many-body effects. J Phys Chem B 113(32):11085–11094

15. Senn HM, Thiel W (2009) QM/MM methods for biomolecular systems. Angew Chem Int Ed 48(7):1198–1229. doi:10.1002/anie.200802019
16. de Vivo M (2011) Bridging quantum mechanics and structure-based drug design. Front Biosci 16(1):1619–1633. doi:10.2741/3809
17. Lodola A, de Vivo M (2012) The increasing role of QM/MM in drug discovery. Adv Prot Chem Struct Biol 87:337–362. doi:10.1016/B978-0-12-398312-1.00011-1
18. Cavalli A, Carloni P, Recanatini M (2006) Target-related applications of first principles quantum chemical methods in drug design. Chem Rev 106(9):3497–3519. doi:10.1021/cr050579p
19. Hu L, Söderhjelm P, Ryde U (2013) Accurate reaction energies in proteins obtained by combining QM/MM and large QM calculations. J Chem Theor Comput 9(1):640–649. doi:10.1021/ct3005003
20. Zalesny R, Papadopoulos MG, Mezey PG, Leszczynski J (eds) (2011) Linear-scaling techniques in computational chemistry and physics. In: Challenges advances computational chemistry physics, vol 13. Springer, Dordrecht. doi:10.1007/978-90-481-2853-2
21. Whitten JL (1973) Coulombic potential energy integrals and approximations. J Chem Phys 58(10):4496–4501. doi:10.1063/1.1679012
22. Baerends EJ, Ellis DE, Ros P (1973) Self-consistent molecular Hartree–Fock–Slater calculations I. The computational procedure. Chem Phys 2(1):41–51. doi:10.1016/0301-0104(73)80059-X
23. Dunlap BI, Connolly JWD, Sabin JR (1979) On some approximations in applications of Xα theory. J Chem Phys 71(8):3396–3402. doi:10.1063/1.438728
24. Dunlap BI, Connolly JWD, Sabin JR (1979) On first-row diatomic molecules and local density models. J Chem Phys 71(12):4993–4999. doi:10.1063/1.438313
25. Feyereisen M, Fitzgerald G, Komornicki A (1993) Use of approximate integrals in ab initio theory. An application in MP2 energy calculations. Chem Phys Lett 208 359–363. doi:10.1016/0009-2614(93)87156-W
26. Eichkorn K, Weigend F, Treutler O, Ahlrichs R (1997) Auxiliary basis sets for main row atoms and transition metals and their use to approximate Coulomb potentials. Theor Chem Acc 97(1-4):119–124. doi:10.1007/s002140050244
27. Weigend F, Häser M, Patzelt H, Ahlrichs R (1998) RI-MP2: optimized auxiliary basis sets and demonstration of efficiency. Chem Phys Lett 294(1–3):143–152. doi:10.1016/S0009-2614(98)00862-8
28. Furche F, Ahlrichs R, Hättig C, Klopper W, Sierka M, Weigend F (2013) Turbomole. WIREs Comput Mol Sci. doi:10.1002/wcms.1162
29. Neese F (2012) The ORCA program system. WIREs Comput Mol Sci 2(1):73–78. doi:10.1002/wcms.81
30. Aguirre D, Chifotides HT, Angeles-Boza AM, Chouai A, Turro C, Dunbar KR (2009) Redox-regulated inhibition of T7 RNA polymerase via establishment of disulfide linkages by substituted Dppz dirhodium(I, II) complexes. Inorg Chem 48(10):4435–4444. doi:10.1021/ic900164j
31. Villar JAFP, Lima FTD, C.L. Veber, A.R.M. Oliveira, A.K. Calgarotto, S. Marangoni, da Silva SL (2008) Synthesis and evaluation of nitrostyrene derivative compounds, new snake venom phospholipase A2 inhibitors. Toxicon 51(8):1467–1478. doi:10.1016/j.toxicon.2008.03.023
32. Parameswari AR, Kumaradhas P (2013) Exploring the conformation, charge density distribution and the electrostatic properties of galanthamine molecule in the active site of AChE using DFT and AIM theory. Int J Quant Chem 113(8):1200–1208. doi:10.1002/qua.24251
33. Silva JRA, Lameira J, Alves CN (2012) Insights for design of *Trypanosoma cruzi* GAPDH inhibitors: a QM/MM MD study of 1,3-bisphospo-D-glyceric acid analogs. Int J Quant Chem 112(20):3398–3402. doi:10.1002/qua.24253
34. Negri M, Recanatini M, Hartmann RW (2011) Computational investigation of the binding mode of bis(hydroxylphenyl) arenes in 17β-HSD1: molecular dynamics simulations, MM-PBSA free energy calculations, and molecular electrostatic potential maps. J Comp Aid Mol Des 25(9):795–811. doi:10.1007/s10822-011-9464-7

35. Arooj M, Thangapandian S, John S, Hwang S, Park JK, Lee KW (2012) Computational studies of novel chymase inhibitors against cardiovascular and allergic diseases: mechanism and inhibition. Chem Biol Drug Des 80(6):862–875. doi:10.1111/cbdd.12006
36. Muzet N, Guillot B, Jelsch C, Howardt E, Lecomte C (2003) Electrostatic complementarity in an aldose reductase complex from ultra-high-resolution crystallography and first-principles calculations. Proc Nat Acad Sci U S A 100(15):8747. doi:10.1073/pnas.1432955100
37. Van Damme S, Bultinck P (2009) Conceptual DFT properties-based 3D QSAR: analysis of inhibitors of the nicotine metabolizing CYP2A6 enzyme. J Comput Chem 30(12):1749–1757. doi:10.1002/jcc.21177
38. Wan J, Zhang LI, Yang G (2004) Quantitative structure-activity relationships for phenyl triazolinones of protoporphyrinogen oxidase inhibitors: a density functional theory study. J Comput Chem 25(15):1827–1832. doi:10.1002/jcc.20122
39. Zhang L, Hao G-F, Tan Y, Xi Z, Huang M-Z, Yang G-F (2009) Bioactive conformation analysis of cyclic imides as protoporphyrinogen oxidase inhibitor by combining DFT calculations, QSAR and molecular dynamic simulations. Bioorgan Med Chem 17(14):4935–4942. doi:10.1016/j.bmc.2009.06.003
40. Lodola A, Capoferri L, Rivara S, Tarzia G, Piomelli D, Mulholland A, Mor M (2013) Quantum mechanics/molecular mechanics modeling of fatty acid amide hydrolase reactivation distinguishes substrate from irreversible covalent inhibitors. J Med Chem 56(6):2500–2512. doi:10.1021/jm301867x
41. Gueto-Tettay C, Drosos JC, Vivas-Reyes R (2011) Quantum mechanics study of the hydroxyethylamines-BACE-1 active site interaction energies. J Comp Aid Mol Des 25(6):583–597. doi:10.1007/s10822-011-9443-z
42. Evin G, Weidemann A (2002) Biogenesis and metabolism of Alzheimer's disease Abeta amyloid peptides. Peptides 23(7):1285–1297. doi:10.1016/S0196-9781(02)00063-3
43. McGeer PL, McGeer EG (2001) Inflammation, autotoxicity and Alzheimer disease. Neurobiol Aging 22(6):799–809. doi:10.1016/S0197-4580(01)00289-5
44. Zhao Y, Truhlar DG (2008) The M06 suite of density functionals for main group thermochemistry, thermochemical kinetics, noncovalent interactions, excited states, and transition elements: two new functionals and systematic testing of four M06-class functionals and 12 other function. Theor Chem Acc 120(1–3):215–241. doi:10.1007/s00214-007-0310-x
45. Xu X, Goddard WA (2004) From the cover: The X3LYP extended density functional for accurate descriptions of nonbond interactions, spin states, and thermochemical properties. Proc Nat Acad Sci U S A 101(9):2673–2677. doi:10.1073/pnas.0308730100
46. Durrington P (2003) Dyslipidaemia. Lancet 362(9385):717–731. doi:10.1016/S0140-6736(03)14234–1
47. Istvan ES, Deisenhofer J (2001) Structural mechanism for statin inhibition of HMG-CoA reductase. Science 292(5519):1160–1164. doi:10.1126/science.1059344
48. Utkov HE, Price AM, Cafiero M (2011) MP2, density functional theory, and semi-empirical calculations of the interaction energies between a series of statin-drug-like molecules and the HMG-CoA reductase active site. Comp Theor Chem 967(1):171–178. doi:10.1016/j.comptc.2011.04.013
49. Leopoldini M, Malaj N, Toscano M, Sindona G, Russo N (2010) On the inhibitor effects of bergamot juice flavonoids binding to the 3-hydroxy-3-methylglutaryl-CoA Reductase (HMGR) Enzyme. J Agr Food Chem 58(19):10768–10773. doi:10.1021/jf102576j
50. Hebert PR, Gaziano JM, Chan KS, Hennekens CH (1997) Cholesterol lowering with statin drugs, risk of stroke and total mortality. An overview of randomized trials. J Am Med Assoc 278(4):313–321. doi:10.1001/jama.1997.03550040069040
51. Leopoldini M, Marino T, Russo N, Toscano M (2009) Potential energy surfaces for reaction catalyzed by metalloenzymes from quantum chemical computations. In: Russo N, Antonchenko VY, Kryachko ES (eds) Self-organization of molecular systems: from molecules and clusters to nanotubes and proteins, NATO Science for Peace and Security Series A: Chemistry and Biology. Springer, New York, p 275–313. doi:10.1007/978-90-481-2590-6_13

52. Antony J, Gresh N, Olsen L, Hemmingsen L, Schofield CJ, Bauer R (2002) Binding of D- and L-captopril inhibitors to metallo-β-lactamase studied by polarizable molecular mechanics and quantum mechanics. J Comput Chem 23(13):1281–1296. doi:10.1002/jcc.10111
53. Antony J, Piquemal J-P, Gresh N (2005) Complexes of thiomandelate and captopril mercaptocarboxylate inhibitors to metallo-beta-lactamase by polarizable molecular mechanics. Validation on model binding sites by quantum chemistry. J Comput Chem 26(11):1131–1147. doi:10.1002/jcc.20245
54. Chen X, Gao F, Zhou Z-X, Yang W-Y, Guo L-T, Ji L-N (2010) Effect of ancillary ligands on the topoisomerases II and transcription inhibition activity of polypyridyl ruthenium(II) complexes. J Inorg Biochem 104(5):576–582. doi:10.1016/j.jinorgbio.2010.01.010
55. Casini A, Edafe F, Erlandsson M, Gonsalvi L, Ciancetta A, Re N, Ienco A, Messori L, Peruzzini M, Dyson PJ (2010) Rationalization of the inhibition activity of structurally related organometallic compounds against the drug target cathepsin B by DFT. Dalton Trans 39(23):5556–5563. doi:10.1039/c003218b
56. Shokhen M, Khazanov N, Albeck A (2011) The mechanism of papain inhibition by peptidyl aldehydes. Proteins 79(3):975–985. doi:10.1002/prot.22939
57. Greig NH, Sambamurti K, Yu Q, Brossi A, Bruinsma GB, Lahiri DK (2005) An overview of phenserine tartrate, a novel acetylcholinesterase inhibitor for the treatment of Alzheimer's disease. Curr Alzheimer Res 2(3):281–290. doi:10.2174/1567205054367829
58. Lahiri DK, Farlow MR, Hintz N, Utsuki T, Greig NH (2000) Cholinesterase inhibitors, beta-amyloid precursor protein and amyloid beta-peptides in Alzheimer's disease. Acta Neurol Scand 102(s176):60–67. doi:10.1034/j.1600-0404.2000.00309.x
59. Leader H, Wolfe AD, Chiang PK, Gordon RK (2002) Pyridophens: binary pyridostigmine-aprophen prodrugs with differential inhibition of acetylcholinesterase, butyrylcholinesterase, and muscarinic receptors. J Med Chem 45(4):902–910. doi:10.1021/jm010196t
60. Yu Q, Holloway HW, Flippen-Anderson JL, Hoffman B, Brossi A, Greig NH (2001) Methyl analogues of the experimental Alzheimer drug phenserine: synthesis and structure/activity relationships for acetyl- and butyrylcholinesterase inhibitory action. J Med Chem 44(24):4062–4071. doi:10.1021/jm010080x
61. Correa-Basurto J, Flores-Sandoval C, Marín-Cruz J, Rojo-Domínguez A, Espinoza-Fonseca LM, Trujillo-Ferrara JG (2007) Docking and quantum mechanic studies on cholinesterases and their inhibitors. Eur J Med Chem 42(1):10–19. doi:10.1016/j.ejmech.2006.08.015
62. Zhang Y, Kua J, McCammon JA (2002) Role of the catalytic triad and oxyanion hole in acetylcholinesterase catalysis: an ab initio QM/MM study. J Am Chem Soc 124(35):10572–10577. doi:10.1021/ja020243m
63. Nascimento ECM, Martins JBL, dos Santos ML, Gargano R (2008) Theoretical study of classical acetylcholinesterase inhibitors. Chem Phys Lett 458(4–6):285–289. doi:10.1016/j.cplett.2008.05.006
64. Khan MAS, Lo R, Bandyopadhyay TT, Ganguly BB (2011) Probing the reactivation process of sarin-inhibited acetylcholinesterase with α-nucleophiles: hydroxylamine anion is predicted to be a better antidote with DFT calculations. J Mol Graph Model 29(8):1039–1046. doi:10.1016/j.jmgm.2011.04.009
65. Khan MAS, Ganguly B (2012) Assessing the reactivation efficacy of hydroxylamine anion towards VX-inhibited AChE: A computational study. J Mol Model 18(5):1801–1808. doi:10.1007/s00894-011-1209-y
66. Shi Y-B, Fu L, Hasebe T, Ishizuya-Oka A (2007) Regulation of extracellular matrix remodeling and cell fate determination by matrix metalloproteinase stromelysin-3 during thyroid hormone-dependent post-embryonic development. Pharmacol Therapeut 116(3):391–400. doi:10.1016/j.pharmthera.2007.07.005
67. Smith MF, Ricke WA, Bakke LJ, Dow MPD, Smith GW (2002) Ovarian tissue remodeling: role of matrix metalloproteinases and their inhibitors. Mol Cell Endocrinol 191(1):45–56. doi:10.1016/S0303-7207(02)00054-0

68. Kawasaki Y, Xu Z-Z, Wang X, Park JY, Zhuang Z-Y, Tan P-H, Gao Y-J, Roy K, Corfas G, Lo EH, Ji R-R (2008) Distinct roles of matrix metalloproteases in the early- and late-phase development of neuropathic pain. Nat Med 14(3):331–336. doi:10.1038/nm1723
69. Egeblad M, Werb Z (2002) New functions for the matrix metalloproteinases in cancer progression. Nat Rev Cancer 2(3):161–174. doi:10.1038/nrc745
70. Noël A, Jost M, Maquoi E (2008) Matrix metalloproteinases at cancer tumor–host interface. Semin Cell Dev Biol 19(1):52–60. doi:10.1016/j.semcdb.2007.05.011
71. Tao P, Fisher JF, Shi Q, Mobashery S, Schlegel HB (2010) Matrix metalloproteinase 2 (MMP2) inhibition: DFT and QM/MM studies of the deprotonation-initialized ring-opening reaction of the sulfoxide analogue of SB-3CT. J Phys Chem B 114(2):1030–1037. doi:10.1021/jp909327y
72. Tao P, Fisher JF, Shi Q, Vreven T, Mobashery S, Schlegel HB (2009) Matrix metalloproteinase 2 inhibition: combined quantum mechanics and molecular mechanics studies of the inhibition mechanism of (4-phenoxyphenylsulfonyl)methylthiirane and its oxirane analogue. Biochemistry 48(41):9839–9847. doi:10.1021/bi901118r
73. Augé F, Hornebeck W, Decarme M, Laronze J-Y (2003) Improved gelatinase a selectivity by novel zinc binding groups containing galardin derivatives. Bioorg Med Chem Lett 13(10):1783–1786. doi:10.1016/S0960-894X(03)00214-2
74. Rouffet M, Denhez C, Bourguet E, Bohr F, Guillaume D (2009) In silico study of MMP inhibition. Org Biomol Chem 7(18):3817–3825. doi:10.1039/b910543c
75. Li D, Zheng Q, Fang X, Ji H, Yang J, Zhang H (2008) Theoretical study on potency and selectivity of novel non-peptide inhibitors of matrix metalloproteinases MMP-1 and MMP-3. Polymer 49(15):3346–3351. doi:10.1016/j.polymer.2008.05.026
76. da Silva SL, Calgarotto AK, Maso V, Damico DC, Baldasso P, Veber CL, Villar JAFP, Oliveira ARM, Comar M, Oliveira KMT, Marangoni S (2009) Molecular modeling and inhibition of phospholipase A2 by polyhydroxy phenolic compounds. Eur J Med Chem 44(1):312–321. doi:10.1016/j.ejmech.2008.02.043
77. Šramko M, Garaj V, Remko M (2008) Thermodynamics of binding of angiotensin-converting enzyme inhibitors to enzyme active site model, J Mol Struct—Theochem 869(1–3):19–28. 10.1016/j.theochem.2008.08.018
78. Opie LH, Gersh BJ (2009) Drugs for the Heart. WB Saunders, Philadelphia
79. Lorthiois E, Bernardelli P, Vergne F, Oliveira C, Mafroud A-K, Proust E, Heuze L, Moreau F, Idrissi M, Tertre A, Bertin B, Coupe M, Wrigglesworth R, Descours A, Soulard P, Berna P (2004) Spiroquinazolinones as novel, potent, and selective PDE7 inhibitors. Part 1. Bioorg Med Chem Lett 14(18):4623–4626. doi:10.1016/j.bmcl.2004.07.011
80. Bernardelli P, Lorthiois E, Vergne F, Oliveira C, Mafroud A-K, Proust E, Pham N, Ducrot P, Moreau F, Idrissi M, Tertre A, Bertin B, Coupe M, Chevalier E, Descours A, Berlioz-Seux F, Berna P, Li M (2004) Spiroquinazolinones as novel, potent, and selective PDE7 inhibitors. Part 2: Optimization of 5,8-disubstituted derivatives. Bioorg Med Chem Lett 14(18):4627–4631. doi:10.1016/j.bmcl.2004.07.010
81. Daga PR, Doerksen RJ (2008) Stereoelectronic properties of spiroquinazolinones in differential PDE7 inhibitory activity. J Comput Chem 29(12):1945–1954. doi:10.1002/jcc.20960
82. Horenstein BA, Schramm VL (1993) Correlation of the molecular electrostatic potential surface of an enzymatic transition state with novel transition-state inhibitors. Biochemistry 32(38):9917–9925. doi:10.1021/bi00089a007
83. Braunheim BB, Miles RW, Schramm VL, Schwartz SD (1999) Prediction of inhibitor binding free energies by quantum neural networks. Nucleoside analogues binding to trypanosomal nucleoside hydrolase. Biochemistry 38(49):16076–16083. doi:10.1021/bi990830t
84. Ehrlich JI, Schramm VL (1994) Electrostatic potential surface analysis of the transition state for AMP nucleosidase and for formycin 5'-phosphate, a transition-state inhibitor. Biochemistry 33(30):8890–8896. doi:10.1021/bi00196a005
85. Debnath AK (2013) Rational design of HIV-1 entry inhibitors. In: Kortagere S (ed) Methods in molecular biology. Springer, Clifton, p 185–204. doi:10.1007/978-1-62703-342-8_13

86. Świderek K, Martí S., Moliner V (2012) Theoretical studies of HIV-1 reverse transcriptase inhibition. Phys Chem Chem Phys 14(36):12614–12624. doi:10.1039/c2cp40953d
87. Liang YH, Chen FE (2007) ONIOM DFT/PM3 calculations on the interaction between dapivirine and HIV-1 reverse transcriptase, a theoretical study. Drug Discov Ther 1(1):57–60.
88. Garrec J, Sautet P, Fleurat-Lessard P (2011) Understanding the HIV-1 protease reactivity with DFT: what do we gain from recent functionals?. J Phys Chem B 115(26):8545–8558. doi:10.1021/jp200565w
89. Garrec J, Cascella M, Rothlisberger U, Fleurat-Lessard P (2010) Low inhibiting power of N···CO based peptidomimetic compounds against HIV-1 protease: insights from a QM/MM study. J Chem Theor Comput 6(4):1369–1379. doi:10.1021/ct9004728
90. Gautier A, Pitrat D, Hasserodt J (2006) An unusual functional group interaction and its potential to reproduce steric and electrostatic features of the transition states of peptidolysis. Bioorg Med Chem 14(11):3835–3847. doi:10.1016/j.bmc.2006.01.031
91. Waibel M, Hasserodt J (2008) Diversity-oriented synthesis of a drug-like system displaying the distinctive N→C=O interaction. J Org Chem 73(16):6119–6126. doi:10.1021/jo800719j
92 Waibel M, Pitrat D, Hasserodt J (2009) On the inhibition of HIV-1 protease by hydrazino-ureas displaying the N→C=O interaction. Bioorg Med Chem 17(10):3671–3679. doi:10.1016/j.bmc.2009.03.059
93. Park C, Koh JS, Son YC, Choi H, Lee CS, Choy N, Moon KY, Jung WH, Kim SC, Yoon H (1995) Rational design of irreversible, pseudo-C2-symmetric hiv-1 protease inhibitors. Bioorg Med Chem Lett 5(16):1843–1848. doi:10.1016/0960-894X(95)00306-E
94. Lee CS, Choy N, Park C, Choi H, Son YC, Kim S, Ok JH, Yoon H, Kim SC (1996) Design, synthesis, and characterization of dipeptide isostere containing cis-epoxide for the irreversible inactivation of HIV protease. Bioorg Med Chem Lett 6(6):589–594. doi:10.1016/0960-894X(96)00087-X
95. Choy N, Choi H, Jung WH, Kim CR, Yoon H, Kim SC, Lee TG, Koh JS (1997) Synthesis of irreversible HIV-1 protease inhibitors containing sulfonamide and sulfone as amide bond isosteres. Bioorg Med Chem Lett 7(20):2635–2638. doi:10.1016/S0960-894X(97)10054-3
96. Kóňa J (2008) Theoretical study on the mechanism of a ring-opening reaction of oxirane by the active-site aspartic dyad of HIV-1 protease. Org Biomol Chem 6(2):359–365. doi:10.1039/b715828a
97. Pommier Y, Johnson AA, Marchand C (2005) Integrase inhibitors to treat HIV/AIDS. Nat Rev Drug Discov 4(3):236–248. doi:10.1038/nrd1660
98. Ingale KB, Bhatia MS (2011) HIV-1 integrase inhibitors: a review of their chemical development. Antivir Chem Chemoth 22(3):95–105. doi:10.3851/IMP1740
99. Messiaen P, Wensing AMJ, Fun A, Nijhuis M, Brusselaers N, Vandekerckhove L (2013) Clinical use of HIV integrase inhibitors: a systematic review and meta-analysis. PloS One 8(1):e52562. doi:10.1371/journal.pone.0052562
100. Liao C, Nicklaus MC (2010) Tautomerism and magnesium chelation of HIV-1 integrase inhibitors: a theoretical study. Chem Med Chem 5(7):1053–1066. doi:10.1002/cmdc.201000039
101. Agrawal A, DeSoto J, Fullagar JL, Maddali K, Rostami S, Richman DD, Pommier Y, Cohen SM (2012) Probing chelation motifs in HIV integrase inhibitors. Proc Nat Acad Sci U S A 109(7):2251–2256. doi:10.1073/pnas.1112389109
102. Thalheim T, Vollmer A, Ebert R-U, Kühne R, Schüürmann G (2010) Tautomer identification and tautomer structure generation based on the InChI code. J Chem Inf Model 50(7):1223–1232. doi:10.1021/ci1001179
103. Nunthaboot N, Pianwanit S, Parasuk V, Kokpol S, Wolschann P (2007) Theoretical study on the HIV-1 integrase inhibitor 1-(5-chloroindol-3-yl)-3-hydroxy-3-(2H-tetrazol-5-yl)-propenone (5CITEP). J Mol Struct 844–845:208–214. doi:10.1016/j.molstruc.2007.06.026

104. Alves CN, Martí S, Castillo R, Andrés J, Moliner V, Tuñón I, Silla E (2007) Calculation of binding energy using BLYP/MM for the HIV-1 integrase complexed with the S-1360 and two analogues. Bioorgan Med Chem 15(11):3818–3824. doi:10.1016/j.bmc.2007.03.027
105. Lespade L, Bercion S (2010) Theoretical study of the mechanism of inhibition of xanthine oxydase by flavonoids and gallic acid derivatives. J Phys Chem B 114(2):921–928. doi:10.1021/jp9041809
106. Lin C-M, Chen C-S, Chen C-T, Liang Y-C, Lin J-K (2002) Molecular modeling of flavonoids that inhibits xanthine oxidase. Biochem Biophys Res Commun 294(1):167–172. doi:10.1016/S0006-291X(02)00442-4
107. Hall LH, Kier LB (1995) Electrotopological state indices for atom types: a novel combination of electronic, topological, and valence state information,. J Chem Inf Model 35(6):1039–1045. doi:10.1021/ci00028a014
108. Fogliani B, Raharivelomanana P, Bianchini J-P, Bouraïma-Madjèbi S, Hnawia E (2005) Bioactive ellagitannins from *Cunonia macrophylla*, an endemic Cunoniaceae from New Caledonia. Phytochemistry 66(2):241–247. doi:10.1016/j.phytochem.2004.11.016
109. Fenton JW, Ofosu FA, Moon DG, Maraganore JM (1991) Thrombin structure and function: why thrombin is the primary target for antithrombotics. Blood Coagul Fibrin 2(1):69–75.
110. Alzate-Morales JH, Contreras R, Soriano A, Tuñon I, Silla E (2007) A computational study of the protein-ligand interactions in CDK2 inhibitors: using quantum mechanics/molecular mechanics interaction energy as a predictor of the biological activity. Biophys J 92:430–439. doi:10.1529/biophysj.106.091512
111. De Oliveira EB, Humeau C, Maia ER, Chebil L, Ronat E, Monard G, Ruiz-Lopez MF, Ghoul M, Engasser J-M (2010) An approach based on Density Functional Theory (DFT) calculations to assess the candida antarctica lipase B selectivity in rutin, isoquercitrin and quercetin acetylation. J Mol Catal B—Enzym 66(3–4):325–331. doi:10.1016/j.molcatb.2010.06.009
112. Benini S, Rypniewski WR, Wilson KS, Ciurli S, Mangani S (2001) Structure-based rationalization of urease inhibition by phosphate: novel insights into the enzyme mechanism. J Biol Inorg Chem 6(8):778–790. doi:10.1007/s007750100254
113. Karplus PA, Pearson MA, Hausinger RP (1997) 70 Years of crystalline urease: what have we learned?. Acc Chem Res 30(8):330–337. doi:10.1021/ar960022j
114. Leopoldini M, Marino T, Russo N, Toscano M (2008) On the binding mode of urease active site inhibitors: a density functional study. Int J Quant Chem 108(11):2023–2029. doi:10.1002/qua.21758
115. Wei D, Lei B, Tang M, Zhan C-G (2012) Fundamental reaction pathway and free energy profile for inhibition of proteasome by epoxomicin. J Am Chem Soc 134(25):10436–10450. doi:10.1021/ja3006463
116. Weigend F (2008) Hartree–Fock exchange fitting basis sets for H to Rn. J Comput Chem 29(2):167–175. doi:10.1002/jcc.20702
117. Neese F, Wennmohs F, Hansen A, Becker U (2009) Efficient, approximate and parallel Hartree–Fock and hybrid DFT calculations. A "chain-of-spheres" algorithm for the Hartree–Fock exchange. Chem Phys 356(1–3):98–109. 10.1016/j.chemphys.2008.10.036
118. Ferrer S, Ruiz-Pernía J, Martí S et al (2011) Hybrid schemes based on quantum mechanics/molecular mechanics simulations goals to success, problems, and perspectives. Adv Protein Chem Struct Biol 85:81–142. doi:10.1016/B978-0-12-386485-7.00003-X

Chapter 8
Molecular Dynamics Simulations of Lipid Bilayers with Incorporated Peptides

Milan Melicherčík, Tibor Hianik and Ján Urban

Abstract Biological membranes are important cell structures that play important role in the transport of the ions and other molecules into and out of the cell and regulate the signaling pathway. They are composed of lipid bilayer, integral and peripheral proteins. The ionic channels, enzymes and most of the membrane receptors belong to integral proteins that span the membrane and contact by their hydrophobic part with hydrophobic interior of the lipid bilayer. These hydrophobic interactions are crucial for the effect of peptide on a lipid bilayer matrix and vice versa. The study of the mechanisms of these interactions is important for understanding the functioning of the peptides in a membrane. However the study of native biomembrane is rather complicated due to its complexity and inhomogeneity. Therefore model lipid bilayers and short peptides can be used as a model for study of the protein–lipid interactions. In this chapter we review the current state of the art in experimental and molecular dynamics simulation study of the short peptide–membrane interactions. As an example we consider in more detail the application of molecular dynamic simulations on the study of interaction of a model lysine-flanked α-helical peptides P_{24}, LA_{12}, L_{24} and its analogues A_{24}, I_{24}, and V_{24} with lipid bilayers composed of dimyristoylphosphatidylcholine (DMPC) and dipalmitoylphosphatidylcholine (DPPC) both in a gel and in a liquid-crystalline state. We have shown that these peptides cause disordering of the lipid bilayer in the gel state and small changes in a liquid-crystalline state. The peptides affect ordering of the surrounding lipids depending on the helix stability, the amount of dihedral angles in *trans* conformation and the number of transitions between *trans* and *gauche* conformation. It has been found the tendency of Lys-flanked peptides to compensate the positive mismatch between peptide and membrane hydrophobic core by tilting.

M. Melicherčík (✉) · T. Hianik · J. Urban
Department of Nuclear Physics and Biophysics, Faculty of Mathematics,
Physics and Informatics, Comenius University, Mlynská dolina,
Bratislava, Slovak Republic 842 48, Europe
e-mail: milan.melichercik@fmph.uniba.sk

L. Gorb et al. (eds.), *Application of Computational Techniques in Pharmacy and Medicine,*
Challenges and Advances in Computational Chemistry and Physics 17,
DOI 10.1007/978-94-017-9257-8_8

In some cases the tilt was replaced by *superhelical* double-twisted structure. The rest of helices were bend or produced kink in addition to the tilt. The lipid structural state around the peptide has been also analyzed.

8.1 Introduction

Lipid–protein interactions are of fundamental importance for understanding both the structural integrity and the functions of biological membranes [1, 2]. In particular, the chemical composition and physical properties of the lipid bilayer membranes (BLMs) can markedly influence the activity, thermal stability, and the location and disposition of a large number of integral membrane proteins in both model and biological membranes [2]. For these reasons, many studies of the interactions of membrane proteins with their host BLM have been carried out, in both biological and reconstituted model systems, employing a wide range of different physical techniques [3–6]. To overcome the problem of the complicated structure of integral proteins and the problems with their isolation and purification, a number of workers have designed and synthesized peptide models of specific regions of natural membrane proteins and have studied their interactions with model lipid membranes of defined composition (see [7, 8]). In particular, the study of the mechanisms of the interactions of peptides with BLM has also very important practical significance for understanding of the mechanism of interaction of e.g. neuropeptides [9] or antimicrobial peptides [10] with membranes. The cell-penetrating-peptides can also transport other macromolecules inside the cell and thus are perspective in drug delivery.

The synthetic peptide acetyl-K_2-G-L_{24}-K_2-A-amide (P_{24}) and its structural analogs, e.g. acetyl-K_2-L_{24}-K_2-amide (L_{24}), have been successfully utilized as a model of the hydrophobic transmembrane α-helical segments of integral proteins [8, 11]. These peptides contain a long sequence of hydrophobic leucine residues capped at both the N- and C-termini with two positively charged lysine residues. The central polyleucine region of these peptides was designed to form a maximally stable α-helix which will partition strongly into the hydrophobic environment of the lipid bilayer core, while the dilysine caps were designed to anchor the ends of these peptides to the polar surface of the BLM and to inhibit the lateral aggregation of these peptides. In fact, circular dichroism (CD) [11] and Fourier transform infrared spectroscopy FTIR [12] spectroscopic studies of P_{24} have shown that it adopts a very stable α-helical conformation both in solution and in lipid bilayers. X-ray diffraction [13], fluorescence quenching [14] and FTIR [12] and deuterium nuclear magnetic resonance (^{2}H-NMR) [15] spectroscopic studies have confirmed that P_{24} and its analogs assume a transbilayer orientation with the N- and C-termini exposed to the aqueous environment and the hydrophobic polyleucine core embedded in hydrocarbon core of the BLM when reconstituted with various phosphatidylcholines (PCs) [16]. ^{2}H-NMR [17] and electron spin resonance (ESR) [18] spectroscopic studies have shown that the rotational diffusion of P_{24} about its long axis perpendicular to the membrane plane is rapid in the liquid-crystalline state of the bilayer.

We applied high-performance liquid chromatography (HPLC), CD, differential scanning calorimetry DSC and attenuated total reflectance ATR-FTIR methods for study studied specially designed α-helical transmembrane peptides (acetyl-K_2-L_m-A_n-K_2-amide, where $m+n=24$) in respect of their solution behavior and interactions with phospholipids [19]. These peptides exhibit strong α-helical conformation in water, membrane-mimetic media and lipid model membranes, however the stability of the helices decreases as the Leu content decreases. Also, their binding to reversed phase high-performance liquid chromatography columns is largely determined by their hydrophobicity and the binding generally decreases with decrease in the Leu/Ala ratio. However, the retention of these peptides by such columns is also affected by the distribution of hydrophobic residues on their helical surfaces, being further enhanced when peptide helical hydrophobic moments are increased by clustering hydrophobic residues on one side of the helix. This clustering of hydrophobic residues also increases peptide propensity for self-aggregation in aqueous media and enhances partitioning of the peptide into lipid bilayer membranes. We also found that the peptides LA_3LA_2 (acetyl-K_2-$(LA_3LA_2)_3LA_2$-K_2-amide) and particularly LA_6 (acetyl-K_2-$(LA_6)_3LA_2$-K_2-amide) associate less strongly with bilayer and perturb the thermotropic phase behavior of phosphatidylcholine bilayers much less than peptides with higher L/A ratios. These results are consistent with free energies calculated for the partitioning of these peptides between water and phospholipid bilayers. This suggests that LA_3LA_2 has an equal tendency to partition into water and into the hydrophobic core of phospholipid model membranes, whereas LA_6 should strongly prefer the aqueous phase. We conclude that for α-helical peptides of this type, Leu/Ala ratios of greater than 7/17 are required for stable transmembrane associations with phospholipid bilayers. Experimental studies have been focused also on the analysis of the effect of substitution of some amino acids in peptides on their properties. Idiong et al. [20] studied α-helical antimicrobial peptides purified from the venom of the Central Asian spider *Lachesana tarabaevi* and showed that replacing the glycine at position 11 with alanine resulted in more rigid peptide structure due to the reduced conformational flexibility.

Detailed DSC, FTIR, NMR and electron paramagnetic resonance (EPR) studies of interaction of P_{24} or L_{24} with BLM [21] have revealed that the results obtained from different physical techniques generally agree well with one another. However, certain discrepancies have been found in comparison of the results obtained by spectroscopic techniques, i.e. FTIR and ^{2}H-NMR. While the ^{2}H-NMR technique indicated that incorporation of P_{24} peptide into the DPPC bilayers resulted in a decrease of the ordering of the membrane in gel state and increase in the liquid crystalline (LC) state, FTIR experiments suggest that peptide induced a decrease of the ordering of the lipid bilayer in both structural state of the membrane [21]. This discrepancy has been explained by different peculiarities of these two methods. While the order parameters in ^{2}H-NMR spectroscopy are primarily sensitive to *trans/gauche* isomerisation, the molecular interpretation of the changes in membrane ordering based on changes in frequency of the methylene stretching modes in IR spectroscopy are likely attributed to the sensitivity of the band position phenomena other than *trans/gauche* isomerisation, such as the interchain coupling and the contribution of peptide in the methylene and methyl stretching region. Interchain coupling is significant enough even in fluid bilayers [21]. Therefore, using exclusively FTIR it is difficult to decide what process is dominant in fluid state—interchain coupling or *trans/gauche* isomerisation.

In contrast with spectroscopic methods that provide information about microscopic changes of the lipid bilayer in close proximity of the protein, macroscopic methods, such are membrane compressibility measurements, are sensitive to changes of large membrane regions. The sensitivity and utility of measurements of volume compressibility has been proved in several studies of the interaction of integral proteins with lipid bilayers, e.g. bacteriorhodopsin [4] or peptides like $ACTH_{24}$ [9] or gramicidin S [10]. In the case of bacteriorhodopsin, it has been shown that one molecule of the peptide is able to change the structural state of the lipid bilayer of whole large unilamellar vesicle (LUV). We applied this method also to the study of the interaction of synthetic α-helical transmembrane peptides like L_{24} with lipid bilayers [22]. We used precise measurement of density and ultrasound velocity to study the physical properties of LUVs composed of a homologous series of n-saturated phosphatidylcholines (PC) containing L_{24}. PCs whose hydrocarbon chains contained from 13 to 16 carbon atoms, thus producing phospholipid bilayers of different thicknesses and gel to liquid-crystalline phase transition temperatures. This allowed us to analyze how the difference between the hydrophobic length of the peptide and the hydrophobic thickness of the lipid bilayer influences the thermodynamical and mechanical properties of the membranes. We showed that the incorporation of L_{24} decreases the temperature and cooperativity of the main phase transition of all LUVs studied. The presence of L_{24} in the bilayer also caused an increase of the specific volume and of the volume compressibility in the gel state bilayers. In the liquid crystalline state, the peptide decreases the specific volume at relatively higher peptide concentration (mole ratio L_{24}:PC = 1:50). The overall volume compressibility of the peptide-containing lipid bilayers in the liquid crystalline state was in general higher in comparison with pure membranes. There was, however, a tendency for the volume compressibility of these lipid bilayers to decrease with higher peptide content in comparison with bilayers of lower peptide concentration. For one lipid composition, we also compared the thermodynamical and mechanical properties of LUVs and large multilamellar vesicles (MLVs) with and without L_{24}. As expected, a higher cooperativity of the changes of the thermodynamical and mechanical parameters took place for MLVs in comparison with LUVs. These results are in agreement with previously reported DSC and ^{2}H NMR spectroscopy study of the interaction of the L_{24} and structurally related peptides with phosphatidylcholine bilayers. An apparent discrepancy between ^{2}H NMR spectroscopy and compressibility data in the liquid crystalline state may be connected with the complex and anisotropic nature of macroscopic mechanical properties of the membranes. The observed changes in membrane mechanical properties induced by the presence of L_{24} suggest that around each peptide a distorted region exists that involves at least two layers of lipid molecules.

Further information on the structure and dynamics of lipid bilayer as well as on the molecular mechanisms of protein–lipid interactions, can be obtained by molecular dynamics simulations (MD). This method is widely used for this purpose. During the last three decades MD method has been applied to many short peptides and larger proteins starting with simple artificial peptides (see e.g. [23–25]). Later, naturally occurred integral proteins such as channel forming peptides gramicidin A [26, 27] and alamethicin [30], or larger transmembrane protein bacteriorhodopsin [28, 29] were analyzed. The specially designed model peptides were also studied. They consist typically from hydrophobic core (usually leucine (Leu) residues [31] or in combination with alanine (Ala) [32]). This alternation of residues decreases the core hydrophobicity and peptide better mimics the natural proteins [33]. These peptides are flanked at both sides by hydrophilic residues stabilizing both ends in headgroup region of the lipid bilayer. As the anchors usually lysine (Lys) [11, 34] or tryptophan (Trp) [35] are used. In a membrane these model peptides form stable α-helix [36, 37], even without polar anchors [38]. The evidence of peptide tilting or changing in membrane thickness is the basis for stating the "hydrophobic matching" theory (more can be found in review published by e.g. Killian [39] or Lee [40]). Among most used are model structures are L_{24}, P_{24}, $WALP_{19}$ and longer $WALP_{23}$ with two more (LA) repeating pairs [41, 42]. But also other peptide lengths were tested—e.g. P_{16} [d43] or different peptides—$KWALP_{23}$: acetyl-G-K-A-L-W-$(LA)_6$-W-L-A-K-A-amid. The peptide–lipid interactions were studied in many lipid bilayers composed of various phosphatidylcholines (PC): dilauroyl PC (DLPC), dimyristoyl PC (DMPC). dipalmitoyl PC (DPPC), dioleoyl PC (DOPC), palmityloleoyl PC (POPC). The MD e.g. Tieleman et al. [30] and NMR studies [42] indicated that although helix is stable in a membrane, some parts of it were bended or even kinked.

There exists a difference in binding to a membrane the peptides flanking with Trp and Lys residues. Maurits et al. [44] studied WALP and KALP peptides of different lengths. The $WALP_{16}$ (acetyl-G-K_2-$(LA)_5$-K_2-A-amid) peptide in DOPC bilayer converts membrane into inverted hexagonal H_{II} phase, while $KALP_{16}$ remains in LC state. In general the behavior of $KALP_{23}$ in a membrane is similar to that of $WALP_{21}$ and consisting in peptide tilt and in inducing changes in the thickness of surrounding lipid layer. This is due to firm interaction of indole group of the peptide with carbonyl group of lipids, while the Lys amino group NH_3^+ lies at the end of long flexible chain. Therefore Lys residue is able to snorkel into (thicker) membrane.

De Jesus and Allen [45, 46] simulated WALP peptides with different number of Trp repeating. The longer Trp parts were used, the higher tilt has been detected (with positive mismatch). The negative mismatch resulted in increases of the membrane deflection, decreases in lipid chain ordering and peptide gets shorter. They also simulated long (92 amino acids) poly-Leu helical peptide with inserted Trp or its analogue 3-methylindol (3-MIND). Trp and 3-MIND lower membrane deflection and interacts with glycerol core, carbonyl oxygens and (farther from membrane center) with phosphate oxygen.

8.1.1 Membrane (Dis-)ordering

Lewis et al. [47] measured energetic effect of transfer the Lys-flanked peptides into the membrane. They found the presence of helix, but also observed the decrease of temperature and enthalpy of phase transition. This suggests that the peptides decrease ordering of membrane in a gel state. Simulations of peptides with membrane in liquid crystalline state were performed by Esteban-MartHn and Salgado [48]. They studied WLP_{23} (acetyl-G-W_2-L_{17}-W_2-A-ethanolamine) and KLP_{23} in DMPC bilayers and observed faster relaxation of Lys-flanked peptides (~10 ns) but lower tilt angle (~20°), while Trp-flanked ones relaxed during 50 ns (in some simulations even till 150 ns), but the final tilt was ~31°. Based on experimental data [49, 50] they stated the hypothesis that the 200 ns simulation is not enough to fully relax the system. Certainly, the measured tilt by ^{2}H-NMR quadrupole splittings is lower than in simulated systems.

Davies et al. studied phase transitions with ^{2}H-NMR quadruple splittings spectra. The longer peptides L_{24} in DPPC membrane in peptide to lipid molar ratio of 100:1 caused 30 % increase of order in lipids acyl chains, while in molar ratio of 43:1 there was only 5 % order increase [11]. In Pan et al. X-ray experiments and MD simulations [51] of alamecithin it has been shown that this peptide decreases thickness of diC22:1-PC membrane and increases its own length in a peptide to lipid molar ratio of 1:10. However, in a DOPC membrane this peptide tilts in approx. ~15°. In both cases the peptide decreased of membrane fluctuations (bending modulus K_c), but the diC22:1-PC is more stabilized (factor ~10) than DOPC (only ~2).

Other peptide, the maculatin 1.1, remains in helical conformation in bilayers of wide range of lipid composition (DHPC to DPPC, POPC, DOPC, DMPA, DMPS, DPPS and DMPG). Only exceptions were the DHPC/LQ and DPPC in liquid crystalline and gel phases, respectively, where there is the too large difference between hydrophobic lengths of membrane and peptide [52].

Hoernke et al. [53] used modified Lys-flanked peptides and those in which Lys has been replaced by ornithine, α,γ-diaminobutyric acid, α,β-diaminopropionic acid in negatively charged phosphatidyl glycerol (PG) membrane. These smaller side-chains caused higher increase of phase transitions temperature and decrease of surface pressure. The long polylysines (more Lys residues at the ends) increase phase transitions temperature, but short ones lower it.

8.1.2 Mutation Studies

Johannson and Lindhal [54] tested systematic mutations of acetyl-G_2PG-A_{19}-GPG_2-amid peptide in DMPC membrane. Some amino acids did dissolve completely in hydrophobic part of the membrane, but caused surface defects and water snorkeling into membrane. This suggests that even for polar/charged residues a large part of solvation cost is due to entropy, not enthalpy losses. Basic side chains cause much less membrane distortion than acidic, since they are able to form hydrogen bonds with carbonyl groups instead of water or other lipid headgroups. This preference is supported by sequence statistics, where basic residues have increased relative occurrence at carbonyl z-coordinates. Snorkeling effects and N-/C-terminal orientation bias are directly observed, which significantly reduces the effective thickness of the hydrophobic core. Aromatic side chains intercalate efficiently with lipid chains (improving Trp/Tyr anchoring to the interface) and Ser/Thr residues are stabilized by hydroxyl groups sharing hydrogen bonds to backbone oxygens.

Similar study has been published by Li et al. [55]. They simulated analogue of Arg side chain—$MguanH^+$. In all membranes tested they detected the ion induced defect. In thinner membranes (DDPC, DLPC) the peptide chain even caused its perforation. The peptide caused increase of lipid area by 0.03 nm^2/lipid for DDPC, while decrease of the area (by 0.03 nm^2/lipid) has been observed for e.g. DLPC or DSPC. This was caused by ordering or disordering of lipid chains, respectively, and agrees with experimental data [56].

Lam et al. [57] studied the antimicrobial peptide protegrin-1 in a membrane by atomic force microscopy (AFM) and MD methods. This 18 amino acid peptide contains six Arg residues and created "edge instabilities" in low concentrations and "wormhole" structure in high concentrations.

MacCallum et al. [58] simulated different amino acid interactions with DOPC membrane. The most hydrophobic amino acids (Ile, Leu, Val, Ala) incorporate into middle part of membrane—lowest energy has been found for Ile (−22 kJ/mol) and highest for Ala (−8 kJ/mol). Cys and Met were located at region between hydrocarbon chains and beginning of choline groups (carbonyl groups)—Region II. Amino acids with aromatic side chains Tyr, Trp, Phe had energetic minimum also in this region. But Tyr has positive energy in Region I (acyl chains). The same holds for Phe. Trp is allowed to be localized in Region I, but with higher energy than in Region II. Polar amino acids Asn, Gln, Ser, Thr have also energetic minimum in Region II, but in Region I they have high positive energy (24–13 kJ/mol). The charged amino acids Arg, Lys, Glu, Asp also prefer Region II. But the negatively charged amino acids show steady increase of energy from water to center of the membrane, while positively charged ones have minimum of the energy in Reg. II, which then increase toward the center. All charged amino acids cause large water defects in a membrane and all but (possibly) Arg lose their charge in the middle of the membrane. In similar work published by Yoo et al. [59] using free energy perturbations method—the pK_a of Arg in DPPC membrane has been estimated. The simulation shows that $pK_a > 7$ (center of membrane has neutral pH = 7). This means, the Arg is probably charged in the middle of membrane.

Daily et al. [47] measured ^{2}H-NMR splittings of $KWALP_{23}$ peptide (acetyl-G-K-A-L-W-$(LA)_6$-x-W-L-A-K-A-amid) in DLPC, DMPC and DOPC membranes. The fitting of splittings of middle six amino acids doesn't fit to helix in the DLPC and DMPC membranes. This suggests creation of kink of 9–13° in this region. The same defect was detected in previous study of $WALP_{23}$ in DLPC bilayers. However significant improvement of previous fitting of the same authors resulted in 15° kink. In the thicker DOPC membranes the peptides didn't exhibited this effect.

Kim and Im [60] used PMF method (potential mean force) in simulations of $WALP_n$ (n = 16, 19, 23, 27) peptides in DMPC or POPC membrane. They also mutated Trp flanking residues with Ala, Lys and Arg. Their results can be summarized as follows: (1) tilting of a single-pass transmembrane (TM) helix is the major response to a hydrophobic mismatch; (2) TM helix tilting up to ~10° is inherent due to the intrinsic entropic contribution arising from helix precession around the membrane normal even under a negative mismatch; (3) the favorable helix–lipid interaction provides additional driving forces for TM helix tilting under a positive mismatch; (4) the minimum-PMF tilt angle is generally located where there is the hydrophobic match and little lipid perturbation; (5) TM helix rotation is dependent on the specific helix-lipid interactions; (6) anchoring residues at the hydrophilic/hydrophobic interface can be an important determinant of TM helix orientation.

Also at large peptide tilt angles the surrounding membrane is even thinner than pure membrane. The tilt angles of different flanking residues depend on its hydrophobicity—lowest tilt angle has RALP peptide in comparison with other peptides: RALP<KALP<WALP<AALP. The authors also compared MD results with previous published experimental data for similar peptides obtained by ^{2}H-NMR splitting measurements (for example the tilt angles have been 4.4° for $KALP_{23}$ and 5.2° for $WALP_{23}$) [61, 62]. They concluded that for correct determination of tilt angles from splittings data, the proper averages of rotation angles is necessary, as it has been done in Ref. [63–65]. For example the florescence spectroscopy determined $WALP_{23}$/DOPC tilt angle 24°±5° [65].

Monticelli et al. [66] published another comparison of ^{2}H-NMR quadrupole splittings and MD simulations. They used $WALP_{23}$ in DMPC and stated that the underestimation of peptide movements can affect the measured tilt angle. This has been also concluded in [63, 64], where authors recognized the problem in GALA method (geometric analysis of labeled alanines) with position averaging. They used nonlinear averaging of goniometric functions and showed that the peptide in membrane tilted by 30° can have the same quadrupolar samplings as motionless peptide with 5° tilt.

In this chapter we show usefulness of the molecular dynamics simulations on the study of model helical peptides composed of acetyl-K_2-A_{24}-K_2-amide (A_{24}), acetyl-K_2-L_{24}-K_2-amide (L_{24}), acetyl-K_2-$(LA)_{12}$-K_2-amide ($(LA)_{12}$), acetyl-K_2-I_{24}-K_2-amide (I_{24}), acetyl-K_2-G-L_{24}-K_2-A-amide (P_{24}) and acetyl-K_2-V_{24}-K_2-amide (V_{24}) incorporated into the phospholipid bilayers (DMPC, DPPC). The behavior of some of these and other peptides in membranes of various lipid compositions has been analyzed by Host and Killian [67]. We have shown that the effect of peptides on the lipid bilayer strongly depends on membrane physical state—gel or liquid crystalline.

8.2 Methods

MD has been applied for the determination of changes of physical properties of lipid bilayers caused by the incorporated peptide as well as for the determination of possible peptide structural alterations. MD were performed under periodic boundary conditions using the GROMACS software [68] and the GROMOS87 [69] forcefield with corrections for lipids [70, 71]. The initial models of transmembrane α-helix peptides have been generated by means of HyperChem [72]. Preequilibrated DMPC and DPPC bilayers with 128 lipid molecules and 3655 molecules of water in L_α liquid-crystalline state published by Tieleman et al. [73] have been used in bilayer modeling. For the simulations with membrane in L_β' gel state, we created the bilayers on the basis of the experimental data (taking into account parameters for the area per lipid and bilayer thickness) [33, 74]. Initial structures were solvated with SPC water model (4764 molecules for DMPC and 4784 for DPPC), energetically minimized and simulated for over 20 ns until the membrane parameters were close to the experimental values. A cylindrical hole has been created in the center of a bilayer by removing four lipids whose atoms were within 0.23 nm of the central axis of a cylinder. The peptide was then inserted into the cavity. The resulting system (peptide, 124 PC molecules, 4 chlorine ions and water) consisted from more than 16,000 atoms for LC and more than 20,000 for gel state of the membrane. The system has been energetically minimized and equilibrated during 0.5 ns while the peptide atoms were fixed. Then MD took place for at least 40 ns at temperatures T=288 and 310 K (bellow and over phase transition) for DMPC and 296/346 K for DPPC bilayer. The temperature of phase transition is 297 K for DMPC and 314 K for DPPC. MD was performed with constant pressure of 1 bar (semi-isotropic barostat), constant temperatures and with the time step of 2 fs. The LINCS algorithm has been used to constrain covalent bond lengths. The used conditions were similar to that reported by Berger et al. [75]. For results confirmation we extended seven simulations to 100 ns.

Trajectories were analyzed from the last 5 ns of the simulations by subroutines (programs) available from Gromacs package. Ramachandran plot were calculated by Procheck software [76].

8.3 Properties of Pure Membranes

Interactions of the peptides with lipid bilayers have been studied in a gel and in a liquid-crystalline state. At the end of the simulation the hydrophobic core of pure lipid bilayers in a gel state is 3.61 nm thick for DPPC and 3.26 nm for DMPC. The area per lipid is 0.4733 nm^2 per lipid for DPPC and 0.4676 nm^2 per lipid for DMPC membrane, respectively. Values determined by experimental methods are 0.479 nm^2/lipid for DPPC and 0.472 nm^2 for DMPC [74, 77]. Membranes with similar difference in parameters have been used in MD simulations by Tu et al. [78] or Tieleman et al. [79].

For the membrane in a gel state most of dihedral angles should be in *trans* conformation [80]. Almost for whole chains it is over 90 %, only on beginnings and ends it goes lower, but it is still more than 80 %. Also deuterium order parameters (S_{CD}) shows highly ordered system. The order parameters are defined as $S_{CD,z} = 3/2\langle\cos^2(\Theta_z)\rangle - 1/2$, where Θ_z is the angle between normal to the membrane and axis of that molecule. The axis is defined as a vector from C_{n-1} to C_{n+1} and angle brackets represents averaging in time. Maximal and minimal values are 1 (parallel to the membrane normal) and $-1/2$ (perpendicular to the membrane) [81]. Order parameters depend on the structural state of lipid membrane—in a gel state it reaches up to value 0.4, but in LC state it is only up to 0.3. Experimental values for gel state reaches maximum 0.35 till 0.4 at 3rd till 4th C^{α} atom and goes down to

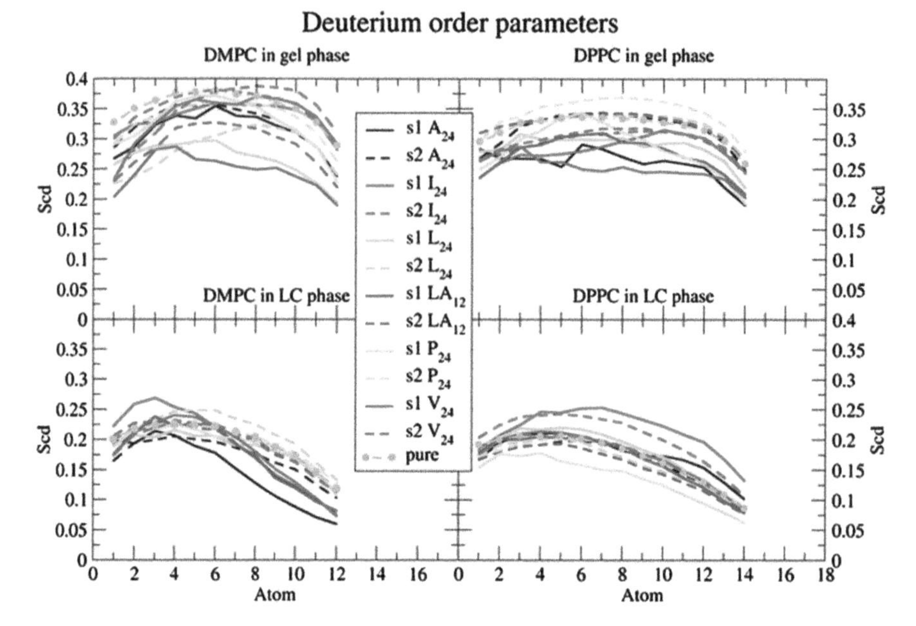

Fig. 8.1 Order parameters from last 5 ns of the simulations of peptides and four types of lipid bilayers. Lipids are separated into two slices around peptide: nearer to the peptide and more influenced as *s1* and farther, less influenced, as *s2*. In gel phase peptide causes disorder by its presence. In PC phase most of peptides relatively keep properties of pure lipid. Only I_{24} increases order in membrane. However more fluctuating peptides disorder membrane in the middle

0.2 in the middle of membrane [82]. The order parameters calculated based on MD simulations of pure membranes are shown on Fig. 8.1 (presented as small spheres). These parameters reach 0.38 for DMPC and 0.34 for DPPC in region between 3rd and 6th atom. The frequency of transitions between *trans* and *gauche* conformation of dihedral angles goes from 7 till 30 changes per ns and lipid. However, as shown below, this parameter depends only on temperature and it is not affected even by presence of peptide.

The corresponding values for lipid bilayers in a LC state indicated in Tieleman's web site are as follows. The hydrophobic thickness of bilayers is 2.97 nm for DPPC and 2.77 nm for DMPC, respectively. The areas per lipid are 0.629 nm^2/DPPC and 0.596 nm^2/DMPC, respectively.

8.4 Changes in the Peptides

Based on molecular modeling studies, it has been estimated, that the hydrophobic length of the α-helix composed of 24 Leu residues is approx. 3.1 nm [83]. Note that this is lower than routinely calculated end-to-end distance of this α-helical peptide, assuming 0.15 nm per amino acid residue × 24 amino acids = 3.6 nm

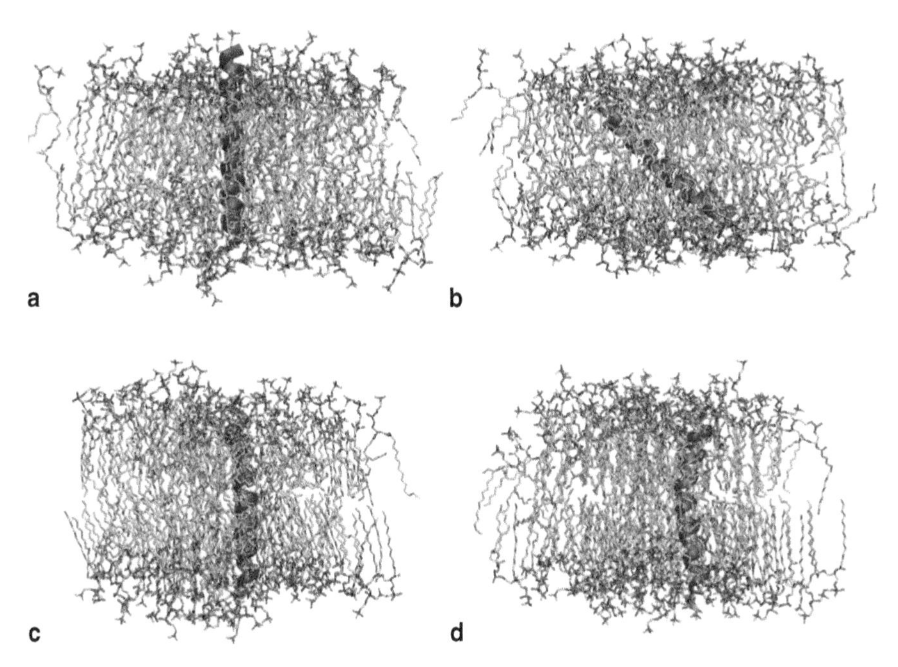

Fig. 8.2 Time development of two systems. On **A** and **B** is linear tilted peptide (L_{24}) in the membrane in LC state; on **C** and **D** is peptide (V_{24}) nearly perpendicular to membrane, but it is bended

[84]). This value is shorter than hydrophobic thickness of the bilayer in a gel state (3.44 for DPPC and 3.2 nm for DMPC), but longer for BLM in a liquid-crystalline state (2.85 nm for DPPC and 2.62 nm for DMPC) [85]. Therefore we have analyzed the geometry of the systems in both gel and LC phases. Two examples of starting state and at the end of simulation (after 40 ns) are presented on Fig. 8.2. On beginning are lipids oriented perpendicular to membrane, but one example showed tilted linear helix, the other tilted but bended helix.

8.4.1 Peptide Tilting

The most basic response of system with positive hydrophobic mismatch is tilting of a peptide. As demonstrated on Fig. 8.3, in most cases the peptide tilts for at least 20° were observed. In some simulations with DPPC membrane in a gel state, the tilt was less than 10°. The difference between thickness of DPPC membrane in gel state and length of the peptide hydrophobic core oriented perpendicular to membrane is very small. The helix changes the conformation of the main chain—the whole helix is twisted again to produce similar configuration like helices in *coiled coil* configuration. This configuration is most visible for LA_{12} in both DMPC and DPPC membranes in a gel state, it was less expressed for P_{24} and V_{24} in DPPC bilayers. This modification shortens effective hydrophobic length of the peptide and equals lengths.

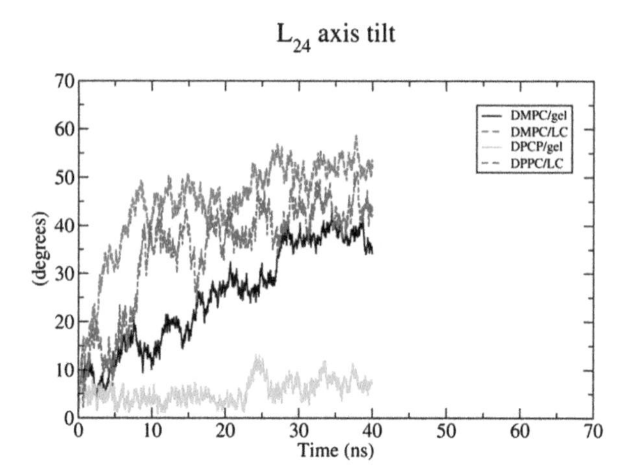

Fig. 8.3 Time development of L_{24} tilt. The peptide is tilted in LC and DMPC/gel simulations. In DPPC/gel was peptide bended (see Table 8.1, angle between 1st and 2nd half of peptide)

In the rest of simulations, the peptide is more or less tilled. In more than half of simulations, the peptide doesn't stay linear, but change its main chain conformation (more to the tilting). The peptide bends like a bow (Barlow et al. called it curved peptide [86]), or even breaks the helical structure (kinked as mentioned by the same authors)—see Fig. 8.4. Amount of bending is different for each simulation, but we were not able to find correlation between composition of system and amount of bending or place of break. But peptide A_{24} keeps best the linear conformation of ideal α-helix and doesn't bend or break (see Table 8.1—the difference between angle of 1st and 2nd peptide half, but helicity analysis is not shown). The peptide V_{24} also keeps linear conformation—exception is DMPC/gel, where the peptides are curved. For providing some quantification of bending we calculated the tilt of whole helix (as axis from first to last four C^{α} atoms) and tilt of separated upper and lower halves (C^{α} from 13th–16th residue)—see Table 8.1. From tilt of the helix and its length we can calculate effective length of peptide in a membrane. The effective length is length of projection of helix to the membrane normal (Table 8.1). In ideal case it should be equal to thickness of membrane to minimize system internal energy.

8.4.2 *Fluctuations of the Peptide*

To check stability of helix itself, we have calculated RMS fluctuations of C^{α} atoms (Fig. 8.5). In general, the highest fluctuations took place at the polar part of the membrane (both ends) and in its central, hydrophobic core. The ends are in hydrophilic regions with many small movable water molecules, while the central part is in region of membrane of lower density. The major fluctuations occur when peptide linear helix bends or breaks (for example I_{24}/LC simulation). Beside of this, the I_{24} is generally of lowest stability and the L_{24} peptide is in contrast the most stable. The V_{24} and $(LA)_{12}$ fluctuate more than L_{24}, but less than I_{24}. The P_{24} fluctuates

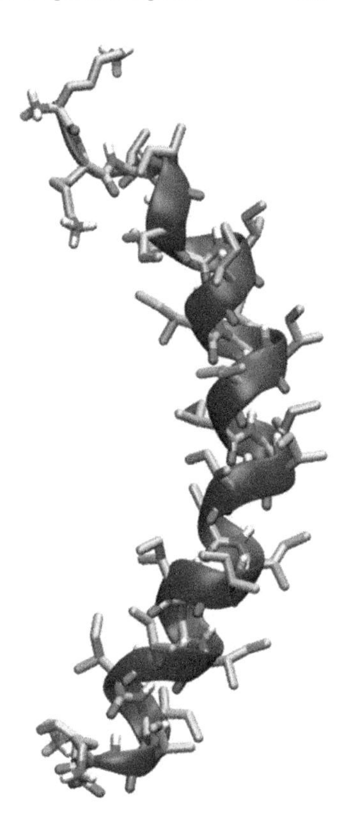

Fig. 8.4 Bended peptide during simulation (I_{24} in DMPC/LC)

similarly like I_{24}, but in DPPC/gel phase fluctuates even more because it is longer due to additional amino acid. But for example the fluctuations of I_{24} in DMPC/LC can be increased also by bending and unbending of peptide, which increase standard deviation from average structure.

The differences in fluctuations of I_{24}, A_{24}, V_{24} and L_{24} are presented on Fig. 8.6, where root mean square fluctuations (RMSF) of C^{β}, C^{γ} and C^{δ} atoms are compared. The fluctuation depends also on temperature and position of C atom in side chain. Because of temperature dependence it is possible only to compare fluctuations from simulations made by same conditions of atoms with same parameters. The Leu side chains are most stable, while the Ile side chains are the most fluctuating. The fluctuation of Val side chains are between that of Leu and Ile. This instability of amino acids side chains affect also surrounding lipids (see below). The fluctuations are affected also by distortions in helical structure—for example C^{δ} in DPPC/gel. In this case in C end of the peptide the fluctuations are similar in Leu and Ile. But the L_{24} peptide formed a kink at this position (aprox. one turn from C end). The stability of side chains has been studied also by Pace et al. [87] and Johanson and Lindahl [54] (determined enthalpy and entropy of exchanging Ala with different amino acids in poly-Ala transmembrane chain) and Barlow et al. [86] (studied entropy of χ angles).

Table 8.1 Angles and lengths of hydrophobic parts—membrane and peptide

System	Angle 1st and 2nd half (nm)	Tilt (°)	Peptide's effective thickness (nm)	Full peptide's length (nm)	Membrane's effective thickness	
					1st shell (nm)	2nd shell (nm)
A_{24}/DMPC/gel	9.33	35.55	2.75	3.38	3.00	2.83
A_{24}/DPPC/gel	6.88	13.9	3.29	3.39	3.33	3.40
A_{24}/DMPC/LC	7.16	50.85	2.15	3.41	2.07	2.34
A_{24}/DPPC/LC	7.16	37.73	2.69	3.41	2.71	2.71
I_{24}/DMPC/gel	4.11	22.43	3.25	3.52	3.08	2.97
I_{24}/DPPC/gel	8.29	18.13	3.31	3.49	3.29	3.33
I_{24}/DMPC/LC	18.46	47.49	2.4	3.56	2.43	2.52
I_{24}/DPPC/LC	8.74	9.19	3.55	3.6	3.06	2.9
L_{24}/DMPC/gel	5.24	37.98	2.67	3.39	2.66	2.74
L_{24}/DPPC/gel	12.66	7.34	3.4	3.43	3.26	3.49
L_{24}/DMPC/LC	10.5	51.92	2.12	3.44	2.4	2.55
L_{24}/DPPC/LC	9.19	43.63	2.49	3.43	2.59	2.72
LA_{12}/DMPC/gel	14.65	16.17	3.26	3.4	2.82	2.93
LA_{12}/DPPC/gel	13.16	5.61	3.38	3.39	3.37	3.43
LA_{12}/DMPC/LC	8.91	50.43	2.18	3.42	2.36	2.27
LA_{12}/DPPC/LC	6.68	41.19	2.58	3.43	2.53	2.74
P_{24}/DMPC/gel	10.42	13.32	3.45	3.54	2.93	2.92
P_{24}/DPPC/gel	5.97	15.69	3.29	3.42	3.4	3.55
P_{24}/DMPC/LC	5.48	50.72	2.28	3.59	2.48	2.56
P_{24}/DPPC/LC	6.12	51.56	2.15	3.45	2.63	2.66
V_{24}/DMPC/gel	19.63	5.24	3.39	3.41	3.24	3.1
V_{24}/DPPC/gel	7.21	3.56	3.48	3.49	3.14	3.05
V_{24}/DMPC/LC	4.86	47.33	2.36	3.48	2.49	2.49
V_{24}/DPPC/LC	7.24	46.66	2.4	3.5	2.58	2.76

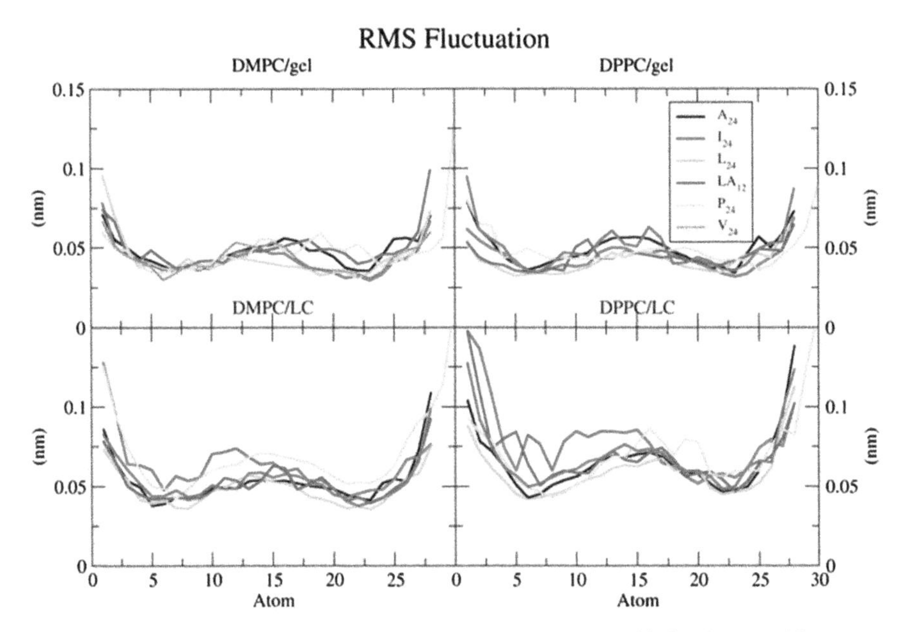

Fig. 8.5 RMS Fluctuations from last 5 ns of simulations. Each peptide has lower stable parts at both ends and in the middle. Stability of peptide is influenced by sidechains—Leu stabilizes helix, Ala, Ile and Val increase the fluctuations

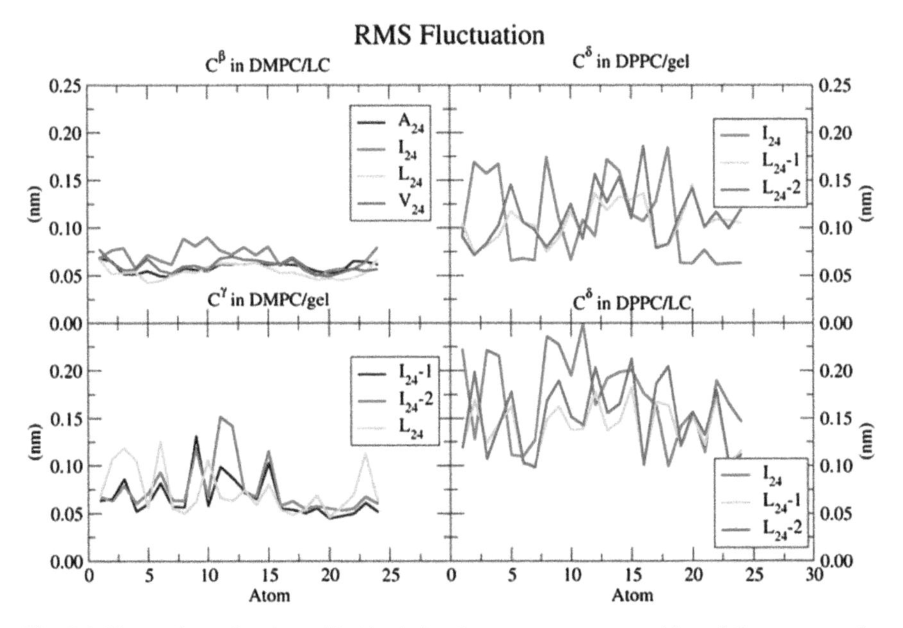

Fig. 8.6 Fluctuations of amino acid side chains. Lys atoms are most stable and Ile atoms are the lowest stable at same temperature and a position on side chain. But the distortions of helical conformation can increase fluctuations (L_{24} at DPPC/gel at C end)

Their results agree with our findings that stability decreases in following order: Ile>Val>Leu>Ala.

8.4.3 Other Helix Parameters

We also calculated (using program *g_helix* from Gromacs package) other helix parameters—rise per residue, twist per residue, helix radius, Φ and Ψ angles, etc. The values computed from simulations are presented in Table 8.2. But the algorithms used by *g_helix* are influenced by changes of dihedral angles and by linearity of analyzed helix. For example on Fig. 8.7 there are two nearly identical structures from I_{24}/DMPC/LC simulation (from 2 ps distant snapshots). Each structure has only 1 amino acid (different in each) outside the favored region for α-helix (Fig. 8.8), but for frame 1 the twist per residue is 100.95° and for frame 2 it is 83.23°. The values of radius of helix are 0.2436 nm (frame 1) and 0.2635 nm (frame 2)—graph not shown. Also the rise per residue and consequently also the helix length are affected (which is computed by multiplying the previous one by number of amino acids). Moreover, it seems that there is correlation between decreasing of the twist and in increasing of the rise per residue.

Problem with calculating of radius of helix is in using single helix axis. Axis is fitted to *z*-axis and using Pythagorean theorem the radius is calculated from *x* and *y* positions of C^{α} atoms. As the helix is bended or kinked, some atoms are moved to one side and other to opposite one. In both cases it causes increasing of helix radius. But in part, which is near to axis, the radius is (relatively) lowered.

There are virtually no changes in Φ and Ψ angles and during the simulation most of the time all angles are in most favorable region for α-helices. Also the averages angle values do not deviate from ideal values for the helix (Table 8.2).

8.5 Changes in the Membrane

The membrane affects the conformation of the peptide due to the difference in the length of their hydrophobic cores and vice versa the peptide affect the membrane structural state. In the case of positive difference, lipids could extend around helix to compensate (at least) part of the difference [8]. We can analyze deuterium order parameters, fraction of dihedral angles in *trans* conformation, frequency of changes between *trans* and *gauche* dihedral angles conformation and thickness of membrane surrounding peptide. All of these parameters are calculated for two cylindrical shells of lipids. First one is up to 0.8 nm and second one lies between 0.8 till 1.6 nm from peptide surface, respectively. Tieleman reported that the effect of peptide on the lipids is negligible at distances surpassing 1.6 nm from the peptide surface [88]. Due to this property we divided lipids into following three groups—1st shell, 2nd shell and the rest of the lipids. We calculated parameters only for first two groups.

Table 8.2 Helix parameters (analyzed last 5 ns of simulations)

System	Φ (°)	Ψ (°)	Φ+Ψ (°)	Radius (nm)	Rise/res. (nm)	Turn/res. (°)
A_{24}&DMPC/gel	−60.83±2.01	−43.16±1.84	−104.00	0.24±0.01	0.15	100.16±1.35
L_{24}&DMPC/gel	−67.25±2.37	−36.32±1.91	−103.57	0.24±0.00	0.15	98.19±0.45
LA_{12}&DMPC/gel	−66.10±2.02	−39.56±1.85	−105.66	0.25±0.01	0.15	98.63±0.54
I_{24}&DMPC/gel	−56.52±1.88	−45.82±1.76	−102.34	0.23±0.00	0.16	101.07±0.99
P_{24}&DMPC/gel	−58.29±1.98	−47.48±1.81	−105.77	0.24±0.00	0.15	99.57±0.46
V_{24}&DMPC/gel	−56.83±1.91	−48.74±1.79	−105.57	0.26±0.01	0.14	96.75±0.29
A_{24}&DMPC/LC	−60.36±1.99	−44.43±1.86	−104.79	0.23±0.00	0.15	99.28±.0.71
L_{24}&DMPC/LC	−58.56±2.01	−45.28±1.88	−103.85	0.24±0.00	0.15	99.24±0.68
LA_{12}&DMPC/LC	−59.24±2.01	−45.12±1.84	−104.36	0.24±0.01	0.15	99.39±0.72
I_{24}&DMPC/LC	−60.60±2.24	−42.58±2.15	−103.18	0.25±0.01	0.16	88.38±5.96
P_{24}&DMPC/LC	−57.76±2.11	−47.44±1.97	−105.20	0.23±0.00	0.15	98.19±1.04
V_{24}&DMPC/LC	−59.38±2.03	−47.41±1.90	−106.79	0.23±0.00	0.15	99.05±0.73
A_{24}&DPPC/gel	−59.13±2.02	−46.27±1.80	−105.40	0.24±0.00	0.15	98.73±0.55
L_{24}&DPPC/gel	−58.60±2.08	−46.82±2.08	−105.41	0.25±0.01	0.15	96.35±0.30
LA_{12}&DPPC/gel	−60.81±2.00	−43.84±1.82	−104.64	0.25±0.00	0.14	99.35±0.47
I_{24}&DPPC/gel	−58.08±1.88	−46.97±1.75	−105.05	0.24±0.00	0.15	99.68±0.88
P_{24}&DPPC/gel	−54.95±2.19	−45.33±1.93	−100.28	0.24±0.00	0.16	97.43±0.87
V_{24}&DPPC/gel	−56.60±2.05	−46.67±1.88	−103.27	0.23±0.00	0.15	99.51±2.38
A_{24}&DPPC/LC	−59.80±2.24	−44.24±2.22	−104.04	0.24±0.00	0.15	99.73±1.00
L_{24}&DPPC/LC	−58.25±2.05	−45.00±1.89	−103.25	0.24±0.00	0.15	99.86±1.04
LA_{12}&DPPC/LC	−62.34±3.14	−46.96±2.17	−109.31	0.24±0.00	0.15	99.02±0.81
I_{24}&DPPC/LC	−63.48±3.57	−40.46±2.56	−103.94	0.23±0.00	0.16	97.61±1.61
P_{24}&DPPC/LC	−58.42±2.22	−44.82±2.08	−103.24	0.23±0.00	0.15	101.12±0.74
V_{24}&DPPC/LC	−61.25±3.16	−41.92±2.12	−103.17	0.24±0.00	0.15	98.33±1.01

Fig. 8.7 Two ns distant frames from trajectory with small differences in atoms positions, but with very changed helix parameters (calculated by g_helix)

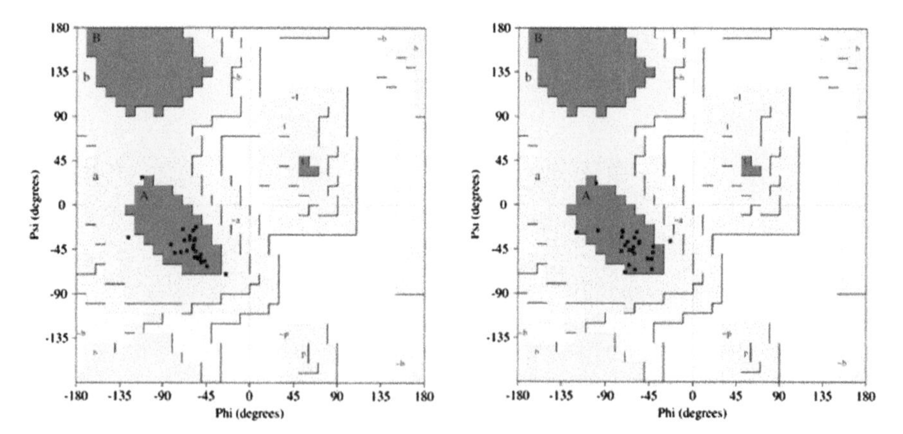

Fig. 8.8 Ramachandran graphs of frames from Fig. 8.7

8.5.1 Order Parameters

In all simulations of the membrane gel phase, the peptides caused changes of the lipid bilayer into less ordered state (lower values of deuterium order parameters)—Fig. 8.1. Each peptide decreases these parameters by different value. The temperature also affected the order parameters. With increasing of the temperature the disorder increases and consequently the order parameters decrease. The decrease of order parameters for membranes in gel state has been confirmed experimentally [16, 21, 89]. This effect is probably correlated with decreasing of membrane thickness (see below).

In simulations with DPPC in LC state, the order parameters increase. The only exception is longer P_{24}. In the DMPC membrane the 1st lipid shell is less ordered than in 2nd shell (with exception of P_{24}). The same holds also for the membrane in a gel state. In experiments with WALP peptides De Planque et al. found that the ordering of membrane can increase and decrease depending mainly on the peptide length [89] (using the same type of the membrane). The order parameters for L_{24} or P_{24} peptides are comparable with that reported by Tieleman et al. [90] and those obtained in experiments [83, 85, 91]. Unfortunately there is no sufficient information for comparison of the behavior of other peptides.

There is also observable effect of different side chains of peptides. I_{24} shows biggest stabilizing effect (higher order parameters) and V_{24} also orders surroundings lipids, but less than I_{24}. The Leu based peptides (L_{24} together with LA_{12} and P_{24}) cause high disordering. But A_{24} induced very small disordering effect. It is probably due to small volume of A_{24} side chains, which are less hydrophobic in comparison with other simulated peptides.

8.5.2 Fraction of Dihedral Angles in Trans Conformation

Another parameter, which can be used for characterizing the phase state of the lipids, is amount of dihedral angles in *trans* conformation (Table 8.3). There is correlation between this parameter and the temperature—the higher temperature, the more energy have atoms and are able to change conformation from nearly *all-trans* state. But there is also influence of peptide on this transition. In a pure gel state the membrane has around 90% of dihedral angles in the *trans* conformation, but in a LC state it is 70–75% [80]. We detected this fraction in 2nd lipid shell to be approx. 85% for gel state and between 72 and 78% for LC membrane state. In 1st shell in a gel phase this fraction decrease is even lower by 2–4°. The lowest decrease of dihedral angles has been caused by I_{24} and V_{24} peptides, while A_{24} and all Leu-based (L_{24}, LA_{12} and P_{24}) produced drop up to 6°. As for LC state, the highest increase in dihedral angles in both shells has been observed for I_{24}. The angle increase has been produced also by V_{24}, LA_{12}, L_{24}. Slight angle decrease is possible for A_{24} or P_{24} peptides. But these changes were lower than 3° for all analyzed peptides. All changes noticed above suggest, that the peptides modify surrounding lipids to produce some type environment. The properties of this environment (order parameters, *trans* fraction, etc.) correspond to the structural state between gel and LC phase, but are shifted nearer to LC.

Table 8.3 Amount of dihedral angles in *trans* conformation

System	% of *trans* dihedral angles		System	% Of *trans* dihedral angles	
	1st shell	2nd shell		1st shell	2nd shell
Pure (DMPC/gel)	87.03 ± 4.75		Pure (DPPC/gel)	90.16 ± 4.88	
A_{24}&DMPC/gel	82.45 ± 3.51	86.04 ± 1.82	A_{24}&DPPC/gel	81.68 ± 4.18	88.64 ± 3.50
L_{24}&DMPC/gel	82.53 ± 1.07	84.10 ± 2.34	L_{24}&DPPC/gel	84.64 ± 2.83	90.61 ± 2.20
$(LA)_{12}$&DMPC/gel	80.58 ± 3.53	83.58 ± 3.33	$(LA)_{12}$&DPPC/gel	85.23 ± 1.90	90.73 ± 2.48
I_{24}&DMPC/gel	86.22 ± 2.73	85.72 ± 2.78	I_{24}&DPPC/gel	85.20 ± 1.98	87.15 ± 2.96
P_{24}&DMPC/gel	84.37 ± 4.71	87.48 ± 2.43	P_{24}&DPPC/gel	83.25 ± 2.57	88.57 ± 2.09
V_{24}&DMPC/gel	85.22 ± 2.48	87.17 ± 1.86	V_{24}&DPPC/gel	84.13 ± 2.01	88.58 ± 2.41
Pure (DMPC/LC)	75.45 ± 2.71		Pure (DPPC/LC)	71.98 ± 2.42	
A_{24}&DMPC/LC	74.23 ± 2.48	75.61 ± 1.43	A_{24}&DPPC/LC	72.30 ± 0.82	72.54 ± 1.27
L_{24}&DMPC/LC	77.06 ± 2.01	77.81 ± 1.89	L_{24}&DPPC/LC	73.05 ± 2.27	72.61 ± 1.05
$(LA)_{12}$&DMPC/LC	75.20 ± 1.36	76.39 ± 2.31	$(LA)_{12}$&DPPC/LC	73.37 ± 1.13	72.08 ± 1.59
I_{24}&DMPC/LC	76.45 ± 2.67	76.01 ± 1.31	I_{24}&DPPC/LC	74.47 ± 2.18	73.76 ± 1.31
P_{24}&DMPC/LC	76.65 ± 1.80	76.53 ± 1.71	P_{24}&DPPC/LC	71.14 ± 1.08	72.65 ± 1.60
V_{24}&DMPC/LC	75.89 ± 2.02	75.97 ± 1.28	V_{24}&DPPC/LC	72.92 ± 1.25	72.86 ± 1.46

Table 8.4 Area per lipid (peptide is not subtracted from total area)

System		Area/lipid (nm^2)	System		Area/lipid (nm^2)
DMPC/gel	A_{24}	0.4942	DPPC/gel	A_{24}	0.4718
	I_{24}	0.4956		I_{24}	0.4730
	L_{24}	0.4948		L_{24}	0.4731
	$(LA)_{12}$	0.4953		$(LA)_{12}$	0.4719
	P_{24}	0.4958		P_{24}	0.4739
	V_{24}	0.4945		V_{24}	0.4751
	Pure	0.4640		Pure	0.4694
DMPC/LC	A_{24}	0.5856	DPPC/LC	A_{24}	0.6207
	I_{24}	0.5881		I_{24}	0.6233
	L_{24}	0.5877		L_{24}	0.6231
	$(LA)_{12}$	0.5872		$(LA)_{12}$	0.6219
	P_{24}	0.5862		P_{24}	0.6206
	V_{24}	0.5873		V_{24}	0.6230
	Pure	0.5860		Pure	0.6213

8.5.3 Area Per Lipid

Table 8.4 shows how incorporated peptide influences the area per membrane lipid. The area occupied by the peptide is not subtracted. The peptides with small side chains (A_{24}) induced lowest area parameter. The highest area can be found in simulations with L_{24} or I_{24}. Area per lipid is increased with increasing of the size of amino acid side chains and for more fluctuating peptide (I_{24}). Because Val has shorter side chains than Leu or Ile, the total area for V_{24} is lower than that for I_{24}, L_{24} or LA_{12} but higher in comparison with A_{24}.

8.5.4 Transitions between Trans and Gauche Conformation

The amount of transitions (Fig. 8.9) between *trans* and *gauche* conformations (per lipid and per ns) depends on the temperature. But also peptides affect the frequency of these changes and amount of transitions. In a gel state all peptides destabilize the surrounding lipids and increase the amount of transitions. There are some but small differences between peptides. A_{24}, L_{24} and LA_{12} provide the highest amount of transitions, while I_{24} the smallest one. The effect of V_{24} lies between these boundaries. In LC state the situation is exactly reversed—the Leu-based peptides modify lipid chains into some conformation, which lies between gel and LC state. The differences in LC state are very small.

In a gel state the changes agrees with previous findings. The side chains of I_{24} and V_{24} shows the highest fluctuations from all amino acids. This movement causes steric clashes with surrounding lipids (1st shell) and pushes lipids away from

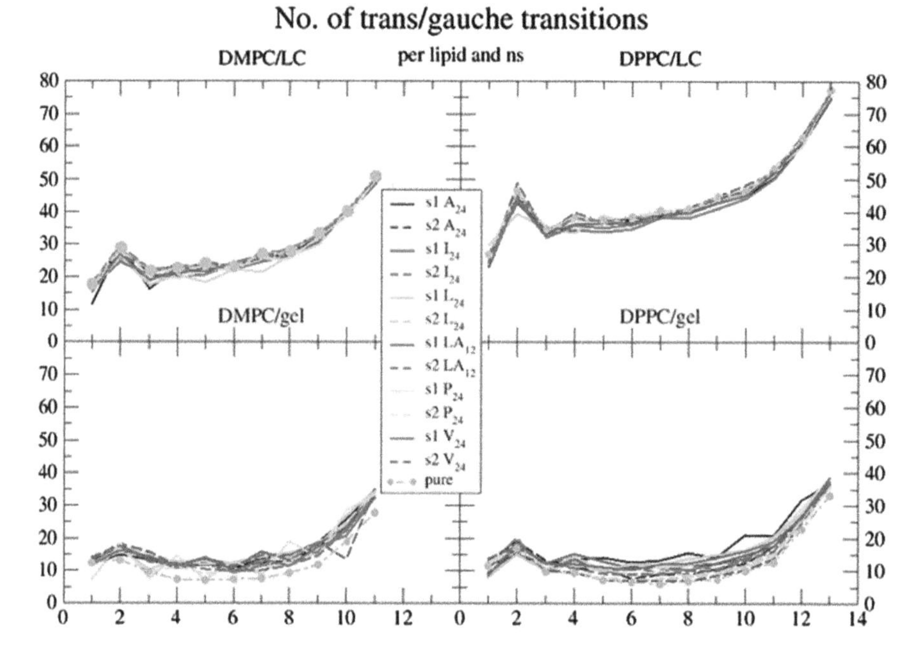

Fig. 8.9 Amount of transitions between *trans* and *gauche* conformations (per lipid and ns)

peptide. This (comparing with other peptides) keeps chains in *trans* conformation of dihedral angles (order parameters) and stabilizes them (amount of transitions). Chains with less *gauche* dihedral angles needs less space (lower area per lipid). V_{24} has shorter side chains (only C^{γ}), which causes lower area per lipid than L_{24}.

8.5.5 Membrane Thickness

We have calculated thickness of membrane for 1st and 2nd shell of lipids around peptide (see Table 8.1). The change of membrane thickness is alternate way (to peptide tilt) to compensate the hydrophobic mismatch. In most of simulations the thickness of 1st shell is lower than 2nd shell. The exceptions are simulations of a gel phase (A_{24}/DMPC, I_{24}/DMPC, P_{24}/DMPC, V_{24}/DMPC, V_{24}/DPPC) and I_{24}/DPPC in LC state. In these 6 cases the peptide did not tilt to compensate the hydrophobic mismatch as mentioned bellow. Above mentioned simulations in a gel state have 1st shell appox. 0.1 nm thicker than 2nd shell and in case of I_{24}/DPPC/LC the difference is 0.15 nm. In the rest of peptides the 1st shell is up to 0.2 nm thinner than 2nd one. In some cases (e.g. A_{24}/DPPC/LC or V24/DMPC/LC) the thickness is virtually the same (the changes are less than 0.01 nm). We did not find any significant influence of amino acid side chains on this parameter.

8.6 Hydrophobic Mismatch

It is possible to calculate peptide's effective thickness from its tilt angle and its length. Effective thickness means length of projection of the peptide to normal of the membrane. This parameter should be equal to length of membrane hydrophobic core to minimize system energy. The effective lengths (perpendicular to membrane surface) of hydrophobic parts of peptides and both lipid shells of the membrane are compared in Table 8.1. When comparing these values, it is important to keep in mind that Lys side chains can flip in or out. This resulted in shortening or prolonging ofeffective length of peptide (max. to 0.4–0.5 nm). Thus, it is not necessary to keep the same effective length of peptide and 1st shell of lipids to satisfy hydrophobic mismatch.

The thickness of hydrophobic part of unmodified membranes has a value of 3.6 nm for DPPC/gel, 3.2 nm for DMPC/gel, 3.0 nm for DPPC/LC and 2.6–2.7 nm for DMPC/LC. In all cases the thickness of the second shell (compared with the first shell) are closer to the unmodified membrane: the average thickness is 3.58 nm for DPPC/gel, 2.93 nm for DPPC/LC, 3.12 nm for DMPC/gel and 2.62 nm for DMPC/LC. Nearly in all cases of simulations, which resulted in the peptide tilt, the membrane is by 0.3 nm thicker than the effective length of the peptide. In those cases the peptide tilt is bigger (according to the simulation results) so its effective length is smaller. But the average membrane thickness does not contain direct information on the orientation of individual lipid chains. Lipid chains can still be longer even in the LC state (higher order parameters, more *trans* conformations of dihedral angles), because they can tilt like the peptide.

The simulations of membrane in a gel state (V_{24}/DPPC, P_{24}/DPPC, LA_{12}/DMPC, and LA_{12}/DPPC) suggest that the membrane affects the peptides conformation. The whole helix is twisted into *superhelical* structure—helix composed from helical chain (see Fig. 8.10). The whole structure resembles single chain from *coiled coil* conformation. This structure is produced only in the gel membrane phase, where there is only small hydrophobic mismatch. The difference of thickness between peptide and membrane in LC state are too large to solve the situation similarly like for gel state.

The I_{24}/DPPC/LC simulations suggest that the average peptide tilt is very small and didn't solve the mismatch. Visual observation shows that at the beginning the peptide tilted (up to approx. 20°), but it didn't stay in this conformation, rather went back nearly into its starting position. After short period of time it tilted again in random direction, but again returned back. During the 40 ns of simulation, whole tilting and returning process is repeated 5 times. The reason of this behavior remains unknown.

Because some membrane phenomena are quite rare, we ran some (I_{24}&DMPC/LC, I_{24}&DPPC/gel, I_{24}&DPPC/LC, LA_{12}&DMPC/gel, L_{24}&DMPC/gel, V_{24}&DPPC/gel, V_{24}/DPPC/LC) simulations to 100 ns. In all cases the conformation didn't differ much from end of original simulations. E.g. tilt in I_{24}&DMPC/LC decreased by 2°, I_{24}&DPPC/LC finally stabilized at average angle 5.15° (effective thickness: 3.2 nm,

Fig. 8.10 Superhelical configuration of LA_{12} peptide at the end of the simulation. *Lime dashed line* is axis connects centers on the beginning and ends of peptide. The peptide tilts around this axis

1st shell 2.87 nm, 2nd shell 2.39 nm). The tilt for V_{24}&DPPC/gel increased to 4.33 nm and for V_{24}/DPPC/LC the tilt decreased to 45.88° (by 1.8°). All these findings agrees with RMS deviation data—the systems reached (at least some meta-) stable state.

Barlow et al. published study of helix conformation from PDB database [86]. Although they studied 48 helices, only 15 % keep conformation near to ideal α-helix. From rest of them 10 % have different lengths, 17 % have been kinked and 58 % were curved. These results are not fully comparable with ours. The reason is that they didn't study helices transmembrane proteins and the amino acid composition was quite different.

Tieleman et al. [73] performed 2 ns MD simulation of α-helix with long hydrophobic segments (Flu_{26} and Flu_{34}) in POPC bilayers. They observed considerable extension of the membrane thickness around Flu_{26} peptide and declination by 10°. At the same time, they did not observe extension of the thickness for the peptide Flu_{34} with longer hydrophobic length, but the peptide molecules declined by 25°. As summarized by Killian [39] from experimental and simulation data, there is change of the membrane thickness near the protein in systems with WALP protein and only a small tilt is created. However Lys flanked peptides such (in his case only L_{24}, $(LA)_{12}$) do not change the membrane thickness so extensively, and rather increase the peptide tilt. This agrees with our results, namely the mismatch of thickness of hydrophobic parts is compensated by peptides tilt.

Petrache et al. [92] also discussed possible drawback of the molecular dynamics simulations. First there are problems connected with the rather short time of the simulations restricted by several ns. At the same time they received similar results with shorter—around 5 ns and longer—around 10 ns simulations. However, it should be noted that characteristic time of relaxation of phospholipid dipole moments following membrane disturbance by voltage jump lies between micro- to mili-second scale. Longer time probably corresponds to the collective movement of lipid clusters [93]. Incorporation of the short peptide influences this relaxation time significantly [92]. We can therefore expect that the relaxation time of the short peptides, like L_{24}, should be comparable or even larger than that for phospholipids. Therefore, in order to receive equilibrium state of the peptide in a membrane, the simulations should last in order of microseconds. Currently it is possible by means of the coarse-graining (CG) models, which however lack the details of atomic resolution. In addition due to less degree of freedom the CG systems move rapidly than atomic models. However, even simple CG models allow to obtain important information on the features of peptide-lipid interactions. This explains growing interest to this method in recent five years, which include also combination of full atomic and CG approach in modeling the peptide/lipid interactions (see Polyansky et al. for recent review [30]). However as stated by Monticelli et al. [95], more shorter simulations can provide better sampling of conformation space than longer one. Despite the large number of limitations, MD represents a useful approach for the study of fast conformational movements of peptides and phospholipids in a membrane, though we cannot be sure whether the model system reached equilibrium or not. But results obtained by MD are consistent with experiments, in respect of inducing hydrophobic mismatch and disordering effect of peptide on the membrane in the gel state.

8.7 Conclusion

Our results confirmed the tendency of Lys-flanked peptides to compensate the positive mismatch between peptide and membrane hydrophobic core by tilting. Some of the peptides, however, produce *superhelical* double-twisted structure. This only occurs in the membrane in the gel phase, where only a small hydrophobic mismatch exists. The peptide also alters certain properties of the surrounding lipids such as membrane ordering, the amount of dihedral angles in *trans* conformation and the number of transitions between *trans* and *gauche* conformation. It is likely that these effects should provide some preferable structural state of the peptides in a membrane. The lipid structural state around the peptide is probably between gel and liquid-crystalline state. This effect depends on peptide amino acid composition. Amino acids with large side chains branched at C^{β} (Ile, Val) produce helix, which has more side chains fluctuates than that of a poly-Leu helix. This holds also for small side chains (Ala).

Acknowledgments The access to the METACentrum and CERIT computing facilities provided under the research intent MSM6383917201 is highly appreciated. This work was also supported by the Slovak Research and Development Agency (Projects No. APVV—0410-10 and LPP-0341-09).

References

1. Gennis RB (1989) Biomembranes: molecular structure and function. Springer, New York
2. Yeagle P (1992) The structure of biological membranes. CRC Press, Boca Raton
3. McElhaney RN (1986) Differential scanning calorimetric studies of lipid–protein interactions in model membrane systems. Biochim Biophys Acta 864(3–4):361–421
4. Hianik T, Passechnik VI (1995) Bilayer lipid membranes: structure and mechanical properties. Kluwer, Netherlands
5. Marsh D, Horvath LI (1998) Structure, dynamics and composition of the lipid–protein interface. Perspectives from spin-labelling. Biochim Biophys Acta 1376(3):267–296
6. Watts A (1998) Solid-state NMR approaches for studying the interaction of peptides and proteins with membranes. Biochim Biophys Acta 1376(3):297–318
7. White SH, Wimley WC (1998) Hydrophobic interactions of peptides with membrane interfaces. Biochim Biophys Acta 1376(3):339–352
8. Killian JA (1998) Hydrophobic mismatch between proteins and lipids in membranes. Biochim Biophys Acta 1376(3):401–415
9. Hianik T, Kaatze U, Sargent DF, Krivanek R, Halstenberg S, Pieper W, Gaburjakova J, Gaburjakova M, Pooga M, Langel U (1997) A study of the interaction of some neuropeptides and their analogs with lipid bilayers. Bioelectrochem Bioenerg 42(2):123–132
10. Krivanek R, Rybar P, Prenner EJ, McElhaney RN, Hianik T (2001) Interaction of the antimicrobial peptide gramicidin S with dimyristoyl-phosphatidylcholine bilayer membranes: a densimetry and sound velocimetry study. Biochim Biophys Acta 1510(1–2):452–463
11. Davis JH, Clare DM, Hodges RS, Bloom M (1983) Interaction of a synthetic amphiphilic polypeptide and lipids in a bilayer structure. Biochemistry 22(23):5298–5305
12. Axelsen PH, Kaufman BK, McElhaney RN, Lewis RNAH (1995) The infrared dichroism of transmembrane helical polypeptides. Biophys J 69(6):2770–2781
13. Huschilt JC, Millman BM, Davis JH (1989) Orientation of α-helical peptides in a lipid bilayer. Biochim Biophys Acta 979(1):139–141
14. Bolen EJ, Holloway PW (1990) Quenching of tryptophan fluorescence by brominated phospholipid. Biochemistry 29(41):9638–9643
15. Morrow MR, Huschilt JC, Davis JH (1985) Simultaneous modeling of phase and calorimetric behavior in an amphiphilic peptide/phospholipid model membrane. Biochemistry 24(20):5396–5406
16. Zhang Y-P, Lewis RNAH, Hodges RS, McElhaney RN (1992) Interaction of a peptide model of a hydrophobic transmembrane α-helical segment of a membrane protein with phosphatidylcholine bilayers. Differential scanning calorimetric and FTIR spectroscopic studies. Biochemistry 31(46):11579–11588
17. Pauls KP, MacKay AL, Soderman O, Bloom M, Taneja AK, Hodges RS (1985) Dynamic properties of the backbone of an integral membrane polypeptide measured by ^{2}H-NMR. Eur Biophys J 12(1):1–11
18. Subczynski W K, Lewis RNAH, McElhaney RN, Hodges RS, Hyde JS, Kusumi A (1998) Molecular organization and dynamics of 1-palmitoyl-2-oleoyl-phophatidylcholine bilayers containing a transmembrane α-helical peptide. Biochemistry 37(9):3156–3164
19. Lewis RNAH, Liu F, Krivanek R, Rybar P, Flach CR, Mendelsohn R, Chen Y, Mant CT, Hodges RS, McElhaney RN (2007) Studies of the minimum hydrophobicity of α-helical peptides required to maintain a stable transmembrane association with phospholipid bilayer membranes. Biochemistry 46(4):1042–1054
20. Idiong G, Won A, Ruscito A, Leung BO, Hitchcock AP, Ianoul A (2011) Investigating the effect of a single glycine to alanine substitution on interactions of antimicrobial peptide latarcin 2a with a lipid membrane. Eur Biophys J 40(9):1087–1100
21. Paré C, Lafleur M, Liu F, Lewis RNAH, McElhaney RN (2001) Differential scanning calorimetry and ^{2}H nuclear magnetic resonance and Fourier transform infrared spectroscopy

studies of the effects of transmembrane α-helical peptides on the organization of phosphatidylcholine bilayers. Biochim Biophys Acta 1511(1):60–73
22. Rybar P, Krivanek R, Samuely T, Lewis RNAH, McElhaney RN, Hianik T (2007) Study of the interaction of an α-helical transmembrane peptide with phosphatidylcholine bilayer membranes by means of densimetry and ultrasound velocimetry. Biochim Biophys Acta 1768(6):1466–1478
23. Damodaran KV, Merz KM Jr (1995) Interaction of the fusion inhibiting peptide carbobenzoxy-d-phe-l-phe-gly with n-methyldioleoylphosphatidylethanolamine lipid bilayers. J Am Chem Soc 117(24):6561–6571
24. Huang P, Loew GH (1995) Interaction of an amphiphilic peptide with a phospholipid bilayer surface by molecular dynamics simulation study. J Biomol Struct Dyn 12(5):937–956
25. Bernéche S, Nina M, Roux B (1998) Molecular dynamics simulations of melittin in a dimyristoylphosphatidylcholine bilayer membrane. Biophys J 75(4):1603–1618
26. Woolf TB, Roux B (1994) Molecular dynamics simulation of the gramicidin channel in a phospholipid bilayer. PNAS 91(24):11631–11635
27. Woolf TB, Roux B (1996) Structure, energetics, and dynamics of lipid–protein interactions: a molecular dynamics study of the gramicidin a channel in DMPC bilayer. Proteins: Struct Funct Gen 24(1):92–114
28. Woolf TB (1997) Molecular dynamics of individual α-helices of bacteriorhodopsin in dimyristoylphosphatidylcholine. I. Structure and dynamics. Biophys J 73(5):2476–2393
29. Woolf TB (1998) Molecular dynamics simulations of individual α-helices of bacteriorhodopsin in dimyristoylphosphatidylcholine. II. Interaction energy analysis. Biophys J 74(1):115–131
30. Tieleman DP, Sansom MSP, Berendsen HJC (1999) Alamethicin helices in a bilayer and in solution: molecular dynamics simulations. Biophys J 76(1):40–49
31. Alter JE, Taylor GT, Sheraga HA (1972) Helix-coil stability constants for the naturally occurring amino acids in water. VI. Leucine parameters from random poly(hydroxypropylglutamine-co-L-leucine) and poly(hydroxybutylglutamine-co-L-leucine). Macromolecules 5(6):739–745
32. Chou PY, Fasman GD (1997) β-Turns in proteins. J Mol Biol 115(2):135–175
33. De Planque MRR, Killian JA (2003) Protein lipid interactions studied with designed transmembrane peptides: role of hydrophobic matching and interfacial anchoring. Mol Membr Biol 20(4):271–284
34. Bloom M, Evans E, Mouritsen OG (1991) Physical properties of the fluid lipid-bilayer component of cell membranes: a perspective. Quart Rev Biophys 24(3):293–397
35. Killian JA, Salemink I, de Planque MRR, Lindblom G, Koeppe II RE, Greathouse DV (1996) Induction of nonbilayer structures in diacylphosphatidylcholine model membranes by transmembrane α-helical peptides: importance of hydrophobic mismatch and proposed role of tryptophans. Biochemistry 35(3):1037–1045
36. Demmers JAA, Haverkamp J, Heck AJR, Koeppe II RE, Killian JA (2000) Electrospray ionization mass spectrometry as a tool to analyze hydrogen/deuterium exchange kinetics of transmembrane peptides in lipid bilayers. PNAS 97(7):3189–3194
37. Demmers JAA, van Duijn E, Haverkamp J, Greathouse DV, Koeppe II RE, Heck AJR, Killian JA (2001) Interfacial positioning and stability of transmembrane peptides in lipid bilayers studied by combining hydrogen/deuterium exchange and mass spectrometry. J Biol Chem 276(37):34501–34508
38. Liu F, Lewis RNAH, Hodges RS, McElhaney RN (2004) Effect of variations in the structure of a polyleucine-based α-helical transmembrane peptide on its interaction with phosphatidylethanolamine bilayers. Biophys J 87(4):2470–2482
39. Killian JA (2003) Synthetic peptides as models for intrinsic membrane proteins. FEBS Lett 555(1):134–138
40. Lee AG (2003) Lipid–protein interactions in biological membranes: a structural perspective. Biochim Biophys Acta 1612(1):1–40
41. De Planque MMR, Bonev BB, Demmers JAA, Greathouse DV, Koeppe II RE, Separovic F, Watts A, Killian JA (2003) Interfacial anchor properties of tryptophan residues in transmem-

brane peptides can dominate over hydrophobic matching effects in peptide–lipid interactions. Biochemistry 42(18):5341–5348
42. Vander Wel PCA, Strandberg E, Killian JA, Koeppe II RE (2002) Geometry and intrinsic tilt of a tryptophan-anchored transmembrane α-helix determined by ^{2}H-NMR. Biophys J 83(3):1479–1488
43. Belohorcová K, Davis JH, Woolf TB, Roux B (1997) Structure and dynamics of an amphiphilic peptide in a lipid bilayer: a molecular dynamics study. Biophys J 73(6):3039–3055
44. Maurits RR, de Planque JA, Kruijtzer W, Rob M, Liskamp J, Marsh D, Greathouse DV, Koeppe II RE, de Kruijff B, Killian JA (1999) Different membrane anchoring positions of tryptophan and lysine in synthetic transmembrane α-helical peptides. J Biol Chem 274(30):20839–20846
45. De Jesus AJ, Allen TW (2013) The role of tryptophan side chains in membrane protein anchoring and hydrophobic mismatch. Biochim Biophys Acta 1828(2):864–876
46. De Jesus AJ, Allen TW (2013) The determinants of hydrophobic mismatch response for transmembrane helices. Biochim Biophys Acta 1828(2):851–863
47. Daily AE, Greathouse DV, van der Wel PCA, Koeppe II RE (2008) Helical distortion in tryptophan- and lysine-anchored membrane-spanning α-helices as a function of hydrophobic mismatch: a solid-state deuterium NMR investigation using the geometric analysis of labeled alanines method. Biophys J 94(2):480–491
48. Lewis RNAH, Zhang YP, Liu F, McElhaney RN (2002) Mechanisms of the interaction of α-helical transmembrane peptides with phospholipid bilayers.Bioelectrochemistry 56(1–2):135–140
49. Fares C, Qian J, Davis JH (2005) Magic angle spinning and static oriented sample NMR studies of the relaxation in the rotating frame of membrane peptides. J Chem Phys 122(19):Art. No. 194908. doi:10.1063/1.1899645
50. Davis JH, Auger M, Hodges RS (1995) High resolution ^{1}H nuclear magnetic resonance of a transmembrane peptide. Biophys J 69(5):1917–1932
51. Pan J, Tieleman DP, Nagle JF, Kučerka N, Tristram-Nagle S (2009) Alamethicin in lipid bilayers: combined use of X-ray scattering and MD simulations. Biochim Biophys Acta 1788(6):1387–1397
52. Sani M-A, Whitwell TC, Separovic F (2012) Lipid composition regulates the conformation and insertion of the antimicrobial peptide maculatin 1.1. Biochim Biophys Acta 1818(2):205–211
53. Hoernke M, Schwieger C, Kerth A, Blume A (2012) Binding of cationic pentapeptides with modified side chain lengths to negatively charged lipid membranes: complex interplay of electrostatic and hydrophobic interactions. Biochim Biophys Acta 1818(7):1663–1672
54. Johansson ACV, Lindahl E (2006) Amino-acid solvation structure in transmembrane helices from molecular dynamics simulations. Biophys J 91(12):4450–4463
55. Li LB, Vorobyov I, Allen TW (2012) The role of membrane thickness in charged protein–lipid interactions. Biochim Biophys Acta 1818(2):135–145
56. Lewis BA, Engelman DM (1983) Lipid bilayer thickness varies linearly with acyl chain-length in fluid phosphatidylcholine vesicles. J Mol Biol 166(2):211–217
57. Lam KLH, Wang H, Siaw TA, Chapman MR, Waring AJ, Kindt JT, Lee KYC (2012) Mechanism of structural transformations induced by antimicrobial peptides in lipid membranes. Biochim Biophys Acta 1818(2):194–204
58. MacCallum JL, Bennett WFD, Tieleman DP (2008) Distribution of amino acids in a lipid bilayer from computer simulations. Biophys J 94(9):3393–3404
59. Yoo J, Cui Q (2008) Does arginine remain protonated in the lipid membrane? Insights from microscopic pK_a calculations. Biophys J: Biophys Lett 94(8):L61–L63. doi:10.1529/biophysj.107.122945
60. Kim T, Im W (2010) Revisiting hydrophobic mismatch with free energy simulation studies of transmembrane helix tilt and rotation. Biophys J 99(1):175–183

61. Ozdirekcan S, Rijkers DTS, Killian JA (2004) Influence of flanking residues on tilt and rotation angles of transmembrane peptides in lipid bilayers. A solid-state ^{2}H NMR study. Biochemistry 44(3):1004–1012. doi:10.1021/bi0481242
62. Strandberg E, Ozdirekcan S, Killian JA (2004) Tilt angles of transmembrane model peptides in oriented and non-oriented lipid bilayers as determined by ^{2}H solid-state NMR. Biophys J 86(6):3709–3721
63. Ozdirekcan S, Etchebest C, Fuchs PF (2007) On the orientation of a designed transmembrane peptide: toward the right tilt angle? J Am Chem Soc 129(49):15174–15181
64. Esteban-Martın S, Salgado J (2007) The dynamic orientation of membrane-bound peptides: bridging simulations and experiments. Biophys J 93(12):4278–4288
65. Holt A, Koehorst RBM, Killian JA (2009) Tilt and rotation angles of a transmembrane model peptide as studied by fluorescence spectroscopy. Biophys J 97(8):2258–2266
66. Monticelli L, Tieleman DP, Fuchs PFJ (2010) Interpretation of ^{2}H-NMR experiments on the orientation of the transmembrane helix WALP23 by computer simulations. Biophys J 99(5):1455–1464
67. Holt A, Killian JA (2010) Orientation and dynamics of transmembrane peptides: the power of simple models. Eur Biophys J 39(4):609–621. doi:10.1007/s00249–009-0567-1
68. Lindahl E, Hess B, van der Spoel D (2001) GROMACS 3.0: a package for molecular simulation and trajectory analysis. J Mol Model 7(8):306–317
69. Van Gunsteren WF, Berendsen HJC (1987) Gromos-87 manual. Biomos BV, Nijenborgh 4, 9747 AG Groningen, Netherlands, p 331–342
70. Van Buuren AR, Marrink SJ, Berendsen HJC (1993) A molecular dynamics study of the decane/water interface. J Phys Chem 97(36):9206–9212
71. Mark AE, van Helden SP, Smith PE, Janssen LHM, van Gunsteren WF (1994) Convergence properties of free energy calculations: α-cyclodextrin complexes as a case study. J Am Chem Soc 116(14):6293–6302
72. HyperChem(TM) Professional 7.51, Hypercube, Inc., 1115 NW 4th Street, Gainesville, Florida 32601, USA
73. Tieleman DP, Marrink SJ, Berendsen HJC (1997) A computer perspective of membranes: molecular dynamics studies of lipid bilayer systems. Biochim Biophys Acta 1331(3):235–270
74. Nagle JF, Tristram-Nagle S (2000) Structure of lipid bilayers. Biochim Biophys Acta 1469(3):159–195
75. Berger O, Edholm O, Jahnnig F (1997) Molecular dynamics simulations of a fluid bilayer of dipalmitoylphosphatidylcholine at full hydration, constant pressure, and constant temperature. Biophys J 72(5):2002–2013
76. Laskowski RA, MacArthur MW, Moss DS, Thornton JM (1993) Procheck—a program to check the stereochemical quality of protein structures. J Appl Cryst 26(2):283–291 doi:10.1107/S0021889892009944
77. Tristam-Nagle S, Liu Y, Legleiter J, Nagle JF (2002) Structure of gel phase DMPC determined by X-ray diffraction. Biophys J 83(6):3324–3335
78. Tu K, Tobias DJ, Blasie JK, Klein ML (1996) Molecular dynamics investigation of structure of a fully hydrated gel-phase dipalmitoylphosphatidylcholine bilayer. Biophys J 70(2):595–608
79. Tieleman DP, Berendsen HJC (1996) Molecular dynamics simulations of fully hydrated DPPC with different macroscopic boundary conditions and parameters. J Chem Phys 105(11):4871–4880
80. Egberts E (1988) Molecular dynamics simulation of multibilayer membranes, Ph.D. Thesis, University of Groningen, Netherlands
81. Nagle JF (1993) Area/lipid of bilayers from NMR. Biophys J 64(5):1476–1481
82. Damodaran KV, Merz KM Jr (1994) A comparison of DMPC- and DLPE-based lipid bilayers. Biophys J 66(4):1076–1067
83. Small DM (1986) The physical chemistry of lipids: from alkanes to phospholipids plenum. Plenum Press, New York

84. Rand RP, Parsegian VA (1989) Hydration forces between phospholipid bilayers. Biochim Biophys Acta 988(3):351–376
85. Cvec G, Marsh D (1987) Phospholipid bilayers: physical principles and models. Wiley, New York
86. Barlow DJ, Thornton JM (1988) Helix geometry in Proteins. J Mol Biol 201(3):601–619
87. Pace CN, Scholtz JM (1998) A helix propensity scale based on experimental studies of peptides and proteins. Biophys J 75(1):422–427
88. Tieleman PD, Forest LR, Samsom MSP, Berendsen HJC (1998) Lipid properties and the orientation of aromatic residues in OmpF, influenza M2, and alamethicin systems: molecular dynamics simulations. Biochemistry 37(50):17554–17561
89. De Planque MRR, Greathous DV, Koeppe II RE, Schäfer H, Marsh D, Killian JA (1998) Influence of lipid–peptide hydrophobic mismatch on the thickness of diacylphosphatidylcholine bilayers. A 2H NMR and ESR study using designed transmembrane α-helical peptides and gramicidin A. Biochemistry 37(26):9333–9345
90. Büldt G, Gally HU, Seelig J (1979) Neutron diffraction studies on phosphatidylcholine model membranes: I. Head group conformation. J Mol Biol 134(4):673–691
91. Smith GS, Sirota EB, Safinya CR, Plano RJ, Clark NA (1990) X-ray structural studies of freely suspended ordered hydrated DMPC multimembrane films. J Chem Phys 92(7):4519–4530. doi:10.1063/1.457764
92. Petrache HI, Zuckerman DM, Sachs JN, Killian JA, Koeppe II REH, Woolf TB (2002) Hydrophobic matching mechanism investigated by molecular dynamics simulations. Langmuir 18(4):1340–1351
93. Sargent DF (1975) Voltage jump/capacitance relaxation studies of bilayer structure and dynamics studies on oxidized cholesterol membranes. J Membr Biol 23(1):227–247
94. Hianik T, Krivánek R, Sargent DF, Sokolíková L, Vinceová K (1996) A study of the interaction of adrenocorticotropin-(1-24)-tetracosapeptide with BLM and liposomes. Progr Coll Polym Sci: Trends Colloid Interface Sci X 100:301–305
95. Monticelli L, Sorin EJ, Tieleman DP, Pande VS, Colombo G (2008) Molecular simulation of multistate peptide dynamics: a comparison between microsecond timescale sampling and multiple shorter trajectories. J Comput Chem 29(11):1740–1752 doi:10.1002/jcc.20935

Chapter 9
Polyphenol Glycosides as Potential Remedies in Kidney Stones Therapy. Experimental Research Supported by Computational Studies

D. Toczek, E. Klepacz, S. Roszak and R. Gancarz

Abstract Polyphenol glycosides are potential compounds in the kidney stones therapy. Experimental data indicate the formation of glycoside complexes with calcium ions. Theoretical studies support experimental finding in elucidating the structures of studied complexes. DFT methods constitutes reasonable approach to investigate the strength and structural properties these complexes. The extraction of main structural factors responsible for complexing activity allows to design new ligands for calcium ions, being helpful in the kidney stones treatment.

9.1 Introduction

Calcium in human body is mostly deposited in skeleton and bones as calcium phosphate. The rest of calcium ions, about 1 % of the whole concentration, play many important functions in the organism. The concentration of Ca^{2+} in blood controls the proper functioning of heart and circulatory system [1]. The Ca^{2+} concentration

D. Toczek (✉) · E. Klepacz · R. Gancarz
Organic and Pharmaceutical Technology Group, Chemistry Department,
Wrocław University of Technology (WUT),
Wyb. Wyspiańskiego str. 27, Wrocław 51-370, Poland
e-mail: dariusz.toczek@pwr.wroc.pl

E. Klepacz
e-mail: ewa.klepacz@pwr.wroc.pl

R. Gancarz
e-mail: roman.gancarz@pwr.wroc.pl

S. Roszak
Institute of Physical and Theoretical Chemistry,
Wrocław University of Technology (WUT),
Wyb. Wyspiańskiego str. 27, Wrocław 51-370, Poland
e-mail: szczepan.roszak@pwr.wroc.pl

L. Gorb et al. (eds.), *Application of Computational Techniques in Pharmacy and Medicine*,
Challenges and Advances in Computational Chemistry and Physics 17,
DOI 10.1007/978-94-017-9257-8_9

controls the blood pressure. The blood clotting is also the Ca^{2+}-dependent process [2]. Moreover, calcium ions, as the second messenger, control processes related to the muscle contraction and nerve impulse transmission [3, 4]. There are hundreds of known proteins binding calcium but the most important is Calmodulin, the special protein dedicated for Ca^{2+}, which after binding with calcium ions changes the conformation and performs different functions. The only available form of Ca^{2+} are dietary sources. The metabolism of Ca^{2+} is regulated by three mechanisms: intestinal absorption, renal reabsorption, and bone turnover [3]. Beside the proper homeostasis of Ca^{2+} in the organism, there are also pathological changes in the calcium economy as osteoporosis or nephrolithiasis. The osteoporosis is related to calcium losses in the biggest reservoir of calcium—bones, while the urolithiasis involves the formation of calcium deposit in kidneys. In almost 80 % cases kidney stones contain calcium oxalate as a fundamental compound. Moreover, nephrolithiasis is the worldwide problem, which is usually more common for men than for women [5]. Lewandowski and Rodger [6] divide factors causing nephrolithiasis on: environmental factors (climate, occupation), physiochemical factors (urine volume, urinary concentration of oxalate and other elements) and dietary patterns and nutrients. Nowadays, medicine offers only few methods of the treatment of kidney stones: traditional-surgery or, less invasive extracorporeal shock wave lithotripsy (ESWL), percutaneous stone removal (PCNL), and endourological stone treatment [7]. The possible treatment constitutes only instrumental techniques and no active drugs or active substances which prevent this disease are available. The only offered drug for patients are those causing increase of urine flow or controlling colic pain [8]. Unfortunately nephrolithiasis is a recurrent disease and about 75 % of patients suffer the recurrence within 10 years. Therefore it is important to develop new more efficient preventative therapies, which can inhibit the formation of kidney stones or possess properties of dissolution of calcium oxalate deposits. The possible approach constitutes modeling the potential inhibitors forming calcium oxalate crystals. On the other hand there are available compounds directly influencing the calcium oxalate formation (inhibitory or dissolution effect) e.g. citrate. It should be noted that compounds possessing dissolution properties of calcium oxalate cannot interfere too much into the whole calcium economy.

Crystallographic data show that the preferred donor for calcium cation are oxygen atoms, while the complexes with nitrogen or sulfur as ligands are a rarity [9]. Hence, potential structures applied for dissolution of kidney stones should contain, in their skeleton, many oxygen atoms. We must remember however that there are many different oxygen atoms with regard to their chemical environment (e.g. hydroxyl, acid, ester, ether, glycosidic and others) and their ability to bind calcium will depend on this factor.

The extract from *Rubia tinctorum L.* is rich in hydroxyanthraquinones and these compounds were used in the kidney stones treatment. Unfortunately it contains also lucidin, one of anthraquinones which exhibit the genotoxic activity [10]. Anthraquinones are polyphenols and they have ability to interact with calcium ions (as will be described later in this chapter). Due to their condensed aromatic polycarbon structure there is a risk of genotoxity. There are some other polyphenols without condensed aromatic polycarbon fragment, which exhibit complexation activity to calcium ions. Polyphenol glycosides constitute the group of such compounds, which beside standard aromatic part substituted by many hydroxyl groups also possess glycon part of the carbohydrate molecule. The number of potential binding sites in these compounds increases due to additional oxygen atom contained in sugar part. The carbohydrate part increases their solubility in water, which is of great advantage if such compounds are used in kidney stone therapy.

9.2 Carbohydrates as Ligands

In this section we try to bring the issue of Ca^{2+} complexation by sugars from the structural point of view. Gyurcsik and Nagy [11] pointed out that although some reviews regarding carbohydrates–metal complexes are available (e.g. "Complexes of natural carbohydrates with metal cation" by Alekseev et al. [12]), no comprehensive monograph which treat carbohydrates as ligands can be found. We also do not pretend to cover the whole topic of the area. Hence, we will mostly focus on the general principles of forming complexes by carbohydrates with a special emphasis to the cases in which calcium constitutes the metal ion and we present selected the most typical structures.

At the beginning, it should be noted that interactions between carbohydrates and metal ions are usually weak. Water strongly binds to hydroxyl groups of carbohydrates, and as a result complexes in water solution and in solid state can be structurally different. The formation of carbohydrate complexes in water requires replacement of water molecules by hydroxyl group from the first coordination sphere. To form complex in water the interaction of the ligand must be stronger than that with water molecules. Therefore there are some complexes which cannot exist in water solution but they exist in crystalline form. Monosaccharides and oligosaccharides as well as polysaccharides possess in their skeleton, as donors, hydroxyl oxygen atoms which are relatively weak ligands. Some sugar derivatives can possess ionizable functional group, which might exhibit stronger interactions with metal ion. Carbohydrates with carboxylic, sulphonic or phosphate groups as an anchoring sites can form much stronger complexes with calcium ions [13–16]. They are not the subject of consideration in this chapter.

In 1961 J.A. Mills studied the acidity of sugars by the paper electrophoresis. He has found that some sugars, even at neutral pH, penetrated toward cathode. This indicated that they are positively charged probably by complexing cations. The extent of migration was a measure of the complexing ability. The only coordination centers in neutral sugars are constituted by oxygen atoms of hydroxyl groups. Since water solvates cations much better, the question that arises is regarding the number hydroxyl groups needed to form a stable complex. It is believed that at least two or three hydroxyl centers in the favorable arrangement are required. There is another question related to their favorable steric arrangements.

The above considerations apply to neutral conditions since in basic environment sugars even possessing only hydroxyl groups can be in the anionic form due to dissociation of hydroxyl groups, and then the situation becomes different. The general rule presented in 70th years of XX century by Angyal [17] is based on his experimental research and defines the preferred spatial arrangement of hydroxyl group in the carbohydrate structure for complex formation. The results of his work indicated that for cyclitols and carbohydrate in the six-membered ring form, the preferred arrangement for the hydroxyl group are 1,3,5-triaxial (*ax–ax–ax*) or 1,2,3-axial–equatorial–axial (*ax–eq–ax*) sequences and also quasi axial–quasi equatorial–quasi axial sequence of three adjacent hydroxyl group for the furanose form. These rules by some are called as Angyal rules[12]. In further Angyal further showed [18] structural arrangement of hydroxyl group for acyclic compounds. Gyurcsik and Nagy [11] collected all possible spatial positions, arranged in the order according to the decreasing ability to form the complex (Fig. 9.1). All above mentioned ligands form 1:1 complexes with metal cations in the hydrophilic solvent [11, 17–19]. Further Angyal studies [20] specified another factor, which leads to the formation of carbohydrate complexes. Namely, it is that the size of the ionic radius of complexed ion. The cation with the ionic radius lower than about 80 pm is better bounded by the *ax–ax–ax* arrangement, whereas the cation with larger ionic radius prefers the *ax–eq–ax* sequence for the coordination. Since the ionic radius of calcium is estimated at 114 pm, the preferable carbohydrate sequence for calcium ions is *ax–eq–ax*. Molecular mechanics computations, carried out by Hancock and Hegetschweiler [21], confirmed such preferences. However, as noted by Alekseev et al. [12] only three complexes in solid state are known, which are in accordance with Angyal rules. On the other hand, there exist carbohydrate complexes with metal ions (e.g. α-D-glucopyranose) without special spatial arrangement required by Angyal rules. Another independent computations carried out by Palma and Pascal [22] have proved that in the gaseous phase general Angyal rules are not fullfilled. It is not surprising that tetra- and pentacoordinated complexes for β-anomers are characterized by the lowest energy. The structural analysis of carbohydrate–metal complexes in water solution is very difficult. The weak complexation results in the equilibrium shifted strongly toward the uncomplexed form. In addition, sugar in water exists in many equilibria, and each conformer can form different complex with the metal cation with the different complex constant. Constants of complexes for simple carbohydrates are in the range between 0.1 and 6.0. Alekseev pointed out that complexes between metal ions of sugar acids are characterized by the one order higher complex constant. In the next part of this section we will present some examples of calcium complexes with carbohydrates and their derivatives.

9.2.1 Calcium–Carbohydrate Complexes in the Crystal Form

Since, as mentioned above, calcium plays very important role in biological systems, our studies are devoted to calcium complexes with polyphenol glycosides. So this review will be devoted mostly to describe the research done on calcium–carbohy-

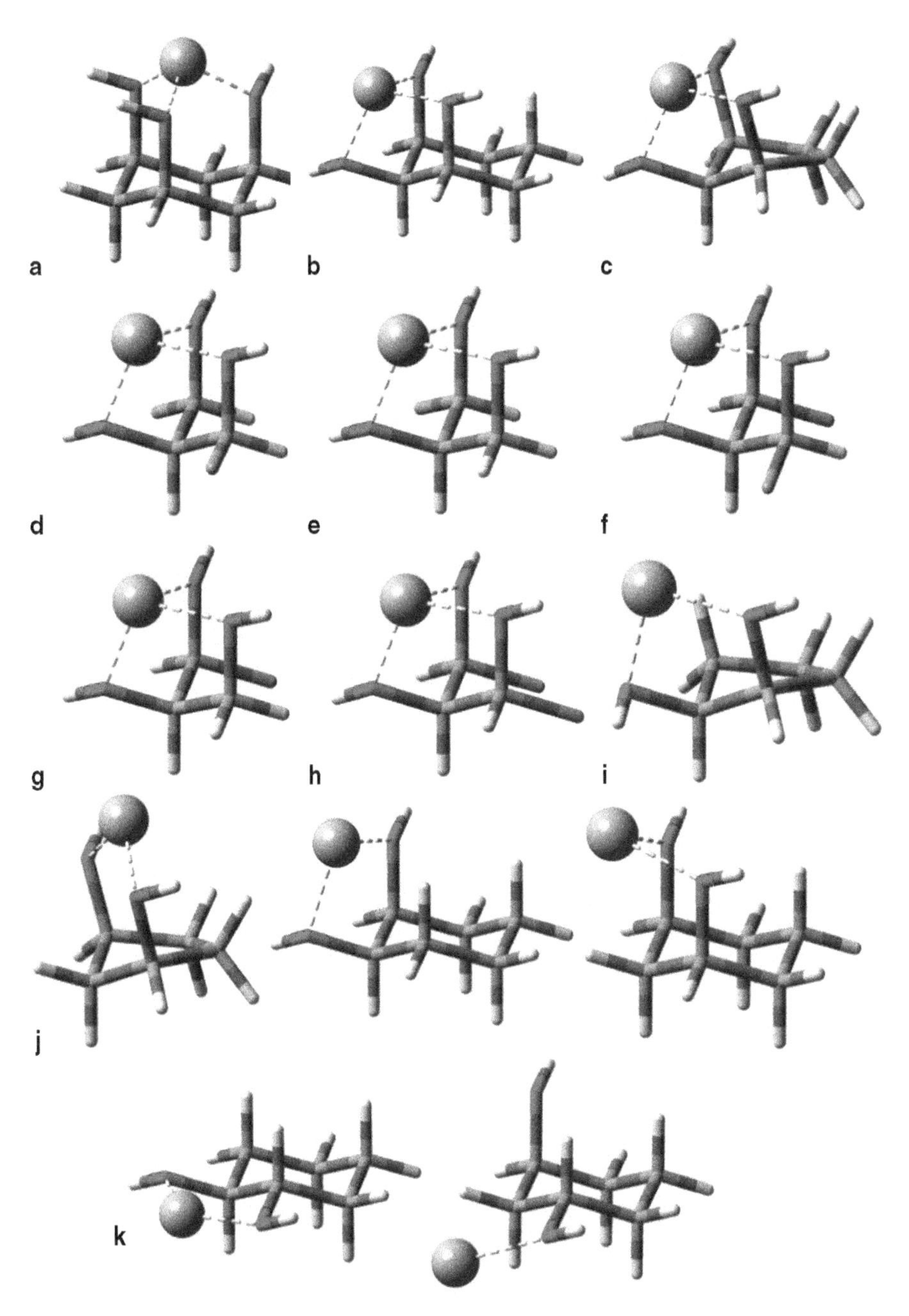

Fig. 9.1 Possible arrangements of hydroxyl groups in six- and five-membered rings and in acyclic form of sugar and their derivatives: (**a**) *ax–ax–ax*, (**b**) *ax–eq–ax*, (**c**) *cis–cis–cis*, (**d**) *threo–threo*-triol, (**e**) *threo*-diol, (**f**) *erythro–threo*–triol, (**g**) *erythro*-diol, (**h**) *erythro–erythro*-triol (**i, j**) *cis*-diol on a five- and six-membered ring, (**k**) *trans*-diol on a six membered rings. (Reprinted from [11], Copyright (2000) with permission from Elsevier)

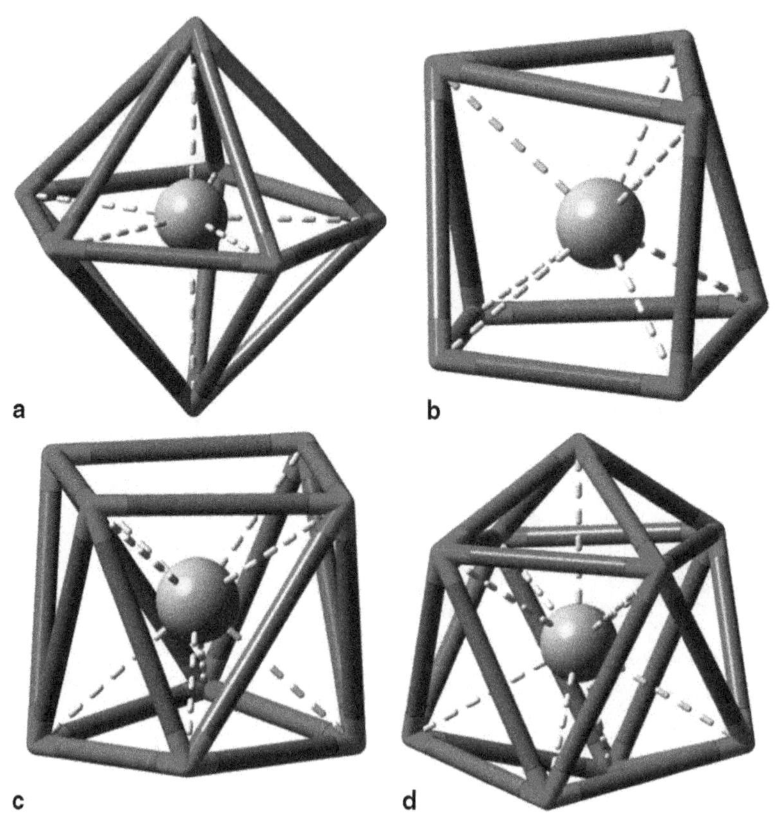

Fig. 9.2 Seven to nine 9-fold coordination geometries: (**a**) bipiramid pentagonal, (**b**) modified square antiprism (**c**) square antiprism, (**d**) nine-fold coordination geometry

drate complexes. Due to variety of coordination numbers, calcium cations shapes are characterized by the great diversity of shapes of its complexes. The majority of calcium complexes with carbohydrates are 7–9 coordinated structures [23]. Typically seven-fold coordination geometry can be described as a pentagonal bipyramid (Fig. 9.2a) or as an alteration of the square antiprism in which one square is replaced by triangle (Fig. 9.2b) and eight coordination as square antiprism (Fig. 9.2c). The centers of squares and triangle in both seven and eight fold geometries are collinear with the calcium ion. The nine coordination geometry (Fig. 9.2d) can also be described as a distorted square antiprism in which one square is significantly larger allowing additional oxygen to interact with calcium in the direction defined by the axis passing through the centers of squares. Polyhedron geometries possesses triangular arrangement of oxygen atoms allowing contacts ranging in length 2.30–2.85 Å, which enables tridendate ligands to bind to calcium ion in the solution. In the crystal structures such arrangements could be the restricted by crystal packing [23].

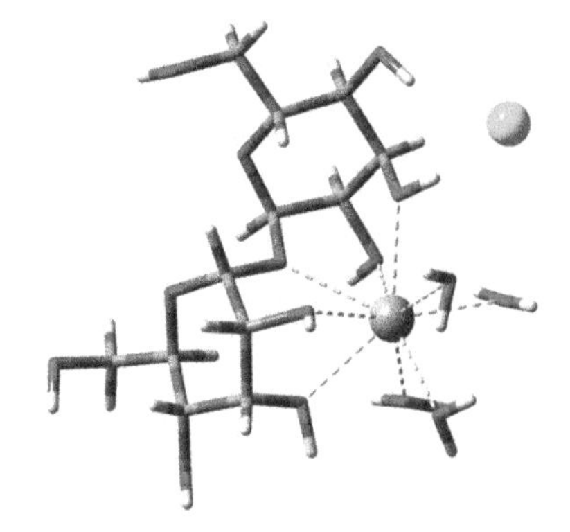

Fig. 9.3 Pentadentate complex of α-allopyranosyl-α-D-allopyranose with $CaCl_2$. (Reprinted from [28], Copyright (1978), with permission from Elsevier)

One of the first historical confirmations of the structure of the carbohydrate complex with metals was the X-ray study of the crystal structure of sucrose $NaBr \cdot 2H_2O$. It has to be mentioned that sucrose does not form the complex in the solution [24].

In 1972 crystal structure of sugar and calcium was determined for D-mannose $CaCl_2 \cdot 2H_2O$. Calcium in this case is coordinated to O1, O2, and O3 of one β-furanose ring and to O5 and O6 of the another mannose molecule [25].

The structure of the sugar unit in complexed and uncomplexed inositol is the same since inositol presents three axial hydroxyl groups in a manner ready for interactions with calcium ion [26]. The only slight change is observed in the distance between *syn*-axial oxygen atoms (2.82 Å in comparison to 2.96 Å in uncomplexed molecule). This reduction in the *gauche* or *syn*-axial interactions between participating oxygen atoms seems to be generally true, providing the driving force toward the complex formation [18].

In 1988 calcium complexes with D-glucose in the solution and in crystal form were investigated by Tajmir-Riahi [27]. He pointed out that there are strong interactions in calcium–D-glucose complexes in the solid and in the non-aqueous solution, while water, as a solvent, causes significant decreasing of these interactions. He also observed that calcium ion complexes in solid state can form two kind of complexes with 1:1 and 1:2 calcium–sugar ratio. In the first one calcium is coordinated by seven oxygen atoms (three oxygen atoms from sugar and four oxygen atoms from water molecules). In the 1:2 complex the calcium atom is coordinated by eight oxygen atoms: four oxygen atoms from water molecules and two hydroxyl oxygen from each D-glucose molecule.

The very interesting case was observed for the disaccharide α-D-allopyranosyl-α-D-allopyranose molecule synthesized intentionally as it has potential pentadendate possibility of binding calcium due to the presence of two ax–eq–ax sequencing. Nuclear magnetic resonance (NMR) studies in the solution did not prove such complexation, however pentadentate interaction was observed in the crystaline $CaCl_2$ complex (Fig. 9.3) [28]. The coordination was found involving O1, O2, O3, O2′, O3′ and four molecules of water. It is worth to point out that isomer α-D-glucopyranosyl-α-D-glucopyranose having no ax–eq–ax sequencing crystallizes in completely different manner [18].

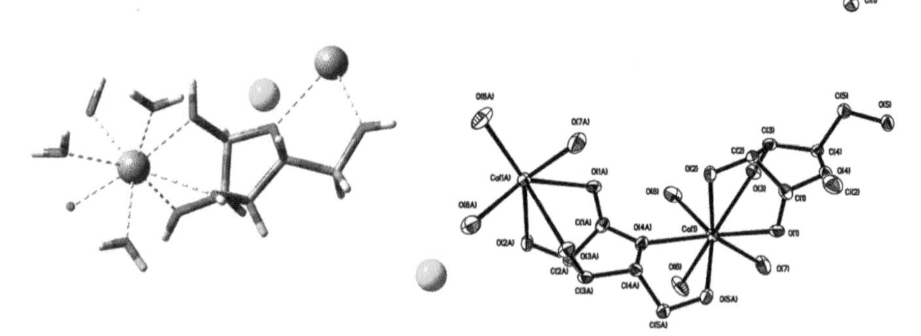

Fig. 9.4 Structure of D-ribose complex with $CaCl_2$. Two molecules of ribose coordinate the calcium cation. The first one is tridentate, the second bidentate. (Reprinted from [29], Copyright (2003), with permission from Elsevier)

Fig. 9.5 Calcium complex with lactose—8 coordination structure. The first coordination sphere is formed by four water molecules, two hydroxyl oxygens from glucose and two hydroxyl oxygen atoms from the galactose part. (Reprinted with permission from [14], Copyright (1973) American Chemical Society)

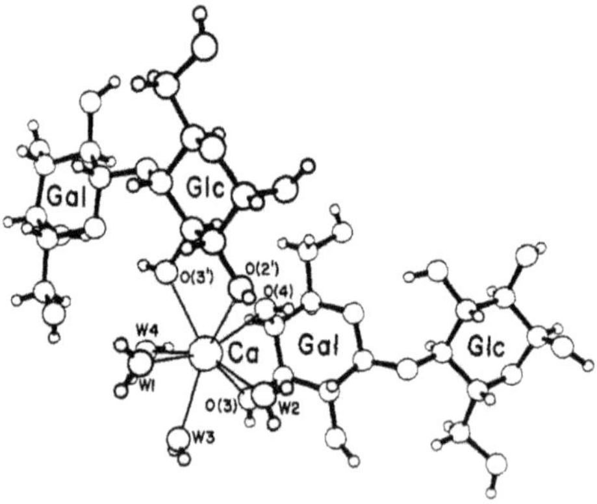

D-Ribose forms a complex in which each calcium ion is shared by two ribose molecules. One ribose molecule represents tridendate ligand binding calcium ion through O1, O2, O3 atoms, the second ribose is a bidendate ligand and binds via O4 and O5 oxygen hydroxyl atoms. The coordination shell is completed by three molecules of water giving the total of eight coordinating oxygens for each calcium (Fig. 9.4). The tridendate binding is due to the fact that arrangement of oxygen atoms O1, O2, O3 in this pentasaccharide is similar to *ax–eq–ax* arrangement in six-membered rings [29].

Bugg [14] analysed the crystal structure of calcium bromide complex of lactose. Lactose, component of milk, is responsible for the absorption of calcium from the gastrointensinal tract due to calcium–metal binding properties. Lactose belongs to disaccharide and its molecule forms the complex in which one lactose molecule is coordinated to the calcium ion through O3 and O4 of its galactose part and through O2′ and O3′ of its glucose moiety. The coordination shell is completed by additional molecules of water (Fig. 9.5). In the same paper another galactose complex is presented in which calcium binds to five hydroxyl groups but from three different carbohydrate molecules; O1 and O2 from the first, O3 and O4 from the second and O6 from the third. The coordination shell is completed by three molecules of water [13].

Takashi et al. [30] and Cook and Bugg [31] studied the complexes of trehalose (disaccharide formed by an α,α-1,1-glucoside bond between two α-glucose units) with calcium chloride and calcium bromide respectively. Both structures were very similar. In each complex the calcium ion is coordinated to one molecule of water and four molecules of trehalose (hydroxyl groups O2 and O3 from two molecules and two O6 hydroxyl oxygen from other molecules. These seven oxygen atoms are arranged in pseudo pentagonal bipyramid—similar to one discussed for the calcium bromide—lactose complex.

Fructose exists in the complex with calcium surrounded by eight oxygen atoms from four molecules of sugar and two molecules of water, which form a distorted square antiprism. Two fructose molecule binds through their O2 and O1 oxygen atoms in bidendate fashion and other two by their O6 oxygen atom in monodentate fashion [32].

The open chain carbohydrate derivatives like alditols form complexes with calcium but their formation strongly depends on their conformation. Angyal [17] based on NMR studies, has proposed that three consecutive hydroxyl groups have to be able to form conformation similar to ax–eq–ax. It is possible when they have the threo–threo configuration. In the case of the erythro–erythro configuration side chains are parallel to each other, thus making such a conformation of relatively higher energy. When three consequitive hydroxyl groups are in threo–erythro configuration two side chains are in the gauche arrangement which makes this isomer less stable than threo–threo but more stable than erythro–erythro. Angyal pointed out, based on electrophoretic mobility, that the ratio of all-threo to all erythro is in the range of 0.24–0.09. The above consideration relates to complexes when three hydroxyl groups from one alditol is complexed to the same calcium atom. If it is not possible the carbohydrate becames bidendate ligand. The situation is less complicated when one or even two side chains are small hydrogen atoms.

Yang et al. [33] have studied the coordination behaviour of neutral erythritol with calcium and lanthanide ions. Erythritol is one of the simplest representative of carbohydrates with four carbons each with the hydroxyl group on it. The hydroxyl group are not in the threo–threo arrangement to form tridentate ligand. They have identified three different metal complexes with the molar ratio of metal ion to erythritol as 2:1, 1:1, and 1:2. In all structures erythritol acts as the bidendate ligand. The structures are shown on Fig. 9.6. In the first structure (Fig. 9.6a) the calcium is surrounded by seven ligands: three chloride ions, two water molecules, and two hydroxyl groups from erithritol, which results in the pentagonal bipyramidal arrangement. In the second structure (Fig. 9.6b), calcium is coordinated to four hydrohyl groups from two erythritol molecules and four molecules of water. Chloride ions do not coordinate to calcium. Calcium is 8-fold coordinated in the bicapped trigonal prism. In the third structure (Fig. 9.6c) the calcium is surrounded by eight hydroxyl groups from four erythritol molecules.

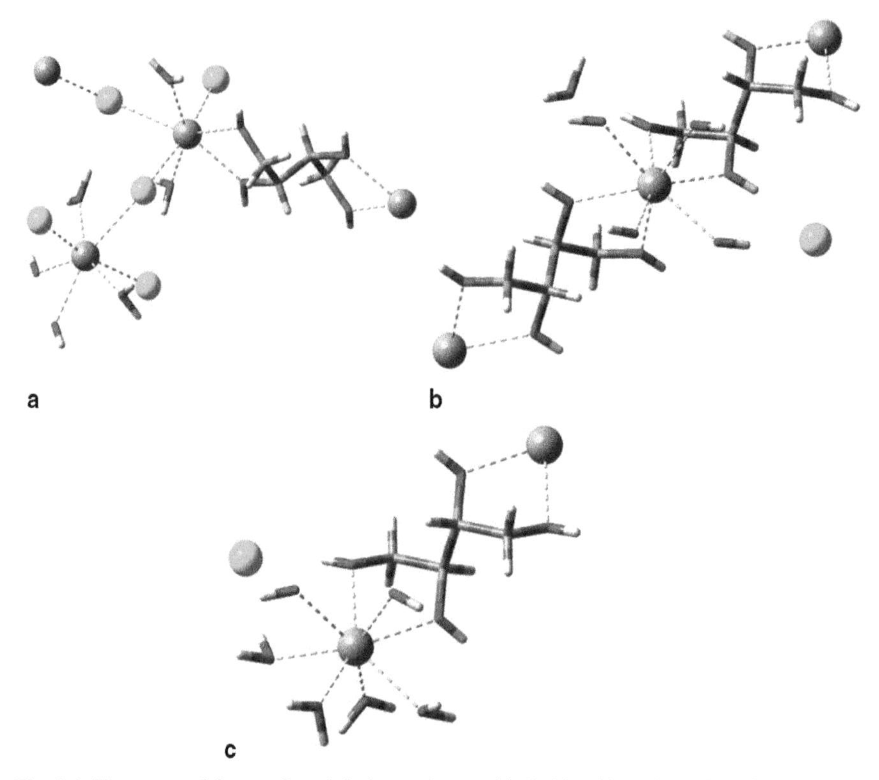

Fig. 9.6 Three crystal forms of erythritol complexes with $CaCl_2$ with molar ratios of metal to erythitol: (**a**) 2:1, (**b**) 1:1, (**c**) 1:2. (Reprinted with permission from [33], Coryright (2003) American Chemical Society)

Another interesting structure of open chain carbohydrate derivative is presented by the galactitol complex with calcium. Galactitol—1,2,3,4,5,6 heksahydroxyhexane is the metabolite of galactose. It lacks the carbonyl group and so it has an the open chain structure. It forms the complex in which the galacitol binds through its four hydroxyl groups to two calcium ions as shown on fig below (Fig. 9.7) [34]. Galactitol in this case is bidentate ligand also, despite that 1,2,3 as well as 4,5,6 hydroxyl groups can arrange in a way to form tridentate ligand with preferred threo–threo geometry.

There is one structure which is worth of analysis despite it is not a neutral polysaccharide. Alginates, b-D-mannopyranosyluronic acids are polysaccharides which form gels with calcium ions. Calcium ion binds to two neighboring molecules of uronic acid to five oxygen atoms: two from hydroxyl oxygens, one from ring oxygen, one from glycosidic oxygen and one oxygen atom from the carboxylic group [17]. More about the structure of alginate gels can be found in Agulhon et al. [35] paper, in which using density functional calculations the interactions were investigated between diuronate units and some divalent cations.

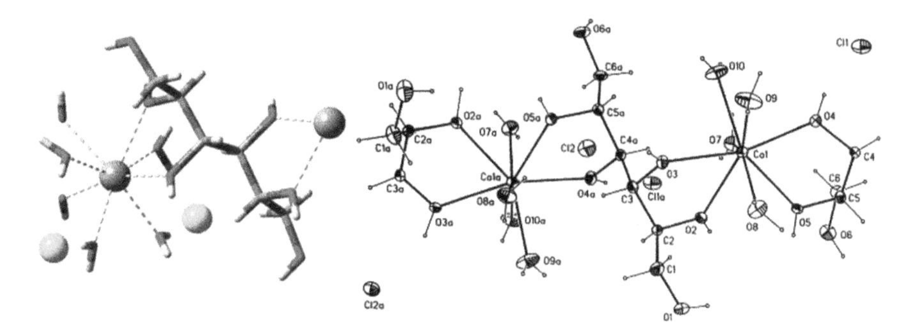

Fig. 9.7 The structure of calcium ions complex with galactitol. (Reprinted from [34], Copyright (2002), with permission from Elsevier)

9.2.2 Calcium–Carbohydrate Complexes in Solution

As mentioned above, simple carbohydrates with s-block metal ions including calcium form complexes in water but their stability is very low. NMR is probably the most suitable method the studies of a complex formation in solution. It has been reported long time ago that the addition of calcium chloride to the solution of epi-inositol and some glycosides in deuterium oxide causes the downfield shift of some or all signals in the ^{1}H-NMR [36, 37]. The shifts are in the same direction, similar to that caused by paramagnetic reagent, however the value of changes are smaller. When the calcium salt is added to the solution of carbohydrates the downfield movement of NMR signals is observed in some or all NMR signals from dipole formed by the metal cation and the coordinated oxygen atom. The shift in hydroxyl–proton resonances in water as a function of the concentration of added calcium chloride yields, in general, has a linear dependence. The larger slope the stronger is the complex formation.

Angyal was also one of the first who studied the complexes of calcium with carbohydrates in the solution by means of NMR. He noticed that addition of paramagnetic ions like europium and praseodium shifts the signals to a greater extend, so that better analysis can be performed [38–40].

Some chosen examples of NMR studies of calcium sugar complexes in the solution are given below.

Similar shifts were observed in 1,2,3,4,5 in the pentahydroxycyclohexanes. In the epi-inositol the interaction was attributed to interaction with O2, O3, and O4 hydroxyl oxygen atoms. The signal shifted the most where the protons are attached to central carbon in the three consecutive carbon atoms with the hydroxyl group in the configuration ax–eq–ax.

α-D-Allopyranose in its more stable conformation contains ax–eq–ax configuration of hydroxyl groups but its β anomer does not. Both anomers are in dynamic equilibrium in water at room temperature [41]. The addition of calcium ions causes a substantial increase in concentration α-D-allopyranose with substantial downfield

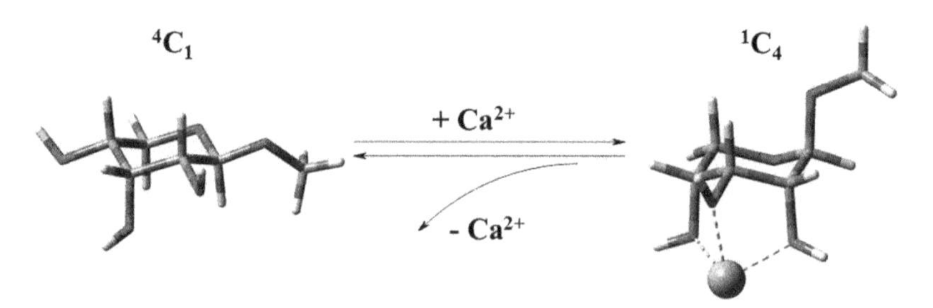

Fig. 9.8 The conformational equilibrium enforced by binding of calcium ions. The nomenclature $^4C_1$ and $^1C_4$ are in accordance with the IUPAC nomenclature [42]. (The letters defining conformations are described by numerals, which are locants of ring-atoms, indicated as superscripts and subscripts)

shift of its anomeric proton. This experiment indicates the formation of complex formation with the α-D-allopyranose only and as a consequence the appropriate equilibrium shift. It is worth to note that the effect after the addition of monovalent sodium ion is small due to weak complexing ability of sodium. The NMR studies also revealed that spectra of α-furanose but not β-furanose was changed indicating that the cis–cis arrangement of five membered α-furanose ring also favors the complex formation. The fact that only α-D-allopyranose and α-D-allofuranose form complexes with calcium and β forms was not independently confirmed by electrophoresis experiments [17].

In the case of 5-O-methyl-D-ribose [17] α-furanose isomer consist of 33 % and β-furanose 67 % in the solution. After the addition of calcium chloride the proportion was changed to 70:30 respectively. It is due to the possibility of α-furanose to adopt conformation in which hydroxyl groups O1, O2, O3 are quasi axial, quasi equatorial, quasi axial which is close to typical arrangement ax–eq–ax.

As expected, α-D-ribopyranose and α-D-ribofuranose [41] signals in NMR are shifted downfield due to the addition of calcium chloride, since they have acquired the arrangement of three hydroxyl groups. In addition β-D-ribopyranose signals has also been shifted due to the change of conformational equilibrium from $^4C_1$ to $^1C_4$ (Fig. 9.8), since only the later has the sequence *ax–eq–ax*.

The conformation change was also observed in the case of D-lyxose. From two possible forms, only β-pyranose has the required ax–eq–ax configuration but α-pyranose does not. The NMR spectrum in the presence of calcium ions [17] has shown that β-anomer consists of 50 % whereas in the solution without calcium chloride its concentration amounts to 28 % only. This is the evidence that equilibrium is shifted due to the complexation with calcium ions.

D-Gulose, D-glycero-D-gulo-heptose are carbohydrates possessing the *ax–ex–ax* sequence in α-pyranose forms. Addition of calcium ions to in the solution also shifts the equilibrium in similar manner to that observed in D-allose. Similar conformational change was observed in the case of apiose, where the α-D-apio-D-furanose signals become the major components of the isomer mixture. This sugar has three cis-hydroxyl groups.

In the contrast, only small changes were observed in the solution of D-mannose, D-glucose, D-xylose, D-arabinose, D-galactose, D-fucose, L-sorbose [17, 43]. These sugars has no *ax–eq–ax* configuration in three subsequent hydroxyl groups. This fact again supports the thesis that such arrangement in six membered cyclitols is required for the complexation with calcium cations. However, some of these sugars can form complexes in bidentate form.

Weak interactions between D-xylose and calcium ions were observed in NMR. None of the forms of D-xylose possesses the required *ax–eq–ax* configuration needed for the strong bonding. Authors [43] attributed the NMR shift change to weak interactions in the bidentate mode. α-anomer in the $^4C_1$ conformation has ax–eq pair and in the $^1C_4$ conformation the *cis ax–ax* pair. Presented results indicate also that from two β-anomers $^4C_1$ and $^1C_4$, only β-anomer in $^1C_4$ conformation has two cis axial OH groups. Since only the β-$^4C_1$ structure dominates, the conclusion is that none of β-anomers form complexes with calcium [44]. The final conclusion was that the chemical shift change was due to the complex formation within the α-$^4C_1$ conformation. The α-$^4C_1$ possess an *ax–eq* pair which form stronger interaction that *ax–ax* in the case of α-$^1C_4$ conformer.

In the case of D-glucose [43], between all possible conformers which are able to bind in the bidentate form, only the α-$^4C_1$ conformer having ax–eq pair (O1, O2 hydroxyl oxygen atoms) is responsible for a weak interaction with calcium. In the absence of calcium the major conformer is constituted of β-$^4C_1$ characterized with all equatorial hydroxyl groups—allowing stronger interactions with water. This conclusion is supported by the fact that 2-deoxyglucose does not bind to the calcium ion.

D-Galactose can form only weak complexes in α-$^4C_1$ and α-$^1C_4$ pyranose conformations due to the presence of *ax–eq* hydroxyl groups and β-$^4C_1$ due to the *ax–eq* pair of the hydroxyl groups arrangement.

D-Fructose and L-sorbose are ketoses. The first one forms weak complexes in the β-pyranose 1C4 conformation with ax–eq sites for binding of calcium and weak complexes by α- and β-furanose isomers. L-Sorbose forms also very weak complexes. It is possible only in the β-4C1 conformer due to two sets of *cis ax–ax*.

9.3 Other Methods Used for Complex Formation Studies in the Case of Polyphenol Glycosides

There are a number of other methods, spectroscopic as infra-red (IR), ultraviolet–visible (UV–Vis), which allow the evaluation of the ability of the Ca^{2+} complex formation. Not all of them are suitable in every case. We have decided to explore the flame photometry and conductometric titration since the NMR changes were too small to perform the correct analysis.

The flame photometry indicates the concentration of calcium in the solution. It could be used for evaluation of dissolving power of the ligand or prevention of crystallization of the solid calcium salts in solution.

9.3.1 *Flame Photometry*

The flame atomic emission spectrometry (FAES) was used by us to measure the complexing ability of tested compound. The flame photometry constitutes the relatively sensitive method which allows the determination of studied element concentrations in the range of 0.1–1000 ppm with a relative small error of 1–3 %. This approach was used for the determination of the calcium concentration in the water solution over solid calcium oxalate i.e. the shift of the solid–liquid interphase equilibrium in and without the presence of synthetic or natural complexing agent [45–47].

9.3.2 *Conductometric Titration*

The conductometric titration is based on the observation of changes in the conductivity of the analyzed solution during the addition of the titrating reagent—(sodium oxalate in this case) to the solution of calcium chloride and the complexing agent. The formation of the complex results in a decreasing of conductivity. Since some of calcium ions are complexed and they are not participating in the conductance. In this reaction, during the addition of Na^+ and $C_2O_4^{2-}$, the insoluble calcium oxalate is formed. The conductivity of such solution decreases until it reaches the end point (EP) of titration, which corresponds to the moment when the total amount of Ca^{2+} ions in the solution is bonded. The conductivity of the solution at this point is minimal. Further addition of sodium oxalate result in increase of conductivity because of the excess of Na^+ and $C_2O_4^{2-}$ ions. If addition the analyzed compound causing that EP is reached sooner than in standard curve, that mean the compound possess complexing ability to calcium ions. The end point of the titration is thus a measure of the degree calcium ions complexing. The greater difference with respect to the calibration curve, the better properties of the inhibition of calcium oxalate formation. When EP values of the standard curve and the curve with analyte are similar, it is considered that the compound does not have the properties of inhibiting of the formation of calcium oxalate. This method was applied to measure the properties of studied compounds [48, 49].

9.4 Results and Discussion

The flame photometry and conductometric titration methods were used to evaluate the complexing property of pure synthetized compounds as well as plant extracts containing polyphenol glycosides [45]. Several glycosides of hydroxyanthraquinones were prepared. The synthesis was conducted according to a four-step method consisting acetylation of sugar compounds, their bromination, coupling with hydroxyanthraquinone, and finally the removal of acetyl groups. Properties of pure

Table 9.1 Results of conductometric titration. (This table was published in Frąckowiak et al. [45],)

Compound	The end of the sample curve (µl)	ΔEP (µl)
None (standard)	270	0
1,2-Dihydroxy-9,10-anthraquinone	230	40
2-(β-D-Glucopyranosyloxy)-1-hydroxy-9,10-anthraquinone	190	80
1,4-Dihydroxy-9,10-anthraquinone	240	30
1-(β-D-Glucopyranosyloxy)-4-hydroxy-1,9-anthraquinone	210	60
1,2,5,8-Tetrahydroxy-9,10-anthraquinone	180	90
2-(β-D-Glucopyranosyloxy)-1,5,8-trihydroxy-9,10-anthraquinone	90	180
2-(β-D-galactopyranosyloxy)-1-hydroxy-9,10-anthraquinone	200	70
2-(β-D-galactopyranosyloxy)-1,5,8-trihydroxy-9,10-anthraquinone	100	170

ΔEP = the end point of the standardization curvex − the end point of the sample curve

hydroxyanthraquinone and their glycosides were compared. The presence of sugar leads to higher change of end point (ΔEP) and the increase of calcium concentration (Δ%) of emission (Tables 9.1, 9.2 and 9.3). The highest dissolving properties were observed for 1,2,5,8-tetrahydroxy-9,10-anthraquinone and its glycosides. It can be expected, that higher the number of hydroxyl groups in the aromatic region the better is complexing properties. The results of flame photometry (Tables 9.2 and 9.3,) indicate that all synthetized glycosides and their aglycones have good properties of coordination to calcium ions and in consequence dissolving properties of model kidney stones. It was also found that the presence of the sugar fragment increases the complexing ability, compared with that of pure aglycones [45]. In the conductometric titration, the best properties were exhibited by 1,2,5,8-tetrahydroxy-9,10-anthraquinone and its glycosides indicating good inhibition of calcium oxalate crystals formation. The analysis of dissolving properties of hydroxyanthraquinone derivatives in the case of natural kidney stones was also performed. The best properties in this case were found for 2-(β-D-glucopyranosyloxy)-1,5,8-trihydroxy-9,10-anthraquinone (Tables 9.2 and 9.3). The effect in the case of real kidney stones is much higher than in calcium oxalate experimental model. The effect is probably due to the different composition and different crystalline structures of kidney stones [45].

Polyphenol- and polyalcohol-glycosides were synthetized according to the same procedure as anthraquinone glycosides.

They have also good calcium oxalate solubility properties, but do not prevent the formation of model kidney stones. Glycosides with the best solubility of calcium oxalate were tested at the case of real kidney stones. In few cases Δ% of emission was negative. It means the reduction of the amount of calcium in the solution, probably by formation of complexes which are not soluble in water. Photometric and mi-

Table 9.2 Results of flame photometry—calcium oxalate test. (This table was published in Frąckowiak et al. [45], Copyright © 2010 Elsevier Masson SAS. All rights reserved)

Compound	Concentration of Ca^{2+} in sample test (10^{-4} M)	ΔCa^{2+} (10^{-4} M)	The increase of calcium concentration (%)
1,2-Dihydroxy-9,10-anthraquinone	9.1	5.4	246
2-(β-D-glucopyranosyloxy)-1-hydroxy-9,10-anthraquinone	22.0	18.3	595
1,4-Dihydroxy-9,10-anthraquinone	7.60	3.9	205
1-(β-D-Glucopyranosyloxy)-4-hydroxy-9,10-anthraquinone	20.0	16.6	541
1,2,5,8-Tetrahydroxy-9,10-anthraquio-none	49.0	45.3	1324
2-(β-D-Glucopyranosyloxy)-1,5,8-trihydroxy-9,10-anthraquinone	56.0	52.3	1513
2-(β-D-Galactopyranosyloxy)-1-hydroxy-9,10-anthraquinone	16.0	12.3	432
2-(β-D-Galactopyranosyloxy)-1,5,8-trihydroxy-9,10-anthraquinone	19.0	15.3	513

Concentration of Ca^{2+} in reference test $= 3.70 \times 10^{-4}$ M. ΔCa^{2+} = Concentration of Ca^{2+} in sample test (2) − Concentration of Ca^{2+} in reference test (1)

Table 9.3 Results of flame photometry—real kidney stones. (This table was published in Frąckowiak et al. [45], Copyright © 2010 Elsevier Masson SAS. All rights reserved)

Compound	Concentration of Ca^{2+} in sample test (10^{-4} M)	ΔCa^{2+} (10^{-4} M)	The increase of calcium concentration (%)
1,2-Dihydroxy-9,10-anthraquinone	8.6	7.4	717
2-(β-D-Glucopyranosyloxy)-1-hydroxy-9,10-anthraquinone	19.0	17.8	1583
1,4-Dihydroxy-9,10-anthraquinone	10.0	8.8	833
1-(β-D-Glucopyranosyloxy)-4-hydroxy-9,10-anthraquinone	24.0	22.8	2000
1,2,5,8-Tetrahydroxy-9,10-anthraquionone	50.0	48.8	4167
2-(β-D-Glucopyranosyloxy)-1,5,8-trihydroxy-9,10-anthraquinone	58.0	56.8	4833
2-(β-D-Galactopyranosyloxy)-1-hydroxy-9,10-anthraquinone	18.0	16.8	1500
2-(β-D-Galactopyranosyloxy)-1,5,8-trihydroxy-9,10-anthraquinone	20.0	18.8	1667

Concentration of Ca^{2+} in reference test $= 1.20 \times 10^{-4}$ M. ΔCa^{2+} = Concentration of Ca^{2+} in sample test (2) − Concentration of Ca^{2+} in reference test (1)

croscopic observations confirm that synthetized glycosides changed the morphology of kidney stones [46]. The similar behavior was observed for the plant extract of *Galium verum*, *Rubia tinctorum*, *Hypericum perforatum* and *Humulus lupulus*, which contain high amount of polyphenol glycosides [47].

Table 9.4 Results of conductometric titration. (Reprinted from [47], Copyright (2010) with permission from Elsevier)

	End point of the sample curve (µl)	End point of the standardization curve (µl)	ΔEP (µl)
Crude extract	250	270	20
Fraction chloroform I	280	280	0
Fraction water II	170	250	80
Fraction soluble in methanol	190	280	90
Fraction insoluble in methanol	90	250	160
Rubinex	270	280	10

Table 9.5 Results of flame photometry for studied preparations—calcium oxalate test. (Reprinted from [47], Copyright (2010) with permission from Elsevier)

	Concentration of Ca^{2+} in sample test (10^{-4} M)	% Of emission in sample test	Concentration of Ca^{2+} in blind test (10^{-4} M)	% Of emission in blind test	ΔCa^{2} (10^{-4} M)	Δ% of emission
Crude extract	14.0	56	13.0	55	−3.50	−14
Fraction water II	–	–	–	–	–	–
Fraction soluble in methanol	9.6	39	0.74	3	5.18	21
Fraction insoluble in methanol	9.9	40	9.9	40	−3.70	−15
Rubinex	1.4	6	3.0	12	−5.10	−21

The plant material was obtained through by the stepwise extraction by the organic solvent: chloroform, methanol, and water. The dry extract was analyzed for their complexing activity. Extracts obtained in each methods were tested by conductometric titration method (Table 9.4) and flame photometry (Table 9.5). Their properties were evaluated as the shift of the end point (EP) in the conductometric titration and the flame photometry results as a *Δ*% of emission (ΔCa^{2+}) as described above. The results are presented in Tables 9.4 and 9.5.

NMR spectra have shown that ethanol–water and water extract at *Galium verum*, *Rubia tinctorum*, *Hypericum perforatum* and *Humulus lupulus* contain: sugar alcohols (myoinositol, arabitol, rybitol, inositol), and monosaccharides (glucose, xylose, ribose, fructose, galactose. Acids, such as citric, malonic, malic, glucuronic are also present. For more details see [47].

9.5 Computational Modeling

Nowadays, computational methods is an inherent and parallel tool for research of new compounds. As mentioned previously, the interactions, between carbohydrates and metal ions usually are very weak. There are several aspects, which can be considered by using computational approaches e.g. structures of complexes, their stability, interaction energies, nature of interactions, dissociation energy, its correlation with complex constants, the contribution of oxygen atoms to interaction due to their different chemical nature, and the influence of solvent model on all mentioned parameters. Beside searching the active site to bind metal cations, relevant observations constitute conformation changes between complexed and non-complexed structures (deformation energy). Hancock and Hegetschweiler [21] by using molecular mechanics calculations proved, above mentioned, Angyal postulates. Using inositols, as compounds structurally similar to sugars, they have presented the correlation between the preferred arrangement of binding and size of metal ionic radius. Because metal ions cannot be readily described by semi-empirical or classical molecular dynamics methods [50], the density functional methods (DFT) [51] are widespread for interaction energy investigations [52]. Moreover, when needed, DFT methods can be easily extended to higher level ab initio methods such as Second-order Møller-Plesset Perturbation Theory (MP2) [53] or the coupled-cluster singles and doubles model (CCSD) [54]. Zheng et al. were investigating interactions between metal cations and inositols by applying the DFT functional. Similar to Hancock and Hegetschweiler, they also used cis-inositols to find possible binding site for Be^{2+}, Mg^{2+}, Ca^{2+} and Li^{+} ions. The computations confirm previous expectations. Additionally, they studied β-D-glucose–calcium complexes and the results show that there are five possible active sites which can bind calcium cation—four bidendate structures and one tridendate. The strongest interaction was observed for structure, which binds the calcium ion by anomeric oxygen atoms, hydroxymethyl and O-ring oxygen atoms [55]. Wong et al. [56] carried out similar studies for mannose complexes with a calcium cation. Their results further indicated several active sites to bind metal cations. Moreover, these two papers unanimously inferred that the stability of complexes and the strength of interactions increase with an increasing coordination number. Our calculations of glucose–Ca complexes confirm those inferences and additionally we obtained four coordinated structures, which according to expectations has the higher interaction energy. Glucose adopts skew-boat—${}^{O}S_2$ conformation, in this four coordinated complex [57]. The result suggests that metal ions could induce conformational interconversions. In 2007 Fabian [58] presented results, which indicated that metal binding might cause the shift of the conformation equilibrium to normally less preferable conformation.

Wong et al. pointed out three types of energy in complexes: binding, deformation, and stabilization energies. Interaction energy (Eq. 9.1) is the only energy between cation and ligand moiety without any additional effect e.g. conformation changes. Deformation energy (Eq. 9.2) is a difference between complexed form of ligand and non-complexed and stabilization energy (Eq. 9.3) is the same like dissociation energy but with the opposite sign. That means that the stabilization energy is a sum of interaction energy and deformation energy (see Fig. 9.9).

$$E_{int} = E_{AB} - E_A - E_B \tag{9.1}$$

$$E_{def} = E_{complex} - E_{non-complex} \tag{9.2}$$

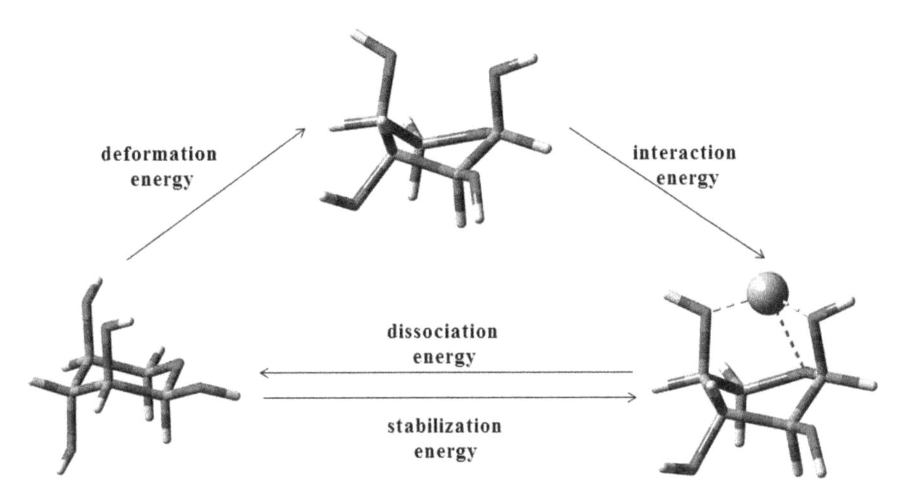

Fig. 9.9 The scheme of defining deformation, interaction, stabilization energies and in complexes

$$E_{stab} = E_{int} + E_{def} \tag{9.3}$$

In our previous study [57] we indicated that not only sugar is able to form complexes with calcium ions but also their derivatives like glycosides could bind metal ions. In this study we further indicated that alizarin glucoside interacts with calcium ion due to aglycon moiety, which possess additional oxygen atom (carbonyl and hydroxyl), there are more potential places to bind calcium. Results show that complexes with higher coordination number are characterized by higher stability than those with lower coordination number (Table 9.6). The presented data in [57] also showed that with increasing coordination number the contribution of single interaction decrease but at the same time total interaction energy increase. Our studies also confirm the above mentioned Fabian conclusion and also experimental findings that metal ions can enforce conformational changes. Our recent studies [59] also support such conclusion. Moreover, apart from the structure, stability and interaction energy in gaseous phase, we also were investigated the influence of solvent on previously mentioned parameters. We calculated all these parameters in two different ways (PCM model or the first shell taken directly into account). In both the cases, results indicate that interaction energy between glycoside and calcium is weaker. The results of the second procedure also indicate that each of water molecule in the first coordination sphere significantly neutralizes the charge on the calcium cation. The results suggest that in water solution glucoside complexes do not form strong complexes. These calculations also confirm the Angyal rule regarding the preferred arrangement of hydroxyl groups. All our computations were carried out applied GAUSSIAN 09 suite of codes [60].

Table 9.6 The number of contacts O–Ca^{2+}·(N), relative interaction energy ($\Delta\Delta E_{int}$), relative enthalpy (ΔH), and relative free enthalpy (ΔG) for glucoside complexes, the atomic charge on Ca and the C=O···Ca^{2+} distance. Energy in kcal/mol, atomic Mulliken charge in electron, distance in angstrom. The reference lowest interaction energy amounts to −233.0 kcal/mol. (Springer and the original publisher/Journal of molecular modeling, Theoretical studies of structure, energetics and properties of Ca^{2+} complexes with alizarin glucoside, Toczek et al. [57], Table 9.3, original copyright (2013) with kind permission from Springer Science and Business Media)

	N	$\Delta\Delta E_{int}$	ΔH	$\Delta\Delta E_{int}-\Delta H$	ΔG	Atomic charge on Ca^{2+}	Distances C=O···Ca^{2+}
a_Gli_1S5_gt_ BEFO1O2_Ca(85)	5	7.8	13.0	−5,2	12.8	1.367	2.28
a_Gli_1S5_gt_ BEFO1O2_Ca(-161)	5	8.4	16.9	−8.5	16.6	1.373	2.30
a_Gli_1C4_gg_ BDFO1O2_Ca	5	9.5	9.1	0.4	10.1	1.374	2.28
a_Gli_1C4_gt_ BDFO1O9_Ca(-85)	5	14.5	23.2	−8.7	23.6	1.356	2.34 (O9)
a_Gli_OS4_tg_BDO1O2_ Ca(75)	4	17.2	14.8	2.4	14.3	1.422	2.27
a_Gli_OS4_tg/ gg_BDO1O2_Ca(77)	4	18.4	12.6	5.8	12.4	1.426	2.27
a_Gli_OS4_gt_BDO1O2_ Ca(72)	4	18.7	18.8	−0.1	17.9	1.423	2.27
a_Gli_25B_gt_DFO1O2_ Ca	4	23.3	25.3	−2.0	24.6	1.409	2.26
a_Gli_1S5_gt_EFO1O9_ Ca	4	24.4	25.9	−1.5	25.2	1.378	2.31 (O9)
b_Gli_OS2_gg_ CEFO1O2_Ca(-95)	5	0.0	0.0	0.0	0.0	1.365	2.28
b_Gli_O3B_gg_ CEFO1O2_Ca(168)	5	2.5	5.9	−3.4	6.5	1.373	2.32
b_Gli_O3B_gg_ CEFO1O9_Ca(-42)	5	9.2	16.1	6.9	16.7	1.335	2.35 (O9)
b_Gli_2S4_gg/ gt_EFO1O2_Ca(-121)	4	14.5	15.5	−1.0	15.7	1.410	2.27
b_Gli_OS4_gt_EFO1O9_ Ca(-75)	4	22.6	26.8	−4.2	25.2	1.374	2.33 (O9)
b_Gli_OS2_gt_CFO1O2_ Ca	4	23.7	24.5	−0.8	23.1	1.395	2.26
b_Gli_2SO_gg_ BEO1O2_Ca	4	29.5	32.3	−2.8	31.7	1.411	2.27
b_Gli_4C1_gt_EFO1O9_ Ca(-59)	4	33.8	31.2	2.6	30.2	1.372	2.33 (O9)
b_Gli_4C1_gt_EFO1O9_ Ca(-83)	4	34.5	25.6	8.9	24.8	1.383	2.32 (O9)

9.6 Concluding Remarks

The above results suggest that polyphenol glycosides possess ability to form complexes with calcium ions contained in calcium oxalate, the main component of kidney stones. Described above experiments and computations indicate that glycosides are characterized by higher ability to form complexes in the comparison to their glycon or aglycon parts alone. The findings indicate that the complexation process proceeds by both part simultaneously. Data for the gas phase show that the interaction energy strongly depends on the coordination number. However computations including the solvent effect indicate the influence of importance of other interactions. The inclusion of the solvent effect increases the contribution of conformational effects of carbohydrate part into the stability of complexes. The studies of glycosides with the potential complexation ability still need optimization of structural parameters responsible for these properties.

Acknowledgments This work was supported by the statutory activity subsidy from Polish Ministry of Science and Technology of Higher Education for the Faculty of Chemistry of Wroclaw University of Technology. The computations were performed in Wroclaw Supercomputing and Networking Center.

References

1. Sejersted OM (2011) Calcium controls cardiac function—by all means! J Physiol 589(12):2919–2920. doi:10.1113/jphysiol.2011.210989
2. Lovelock JE, Porterfield BM (1952) Blood clotting: the function of electrolytes and of calcium. Biochemistry 50:415–420
3. Peacock M (2010) Calcium metabolism in health and disease. Clin J Am Soc Nephrol 5:S23–S30. doi:10.2215/CJN.05910809
4. Endo M (2006) Calcium ion as a second messenger with special reference to excitation–contraction coupling. J Pharmacol Sci 100:519–524. doi:10.1254/jphs.CPJ06004X
5. Moe OW (2006) Kidney stones: pathophysiology and medical management. Lancet 367:333–344. doi:10.1016/S0140-6736(06)68071-9
6. Lewandowski S, Rodgers AL (2004) Idiopathic calcium oxalate urolithiasis: risk factors and conservative treatment. Clin Chim Acta 345:17–34. doi:10.1016/j.cccn.2004.03.009
7. Coe FL, Evan A, Worcester E (2005) Kidney stone disease. J Clin Invest 115(10):2598–2608. doi:10.1172/JCI26662
8. Hall PM (2009) Nephrolithiasis: treatment, causes, and prevention. Cleve Clin J Med 76(10):583–591. doi:10.3949/ccjm.76a.09043
9. Kaufman Katz A, Glusker JP, Beebe SA, Bock CW (1996) Calcium ion coordination: a comparison with that of beryllium, magnesium, and zinc. J Am Chem Soc 118:5752–5763. doi:10.1021/ja953943i
10. Nakanishi F, Nagasawa Y, Kabaya Y, Sekimoto H, Shimomura K (2005) Characterization of lucidin formation in *Rubia tinctorum* L. Plant Physiol Biochem 43:921–928. doi:10.1016/j.plaphy.2005.08.005
11. Gyurcsik B, Nagy L (2000) Carbohydrates as ligands: coordination equilibria and structure of the metal complexes. Coord Chem Rev 203:81–149. doi:10.1016/S0010-8545(99)00183-6
12. Alekseev YE, Garnovskii AD, Zhdanov YA (1998) Complexes of natural carbohydrates with metal cations. Russ Chem Rev 67(8):649–669.
13. Bugg CE, Cook WJ (1972) Calcium ion binding to uncharged sugars: crystal structures of calcium bromide complexes of lactose, galactose, and inositol. J Chem Soc Chem Commun 12:727–729. doi:10.1039/C39720000727
14. Bugg CE (1973) Calcium binding to carbohydrates. Crystal structure of a hydrates calcium bromide complex of lactose. J Am Chem Soc 95:908–913. doi:10.1021/ja00784a046

15. Saladini M, Menabue L, Ferrari E (2001) Sugar complexes with metal^{2+} ions: thermodynamic parameters of associations of Ca^{2+}, Mg^{2+} and Zn^{2+} with galactaric acid. Carbohydr Res 336(1):55–61. doi:10.1016/S0008-6215(01)00243-9
16. Yang L, Su Y, Liu W, Jin X, Wu J (2002) Sugar interaction with metal ions. The coordination behavior of neutral galactitol to Ca(II) and lanthanide ions. Carbohydr Res 337(14):1485–1493. doi:10.1016/S0008-6215(02)00130-1
17. Angyal SJ (1973) Complex formation between sugars and metal ions. Pure Appl Chem 35(2):131–146. doi:10.1351/pac197335020131
18. Angyal SJ (1980) Haworth memorial lecture. Sugar–cation complexes—structure and applications. Chem Soc Rev 9(4):415–428. doi:10.1039/CS9800900415
19. Alvarez AM, Morel-Desrosiers N, Morel J-P (1987) Interactions between cations and sugars. III. Free energies, enthalpies, and entropies of association of Ca^{2+}, Sr^{2+}, Ba^{2+}, La^{3+}, Gd^{3+} with D-ribose in water at 25 °C. Can J Chem 65(11):2656–2660. doi:10.1139/v87–439
20. Angyal SJ (1974) Complexing of polyols with cations. Tetrahedron 30(12):1695–1702. doi:10.1016/S0040-4020(01)90691-X
21. Hancock RD, Hegetschweiler K (1993) A molecular mechanics study of the complexation of metal ions by inositols. J Chem Soc Dalton Trans 2137–2140. doi:10.1039/DT9930002137
22. Palma M, Pascal YL (1995) Étude théorique de la complexation des cations Pb^{2+} et Hg^{2+} par le D-talose. Can J Chem 73(1):22–40. doi:10.1139/v95-005
23. Dheu-Andries ML, Perez S (1983) Geometrical features of calcium–carbohydrate interactions. Carbohydr Res 124(2):324–332. doi:10.1016/0008-6215(83)88468-7
24. Beevers CA, Cochran W (1947) The crystal structure of sucrose sodium bromide dihydrate. Proc R Soc London Ser A 190:257–272.
25. Craig DC, Stephenson NC, Stevens JD (1972) An X-ray crystallographic study of β-D-mannofuranose-$CaCl_2 \cdot 4H_2O$. Carbohydr Res 22(2):494–495. doi:10.1016/S0008-6215(00)81309-9
26. Jeffrey GA, Kim HS (1971) The crystal and molecular structure of epinositol. Acta Crystallogr Sect B: Struct Crystallogr Cryst Chem 27(9):1812–1817. doi:10.1107/S0567740871004837
27. Tajmir-Riahi H-A (1988) Interaction of D-glucose with alkaline-earth metal ions. Synthesis, spectroscopic, and structural characterization of Mg(II)- and Ca(II)-D-glucose adducts and the effect of metal-ion binding on anomeric configuration of the sugar. Carbohydr Res 183:35–46. doi:10.1016/0008-6215(88)80043-0
28. Ollis J, James VJ, Angyal SJ, Pojer PM (1978) An X-ray crystallographic study of α-D-allopyranosyl α-D-allopyranoside·CaCl2$\cdot 5H_2O$ (a pentadentate complex). Carbohydr Res 60(2):219–228. doi:10.1016/S0008-6215(78)80029-9
29. Lu Y, Deng G, Miao F, Li Z (2003) Sugar complexation with calcium ion. Crystal structure and FT-IR study of a hydrated calcium chloride complex of D-ribose. J Inorg Biochem 92:487–492. doi:10.1016/S0162-0134(03)00251-4
30. Takashi F, Kazuyuki O, Mitsuru T, Tomoya M (2006) Crystal structure of α,α-trehalose–calcium chloride monohydrate complex. J Carbohydr Chem 25(7):521–532. doi:10.1080/07328300600966414
31. Cook WJ, Bugg CE (1973) Calcium interactions with D-glucans: crystal structure of α,α-trehalose–calcium bromide monohydrate. Carbohydr Res 31:265–275. doi:10.1016/S0008-6215(00)86191-1
32. Guo J, Lu Y, Whiting R (2012) Metal–ion interactions with sugars. The crystal structure of $CaCl_2$–fructose complex. Bull Korean Chem Soc 33(6):2028–2030. doi:10.5012/bkcs.2012.33.6.2028
33. Yang L, Su Y, Xu Y, Wang Z, Guo Z, Weng S, Yan C, Zhang S, Wu J (2003) Interactions between metal ions and carbohydrates. Coordination behavior of neutral erythritol to Ca(II) and lanthanide Ions. Inorg Chem 42:5844–5856. doi:10.1021/ic0300464
34. Yang L, Su Y, Liu W, Jin X, Wu J (2002) Sugar interaction with metal ions. The coordination behavior of neutral galactitol to Ca(II) and lanthanide ions. Carbohydr Res 337(16):1485–1493. doi:10.1016/S0008-6215(02)00130-1

35. Agulhon P, Markova V, Robitzer M, Quignard F, Mineva T (2012) Structure of alginate gels: interaction of diuronate units with divalent cations from density functional calculations. Biomacromol 13(6):1899–1907. doi:10.1021/bm300420z
36. Angyal SJ, Davies KP (1971) Complexing of sugars with metal ions. Chem Commun 500–501. doi:10.1039/C29710000500
37. McGavin DG, Natusch DFS, Young JD (1969) Complexes of sugars with metalions. In: Proceedings of the Xllth International Conference on Coordination Chemistry Sydney 134–5 Science Press Marrickville New South Wales Australia
38. Angyal SJ, Greeves D (1976) Complexes of carbohydrates with metal cations. VII. Lanthanide-induced shifts in the P.M.R. spectra of cyclitols. Aust J Chem 29:1223–1230. doi:10.1071/CH9761223
39. Angyal SJ, Greeves D, Mills JA (1974) Complexes of carbohydrates with metal cations. III. Conformations of alditols in aqueous solution. Aust J Chem 27:1447–1456. doi:10.1071/CH9741447
40. Angyal SJ (1973) Shifts induced by lanthanide ions in the N.M.R. spectra of carbohydrates in aqueous solution. Carbohydr Res 26(1):271–273. doi:10.1016/S0008-6215(00)85054-5
41. Angyal SJ (1972) Complexes of carbohydrates with metal cations. I. Determination of the extent of complexing by N.M.R. spectroscopy. Aust J Chem 25:1957–1966. doi:10.1071/CH9721957
42. Angyal SJ (1981) Conformational nomenclature for five- and six-membered ring forms of monosaccharides and their derivatives. Pure & Appl Chem 53(10):1901–1905. doi:10.1351/pac198153101901
43. Symons MCR, Benbow JA, Pelmore H (1984) Interactions between calcium ions and a range of monosaccharides studied by hydroxy-proton resonance spectroscopy. J Chem Soc Faraday Trans 80:1999–2016. doi:10.1039/F19848001999
44. Angyal SJ (1969) The Composition and conformation of sugars in solution. Angew Chem Int Ed Engl 8(3):157–166. doi:10.1002/anie.196901571
45. Frąckowiak A, Skibiński P, Gaweł W, Zaczyńska E, Czarny A, Gancarz R (2010) Synthesis of glycoside derivatives of hydroxyanthraquinone with ability to dissolve and inhibit formation of crystals of calcium oxalate. Potential compounds in kidney stone therapy Eur J Med Chem 45(3):1001–1007. doi:10.1016/j.ejmech.2009.11.042
46. Kubas K (2012) Glycosides of polyphenols, hydroxyketones and their derivatives. Sythesis and evaluation their complexing ability toward calcium ions. Doctoral dissertation Wroclaw University of Technology (in polish)
47. Frąckowiak A (2010) Solubility, inhibition of crystallization and microscopic analysis of calcium oxalate crystals in the presence of fractions from *Humulus lupulus* L. J Cryst Growth 31(23):3525–3532. doi:10.1016/j.jcrysgro.2010.09.040
48. Das I (2005) In vitro inhibition and dissolution of calcium oxalate by edible plant Trianthema monogyna and pulse Macrotyloma uniflorum extracts. J Cryst Growth 273(3–4):546–554. doi:10.1016/j.jcrysgro.2004.09.038
49. Das I (2004) Inhibition and dissolution of calcium oxalate crystals by Berberis Vulgaris-Q and other metabolites. J Cryst Growth 267(3–4):654–661. doi:10.1016/j.jcrysgro.2004.04.022
50. Sillanpaa AJ, Aksela R, Laasonen K (2003) Density functional complexation study of metal ions with (amino) polycarboxylic acid ligands. Phys Chem Chem Phys 5:3382–3393. doi:10.1039/B303234P
51. Parr RG, Yang W (1995) Density-functional theory of the electronic structure of molecules. Annu Rev Phys Chem 46:701–728. doi:10.1146/annurev.pc.46.100195.003413
52. Suarez D, Rayon VM, Diaz N, Valdes H (2011) Ab initio benchmark calculations on Ca(II) complexes and assessment of density functional theory methodologies. J Phys Chem A 115:11331–11343. doi: 10.1021/jp205101z
53. Head-Gordon M, Pople JA, Frisch MJ (1988) Chem Phys Lett 153:503–506
54. Purvis GD, Bartlett RJ, (1982) A full coupled-cluster singles and doubles model: the inclusion of disconnected triples. J Chem Phys 76:1910–1918. doi: 10.1063/1.443164

55. Zheng YJ, Ornstein RL, Leary JA (1997) A density functional theory investigation of metal ion binding sites in monosaccharides. J Mol Struc (Theochem) 389(3):233–240. doi:10.1016/S0166-1280(96)04707-0
56. Wong CHS, Siu FM, Ma NL, Tsang CW (2001) Interaction of Ca2 + with mannose: a density functional study. J Mol Struc (Theochem) 536(2):227–234. doi:10.1016/S0166-1280(00)00634-5
57. Toczek D, Kubas K, Turek M, Roszak S, Gancarz R (2013) J Mol Model. 19:4209–4214. doi:10.1007/s00894-013-1841-9
58. Fabian WMF (2007) Metal binding induced conformational interconversions in methyl ß-D-xylopyranoside. Theor Chem Acc 117:223–229. doi:10.1007/s00214-006-0130-4
59. Toczek D, Kubas K, Roszak S, Gancarz R (in the preparation)
60. Gaussian 09, Revision A.1(2009), Frisch MJ, Trucks GW, Schlegel HB, Scuseria GE, Robb MA, Cheeseman JR, Scalmani G, Barone V, Mennucci B, Petersson GA, Nakatsuji H, Caricato M, Li X, Hratchian HP, Izmaylov AF, Bloino J, Zheng G, Sonnenberg JL, Hada M, Ehara M, Toyota K, Fukuda R, Hasegawa J, Ishida M, Nakajima T, Honda Y, Kitao O, Nakai H, Vreven T, Montgomery Jr JA, Peralta JE, Ogliaro F, Bearpark M, Heyd JJ, Brothers E, Kudin KN, Staroverov VN, Kobayashi R, Normand J, Raghavachari K, Rendell A, Burant JC, Iyengar SS, Tomasi J, Cossi M, Rega N, Millam NJ, Klene M, Knox JE, Cross JB, Bakken V, Adamo C, Jaramillo J, Gomperts R, Stratmann RE, Yazyev O, Austin AJ, Cammi R, Pomelli C, Ochterski JW, Martin RL, Morokuma K, Zakrzewski VG, Voth GA, Salvador P, Dannenberg JJ, Dapprich S, Daniels AD, Farkas Ö, Foresman JB, Ortiz JV, Cioslowski J, Fox DJ Gaussian, Inc., Wallingford

Chapter 10
Quantum-Chemical Investigation of Epoxidic Compounds Transformation. Application for *In Vitro* and *In Vivo* Processes Modeling

Sergiy Okovytyy

Abstract Transformation of epoxides is a key step for numerous processes important both for synthetic organic chemistry and biochemistry. Since experimental methods are restricted by the fixation of source compounds, intermediates and products of reactions, quantum chemical calculations serve as the only direct approach for prediction of the structure and energy of transition states thus clarifying detailed mechanisms of chemical reactions. This chapter summarizes results of quantum chemical investigation of epoxides transformation mechanisms in alkaline, neutral and acidic environments. Special attention has been paid to stereo- and regiochemistry of the processes, influence of solvation effects and nature of catalytic action of mono- and bidentate acids.

10.1 Introduction

Epoxides play important role in the synthesis of wide row of chemicals and pharmaceutical intermediates [1]. Owing to the presence of heteroatom and high strain energy of three-membered ring (119 kJ/mol [2]) oxiranes possess much higher reactivity if compare to acyclic analogs which results in easy C–O bond rupture [3–6] through where disadvantage of a highly basic leaving group is compensated by significant decreasing of oxirane ring strain during reaction [7]. In addition to nucleophilic opening reactions epoxides can undergo elimination resulted in formation of allylic alcohols [8, 9]. In acidic medium or in the presence of Lewis acids ring opening reaction can be accompanied by rearrangement to carbonyl compounds [10].

S. Okovytyy (✉)
Department of Organic Chemistry, Oles Honchar Dnipropetrovsk National University,
Dnipropetrovsk 49010, Ukrain
e-mail: sokovyty@icnanotox.org

L. Gorb et al. (eds.), *Application of Computational Techniques in Pharmacy and Medicine,*
Challenges and Advances in Computational Chemistry and Physics 17,
DOI 10.1007/978-94-017-9257-8_10

Reactions of epoxides are not restricted by *in vitro* transformations but also have medical relevance since epoxides can be formed via endogenous biochemical pathways or oxidation of xenobiotics, for example, by cytochrome P450 [11–13]. This process is important in drug metabolism, as epoxides formed during the oxidation of drug molecules [14] can undergo ring opening *in vivo* by biological nucleophiles, such as DNA/RNA bases [15, 16], causing damage to cells. Another direction of epoxides transformation vital for organisms is hydrolysis facilitated by epoxide hydrolase (EH) [15, 16]. Epoxides that are poor EH substrates tend to be highly carcinogenic [17, 18]. Thus *in vivo* toxicity of epoxides stated to be correlated with their alkylation rate [19–23].

In this chapter we summarize the results of computational investigation of epoxide ring opening reactions modeling the main features of the processes such as the nature of medium effects, solvent, attacking nucleophile and substitutes on the mechanisms of reactions. Special attention has been paid to study of stereo- and regiochemical peculiarities of the processes; as example of epoxycycloalkanes and spirooxiranes relationship between reactivity of oxiranes and their strain energy has been investigated. In particular, implicit and explicit consideration of solvation effects on epoxide ring opening has been performed. Investigation of the oxiranes transformation in the presence of electrophilic activators includes consideration of the following issues: relation between S_N1 and S_N2 mechanisms, and nature of catalytic action of mono- and bidentate acids.

10.2 Interaction of Epoxides with Anionic Nucleophiles

Among indexes of reactivity of epoxides with nucleophiles the most useful are parameters of electronic density distribution such as atomic charges, C–O bond orders, energy of LUMO and hybridization of atomic orbitals of oxirane ring [24–30]. However it should be noted that afore mentioned parameters of epoxides are not always in a good agreement with the results of kinetic studies [27, 28, 30], especially in those cases when steric effect in the course of reaction overcomes electronic. By these reasons, predictions of reactivity of epoxides has been mostly based on the analysis of potential energy surfaces (PES) of corresponding reactions with localization of transition states (TS) and prereactive complexes and further calculations of activation barriers.

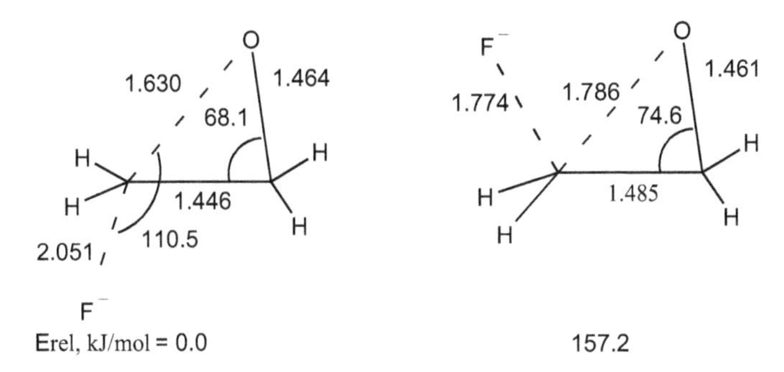

Fig. 10.1 Structure of transition states and values of relative energies (E_{rel}) for reaction of oxirane with fluoride ion [31]

One of the first theoretical study of the nucleophilic bimolecular ring-opening of ethylene oxide has been performed by Fujimoto and co-workers using semiempirical approach as an example of oxirane (**1**) interaction with hydride ion [25]. It has been shown that oxirane contributes to the interaction not only LUMO but also next unoccupied molecular orbital. The authors also stated the preference of reaction with inversion of epoxide carbon atom. Later on *ab initio* study (at the HF/3-21G level) performed by this group confirmed the preference of back-side attack of nucleophile (by fluoride ion) (see Fig. 10.1) [31]. Destabilization of transition state for front-side attack has been explained by strong repulsion of oxygen-atom and fluoride ion due to antibonding overlap of their orbitals and significant deformation of three-membered ring in a tighter transition state.

Investigation of potential energy surface for alkaline hydrolysis of oxirane performed by Lundin and co-workers at B3LYP/6-311+G(d, p) level of theory has also shown that *trans* S_N2 reaction is strongly favored as compare to the corresponding *cis* reaction (see Fig. 10.2) [32].

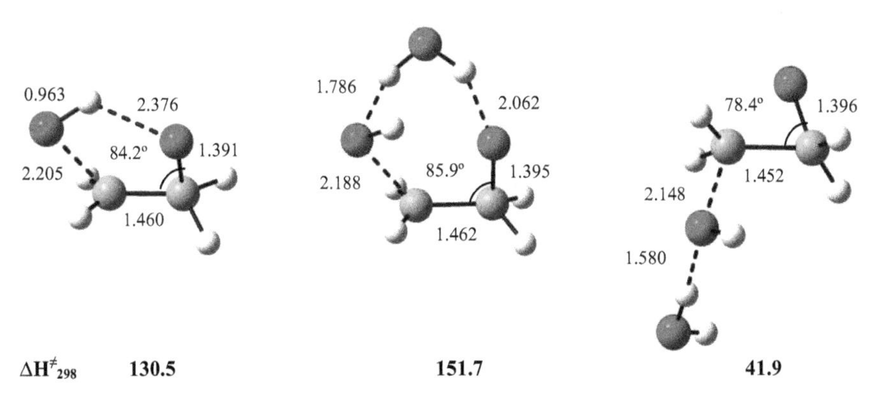

Fig. 10.2 Structure of transition states and values of activation enthalpies ($\Delta H^{\neq}_{298}$, kJ/mol) for alkaline hydrolysis of oxirane. (Adopted from [32])

Table 10.1 Geometrical parameters and values of SKIE calculated at MP2/6-31+G(d) level of theory [33]

Nu	C_1–O	C_1–Nu	C_2–O	C_1C_2O	NuC_1C_2	SKIE
H^-	1.654	2.148	1.436	70.2	106.7	1.058
NH_2^-	1.708	2.307	1.423	73.1	113.5	1.011
OH^-	1.753	2.129	1.420	75.4	112.1	0.981
F^-	1.831	1.922	1.413	79.7	111.7	0.940
SH^-	1.861	2.463	1.402	81.3	110.9	1.049
Cl^-	1.988	2.212	1.388	88.2	113.6	0.999
NH_3	2.075	1.763	1.361	93.2	114.8	0.916

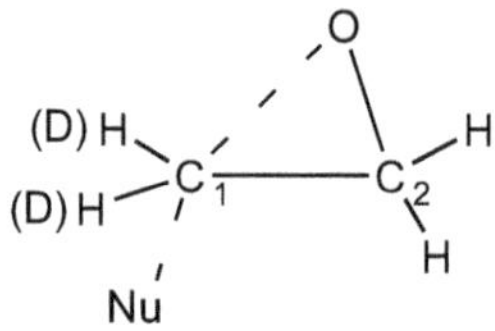

Interesting from methodological point of view investigations have been performed by Glad and Jensen [33]. They have calculated deuterium α-secondary kinetic isotope effects (SKIE) as an example of oxirane interaction with nucleophiles and compared them with optimized transition state geometries (Table 10.1).

In the row of H^-...NH_3 when nucleophilicity decreases transition states becomes to have more pronounced late character. All calculated at MP2/6-31+G(d) level values of SKIE lie between 0.91 and 1.06 in absolute value, typical for an S_N2 reaction [34]. Analysis of changes of SKIE as function of broken C_1–O bond length confirms relationship between TS geometry and KIE. For second-row nucleophiles and hydride ion a difference in SKIE of 0.02 (a typical experimental uncertainty) corresponds to geometry changes at the TS of ~0.03 Å in bond length (R_{CO}) and ~3^0 difference in torsional angle (Θ_{HCCO}), i.e. isotope effects are quite sensitive to the TS geometry.

To discuss the early/late and loose/tight features of the TSs, degrees of forming and breaking of bonds ($n^{\neq}$) have been calculated according to (Eq. 10.1)

$$n^{\neq} = \exp\left[(r^0 - r^{\neq})/0.6 \right], \quad (10.1)$$

where r^0—bond length in oxirane, $r^{\neq}$—corresponding parameters in transition state.

Analysis of two-dimensional More O'Ferrall-Jencks diagram, drawn as function of degrees of formation of C–Nu bond on $(1-n^{\neq}_{C1-O})$ value, has shown that the main determining factor for the SKIE is the loose/tight feature of the TS rather than the early/late position, i.e. the lower KIE for the more product-like TSs is primarily due to the increased tightness.

Table 10.2 Values of E_{act} (kJ/mol) for epoxide ring opening reactions[a]

Epoxide	Nucleophile	Level of theory	E_{act}	
			C_1	C_2
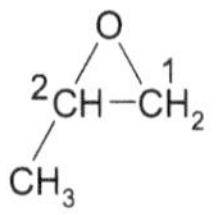 **2**	OH^- [7]	HF	58.6	70.7
		MP2	39.3	50.2
	$HCOO^-$ [35]	B3LYP	64.6	76.5
		MP2	83.3	97.3
		MP4(SDQ)//MP2	83.0	-
		B3LYP//HF + IPCM	68.6	79.3
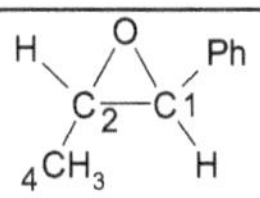	CH_3COO^- [36]	HF	136.4	120.5
		B3LYP	83.7	87.9
		MP2//HF	86.6	88.7
		COSMO+B3LYP//B3LYP	125.1	136.0

[a] Calculation of E_{act} values has been performed relatively to prereaction complexes using 6–31+G** basis set for geometry, optimized with 6–31G* basis set for OH^- and 6–31+G** for the rest cases

Regiochemistry of basic-catalyzed epoxide ring opening has been studied for the case of interaction of methyloxirane with hydroxyl- [7] and formiat-anion [35], and 1S,2S-*trans*—2-methylstyrene oxide (tMSO) with acetate [36].

According to Krasusky rule in case of methyloxirane (**2**) more energetically preferable is attack of nucleophile on sterically more accessible primary carbon atom of epoxide ring. In case of tMSO lower activation barrier corresponds to phenyl side attack by acetate (Table 10.2).

As could be seen from Table 10.2, Hartree-Fock approach significantly (by almost 50 %) overestimates E_{act} values if compared to that predicted by correlated approaches such as B3LYP and MP2. Forth-order Meller-Plesset perturbation theory at MP4(SDQ) level predicts virtually the same values of activation energy [35].

Reaction of methyloxirane with strong nucleophile OH^- is characterized by lower activation barriers if compare to formiate; transtition state in the latter case has rather late nature. Transition states corresponding to nucleophile attack on the primary carbon atom of epoxide cycle for both nucleophiles are tighter (Fig. 10.3).

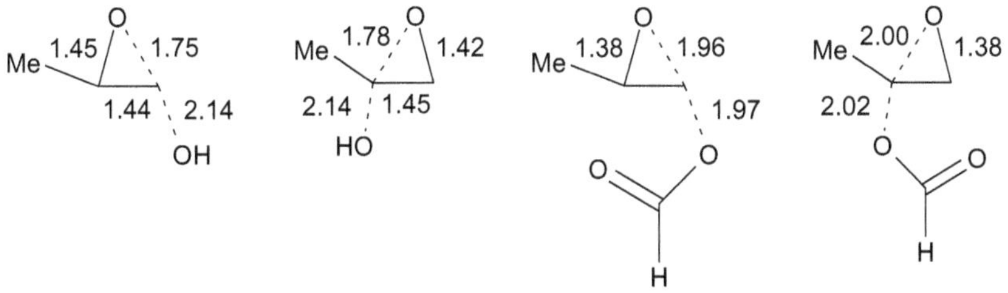

Fig. 10.3 Structure of transition states of epoxide ring opening of methyloxirane (**2**) with hydroxyl anion (MP2/6-31G* level of theory [7]) and formiate (B3LYP/6-31G* [35]) level of theory)

To assess the effect of ring strain Gronert and co-workers have compared interaction of HO^- with methyloxirane and corresponding acyclic analog $CH_3CH_2OCH_3$. According to calculations in the case of methylethyl ether activation barrier is by 87.0 kJ/mol higher if compared to that for methyloxirane so from ~113.0 strain energy of oxirane ring ~75 % is released at the transition state [7].

A specific group of epoxy derivatives consists of compounds in which the oxirane ring is fused to an alicyclic fragment (**3–9**). Kinetic study of alkaline methanolysis reaction for this row of compounds (hexane-1 oxide **10** has been studied for comparison) has shown the relation between their strain and reactivity [27, 37]. It has also been shown that stereoisomeric epoxynorbornanes possess isomers having different reactivity: *exo*-isomer (**8**) is stable to sodium methoxide ($[MeO^-]=6.21$ M) for 20 h at 60° C, whereas under the same conditions *endo*-isomer (**7**) undergoes slow methanolysis at a rate of 0.036 l mol^{-1} s^{-1} [27].

O α O O O O O O O

3 **4** **5** **6** **7** **8** **9** **10**

As has been shown in [27], increase in the strain energy of epoxycycloalkanes is accompanied by increase in the LUMO energy and the order of the C–O bond (Table 10.3). Obtained therein parameters of electron density distribution evidence that in the case of epoxides (**3–5**) and epoxybicycle[2.2.2]octane (**9**) values of

Table 10.3 Calculated parameters of oxiranes (**1, 3–10**) and relative values of rate constants (k_{rel}) for their alkaline methanolysis reaction [27]

Epoxide	Strain energy (kJ/mol)[a]	E_{LUMO} (eV)[b]	Angle α (deg.)[b]	C–O bond order[b]	k_{rel}
1		2.3862		0.9705	–
3	224.86	2.1415	90.6	0.9650	–
4	134.61	2.3628	108.3	0.9603	0.41
5 C_1–O	126.78	2.3591	120.8	0.9583	1.00
C_2–O				0.9595	
6	142.68	2.4061	121.6	0.9605	15.5×10^{-3}
7	206.32	2.3392	104.5	0.9654	Does not react
8	196.83	2.4188	104.6	0.9638	Does not react
9	173.91	2.4857	111.0	0.9630	5.52×10^{-3}
10		2.3628	–	0.9703	5.96

[a] Method MM2E [38]
[b] Method PM3 [39]

E_{LUMO} and C–O bond orders are in a good agreement with the kinetic study [27], which reveals the following order of reactivity changes:**5**>**4**>**8**>**9**. At the same time, results of calculation of bond orders and values of E_{LUMO} do not reflect any particularities for methanolysis of hexane-1 oxide (**10**) and steroisomeric epoxynorbornanes (**7, 8**). In fact, for epoxide (**10**), which is characterized by the highest reactivity, calculations predict the highest values of breaking C–O bond order (0.9703). For epoxides (**7, 8**) this parameter is equal to 0.9654 and 0.9638 correspondingly.

Using semiempirical method PM3 transition states and corresponding prereaction complexes have been localized for interaction of epoxides (**1**, **3–10**) with methoxy-anion modeling gas-phase conditions and taking into account solvent effects using macroscopic and supermolecular approach with explicit consideration of solvent (methanol) molecules [30]. Activation barriers calculated for reaction *in vacuo* (Table 10.4) are in a good agreement with such characteristics of epoxides as strain energy and the C–O bond orders: increasing strain in the alicyclic fragment is accompanied by increase in the enthalpy of activation.

Among geometric parameters of transition states the most illustrative are the bond angles O_1CO_2 (β) and O_1CCO_2 (γ), where O_1 is the oxirane oxygen atom, and O_2 is the oxygen atom of methoxide ion. The first of these angles decreases as $\Delta H^{\neq}$ rises: from 167.9° for oxirane (**1**) to 144.0° for the most strained endo-epoxynorbornane (**7**):

$$\Delta H^{\neq} = -1.35 \times \beta + 331.72; \quad r = 0.95 \quad n = 10.$$

An analogous correlation is observed between $\Delta H^{\neq}$ and γ. The existence of such correlations is closely related to the S_N2 character of the reaction under study: the corresponding angles in a classical bimolecular substitution reaction approach 180°. Deviation from this value reduces overlap of molecular orbitals, and the activation barrier increases (Fig. 10.4).

Epoxy compounds (**4–6, 9, 10**) show a satisfactory correlation between the calculated values of $\Delta H^{\neq}$ (Table 10.4) and logarithms of the rate constants given in [2]:

$$\lg k_{rel} = -0.31 \times \Delta H^{\neq} + 37.42; \quad r = 0.95 \quad n = 5.$$

However, the calculations performed for the gas phase incorrectly predict greater reactivity of exo-epoxynorbornane (**8**) relative to its endo-isomer (**7**). This may be due to underestimation of steric factor whose contribution considerably increases in reactions of epoxy derivatives with solvated methoxide ion. Taking into account that experimental data on alkaline methanolysis of epoxycycloalkanes were obtained in methanol which is a fairly polar solvent capable of forming hydrogen bonds, we performed a theoretical study of the solvent effect on the process.

Transition states, optimized at macroscopic approximation using the COSMO procedure [40] are characterized by lower degree of O_2–C bond formation and greater degree of the C–O_1 bond cleavage, as compared with the gas-phase calculations, i.e. the transition states are looser in the former case. In this case, the endo isomer of epoxynorbornanes turns out to be more reactive; however, the variation of

Table 10.4 Calculated geometric parameters of transition states for opening of the oxirane ring and corresponding activation barriers for compounds (**1, 3–10**) [30]

Compound no.	Bond length (Å)		Angle (deg.)			ΔH≠ (kJ/mol)
	CO_1	CO_2	CCO_1	β	γ	
In vacuo						
1	1.740	1.995	74.5	167.9	175.6	108.74
3	1.779	2.071	76.1	147.8	−157.6	133.55
4	1.775	2.005	76.6	150.9	−158.5	129.91
5[a]	1.787	2.000	76.8	152.5	159.6	120.58
6	1.837	2.036	79.3	148.8	162.8	134.08
7	1.854	2.058	80.3	144.0	−142.7	141.24
8	1.876	2.064	81.5	144.4	−144.9	135.60
9	1.867	2.043	81.2	148.0	−148.5	136.20
10	1.737	1.982	74.1	159.3	175.1	115.23
Macroscopic approximation						
1	1.772	2.065	76.1	154.7	179.5	128.24
3	1.841	2.193	79.0	142.3	−153.1	149.66
4	1.832	2.142	79.1	145.8	−157.6	156.98
5[a]	1.826	2.101	78.7	148.1	−151.7	150.37
6	1.873	2.124	81.3	147.8	−151.5	127.32
7	1.893	2.155	82.2	139.3	−137.2	169.74
8	1.981	2.292	87.0	137.0	−136.8	172.05
9	1.903	2.152	82.8	144.0	−144.5	132.97
10	1.775	2.014	75.9	159.5	176.1	149.62
Supermolecular approximation						
1	1.862	1.991	81.3	156.0	179.6	141.46
3	1.950	2.088	85.5	147.5	16.1	179.79
4	1.890	2.046	82.2	145.0	157.8	163.43
5[a]	1.892	2.030	82.3	143.3	155.5	188.28
6	2.046	2.298	91.1	143.7	138.8	205.23
7	2.078	2.205	92.8	134.5	124.3	214.85
8	2.015	2.190	89.4	138.4	134.3	198.76
9	1.871	1.993	81.1	158.1	174.8	148.82

[a] For the reaction at C_1 (favorable attack); the value of ΔH≠ for gas-phase attack on C_2 is equal to 129.12 kJ/mol

the activation barrier is as small as 2.38 kJ/mol, which is not quite consistent with the experimental data [27]. In addition, the ΔH≠ values for compounds (**9, 10**) are considerably underestimated.

The supermolecular approximation provides a more appropriate description of specific solvation and reaction mechanisms, where solvent molecules are considered to be reagents. Successful application of the supermolecular approach requires

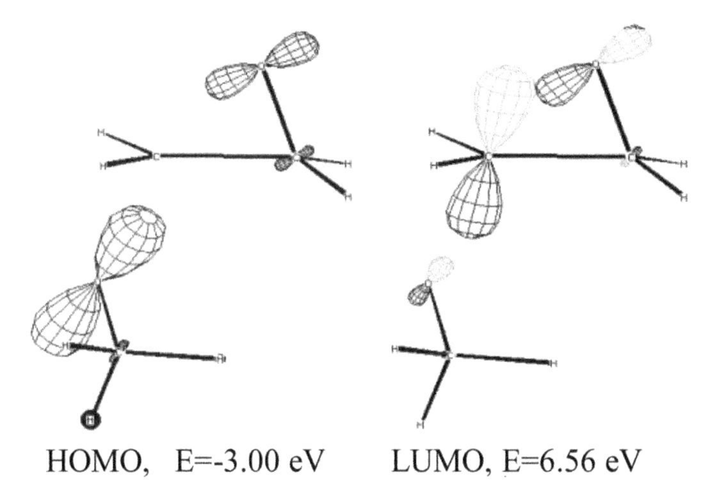

Fig. 10.4 Structure and energies of frontier molecular orbitals of the transition state in alkaline methanolysis of oxirane, calculated by PM3 approach

that the necessary and sufficient number of solvent molecules be adjusted. While studying the reaction of oxirane (**1**) with methanol [41], the most acceptable results were obtained when the first solvate shell of methoxide ion contained four solvent molecules (methanol) and the less basic oxirane oxygen atom was solvated with one methanol molecule, i.e., the model $CH_3O_3(CH_3OH)_n$ + oxirane$(CH_3OH)_m$ (n = 4, m = 1) has been used. The same model has been applied to examine alkaline methanolysis of epoxy derivatives (**3–10**).

Localized therein transition states have more pronounced loose character if compare to calculations *in vacuo* and using COSMO approximation (Table 10.4). Calculated values of $\Delta H^{\neq}$ are in good agreement with logarithms of the experimental relative rate constants.

$$\lg k_{rel} = -0.14 \times \Delta H^{\neq} + 23.14 \quad r = 0.96 \quad n = 5$$

It should be noted that the activation barrier for the reaction of *endo* isomer (**7**), calculated in the supermolecular approximation, is lower by 9.62 kJ/mol than the corresponding barrier calculated for *exo* isomer (**8**). Comparison of these results with those obtained by *in vacuo* calculations shows that steric factor is actually determinative for the reactivity of strained epoxynorbornanes (**7, 8**). Only this factor is taken into account in terms of the supermolecular approach, where solvent molecules are included in the explicit form. As a result, the effective volume of the reagent considerably increases (Fig. 10.5).

Such supermolecular model has been successfully applied for investigation of chemo-, region- and stereoselectivity of dicyclopentadiene diepoxide (**11**) alkaline methanolysis [42, 43] and reaction of spirooxiranes (**12–19**) with methoxy anion [44].

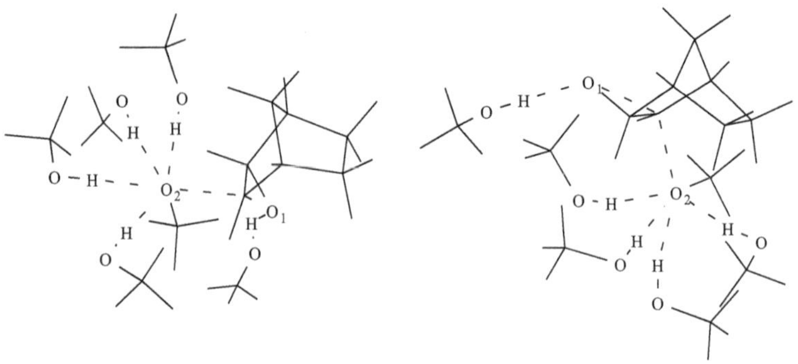

Fig. 10.5 Structure of transition states in the methanolysis of stereoisomeric epoxynorbornanes (**7, 8**), calculated in the supermolecular approximation (n=4, m=1)

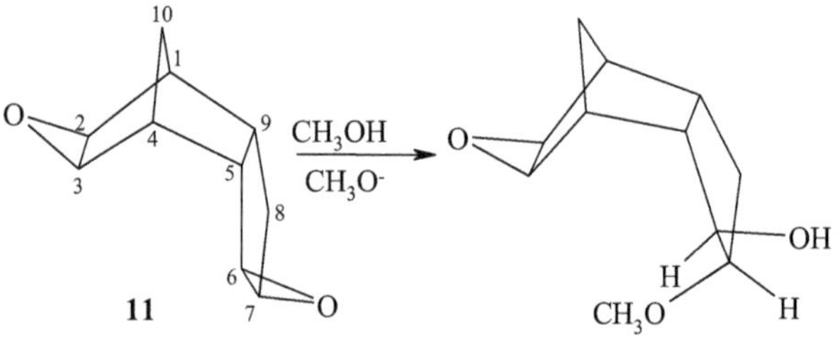

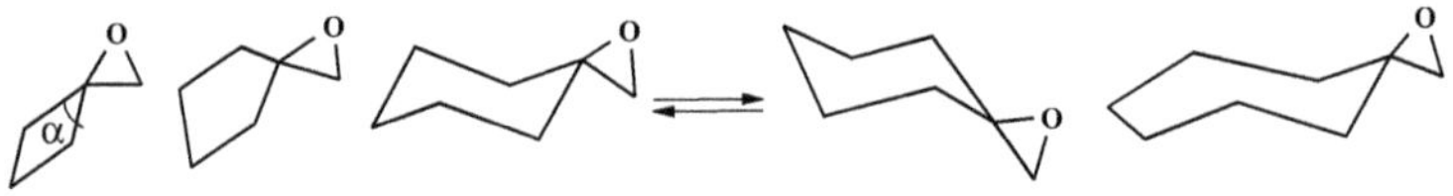

	12	13	14A	14B	15
$\Delta H^{\neq}_{supermolecular}$, kJ/mol [44]	124.47	139.08	137.82	147.82	138.53
K_{rel} [37]	9.19	-	5.50	-	3.13

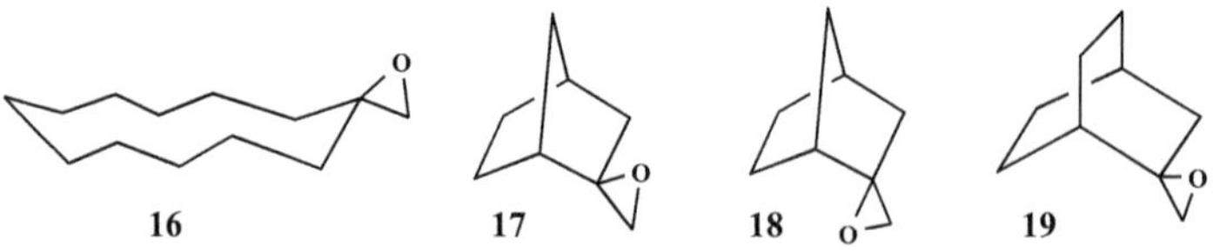

	16	17	18	19
$\Delta H^{\neq}_{supermolecular}$, kJ/mol [44]	140.12	149.79	136.65	151.04
K_{rel} [37]	1.00	-	5.75	2.63

Calculated values of $\Delta H^{\neq}$ for epoxides (**12, 14, 15, 18, 19**) are in a good agreement with kinetic studies:

$$\lg k_{rel} = -16.86 \times \Delta H^{\neq} + 163.94, \quad r = 0.90, \quad n = 5$$

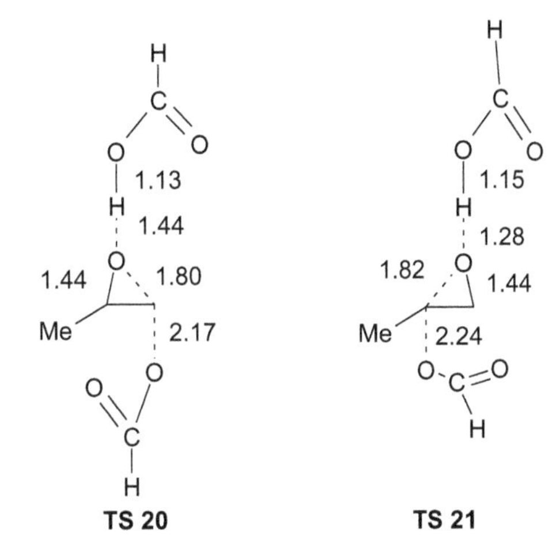

Fig. 10.6 Structure and some geometrical parameters (Å) of transition states for opening of methyloxirane epoxide ring by formiat-anion in the presence of formic acid as catalyst (B3LYP/6-31G* level of theory [45])

In the row of conformationally flexible spirooxiranes (**14–16**) conformers with pseudo equatorial orientation of methylene group (conformer A) possess higher reactivity. Worthwhile mentioning, that no one calculated parameter of initial epoxides as well as *in vacuo* $\Delta H^{\neq}$ values, does not correlate with experimental kinetic results [44].

In vacuo reactions of formiat-anion with complex "methyloxiran—formic acid", which correspond to nucleophile attack of primary (**TS 20**) and secondary (**TS 21**) carbon atoms of epoxide ring are characterized by E_{act} values equal to 28.0 and 36.0 kJ/mol correspondingly (calculated at B3LYP/6-31+G(d, p) level of theory) (see Fig. 10.6) [45]. Taking into account solvation effects at B3LYP/6-31+G(d, p)/IPCM-HF/6-31+G(d, p) level results in E_{act} values equal to 40.5 and 56.3 kJ/mol. Thus, activation of epoxidic ring in methyloxirane by formic acid does not change regiochemistry which corresponds to Krasusky rule. Acid catalysis decreases activation barriers for alternative reactions by 37 and 40 kJ/mol for reaction *in vacuo* and by 28 and 23 kJ/mol for reaction in solution.

Adding of phenol as catalyst of trans-methylstirole epoxide acidolysis also leads to decreasing of $E_{акт}$ by about 40 kJ/mol if compared to uncatalyzed reaction [36]. Calculated at MP2/6-31+G(d, p)//HF/6-31+G(d, p) and B3LYP/6-31+G(d, p)//HF/6-31+G(d, p) levels of theory values of E_{act} are equal to 43.5 and 38.5 kJ/mol respectively, for attack on benzilic and 47.3 and 45.2 kJ/mol, respectively, for attack on the secondary carbon atoms.

Modeling the inhibition activity of peptides and peptidomimetics containing epoxide ring against the cysteine protease Helter and co-workers explored the potential energy surface for interaction of oxirane (**1**), α,β-epoxy carbonyl compounds (**22, 23**) with methylthiolate-anion at BLYP/6-311+G(d) and BLYP/TZV+P levels of theory [46, 47].

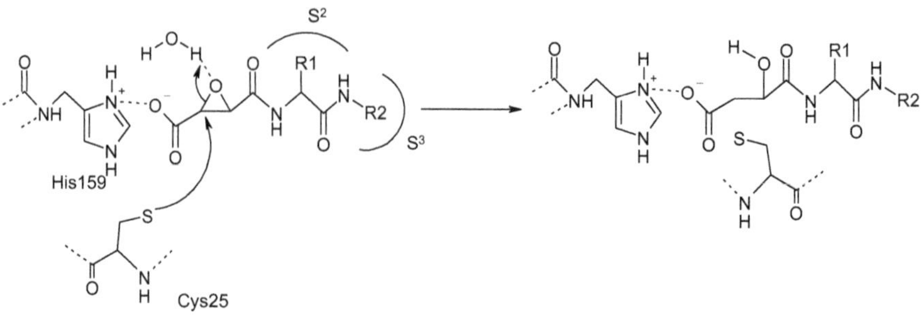

Fig. 10.7 Proposed inhibition mechanism of cysteine protease inhibitors containing three-membered heterocycles (His159, Cys25: active site diad, papain numbering) [47]

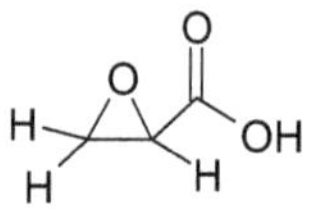

22 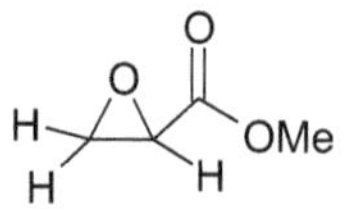 23

Their inhibition potency of peptides and peptidomimetics is usually characterized by two inhibition constants (Eq. 10.2, minimal two-step mechanism for irreversible enzyme inhibition where E = enzyme, I = inhibitor, EI = noncovalent enzyme-inhibitor complex, E–I = inactivated enzyme): the dissociation constant Ki, characterizing the preliminary reversible complexation step, and the first-order rate constant of inhibition ki, ascribing the rate of irreversible enzyme alkylation (Fig. 10.7) [46].

$$E + I \underset{}{\overset{Ki}{\rightleftarrows}} EI \xrightarrow{Ki} E - 1 \tag{10.2}$$

In model system, the attacking cysteine is mimicked by a methyl thiolate while oxirane was considered as inhibitor.

The effect of a decreasing pH value on the reaction profile was captured by a series of model systems in which solvent molecules with increasing proton donor ability are placed in the vicinity of the heteroatom of the oxirane and in the vicinity of the methyl thiolate. Water molecules were employed to mimic environments with weak proton donor ability (pK_a = 15.74), while NH_4^+(pK_a ≈ 9.3) and HCO_2H (pK_a ≈ 3.8) molecules were used to simulate environments with higher proton donor abilities (Fig. 10.8). According to calculations the electrophylic activation of oxirane ring by water molecule decreases activation barrier by 10.46 kJ/mol, while for a stronger proton donor such as $NH_4^+$and HCO_2H this effect is more pronounced (activation barrier decrease is 15.06 and 20.08 kJ/mol, respectively). There is one more reaction pathway found on the potential energy surface for reaction in the presence of formic acid, which corresponds to interaction of activated oxirane with compex "methyl mercaptan—formiat anion". This pathway is characterized by significantly higher activation barrier (71.13 kJ/mol). Obtained results explain the experimental observations of Meara and Rich [48] who found a strong decrease of

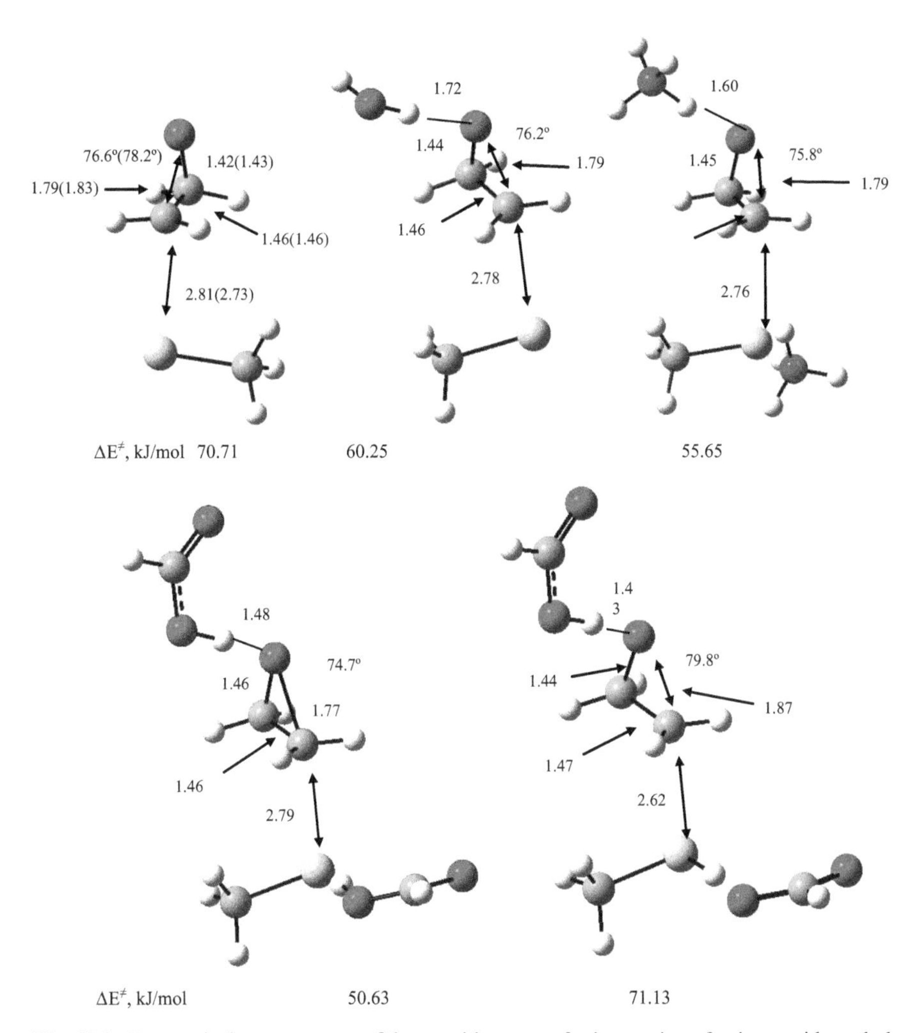

Fig. 10.8 Geometrical arrangements of the transition states for interaction of oxirane with methyl thiolete with explicit consideration of proton donors H_2O, $NH_4^+$and HCO_2H and corresponding values of activation barriers calculated at COSMO/BLYP/6-311+G(d) level of theory (the geometrical values obtained if only bulk effects of water as solvent are considered (given in parenthesis)) [46]

*k*i at decreasing pH values and assigned it to the protonation of an acidic group possessing pK_a of about 4.

Comparison of activation energy values for interaction of methyl thiolate with oxirane and epoxides (**24, 25**) shows decreasing of reaction rate for epoxyacid (Fig. 10.9) [47], while increasing of inhibition activity has been found experimentally for acid-substituted oxiranes [48]. According to [47] increased activity of epoxide (**24**) arises from interaction of carboxylate with imidazolium ion stabilizes the noncovalent enzyme inhibitor complex and thus improves Ki.

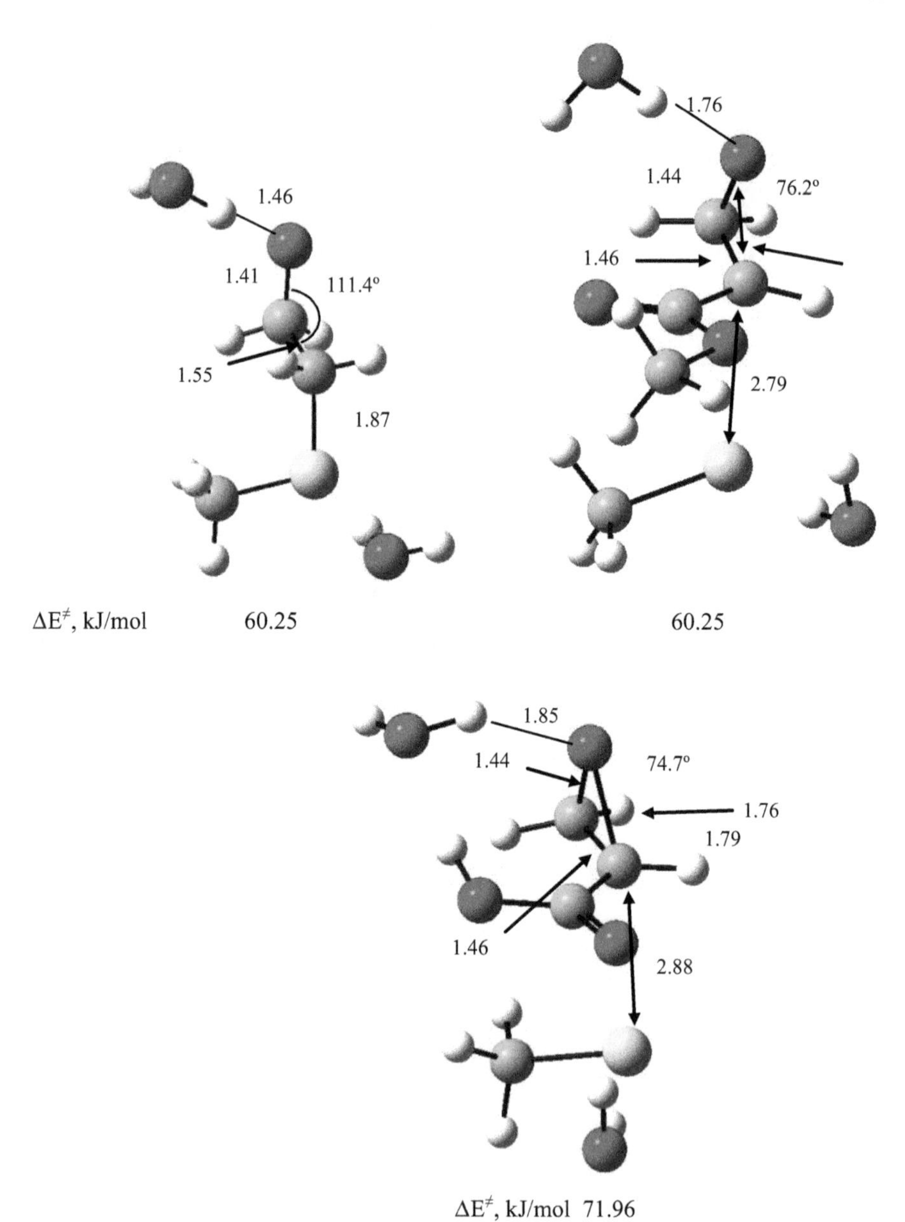

Fig. 10.9 Geometrical arrangements of the transition states for interaction of oxiranes (**1, 24, 25**) with methyl thiolate with explicit consideration of water molecule and corresponding values of activation barriers calculated at COSMO/BLYP/TZV+P level of theory COSMO/BLYP/TZV+P [47]

10.3 Transformation of Oxiranes in Neutral Environment

Among uncatalyzed reactions of oxiranes in neutral environment theoretical investigations mostly focused on interaction with HF [49], NH_3 [33, 50–52], H_2O [32, 53], N-(3-chlorophenyl)piperazine molecules [54]. In early study Alagona and co-workers optimized geometry of transition states for alternative pathways of HF interaction with oxirane with the retention and inversion of carbon atom [49]. According to their calculations the activation energy for pathway with inversion of configuration is equal to 136.6 and 141.7 kJ/mol at CI/STO-3G and HF/4-31G levels of theory. Reaction with retention of configuration is characterized by significantly lower values of activation energy (67.5 and 84.0 kJ/mol). The authors concluded that significant decreasing of activation energy observed at CI level could be explained by pronounced contribution of excited states to wave function of transition states.

Mechanism of *in vacuo* oxirane aminolysis has been studied by numerous quantum-chemical approaches (MINDO/3 [50, 52], B3LYP/6-31G(d) [51], HF/6-31+G(d) [33] and MP2/6-31+G(d) [33]). In all cases transition states are localized close to typical S_N2-type transition states where nucleophile is located in the plane of oxirane ring and bond angle between forming and breaking bonds is close to 180° (Fig. 10.10).

It should be noted that in spite of some difference in transition states geometries, all used approached give close values of E_{act}. In contrast to reaction with HF the alternative reaction with front-side approach of ammonia molecule to oxirane energetically is less favorable (E_{act} = 200.7 kJ/mol (B3LYP/6-31G(d) [51])).

More complicated aminolysis reaction has been studied theoretically for cyclohexadiene monoepoxide interaction with N-(3-chlorophenyl)piperazine [54]. Both pathways corresponding to front- and rear-side attacks have been found. The first pathway is one-step reaction where breaking of C–O bond, forming of C–N bond and proton transfer between N and O atoms occur simultaneously (Fig. 10.11). In contrast, at the first step of rear-side attack zwitterionic intermediate is formed

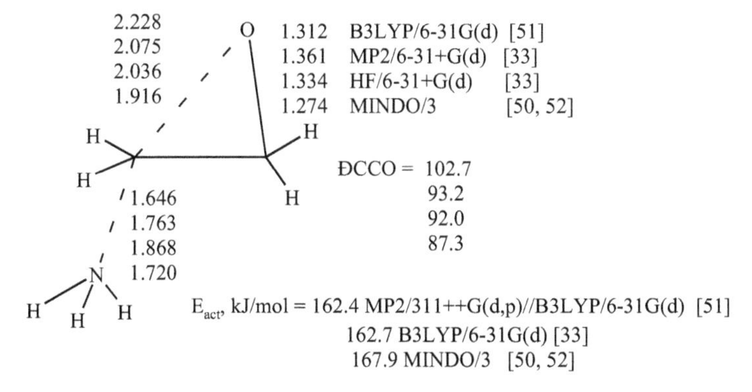

Fig. 10.10 Structure of transition state for oxirane aminolysis and corresponding values of activation energy (E_{act})

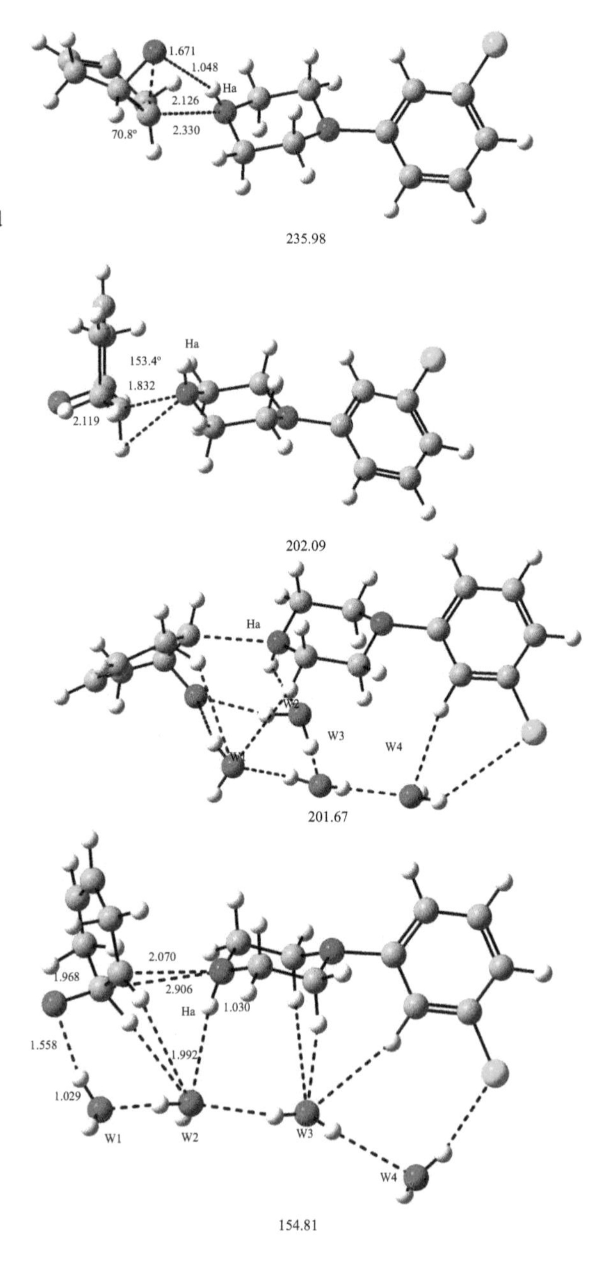

Fig. 10.11 Optimized structures and relative free energies for the pathways of the neat and water-catalyzed mechanism, calculated at the B3LYP/6-311++G(d, p) level (bond lengths in Å and energies in kJ/mol as adopted from [54])

under the neat conditions and another amine molecule is required for intermolecular proton transfer on the next step. Comparison of activation parameters shows the preference of rear-side attack. In order to model process of nucleophilic opening of epoxide at the "oil" droplet/water interface Zheng and Zhang have studied aforementioned reaction with explicit consideration of four-water cluster which consid-

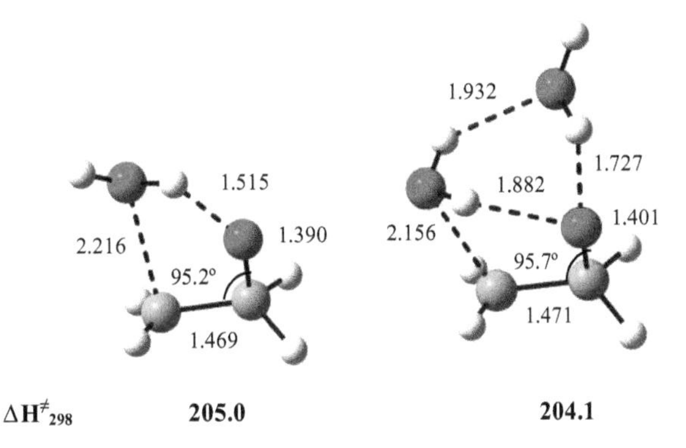

Fig. 10.12 Geometrical parameters of transition states for neutral hydrolysis of oxirane and corresponding values of activation enthalpies (kJ/mol) calculated at B3LYP/6-311++G(d, p) level of theory

ered as a key fragment in the reaction surface. On the contrary to reaction under neat conditions, both front- and rear-side pathways therein are one-step processes at that water assistance results in decreasing of $\Delta G^{\neq}$ values on 8.2 and 11.1 kJ/mol correspondingly (Fig. 10.11). The performed NPA analysis revealed that the water cluster accelerated the reaction not only by assisting the proton transfer, but also by strengthening both the entering and the leaving groups through a charge-transfer process induced by different strengths between the two proton-transfer processes. The enhancement of the entering- and leaving-group effects were qualitatively supported by the evaluation of the nucleophilicity index and the stabilization energy, respectively [54].

For neutral hydrolysis of oxirane Lundin and co-workers have been investigated two mechanisms, one where OH^- and $H^+$are formed as a results of hydrolysis reaction, and a second where the heterolytic decomposition of the epoxide occurs in concert with protolysis of water which is the limiting bare reaction [32]. In contrast to aminolysis reactions there is no transition state on potential energy surface which correspond to *transoid* hydrolysis and formation of ions OH^- and H^+. Transition state for *cisoid* opening for bare reaction is shown in Fig. 10.12, adding a second water molecule does not effect on activation energy since, both transition states are associated with the protolysis of water. Both reactions in neutral environment are characterized by significantly greater activation enthalpy than those under acidic and alkaline conditions, which agrees well with experimental results [55–59].

Comparison of activation enthalpies for neutral hydrolysis of substituted oxiranes (propene oxide and butane oxides) [53] clearly demonstrates the preference of Beta pathways corresponding to the attack of water molecule on more substituted carbon atom (Fig. 10.13). Activation enthalpies lowering has been ascertained in the series: ethene oxide > trans-2-butene oxide ≈ cis-2-butene oxide ≈ propene oxide > isobutene oxide.

A special role in the chemistry of epoxidic compounds have 2-oxabicyclobutane. Its derivatives have been postulated as intermediates in various thermal, photochemical and chemical reactions which have as a final product different aldehydes

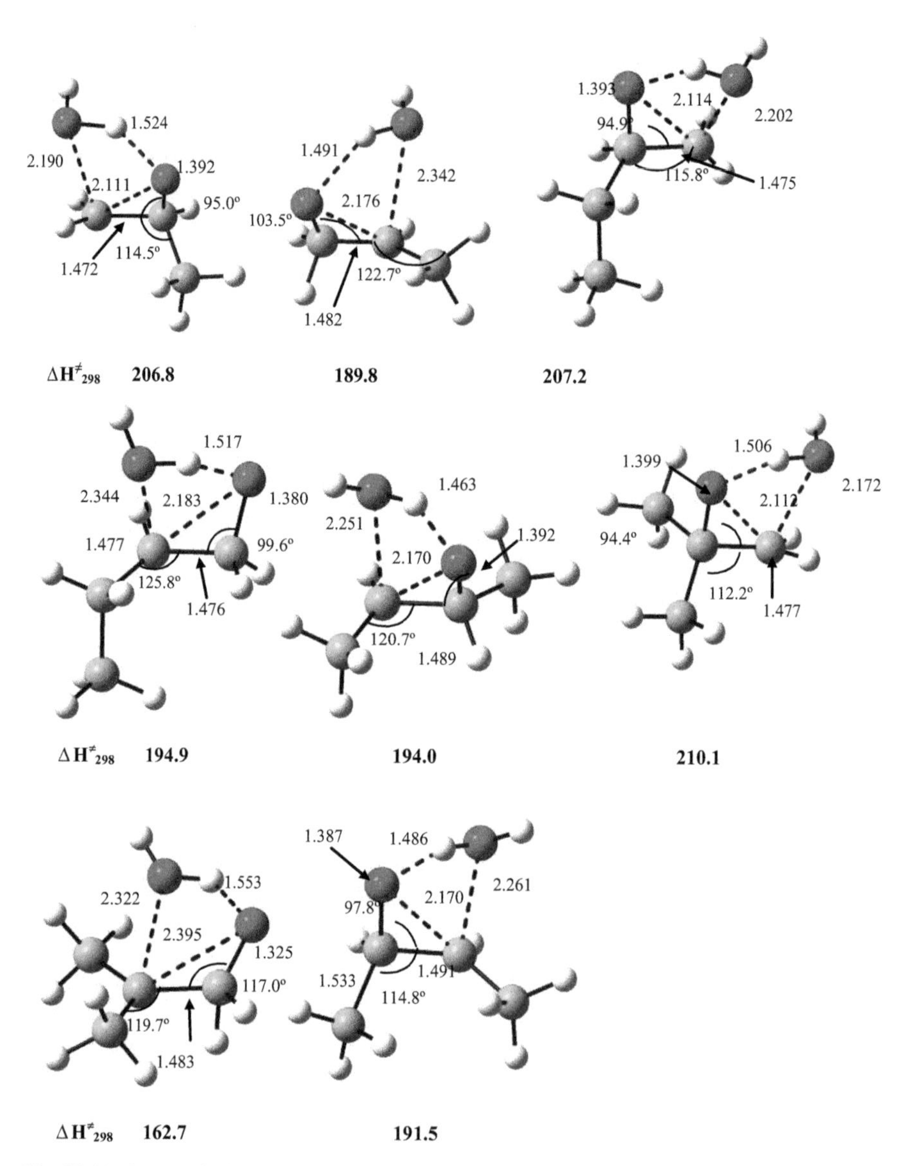

Fig. 10.13 Geometrical parameters of transition states for neutral hydrolysis of oxiranes and corresponding values of activation enthalpies (kJ/mol) calculated at B3LYP/6-311++G(d, p) level of theory. (Adapted from [53])

and ketones. However, all attempts to isolate or even spectroscopically detect oxabicyclobutanes were unsuccessful.

The analysis of the potential energy surface for unimolecular fragmentation of 2-oxabicyclobutane at CASSCF(10,10)/6-31G(d) and UQCISD(T)/6-31G(d) levels of theory has shown that reaction may occur through two pathways (see Fig. 10.14)

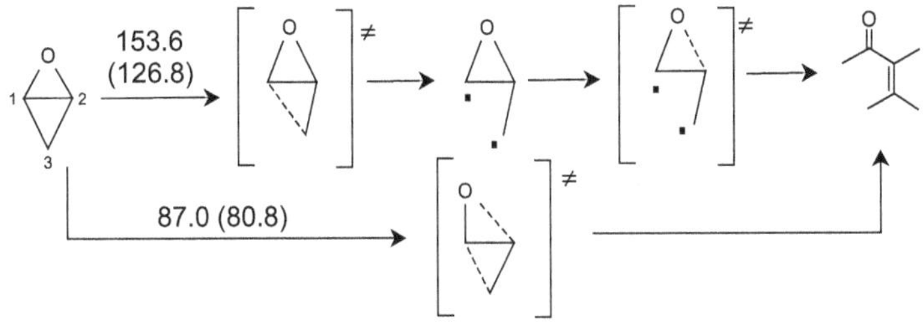

Fig. 10.14 Pathways for transformation of 2-oxabicyclobutane and values of activation barriers for rate-limiting stages, calculated at UQCISD(T)//6-311++G(d, p)//UQCISD/6-31G(d) and CASSCF(10,10)/6-31G(d) (in parenthesis), kJ/mol

[60]. The first one is an asynchronous concerted transformation with a prior breaking of the C_2–O bond of the epoxidic cycle in the transition state. The second mechanism involves stepwise nonconcerted transformation with breaking during the first (rate-determining) stage of the C_1–C_3 bond and the formation of the biradical intermediate which transforms to acrolein with a very small barrier. As could be seen from Fig. 10.14, transformation of 2-oxabicyclobutane through both pathways is characterized by high values of activation barriers thus one may conclude that interaction of cyclopropene with epoxidation reagents most probably proceeds via routes which exclude formation of epoxycyclopropanes as intermediates [60].

10.4 Transformation of Epoxides in Acidic Environment

A number experimental and theoretical studies confirm significant increasing of epoxides reactivity in the presence of electrophylic catalysts [27, 61–64]. Along with increasing of reaction rate the activation of epoxide ring leads in some cases to alteration of stereo- and regiochemistry of the processes.

The simplest system modeling reaction in the presence of electrophylic catalyst is the nucleophiles interacting with protonated epoxide. Exploring equilibrium structures “oxirane-proton” using CNDO/2 approach Kretov and co-workers have located stable forms, which correspond to O-protonation and proton coordination of C–C bond [29, 65, 66]. More precise analysis of potential energy surface for $C_2H_5O^+$ system at HF/4-31G has shown, that O-protonated oxirane is the only structure which corresponds to minimum. Protonation at C–C and C–O bonds leads to barrierless transformation to isomeric structures [67]. O-protonated form of oxirane also has been confirmed at MP2 [68–70], CCSD [69, 70] and DFT levels [60, 70]. Among mono- and dimethyl substituted oxiranes analyzed in [60, 70] 2-methyl-1,2-epoxypropane has been found to be a challenging problem for density functional theory. Numerous functionals including popular B3LYP fail in predicting the structure of protonated 2-methyl-1,2-epoxypropane while the functionals M05 and M05-2X recently proposed by Truhlar and co-workers give a good correspondence with CCSD and MP2 results (Fig. 10.15) [69].

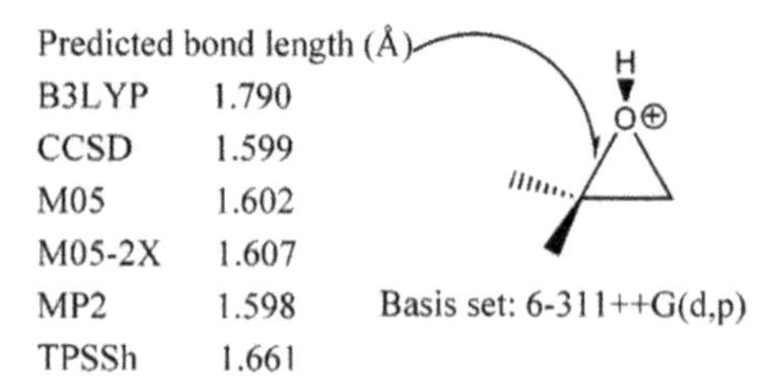

Fig. 10.15 Structure of protonated 2-methyl-1,2-epoxypropane [69]

In contrast to neutral oxiranes molecule their protonated form characterized by shorter C–C and remarkable longer C–O bond length [68–70]. In addition, due to increasing of π-character of protonated epoxides flattening of hydrogen or substitutes part of molecules takes place [27, 68–70], which results in lighter access of nucleophiles to reaction center.

Along with steric factor, stereoelectronic factor plays significant (and, probably, decisive) role in determination of protonated oxiranes reactivity with nucleophiles. This statement is supported by increasing of positive charge on carbon atoms [27], decreasing of C–O bond orders and energy [26, 27, 66], and significant decreasing of the lowest unoccupied molecular orbitals [71]. It should be noted that formally in the case of protonated oxiranes neutral hydroxyl group serves as leaving group instead of highly unstable O^- anion.

Aforementioned changes of structural and electronic characteristics of protonated epoxides naturally leads to alteration of their transformation mechanism. Depending on epoxide substituent character and strength of attacking nucleophile the mechanism may change from borderline S_N2' to S_N1-like mechanism (Fig. 10.16). The possibility of realization of the S_N1-like mechanism has been confirmed by formation of structures with retention of epoxidic carbon configuration [72]. Exploring of potential energy surface for monomolecular opening of protonated oxirane in vacuo at MP2/6-31G(d, p)//HF/6-31G(d) [73] and MP2/6-31G(d, p)//MP2/6-31G(d) [68] levels of theory has shown that reaction in one stage leads to protonated acetic aldehyde with activation barrier of 102.9 and 115.9 kJ/mol, correspondingly. Monomolecular transformation of protonated propylene oxide to protonated propanale is characterized by lower value of E_{act} (74.1 and 76.8 kJ/mol at MP2/6-31G(d) and MP2/6-311++G(d, p) levels, correspondingly) and also occurs in one stage [72, 74]. Detailed analysis of intrinsic reaction coordinate paths has shown that the lowest energy pathway involves two distinct steps. The first step, rupture of the oxirane ring, is followed by a second step, hydride migration, a process not commenced until breaking of the C–O bond is complete. The combination of these two steps defines a concerted asynchronous pathway. Although the carbocation was identified on the of potential energy surface it was not characterized as a minimum at the MP2 level which contradicts the S_N1 mechanism (reaction steps) where carbocation is the reaction coordinate [72, 74]. Thus for investigation of "classic" epoxide ring-opening reaction in acidic environment nucleophile molecule has to be involved from the first stage.

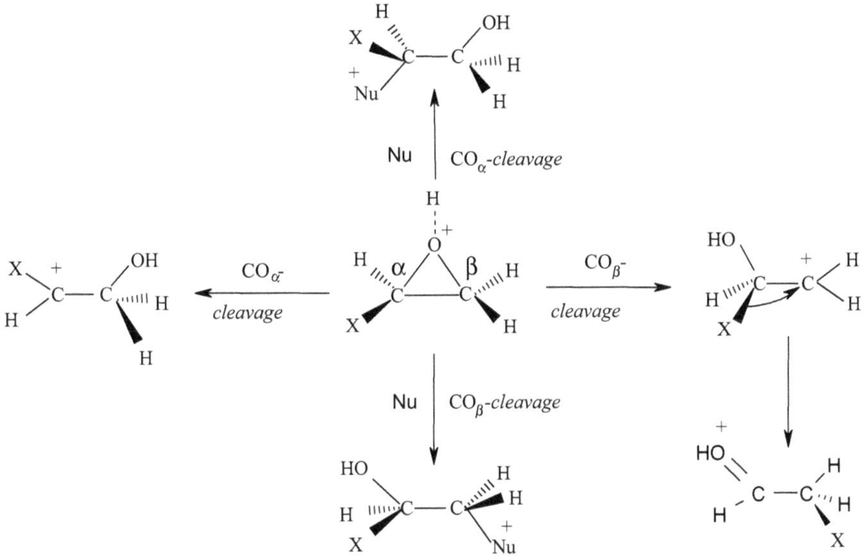

Fig. 10.16 Reactions of mono- and bimolecular opening of protonated oxiranes. (Adapted from [72])

Comparison of activation barriers values for hydrolysis of oxirane, catalyzed by oxonium ion, clearly demonstrates the preference of rear-side attack of nucleophile if compared with front-side attack, where transition state destabilized by Coulomb and Pauli repulsion between the electron rich $OH^{\delta-}$ and $CH_2O^{\delta-}$ fragments (Fig. 10.17) [32].

For unsymmetrical epoxides using isotopic labeling in the $H_2{}^{18}O$ molecule Long and Pritchard [56] showed that hydrolysis reaction was regioselective with formation of so-called abnormal products resulted from nucleiphilic attack on the more substituted carbon atom [56]. As it could be seen from Fig. 10.18, for hydrolysis of protonated propylene oxide *in vacuo* formation of abnormal product is in 5.02 kJ/mol favored relative to attack on the less substituted carbon atom [72].

In contrast to oxirane and methyloxirane, corresponding carbocations have been located at MP2/6-31G(d) level of theory on the potential energy surface of protonated epoxide derivatives of benzene (**26**), and naphthalene (**27**). Epoxide (**26**) transformation to carbonium ion requires activation energy 8 kJ/mol, in case of naphthalene oxide (**27**) pathways for C_a–O and C_b–O bond cleavage are characterized by values of activation energy equal to 1 and 6 kJ/mol, correspondingly. In all cases carbonium ions formed are more stable if compared to protonated epoxide (on 52 kJ/mol for **26** and −53 and 29 kJ/mol for C_a–O and C_b–O bond cleavage in **27**) [75].

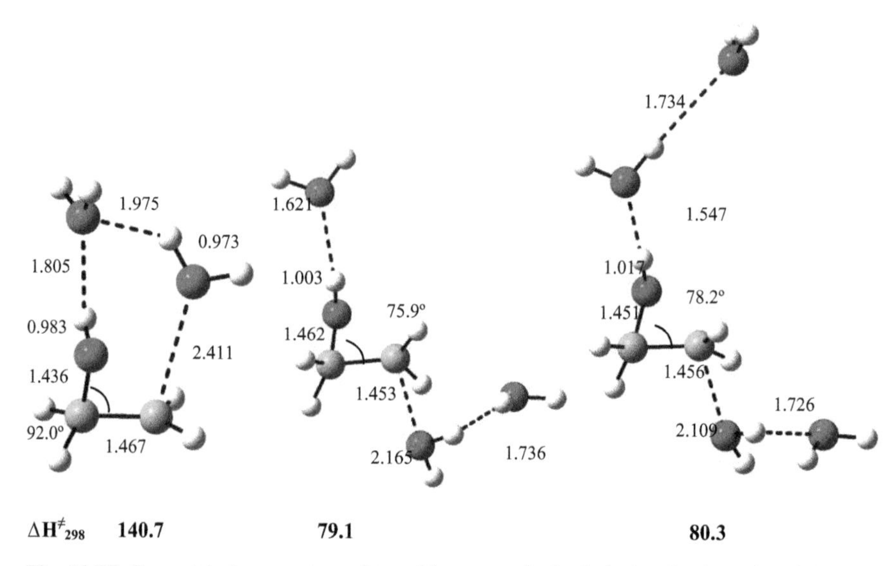

Fig. 10.17 Geometrical parameters of transition states for hydrolysis of oxirane in acidic environment and corresponding values of activation enthalpies (kJ/mol) calculated at B3LYP/6-311++G(d, p) level of theory. (Adopted from [32])

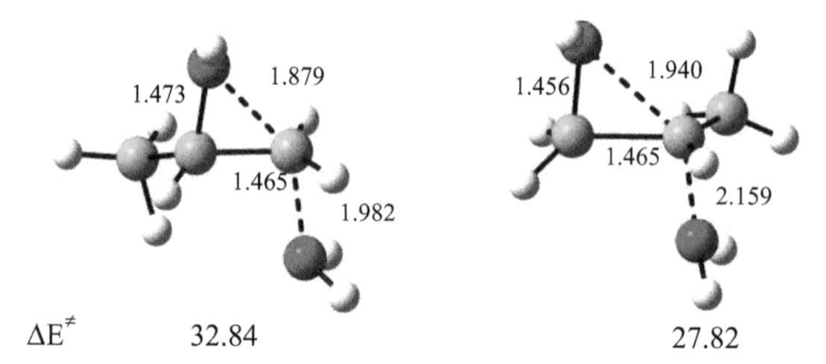

Fig. 10.18 Geometrical parameters of transition states for *in vacuo* hydrolysis of oxirane in acidic environment and corresponding values of activation enthalpies (kJ/mol) calculated at MP2/6-311++G(d, p) level of theory. (Adopted from [72])

26 **27** **28** **29**

Table 10.5 Calculated activation energy (E_{act}) for the reaction of oxirane (**1**) and complexes (**30–36**) with ammonia (in kJ/mol)

Complex	**1**	**30**	**31**	**32**	**33**	**34**	**35**	**36**
E_{act} B3LYP/6-31G(d)	162.6	58.7	93.1	87.0	81.4	82.9	107.9	109.5
E_{act} MP2/6-311++G(d, p)// B3LYP/6-31G(d)	162.3	84.0	115.5	111.7	108.1	108.9	130.8	133.6

In case of partially saturated systems the compounds (**28, 29**), do not appear as stable species on the MP2/6-31G(d) potential energy surface. The approach of proton leads to spontaneous ring opening *via* C_a–O and C_b–O bond breaking for compounds (**28**) and (**29**), respectively [75]. Formation of carbonium ion intermediates also has been shown for monomolecular transformation of fluorooxirane [73], chlorooxirane [76], and styrene oxide [77].

Investigation of kinetic particularities of epoxide interaction with carboxylic acids [62–64, 78] showed that proton transfer with formation of protonated epoxide takes place only in case of reaction with strong acids such as trifluoroacetic acid. Interaction with relatively weak dichloroacetic acid leads to formation of complex without proton transfer and involvement of the second molecule of acid is required for epoxide ring opening which corresponds to experimentally determined second order of reaction [63].

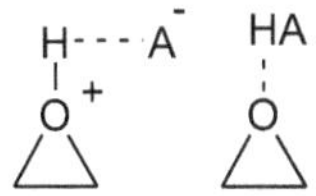

H- - - A⁻ HA

Complexes of such kind have been used by Omoto and Fujimoto for investigation of catalytic strength of 1,8-biphenyldiole (**30**) and number of monodentate catalysts (complexes **31–36**) [51].

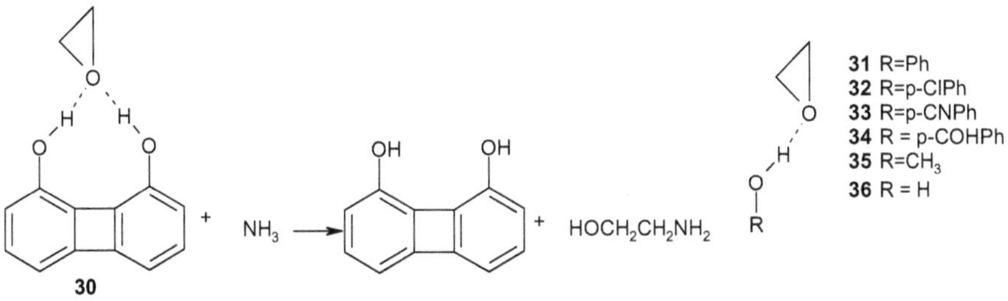

+ NH_3 → OH OH + $HOCH_2CH_2NH_2$

30

31 R=Ph
32 R=p-ClPh
33 R=p-CNPh
34 R = p-COHPh
35 R=CH_3
36 R = H

As could be seen from Table 10.5 the least effective activators are water and methanol molecules. Among phenols more effective are compounds with electron-attracting groups in *para*-position of benzene ring. It should be mentioned that

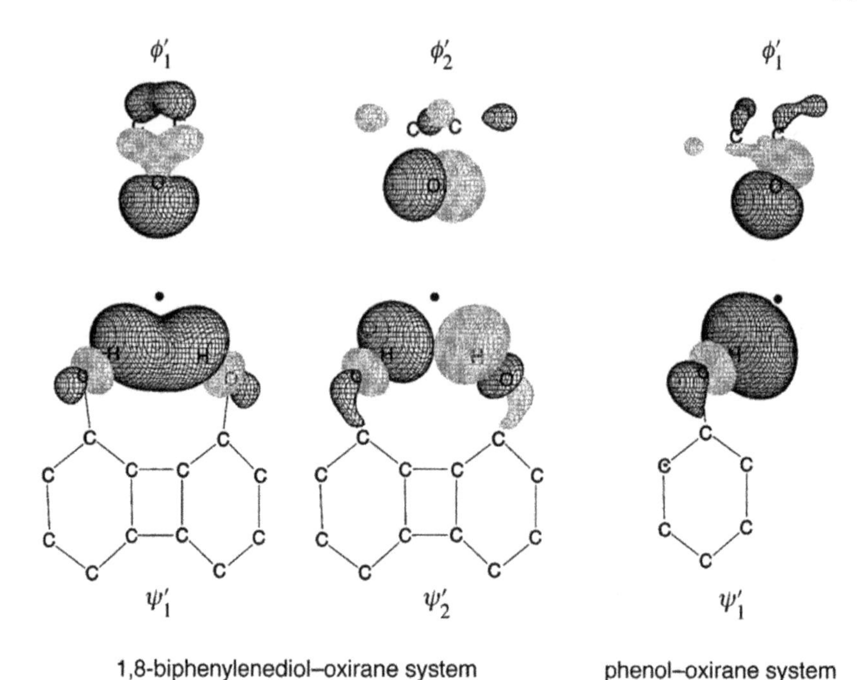

Fig. 10.19 Interaction orbitals responsible for electron delocalization from the oxirane part to the acid part in the 1:1 complex of oxirane and an acid. The orbital ϕ's are given by the combination of the occupied MOs of the oxirane part, and Ψ's are given by the combination of the unoccupied MOs of the acid part. Two orbital pairs on the left-hand side are for the (oxirane + 1,8-biphenylene-diol) system and a pair on the right-hand side is for the (oxirane + phenol) system [51]

1,8-biphenyldiole catalyzes reaction in greater extent as it could be expected from Brønsted dependence [35, 36]. Similar influence of bidentate phenols has been also shown in some experimental investigations devoted to interaction of phenyl glycidyl ether with diethylamine [79], and for Diels-Alder reaction with α,β-unsaturated ketones and aldehydes [80].

The reason of such high catalytic activity of 1,8-biphenylenediol has been studied in [51] by the method of paired interaction orbitals [81–84]. As could be seen from Fig. 10.19, orbitals Ψ_1' and Ψ_2' are localized completely on the two O–H bonds of the diol and overlaped *in-phase* with the orbitals ϕ_1' and ϕ_2', for the lone pairs of electrons of oxirane, respectively. By means of these two pairs of orbitals, the diol can accept the electronic charge from the oxygen, and the two hydrogen bonds are formed. In the system "phenole-oxiran" interaction between molecular orbitals is not so effective. Two effects resulted from addition of acidic catalyst have been derived from the detailed analysis of electron density distribution in complexes (**30–36**) and transition states of the corresponding reactions. One is to enhance electrophilicity of the C–O bond connected with removing the electronic charge from oxirane before the attack of a nucleophile and the second is stabilization of transition states due to electronic charge shifting from the attacking nucleophile to the oxirane ring further onto the acid framework. That is, acids serve as temporary reservoir of electronic charge during the reaction to maintain high nucleophilicity of oxirane and to reduce overlap repulsion between the substrate and the attacking nucleophile.

10.5 Conclusion

In this brief review, there are summarized the results of quantum chemical investigation of epoxides transformation in neutral, alkaline and acidic environment. It has been shown that in the presence of both basic and acid catalysts back-side attack of nucleophile is more preferable if compared to front-side approach due to strong repulsion between nucleophile and oxygen atom of epoxide. Having the same stereochemistry the reactions in alkaline and acid medium possess opposite regiochemisty—protonation facilitates nucleophile attack on the less substituted carbon atom resulted in formation of abnormal product while in alkaline medium nucleophile forms bond with more substituted carbon atom. Uncatalyzed transformation of epoxides is characterized by high values of activation energy. Modeling hydrolysis of oxirane with explicit consideration of one and two water molecules showed the preference of front-side attack. For neutral aminolysis, on the other hand, back-side attack is more favorite where four-water cluster assists the proton transfer, and strengthens both the entering and the leaving groups through a charge-transfer process induced by different strengths between the two proton-transfer processes.

Detailed analysis of epoxides model reactions in different environment built background for theoretical modeling of large-scaled biologically valuable processes.

References

1. Yudin AK (2006) Aziridines and epoxides in organic synthesis. Wiley, Weinheim
2. Stirling CJM (1985) Evaluation of the effect of strain upon reactivity. Tetrahedron 41(9):1613–1666
3. Bartok M, Lang KL (1980) The chemistry of functional groups. Supplement E. In: Patai S (ed) The chemistry of ethers, hydroxyl groups and their sulfur analogues. Part 2. Willey, New York, pp 609–682
4. Lewars EG (1984) Structure of small and large rings. In: Katritzky AR, Rees CW, Lwowski W (eds) Comprehensive heterocyclic chemistry, vol 7. Pergamon, New York, pp 95–130
5. Dittmer DC (1984) Thiiranes and Thiirenes. In: Katritzky AR, Rees CW, Lwowski W (eds) Comprehensive heterocyclic chemistry, vol 7. Pergamon, New York, pp 131–184
6. Crandall JK, Lin L-HC (1968) Base-promoted reactions of epoxides. V. 1-Alkylcycloalkene oxides. J Org Chem 33(6):2375–2378
7. Gronert S, Lee JM (1995) Gas phase reactions of methyloxirane with HO^- and methylthiirane with HO^- and HS^-. An ab initio study of addition and elimination. J Org Chem 60(14):4488–4497

8. Apparu M, Barrelle M (1978) Reactivite des epoxydes—III: Methodes de synthese d'hydroxyalkyl-1 vinyl-2 cyclopropanes: reactions d'epoxydes γ,δ-ethyleniques avec les amidures de lithium dans l'HMTP. Tetrahedron 34(11):1691–1697
9. Rickborn B, Thumell RP (1969) Stereoselectivity of the base-induced conversion of epoxides to allylic alcohols. J Org Chem 34(11):3583–3586
10. Norman ROC, Coxon JM (1993) Principles of organic synthesis, 3rd edn. Chapman and Hall, Blackie
11. Chacos N, Capdevila J, Falck JR, Manna S, Martin-Wixtrom C, Gill SS, Hammock BD, Estabrook, RW (1983) The reaction of arachidonic acid epoxides (epoxyeicosatrienoic acids) with a cytosolic epoxide hydrolase. Arch Biochem Biophys 223:639–648
12. Yu Z, Xu F, Huse LM, Morisseau C, Draper AJ, Newman JW, Parker C, Graham L, Engler MM, Hammock BD, Zeldin DC, Kroetz DL (2000) Soluble epoxide hydrolase regulates hydrolysis of vasoactive epoxyeicosatrienoic acids. Circ Res 87:992–998
13. Lonsdale R, Harvey JN, Mulholland AJ (2010) Compound I reactivity defines alkene oxidation selectivity in cytochrome P450cam. J Phys Chem B 114(2):1156–1162
14. Jean DJ St Jr, Fotsch C (2012) Mitigating heterocycle metabolism in drug discovery. J Med Chem 55(13):6002–6020
15. Koskinen M, Plna K (2000) Specific DNA adducts induced by some mono-substituted epoxides *in vitro* and *in vivo*. Chem Biol Interact 129(3):209–229
16. Suresh CH, Vijayalakshmi KP, Mohan N, Ajitha M (2011) Mechanism of epoxide hydrolysis in microsolvated nucleotide bases adenine, guanine and cytosine: a DFT study. Org Biomol Chem 9:5115–5122
17. Oesch F (1973) Mammalian epoxide hydrases. Inducible enzymes catalyzing inactivation of carcinogenic and cytotoxic metabolites derived from aromatic and olefinic compounds. Xenobiotica 3:305–340
18. Oesch F, Hengstler JG, Arand M (2004) Detoxification strategy of epoxide hydrolase—the basis for a novel threshold for definable genotoxic carcinogens. Nonlinearity Biol Toxicol Med 2(1):21–26
19. Jones RB, Mackrodt WC (1983) Structure–genotoxicity relationship for aliphatic epoxides. Biochem Pharmacol 32:2359–2362
20. Ross WC (1950) The reactions of certain epoxides in aqueous solutions. J Chem Soc 2257–2272
21. Jones RB, Mackrodt WC (1982) Structure–mutagenicity relationships for chlorinated ethylenes: a model based on the stability of the metabolically derived epoxides. Biochem Pharmacol 31:3710–3712
22. Kirkovsky LI, Lermontov SA, Zavorin SI, Sukhozhenko II, Zavelsky VI, Thier R, Bolt HM (1998) Hydrolysis of genotoxic methyl-substituted oxiranes: experimental kinetic and semiempirical studies. Environ Toxicol Chem 17:2141–2147
23. Gervasi PG, Citti L, Delmonte M, Longo V, Benetti D (1985) Mutagenicity and chemical-reactivity of epoxidic intermediates of the isoprene metabolism and other structurally related-compounds. Mutat Res 156:77–82
24. Polansky OE, Fratev F (1976) Pars orbital method: character orders in CNDO calculations and a new definition for bond orders in all valence electron methods. Chem Phys Lett 37(3):602–607
25. Fujimoto H, Katata M, Yamabe S, Fukui K (1972) An MO-theoretical interpretation of the nature of chemical reactions III. Bond interchange. Bull Chem Soc Jpn 45(5):1320–1324
26. Kas'yan LI (1999) Steric strain and reactivity of epoxynorbornanes (3-oxatricyclo[3.2.1.02,4] octanes). Rus J Org Chem 35(5):635–665
27. Kas'yan LI, Gorb LG, Seferova MF, Chernousov DA, Svyatkin VA, Boldeskul IE (1990) Quantum-chemical investigation of alicyclic epoxide compounds. Zh Org Khim 26(1):3–10
28. Kas'yan LI, Gorb LG, Seferova MF, Dryuk VG (1990) Quantum-chemical study of electronic structure and reactivity of spirooxiranes. Ukr Khim Zhurn 56(10):1071–1076
29. Kretova EN, Podgornova VA, Shpekht NA, Ustavshchikov BF (1983) Theoretical and experimental studies of the reactivity of alkene oxides with acids. I. Calculation of the electronic

structure of a series of alkene oxides and mechanism of cleavage of the epoxy ring in acid media. Osnov Organ Sintez i Neftekhimiya. Yaroslavl 19:82–87
30. Okovityi SI, Platitsyna EL, Kas'yan LI (2001) Theoretical study of alkaline methanolysis of alicyclic epoxy derivatives. Rus J Org Chem 37(3):345–350
31. Fujimoto H, Hataue S, Koga N, Yamasaki J (1984) Ring-opening of oxirane by nucleophilic attack. Transition states and paired Interacting Orbitals. Tetrahedron Lett 25(46):5339–5342
32. Lundin A, Panas I, Ahlberg E (2007) A mechanistic investigation of ethylene oxide hydrolysis to ethanediol. J Phys Chem A 111(37):9087–9092
33. Glad SS, Jensen F (1994) *Ab initio* study of the nucleophilic ring opening of ethylene oxide. Connection between secondary isotope effects and transition structures. J Chem Soc Perkin Trans II 4:871–876
34. Scheppele SE (1972) Kinetic isotope effects as a valid measure of structure–reactivity relations. Isotope effects and nonclassical theory. Chem Rev 72(5):511–532
35. Williams IN (1984) Theoretical modelling of compression effects in enzymic methyl transfer. J Am Chem Soc 106(23):7206–7212
36. Lau EY, Newby ZE, Bruice TC (2001) A theoretical examination of the acid-catalyzed and noncatalyzed ring-opening reaction of an oxirane by nucleophilic addition of acetate. Implications to epoxide hydrolases. J Am Chem Soc 123(14):3350–3357
37. Kas'yan LI, Stepanova NV, Belyakova TA, Kunanets VK, Lutsenko AI, Zefirov NS (1984) Methanolysis of cyclic epoxide compounds. Zh Org Khim 20(11):2295–2301
38. Podlogar BL, Raber DJ (1989) Molecular mechanics calculations of epoxides. Extension of the MM2 force field. J Org Chem 54(21):5032–5035
39. Stewart JJP (1989) Optimization of parameters for semiempirical methods. I. Method. J Comput Chem 10(2):209–220
40. Klamt A (1995) Conductor-like screening model for real solvents: a new approach to the quantitative calculation of solvation phenomena. J Phys Chem 7(99):2224–2235
41. Okovytyy SI, Platicina EL, Kasyan LI (1998) Theoretical investigation of the solvent influence on the oxirane alkaline methanolysis mechanism. Visn dnipropetr univ: Khim 2:132–135
42. Kas'yan LI, Gapanova RG, Okovityi SI (1994) Effect of medium on methanolysis of dicyclopentadiene diepoxide. Zh Org Khim 30(5):692–698
43. Okovytyy SI, Platicina EL, Kasyan LI (2000) Quantum chemical investigation of the environment influence on diepoxide dicyclopentadiene methanolysis. Visn Dnipropetr Univ: Khim 4:42–47
44. Okovytyy SI, Platicina EL, Seferova MF, Kasyan LI (2001) Theoretical investigation of the mechanism of spirooxiranes interaction with methanole. Visn Dnipropetr Univ: Khim 6:46–49
45. Laitinen T, Rouvinen J, Peräkylä M (1998) Ab initio quantum mechanical and density functional theory calculations on nucleophile- and nucleophile and acid-catalyzed opening of an epoxide ring: a model for the covalent binding of epoxyalkyl inhibitors to the active site of glycosidases. J Org Chem 63(23):8157–8162
46. Helten H, Schirmeister T, Engels B (2004) Model calculations about the influence of protic environments on the alkylation step of epoxide, aziridine, and thiirane based cysteine protease inhibitors. J Phys Chem A 108(38):7691–7701
47. Helten H, Schirmeister T, Engels B (2004) Theoretical studies about the influence of different ring substituents on the nucleophilic ring opening of three-membered heterocycles and possible implications for the mechanisms of cysteine protease inhibitors. J Org Chem 70(1):223–237
48. Meara JP, Rich DH (1996) Mechanistic studies on the inactivation of papain by epoxysuccinyl inhibitors. J Med Chem 39(17):3357–3366
49. Alagona G, Scrocco E, Tomasi J (1979) Theoretical *ab initio* study of the reaction of formation of 2-fluorethanol. Theor Chim Acta 51(1):11–35
50. Bobylev VA, Koldobskii SG, Tereshchenko GF, Gidaspov BV (1988) Khim Geterotsikl Soedin 9:1155–1168

51. Omoto K, Fujumoto H (2000) Theoretical study of activation of oxirane by bidentate acids. J Org Chem 65(8):2464–2471
52. Shibaev AYu, Astrat'eva NV, Tereshchenko GF (1984) Quantum-chemical study of ethylene oxide amination. Zh Obsh Khim 54(12):2744–2747
53. Lundin A, Panas I, Ahlberg E (2009) Quantum chemical modeling of propene and butene epoxidation with hydrogen peroxide. J Phys Chem A 113(1):282–290
54. Zheng Y, Zhang J (2010) Interface water catalysis of the epoxide opening. Chem Phys Chem 11(1):65–69
55. Long FA, Pritchard JG (1956) Hydrolysis of substituted ethylene oxides in H_2O^{18} solutions. J Am Chem Soc 78(12):2663–2667
56. Parker RE, Isaacs MS (1956) Mechanisms of epoxide reactions. Chem Rev 59(4):737–799
57. Pritchard JG, Long FA (1956) Kinetics and mechanism of the acid-catalyzed hydrolysis of substituted ethylene oxides. J Am Chem Soc 78(12):2667
58. Taft RW (1952) The Dependence of the rate of hydration of isobutene on the acidity function, H_0, and the mechanism for olefin hydration in aqueous acids. J Am Chem Soc 74(21):5372–5376
59. Wohl RA (1974) The Mechanism of the acid-catalized king opening of epoxides. A reinterpretative review. Chimia 28(1):1–5
60. Okovytyy S, Gorb L, Leszczynski J (2001) New insight on the mechanism of 2-oxabicyclobutane fragmentation. A high-level ab initio study. Tetrahedron 57(8):1509–15013
61. Lebedev NN, Guskov KA (1963) Reaction of α-oxides. II. Kinetics of the reactions of ethylene oxides with acetic and monochloroacetic acids. Kinet Katal 4(1):116–127
62. Lebedev NN, Kozlov VM (1966) Kinetics and mechanism of reaction of ethylene oxide and acids. Zh Obsh Khim 2(2):261–265
63. Kakiuchi H, Tijima T (1980) The ring-opening reactions of propylene oxide with chloroacetic acids. Tetrahedron 36(8):1011–1016
64. Stepanov EG, Podgornova VA, Ustavshchikov BF (1976) Effect of the structure of the carboxylic acid on the reactivity in the reaction with propylene oxide. Osnovn Organ Sintez i Neftekhimiya 6:29–31
65. Kretova EN, Podgornova VA, Bastrakova ZA, Chagina NT (1984) Zh Fizich Khim 58(11):2885–2887
66. Kretova EN (1985) Calculation of the electronic structure of protonated ethylene oxide by the SCF MO LCAO CNDO/BW method. Zh Strukt Khim 26(1):133–135
67. Nobes RH, Rodwell WR, Bouma WJ, Radom L (1981) The oxygen analog of the protonated cyclopropane problem. A theoretical study of the $C_2H_5O^+$ potential energy surface. J Am Chem Soc 103(8):1913–1922
68. Bock CW, George P, Glusker JP (1993) Ab initio molecular orbital studies on $C_2H_5O^+$ and $C_2H_4FO^+$: oxonium ion, carbocation, protonated aldehyde, and related transition-state structures. J Org Chem 58(21):5816–5825
69. Zhao Y, Truhlar DG (2007) How well can new-generation density functionals describe protonated epoxides where older functionals fail? J Org Chem 72(1):295–298
70. Carlier PR, Deora N, Crawford TD (2006) Protonated 2-methyl-1,2-epoxypropane: a challenging problem for density functional theory. J Org Chem 71(4):1592–1597
71. Frenking G, Kato H, Fukui K (1975) An MO-theoretical treatment of the cationic ring-opening polymerisation. I. Ethylene oxide. Bull Chem Soc 45(1):6–12
72. F de Sousa (2010) Theoretical study of acid-catalyzed hydrolysis of epoxides. J Phys Chem A 114(15):5187–6194
73. Ford GP, Smith CT (1987) Gas-phase hydrolysis of protonated oxirane. Ab initio and semiempirical molecular orbital calculations. J Am Chem Soc 109(5):1325–1331
74. Coxon JM, Maclagan DGAR, Rauk A, Thorpe AJ, Whalen D (1997) Rearrangement of protonated propene oxide to protonated propanal. J Am Chem Soc 119(20):4712–4718
75. Korzan R, Upton B, Turnbull K, Seybold PG (2010) Quantum chemical study of the energetics and directionality of acid-catalyzed aromatic epoxide ring openings. Int J Quant Chem 110(15):2931–2937

76. Shinoda H, Mori Y, Mizuguchi M (2006) Reinvestigation into the ring-opening process of monochloroethylene oxide by quantum chemical calculations. Int J Quant Chem 106(4):952–959
77. George P, Bock CW, Glusker JP (1992) Protonation of monosubstituted oxiranes: a computational molecular orbital study of oxonium ion versus carbonium ion formation. J Phys Chem 96(9):3702–3708
78. Schlegel HB, Mislow K, Bernardi F, Bottoni A (1977) An ab initio investigation into the S_N2 reaction: frontside versus backside attack in the reaction of F^- with CH_3F. Theor Chim Acta 44:245–256
79. Partansky AM (1970) A study of accelerators for epoxy–amine condensation reaction. In: Lee H (ed) Epoxy resins. Am Chem Soc Adv Chem Ser 92:29–47
80. Kelly TR, Meghani P, Ekkundi VS (1990) Diels–Alder reactions: rate acceleration promoted by a biphenylenediol. Tetrahedron Lett 31(24):3381–3384
81. Fujimoto H, Koga N, Fukui K (1981) Coupled fragment molecular orbital method for interacting systems. J Am Chem Soc 103(25):7452–7457
82. Fujimoto H, Koga N, Hataue I (1984) Orbital transformations in configuration analysis. A simplification in the description of charge transfer. J Phys Chem 88(16):3539–3544
83. Fujimoto H, Yamasaki T, Mizutani H, Koga N (1985) A theoretical study of olefin insertions into titanium–carbon and titanium–hydrogen bonds. An analysis by paired interacting orbitals. J Am Chem Soc 107(22):6157–6161
84. Fujimoto H, Yamasaki T (1986) A theoretical analysis of catalytic roles by paired interacting orbitals. Palladium(II)-catalyzed nucleophilic additions to carbon–carbon double bonds. J Am Chem Soc 108(4):578–581

Chapter 11
Computational Toxicology in Drug Discovery: Opportunities and Limitations

Alexey Zakharov and Alexey Lagunin

Abstract Different methods of computational toxicology are used in drug discovery to reveal toxic and dangerous side effects of drug candidates on early stages of drug development. Information about chemoinformatic, toxicogenomic and system biological approaches, commercial and freely available software and resources with data about toxicity of chemicals used in computational toxicology are represented. General rules and key components of QSAR modeling in respect to opportunities and limitations of computational toxicology in drug discovery are considered. The questions of computer evaluation of drug interaction with antitargets, drug-metabolizing enzymes, drug-transporters and related with such interaction toxic and side effects are discussed in the chapter. Along with an overview of existing approaches we give examples of the practical application of computer programs GUSAR, PASS and PharmaExpert to assess the general toxicity and toxic properties of individual drug-like compounds and drug combinations.

11.1 Introduction

The practical use of computer technology to predict the effects of chemicals on the environment and human health, preclinical evaluation of toxicity, side effects and metabolism of drug candidates is of great interest to the scientific community and human society [1]. Currently, the benefits of using computational methods to predict the toxicity of compounds are properly recognized by members of the business

A. Zakharov (✉)
National Institutes of Health, National Cancer Institute, Chemical Biology Laboratory,
376 Boyles St., Frederick, MD 21702, USA
e-mail: alexey.zakharov@nih.gov

A. Lagunin
Orechovich Institute of Biomedical Chemistry of Russian Academy of Medical Sciences,
Laboratory of Structure-Function Based Drug Design,
Pogodinskaya St. 10/7, Moscow 119121, Russia
e-mail: alexey.lagunin@ibmc.msk.ru

L. Gorb et al. (eds.), *Application of Computational Techniques in Pharmacy and Medicine,*
Challenges and Advances in Computational Chemistry and Physics 17,
DOI 10.1007/978-94-017-9257-8_11

community and the public authorities responsible for safety of the environment and human health. Pharmaceutical companies use computer predictions of toxicity at the design stage to identify lead compounds with low toxic properties, as well as in selection of candidates at the optimization stage of potential drugs [2]. An important priority of pharmaceutical companies during the drug design programs and safety assessment is an early detection of dangerous toxic effects before significant time and financial resources will be spent for new drugs at the latest stages of clinical trials.

Computational prediction of toxicity is a significant part of a more general field of science—Computational Toxicology. Definition of Computational Toxicology was given by U.S. Environmental Protection Agency (EPA): an integration of modern computational and information technology with molecular biology, which is aimed to improve the prioritization and risk assessment of chemicals [3]. Thus, we have the following definition of toxicology: Toxicology (from the Greek τοξικος—poison and λογος—science, that is τοξικολογία—the science of poisons) means the science that studies poisonous, toxic and harmful substances, a potential risk of their effects on organisms and ecosystems, mechanisms of toxicity, and methods of diagnosis, prevention and treatment of emerging diseases as the result of such exposure. Therefore, computational prediction of toxicity can be characterized as prediction of the effects provided by chemical compounds on biological organisms and ecosystems which is based on the analysis of structure-activity relationships using the modern computational and informational technology.

The National Research Council of the United States (NRC) has recently published a basic report entitled "Toxicity Testing in the twenty-first Century: A Vision and Strategy" [4]. The report is devoted to a well-established methodology in the toxicity study and discussion on the use of alternative methods, strategies to increase effectiveness and appropriateness of toxicity tests for the risk assessment of chemicals. According to NRC, the application of system biology, high-throughput screening and computational technology will be increased significantly in future, together with other toxicological tests that generate a huge amount of the biological data (toxicogenomics, proteomics, metabolomics, etc.). The progress in a bioinformatics field, system biology, omics and computational toxicology can transform and change the animal toxicity testing to the alternative testing methods.

As a practical development and promotion of the computational toxicity prediction for the risk assessment of chemicals in industry, the European Community has adopted a special law—Registration, Evaluation, Authorization and Restriction of Chemicals (REACH). REACH provides the basis for a regular use of quantitative/ qualitative analysis of structure—activity (Quantitative Structure-Activity Relationships—(Q)SAR analysis) in the European Community. The aim of REACH is to improve the protection of humans and the environment through the better and earlier identification of the toxic properties of compounds [5]. The effect of 60,000 compounds on humans and the environment, which are produced in the EU in amounts of more than 1 ton per year, will be evaluated by REACH. Examples of QSAR practice in REACH are given in the following review [6, 7].

The advantage of computational prediction methods in comparison with the experimental biological toxicity is their lower cost and time, high efficiency and reproducibility using the same model. (Q)SAR models have no restrictions related to chemical synthesis, they can be continuously improved (allow adding important properties, descriptors, and expansion of the chemical space), and can also help to reduce the number of experimental animals. It is important to understand that, in spite of all possibilities for computation prediction methods, they cannot be used separately from experimental studies and cannot fully replace biological experiments designed to determine the toxicity of compounds. In most cases the computational prediction of toxicity is an effective tool to support the decision about experimental testing of compounds. In addition, an ability of computational methods to predict ADME (Adsorption, Distribution, Metabolism and Excretion) properties for virtual structures allows investigating the chemical space without chemical synthesis and experimental testing of compounds.

11.2 General Principles of the Computational Prediction of Toxicity

Toxicity prediction is based on the assumption that an activity of chemicals depends on their structures. This statement is valid both for the creation of (Q)SAR models and for calculation of risk assessment using expert rules, which allow detecting the toxic compounds based on the so-called alerts, simple structural components associated with the manifestation of toxicity [8, 9, 10].

For the construction of any toxicological model the three key components are used:

1. **Data** of chemical compounds (structure, physicochemical and biological properties) and biological experimental systems (species, strain, sex, clinical characteristics, gene expression and protein synthesis);
2. **Descriptors** are used for description of chemical structures (constitutional descriptors, topological, electro-topological, quantum-chemical, structural fragments, fingerprints, physicochemical descriptors);
3. **Mathematical methods** are used to identify the relationship between descriptors and the biological effects (multiple linear regressions, neural networks, nearest neighbors, support vector machine, random forest, etc.).

In 2002, the Organization for Economic Cooperation and Development (OECD) had developed and introduced the guidelines for creation of predictive (Q)SAR models [11]. The guidelines are that a valid QSAR/QSPR should have:

1. A defined endpoint. An independent variable, which is used for modeling has to be well defined;
2. An unambiguous algorithm. It is a necessary to use an unambiguous algorithm for the model building;

3. A defined domain of applicability. It is a necessary to estimate an applicability domain for each developed model;
4. Appropriate measures of goodness for fit, robustness and predictivity. It is a necessary to validate the developed models according to different types of validation procedure;
5. A mechanistic interpretation, if possible. Each model, if possible, should have a mechanistic interpretation;

The analysis of key components in QSAR modeling procedure, which includes common mistakes, is given below.

11.2.1 Data

A considerable attention should be given to data, which are used for the models construction. The following criteria are used for this purpose:

1. **Reliability**: accurate and complete description of the experimental values;
   Data should be obtained by the same experimental protocol (using the same type and sex of the animal, route of administration, time of exposure);
   For quantitative parameters (LD_{50}, EC_{50}, etc.) the molar concentrations (mmol/kg) should be used instead of weight (mg/kg) or volume (ppm). The following equations are used for conversion of weight and volume values to the molar concentrations:

$$mMole / kg = \frac{mg / kg}{g / mole}, \quad mMole / m^3 = \frac{ppm}{24.25}$$

   The structure of compounds should be a single-component and presented in a neutral form;
   References or literature sources should be provided for each type of data;
2. **Consistency**: experimental results must be reproducible with low error;
   For qualitative data (for example, a carcinogen and not a carcinogen) should not be contradictions;
   Classification of the compounds should be made according to the same criteria;
3. **Reproducibility**: experimental results should be reproducible in different laboratories.

In addition, it is a necessary to take into account the range of a dependent variable during the construction of (Q)SAR models. Gedeck et al. has showed on different datasets that the prediction accuracy is significantly increased with magnification of a value range in the dependent variable [12]. The authors have proved that to develop a good model, it is required to use the data, which has the range of dependent values as at least one logarithm (the value of Y (e.g., LD_{50}) varies by 10 times).

Some database may contain information about the weak activity of compounds that are labeled like $LD_{50}>300$ mg/kg. This type of data should be censored. There are different methods of using this data for QSAR modeling [13, 14], but it is better to avoid using them. Especially, these values could not be assigned for a specific threshold (for example, $LD_{50}>300$ mg/kg cannot be assigned to the LD_{50} 300 mg/kg or 600 mg/kg or 900 mg/kg). Quality and reproducibility of the data used for the toxicity modeling is one of the most important issues for creation of (Q)SAR models. For example, can rodent's experimental test results be used for the (Q)SAR modeling? Although, currently, there are more than 5000 experimental results of carcinogenicity in rodents for compounds, most of them are not public available (e.g., private and proprietary archival research) [15]. As well as quality of data, the transparency of standardized bioassays using specific protocols, such as NTP (National Toxicology Program: http://ntp.niehs.nih.gov/), is important. It is considered that the usage of rodent's carcinogenicity data is efficient for (Q)SAR modeling. However, it was found that the data of experimental protocols are varied depending on the sources. For some compounds contradictions in the measurement data were found in the published reports [16], while for other compounds results were well reproducible [17].

Since the correlation strategy of toxicity values with molecular structures of the training set is used during *in silico* modeling, an inability of software to process the mixture of compounds with a small amount of salts containing ions such as HCl- or SO_4 or hydrated condition indicates that the wrong representation of endogenous molecules can result in inaccurate predictions [18]. Some (Q)SAR developers ignore the small molecule ions, but they can affect the pKa of the molecule, which depends on various physiological conditions and may affect the behavior of the whole molecule, such as absorption. Therefore, this ignoring leads to errors in the model as well as prediction results.

11.2.2 Databases with Experimental Toxicity Data of Compounds

Toxicity databases are widely used for developing models, prediction of undesirable drug effects, safety assessment of different xenobiotics, selection of promising compounds and, ideally, estimation of the risk assessment of compounds. The main aim in the design of such database is aggregation of the acceptable scientific data from different toxicity studies for constructing an electronic resource that can be used for search of chemicals, for model developing and for read-across strategy of structurally similar compounds. OECD has published the guidance on the quantitative and qualitative read-across approach, which can be used to fill the data gaps in the risk assessment of chemicals [19]. Computer toxicology databases are often used by regulatory agencies and industry for the safety assessment and predictions of xenobiotics side effects [20].

Definition of toxicology databases is varied as the definition of computational toxicology. The most common definition of toxicology databases is a set of electronic information that can be related to the toxicity of compounds, which is organized by the certain computer software and is used for the safety assessment and risk analysis of industrial compounds, research and development products in the field of biomedical and toxicological sciences [1]. Therefore, toxicology databases have to be constantly updated by cheminformatic resources, which are useful for the creation of secondary data sets, e.g. training sets for the (Q)SAR modeling of toxicity.

A huge amount of information from toxicity databases has become freely available recently. It has played an important role in the development of (Q)SAR models and computer programs for the toxicity prediction. Unfortunately, the content of freely available databases is still different from compound libraries used for development of drugs and from industrial compounds. Recent initiatives of regulatory agencies require to develop the toxicology database with a free access and promote the usage of a computer technology [6]. Table 11.1 shows the list of publicly available toxicity databases, which describe the effect of substances on the human health, and electronic resources, which are useful for risk assessment and safety of chemical compounds [1].

The private toxicity databases offer more accurate toxicity data and the extended chemical space of representative structures in comparison with public databases. Despite expansion of the chemical space and various numbers of proposed descriptors, the private databases have limitations related to the models selection, types of algorithms and the content of data, which is probably a part of confidential business information, such as a proprietary structure of pharmaceutical molecules. The private toxicity databases may also provide internal systems created by industry or government agencies [21]. These databases may not be suitable for a commercial usage, but are useful for the internal analysis. Therefore, publication of the scientific research based on these data is often difficult to evaluate independently. The most known commercial databases associated with toxicity are Accelrys Toxicity Database (contains information about the structure and different types of toxicity for more than 150,000 compounds from RTECS and other sources) and Leadscope Toxicity Database. Standardization of toxicity databases is designed to facilitate integration between different sources and to provide their quality. Since databases are often not compatible with each other, standardization initiatives (e.g., controlled vocabularies) can help to combine their data [22].

11.2.3 Descriptors

Appropriate description of the chemical structure is a major component and limitation for creation of high-quality (Q)SAR models. Molecular descriptors are important for the toxicity modeling technology because their numeric representation is the basis for construction of structure-activity relationships by computational models. Therefore, if the selected descriptors do not reflect aspects influencing on manifestation of the molecule toxicity, the developed model may show a poor accuracy. There are both commercial and public software which allow generating different

Table 11.1 Publicly available toxicity databases (DB) and resources

Database	Definition
ACToR	ACToR (Aggregated Computational Toxicology Resource) is EPA's online warehouse of all publicly available chemical toxicity data. Publicly available data produced by the industry in the medium and large amounts, includes information about the components of drinking water. Database has a searching system, which includes a chemical structure, physicochemical properties, in vitro and in vivo experimental toxicity data: http://actor.epa.gov/actor/faces/ACToRHome.jsp
AffyTrack	FDA Toxicogenomics-centric gene microarray expression database: http://www.fda.gov/ScienceResearch/BioinformaticsTools/Arraytrack/default.htm
Cal/EPA	The toxicity database includes the standard criteria levels of chronic exposure and carcinogenic potential. Developed by the U.S. Agency for Environmental Protection California: http://www.oehha.ca.gov/risk/ChemicalDB/index.asp
CCRIS	Chemical Carcinogenesis Research Information System (CCRIS). System includes carcinogenicity and mutagenicity test results for over 8000 chemicals: http://toxnet.nlm.nih.gov/cgi-bin/sis/htmlgen?CCRIS
CEBS	US NIH/NIEHS Chemical Effects in Biological Systems Knowledgebase; integrates genomic and biological data, including dose-response studies in toxicology and pathology: http://www.niehs.nih.gov/research/resources/databases/cebs/index.cfm
CEDI/ADI Database	US FDA/CFSAN Cumulative Estimated Daily Intake/Acceptable Daily Intake Database of food compounds: http://www.fda.gov/Food/IngredientsPackagingLabeling/PackagingFCS/CEDI/default.htm
CERES	Chemical Evaluation and Risk Estimation System (CERES). US FDA/CFSAN develops a knowledge base of nutritional supplements. Knowledge base includes a toxicity database of food ingredients, pharmaceuticals, agricultural and industrial chemicals, structural alerts and QSAR-based toxicity prediction: http://www.accessdata.fda.gov/FDATrack/track-proj?program=cfsan&id=CFSAN-OFAS-Chemical-Evaluation-and-Risk-Estimation-System
ChEMBLdb	ChEMBLdb contains freely available data on cytotoxicity and interaction with more 9000 targets including off-targets, transporters and drug-metabolizing enzymes: https://www.ebi.ac.uk/chembldb/
ChemIDPlus	It contains information about the structure, name, physical-chemical properties, biological activity and toxicity for 139,354 compounds: http://chem.sis.nlm.nih.gov/chemidplus/
CPDB	Carcinogenic Potency DataBase (Database of carcinogenic potential of chemical compounds supported by the University of California, Berkeley), includes the results collected from the literature of 2-year carcinogenicity tests for different species of mammals: http://potency.berkeley.edu/
CTD	Comparative Toxicogenomics Database contains the data on changes of gene expression, gene ontology, associations and relationships of genes with diseases and biological pathways that give an insight into the mechanisms of influence of chemical compounds on human health: http://ctdbase.org/
Danish (Q)SAR Database	Database of Danish Agency for Environmental Protection includes more than 70 QSAR models and the information about biological effects for 166,072 chemical compounds: http://qsar.food.dtu.dk/
DITOP	Drug-Induced Toxicity-Related Protein Database provides information about drug-induced toxicity associated with proteins. It includes data overdose toxicity, idiosyncratic toxicity, drug-drug interactions and genotoxicity: http://bioinf.xmu.edu.cn/databases/DITOP/

Table 11.1 (continued)

Database	Definition
DRAR-CPI	Prediction of Drug Repositioning and Adverse Reaction via Chemical-Protein Interactome: http://cpi.bio-x.cn/drar/
Drugs@FDA	Information about the US FDA approved drug products: http://www.fda.gov/Drugs/InformationOnDrugs/ucm135821.htm
DSSTox	Distributed Structure-Searchable Toxicity Database Network. Distributed toxicity databases, which is focused on structure search, upload and the standardization of the structural information associated with toxicity data: http://www.epa.gov/ncct/dsstox/index.html
DTome	Information on adverse drug–drug interactions: http://bioinfo.mc.vanderbilt.edu/DTome/
EAFUS	US FDA/CFSAN Everything Added to Food Database. It contains data about ingredients, which are added to food: http://www.accessdata.fda.gov/scripts/fcn/fcnnavigation.cfm?rpt=eafuslisting
ECETOC	European Centre for Ecotoxicology and Toxicology of Chemicals: http://www.ecetoc.org/
ECOTOX	US EPA toxicity database includes information about water and soil organisms: http://cfpub.epa.gov/ecotox/
ESIS	European chemical Substances Information System provides information about the risks and safety of chemical compounds: http://esis.jrc.ec.europa.eu/
eTOX Library	Contains links to articles of toxicological relevance (data that can be used for modeling purposes, computational models, and toxicity mechanisms), public databases, standardized vocabularies and modeling tools: http://cadd.imim.es/etox-library/
EXTOXNET	The database provides information about toxicity of pesticides: http://extoxnet.orst.edu/ghindex.html
FAERS	The FDA Adverse Event Reporting System (FAERS) is a database that contains information on the adverse event and medication error reports submitted to FDA. The database is designed to support the FDA's post-marketing safety surveillance program for drug and therapeutic biologic products: http://www.fda.gov/Drugs/GuidanceComplianceRegulatoryInformation/Surveillance/AdverseDrugEffects/default.htm
FDA Poisonous Plant Database	Database US FDA/ CFSAN with references to the scientific literature describes the study of toxic properties of plants: http://www.accessdata.fda.gov/scripts/plantox/index.cfm
GAC	US NIH/NIEHS Genetic Alterations in Cancer database; Quantitatively described mutations found in the tumors induced by the compounds present in the environment: http://www.niehs.nih.gov/research/resources/databases/gac/index.cfm
Gene-Tox	Expertly curate data of genetic toxicity tests for more than 3000 chemical compounds of US NLM: http://toxnet.nlm.nih.gov/cgi-bin/sis/htmlgen?GENETOX
HERA	Human and Environmental Risk Assessment on Ingredients and Household Cleaning Products. It contains toxicological information about ingredients supplied by European manufacturers: http://www.heraproject.com/RiskAssessment.cfm
Household Products Database	Database provided by Department of Health and Human Services with Material Safety Data Sheets. It includes products used in the household with the assessment of their effects on human health and chemicals, which are presented in these products: http://hpd.nlm.nih.gov/

Table 11.1 (continued)

Database	Definition
IARC Monograph	Monograph of International Agency for Research on Cancer (IARC) describes human carcinogenic risks assessment: http://monographs.iarc.fr/
IRIS	Integrated Risk Information System (IRIS) includes electronic reports on the compounds available in the environment and their potential to cause effects on the health: http://cfpub.epa.gov/ncea/iris/index.cfm
ITER	The database of the human health risk values and the cancer classification for over 680 chemical compounds from the environment: http://www.tera.org/iter/
JECDB	Database of chemical toxicity provided by Ministry of Health, Labour and Welfare, Japan. It contains reports of toxicological tests of compounds in the environment: http://dra4.nihs.go.jp/mhlw_data/jsp/SearchPageENG.jsp
LAZAR	The resource allows to predict hepatotoxicity, mutagenicity and carcinogenicity: http://www.in-silico.de/
MRL	Database provides Minimal Risk Levels for compounds. Supported by US DHHS/ATSDR: http://www.atsdr.cdc.gov/mrls/index.html
MRTD	The database includes maximum recommended therapeutic doses of 1220 drugs. Developed for US FDA/CDER/OPS/SRS/ICSAS: http://www.fda.gov/AboutFDA/CentersOffices/OfficeofMedicalProductsandTobacco/CDER/ucm092199.htm
NPIC	The National Pesticide Information Center at Oregon State University and US EPA, provides information about pesticides and toxicity of compounds: http://npic.orst.edu/
NTP	The National Toxicology Program of US NIH/NIEHS gives information about the compounds that are registered in the U.S. and providing the public interest in terms of health: http://ntp.niehs.nih.gov/
OpenTox	Web resource provides the ecotoxicity data of chemical compounds. It has standardized data exchange, allows to calculate descriptors on-line and to apply of mathematical methods for QSAR modeling: http://www.opentox.org
PAN Pesticide	Pesticide Action Network North America includes data on 6500 pesticides, insecticides and herbicides, water pollution, ecological toxicity and regulatory status: http://www.pesticideinfo.org/
Pesticide Database	Database of pesticides. Supported by Toyohashi University of Technology, Japan: http://chrom.tutms.tut.ac.jp/JINNO/PESDATA/00alphabet.html
PubChem	PubChem provides information on the biological activities of small molecules including toxicity data: http://pubchem.ncbi.nlm.nih.gov/
RAIS	Risk Assessment Information System provides values of specific toxicity of chemical compounds: http://rais.ornl.gov/
RITA	Registry of Industrial Toxicology Animal-data. Developed for comparison and interpretation carcinogenicity studies: http://reni.item.fraunhofer.de/reni/public/rita/
SIDER	Information on marketed medicines and their recorded adverse drug reactions (ADRs): http://sideeffects.embl.de/
STITCH	Search Tool for Interactions of Chemicals (STITCH). STITCH contains interactions between 300,000 small molecules and 2.6 million proteins from 1133 organisms: http://stitch.embl.de/
TEXTRATOX	Agricultural Institute of the University of Tennessee. The TETRATOX database is a collection of toxic potency data for more than 2400 industrial organic compounds of which more than 1600 have been published: http://www.vet.utk.edu/TETRATOX/index.php

Table 11.1 (continued)

Database	Definition
ToxML Editor	Enter toxicity data with a standardized vocabulary for data exchange and integration: http://www.leadscope.com/product_info.php?products_id=51
TOXNET	Databases on toxicology, hazardous chemicals, environmental health, and toxic releases: http://toxnet.nlm.nih.gov/
ToxRefDB	Toxicity Reference Database captures standard toxicological studies of pesticides and other chemical compounds present in the environment, including acute, subacute, chronic toxicity and influence on the development of the body. Support by ToxCast program: http://www.epa.gov/ncct/toxrefdb/
Toxtree	Open source application, which is able to estimate toxic hazard by applying a decision tree: http://toxtree.sourceforge.net/
USGS	The results of aqueous toxicity tests collected by US Geological Survey, Columbia Environmental Research Center: http://137.227.231.90/data/acute/acute.html

types of molecular descriptors. The most popular programs are public Mol2d, which is available from FDA's National Center for Toxicological Research [23] and commercial Dragon, provided by Italian company Talete. Several descriptor and fingerprint generators are also available in KNIME and CHEMBENCH. The list of mostly known generators of molecular descriptors is presented in Table 11.2.

It is necessary to emphasize that sometimes different descriptors generators provide the same descriptors, but called in different ways. It is considered that one should avoid the use of collinear descriptors, when correlation between the descriptors for the compounds in a training set is close to 100% [24]. In addition to descriptors directly calculated from the molecular structure (the topology of a molecule, the number of acceptors, etc.), there are experimentally received descriptors or based on computer predictions (both biological (e.g., absorption, resulting line CaCo-2 cells) and physical-chemical (e.g., LogP)). Currently, 20 computer programs calculates the LogP values and in most cases the prediction results are not the same [24]. The usage of these descriptors may provide an error and noise into developed model. In these cases it is necessary to pay attention to coefficients for these descriptors in QSAR equation. If the descriptor coefficient is equal to or less than the descriptor's error, then the descriptor should be deleted from the equation, since it would result in less accuracy of prediction for compounds from the test set. Another important feature of descriptors is their intervals. For example, for some training sets one descriptor (LogD) can range from −3.572 to 3.773, second descriptor (F) can range from −0.04 to 0.67 and the third (dCox) has values from 0.0000 to 0.0646 [25]. Therefore, for all descriptors which will be used for the model building, it is necessary to carry out the procedure of auto-scaling when the value of all used descriptors vary in the same range, for example from 0 to 1.

Many systems use computer predictions of toxicity based on a two-dimensional representation of the chemical structure in the training set. However, the three-dimensional representation of the molecular structure can be sometimes more reasonable and demonstrative regarding to manifestations of toxicological and pharmaco-

Table 11.2 Generators of molecular descriptors

Name	Source	Definition
AFGen	University of Minnesota	It generates the set of fragment-based descriptors with three different types of topologies: paths, acyclic subgraphs, and arbitrary topology subgraphs: http://glaros.dtc.umn.edu/gkhome/afgen/overview
CDK	Independent developers	The Chemistry Development Kit (CDK) is a Java library for structural chemo- and bioinformatics. Calculates 260 types of descriptors: http://cdk.sourceforge.net/
CODESSA	University of Florida	Program automatically calculates more than 500 types of descriptors (Constitutional, Topological, Geometrical, Electrostatic, Thermodynamic, Quantum-chemical): http://www.codessa-pro.com/index.htm
DRAGON	Talete s.r.l.	Calculates 4885 types of descriptors (represented by all classes of descriptors): http://www.talete.mi.it/index.htm
E-DRAGON	VCCLAB	E-DRAGON remote version of the DRAGON, which is an application for the calculation of molecular descriptors developed by the Milan group of Prof. Todechini: http://www.vcclab.org/
ISIDA	Université de Strasbourg	Software for calculation of Substructural Molecular Fragments (SMF) as well as ISIDA Property-Labeled Fragments (IPLF) descriptors: http://infochim.u-strasbg.fr/spip.php?rubrique49
MODEL	National University of Singapore	Software for a non-commercial use that calculates about 4000 molecular descriptors based on 3D structure of a molecule: http://jing.cz3.nus.edu.sg/cgi-bin/model/model.cgi
MOE	Chemical Computing Group	Calculates over 600 molecular descriptors including topological indices, structural keys, E-state indices, physical properties (such as LogP, molecular weight and molar refractivity), topological polar surface area (TPSA) and CCG's VSA descriptors: http://www.chemcomp.com/software-chem.htm
Mol2D	FDA	Freely available software that calculates more than 700 descriptors types based on a two-dimensional structure of molecules. http://www.fda.gov/ScienceResearch/BioinformaticsTools/Mold2/default.htm
MOLGEN	MOLGEN	Freely available web-service that calculates 708 arithmetical, topological and geometrical descriptors: http://molgen.de/
PreADMET	PreADMET	Calculates more than 2000 descriptors, including both 2D and 3D descriptors: http://preadmet.bmdrc.org/
PaDEL	National University of Singapore	The software currently calculates 863 descriptors (729 1D, 2D descriptors and 134 3D descriptors) and 10 types of fingerprints: http://padel.nus.edu.sg/software/padeldescriptor/
PCLIENT	VCCLAB	PCLIENT (Parameter Client) provides an interface for different programs, which calculate several groups of indices. In total, more than 3000 kinds of descriptors: http://www.vcclab.org/
QSARpro	Vlife	Calculation of over 1000 molecular descriptors of various classes: http://www.vlifesciences.com/products/QSARPro/Product_QSARpro.php

logical effects of some molecules in the biological system, e.g. the ligand-receptor interaction has a distinct isomeric specificity. Influence of three-dimensional characteristics of ligand-receptor interactions, such as dimensional alignment of molecules and fitting electronic properties of molecules based on their surfaces can be very significant in pharmacology [26]. For example, the interaction of ligands with CYP2C9 was extended to the fourth dimension (conformer's analysis) and led to a 4-D classification of drugs [27]. The multi-dimensional relationships between precursors of anabolic steroids and mineral corticoid receptor were modeled and it was shown how these substances disrupted the endocrine system [28].

11.2.4 Mathematical Methods

There are several handbooks [29, 30, 31] with descriptions of mathematical methods (machine learning techniques) used in QSAR modeling. The most known methods are: Naïve Bayes, Decision Trees, Fuzzy Logic, Genetic Algorithms, Multiple Regressions, Neural Networks, Partial Least Squares, Radial Basis Function, Support Vector Machines. Here, we describe the most well-known computer programs, providing the algorithms, used to build (Q)SAR models (Table 11.3).

Currently, the main tendency in the development of QSAR modeling is the use of consensus models. When several models with different machine learning techniques and/or different sets of descriptors are developed based on the same training set. The consensus model results in aggregation of predictions from all developed models. Predictions from the models can be arithmetically averaged (simple unweighted consensus) or can be averaged with some weights for each model (weighted consensus) [32, 33]. It is considered that the use of the consensus model reduces the variability of the individual models, which leads to more reliable predictions [34]. These statements are valid for both QSAR, and SAR models.

11.3 The Practical Use of the Methods for Computational Toxicity Prediction

Toxicological (Q)SAR models are essentially used for the toxicity prediction of new compounds. These models are developed mainly from the training set of compounds with a known activity. If the training set is large and diverse (chemically heterogeneous), then (Q)SAR models based on this set are considered to be global. If the training set consists of a homogeneous compound, the (Q)SAR models, based on this set, are called as local. The assessment of existing (Q)SAR methods to predict most significant toxicological values was performed during preparation of the computational toxicology report by the European Commission under REACH development (early 2000's). According to this report the accurate (Q)SAR models were developed based on non-heterogeneous data (local (Q)SAR models). Nowadays the situation has been considerably changed. At the present time a reasonable

Table 11.3 Computer programs used to create the (Q)SAR models

Name	Source	Definition
ASNN	VCCLAB	Creates a non-linear model based on neural network. Freely available: http://www.vcclab.org/
CHEMBENCH	Carolina Exploratory Center for Chemin-formatics Research	A freely available portal providing cheminformatics research support to molecular modelers, medicinal chemists and quantitative biologists by integrating robust model builders, property and activity predictors, virtual libraries of available chemicals with predicted biological and drug-like properties: http://chembench.mml.unc.edu/
GUSAR	GeneXplain	Software for creation of (Q)SAR models on the basis of atom-centric QNA and MNA descriptors: http://www.genexplain.com/
KNIME	KNIME.com AG	A graphical workbench for the analysis process: data transformation, predictive analytics, visualization and reporting. It includes plug-ins for descriptors generation, creation of QSAR models: http://www.knime.org/
MATLAB	The MathWorks	Interactive environment, using its own language, and includes almost all of the most commonly used machine learning methods in QSAR: http://www.mathworks.com/
MOE	Chemical Computing Group	Construction of QSAR/QSPR models using probabilistic methods and decision trees, PCR and PLS methods: http://www.chemcomp.com/software-chem.htm
PLS	VCCLAB	Partial Least Squares (PLS). The original two-step descriptors selection procedure: http://www.vcclab.org/
PNN	VCCLAB	Polynomial Neural Network (PNN): http://www.vcclab.org/
R	R-project	A freeware product for statistical calculation and graphics creation. R provides a wide range of tools (linear and nonlinear modeling, classical statistical tests, consistent analysis, classification, clustering): http://www.r-project.org/
Small-molecule drug discovery suite	Schrodinger	2D/3D QSAR with a large selection of fingerprint options, shape-based screening, with or without atom properties, ligand-based pharmacophore modelling, R-group analysis: http://www.schrodinger.com/productsuite/1/
StarDrop	Optibrium	The software for QSAR modeling, data analysis and structures optimization: http://www.optibrium.com/
Statistica	StatSoft	Data processing environment includes almost all of the most frequently used mathematical methods in QSAR: http://www.statsoft.com/
Discovery-studio	Accelrys	The software package for QSAR modeling, for data analysis and structures optimization: http://accelrys.com/products/discovery-studio/
WEKA	University of Waikato	A collection of machine learning algorithms for data analysis. It contains tools for data pre-processing, classification, regression, clustering and visualization: http://www.cs.waikato.ac.nz/ml/weka/

Table 11.4 The applicability of (Q)SAR methods

Type of activity	2003	2013
Acute toxicity, fish	General	General
Acute toxicity, algea	Local	Local
Acute toxicity, daphnia	Local	General
Acute toxicity, rodents	Local	General
Human chronic toxicity	Local	Local
Skin irritation	Local	General
Eye irritation	Local	General
Skin sensitization	Local	General
Mammalian chronic toxicity	Local	Local
Mutagenicity in vitro	Local—general	General
Carcinogenicity	Local	Local
Teratogenicity	Local	General

local QSAR models created for non-heterogeneous data; *general* QSAR models created for heterogeneous data

robust classification or continuous global QSAR models on heterogeneous data have been created for many toxicological endpoints. Nevertheless for some, mostly chronic toxic effects only reasonable local QSAR models exist (Table 11.4). The main reason of that is an insufficient size of accurate and heterogeneous experimental data.

There are software products providing an opportunity for prediction of different toxic effects on the basis of already existing (Q)SAR models. Most of them are based on the heterogeneous data of toxicity effects mentioned in Table 11.4. Table 11.5 shows the well known software products which are used for prediction of toxic effects.

The parameters of some software presented in Table 11.5 have been compared in Table 11.6.

The Table 11.6 shows that most programs do not provide the access to the training set, but some programs may be modified (adding the data and retraining of the models). Also, for some programs, there is no information on the external validation of accuracy. It does not allow estimating in advance the accuracy of these programs.

It is a necessary to emphasize that some developers provide a free access to the created models. For instance, GUSAR development team provides on-line web service, which allows predicting the acute rat toxicity for four type of administration: oral, intraperitoneal, intravenous and subcutaneous. Authors used in-house database (more than 10,000 compounds) prepared on the basis of data from SYMYX MDL Toxicity Database (now Accelrys Toxicity Database) for models developing and validation. The models were created using GUSAR program. The developed models were compared to prediction results of the acute rodent toxicity for non-congeneric sets made by ACD/Labs Inc. It was shown [35] that the consensus prediction results obtained by GUSAR models on the test sets were equal or higher than those achieved by ACD/Labs models. In addition, the results of GUSAR predictions were also compared ones given by the T.E.S.T. program (Toxicity Estimation Software Tool) Version 3.0, developed by U.S. Environmental Protection Agency, 2008 on

Table 11.5 Software for prediction of toxicity and adverse side effects

Name	Definition
ACD/Percepta	It predicts ADME, toxicological, and physicochemical property endpoints: http://www.acdlabs.com/products/percepta/
Admensa Interactive	Computer system of QSAR modeling is developed for ADME optimization: http://www.bioportfolio.com/biotech_news/Inpharmatica_3.htm
ASTER	ASTER (Assessment Tools for the Evaluation of Risk) was developed in US EPA. It is a database of toxic effects AQUIRE (AQUatic toxicity Information Retrieval) integrated with an expert system QSAR and valuation techniques for activity prediction: http://www.epa.gov/med/Prods_Pubs/aster.htm
CATABOL	A hybrid expert system for predicting the biotransformation pathways, works together with model which calculates the probabilities of individual transformation of molecule: http://oasis-lmc.org/?section=software&swid=1
Cerius^2	Molecular modeling package providing computer models for predicting ADME and toxic properties of compounds: http://accelrys.com/products/cerius2/
DEREK	DEREK (Deductive Estimation of Risk from Existing Knowledge). An expert system is based on rules. It identifies the so-called toxicophores fragments of the molecule associated with the corresponding activity) and provides the related commentary for them and references to the available information: http://www.lhasalimited.org/
DIGEP-Pred	Web-service for prediction of drug-induced changes in the gene expression profile based on the structural formula of drug-like ompounds: http://www.way2drug.com/GE
DISCAS	The cascade model with an ability to analyze local correlations in the training sets with a large number of variables: http://www.clab.kwansei.ac.jp/mining/discas/discas.html
DvD	An R/Cytoscape plug-in assessing system-wide gene expression data to predict drug side effects and drug repositioning: http://www.ebi.ac.uk/saezrodriguez/DVD
ECOSAR	Ecological Structure Activity Relationships (ECOSAR)—a computer prediction system for assessing the aquatic toxicity of industrial compounds. It was developed in 1979 by US EPA. The program is based on the SAR calculates acute and chronic toxicity for aqueous organisms (fish, aquatic invertebrates and plants): http://www.epa.gov/oppt/newchems/tools/21ecosar.htm
GUSAR	Software for modeling of any quantitative and qualitative relationships based on self-consistent regression and neighborhoods of atoms descriptors (QNA, MNA). GUSAR provides prediction of rat acute toxicity (LD50), ligand interaction with several antitargets and some ecotoxicological end-points: http://www.way2drug.com/GUSAR
HazardExpert	An expert system is based on rules. The program had been developed in 1992 by Smithing Darvas. Program identifies toxicophores (fragments of the molecule associated with the corresponding activity). It is based on US EPA database: Toxic Fragments Knowledge Base: http://www.compudrug.com/
Lazar	Lazy Structure-Activity Relationships. Gets the prediction from the toxicity data by searching similar compounds in the database which are associated with a given toxic activity: http://in-silico.de/
MCASE	MCASE (Multiple Computer Automated Structure Evaluation). Program automatically generates its own descriptors (Biophores) from the training set, which are related to the activity. Biophores may be a group of atoms and distances between them, the physico-chemical parameters. The program estimates the statistical significance of each Biophore for active and inactive compounds, and on the basis of these estimations it makes a prediction: http://www.multicase.com/

Table 11.5 (continued)

Name	Definition
MetaDrug	Toxicity assessment by the generation of networks around the proteins and genes (toxicogenetics platform): http://www.genego.com/metadrug.php
OASIS	A computer system is designed for modeling acute and chronic toxicity, for screening and prioritization of compounds: http://www.oasis-lmc.org/
OncoLogic	An expert system which was developed by US EPA to predict carcinogenicity in rodents. Chemicals are divided into 4 groups: fibers, polymers, metals and organic compounds. OncoLogic makes prediction for each group based on the rules: http://www.epa.gov/oppt/newchems/tools/oncologic.htm
PASS	The program predicts the biological activity spectrum of chemical compounds on the basis of their structure. A freely available web-site provides prediction for several thousand types of biological activity, including pharmacological effects, mechanisms of action, adverse or toxic effects, interaction with metabolic enzymes, transporters, and influence on the gene expression: http://way2drug.com/PASS
PreADMET	The calculation of the important descriptors and neural network to create QSAR models: http://preadmet.bmdrc.org/
SADR-Gengle	PubMed records text mining-based data on 6 serious adverse drug reaction: http://gengle.bio-x.cn/SADR
SePreSA	Binding pocket polymorphism-based serious adverse drug reaction predictor: http://sepresa.bio-x.cn
SRC EPIWIN (EPI Suite)	Package of freely downloadable models from the site of Syracuse Research Corporation. It calculates physicochemical properties and predicts bioconcentration factor (BCF) based on linear regression, log octanol/ water partition (LogKow) and taking into account type of substances (ionic and non-ionic): http://www.epa.gov/oppt/exposure/pubs/episuite.htm
TOPKAT	TOPKAT (Toxicity Prediction by Komputer Assisted Technology) uses multiple linear regression equation (quantitative prediction) or two-group discriminant function for qualitative prediction of different effects: mutagenicity, carcinogenicity and teratogenicity. It uses substructural, electro-topological descriptors and bonds between the atoms from the library containing about 3000 molecular fragments: http://accelrys.com/products/discovery-studio/toxicology/

the available training and test sets with data on oral acute toxicity measured in LD_{50} (mmol/kg) values for 7286 compounds. It was demonstrated that GUSAR models had the highest accuracy in comparison with the models from T.E.S.T. program and provided the highest speed of prediction (18 times faster).

The developed models are freely available on the web site: http://www.way2drug.com/gusar/acutoxpredict.html. The characteristics of the models are given in Table 11.7. The service includes an on-line chemical editor [36] for drawing the studied structure. It provides acute toxicity prediction results in the different units, but does not support the batch prediction mode.

Table 11.6 Comparison of software for the toxicity prediction

	ACD/ Percepta	ADMET predictor	CASE Ultra	Meteor/ Derek	GUSAR	PASS	StarDrop	TOPKAT
Modification	–	–	+	+	+	+	+	–
Availability of training set	+[a]	–	+	–	–	–	–	+
Method description	+	+	+	+	+	+	+	+
Internal validation	+	+	+	+	+	+	+	+
External validation	+	+	–	+[a]	+	+	+	+[a]
Applicability domain	+	–	–	+	+	+	+	+
Number of end-points (ADME/T)	30/26	37/30	6/471	dozens/40	0/44	282/638	11/40	7/16
Source	ACD/ Labs	Simulations Plus, Inc.	Multicase Inc.	Lhasa Ltd.	GeneXplain	GeneXplain	Optibrium, Ltd.	Accelerys

[a] for some models

Table 11.7 Characteristics of GUSAR models for the rat acute toxicity predictions

Administration	N_{train}	N_{test}	N models	R^2	Q^2	R^2_{test}	$RMSE_{test}$	Coverage, %
Oral	6280	2692	40	0.61	0.57	0.59	0.57	97.5
Intraperitoneal	2480	1065	68	0.66	0.56	0.57	0.57	96.1
Intravenous	920	394	50	0.73	0.66	0.63	0.62	99.2
Subcutaneous	759	325	7	0.69	0.59	0.50	0.69	92.0

N_{train} number of compounds in the training set, N_{test} number of compounds in the test set, R^2 average R^2 of the models calculated for the appropriate training set, Q^2 average Q^2 of the models calculated for the appropriate training set, *Coverage* percent of the compounds from the test set fell in the Applicability Domain

11.4 Development of New Medicines

Availability of side effects in drugs is one of the major problems in clinical practice. Serious pathologies that affect the number of patients due to taking of some drugs are the cause for rejection of using these drugs in the clinic. Approximately 35–40 % failures of the drug application fall on these cases [37, 38, 39]. This leads to withdrawal of the drug from the market and increasing costs, since the usage of these drugs results in severe injuries in various organs and tissues, which may cause the patient death. The central nervous system, cardiovascular system and liver (9–26 % of all cases of stopping treatment) are affected in the most commonly cases [40, 41].

Table 11.8 Modified OECD project of toxicity classification of chemicals

Parameter	1 class	2 class	3 class	4 class	5 class	Low toxic
LD_{50}, mg/kg (oral)	≤5	(5:50]	(50:300]	(300:2000]	(2000:5000]	>5000
LD50, mg/kg (i. v.)	≤0.7	(0.7:7]	(7:40]	(40:300]	(300:700]	>700
LD50, mg/kg (i. p.)	≤1	(1:10]	(10:75]	(75:500]	(500:1250]	>1250
LC50, ppm (inhalation)	≤100	(100:500]	(500:2500]	(2500:5000]		>5000
LD50, mg/kg (s. c.)	≤2	(2:20]	(20:150]	(150:1000]	(1000:2500]	>2500
LD50, mg/kg (skin)	≤50	(50:200]	(200:1000]	(1000:20000]		>20000

There are several indexes which are useful to assess the general drug toxicity in clinical studies or to compare drug-candidates on the basis of experimental or predicted data:

1. **Therapeutic index (TI):** $TI = LD_{50}/ED_{50}$ or $TI = LD_{10}/ED_{90}$. The safest drug has the highest index. Review of QSAR models for prediction of rodent LD_{50} values is presented above.
2. **Maximum recommended therapeutic dose (MRTD)**. The MRDD is essentially equivalent to the NOAEL (no observed adverse effect level) in humans, a dose beyond which adverse (toxicological) or undesirable pharmacological effects are observed. The correlation between MRDD and LD_{10} is about 94%. FDA's Center for Drug Evaluation and Research provides MRTD database [42] with information on MRTD for 1220 drugs. QSAR models for prediction of MDTD are provided by Simulation Plus and ACD/Labs. QSAR models for mice and rat maximum tolerated dose are provided by Accelrys.

There is a classification of hazard classes to humans based on LD_{50} values for rats and mice (OECD Project of Toxicity Classification of Chemicals). It is presented in Table 11.8.

As stated above, assessment of the specific drug toxicity is very important during the drug development process. For instance, it is a well known that the non-steroidal anti-inflammatory drug (NSAID) Vioxx (rofecoxib) produces the side effects. It is a selective inhibitor of cyclooxygenase 2, which has been developed for the treatment of arthritis. The basic idea to create such NSAIDs is the reduction of ulcerogenic effect due to affinity decreasing of cyclooxygenase 1 (COX 1), towards a greater selectivity for cyclooxygenase 2 (COX 2). COX 1 is a conservative enzyme in contrast to COX 2, which is synthesized only during inflammation and is involved in the synthesis of prostaglandins, providing a protective effect on the gastric mucosa. In clinical trials of VIGOR (VIOXX Gastrointestinal Outcomes Research), a low ulcerogenic effect of rofecoxib compared to naproxen was obtained, but, in addition, toxic effects on the cardiovascular system were also found. However, further clinical trials have not revealed the increase in risk of these side effects, which became the basis for the product approval by regulatory authorities and its release

Table 11.9 The most frequent side effects of drugs

Localization	Effect
Gastrointestinal tract	Hepatitis and/or hepatocellular damage; Constipation; Diarrhea; Nausea and/or vomiting; Ulceration; Pancreatitis; Dry mouth
Blood	Agranulocytosis; Hemolytic anemia; Pancytopenia; Thrombocytopenia; Megaloblastic anemia; Clotting and/or bleeding; Eosinophilia
Cardiovascular system	Arrhythmias; Hypotension; Hypertension; Congestive heart failure; Angina and/or chest pain; Pericarditis; Cardiomyopathy
Integuments	Erythemas; Hyperpigmentation; Photodermatitis; Eczema; Urticaria; Acne; Alopecia
Metabolism	Hyperglycemia; Hypoglycemia; Hyperkalemia; Hypokalemia; Metabolic acidosis; Hyperuricemia; Hyponatremia
Respiratory System	Airway obstruction; Pulmonary infiltrates; Pulmonary edema; Respiratory depression; Nasal congestion
Musculoskeletal system	Myalgia and/or myopathy; Rhabdomyolysis; Osteoporosis
Urogenital System	Nephritis; Nephrosis; Tubular necrosis; Renal dysfunction; Bladder dysfunction; Nephrolythiasis
Endocrine system	Thyroid dysfunction; Sexual disfunction; Gynecomastia; Addison syndrome; Galactorrhea
Nervous system	Seizures; Tremor; Sleep disorders; Peripheral neuropathy; Headache; Extrapyramidal effects
Central nervous system	Delirium, confusion; Depression; Hallucination; Drowsiness; Schizophrenic and/or paranoid reactions; Sleep disturbances
Eye	Disturbed color vision; Cataract; Optic neuritis; Retinopathy; Glaucoma; Corneal opacity
Hearing	Deafness; Vestibular disorders

to the market. Then a prospective, multicenter, randomized, placebo-controlled, double-blind clinical trial APPROV (Adenomatous Polyp Prevention on VIOXX) was performed. Its purpose was to evaluate the effectiveness for prevention of colon polyps in patients with colorectal adenomatous polyps in the medical history. An increasing risk of cardiovascular events such as myocardial infarction and stroke was found. Therefore, clinical trials (APPROV) were terminated and the drug Vioxx was withdrawn from the market [43].

Along with these critical side effects, like VIOXXa, there are less serious side effects, such as nausea, dizziness, dry mouth, skin rash, etc. They are significant only with the long-term usage of drug compounds. Table 11.9 shows the most frequent adverse effects of therapeutic compounds [44].

Traditionally, the drug development process starts with definition of the target protein. An interaction of the drug with this target leads to the manifestation of a therapeutic effect. It is also well known, that drugs in most cases interact with different additional targets (off-targets). Availability of these additional mechanisms of action provides the emergence of side effects. The proteins which led to the appearance of side effects are known as antitargets. The list of antitargets associated with the cardiovascular system was published by Whitebread with co-authors [44]. The antitarget interactions predicted by GUSAR Online web-service [http://www.way2drug.com/GUSAR] and associated with the toxic and side effects are shown in Table 11.10.

Table 11.10 Relationships between drug action on antitargets and adverse side effects from PharmaExpert

Action on antitarget	Related side effects
5HT 2A receptor antagonist	Hypnotic; Sedative
5-HT 2C receptor antagonist	Obesity; Weight loss
Alpha 1a adrenergic receptor antagonist	Dizziness; Flushing; Impotence; Nasal congestion; Postural (orthostatic) hypotension; Tachycardia; Weakness
Alpha 1b adrenergic receptor antagonist	Ejaculation dysfunction
Alpha 2a adrenergic receptor antagonist	Anxiety; Depression
Androgen receptor antagonist	Virilization; gynecomastia; hepatic pelioza; hepatoma
Carbonic anhydrase inhibition	Acidosis, metabolic; Alopecia (hair loss); Anaphylaxis; Anemia, aplastic; Anxiety; Bone marrow suppression; Chronic fatigue syndrome; Confusion; Corneal edema; Depression; Dyspepsia; Dysphagia; Impotence; Irritation; Keratitis; Malaise; Nausea; Nephrotoxic; Neuroprotector; Paralysis; Pulmonary edema; Renal tubular acidosis; Stevens-Johnson syndrome; Taste disturbance; Thrombocytopenia; Urinary stone; Vision blurring; Weight loss
Estrogen receptor antagonist	Depression; Headache; Obesity; Sickness; Hot flashes; Puffiness
Delta-type opioid receptor antagonist	Reverse analgesia; Opioid withdrawal symptoms
MAO A inhibitor	Blood pressure lability; Bradycardia; Chorea; Convulsions; Delirium; Diarrhea; Hepatotoxicity; Drowsiness
Mu-type opioid receptor antagonist	Laxative
Sodium- and chloride-dependent GABA transporter 1 inhibitor	Neurotoxicity
Sodium-dependent dopamine transporter inhibitor	Neurotoxicity
Sodium-dependent serotonin transporter inhibitor	Acute respiratory distress syndrome (ARDS); Agitation; Akathisia; Anorgasmia; Antialcoholic; Constipation; Convulsant; Diarrhea; Dizziness; Drowsiness; Dysesthesia; Dyskinesia; Dystonia; Ecchymosis; Ejaculation dysfunction; Emetic; Epistaxis; Exanthema; Extrapyramidal effect; Glaucoma; Hallucination, visual; Headache; Hemorrhage; Hepatotoxic; Hormone secretion (SIADH); Hot flush; Hyperprolactinemia; Hypersexuality; Hypertensive; Hypoglycemic; Hypomania; Hyponatremia; Hypothermic; Insomnia; Ischemic colitis; Mania; Menorrhagia; Myoclonus; Nausea; Neuroleptic malignant syndrome; Panic; Parkinsonism; Purpura; QT interval prolongation; Serotonin syndrome; Sexual dysfunction; Shivering; Sleep disturbance; Stevens-Johnson syndrome; Sweating; Syncope; Taste disturbance; Thrombocytopenia; Tremor; Weight gain; Weight loss; Xerostomia

PharmaExpert is a commercial software providing data on relationships between drug interactions with antitargets and specific toxicity [45]. The part of known antitargets associated with specific toxicity and side effects are freely available in Drug Adverse Reaction Target [http://bidd.nus.edu.sg/group/drt/dart.asp] and Drug-Induced Toxicity-Related Protein [http://bioinf.xmu.edu.cn/databases/DITOP/] databases. Interaction with a primary target may also result in adverse reactions due to no local distribution of a target in the body and its multiple functions.

All side effects are usually divided into five classes [46, 47]:

a. Includes dose-dependent side effects, which are often found in preclinical studies.
b. Includes side effects, the frequency of which is not dose-dependent, they are often detected in observation of marketed drugs.
c. Includes adaptive functional changes in the body during a long-term drug usage. They are identified in the measurement of functional parameters during the long-term studies.
d. Includes delayed side effects such as carcinogenicity and teratogenicity.
e. Includes those side effects, which can lead to refuse of the drug.

Side effects of B, C, D and E classes are of considerable interest due to the difficulty of timely registration.

Currently, QSAR (quantitative structure-activity relationship) and SAR (structure-activity relationship) methods are widely used for computational prediction of different toxicity types, such as cardio-, hepato-, renal toxicity, teratogenicity, and carcinogenicity.

In addition to traditional SAR methods used in DEREK, TOPKAT, MCASE there are examples of using the method of molecular docking to predict side effects of drugs. For example, Ji and co-authors described the search of targets associated with side effects for various anti-HIV drugs available on the market [48]. For the docking procedure, the authors used the docking program INVDOCK [49]. This program was designed for an automated search of targets for low-weight ligands, by attempting to integrate them into "cavities" of each protein, e.g. search of the corresponding binding sites. Three-dimensional protein structures, data about inhibitors, activators, agonists, antagonists, and the toxic side effects data caused by interaction with targets were obtained from the DART database [50]. As a result, for 11 anti-HIV drugs, which are inhibitors of HIV protease and reverse transcriptase, nonviral target molecule interactions have been found which lead to the side effects. Existence of two targets was shown for delavirdine: DNA polymerase beta and DNA topoisomerase I. The action on these targets causes pancreatitis, nausea, vomiting, leukopenia, peripheral neuropathy, and abdominal pain.

This method also has significant drawbacks. Firstly, it requires a lot of computational power and time. Secondly, a clear correlation between the target and side effects is not always established, and mechanisms of adverse drug reactions are not always known. Thirdly, three-dimensional structures are known only for a limited number of proteins. All these drawbacks limit the application of the method.

Current methods based on "structure–activity" relationships, have also some disadvantages. Firstly, they are not applicable to predict biological activities of polymers, especially proteins, and inorganic compounds that can act as drugs. Secondly, many experimental data is obtained from animals. Therefore predictions based on these end-points cannot be always extrapolated to humans. Thirdly, predictions were made for the substance itself, but toxic or side effects might be provided by its metabolites. Fourthly, most of the methods allow predicting only one type of toxicity or one mechanism of side effects.

Thus, the desirable method is to predict the full range of side effects, without a three-dimensional structure of the target, without spending a lot of time and computing resources. These capabilities are presented in the program PASS—Prediction of Activity Spectra for Substances (www.way2drug.ru/PASS). Its algorithm is based on MNA descriptors for describing the structure of compounds and modified Bayesian approach for prediction of biological activities [31]. The program allows predicting both the mechanisms of toxic effects (interaction with antitargets) and main side effects of compounds [51]. Table 11.11–11.12 shows the main toxic side effects and interaction with antitargets predicted by PASS 2012, the number of active compounds in the training set and the prediction accuracy. An average accuracy of prediction calculated by leave-one-out cross-validation is 87.7% for side and toxic effects and 96.7% for interaction with antitargets.

GUSAR software is another method corresponded to above mentioned requirements. It is based on Multilevel and Quantitative Neighbourhoods of Atoms (MNA, QNA) descriptors [52, 53] and the self-consistent regression (SCR) algorithm [52, 54]. It was shown that GUSAR may successfully be applied for multiple QSAR tasks [35, 52, 54, 55, 56].

A freely available on-line service for the simultaneous prediction of thirty two antitarget end-points (IC_{50}, K_i and K_{act}) has been developed on the basis of GUSAR [http://www.way2drug.com/GUSAR/Antitargets/]. These antitarget end-points are related to 18 proteins: 13 receptors, 2 enzymes and 3 transporters. The relationships between predicted drug interactions with antitargets and adverse side effects are represented in Table 11.10. The accuracy of end-point predictions, calculated for the appropriate external test sets was typically in the range of R^2_{test}=0.6–0.9. This service provides a reasonable computational speed (about 2 compounds per second for the simultaneous prediction of 32 antitarget end-points).

In addition, the web service allows calculating the total number of targets for which the input compound has been predicted to be active. This can be useful for selection and prioritization of compounds during the drug discovery process. A particular compound can be considered as a potential source of adverse drug reactions (ADRs) if interactions with three or more antitargets have been predicted and exceed the cut-off value (1 μM). Compounds for which antitargets are not predicted can be selected for further development as potential drugs. The service can also help medical chemists to determine targets (molecular mechanism of toxicity) on which a particular compound should be tested experimentally, to avoid ADRs.

Fourteen known drugs, which had been withdrawn from the market, were analyzed by Zakharov with co-authors using GUSAR Online web-service [57]. In addi-

Table 11.11 Main adverse and toxic effects predicted by PASS 2012

Activity name	*N*	AUC, %
Abortion inducer	38	94.9
Agranulocytosis	408	84.1
Allergic reaction	789	81.9
Anaphylaxis	392	83.0
Anemia	228	84.8
Arrhythmogenic	481	81.2
Carcinogenic	1813	93.3
Carcinogenic, group 1	23	86.9
Carcinogenic, group 2A	36	95.5
Carcinogenic, group 2B	194	96.0
Carcinogenic, group 3	387	94.8
Carcinogenic, mouse	456	92.1
Carcinogenic, rat	603	93.1
Cardiodepressant	160	91.5
Cardiotoxic	982	82.4
Cataract	115	82.0
Coma	311	82.6
Convulsant	890	85.3
Cytotoxic	721	93.1
DNA damaging	466	96.3
Dependence	70	84.1
Depression	125	81.7
Embryotoxic	1743	90.7
Emetic	788	82.0
Endocrine disruptor	223	85.2
Eye irritation, high	528	93.7
Eye irritation, moderate	229	96.7
Eye irritation, weak	348	96.8
Genotoxic	146	94.9
Hematotoxic	1142	82.7
Hepatotoxic	1033	83.7
Hypercalcaemic	19	84.7
Hypercholesterolemic	10	85.1
Hyperglycemic	71	80.1
Hypertensive	736	81.5
Hypertensive, ophthalmic	10	95.8
Hypocalcaemic	29	85.4
Hypoglycemic	413	91.9
Hypothermic	295	84.4
Immunotoxin	43	79.1
Leukopenia	542	82.0
Mutagenic	3542	97.1
Nephrotoxic	787	84.0
Neurotoxic	638	85.8
Ocular toxicity	639	82.1
Ototoxicity	43	87.4
Pneumotoxic	17	95.5
QT interval prolongation	117	86.2

Table 11.11 (continued)

Activity name	*N*	AUC, %
Reproductive dysfunction	219	79.2
Respiratory failure	323	83.4
Sedative	1223	91.8
Sensitization	161	92.5
Skin irritative effect	1117	95.9
Teratogen	1552	90.5
Thrombocytopenia	494	81.6
Thrombocytopoiesis inhibitor	11	84.0
Thrombophlebitis	146	80.2
Torsades de pointes	90	82.2
Toxic, respiratory center	15	92.7
Ulceration	53	90.7

N number of active compounds in the training set *AUC* Area Under Curve, calculated by leave-one-out cross-validation procedure

Table 11.12 Antitarget activities predicted by PASS 2012

Activity name	*N*	AUC, %
11-Beta-hydroxysteroid dehydrogenase 2 inhibitor	253	99.8
5 Hydroxytryptamine 1A agonist	1131	98.8
5 Hydroxytryptamine 2A agonist	59	95.0
5 Hydroxytryptamine 2B agonist	9	94.7
5 Hydroxytryptamine 2C antagonist	1781	98.5
5 Hydroxytryptamine 3 agonist	73	96.9
5 Hydroxytryptamine uptake inhibitor	4311	98.5
ATPase inhibitor	150	93.3
Acetylcholine M1 receptor antagonist	1263	98.8
Acetylcholine M2 receptor agonist	39	97.1
Acetylcholine M2 receptor antagonist	1374	98.9
Acetylcholinesterase inhibitor	1646	97.5
Aconitate hydratase inhibitor	22	99.2
Acyl-CoA dehydrogenase inhibitor	15	99.4
Adenosine deaminase inhibitor	165	98.4
Adenylate cyclase I inhibitor	54	99.7
Adenylate cyclase inhibitor	174	95.0
Adenylate kinase inhibitor	23	93.5
Adrenaline uptake inhibitor	2366	98.8
Alcohol dehydrogenase inhibitor	119	95.7
Aldosterone antagonist	58	87.2
Alkaline phosphatase inhibitor	336	95.7
Alpha 1a adrenoreceptor agonist	48	100.0
Alpha 1a adrenoreceptor antagonist	3142	98.2
Alpha 2a adrenoreceptor antagonist	629	98.1
Alpha galactosidase inhibitor	16	100.0
Alpha-mannosidase inhibitor	59	98.5
Aminopeptidase A inhibitor	55	99.9
Aminopeptidase N inhibitor	183	98.2
Androgen agonist	219	98.5
Arginase inhibitor	34	99.3

Table 11.12 (continued)

Activity name	*N*	AUC, %
Argininosuccinate synthase inhibitor	21	98.9
Aryl hydrocarbon receptor agonist	12	92.1
Beta 1 adrenoreceptor agonist	49	97.6
Beta 1 adrenoreceptor antagonist	677	98.9
Beta 2 adrenoreceptor agonist	683	99.5
Beta 2 adrenoreceptor antagonist	670	99.0
Butyrylcholinesterase inhibitor	952	98.9
CYP1A2 inhibitor	412	89.0
CYP2C9 inhibitor	748	90.1
CYP2D6 inhibitor	994	93.6
CYP3A4 inhibitor	1396	91.1
Ca2+-transporting ATPase inhibitor	12	86.1
Carbamoyl-phosphate synthase (ammonia) inhibitor	15	99.0
Carbonic anhydrase I inhibitor	812	99.9
Carbonic anhydrase II inhibitor	1211	99.8
Carbonic anhydrase inhibitor	1344	99.8
Catalase inhibitor	27	96.1
Catechol O methyltransferase inhibitor	58	98.2
Cyclooxygenase inhibitor	4216	96.6
Cystathionine beta-synthase inhibitor	12	99.0
Cytochrome oxidase inhibitor	11	99.9
DNA polymerase beta inhibitor	31	99.6
DOPA decarboxylase inhibitor	10	98.5
Diamine oxidase inhibitor	32	99.0
Dihydrofolate reductase inhibitor	1812	99.3
Dipeptidyl peptidase IV inhibitor	1195	99.0
Dopamine D2 antagonist	4605	98.1
Electron transport complex I inhibitor	26	98.5
Estrogen agonist	734	98.0
Excitatory amino acid transporter 2 inhibitor	44	99.9
Fumarate hydratase inhibitor	24	99.6
GABA A receptor agonist	231	94.5
GABA A receptor antagonist	2765	98.7
GABA aminotransferase inhibitor	14	85.3
GABA transporter 1 inhibitor	150	99.6
Glucocorticoid agonist	307	99.4
Glucose-6-phosphate isomerase inhibitor	22	99.9
Glutamate dehydrogenase inhibitor	34	96.0
HERG channel blocker	1207	95.2
HMG CoA reductase inhibitor	484	99.2
Heat shock protein 70 antagonist	23	96.4
Hexokinase inhibitor	193	95.6
Histamine H1 receptor antagonist	858	97.7
Histamine H2 receptor antagonist	375	98.2
Hypoxanthine phosphoribosyltransferase inhibitor	37	99.0
Insulin and insulin analogs	10	78.7
Insulin antagonist	339	96.7
Luteinizing hormone-releasing hormone antagonist	969	98.8

Table 11.12 (continued)

Activity name	*N*	AUC, %
Lysine carboxypeptidase inhibitor	35	99.8
MAO A inhibitor	463	98.2
MAO inhibitor	1124	97.1
Na+K+ transporting ATPase inhibitor	165	98.5
Neutral endopeptidase inhibitor	698	99.5
Opioid delta receptor antagonist	2044	99.0
Opioid mu receptor agonist	133	95.7
Ornithine carbamoyltransferase inhibitor	40	99.0
Peroxidase inhibitor	16	93.1
Phenylalanine 4-hydroxylase inhibitor	30	97.9
Phosphodiesterase inhibitor	5905	97.1
Phosphofructokinase-1 inhibitor	33	98.5
Phosphoglycerate kinase inhibitor	189	99.6
Phospholipase A2 inhibitor	541	96.5
Phospholipase C inhibitor	51	92.8
Phosphorylase inhibitor	279	98.5
Platelet activating factor antagonist	2431	97.2
Prostaglandin F2 alpha agonist	26	100.0
Protein kinase C stimulant	61	98.8
Pyruvate kinase inhibitor	38	97.3
Retinoic acid alpha receptor agonist	27	99.7
S-adenosyl-L-homocysteine hydrolase inhibitor	69	100.0
Sarcoplasmic reticulum calcium ATPase inhibitor	11	94.3
Sodium channel blocker	1004	93.8
Succinate dehydrogenase inhibitor	14	95.7
Superoxide dismutase inhibitor	15	88.4
Thyroid hormone agonist	104	98.6
Thyroid hormone beta agonist	74	100.0
Topoisomerase I inhibitor	478	96.3
Topoisomerase II alpha inhibitor	44	96.7
Triose-phosphate isomerase inhibitor	23	98.6
Tyrosine 3 hydroxylase inhibitor	12	88.4
UDP-glucose 4-epimerase inhibitor	28	97.2
Ubiquinol-cytochrome-c reductase inhibitor	10	81.6

N number of active compounds in the training set *AUC* Area Under Curve, calculated by leave-one-out cross-validation procedure

tion to the withdrawn drugs, seven currently marketed drugs were also analyzed to find out the difference in the number of predicted antitargets. The prediction results are presented in Table 11.13.

The results have shown that the number of predicted antitargets for withdrawn drugs is considerably higher than one for marketed drugs. Thus, this service can successfully be applied for the selection and prioritization of safe compounds during the drug discovery process.

Table 11.13 Prediction results for withdrawn and marketed drugs

Drug name	State	The number of predicted antitargets
Amineptine	Withdrawn	13
Duract	Withdrawn	8
Vioxx	Withdrawn	7
Astemizole	Withdrawn	17
Cerivastatin	Withdrawn	8
Chlormezanone	Withdrawn	10
Fenfluramine	Withdrawn	11
Flosequinan	Withdrawn	11
Glafenine	Withdrawn	14
Grepafloxacin	Withdrawn	12
Mibefradil	Withdrawn	16
Rofecoxib	Withdrawn	7
Troglitazone	Withdrawn	14
Ximelagatran	Withdrawn	14
Aspirin	Marketed	2
Ibuprofen	Marketed	2
Valtrex	Marketed	3
Microzide	Marketed	3
Neurontin	Marketed	3
Enoxaparin	Marketed	2
Lyrica	Marketed	2

Typically, the affinity of a pharmaceutical agent to the drug target should exceed the affinity to antitargets for at least one to two orders of magnitude. The medium affinity of current drugs to drug targets is about 16 nM, ranging from 16 mM to 1.6 pM [58]. Therefore, GUSAR prediction of interaction with antitarget(s) should be carefully considered in each individual case taking into account the predicted/measured affinity of an analyzed compound to the drug target. A particular attention should be paid to compounds, for which the predicted affinity of interaction with four or more antitargets exceeded 1 μM.

One of the main limitations for *in silico* assessment of toxic/side effects based on prediction of ligand interaction with antitargets is an insufficient knowledge about "target-side effects" relationships. The computer evaluation of relationships between the predicted targets or drug-target interactions and known side effects is carried out using statistical methods of disproportionality analysis. The prediction of drug-target interactions were made by a similarity assessment with the compounds from ChEMBL database [59], PASS prediction of biological activity spectra [60] or docking [61]. Another method for revealing associations between targets and side effects is based on creation of SAR models for compounds interacted with each target and causing the side effect. For example, if the prediction of interactions with targets and prediction of side effects carried out based on the Bayesian approach, the relationships between targets and adverse effects can be calculated by the Pearson correlation coefficient between the conditional probability P (A|Di) and P (M|Di) models (A—a side effect, M—target, Di—descriptor) [62].

Biological pathways and biological processes can characterize mechanisms of drug side effects at more general level than the target molecules. Their assessment is based on prediction of drug interactions with the targets and analysis of drug-induced changes in gene expression profiles. In the first case, assessment of profiles in drug interactions with proteins is made for a set of compounds, some of which cause, and others do not cause side effects. Each of the predicted protein is associated with the biological pathway (signaling, metabolic, and regulatory). Every biological pathway receives a score, which is calculated as the sum of probable interactions with proteins that are a part of the way for all compounds causing an appropriate side effect. The same estimation is calculated for the compounds that do not cause this side effect. After that the ratio of estimates is calculated. Pathways are considered associated with a side effect if the ratio is greater than 1, or the way, the interaction with which is predicted only for compounds that cause the side effect [63].

Drug-induced changes in gene expression profiles can be used for searching of the biological processes associated with side effects by the gene set enrichment analysis [64]. In this method a set of genes involving in a biological process is created. Then, a list of estimates for changes in the gene expression is calculated based on comparison of gene expression after the compound action and in the norm. The list is sorted by ascending or descending the estimates subject to direction in which the gene expression is changed (hyper-or hypo-expression). The basic analysis hypothesis is that the genes involved in the same biological process should be clustered mainly on the top or bottom of the sorted list.

In 2006, Lamb and co-authors introduced Connectivity Map (CMap) as a phenotypic-based drug discovery approach based on comparison of the disease gene signature and drug-induced changes in gene expression profiles [65]. It was shown that CMap approach can be used to reveal side effects of drugs [66]. The constructed multigene expression signature was used to predict future onset of the proximal tubular injury in rats [67]. CMap approach has a limitation due to its applicability only for drugs having experimentally determined drug-induced changes of gene expression and it cannot be used for new drugs or new drug-candidates. This limitation may be partly overcome by prediction of possible drug-induced changes in the gene expression for new drug-like compounds on the basis of existed experimental microarray data. Such possibility is realized on a freely available DIGEP-Pred web-service (http://www.way2drug.com/GE). It also provides the links between gene names in the predicted drug-induced changes in the gene expression and Comparative Toxicogenomics Database [68] which simplifies interpretation of predicted results due to the access to relationships of genes with diseases, side effects and biological pathways [69].

One of most important parameters considered during the drug development is assessment of a potential participation of drugs in drug-drug interactions. Usage of multiple drugs is a common practice in the treatment of many diseases, including cardiac insufficiency (diuretics and vasodilators), malignant neoplasms and some infectious diseases (HIV, hepatitis C, etc.). Despite a positive effect of the simultaneous usage for several drugs, there is a risk of negative influence on each other and on the human body. On the average the frequency of adverse drug interactions is between 3–5 to 20 %, if patients take simultaneously from 2 to 10 drugs [70]. The possible estimation of those interactions will increase the safety of drug therapy.

Several computational studies of drug-drug interactions depending on the molecular mechanisms of action and biotransformation pathways were performed [71, 72]. They are mainly related to the analysis of drug metabolizing enzymes, such as isoforms of cytochrome P450, and are restricted by narrow classes of chemical compounds.

Drug-drug interactions can be a physical (e.g., changing the pH, which depends on the absorption of these compounds as ketoconazole and glipizide), chemical (e.g., ciprofloxacin is a chelator of cations such as aluminum, magnesium and iron), and biological, which depends on interactions with human proteins. The last type of interaction is of great interest for computational predictions.

There are many mechanisms of the drug-drug interactions in humans. They may be divided into two large groups: pharmacokinetic and pharmacodynamic drug interactions. Pharmacokinetic drug interactions include cases where one drug affects the absorption, distribution, metabolism and excretion of another drug. Pharmacodynamic drug interactions include cases where drugs have additive or antagonistic pharmacological effects.

11.4.1 Pharmacokinetic Drug-Drug Interactions

The key molecular mechanisms were identified during the detailed analysis of pharmacokinetic drug-drug interactions. Transport proteins play an important role for manifestation of negative drug interactions during absorption and excretion [73]. The reason for changing the drug absorption and excretion can be the direct competition for reaction with transport protein (compounds are substrates of the same transporter enzyme) and the influence of one drug on the activity (inhibition) or the amount (induction of expression) of transport protein, while the other drug is a substrate of this transport protein. Therefore, the computer prediction of interaction for chemical compounds with transporter proteins can be used to assess possible drug-drug interactions. The QSAR models for prediction of drug interaction with transporters are provided by Simulation Plus Inc. (OATP1B1 transporter), ACD/Labs (P-glycoprotein), Optibrium (P-glycoprotein). QSAR modeling for the several major transporters including MDR1, BCRP, MRP1–4, PEPT1, ASBT, OATP2B1, OCT1, and MCT1 was made by Sedykh with co-authors [74]. The most representative profile of interaction with transporters is calculated by PASS software (Table 11.14).

Drug distribution in the body depends on several factors: the total amount of extracellular liquid, the percentage of adipose tissue and an ability to bind with plasma proteins, which depends on a structural formula of the compound. The last factor plays a significant role in process of drug-drug interactions during distribution. Albumin and alpha-1 glycoprotein plasma proteins are responsible for the transfer of major drugs and undesirable interactions between these drugs may occur due to

Table 11.14 Interactions with transport proteins and transport systems predicted by PASS 2012

Activity name	*N*	AUC, %
5 Hydroxytryptamine uptake inhibitor	4311	98.5
ATP-binding cassette G2 inhibitor	44	96.9
Adenosine uptake inhibitor	55	100.0
Adrenaline uptake inhibitor	2366	98.8
Amino acid-polyamine-organocation (APC) transporter antagonist	82	100.0
Apical sodium codependent bile acid transporter inhibitor	16	100.0
Cationic amino acid transporter 4 inhibitor	32	100.0
Dopamine transporter inhibitor	1221	99.1
Dopamine uptake inhibitor	1848	99.2
Electron transport complex I inhibitor	26	98.5
Endocannabinoid uptake inhibitor	29	99.2
Equilibrative nucleoside transport protein 1 inhibitor	97	99.5
Equilibrative nucleoside transport protein inhibitor	127	99.5
Excitatory amino acid transporter 1 inhibitor	33	99.9
Excitatory amino acid transporter 2 inhibitor	44	99.9
Excitatory amino acid transporter 3 inhibitor	10	100.0
Fatty acid transport protein 4 inhibitor	19	00.0
GABA transporter 1 inhibitor	150	99.6
GABA transporter 2 inhibitor	11	98.9
GABA transporter inhibitor	157	99.4
GABA uptake inhibitor	68	99.3
Glycine transporter 1 inhibitor	466	99.7
Glycine transporter 2 inhibitor	74	98.4
Glycine transporter inhibitor	550	99.2
Ileal bile acid transport inhibitor	144	99.8
Monoamine uptake inhibitor	41	91.6
Multidrug resistance-associated protein 1 inhibitor	223	96.7
Multidrug resistance-associated protein inhibitor	224	96.2
Neuronal K-Cl cotransporter inhibitor	17	91.1
Nucleoside transporters inhibitor	18	99.9
P-glycoprotein 1 inhibitor	450	93.6
P-glycoprotein 3 inhibitor	62	92.2
P-glycoprotein inhibitor	644	93.4
P-glycoprotein substrate	59	85.1
P2 nucleoside transporter inhibitor	11	100.0
Phosphate transporter inhibitor	26	98.1
Proline transporter inhibitor	23	100.0
Sodium/bile acid cotransporter inhibitor	16	97.0
Sodium/calcium exchanger 1 inhibitor	25	97.3
Sodium/calcium exchanger inhibitor	109	98.7
Sodium/glucose cotransporter 1 inhibitor	180	99.2
Sodium/glucose cotransporter 2 inhibitor	425	99.8
Sodium/hydrogen exchanger 1 inhibitor	296	99.9
Sodium/hydrogen exchanger 2 inhibitor	157	100.0
Sodium/hydrogen exchanger 3 inhibitor	220	99.9
Solute carrier family 22 member 12 inhibitor	19	100.0
Tetracycline antiport transporter antagonist	36	100.0

Table 11.14 (continued)

Activity name	*N*	AUC, %
Urate transporter 1 inhibitor	29	97.9
Vesicle monoamine transporter 2 inhibitor	13	100.0
Vesicle monoamine transporter inhibitor	14	99.6
Vesicular acetylcholine transporter inhibitor	75	99.8

N number of active compounds in the training set *AUC* Area Under Curve, calculated by leave-one-out cross-validation procedure

the competition with their binding center, which lead to increase in concentrations of the drugs in blood and cause adverse and toxic effects. It is considered that the clinically significant binding of drugs with plasma proteins is 95 %. [75]. Currently, several studies of the plasma proteins binding predictions were performed [76, 77]. QSAR models for assessment of the plasma proteins binding are provided by Accelrys, Simulation Plus Inc., ACD/Labs, PreADMET, Optibrium.

Metabolic change of one drug by another is one of the most significant causes for drug-drug interactions. Thus, the study of drug interactions with enzymes involved in their metabolism is very popular. There are four types of adverse drug reactions arising from influence on metabolic pathways of drugs.

1. Inhibition of the enzyme increases toxicity of the drug compound, which is a substrate of this enzyme. Most drugs are metabolized into inactive or less active metabolites by the liver and intestine enzymes. Inhibition of metabolism can increase the compound concentration and, consequently, enhance its effect. If the increasing of concentration is significant, it may cause toxicity. It is one of the most common and important interaction mechanisms of drug compounds in the clinic. The limited number of cytochrome P450 isoforms is involved in metabolism of drugs. Thus, competition between two drugs for these isoforms is possible. For example, inhibitors of CYP1A2 (cytochrome P450 isoform) may increase the risk of toxicity for clozepine and theophylline. Inhibitors of CYP2A9 may cause toxicity of phenytoin, tolbutamide and oral anticoagulants (e.g. warfarin). Inhibitors of CYP3A4 (e.g., phenytoin) increase the risk of toxicity of many drugs, including carbamazepine, ciclosporin, lovastatin, protease inhibitors, rifabutin, simvastatin and vinca alkaloids.
2. Enzyme inhibition reduces a therapeutic effect of the drug compound, an active form of which is produced by metabolism of an initially inactive or low active compound (prodrug). Inhibition of prodrug metabolism may reduce the amount of active forms and hence reduce the therapeutic effect. For example, analgesic and toxic effects of codeine occur as a result of its transformation into morphine by CYP2D6. Thus, CYP2D6 inhibitors can reduce the therapeutic effect of codeine.
3. The enzyme induction reduces therapeutic effects of its substrates. Some drugs, which are enzyme inductors, are able to increase the activity of drug metabolizing enzymes, which lead to reduce the therapeutic effect of other drugs. For example, the most common enzyme inducers are aminoglutethimide, barbitu-

rates, carbamazepine, glyutetimid, griseofulvin, primidone, finitoin, rifabutin, rifampin and troglitazone. Some drugs, as ritonavir, may act both inducers of enzymes and their inhibitors. It is considered that drugs metabolized by CYP3A4 and CYP2A9 are especially sensitive to enzyme induction.

4. The enzyme induction may increase toxic metabolites because of some drugs are transformed into toxic metabolites. For example, analgesic acetaminophen is mainly converted into non-toxic metabolites, but its small amount is transformed by CYP2E1 into a cytotoxic metabolite N-acetyl-p-benzoquinone imine. Enzyme inductors may increase the formation of toxic metabolite and increase the risk of hepatotoxicity and damage of other organs.

Obviously, it is necessary to create models to predict an interaction of compounds with the most important drug metabolizing enzymes and to develop an algorithm for analyzing results of these predictions. QSAR models for estimation of drug interactions with drug-metabolizing enzymes are provided by Simulation Plus, ACD/Labs, GeneXplain, Accelris, Optibrium, Lhasa and Multicase. The excellent review of software and *in silico* methods for evaluation of possible ligand interactions with drug-metabolizing enzymes and prediction of possible metabolites was recently published [78].

11.4.2 Pharmacodynamic Drug Interactions

Pharmacodynamic drug interactions include the cases where drugs show additive or antagonistic pharmacodynamic effects.

1. ***Antagonistic pharmacodynamic effects.*** This group of interactions includes the cases where drugs have the opposite pharmacodynamic effects, leading to a decrease in the exposure of one or both drugs. For example, compounds, which have a tendency to increase the blood pressure (as a non-steroidal anti-inflammatory compound), can inhibit an antihypertensive effect of angiotensin-converting enzyme inhibitors. Decrease of benzodiazepine effects by theophylline is another example.
2. ***Additive pharmacodynamic effects.*** In the case when two or more drugs exhibit similar pharmacodynamic effects it may produce an excessive manifestation of toxicity. It could be compounds whose combination may cause QT interval prolongation, leading to ventricular arrhythmia, as well as compounds that increase the concentration of potassium in blood and lead to hyperkalemia. An additive pharmacodynamic effect is also used for therapeutic purposes, so diuretics and angiotensin-converting enzyme inhibitors cause the blood pressure reduction.

To reveal these types of interaction, different types of prediction results are required. It is necessary to take into account the interaction of compounds with proteins, as well as information on the relationship between molecular mechanisms of action and biological effects. This analysis can be done with a computer program PASS (Prediction of the Activity Spectrum of Substance) and PharmaExpert (analysis of

PASS prediction results, which is based on the knowledgebase of "mechanism-effect" relationships) [45]. The current version of PASS predicts the biological activity spectrum, including molecular mechanisms of action, pharmacological effects and toxicity, interaction of compounds with transport proteins and drug metabolizing enzymes. Analysis of the biological activity spectrum prediction allows identifying potential pharmacokinetic and pharmacodynamic drug interactions by PharmaExpert. Pharmacokinetic interactions are detected based on prediction results of compounds interactions with drug metabolizing enzymes and transport proteins involved in absorption, distribution and excretion. Pharmacodynamic interactions are identified based on analysis of prediction results for the molecular mechanisms of action, pharmacological effects and toxicity. Overall flowchart identification of pharmacokinetic drug interactions is shown in Fig. 11.1.

The algorithm of the pharmacokinetic drug interaction identification is based on comparison of biological activity spectra in compounds (in the block diagram they are labeled as Compound 1 and Compound 2), predicted by PASS. It is necessary to determinate whether compounds are predicted as substrates of the same transporter

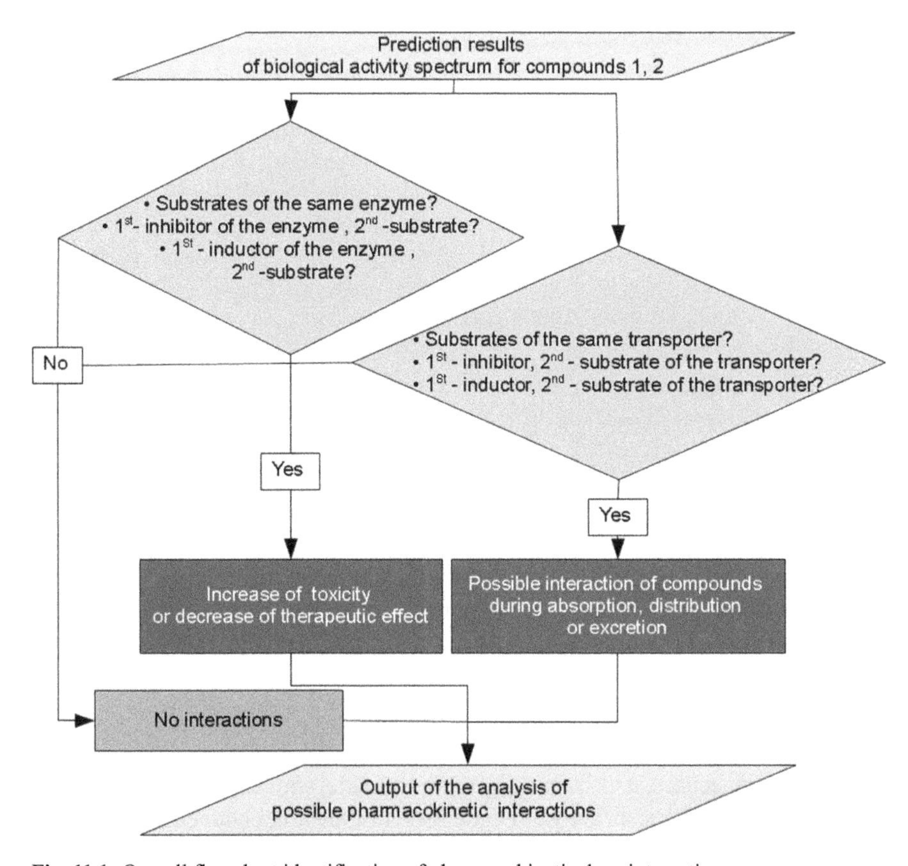

Fig. 11.1 Overall flowchart identification of pharmacokinetic drug interactions

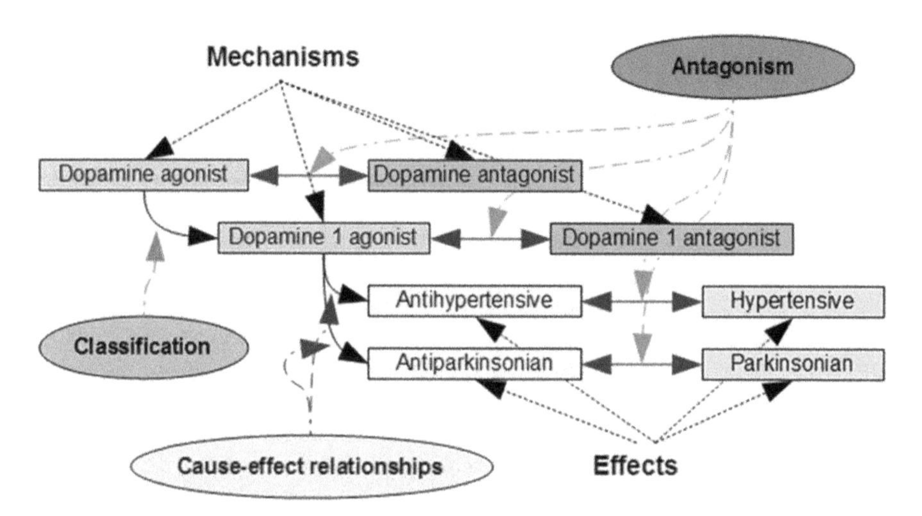

Fig. 11.2 The "activity-activity" relationships in PharmaExpert knowledge base. Mechanisms and Effects are classes of biological activities. Classification, Cause-effect relationships and Antagonism are classes of "activity-activity" relationships

or biotransformation enzyme, or compounds are predicted as an inhibitor or inducer of another transporter or drug metabolizing enzyme. The possible pharmacokinetic drug interactions are concluded in accordance with the result of comparison.

The analysis of potential pharmacodynamic drug interactions is based on comparison of the biological activity spectra in compounds and information from the database of "mechanism-effect" relationships provided by PharmaExpert. The knowledge base contains data of "cause-effect" relationships, classifications and antagonism of biological activities (Fig. 11.2).

Currently, the knowledge base contains information about 6233 mechanisms of action, 707 pharmacological and 996 side effects, and 12,785 relationships between them (PharmaExpert 2012). Mechanisms of action which may cause additive and antagonistic effects associated with drug interactions can be easily determined by using these data. The overall flowchart for identification of the pharmacodynamic drug interactions is shown in Fig. 11.3.

Algorithm for identification of pharmacodynamic drug interactions is based on comparison of biological activity spectra of the compounds (in the block diagram they are labeled as compound 1 and compound 2), predicted by PASS, together with information from the PharmaExpert knowledge base ("activity-activity" relationships). Antagonistic pharmacodynamic effects are determined by:

1. Biological effects with an opposite action. For example, hypertensive and antihypertensive effects, antispasmodic and spasmogenic effects, etc.
2. Antagonistic action on the same target. For example, stimulation and blocking of beta adrenoreceptors, an activator and blocker of the potassium channel, etc.

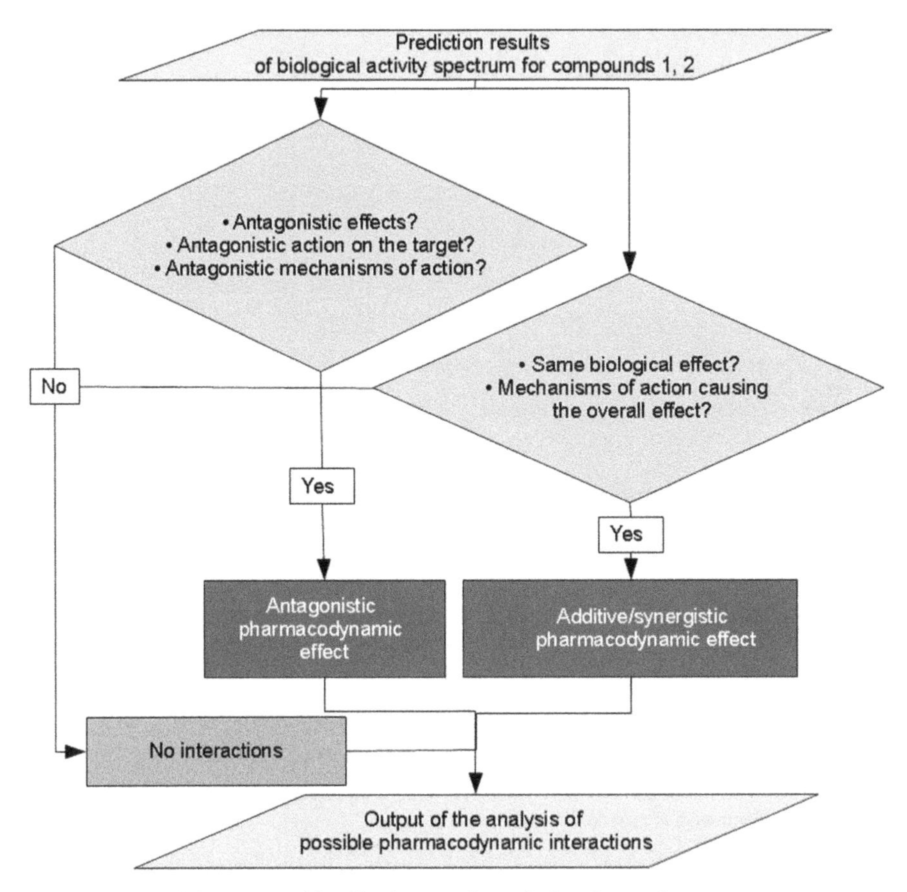

Fig. 11.3 Overall flowchart to identify pharmacodynamic drug interactions

3. Molecular mechanisms causing the opposite biological effects. For example, an alpha adrenergic agonist increases the blood pressure, and angiotensin converting enzyme inhibitor reduces the blood pressure.

Additive pharmacodynamic effects are determined by:

1. Identical biological effects, if these compounds provide the same biological effect. For example, the blood pressure reduction or increase of QT interval.
2. Mechanisms of action providing the common biological effect, if these compounds have different molecular mechanisms of action, causing the common biological effect. For example, a diuretic and angiotensin-converting enzyme inhibitors decrease the blood pressure, and inhibitors of carbamoyl phosphate synthetase and acyl-CoA dehydrogenase induce hyperammonemia.

The algorithms for the analysis of pairwise drug-drug interactions are also useful for the analysis of interactions between several substances. The example of evalu-

Drug-drug interaction between compounds with ID: Efavirenz (1), Emtricitabine (2), Tenofovir disoproxil (3)

Synergistic therapeutic effects and actions

Antiviral (HIV) – *name of synergistic effect*
Antiviral (HIV) (3):1,2,3 – *name of synergistic effect (number of compounds with the effect): Structure identifications*
HIV-1 reverse transcriptase inhibitor (2): 1,2
RNA directed DNA polymerase inhibitor (3):1,2,3
Antiviral (HIV) 0.926 0.003 ID:1 Antiviral (HIV) 0.501 0.004 ID:3
Name of effect Pa Pi ID Name of effect Pa Pi ID
Antiviral (HIV) 0.926 0.003 ID:1 Antiviral (HIV) 0.995 0.002 ID:2
HIV-1 reverse transcriptase inhibitor 0.930 0.001 ID:1 HIV-1 reverse transcriptase inhibitor 0.849 0.002 ID:2
Antiviral (HIV) 0.926 0.003 ID:1 RNA directed DNA polymerase inhibitor 0.837 0.002 ID:2

Synergistic toxic and side effects

Hepatotoxic
Hepatotoxic (2): 1,2
Hepatotoxic 0.853 0.011 ID:1 Hepatotoxic 0.945 0.003 ID:2

Emetic
Emetic (2): 1,2
Emetic 0.948 0.006 ID:1 Emetic 0.981 0.003 ID:2

Nephrotoxic
Nephrotoxic (2): 1,3
Nephrotoxic 0.710 0.021 ID:1 Nephrotoxic 0.768 0.014 ID:3

Metabolic drug-drug interactions
No interactions

Target-level drug-drug interactions

RNA directed DNA polymerase inhibitor (3):1,2,3
HIV-1 reverse transcriptase inhibitor (2): 1,2

Fig. 11.4 The results of PharmaExpert analysis of drug-drug interaction between the three drugs from Atripla. Comments are marked in italics

ation of possible interaction between three drugs (HIV reverse transcriptase inhibitors—efavirenz, emtricitabine, and tenofovir disoproxil) using in the treatment of HIV, which are included into the medicinal preparation Atripla [79], at a threshold of Pa > 0.5, is represented at the Fig. 11.4.

Figure 11.4 shows that in addition to correctly predicted antiviral effect, the known side effects given from Drugs.com (http://www.drugs.com/sfx/atripla-side-effects.html): hepatotoxic, nephrotoxic and emetic were also predicted by PASS/PharmaExpert as possible additive/synergistic side effects. None of additive/synergistic/antagonistic effects of the test compounds by reaction with drug metabolism enzyme or transporter protein was predicted. The presence of such interactions on the website Drugs.com information is also not shown. The known interactions of the compounds with HIV reverse transcriptase (a known target of the studied drugs) and RNA-dependent DNA polymerase (the name of the general class of enzymes including HIV reverse transcriptase) were also predicted.

The another example of the interaction analysis between compounds from Saint John's wort by PharmaExpert was described by Lagunin and co-authors [80].

11.5 A Critical Assessment of Computational Approaches for Toxicity Prediction

For a better understanding of the possible toxicity prediction it is necessary to pay attention to a number of limitations that may occur using modern computer tools for creation of (Q)SAR models. The major limitation factors are the guarantee of high-quality experimental data used for creation of the training sets and understanding of what is exactly modeled by the user. When mistakes occur (e.g., an incorrect structure of the molecule or incorrect data from toxicological studies) in the training set, it leads to the wrong model, which provides incorrect predictions. Therefore, considerable efforts should be made for the appropriate high-quality selection of experimental data, which will be used for creation of the model. Some recommendations for data curation in cheminformatics and QSAR modeling were published by Fourches and co-authors [81]. Since the concept of the "most suitable" and "quality" is subjective, even among experts, determination of the quality data can be done in several ways. FDA considers prospects of the evidence base for study and approval of products. This includes the standard of proof, performed by regulations, recommendations, guidelines, GLP (Good Laboratory Practice), GCP (Good Clinical Practice), under accurate and standardized research protocols. It can be used for well-defined parameters of the compounds studied in toxicity tests and required for risk assessment and design of experiments [18]. But there are some factors that cannot be clearly defined. For example, there is no sufficient information on nature of the liver damage which occurs at the hepatic toxicity. A priori it is not clear which factors should be taken into account for the dose modeling of this damage. From a pathology point of view it may be necrosis, fibrosis, inflammation, etc. Which of these data should be used for QSAR modeling? It is necessary to check carefully the data sources, how the data were obtained and to use methods for ranking the data quality before applying it for the toxicity modeling. Therefore, the model built for highly specialized mechanisms or clinical measurements (e.g., the prediction of transaminases or bilirubin increase in the blood plasma) can be more accurate and useful. However, in the computational prediction of toxicity, these cases are rare and it is usually required to predict more uncertain parameters. Other limitations of modern QSAR methods are difficult to build models for organometallic compounds, complex mixtures (e.g., plant extracts), and macromolecular compounds as polymers.

Another problem is to assess the safety of polypharmacological compounds acting on multiple targets. Also, there are significant limitations for the models building of carcinogenicity in rodents due to many existing mechanisms which may be caused by this effect. More important is how to interpret the data obtained in the carcinogenicity testing of drug compounds for rodents to humans. Some animal tumors have no analogs in humans [80]. If a molecule acts as a prooncogen, then it is difficult to estimate an activity dose and tumor tissue. It is considered that creation of the model describing toxicity or carcinogenicity based on various mechanisms as QSAR equation is much more complicated. However, there are several computer programs designed to solve those problems [82].

Another limitation is the difficulty of combining biological data generated in real time with computer predictions. Nevertheless, there are few successful examples of this approach, such as association of structure metabolites predictions with the spectra of liquid chromatography in mass spectrometry [83, 84].

It is considered to be the fundamental concept that the computer prediction of toxicity applied to the analysis of preclinical drug compounds, in fact, is a prediction of the prediction. It is necessary to take into account that most preclinical parameters (e.g., carcinogenicity, genetic toxicity and teratogenicity) are predictions of human toxicity, which help to establish the safety of drugs before their clinical trials. Thus, the creation of models based on other models, can only result in uncertainty. In reality, (Q)SAR is a theoretical analysis, based on the modeling of the chemical space and data from human toxicity models [1]. Unfortunately, uncertainty is still inevitable, because another layer of the modeling is added for the safety assessment. QSAR predictions have to be based on the same type of features among many important pieces of information (e.g., duration and level of exposure, confounding factors, and a risk/benefit ratio) in the general risk analysis. An applicability domain of the model is also the main criterion and restricting factor to use (Q)SAR models in toxicology and pharmacology [85, 86]. If a compound does not fall into an applicability domain of the model during *in silico* screening the prediction is considered to be incorrect. At the same time, it can be expected that the development of new chemical entities requires moving towards new chemical spaces, since they are created for new therapeutic targets. Thus, the domain of applicability for new QSAR models is needed to be extended for new molecules. This will lead to overcoming limitations and thus provide more accurate and acceptable predictions.

Predictions of a specific toxicity (e.g., carcinogenicity) based on QSAR models associated with the alerts classification schemes (e.g. Ashby-Tennant alerts) or expert rules, also have their own limitations. A significant part of used drugs shows a positive result in the rodent's carcinogenicity and negative results for genotoxicity [87]. It leads to the dilemma how to predict non genotoxic carcinogens, can it be based either on the structure formula or on the structural alerts associated with the inducing of DNA damage? Therefore, the developments, which are devoted to this problem, and the predictions of epigenetic mechanisms with a carcinogens action have the primary importance. In addition, recently it has been proposed that there is a new research area for computer predictions of carcinogenicity: creating computer models to predict carcinogens acting through inhibition of protein kinase networks.

The expert system based on rules has limitations in an ability to identify structural alerts, which lead to the manifestation of activity and does not have built-in rules for the estimation of compounds, which includes two or more structural alert or deactivating fragments in the molecule. Moreover, the "negative" prediction of these programs means that nothing has been found and prediction could not indicate the loss of toxicity. Finally, there is a human factor underlying this approach. It is represented by the consensus opinion or expert opinion which are prone to subjectivity and can result in incorrect or inaccurate prediction rules [88].

Another important factor that influences on the accuracy of models is a metabolic pathway or metabolic activation of the parent compound which is not usually considered. It would be reasonable to predict the first reactive metabolite or find an actual proof that it is produced, and then make a separate prediction for the metabolite of interest. Several computer programs provide either the prediction of human metabolism or common metabolic pathway based on the rules extracted from several species of mammals (rat, mouse, human, hamster), using a mixture of *in vitro* and *in vivo* metabolism data of compounds [68, 89]. However, in these programs, predictions made for the possible reactive metabolites do not affect the prediction for a parent compound.

References

1. Valerio LG Jr (2009) In silico toxicology for the pharmaceutical sciencesâ. Toxicol Appl Pharmacol 241:356–370. doi:10.1016/j.taap.2009.08.022
2. Snyder RD (2009) An update on the genotoxicity and carcinogenicity of marketed pharmaceuticals with reference to in silico predictivity. Environ Mol Mutagen 50:435–450. doi:10.1002/em.20485
3. Kavlock RJ, Ankley G, Blancato J et al (2008) Computational toxicology—a state of the science mini review. Toxicol Sci Off J Soc Toxicol 103:14–27. doi:10.1093/toxsci/kfm297
4. (2007) Tools and technologies Chapter 4. Toxic. Test. 21st Century Vis. Strategy. National Academies Press, Washington, DC, pp 98–119
5. (2006) EU. Official. J. Eur. Union. L396
6. Lahl U, Gundert-Remy U (2008) The use of (Q)SAR methods in the context of REACH. Toxicol Mech Methods 18:149–158. doi:10.1080/15376510701857288
7. Benfenati E, Diaza RG, Cassano A et al (2011) The acceptance of in silico models for REACH: requirements, barriers, and perspectives. Chem Cent J 5:58. doi:10.1186/1752-153X-5-58
8. Ashby J (1985) Fundamental structural alerts to potential carcinogenicity or noncarcinogenicity. Environ Mutagen 7:919–921
9. Ashby J, Tennant RW (1988) Chemical structure, Salmonella mutagenicity and extent of carcinogenicity as indicators of genotoxic carcinogenesis among 222 chemicals tested in rodents by the U.S. NCI/NTP. Mutat Res 204:17–115
10. Ashby J, Tennant RW (1991) Definitive relationships among chemical structure, carcinogenicity and mutagenicity for 301 chemicals tested by the U.S. NTP. Mutat Res 257:229–306
11. OECD principles. OECD Princ. http://www.oecd.org/env/ehs/risk-assessment/37849783.pdf. Accessed 7 May 2013
12. Gedeck P, Rohde B, Bartels C (2006) QSAR—how good is it in practice? Comparison of descriptor sets on an unbiased cross section of corporate data sets. J Chem Inf Model 46:1924–1936. doi:10.1021/ci050413p
13. Borth DM (1996) Optimal experimental designs for (possibly) censored data. Chemom Intell Lab Syst 32:25–35. doi:10.1016/0169-7439(95)00057-7
14. Borth DM, Wilhelm MS (2002) Confidence limits for normal type I censored regression. Chemom Intell Lab Syst 63:117–128. doi:10.1016/S0169-7439(02)00019-9
15. Matthews EJ, Kruhlak NL, Benz RD et al (2008) Combined use of MC4PC, MDL-QSAR, BioEpisteme, Leadscope PDM, and Derek for Windows Software to Achieve High-performance, high-confidence, mode of action-based predictions of chemical carcinogenesis in rodents. Toxicol Mech Methods 18:189–206. doi:10.1080/15376510701857379
16. Gottmann E, Kramer S, Pfahringer B, Helma C (2001) Data quality in predictive toxicology: reproducibility of rodent carcinogenicity experiments. Environ Health Perspect 109:509–514

17. Gold LS, Wright C, Bernstein L, deVeciana M (1987) Reproducibility of results in "near-replicate" carcinogenesis bioassays. J Natl Cancer Inst 78:1149–1158
18. Helma C, Kramer S, Pfahringer B, Gottmann E (2000) Data quality in predictive toxicology: identification of chemical structures and calculation of chemical properties. Environ Health Perspect 108:1029–1033
19. (2007) Chapter 3: Guidance on grouping of chemicals. Approaches data gap fill. Chem. Categ. OECD, pp 30–41
20. Arvidson KB (2008) FDA toxicity databases and real-time data entry. Toxicol Appl Pharmacol 233:17–19. doi:10.1016/j.taap.2007.12.033
21. Benz RD (2007) Toxicological and clinical computational analysis and the US FDA/CDER. Expert Opin Drug Metab Toxicol 3:109–124. doi:10.1517/17425255.3.1.109
22. Richard AM, Yang C, Judson RS (2008) Toxicity data informatics: supporting a new paradigm for toxicity prediction. Toxicol Mech Methods 18:103–118. doi:10.1080/15376510701857452
23. National Center for Toxicological Mold2. NCTR. http://www.fda.gov/ScienceResearch/BioinformaticsTools/Mold2/default.htm. Accessed 7 May 2013
24. Dearden JC, Cronin MTD, Kaiser KLE (2009) How not to develop a quantitative structure-activity or structure-property relationship (QSAR/QSPR). SAR QSAR Environ Res 20:241–266. doi:10.1080/10629360902949567
25. Dearden JC, Cronin MTD, Schultz TW, Lin DT (1995) QSAR study of the toxicity of nitrobenzenes to tetrahymena pyriformis. Quant Struct-Act Relatsh 14:427–432. doi:10.1002/qsar.19950140503
26. Kortagere S, Krasowski MD, Ekins S (2009) The importance of discerning shape in molecular pharmacology. Trends Pharmacol Sci 30:138–147. doi:10.1016/j.tips.2008.12.001
27. Potemkin V, Grishina M (2008) Principles for 3D/4D QSAR classification of drugs. Drug Discov Today 13:952–959. doi:10.1016/j.drudis.2008.07.006
28. Peristera O, Spreafico M, Smiesko M et al (2009) Mixed-model QSAR at the human mineralocorticoid receptor: predicting binding mode and affinity of anabolic steroids. Toxicol Lett 189:219–224. doi:10.1016/j.toxlet.2009.05.025
29. Rose JR (2008) Machine learning techniques in chemistry. In: Gasteiger J (ed) Handbook of chemoinformatics. Wiley-VCH Verlag GmbH, Weinheim, Germany, pp 1082–1097
30. Martin S (2010) Machine learning-based bioinformatics algorithms: application to chemicals. Handbook of chemoinformatics algorithms. Chapman and Hall/CRC, Boca Raton, Florida, USA, pp 383–398
31. Filimonov D, Poroikov V (2008) Probabilistic approach in activity prediction. In: Varnek A, Tropsha A (eds) Chemoinformatics approaches virtual screen. RSC Publishing, Cambridge, pp 182–216
32. Hewitt M, Cronin MTD, Madden JC et al (2007) Consensus QSAR models: do the benefits outweigh the complexity? J Chem Inf Model 47:1460–1468. doi:10.1021/ci700016d
33. Lagunin A, Filimonov D, Zakharov A et al (2009) Computer-aided prediction of rodent carcinogenicity by PASS and CISOC-PSCT. QSAR Comb Sci 28:806–810. doi:10.1002/qsar.200860192
34. Geman S, Bienenstock E, Doursat R (1992) Neural networks and the bias/variance dilemma. Neural Comput 4:1–58. doi:10.1162/neco.1992.4.1.1
35. Lagunin A, Zakharov A, Filimonov D, Poroikov V (2011) QSAR modelling of rat acute toxicity on the basis of PASS prediction. Mol Inform 30:241–250. doi:10.1002/minf.201000151
36. ChemAxon Marvin Sketch. http://www.chemaxon.com/marvin/help/index.html. Accessed 8 May 2013
37. Kennedy T (1997) Managing the drug discovery/development interface. Drug Discov Today 2:436–444. doi:10.1016/S1359-6446(97)01099-4
38. Lasser KE, Allen PD, Woolhandler SJ et al (2002) Timing of new black box warnings and withdrawals for prescription medications. JAMA J Am Med Assoc 287:2215–2220
39. Kola I, Landis J (2004) Can the pharmaceutical industry reduce attrition rates? Nat Rev Drug Discov 3:711–715. doi:10.1038/nrd1470

40. Fung M, Thornton A, Mybeck K et al (2001) Evaluation of the characteristics of safety withdrawal of prescription drugs from worldwide pharmaceutical markets-1960 to 1999*. Drug Inf J 35:293–317. doi:10.1177/009286150103500134
41. Stephens MDB (2004) Introduction. In: Talbot J, Waller P (eds) Stephens detect. New adverse drug react., 5th edn. Wiley, Chichester, pp 1–91
42. MRTD database. http://www.epa.gov/ncct/dsstox/sdf_fdamdd.html. Accessed 8 May 2013
43. Mitchell JA, Warner TD (2005) Discontinuation of Vioxx. The Lancet 365:27–28. doi:10.1016/S0140-6736(04)17659-9
44. Whitebread S, Hamon J, Bojanic D, Urban L (2005) Keynote review: in vitro safety pharmacology profiling: an essential tool for successful drug development. Drug Discov Today 10:1421–1433. doi:10.1016/S1359-6446(05)03632-9
45. Lagunin A, Filimonov D, Poroikov V (2010) Multi-targeted natural products evaluation based on biological activity prediction with PASS. Curr Pharm Des 16:1703–1717
46. Breckenridge A (1996) A clinical pharmacologist's view of drug toxicity. Br J Clin Pharmacol 42:53–58
47. Lazarou J, Pomeranz BH, Corey PN (1998) Incidence of adverse drug reactions in hospitalized patients: a meta-analysis of prospective studies. JAMA J Am Med Assoc 279:1200–1205
48. Ji ZL, Wang Y, Yu L et al (2006) In silico search of putative adverse drug reaction related proteins as a potential tool for facilitating drug adverse effect prediction. Toxicol Lett 164:104–112. doi:10.1016/j.toxlet.2005.11.017
49. Chen YZ, Zhi DG (2001) Ligand-protein inverse docking and its potential use in the computer search of protein targets of a small molecule. Proteins 43:217–226
50. Ji ZL, Han LY, Yap CW et al (2003) Drug Adverse Reaction Target Database (DART): proteins related to adverse drug reactions. Drug Saf Int J Med Toxicol Drug Exp 26:685–690
51. Poroikov V, Filimonov D, Lagunin A et al (2007) PASS: identification of probable targets and mechanisms of toxicity. SAR QSAR Environ Res 18:101–110. doi:10.1080/10629360601054032
52. Filimonov DA, Zakharov AV, Lagunin AA, Poroikov VV (2009) QNA-based "Star Track" QSAR approach. SAR QSAR Environ Res 20:679–709. doi:10.1080/10629360903438370
53. Sadym A, Lagunin A, Filimonov D, Poroikov V (2003) Prediction of biological activity spectra via the Internet. SAR QSAR Environ Res 14:339–347. doi:10.1080/10629360310001623935
54. Lagunin AA, Zakharov AV, Filimonov DA, Poroikov VV (2007) A new approach to QSAR modelling of acute toxicity. SAR QSAR Environ Res 18:285–298. doi:10.1080/10629360701304253
55. Kokurkina GV, Dutov MD, Shevelev SA et al (2011) Synthesis, antifungal activity and QSAR study of 2-arylhydroxynitroindoles. Eur J Med Chem 46:4374–4382. doi:10.1016/j.ejmech.2011.07.008
56. Zakharov AV, Peach ML, Sitzmann M et al (2012) Computational tools and resources for metabolism-related property predictions. 2. Application to prediction of half-life time in human liver microsomes. Future Med Chem 4:1933–1944. doi:10.4155/fmc.12.152
57. Zakharov AV, Lagunin AA, Filimonov DA, Poroikov VV (2012) Quantitative prediction of antitarget interaction profiles for chemical compounds. Chem Res Toxicol 25:2378–2385. doi:10.1021/tx300247r
58. Overington JP, Al-Lazikani B, Hopkins AL (2006) How many drug targets are there? Nat Rev Drug Discov 5:993–996. doi:10.1038/nrd2199
59. Lounkine E, Keiser MJ, Whitebread S et al (2012) Large-scale prediction and testing of drug activity on side-effect targets. Nature 486:361–367. doi:10.1038/nature11159
60. Ivanov SM, Lagunin AA, Zakharov AV et al (2013) Computer search for molecular mechanisms of ulcerogenic action of non-steroidal anti-inflammatory drugs. Biochem Mosc Suppl Ser B Biomed Chem 7:40–45. doi:10.1134/S199075081301006X
61. Yang L, Wang K, Chen J et al (2011) Exploring off-targets and off-systems for adverse drug reactions via chemical-protein interactome-clozapine-induced agranulocytosis as a case study. PLoS Comput Biol 7:e1002016. doi:10.1371/journal.pcbi.1002016

62. Bender A, Scheiber J, Glick M et al (2007) Analysis of pharmacology data and the prediction of adverse drug reactions and off-target effects from chemical structure. ChemMedChem 2:861–873. doi:10.1002/cmdc.200700026
63. Scheiber J, Chen B, Milik M et al (2009) Gaining insight into off-target mediated effects of drug candidates with a comprehensive systems chemical biology analysis. J Chem Inf Model 49:308–317. doi:10.1021/ci800344p
64. Subramanian A, Tamayo P, Mootha VK et al (2005) Gene set enrichment analysis: a knowledge-based approach for interpreting genome-wide expression profiles. Proc Natl Acad Sci U S A 102:15545–15550. doi:10.1073/pnas.0506580102
65. Lamb J, Crawford ED, Peck D et al (2006) The Connectivity Map: using gene-expression signatures to connect small molecules, genes, and disease. Science 313:1929–1935. doi:10.1126/science.1132939
66. Toyoshiba H, Sawada H, Naeshiro I, Horinouchi A (2009) Similar compounds searching system by using the gene expression microarray database. Toxicol Lett 186:52–57. doi:10.1016/j.toxlet.2008.08.009
67. Minowa Y, Kondo C, Uehara T et al (2012) Toxicogenomic multigene biomarker for predicting the future onset of proximal tubular injury in rats. Toxicology 297:47–56. doi:10.1016/j.tox.2012.03.014
68. Davis AP, Murphy CG, Johnson R et al (2013) The comparative toxicogenomics database: update 2013. Nucleic Acids Res 41:D1104–1114. doi:10.1093/nar/gks994
69. Lagunin A, Ivanov S, Rudik A et al (2013) DIGEP-Pred: web-service for in silico prediction of drug-induced gene expression profiles based on structural formula. Bioinforma Oxf Engl. doi:10.1093/bioinformatics/btt322
70. Hardman JG, Limbird LE, Gilman AG (2001) Goodman & Gilman's the pharmacological basis of therapeutics, 10th edn. The McGraw-Hill, NY, USA
71. Manga N, Duffy JC, Rowe PH, Cronin MTD (2005) Structure-based methods for the prediction of the dominant P450 enzyme in human drug biotransformation: consideration of CYP3A4, CYP2C9, CYP2D6. SAR QSAR Environ Res 16:43–61. doi:10.1080/10629360412331319871
72. Jonker DM, Visser SAG, van der Graaf PH et al (2005) Towards a mechanism-based analysis of pharmacodynamic drug-drug interactions in vivo. Pharmacol Ther 106:1–18. doi:10.1016/j.pharmthera.2004.10.014
73. Han H-K (2011) Role of transporters in drug interactions. Arch Pharm Res 34:1865–1877. doi:10.1007/s12272-011-1107-y
74. Sedykh A, Fourches D, Duan J et al (2013) Human intestinal transporter database: QSAR modeling and virtual profiling of drug uptake, efflux and interactions. Pharm Res 30:996–1007. doi:10.1007/s11095-012-0935-x
75. Manzi SF, Shannon M (2005) Drug interactions—a review. Clin Pediatr Emerg Med 6:93–102. doi:10.1016/j.cpem.2005.04.006
76. Hou T, Wang J (2008) Structure-ADME relationship: still a long way to go? Expert Opin Drug Metab Toxicol 4:759–770. doi:10.1517/17425255.4.6.759
77. Metcalfe PD, Thomas S (2010) Challenges in the prediction and modeling of oral absorption and bioavailability. Curr Opin Drug Discov Devel 13:104–110
78. Kirchmair J, Williamson MJ, Tyzack JD et al (2012) Computational prediction of metabolism: sites, products, SAR, P450 enzyme dynamics, and mechanisms. J Chem Inf Model 52:617–648. doi:10.1021/ci200542m
79. Deeks E, Perry C (2010) Efavirenz/emtricitabine/tenofovir disoproxil fumarate single-tablet regimen (Atripla®): a review of its use in the management of HIV infection. Drugs 70:2315–2338
80. Jacobs A (2005) Prediction of 2-year carcinogenicity study results for pharmaceutical products: how are we doing? Toxicol Sci Off J Soc Toxicol 88:18–23. doi:10.1093/toxsci/kfi248
81. Fourches D, Muratov E, Tropsha A (2010) Trust, but verify: on the importance of chemical structure curation in cheminformatics and QSAR modeling research. J Chem Inf Model 50:1189–1204. doi:10.1021/ci100176x
82. SYMMETRY®. http://www.prousresearch.com/Technology/SYMMETRY.aspx. Accessed 7 May 2013

83. Stranz DD, Miao S, Campbell S et al (2008) Combined computational metabolite prediction and automated structure-based analysis of mass spectrometric data. Toxicol Mech Methods 18:243–250. doi:10.1080/15376510701857189
84. Pelander A, Tyrkkö E, Ojanperä I (2009) In silico methods for predicting metabolism and mass fragmentation applied to quetiapine in liquid chromatography/time-of-flight mass spectrometry urine drug screening. Rapid Commun Mass Spectrom 23:506–514. doi:10.1002/rcm.3901
85. Dimitrov S, Dimitrova G, Pavlov T et al (2005) A stepwise approach for defining the applicability domain of SAR and QSAR models. J Chem Inf Model 45:839–849. doi:10.1021/ci0500381
86. Weaver S, Gleeson MP (2008) The importance of the domain of applicability in QSAR modeling. J Mol Graph Model 26:1315–1326. doi:10.1016/j.jmgm.2008.01.002
87. Snyder RD, Green JW (2001) A review of the genotoxicity of marketed pharmaceuticals. Mutat Res 488:151–169
88. Guzelian PS, Victoroff MS, Halmes NC et al (2005) Evidence-based toxicology: a comprehensive framework for causation. Hum Exp Toxicol 24:161–201
89. Klopman G, Dimayuga M, Talafous J (1994) META. 1. A program for the evaluation of metabolic transformation of chemicals. J Chem Inf Comput Sci 34:1320–1325

Chapter 12
Consensus Drug Design Using IT Microcosm

Pavel M. Vassiliev, Alexander A. Spasov, Vadim A. Kosolapov, Aida F. Kucheryavenko, Nataliya A. Gurova and Vera A. Anisimova

Abstract This chapter discusses Microcosm, an information technology package for predicting the pharmacological activity of chemical compounds. This technology is based on a complex prediction methodology with a consensus approach to prediction as its central component. The complex methodology of prediction in IT Microcosm is essentially different from that of other QSAR approaches in that it employs a redundant multi-descriptor, multi-level representation of the structure of chemical compounds by an aggregate of fragment descriptors with different physicochemical meanings and varying extents of complexity. The methodology also includes several classification methods that differ in their mathematical formalisms and several decision making circuits that are conceptual in the results they yield. At the same time, no feature space reductions are made, and no significant variables are isolated; all of the parameters of description are used in the construction of the prediction regularities. The integral decision rules are constructed by generalizing the spectrum of primary prediction estimates using different levels and types of consensus. In this chapter, we describe the paradigm of IT Microcosm, including its theoretical concepts, a specialized QL language for chemical structure representation, and prediction methods and strategies using the package. The adequacy, validity and high accuracy of IT Microcosm are demonstrated via sample predictions of the various pharmacological activities of structurally similar and structurally diverse organic compounds, complex organic salts, supramolecular complexes and substance mixtures, accounting for the synergy between the individual components of mixtures. The authors also present the results of a successful application of IT Microcosm, along with *in vivo* and *in vitro* experimental methods for (1) the search for novel potent antioxidants, antiarrhythmics and antiplatelet agents; (2) the optimization of the composition of supramolecular complexes with antioxidant and

P. M. Vassiliev (✉) · A. A. Spasov · V. A. Kosolapov · A. F. Kucheryavenko · N. A. Gurova
Volgograd State Medical University (VSMU), Pavshikh Bortsov Sq. 1,
Volgograd 400131, Russian Federation
e-mail: pmv@avtlg.ru

V. A. Anisimova
Institute of Physical and Organic Chemistry at Southern Federal University (IPOC SFU),
Stachka Ave. 194/2, Rostov-on-Don 344090, Russian Federation
e-mail: anis39@mail.ru.

L. Gorb et al. (eds.), *Application of Computational Techniques in Pharmacy and Medicine*,
Challenges and Advances in Computational Chemistry and Physics 17,
DOI 10.1007/978-94-017-9257-8_12

antiarrhythmic activity; (3) the evaluation of the spectrum and the extent of pharmacological effects and the optimization of the composition of naturally occurring multicomponent drugs; and (4) the evaluation of the synergistic effects of mixtures of drug substances. IT Microcosm consists of a package of 20 computer programs; there is a separate free Microcosm White computer program.

Abbreviations

5-HT	5-Hydroxytryptamine
BA	Bayesian approach
CS	Conservative strategy
DLOOCV	Double leave-one-out cross-validation
DM	Distance method
GA	Glycyrrhizinic acid
H	Histamine
in silico	research of biologically active substances that is performed using a computer or via computer simulation
IT	Information Technology
LBDD	Ligand-Based Drug Design
LDM	Local distribution method
LOOCV	Leave-one-out cross-validation
LP	Lipid peroxidation
NNM	Nearest neighbor method
NS	Normal strategy
QSAR	Quantitative Structure-Activity Relationships
QSPR	Quantitative Structure-Property Relationships
RS	Risk strategy
SBDD	Structure-Based Drug Design
SHCV	Split-half cross-validation
ST	Self-testing

12.1 Introduction

In silico methods are currently common practice in drug discovery [61, 72]. The methods for *in silico* drug design are traditionally divided into the following types: 2D and 3D approaches based on the way that a chemical structure is represented [27] or SBDD and LBDD based on the information that is available about the structure of pharmacologically relevant target proteins [61]. The use of SBDD methods is standard under several circumstances, including cases where a 3D model is not informative enough but there is sufficient information about the activity of the studied compounds or where no structural data are available for a given target, such as when the desired effect is of a systemic nature. For this reason, QSAR appears to be the most common method of *in silico* drug discovery [27]. MFTA [73], HIT QSAR [51], PASS [70], ISIDA [92], NASAWIN [8], and CORAL [89] are QSAR software packages that have been successfully utilized for the prediction of a wide variety of biological activities.

The fragmental approach is one the most productive QSAR methods [43, 111, 139]; this approach is based on the idea that a chemical structure is a set of sub-structural fragments that can be isolated from the structural formulas of compounds according to certain rules, together with various parameters that characterize these fragments. In a certain sense, the classical structural formula is superior to any 3D model with respect to the reliability of information about a compound; the reliability of 3D models depends on the methods and conditions of their development. An adequate methodology of extracting information contained in the structural formula yields up to 90 % of for information about the properties of a given substance, even without resorting to 3D modeling [104, 110].

Any parametric description of a chemical structure, including a fragment-based description, destroys the pattern of a compound as a whole [27, 43], which can lead to the loss of information about the substance as a unique object at an over-cybernetic level of organization. Therefore, an adequate representation of a chemical structure, without the loss of information specific to any compound, is of the utmost importance in QSAR. This is especially relevant for highly active compounds, which commonly show chemical novelty and can be termed "upstarts" according to their characteristics. Such compounds do not follow the conventional regularities describing compounds with medium activity [135]. One rewarding way to overcome the loss of information about the integrity of a compound is to resort to a multilevel hierarchical description of the chemical structure by a set of substructural descriptors with increasing complexity [51, 107].

A wide range of methods for restoring empirical regularity are used within the framework of the fragmental approach for the calculation of structure-activity relationships: regression [51, 137], pattern recognition [23, 57], artificial neural networks [9], and machine learning [23, 31]. Meanwhile, the relationship between the pharmacological activity and structure of a chemical compound is not originally continuous in nature because it includes a multiplicity of discrete components such as pleiotropic effects [71] (i.e., multiple physiological mechanisms of action [25] or interactions with several biological targets [62]), selective complementarity to certain pockets of binding sites (the privileged molecule phenomenon) [21], selective transport by specific proteins [7], synergy with other compounds [86], and so on. In addition, the variability of pharmacological data is very high, due to the extreme complexity of higher animals; for example, the parameter dispersion of techniques for studying behavioral performance can be as high as dozens of percent [14]. Therefore, the use of methods for restoring smooth continuous dependencies in QSAR analysis of pharmacological activity is considerably limited.

The vast majority of QSAR studies depend on a rather controversial working concept: the choice of a "better equation" for the prediction of the studied activity.

These "better equations" are always context-dependent; minimally, they depend on the way that the compound structure is represented and on the computational method for predictive relation. It was shown as early as 1972 that it is impossible to choose a single regularity that provides an adequate description of a predicted biological activity when you are faced with a number of QSAR models with comparable accuracy [18]. On the other hand, if one simultaneously uses several equations for prediction and the calculated estimates of the activity of a certain compound coincide, the prediction error is considerably lower. Mathematical tools for object classification that simultaneously use several decision rules began to be developed heavily starting in the 1990s [65]. Within the framework of *in silico* drug discovery, this approach is referred to as a consensus, ensemble, or committee approach by different authors, and methods using these approaches were first used in the 2000s [10]. At present, the consensus approach to prediction based on the synthesis of data obtained from several QSAR dependencies is at the peak of popularity [42].

The process of selecting the so-called significant variables is an indispensable stage of virtually all QSAR analysis methods [9, 27, 33, 72]. Despite its apparent clarity and attractiveness, this approach often results in the generation of artifact dependencies, and this effect has been noted by several authors on many occasions [18, 39]. The equations calculated in this manner, though simple and clear at first glance, do not give a fair representation of the individual features of the predicted compounds; they are therefore only marginally suitable for the design of intriguing, highly active substances that show mostly nonstandard characteristics, which distinguishes them from other compounds. The transformation of a primary parameter set into latent variables, as in PLS-regression [24] or in multilayer artificial neural networks [40], does not settle the issue because primary variable weighting (the determination of significance) is also performed in these cases. On the other hand, the mathematical methods themselves often contain limitations that preclude the simultaneous use of a large number of variables in model construction. Thus, there should be no less than three observations for each variable in a regression analysis [22], and no less than two observations in artificial neural network modeling [9]. QSAR approaches that consider the use of all of the available variables in making decision rules (for example, the support vector machine (SVM) method) have only recently appeared [91] in conjunction with kernel function use [13, 58].

The concept of significant variables presumes their independence. The methods for restoring empirical regularities that are used in QSAR are also intended to be only used in a Euclidean space [1]. However, all of the parameters of a compound description are generated from the same object (the chemical structure of the compound), which is why all of the obtained variables are always inter-dependent. Furthermore, if the feature space is also nonlinear, then we must at least face the task of adapting the existing QSAR methods to such spaces.

Taking into account all of the peculiarities of the chemical-biological universum discussed above, and in order to overcome the deficiencies of the existing QSAR systems, a **complex** methodology to predict the properties of organic compounds was developed [93, 96, 102, 116] and served as a base for the development of IT Microcosm [45, 98, 103, 105,] in the form of a software package [113]. This software package is designed to aid in the *in silico* discovery of chemical compounds with a desired pharmacological activity, which can be defined as "high" in case of positive effects, and "minimal" in case of adverse effects.

This complex prediction methodology is dramatically different from other approaches to structure-activity analysis in that it simultaneously employs:

- a redundant multidescriptor and multilevel representation of the structure of chemical compounds as a set of fragment descriptors with different physicochemical meanings and varying extents of complexity;
- several classification methods that differ considerably in their mathematical formalism; and
- several decision-making circuits that are conceptual in the results that they yield.

An additional point to emphasize is that in calculating the prediction dependencies, no feature space reduction is carried out, and no significant variables are selected; all of the parameters of the object area description are used in the construction of the separating functions.

A complex methodology of prediction allows the formation of integral consensus decision rules that are context-independent and work stably in extra-large-dimension correlated spaces. The current version of IT Microcosm 5.1 [113] for the calculation of QSAR dependencies uses a fourth-order consensus.

The adequacy, validity and high accuracy of IT Microcosm have been demonstrated on multiple occasions when predicting various pharmacological activities of "conventional" structurally diverse [109, 117, 121, 140] and structurally similar [80, 127, 131, 132, 130] organic compounds as well as "nonstandard" chemical systems, complex organic salts [105, 123], supramolecular complexes [128, 129] and substance mixtures [35, 74, 75, 97, 112, 111, 114, 119, 120], including cases where the synergy of the components of a mixture was taken into consideration [74, 75, 112, 111, 114].

12.2 Theoretical Basis of IT Microcosm

Information technology, in the general meaning of the word, is a package of technological components (devices or methods, for example) that are used by people to manage information [20].

Information technology Microcosm for predicting the properties of organic compounds is a package of original theoretical concepts, mathematical methods, and rules driving computer algorithms and software that allow a calculated estimate of the properties of a chemical compound based on its structural formula, with the help of multilevel consensus classification QSAR dependencies [105].

In IT Microcosm, predictions are based on the task of classifying compounds into two classes: active compounds and inactive compounds. Compounds are called active if they show a certain level of a given biological activity; this level is pre-set by the researcher. Inactive compounds do not meet the requirement of this pre-set activity level. For example, activity may be described as "absent-present," "high-low" (in the opinion of an expert pharmacologist), or "fitting-missing" (when we refer to a range of some quantitative evaluation, such as $100 < LD_{50} < 500$ mg/kg). For a quantitative evaluation, this prediction is carried out according to several binary oppositions describing the activity; for example, "high-other," "moderate-other," or "low-other." To enhance the adequacy and consistency of the prediction, associated classes of activity are formed, such as "high or moderate-other", "high or moderate or low-other" ("active-inactive"). This delineation usually suffices for practical purposes because the experimenting pharmacologist is typically interested in highly active compounds, which he or she will test on a first-priority basis.

In IT Microcosm, prediction is made on 11 levels by describing the chemical structure by various types of descriptors using four mathematical methods; the spectrum of intermediate prediction estimates is generalized on the basis of three voting strategies, and the prediction results are generalized for all strategies, while the spectrum of prediction estimates is simultaneously checked for noncontradiction [96, 102, 103, 105, 110, 127].

In a training set, the classification dependences are established by methods of pattern recognition and machine learning. These regularities unite various activity levels that are pre-set in the form of semiquantitative gradations, and the structure of a given compound is represented as a matrix of structural descriptors. The chemical structure is represented by 11 types of descriptors in a specialized hierarchic multilevel language, QL [107, 110]. These descriptors form generalized patterns of classes of active/inactive compounds, represented as matrices of structural descriptors [109] within the framework of a generalized pattern of a compound class with the desired properties [94, 101, 105].

When predicting the presence or extent of a desired pharmacological activity, the structural formula of an untested compound represented as a standard connection table is transformed by the translator program into descriptors in the working language. By comparing the obtained pattern with models of class patterns using four prediction methods that are essentially different in their mathematical formalism, a spectrum of 44 intermediate prediction estimates (11 for each type of QL descriptor) is calculated for each type of activity. This spectrum is then generalized on the basis of one of three strategies, and a final estimate of the predicted compound activity is calculated [103, 105, 109].

At the final stage, the prediction results are generalized in relation to all strategies, and the spectrum of prediction estimates is checked for noncontradiction [102]. To enhance the reliability of this method, one can consolidate the prediction results in relation to several levels of a predicted pharmacological activity [133].

The integral decision rules that are generated by IT Microcosm are consensus QSAR regularities of the fourth level; the first consensus level relates to the 11 types of QL descriptors, the second level relates to the four prediction methods, the third level consists of the three prediction strategies, and the fourth level is based on the levels of the predicted activity [134].

12.2.1 Concepts

The IT Microcosm paradigm goes as follows: The biological activity of a chemical compound is determined by the complex effect of this compound as a whole on all of the components of a biological system by multiple characteristics of its structures [96, 105, 99, 110]. By extension, we consider the properties of a compound instead of its biological activity and a high-complexity dynamic chemical system instead of a biological system.

The semantic content of the IT Microcosm paradigm is a complex methodology for the computer prediction of the properties of a chemical compound. This methodology is founded on uniting the following for calculations: various ways of representing the chemical structure that differ in their physicochemical meaning, various levels of representing the chemical structure that differ in their complexity, redundant representations of the chemical structure that expand its parameters, classification methods that vary in their mathematical formalism, and decision circuits that are conceptual in the results they yield [105, 122]. Because it synthesizes all of the above mentioned components, the complex methodology produces context-independent decision rules (i.e., rules that do not depend on the composition of the training set, the ways of representing compound structures, or the methods of recognizing regularities) and predictive estimates of biological activity founded on these rules. The IT Microcosm paradigm comprises several basic theoretical concepts: high-complexity dynamic chemical systems, a generalized pattern of a class of compounds with desired property, a multidescriptor hierarchic multilevel representation of the structure of a chemical compound, mega-dimensional spaces, complementarity of the decision rules for the computer prediction of chemical compound properties, and strategies for the computer prediction of chemical compound properties.

A high-complexity dynamic chemical system (hereafter referred to as a complex chemical system) [95] is defined as an entire system containing a large number of individual chemical compounds contained in a space with a limited volume; these compounds interact with one another and the external environment and are separated from the external environment and one another by one or more semipermeable surfaces.

The activity of a chemical compound can be defined as the ability of this compound to cause a change in the value of one or more external parameters in a complex chemical system by a system interaction. The methods that were developed to predicting the activity of a compound in a certain complex chemical system can be applied to other complex chemical systems with minor adjustments. A complex chemical system recognizes a chemical compound through the sum of its characteristics, where each characteristic in isolation is of no great consequence. An active compound interacting with a complex chemical system is regarded as an object with a large number of degrees of freedom. An adequate description of such an object is only possible using a huge number (ideally, an infinite number) of parameters that differ in their physical and chemical meanings. The complex chemical system responds to components with varying degrees of complexity and at different stages.

Therefore, a representation of a chemical structure should be multilevel, and the variables should reflect both the local and integral properties of the compound.

The generalized pattern of a class of compounds with a desired property is a set of all of the compounds that showing this property described by the set of parameters that characterize this compound [94, 101, 110]. The cardinality of this generalized pattern goes to infinity because its elements include both synthesized (tested) and nonsynthesized (untested) active compounds when the number of parameters is not limited. The more compounds in the training set and the greater number of contrast variables of varying degrees of complexity describing their structure, the more adequate the model of the generalized pattern.

If we unite these concepts, three important consequences ensue:

1. the biological activity shown by a chemical compound is not necessarily related to its interaction with a specific biological target;
2. the chemical compound is regarded as a whole. There are no "significant" or "insignificant" fragments in its structure; likewise, there are no "significant" or "insignificant" variables describing this structure; and
3. a parametric description of the model of a generalized pattern is context-independent from the method of data analysis because it is redundant; the model also does not presuppose the involvement of any procedures for detection of "informative" variables.

Thus, the parameter space of the models of generalized pattern of a class of compounds with a desired property is of extremely great dimension; it is not divided into "informative" and "noninformative" subspaces. With this method of representation, no information that determines the individual specifics of the chemical structures to be recognized is lost, and this retention of information allows an effective extrapolation of the obtained QSAR regularities to the area of new or poorly studied compounds with nontrivial specifics of action.

A multidescriptor, hierarchical, multilevel representation of the structure of a chemical compound consists of description methods that vary in their complexity and physicochemical meaning; in other words, groups of parameters that are simultaneously divided into several levels of chemical structure representation with increasing complexity, where each subsequent level of greater complexity is generated by the preceding one. The redundancy of description expansion in the parameters provides for an increased cardinality and improves the resulting model of a generalized pattern of a compound class with the desired properties. This concept was implemented when developing QL [110], a specialized language that describes the structure of chemical compounds using 11 groups of QL descriptors at four levels of complexity, with each of the descriptors varying in their physicochemical meaning. The higher-ranking descriptor is formed as a combination of the preceding rank and the elementary QL descriptors.

The mega-dimensional space is a nonlinear space with a variable curve of extra-large dimensionality [105]. As a consequence, it is strongly correlated while it is neither orthogonal nor normed. Generally speaking, such a space is mixed and discrete-continuous. Mega-dimensional spaces are, in fact, an "impression" of real-world spaces representing real-world objects where there are no rough or postulative assumptions about the linearity or orthogonality of these spaces; the spaces are not divided into "informative" or "noninformative" subspaces. An object description, as a totality of the values of all of the variables in a mega-dimensional space, is context-independent because it makes no arbitrary assumptions about the properties of the space nor does it include the results of any preliminary analyses in any manner, which provides for a high adequacy of this description.

Complementarity of Decision Rules in the Computer Prediction of the Chemical Properties of a Compound. As a rule, making a prediction algorithm for a high-complexity dynamic chemical system using a single mathematical method is based on a series of assumptions that do not prove to be true. When comparing prediction relationships that are comparable in their accuracy, one cannot unambiguously determine which of them are more adequate. These contradictions in predicting the activity of the same compound can be resolved by applying several approaches that differ in their mathematical content as much as possible. For each level of description and each parameter group, several methods are used to calculate several classification rules including all of the variables of the given local space. An integral multimodel decision rule is developed by generalizing the obtained spectrum of intermediate prediction estimates. This suggests that the final estimate that is calculated in this way is a reliable indicator of the actual activity of the predicted compound; it also gives an adequate idea of the specifics of its behavior in the given biological system.

The generally accepted classification system [1] that is applied to the prediction of biological activity consists of the following stages:

1. developing a training set from active and inactive compounds;
2. building a primary space that describes the compound structure according to one of the classical QSAR paradigms on the basis of a simple, well-known model;
3. selecting "significant" variables using a procedure that is correlated with the model; and
4. calculating several activity-structure relationships using the "significant" variables set and choosing the most appropriate one.

This scheme of decision rule development is context-dependent; thus, "the most appropriate" QSAR regularities built from the "significant" variables are mostly artificial dependencies.

IT Microcosm produces context-independent classification rules; it has the following features [105] that distinguish it from the classical scheme above:

1. When the training set is developed, no primary space for the description is built; rather, working models of generalized patterns of active/inactive compounds in extra-large dimensions are developed without defining the "significant" and "insignificant" variables. The parametric description of models and generalized patterns is context-independent from the methods of data analysis because it is extremely redundant and does not suggest any procedure for the detection of "informative" features.

2. No reduction of the dimensionality of mega-dimensional space of description is carried out; the totality of the parameters is assumed to be informative. As a consequence, no information determining the specifics of the objects to be recognized is lost, which allows for a more effective extrapolation of the obtained regularities to new or poorly studied compounds with nontrivial specifics of action.
3. A higher adequacy and enhanced prognostic capacity of the decision rules is achieved by expanding the feature space by adding new parameter groups and new levels of compound structure representation, rather than selecting "significant" variables or reducing the feature space.
4. This approach develops an ensemble of decision rules that are based on several classification methods that differ essentially in their mathematical formalism instead of a "better" prediction regularity. An independent classification of the properties of the predicted object is done by each method for each description level of each parameter group. The obtained spectrum of prediction estimates of chemical compound activity is then used in final classification procedures.
5. The context-independence of the integral decision rule is a result of the generalization of the spectrum of intermediate prediction estimates of activity by the methods of decision-making theory. As a result, primary decision rules that differ in their mathematical meaning and chemical structure representation complement each other; in particular, prediction errors are mutually compensated.

Strategies for the Computer Prediction of the Properties of a Chemical Compound. A strategy is an integral decision-making rule regarding the final activity of a chemical compound that is based on a set of intermediate prediction estimates of its activity [96]. Methods of decision-making theory [54], particularly various voting procedures [136], are used to develop such an integral decision rule. When classifying into two classes, a single activity estimate generated using a separate method to describing a certain parameter group at a given level is a binary variable. Its meanings correspond to one of alternate possibilities: "pro" or "contra." Because there are different classification methods, description levels and parameter groups, the use of all of the intermediate calculated estimates of activity in the final vote mimics an objective decision made by an independent expert group. The outcome of this voting is context-independent with respect to both the method of structure representation and the methods of intermediate prediction estimates construction, and we can therefore regard these strategies as reliable tools for the evaluation of untested compound activity.

Three prediction strategies are defined in IT Microcosm: a conservative strategy that is based on a model of general unweighted consensus, a normal strategy that uses a model of selective weighted consensus, and a risk strategy that implements a model of supremum consensus.

The combined use of several different strategies during prediction constitutes a universal hierarchic multistage final voting procedure; it mimics an objective decision by several independent expert groups. The ultimate integral decision rule is, in essence, a multilevel consensus QSAR model.

A Complex Methodology for the Computer Prediction of the Properties of a Chemical Compound. This is the central concept of IT Microcosm [93, 105, 116] and features a synthesis of the fundamental theoretical concepts that underlie the development of rules and principles for generating the applied components of the technology: algorithms and programs that predict compound activity.

An adequate prediction of chemical compound activity is only possible through the generalization of a spectrum of prediction estimates that are obtained by several methods of classification that differ in their mathematical formalism; these methods are applied to levels of varying complexity and methods of structure representation with a varying physicochemical meaning. We use every available description variable and the redundancy of this description when expanding parameters, on the basis of several decision-making circuits that are conceptual in the results they yield.

The following decision making circuit conforms to the concept described above:

1. constructing a representative training set that includes the structures of reliably active and inactive tested substances;
2. constructing models of the generalized patterns of active/inactive compound classes on the basis of a mega-dimensional multilevel description of their structure with parameter groups of different physicochemical meaning;
3. establishing a set of decision rules using several essentially different classification methods; each method is used separately for each description level of each parameter group;
4. calculating the spectrum of prediction estimates of compound activity in the training set using all of the developed classification rules; and
5. constructing integral multimodel consensus decision-making rules using strategies with different spectra of prediction estimates, and evaluating the prognostic ability of these decision rules.

Utilizing this complex methodology, we obtain decision rules and results of activity prediction that are context-independent from the training set composition, the methods of compound structure representation, and the methods of regularity detection.

The following principles of constructing the applied components of IT Microcosm arise from the summation of these theoretical concepts:

1. a compound structure should be described by the maximum possible number of parameters;
2. the structure representation should be multilevel;
3. the groups of description parameters should differ in their physicochemical meaning;
4. the decision rules for activity prediction should include all of the parameters of structure representation;
5. the prediction methods should be adapted for use in nonlinear spaces with extra-large dimensions;
6. the compound activity prediction should be performed by several methods simultaneously;
7. the prediction methods should differ in their mathematical formalism; and

8. the final estimate of compound activity should be made by generalizing the prediction results obtained from different methods.

These principles underlie the means for processing information about chemical structure and biological properties that are discussed below.

12.2.2 QL Language for Fragment Coding of Chemical Compound Structure

QSAR Language is a specialized multidescriptor, hierarchical, multilevel language for the description of the structure of a chemical compound with fragmental substructural notation [107, 110]. It is the working language of IT Microcosm and provides an adequate representation of a full molecular graph.

Substructural descriptors are the key elements of the language; a substructural descriptor is a chemical structure fragment of a varying degree of complexity, isolated from the structural formula of a compound according to the rules and postulates of the QL language. The simplest fragments are called elementary descriptors; they constitute the QL alphabet. All of the other descriptors are a combination of several elementary descriptors; in relation to the elementary descriptors, they are composite descriptors. Eleven descriptor types of the first four ranks would be sufficient for an unambiguous representation of a molecular graph in QL.

The QL alphabet (a list of first-rank descriptors) is postulated; it is defined by three types of elementary descriptors.

A Structural Descriptor (SD) is a fragment of a compound structure with a sufficiently labile electron system constituting a non-hydrogen atom or a group of atoms with the immediate environment taken into consideration. The SD subalphabet in QL is postulated; it contains 4352 types of descriptors for all of the elements in the periodic table, including 378 types of heteroatom-containing SDs, 11 types of carbon-containing SDs, 3963 types of cyclic SDs (of three classes—nonconjugated, conjugated nonaromatic and aromatic rings smaller than 99 atoms) (Table 12.1). If a SD shows chirality, its symbol bears an R or S in accordance with the Cahn-Ingold-Prelog priority rules [82].

A Length Descriptor (LD) is the length in bonds of the shortest path along the carbon chain between two SDs (values ranging from 1 to 99) or the number of total atoms when one SD is superposed immediately over another (the occurrence index has a negative value in this case). When looking for the shortest path, heteroatomic SDs do not cross.

A Bond Descriptor (BD) characterizes the type of electron system in an LD and reflects the types of bonds, their number and the presence of conjugation. The BD subalphabet is postulated; it consists of 54 types of descriptors, where each one is formed from a combination of 4 binary indices: the presence of one or several multiple bonds (p, P), aromatic (a, A) and noncovalent bonds (n, N), and conjugation index (0, 1) (Table 12.2).

Table 12.1 Structural descriptors in the QL language

ID	SD	ID	SD	ID	SD	ID	SD	ID	SD
Nitrogen		35	–P<	69	–B<	Arsenic		132	>Se=
1	–NH2	36	–P=	70	–B=	99	–AsH2	133	=>Se=
2	>NH	37	–PH2=	71	>B′<	100	>AsH	134	–Se+<
3	=NH	38	>PH<	11072	>B+	101	–As<	135	–Se′
4	–N<	39	>PH=	Silicon		102	–As=	136	Se+n
5	–N=	40	–>P<	72	–SiH3	103	–>As<	Tellurium	
6	#N	41	–>P=	73	>SiH2	104	–>As=	137	–TeH
7	–=N=	42	–=P=	74	–SiH<	105	>As+<	138	>Te
8	–NH3+	43	–PH3+	75	>Si<	106	>As′	139	=Te
9	>NH2+	44	>PH2+	76	>Si=	107	>>As′<	141	>Te<
10	=NH2+	45	–PH+<	77	Si′n	Antimony		142	>Te=
11	–NH+<	46	–PH+=	Germanium		108	–SbH2	143	=>Te=
12	–NH+=	47	>P+<	78	–GeH3	109	>SbH	144	–Te+<
13	>N+<	48	>P+=	79	>GeH2	110	–Sb<	145	–Te′
14	>N+=	49	>P′	80	–GeH<	111	–Sb=	146	Te+n
15	>N′	50	>P′<	81	>Ge<	112	–>Sb<	Metals	
Oxygen		51	>P′=	82	>Ge=	113	–>Sb=	147-301	Mt
16	–OH	52	>>P′<	83	Ge′n	114	>Sb+<	148-302	Mt+n
17	>O	Halogens		84	Ge+n	115	>Sb′	11159-11303	Mt′n
18	=O	53	–F	Tin		116	Sb′n	Carbon	
20	–O+<	54	–F′	85	–SnH3	117	Sb+n	303	–C+<
21	–O+=	55	–Cl	86	>SnH2	Bismuth		304	–C′<
22	–O′	56	–Cl′	87	–SnH<	118	–BiH2	305	–CH3
Sulfur		57	Cl+n	88	>Sn<	119	>BiH	306	=CH2
23	–SH	58	–Br	89	>Sn=	120	–Bi<	307	#CH
24	>S	59	–Br′	90	Sn′n	121	–Bi=	308	–CH=
25	=S	60	Br+n	91	Sn+n	122	–>Bi<	309	–C#
27	>S<	61	–I	Lead		123	–>Bi=	310	>C=
28	>S=	62	–I′	92	–PbH3	124	>Bi+<	311	=C=
29	=>S=	63	I+n	93	>PbH2	125	Bi′n	312	>C(<)
30	–S+<	64	–At	94	–PbH<	126	Bi+n	315	–C(Ar)<
31	–S′	65	–At′	95	>Pb<	Selenium		Cycles	
Phosphorus		66	At+n	96	>Pb=	127	–SeH	5	Cycnn
32	–PH2	Boron		97	Pb′n	128	>Se	600-10302	Cycnnkk
33	>PH	67	–BH2	98	Pb+n	129	=Se	10303-10399	CycArnn
34	=PH	68	>BH			131	>Se<		

For each structural descriptor, its digital code and symbol are indicated.
Special symbols: «#»—triple bond; «‘»—negative charge; «Ar»—aromatic.
For “Metals” descriptors Mt is a metal symbol in the Periodic Table.
For “Rings” descriptors: nn—ring size; kk—the number of electrons in a conjugated system.

Different types of elementary descriptors have different physicochemical meanings. Thus, an SD shows the local geometric, electron and lipophilic characteristics, an LD indicates the integral geometric characteristics, and a BD shows the integral electron characteristics.

Table 12.2 Bond descriptors in the QL language

ID	BD	ID	BD	ID	BD	ID	BD
Single-index		5	. . N 1	23	p . N 1	45	P a n 1
0	. . . 0	Three-index		40	P . N 0	32	p A n 0
1	. . . 1	24	p a . 0	41	P . N 1	33	p A n 1
Two-index		25	p a . 1	8	. a n 0	50	P A n 0
18	p . . 0	42	P a . 0	9	. a n 1	51	P A n 1
36	p . . 0	43	P a . 1	14	. A n 0	28	p a N 0
19	p . . 1	30	p A . 0	15	. A n 1	29	p a N 1
37	P . . 1	31	p A . 1	10	. a N 0	46	P a N 0
6	. a. 0	48	P A . 0	11	. a N 1	47	P a N 1
12	. A. 0	49	P A. 1	16	. A N 0	34	p A N 0
7	. a . 1	20	p . n 0	17	. A N 1	35	p A N 1
13	. A . 1	21	p . n 1	Four-index		52	P A N 0
2	. . n 0	38	P . n 0	26	p a n 0	53	P A N 1
4	. . N 0	39	P . n 1	27	p a n 1		
3	. . n 1	22	p . N 0	44	P a n 0		

For each bond descriptor, its digital code and symbol are indicated

Elementary descriptors are simple first-rank descriptors. They generate composite descriptors in QL. Simple composite descriptors of rank 2–4 consist of 2–4 elementary descriptors. Second-rank descriptors are of four types (SD-LD, SD_1-SD_2, SD-BD and LD-BD), third-rank descriptors are of three types (SD_1-LD-SD_2, SD_1-LD-BD and SD_1-SD_2-BD), and fourth-rank descriptors are of one type (SD_1-LD-SD_2-BD). In all, 11 types of simple descriptors of rank 1–4 are defined in QL. Upon translation, a primary QL representation of a structure is constructed as a list of fourth-rank descriptors where all of the SDs are numbered. Later on, these descriptors generate the descriptors of all other ranks; therefore, the fourth-rank descriptors are also referred to as basic descriptors. Descriptors of rank 5 and higher consist of two or more basic descriptors; they are referred to as complex descriptors.

As the rank increases, the dispersion of the compound properties in the descriptors also grows. The simpler the descriptor, the better its extrapolation (prediction) ability; more complex descriptors show a better interpolation (recognition) ability. For example, using a CH_3 group permits a prediction of the activity of many organic compounds. However, for this prediction to be sufficiently accurate, one should take into consideration more complex structural fragments, too.

In the QL dictionary, all of the elementary descriptors are independent words. Basic syntax contains two rules only: (1) each simple descriptor of a higher rank (ranks 2–4) is generated by the combination of a descriptor of a lower rank and one elementary descriptor, and (2) each complex descriptor of a higher rank at rank 5 and higher is generated by the combination of a descriptor of a lower rank and one basic descriptor through a common SD of the same numerical order in both descriptors. Because the SDs in the basic descriptors are numbered, the rule governing the generation of complex descriptors allows the complete reconstruction of the structural formula of a compound. Thus, QL is a one-to-one language.

A study of information redundancy of QL showed that a set of simple descriptors of ranks 1–4 with an indication of their number is almost identical to the molecular graph; therefore, only these 11 types of QL descriptors are used in IT Microcosm for the calculation of decision rules and the prediction of activity.

The meaning of a QL-based description is easily derived from the symbols of descriptors. For instance:

{–CH3 2 ○ ○} This is a methyl group in a two-bond chain with arbitrary type bonds, with conjugation present or absent ("○" stands for any elementary descriptor);

{–N< 5 –CH3 ○} This is a tertiary amino group bonded to a methyl group by a five-bond chain with arbitrary type bonds, with conjugation present or absent;

{–N=–1 CycAr05 …1} A secondary imino group included into a five-membered aromatic ring.

The structure of an organic compound of medium complexity is usually described by 50-1,000 types of QL descriptors.

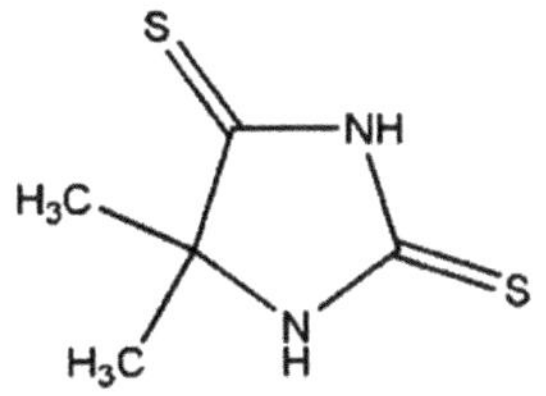

For example, the QL description of the structure of Thiomedan, an antiepileptic drug, includes 165 types of QL descriptors.

For a desired activity type, models of generalized patterns of active/inactive compounds are constructed from a training set in the form of a substructural descriptor matrix, which is a table where the lines feature the symbols for unique QL descriptors of 11 types of the first four ranks and the columns indicate the numbers of compounds. Each table cell shows the number of QL descriptors of this type in the structure of a certain compound. In the matrix, the descriptors are placed in the order of increasing rank in lexicographic order. The activity of the compounds in the training set is annotated in a separate file.

12.2.3 Prediction Methods

To obtain a spectrum of intermediate prediction estimates, IT Microcosm utilizes four original classification methods that show a consistent performance in megadimensional spaces. The prediction estimates are binary variables; they can only assume two values: A or N (in numerical expression, 1 or 0, correspondingly). Each method yields 11 prediction estimates (according to the number of QL descriptor types). All four methods are essentially different in the way that their decision rule is constructed. This suggests that a spectrum of 44 prediction estimates gives a fairly accurate idea of the various specifics of a generalized pattern of active/inactive compound class, which hereafter allows for a reliable evaluation of the total activity of the predicted compound [102, 105, 109].

We will now introduce the following designations:

i QL descriptor type, $i=1,\ldots, 11$;
j type of i-type descriptor in QL matrix, $j=1,\ldots, d_i$;
d_i number of types of unique descriptors of i-type in QL matrix;
ij j^{th} descriptor of i-type;
a active compound class;
n inactive compound class;
k compound class, $k=a, n$;
N the number of compounds in a training set;
β 0.001– unbiasedness parameter.

The Bayesian approach is one of the probabilistic central parametric classification methods; it is based on the consistent application of the classic Bayes equation (also known as "the naïve Bayes classifier") for conditional probability [34] to construct a decision rule; a modified algorithm is explained in references [105, 109, 121]. In this approach, a chemical compound C, which can be specified by a set of probability features $(c_1,\ldots,c_m)$ whose random values are distributed through all classes of objects, is the object of recognition. The features are interpreted as independent random variables of an m-dimensional random variable. The classification metric is an *a posteriori* probability that the object in question belongs to class k. Compound C is assigned to the class where the probability of membership is the highest.

The logarithm of the probability that the predicted compound C belongs to class k, given it has d_i QL descriptors of ij-type B_{ij}

$$\Pr_i (C \in k) = \log\left[P_0(C \in k)\right] + \sum_{j=1}^{d_i} \log\left[P(B_{ij} \mid C \in k)\right], \qquad (12.1)$$

where $P_0(C \in k) = 0,5$ is the *a priori* probability that compound C belongs to class k before the analysis is started;

$P(B_{ij} \mid C \in k)$ is the *a priori* probability that descriptor B_{ij} occurs in class k.

In the formula (1)

$$P(B_{ij} \mid C \in k) = \frac{S_{ijk} + \beta}{S_{ik} + d_i \cdot \beta}, \qquad (12.2)$$

where S_{ijk} is the number of ij descriptors in the QL matrix for class k; and

S_{ik} is the total number of all QL descriptors of i-type for class k.

Compound C is defined as active for the i-type descriptor, if

$$\Pr_i (C \in a) \geq \Pr_i (C \in n); \qquad (12.3)$$

otherwise it is classified as inactive.

The boundary conditions of the Bayesian formula are not satisfied, as a rule, in cases of real-world training sets. The set of events may be neither complete nor mutually exclusive, while QL descriptors of different *ij* types are virtually always interdependent. Therefore, the value $P_i(C \in k)$ calculated from the training set is not, in fact, an indicator of probability, and the aggregate of the tested compounds is a fuzzy set. The fuzzy set *A* of universe *X* is characterized by the membership function $\mu_A : X \to [0, 1]$, which places each element $x \in X$ in correspondence with the number $\mu_A(x)$ from an interval [0, 1] describing the degree of membership of element *x* in set *A* [2]. The membership function in IT Microcosm expresses the degree to which the model of the tested compound *C* pattern corresponds to the generalized pattern of class *k* compounds with the desired activity.

In the Bayesian approach, the membership function of compound *C* belonging to activity class *k* for *i*-type descriptor is

$$Fb_i(C \in k) = \frac{\left|\Pr_i(C \in k)\right| + \beta}{\left|\Pr_i(C \in a)\right| + \left|\Pr_i(C \in n)\right| + 2 \cdot \beta}. \tag{12.4}$$

The distance method is a geometric central parametric method; its modified algorithm is described in references [105, 109]. In this case, the object of classification (chemical compound *C*) is defined by a set of determined features $(c_1,\ldots,c_m)$ whose values are interpreted as coordinates of a point in a multidimensional space of *m* dimension. The classification metric is the distance from the object in question to the geometric center of class *k*. Compound *C* belongs to the class placed at a shorter distance.

A Pearson weighted L_1-distance is calculated in the space of QL descriptors of *i*-type from the predicted compound *C* to the center of class *k*

$$D_{ik} = \sum_{j=1}^{d_i} w_{ij} \cdot \left|c_{ij} - z_{ijk}\right|, \tag{12.5}$$

where c_{ij} are the coordinates of compound *C* for the descriptor *ij*;

z_{ijk} are the coordinates of the center of class *k* for descriptor *ij*; and

$w_{ij}=(z_{ija}+z_{ijn})^{-1}$ is the weighting coefficient for descriptor *ij*.

Compound *C* is regarded active for descriptor of *i*-type, if

$$D_{ia} \le D_{in}; \tag{12.6}$$

otherwise it is classified as inactive.

In the distance method, the membership function of compound *C* belonging to activity class *k* for descriptor of *i*-type appears as

$$Fb_i(C \in k) = 1 - \frac{D_{ik} + \beta}{D_{ia} + D_{in} + 2 \cdot \beta}. \tag{12.7}$$

The nearest neighbor method is a local geometric nonparametric method; it is useful for the calculation of a piecewise linear separating function [32]. The basic algorithm was described in reference [34] and adapted to extra-large dimensional spaces as described in [105, 109]. The classification metric is the distance in the *m*-dimensional feature space between the object of interest and the object of class *k*. Compound *C* belongs to the class where its nearest neighbor is located.

The squared Euclidean distance in the space of QL descriptors of *i*-type from the predicted compound *C* to each compound H_l in the training set is

$$D_i^2(H_l) = \sum_{\substack{j=1 \\ j \in C \cup H_l}}^{d_i} (c_{ij} - h_{ijl})^2,\ l = 1,\dots,N, \tag{12.8}$$

where c_{ij} are coordinates of compound *C* for descriptor *ij*; and

h_{ijl} are coordinates of compound H_l for descriptor *ij*.

Therefore, in the space of *i*-type descriptors, the distance from compound *C* to its nearest neighbor of class *k* is

$$D_{ik}^2 = \min_{\substack{l=1 \\ H_l \in k}}^{N} \{D_i^2(H_l)\} \tag{12.9}$$

Compound *C* is defined as active for a descriptor of *i*-type, if

$$D_{ia}^2 \le D_{in}^2; \tag{12.10}$$

otherwise it is classified as inactive.

In the nearest neighbor method, the membership function of compound *C* belonging to activity class *k* for descriptor of *i*-type is

$$Fb_i(C \in k) = 1 - \frac{D_{ik}^2 + \beta}{D_{ia}^2 + D_{in}^2 + 2 \cdot \beta}. \tag{12.11}$$

The local distribution method is one combination method using the geometric local nonparametric method in parallel to a probabilistic central parametric method for decision rule construction. The algorithm was first described in [109] and later modified as described in [105]. Two metrics serve as classification metrics: the similarity coefficient of the features of the object to be predicted and class *k* objects in *m*-dimensional space, and the probability that the object of interest belongs to the subclass of similar objects in class *k*. Compound *C* is assigned to the class with the greatest local probability that the compound belongs to the structurally similar subclass.

The similarity coefficient of the predicted compound *C* and each compound H_l in the training set in the space of *i*-type descriptors is

$$Q_i(H_l) = \frac{1}{d_i} \cdot \sum_{\substack{j=1 \\ j \in C \cup H_l}}^{d_i} \frac{\min\{B_{ijC}, B_{ijl}\}}{\max\{B_{ijC}, B_{ijl}\}}, \; l = 1, ..., N, \tag{12.12}$$

where B_{ijC} is the number of descriptors ij in compound C; and

B_{ijl} is the number of descriptors ij in compound H_l.

Therefore, the subclass of class k compounds that is structurally similar to compound C in the space of i-type descriptors appears as

$$k_i = \{H'_1, H'_2, H'_3, H'_4\} \cup \{H_l \in k \mid Q_i(H_l) = Q_i(H'_4)\} \cup \{H_l \in k \mid Q_i(H_l) \geq 0.8\}, \tag{12.13}$$

where $\{H'_1, H'_2, H'_3, H'_4\}$ are the first four members of a set of compounds H_l arranged in descending order of the $Q_i(H_l)$ value, $l = 1, \dots N$.

In other words, four structures that are similar to the predicted compound C, compounds with a minimal value of $Q_i(H_l)$, and compounds with a similarity coefficient for compound C of no less than 0.8 are selected in each class a and n.

Applying the formulas of the Bayesian approach (1–4) to the constructed local training set $k_i = \{a_i, n_i\}$, we obtain

$$\Pr_i(C \in k_i) = \log\left[P_0(C \in k_i)\right] + \sum_{j=1}^{d_i} \log\left[P(B_{ij} \mid C \in k_i)\right]. \tag{12.14}$$

Compound C is defined as active for the i-type descriptor if

$$\Pr_i(C \in a_i) \geq \Pr_i(C \in n_i); \tag{12.15}$$

otherwise it is classified as inactive.

In the local distribution method, the membership function for compound C belonging to activity class k for the i-type descriptor is

$$Fb_i(C \in k) = \frac{\left|\Pr_i(C \in k_i)\right| + \beta}{\left|\Pr_i(C \in a_i)\right| + \left|\Pr_i(C \in n_i)\right| + 2 \cdot \beta}. \tag{12.16}$$

The classification metric (14) and membership function (16) for compound C are only calculated for compounds of the local training set.

The central parametric methods of classification are based on generalized information about all of the objects in the training set, so when applying these methods, one generally takes the most significant regularities that are typical of this type of activity into consideration. Conversely, local nonparametric methods consider the characteristics of objects that are closest to the predicted structure, so they mostly reflect the fine specific regularities, the individual peculiarities of the predicted compound, and the degree of its novelty.

The four classification methods described above are essentially different in the way the decision rule is constructed; when applied jointly, they compensate for each other's errors.

12.2.4 Prediction Strategies

IT Microcosm implements three decision strategies that classify a compound as either active or inactive on the basis of a prediction estimate spectrum obtained by different methods [96, 105]. Each strategy is a method of constructing an integral decision rule in the form of a consensus QSAR model; different types of consensus are employed.

The conservative prediction strategy uses a simple vote procedure. Here a general nonweighted consensus model is implemented, all 44 prediction estimates are deemed to be equally significant irrespective of the QL representation level and prediction method, and a decision is made according to the majority of coinciding estimates.

For example, if we set θ as the number of positive prediction estimates "A" that the predicted compound *C* belongs to class *a* of active compounds. Within the framework of the conservative strategy, compound *C* is considered to be active if $\theta \geq 27$, and defined as inactive if $\theta \leq 17$ (95 % is the confidence interval for the median in binomial distribution [41]). If the $17 < \theta < 27$ prediction is discontinuous, compound *C* can be considered conditionally active if $22 \leq \theta < 27$, and conditionally inactive if $17 < \theta < 22$.

In the conservative strategy, the membership function for compound *C* belonging to the activity class *k* is

$$Fb(C \in k) = \frac{\max\{\theta, (44 - \theta)\} + \beta}{44 + 2 \cdot \beta}. \tag{12.17}$$

The conservative strategy takes into consideration the most stable, standard, and perhaps even trivial regularities that are typical of this activity type, so with this strategy, prediction is notoriously reliable. The conservative strategy can be employed in searches for novel but typical active compounds or to improve the parameters of preexisting standard substances, such as for interpolation or placement into the "center" of the class.

The normal strategy implements a model of selective weighted consensus. This implies selecting a method with superior accuracy out of four prediction methods, based on the results of a leave-one-out cross validation of the training set, with the generalization of 11 prediction estimates calculated within the framework of each method with the help of weighted voting.

The Bayesian binary classifier serves for the voting procedure [32].

$$L_l(C) = w_{l0} + \sum_{i=1}^{11} w_{li} \cdot \delta_{Cli}, \quad l = 1, ..., 4 \qquad (12.18)$$

The weighting coefficients are

$$w_{l0} = \sum_{i=1}^{11} \log \frac{1 - p_{ali}}{1 - p_{nli}} \quad w_{li} = \log \frac{P_{ali} \cdot (1 - p_{nli})}{P_{nli} \cdot (1 - p_{ali})}, \quad l = 1, ..., 4 \qquad (12.19)$$

where

$p_{kli} = (n_{kli} + 1)/(n_{ali} + n_{nli} + 2)$ is an *a priori* probability of classifying compound C for the i-type descriptor into class $k = a, n$ by method l;

n_{kli} is the number of compounds in class k of the training set that are classified as active for the i-type descriptor by method l; and

δ_{Cli} is the result of classifying compound C into classes a ($\delta_{Cli} = 1$) or n ($\delta_{Cli} = 0$) for the i-type descriptor by method l.

Within the framework of method b (with superior accuracy), compound C is considered to be active in the whole of the QL representation if the separating function (calculated as in (18)) $L_b(C) \geq 0$; otherwise, the compound is classified as inactive.

In the normal strategy, the membership function of compound C belonging to the activity class k is

$$Fb(C \in k) = \max \left\{ \left[\left(\frac{L_b(C) - L_{b,low} + \beta}{L_{b,high} - L_{b,low} + 2 \cdot \beta} \right), \left(1 - \frac{L_b(C) - L_{b,low} + \beta}{L_{b,high} - L_{b,low} + 2 \cdot \beta} \right) \right] \right\}, \qquad (12.20)$$

where

$L_b(C)$ is the meaning of the separating function (18) for compound C in method b of superior accuracy;

$L_{b, low}$ is the minimal value of the separating function (18) in method b of superior accuracy (a sum of w_0 and all negative w_i); and

$L_{b, high}$ is the maximum value of the separating function (18) in method b of superior accuracy (a sum of w_0 and all positive w_i).

When using the normal strategy of prediction, we consider both standard and non-standard regularities for this type of property. Such an approach is advantageous when modifying the structures of atypical compounds; it expands the prediction area to the nearest boundary of the training set, but it also increases the risk of artifacts.

The risk strategy is also based on selecting the method of superior accuracy out of four methods according to the results of a leave-one-out cross-validation of the training set, but it is performed for each one of 11 levels of QL representation separately, with consideration to the descriptor type. This strategy implements a model of supremum consensus. There is no final voting procedure because each one of 44 prediction sets is treated as an independent information space.

Within the risk strategy, the classification metric and membership function for the predicted compound C are calculated from the formula pairs (1, 4), (5, 7), (8, 11), (14, 16) corresponding to the selected prediction method and QL description level.

The risk strategy permits the indirect consideration of the peculiarities associated with possible effect mechanisms, assigns more weight to the novelty of the predicted structure than the two other strategies, and offers a broad range of extrapolation possibilities. However, this strategy requires great caution because there is typically a high probability of erroneous results.

12.2.5 Evaluation of Prediction Accuracy

The accuracy of the consensus prediction regularities obtained by these three strategies is evaluated according to four indicators of the recognizing and predicting abilities of the integral decision rule, i.e., the results of self-prediction, leave-one-out cross-validation, split-half cross-validation, and double leave-one-out cross-validation.

Self-Prediction. The activity of each one of N compounds in the training set is calculated without any changes in the QL matrix or recalculation of the decision rules.

Leave-One-Out Cross-Validation. Each compound in the training set is in turn excluded from the QL matrix. In the changed set, new decision rules are calculated for $N1$ compounds, and the excluded compound is used as an independent testing object. The procedure is repeated N times.

Split-Half Cross-Validation. The working QL matrix and new decision rules are calculated with the odd-numbered compounds in the training set, and the even compounds serve as an independent testing set. A reverse procedure is then carried out, and the classification regularities are calculated with the even compounds, while the odd ones serve as a testing set. The results of testing are averaged out.

Double Leave-One-Out Cross-Validation. One compound is excluded from the training set. The procedure of leave-one-out validation is performed with the remaining $N1$ compounds. On the basis of the leave-one-out validation results, the parameters of the final decision rule are calculated for the Bayesian binary classifier. This decision rule is used to classify the excluded compound, and the procedure is repeated N times. This testing method is only used in the normal strategy.

Prediction Accuracy Metrics. In all testing methods, the following indicators are worked out (and measured as percentages):

F_0 the proportion of correctly classified compounds irrespective of the activity class (accuracy);
F_a the proportion of correctly classified active compounds (sensitivity);
F_n the proportion of correctly classified inactive compounds (specificity); and
F_u the proportion of compounds that were rejected from prediction (this is only calculated in the conservative strategy).

12.2.6 Noncontradiction Check of the Prediction

Substances that showed the coincidence of two or three calculated estimates of activity obtained with different strategies are of the greatest interest for experimental study because in this case we take into consideration both standard and nonstandard QSAR regularities and the novelty of the chemical structure; the reliability and adequacy of the predictions are therefore greatly increased.

When selecting such structures, IT Microcosm allows a noncontradiction check of the prediction estimate spectrum in case of semiquantitative gradations of activity [102]. Each activity level is matched to a set of correct evaluations of other activity levels referred to as a template. For example, if the activity can be scored as "high," "moderate," "low," "high or moderate," "active," and "inactive," the template set appears as ANNAAN, NANAAN, NNANAN, AANAAN, AAAAAN, NNNNNA. If one decision rule assigns a compound to the "high" class, the classification of this compound as "moderate", "low" or "inactive" in other decision rules is logically incorrect.

According to the prediction results, we select compounds that received an "A" score for the target level (type) of activity in at least one strategy. In this case, for each such compound *C*, the conformity coefficient (ranging from 0 to 1) is calculated from the spectrum of activity prediction estimates as a ratio of the number of estimates that logically predict activity to the number of total estimates

$$K_{Cq} = \frac{1}{3 \cdot V_q \cdot (G-1)} \cdot \sum_{s=1}^{3} \sum_{t=1}^{G} \omega_{Cts}, \tag{12.21}$$

where

G is the number of gradations (levels) of activity, $G \geq 3$;
q is the index of activity gradation, $q = 1, \ldots, G$;
s is the index of the prediction strategy, $s = 1, 2, 3$;
V_q is the number of templates of gradation activity q; and

ω_{Cts} is the conformity index {0, 1} for compound C based on the results of activity gradation t prediction in strategy s, and the corresponding value in the template of gradation activity q.

To determine the order of experimental testing for a high level of activity, the compounds are first ranked in decreasing order by the number of positive final prediction estimates and then ranked in descending order of conformity coefficient (21). When the prediction is refined for a certain compound, a full spectrum of prediction estimates and the values of membership functions are used in all three strategies.

12.2.7 Levels and Types of Consensus

Let us define the level of consensus as the step number in a consecutive hierarchical generalization of classification estimates.

In IT Microcosm, the chemical structure is represented by 11 types of QL descriptors. For each i level of QL description, separate decision rules are calculated including all of the i-type descriptors. Thus, each primary prediction estimate of an activity obtained by one of the four prediction methods for the given level of QL description is the result of a first-level consensus. All of the prediction methods implement a procedure of general weighted consensus, and the contribution of each descriptor of type ij to the final estimate is symbatic with its number in the structure of the predicted compound.

On the basis of a generalized spectrum of 44 primary prediction estimates, the integral decision rules are constructed with the help of three prediction strategies. Each of the three strategies models decision-making by an independent expert group with a unique method. This procedure may be referred to as second-level integral consensus. As stated above, the conservative strategy implements a model of general nonweighted consensus, the normal strategy implements a model of selective weighted consensus, and the risk strategy implements a model of supremum consensus.

The generalization of prediction results by three strategies corresponds to decision-making by three expert groups employing different methods to determine the "quality" of the predicted object. This type of a QSAR model is something akin to a "consensus of consensuses;" it is a third-level integral consensus.

To enhance the reliability of predictions and determine the order of experimental testing of compounds, IT Microcosm uses a fourth-level integral consensus in the generalization of prediction estimates derived for several activity levels.

12.2.8 Pharmacophore Analysis

The theoretical concepts of IT Microcosm do not include the basic concept of a pharmacophore; pharmacophores are merely considered to be extreme particular manifestations of a generalized pattern of compound showing a desired biological activity [94, 101, 105]. Nevertheless, the concept of pharmacophore analysis is preserved in IT Microcosm because it is the conventional approach to structure-activity tasks and it is notable for its demonstrativeness and simplicity. Within the framework of this approach, IT Microcosm defines a new class of QSAR objects, pharmacophorepatterns, each including multiple conjunctions of "conventional" pharmacophores, i.e., fixed fragments of the chemical structure. Compared to conventional pharmacophores, pharmacophorepatterns show a much greater predictive ability; however, this predictive ability is greatly inferior to that of the generalized pattern. The procedure of constructing pharmacophorepatterns is described in detail in reference [79]; it consists of two stages.

Stage one entails calculating the lists of the primary statistically significant QL features of the desired type (level) of activity.

The Bayesian frequency of occurrence of descriptor *ij* in class $k=a, n$

$$P_{ijk} = \frac{S_{ijk}+1}{S_{ik}+d_i}, \quad j=1,\ldots,d_i, \quad i=1,\ldots,11 \tag{12.22}$$

where

S_{ijk} is the number of descriptors of *ij*-type in class *k*; and
S_{ik} is the total number of descriptors of *i*-type in class *k*.

For each descriptor *ij*, we evaluate the statistical significance of the Bayesian frequency of the occurrence of this descriptor in classes *a* and *n* according to the hypergeometric test

$$\mathrm{Pr}_{ij} = \frac{\begin{pmatrix} (S_{ija}+S_{ijn}+2) \\ (S_{ija}+1) \end{pmatrix} \times \begin{pmatrix} (S_{ia}+S_{in}+2d_i)-(S_{ija}+S_{ijn}+2) \\ (S_{ia}+d_i)-(S_{ija}+1) \end{pmatrix}}{\begin{pmatrix} (S_{ia}+S_{in}+2d_i) \\ (S_{ia}+d_i) \end{pmatrix}}, \tag{12.23}$$

where $\begin{pmatrix} R \\ r \end{pmatrix} = \dfrac{R!}{r!(R-r)!}$ is the combination.

The QL descriptor of *ij*-type is a statistically significant feature of activity if

$$P_{ija} \geq P_{ijn} \text{ and } \mathrm{Pr}_{ij} \leq \mathrm{Pr}_0, \tag{12.24}$$

where Pr_0 is the statistical significance threshold (0.05, 0.01 or 0.001) selected in relation to the reliability extent desired by the researcher.

Stage two entails the construction of pharmacophorepatterns according to the lists of statistically significant QL features. By the pair-wise comparison of all of the features on the list, the pairs of QL descriptors that have coinciding or structurally similar features are selected. The selected feature pairs are united into intermediate "integral" features. The obtained "integral" features are compared against each other and against the QL features that remain on the list, and the resulting analogous pairs are united again. This procedure is repeated until the possibility of feature pairing of both primary and intermediate features is completely exhausted.

12.2.9 IT Microcosm Software Complex

All of the theoretical concepts, methods and rules discussed above are implemented in IT Microcosm software package for Windows [113], which includes 20 applications and a total of over 58,000 lines of source code.

The IT Microcosm software package makes it possible to solve the following tasks associated with the calculation of prediction regularities:

- Notation of compound activity levels;
- Translation of the structure of a compound into QL representation and the construction of generalized classes of compounds with the desired activity;
- The calculation of first-level consensus decision rules using four prediction methods;
- The construction of second-level integral consensus decision rules using three prediction strategies;
- The construction of third-level integral consensus decision rules by generalizing the prediction results for all strategies;
- The validation of the obtained decision rules;
- Making mixtures of compounds with any composition in the form of a set of structural files of a standard format;
- Revealing pharmacophorepatterns that are associated with high activity.

The IT Microcosm software package offers researchers the following possibilities that are associated with the prediction of the presence or extent of pharmacological and biological activity of various types in organic compounds with any chemical structure:

- Prediction of the activity of untested compounds;
- Directed search for highly active drug substances;
- Directed design of novel, highly active drug substances;
- Directed search for highly active multi-target compounds that show several mechanisms of action at the same time;
- Prediction of the activity of organic compound salts taking the mutual effects of their components into consideration;
- Optimization of qualitative and quantitative composition of organic compound salts with the aim of achieving the maximum pharmacological effect;
- Predicting the activity of supramolecular complexes that are formed as a result of a noncovalent intermolecular interaction between several compounds;
- Optimization of qualitative and quantitative composition of supramolecular complexes with the aim of achieving the maximum pharmacological effect;

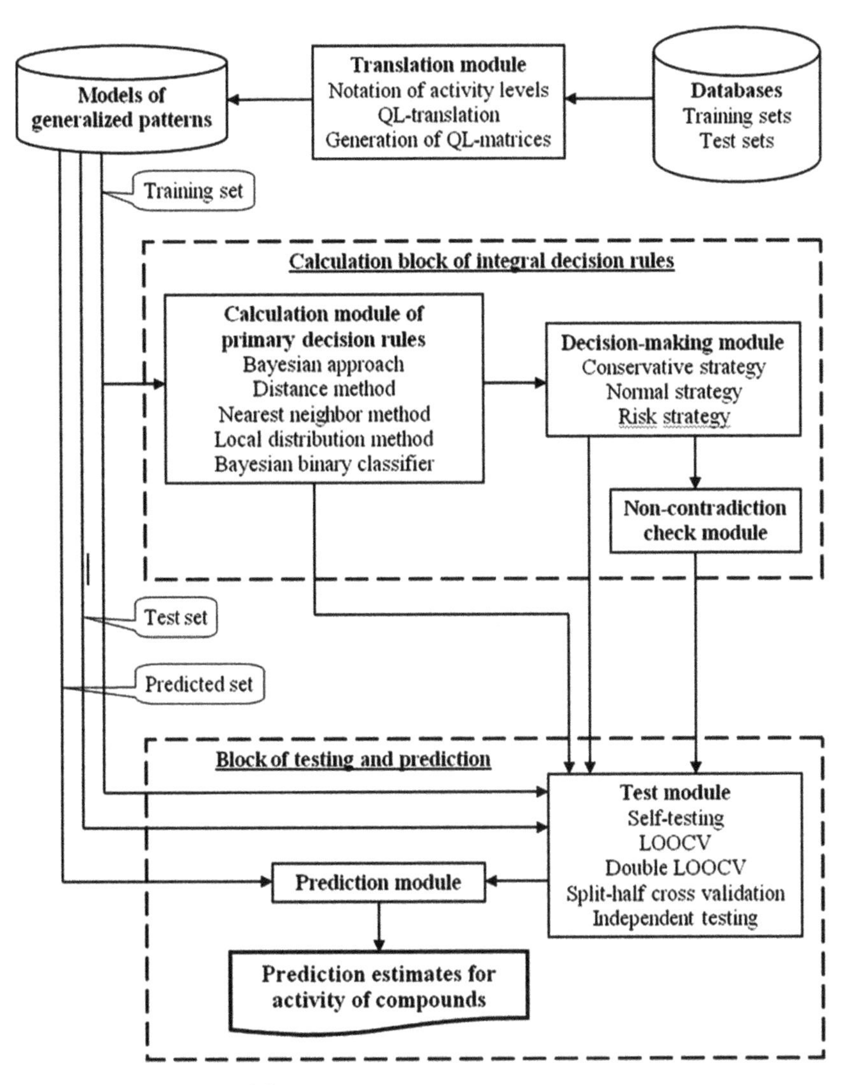

Fig. 12.1 IT Microcosm workflow

- Predicting the activity of mixtures that consist of several individual compounds accounting for their quantitative composition and component synergism;
- Optimization of the qualitative and quantitative composition of compound mixtures with the aim of achieving the maximum pharmacological effect;
- Directed search for compounds that are highly selective for a certain biological target subtype compared to other subtypes of the same target.

A general workflow of data processing in IT Microcosm is shown in Fig. 12.1. The IT Microcosm software is available upon request. IT Microcosm White is a freely distributed version with limited functionality, for the prediction of individual organic compound activity.

12.3 Predictive Power of IT Microcosm

The adequacy, validity and accuracy of IT Microcosm were analyzed by testing the training sets for the structure and pharmacological activity of structurally diverse and structurally similar chemical compounds. It is well known that lead generation methods are intended to represent searches for structures with high novelty, parent compounds of new chemical classes with the desired pharmacological activity, and new chemical entities; these methods should provide for an adequate prediction of activity in structurally diverse compounds. Lead optimization methods are aimed at revealing the most active compounds in a series of structurally similar compounds; these approaches are expected to produce stable results and to predict the pharmacological activity of one class of chemical derivatives. The theoretical concepts of IT Microcosm are of a universal nature; therefore, the technology permits an equally successful prediction of the presence and extent of pharmacological activity in both structurally diverse and structurally similar compounds, including chiral compounds.

12.3.1 Structurally Diverse Compounds

The training sets were constructed from available reference literature and information from the internet. The sets include structures of compounds that reliably show the predicted activity and compounds that reliably show no activity. The total size of databases for 34 types of activity is 10,703 structures of known drugs and biologically active substances. For 19/34 activity types, the active compounds were divided into highly active and moderately active classes using the expert method. The selection, verification and primary processing of information in relation to structure and activity were performed by competent experts: chemists and doctoral-level pharmacologists with considerable experience in the corresponding fields. The method of training set construction is discussed in detail in references [105, 109].

The sizes of the training sets ranged from 30 to 1140 compounds. Depending on the type of activity, the indices of chemical diversity varied within the following limits: the dimensionality of the object domain description ranged from 1798 to 22,461 variables, and the mean number of unique features per compound ranged from 14 to 149 QL descriptors.

A summary of the analysis of the adequate decision rules when predicting the presence/absence of an activity in structurally diverse compounds is shown in Table 12.3. A decision rule was deemed to be adequate if the values of all of the prediction accuracy indices F_0, F_a and F_n in all testing methods were at least 60 %, which corresponds to a confidence level of $p \geq 0.9$ with a training set size of $N \geq 30$.

Table 12.3 General indices of prediction accuracy for the activity of structurally diverse compounds

Strategy	ST, %		LOOCV, %		SHCV, %		DLOOCV, %	
	Min	Max	Min	Max	Min	Max	Min	Max
Accuracy F_0								
Conservative	97	100	77	100	70	100	–	–
Normal	81	100	69	100	65	95	69	100
Risk	85	100	67	100	65	97	–	–
Sensitivity F_a								
Conservative	88	100	67	100	67	100	–	–
Normal	83	100	63	100	62	97	63	100
Risk	83	100	67	100	63	100	–	–
Specificity F_n								
Conservative	96	100	68	100	60	100	–	–
Normal	79	100	69	100	64	96	69	100
Risk	85	100	67	100	60	100	–	–

DLOOCV is not used in the conservative or risk strategy

Altogether, in all strategies and methods of testing, the maximum value of all three indices F_0, F_a and F_n amounts to 100%.

The estimates of prediction accuracy for the best decision rules (for the F_0 value sum for ST, LOOCV and SHCV) are shown in Table 12.4. In this case, the conservative strategy proved to be the best; optimal decision rules were obtained for 26 activities out 34 (76%), mainly for systemic pharmacological effects. This is not surprising because the presence/absence of activity is mostly determined by “standard” regularities, and the structural diversity of compounds underlies the multiplicity of action mechanisms, especially when dealing with therapeutic activities. The normal and risk strategies yielded 4 optimum decision rules (12% each). Notably, the risk strategy produced decision rules for only the receptor activity type, where subtle interaction mechanisms between the ligand and receptor site play a key role.

A summary of testing adequate decision rules in predicting a high-level activity in structurally diverse compounds is shown in Table 12.5. In this case, the maximum values F_0, F_a and F_n in all strategies only amount to 100% in the self-prediction model. In the leave-one-out and split-half cross-validations, the maximum values of F_0, F_a, and F_n were 99%, 95%, and 100%, respectively.

The estimates of the best decision rules accuracy in predicting a high activity are shown in Table 12.6. In this case, the conservative strategy again gave the best results; optimum decision rules were obtained for 7 activities out of 16 (44%). However, the normal (5 optimum regularities, 31%) and risk strategies (4 optimum regularities, 25%) were used more often. This outcome is also understandable because highly active compounds often show a “nonstandard” chemical structure and as a result tend to have unconventional mechanisms of action.

Table 12.4 Accuracy of the best strategy for predicting the presence of an activity in structurally diverse compounds

Activity	N	Better strategy	ST F_0, %	LOOCV F_0, %	SHCV F_0, %
Neuroleptic	645	Conservative	99	97	97
Tranquilizer	532	Conservative	100	89	89
Antidepressant	628	Conservative	100	91	90
Nootropic	420	Conservative	99	77	70
Analgesic narcotic	320	Conservative	100	97	95
Anesthetic local	324	Conservative	100	91	91
Spasmolytic	804	Conservative	99	81	82
Antianginal	410	Conservative	99	81	79
Cardiotonic	304	Conservative	100	89	89
Cardiac stimulant	233	Conservative	100	96	95
Hypoglycemic	230	Conservative	100	82	82
Anabolic	184	Normal	99	91	90
Antiseptic	494	Conservative	100	87	86
Leprostatic	52	Conservative	100	79	80
Tuberculostatic	386	Conservative	100	93	92
Antifungal	471	Conservative	100	83	80
Anti-HIV	1140	Conservative	97	80	80
Antiherpetic	412	Normal	90	69	68
Anti-paramixovirus	54	Conservative	98	96	94
Anti-picornavirus	512	Conservative	97	87	88
Anti-rheovirus	30	Normal	93	77	76
Anti-orthovirus	72	Conservative	97	92	92
Antileukemic	252	Conservative	98	81	79
Antineoplastic	821	Conservative	100	89	87
Antioxidant	82	Conservative	100	86	85
K-opioid agonist	74	Conservative	100	95	83
5-HT_2 antagonist	33	Conservative	100	86	93
5-HT_3 antagonist	47	Risk	87	87	81
H_1 antagonist	66	Normal	100	89	89
H_2 antagonist	50	Risk	100	92	92
H_3 antagonist	46	Risk	100	98	96
H_3 agonist	47	Conservative	100	100	100
$P2Y_1$ antagonist	36	Risk	100	100	97
Carcinogenic	492	Conservative	100	87	86

N is the number of compounds in a training set

12.3.2 Structurally Similar Compounds

Similar testing was performed to predict the levels of 28 types of pharmacological activity in structurally similar condensed azole derivatives of the following chemical classes: imidazoles, 1,2,4-triazoles, indoles, purines, benzimidazoles, imidazo[1,2-a]benzimidazoles, pyrimido[1,2-a]benzimidazoles, pyrazolo[1,5-a]benzimidazoles, pyrrolo[1,2-a]benzimidazoles, 1,2,4-triazolo[1,5-a]benzimidazoles, thiazolo[2,3-a]

Table 12.5 General indices of prediction accuracy for high-level activity in structurally diverse compounds

Strategy	ST, %		LOOCV, %		SHCV, %		DLOOCV, %	
	Min	Max	Min	Max	Min	Max	Min	Max
Accuracy F_0								
Conservative	97	100	84	99	85	99	–	–
Normal	81	100	72	97	74	96	72	97
Risk	76	99	73	99	97	97	–	–
Sensitivity F_a								
Conservative	88	100	74	94	72	93	–	–
Normal	89	100	66	91	63	90	66	91
Risk	72	100	69	95	64	91	–	–
Specificity F_n								
Conservative	97	100	89	100	87	100	–	–
Normal	78	100	73	98	74	98	74	98
Risk	75	100	73	100	77	100	–	–

DLOOCV is not used in the conservative or risk strategy

benzimidazoles, 1,2,4-triazino[2,3-a]benzimidazoles, 1,3,4-thiadiazino[3,2-a]benzimidazoles, thiazolo[2,3f]purines, and oxazolo[2,3-f]purines [105]. The sizes of the training sets varied from 17 to 459 compounds. A model of a generalized pattern of 1312 condensed azole derivatives is described by 8615 types of QL descriptors (seven descriptors per compound on average), which testifies to their great structural similarity. Quantitative data for all 28 types of pharmacological activity of these compounds underwent cluster analysis [59], and four classes of activity were detected: high, moderate, low, and inactive. In each case, the class borders were set using the numerical values of one or several indices of the tested activity.

In the first series of tests, we evaluated the prediction accuracy of expressed activity, which corresponds to the combined gradation of "high or moderate" versus "low or inactive" compounds. The choice of gradations was dictated by the fact that in an experimental screening of structurally similar compounds, substances with low or no activity were immediately eliminated from further studies.

A summary of the adequacy of decision rules in predicting expressed activity of structurally similar compounds is shown in Table 12.7. In this case, the maximum values F_0, F_a and F_n in all strategies amount to 100% only in the self-prediction model. In leave-one-out and split-half cross-validations the maximum values of F_0, F_a, and F_n were 91%, 100%, and 96%, respectively.

The estimates of the accuracy of the best decision rules in predicting an expressed activity are shown in Table 12.8. In this case, the risk strategy showed the best results; optimum decision rules were obtained for 10 activities out of 21 (48%).

The summarized results of testing the adequate decision rules in predicting a high activity level in structurally similar condensed azole derivatives are shown in Table 12.9. In this case, too, the maximum values for F_0, F_a and F_n in all strategies amount to 100% only in self-prediction. In leave-one-out and split-half cross-validations the maximum values of F_0, F_a, and F_n were 93, 100, and 97%, respectively.

Table 12.6 Accuracy of prediction of the best strategy for a high activity in structurally diverse compounds

Activity	N	Better strategy	ST F_0, %	LOOCV F_0, %	SHCV F_0, %
Neuroleptic	645	Conservative	99	95	94
Tranquilizer	532	Conservative	99	98	97
Antidepressant	628	Risk	77	76	79
Analgesic narcotic	320	Normal	100	97	96
Antianginal	410	Conservative	100	99	99
Cardiotonic	304	Normal	100	95	96
Hypoglycemic	230	Risk	87	77	77
Antiseptic	494	Normal	86	81	84
Tuberculostatic	386	Conservative	100	97	95
Anti-HIV	1140	Normal	99	80	80
Anti-paramixo-virus	54	Conservative	100	93	92
Anti-picornavirus	512	Conservative	97	89	88
Anti-orthovirus	72	Risk	99	99	97
Antileukemic	252	Conservative	98	84	85
Antineoplastic	821	Normal	90	83	81
Antioxidant	82	Risk	96	90	91

N is the number of compounds in a training set

Table 12.7 General indices of prediction accuracy for expressed activity in structurally similar condensed azole derivatives

Strategy	ST, %		LOOCV, %		SHCV, %		DLOOCV, %	
	Min	Max	Min	Max	Min	Max	Min	Max
Accuracy F_0								
Conservative	88	100	67	86	68	87	–	–
Normal	81	100	67	90	69	86	67	90
Risk	72	100	67	87	69	91	–	–
Sensitivity F_a								
Conservative	89	100	64	100	66	89	–	–
Normal	80	100	66	100	60	83	66	100
Risk	65	100	60	92	60	100	–	–
Specificity F_n								
Conservative	81	100	60	90	61	96	–	–
Normal	74	100	67	86	61	92	67	86
Risk	74	100	64	92	67	92	–	–

DLOOCV is not used in the conservative or risk strategy

The estimates of the accuracy of the best decision rules in predicting a high activity are shown in Table 12.10. Here, too, the risk strategy proved to be the best. Notably, it showed better results compared to the previous example of an expressed activity; optimum decision rules were obtained for 14 activities out of 22 (64%). This result is quite understandable; it is due to the novelty of the chemical structures

Table 12.8 Accuracy of the best strategy for predicting an expressed activity in structurally similar condensed azole derivatives

Activity	N	Better strategy	ST F_0, %	LOOCV $_0$, %	SHCV F_0, %
Antioxidant	325	Conservative	97	80	82
Antiradiomimetic	73	Risk	85	74	74
PDE cAMP inhibitor	41	Conservative	97	86	87
Anti-calmodulin	23	Risk	100	87	82
H_1 antagonist	62	Risk	84	77	77
$P2Y_1$ antagonist	56	Normal	88	73	73
K-opioid agonist	91	Conservative	99	84	86
Ca^{+2} channel blocker	69	Normal	84	71	70
Antiplatelet	312	Risk	100	72	78
Hemorheologic	160	Conservative	98	74	68
Spasmolytic	170	Normal	100	67	69
Antiarrhythmic	305	Conservative	94	75	78
Anesthetic local, surface	459	Risk	85	84	86
Anesthetic local, infiltration	459	Risk	87	85	88
Anesthetic local, conductive	459	Normal	81	79	79
Hypotensive	336	Normal	83	77	77
Hypoglycemic	125	Risk	95	67	69
Antisecretory	73	Conservative	93	76	81
Cerebroprotective	36	Risk	83	81	81
Anti-hypoxic	17	Risk	100	69	82
Actoprotective	32	Risk	72	67	69

N is the number of compounds in a training set

and the nonstandard mechanisms of action of highly active compounds. The conservative strategy yielded 3 optimum regularities (13 %), and the normal strategy yielded 5 (23 %).

12.3.3 Chiral Compounds

Different streoisomers of the same medicinal substance differ in the range of pharmacological effects and the extent of their manifestation.

Three training sets demonstrated the possibility of predicting the pharmacological activity of chiral compounds [82].

1. The data for the structure and activity of 26 structurally diverse dopamine D_2 receptor agonists (7 diastereomers, 15 enantiomers, 4 achiral compounds) [87] were clustered into three activity classes: "high," "moderate," and "low."

Table 12.9 General indices of prediction accuracy for high activity in structurally similar condensed azole derivatives

Strategy	ST, %		LOOCV, %		SHCV, %		DLOOCV, %	
	Min	Max	Min	Max	Min	Max	Min	Max
Accuracy F_0								
Conservative	89	100	64	93	64	90	–	–
Normal	73	100	67	90	64	89	67	90
Risk	71	100	64	90	66	91	–	–
Sensitivity F_a								
Conservative	91	100	64	100	61	96	–	–
Normal	79	100	64	86	63	91	64	86
Risk	71	100	60	92	60	100	–	–
Specificity F_n								
Conservative	85	100	60	97	66	95	–	–
Normal	69	100	64	93	61	92	64	93
Risk	65	100	61	94	68	93	–	–

DLOOCV is not used in the conservative or risk strategy

2. β-Adrenoblocking activity and four pharmacokinetic indices of twelve *S*- and *R*-isomers of six known structurally similar β-adrenoblockers, 1-aryloxy-3-alkylamino-propane-2-ole derivatives (acebutolol, atenolol, carvedilol, metoprolol, pindolol, sotalol) [60]. The compounds were classified into two classes based on each parameter; 1) the enantiomer activity was higher than that of a racemate, or 2) the enantiomer activity was lower than that of a racemate.
3. The values for the β_1-adrenoblocking activity of ten optical isomers of a single drug, nebivolol, which has four chiral centers [77], were divided into two classes: highly active ones and those with a low activity.

The results of testing the best decision rules are shown in Table 12.11. The normal strategy was the best in all of these cases, which corresponds to the characteristics of chiral compounds; the basal activity level of chiral compounds is established by the chemical structure, and specific effects are determined by chiral structural fragments.

In summary, IT Microcosm allows an accuracy approaching 100 % in predicting the presence and extent of a pharmacological activity in structurally diverse and structurally similar compounds and their stereoisomers.

12.4 IT Microcosm in the Search for Novel Drugs

The fundamental nature of the theoretical concepts behind IT Microcosm makes it possible to use this technology to successfully solve various QSAR/QSPR optimization tasks, such as searching for novel compounds with high pharmacological

Table 12.10 Accuracy of the best strategy for predicting high activity in structurally similar condensed azole derivatives

Activity	N	Better strategy	ST F_0, %	LOOCV F_0, %	SHCV F_0, %
Antiradical	36	Risk	100	81	81
Antioxidant	310	Conservative	98	93	90
Antiradiomimetic	73	Conservative	89	72	75
PDE cAMP inhibitor	109	Normal	100	83	85
Anti-calmodulin	23	Risk	100	83	91
5-HT_2 antagonist	85	Risk	72	65	69
5-HT_3 antagonist	98	Risk	73	68	71
H_1 antagonist	62	Normal	94	85	84
$P2Y_1$ antagonist	56	Risk	80	73	71
K-opioid agonist	91	Normal	100	75	67
Ca^{+2} channel blocker	69	Risk	100	80	71
Hemorheologic	160	Risk	71	71	72
Spasmolytic	170	Normal	85	74	76
Antiarrhythmic	305	Conservative	98	87	89
Anesthetic local, surface	459	Normal	89	88	88
Anesthetic local, infiltration	459	Risk	82	82	84
Anesthetic local, conductive	459	Risk	85	83	85
Hypotensive	336	Risk	75	71	74
Hypoglycemic	125	Risk	100	73	71
Anti-ulcerogenic	77	Risk	78	68	66
Cerebroprotective	36	Risk	83	81	81
Anti-hypoxic	17	Risk	100	82	82

N is the number of compounds in a training set

[19, 37, 38, 47, 48, 49, 79, 80, 82, 1, 122, 127, 130–134 140] or biological [15–17, 45, 63, 100] activities, making an assessment of the carcinogenic potential of substances [67, 84, 117, 121], developing new effective polymer composite additives [29, 30, 64, 68] and rubber mixture additives [84, 118, 121], and predicting the environmental hazards of chemical production plants [67, 108, 115].

The effectiveness of IT Microcosm in the search for novel drugs with high antioxidant, antiarrhythmic and antiplatelet activities among condensed azole derivatives is shown below; the general formulas of these compounds are given in Fig. 12.2. These compounds satisfy Lipinski's rules [56] and the order of priority for cyclic and heterocyclic structures [36, 11]; some of them are so-called privileged structures [21].

Table 12.11 Indices of prediction accuracy for the best strategy of chiral compound activity

Activity level	Better strategy[a]	F_0, %	F_a, %	F_n, %
D_2 agonists				
High	Normal	89	83	90
High or moderate	Normal	73	73	73
β-Adrenoblockers				
High adrenoblocking	Normal	100	100	100
High C_{max}[b]	Normal	100	100	100
High AUC[c]	Normal	100	100	100
High Cl_R[d]	Normal	100	100	100
High $t_{1/2}$[e]	Normal	100	100	100
Nebivolol				
High $β_1$-adrenoblocking	Normal	100	100	100

[a] In LOOCV
[b] Maximum blood plasma concentration
[c] Area under plasma concentration-time curve
[d] Renal clearance
[e] Half-life

12.4.1 *Antioxidant Activity*

The training set was constructed according to the results of experimental studies of 325 condensed azole derivatives of four classes (Fig. 12.2): 232 imidazo[1,2-a]benzimidazoles (IV), 34 benzimidazoles (II), 33 pyrimido[1,2-a]benzimidazoles (VIII) and 26 pyrrolo[1,2-a]benzimidazoles (V) [46–48, 105, 122, 124].

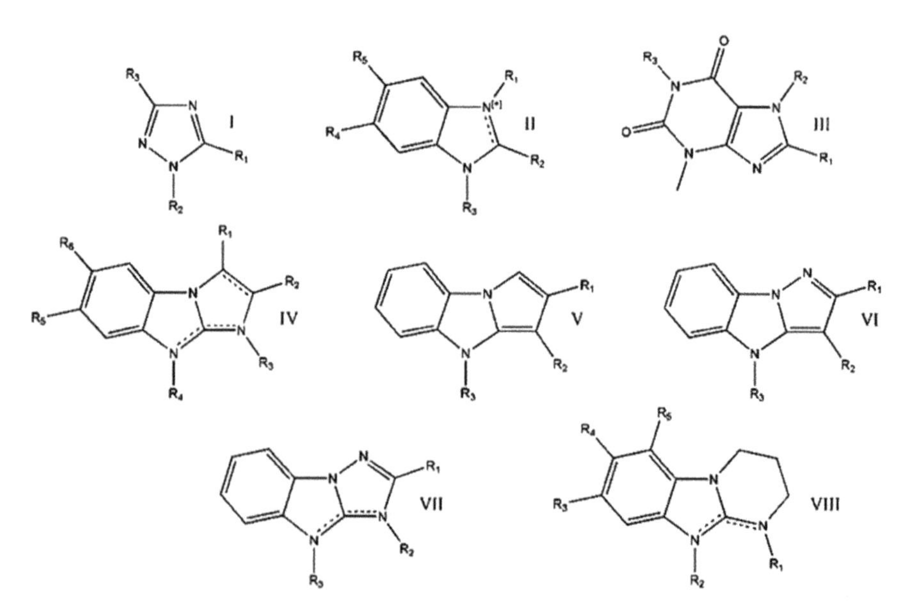

Fig. 12.2 Condensed azole derivatives tested for antioxidant, antiarrhythmic and antiplatelet activity (the interrupted line indicates a possible double bond)

The substances were studied *in vitro* on a model of ascorbate-induced lipid peroxidation (LP) in the homogenate of rat liver [53]. The indices of the antioxidant activity were as follows: EC_{50}–compound concentration (mol/L) inhibiting LP by 50%; $\Delta(10^{-6})$—percentage of LP inhibition at a concentration $1 \cdot 10^{-6}$ M of the studied substance; Ind_{10}—the module of the exponent of the measure of the substance concentration inducing LP inhibition by 10% (using the value of 2 to 7 points). Trolox C (CAS 53188-07-1) was studied as the comparison drug, and it yielded EC_{50} values of $2.76 \cdot 10^{-6}$ M, $\Delta(10^{-6}) = 36.9\%$, and $Ind_{10} = 7$.

Based on the results of cluster analysis in combination with an expert assessment, the following activity classes were distinguished:

- high—$EC_{50} < 5.0 \cdot 10^{-6}$ M or $\Delta(10^{-6}) > 20.0\%$ (78 compounds), or $Ind_{10} = 6, 7$ (86 compounds);
- moderate—$5.0 \cdot 10^{-6} \leq EC_{50} < 1.0 \cdot 10^{-4}$ M or $10.0 < \Delta(10^{-6}) \leq 20.0\%$ (69 compounds), or $Ind_{10} = 5$ (59 compounds);
- high or moderate—$EC_{50} < 1.0 \cdot 10^{-4}$ M or $\Delta(10^{-6}) > 10.0\%$ (147 compounds), or $Ind_{10} = 5, 6, 7$ (145 compounds); and
- low—$EC_{50} \geq 1.0 \cdot 10^{-4}$ M or $\Delta(10^{-6}) \leq 10.0\%$ (178 compounds), or $Ind_{10} = 2, 3, 4$ (165 compounds).

The results of computational accuracy testing for the prediction of the extent of the antioxidant activity of the condensed azole derivatives are shown in Table 12.12. All of the obtained decision rules proved inadequate for the moderate activity; therefore, they were excluded from further prediction.

Substances with a high antioxidant activity were sought among 721 novel, untested condensed azole derivatives of 15 chemical classes (see Sect. 3.2) using several consensus approaches to the selection of the most promising compounds, including a method of using three strategies in combination, and testing the spectrum of the predicted estimates for noncontradiction.

The key criteria for selecting experimental-study candidates were as follows ("A" stands for a positive prediction estimate of a compound activity as "high"):

- 2A or 3A in all six decision rules, with a conformity coefficient of the prediction-estimate spectrum (21) $K_{High} > 0.5$;
- 3A for the index EC_{50} and 3A for Ind_{10} simultaneously;
- 3A for EC_{50} or 3A for Ind_{10}, while $K_{High} = 1.0$; and
- 3A for EC_{50} and $K_{High} > 0.8$.

Altogether, 41 substances were selected according to these criteria and studied experimentally.

Twenty-eight compounds (68.3%) with an expressed antioxidant activity $\Delta(10^{-6}) > 10.0\%$ were found; 17 of them (41.5%) were compounds showing a high activity $\Delta(10^{-6}) > 20.0\%$. Of 17 highly active compounds, 13 compounds (76.5%) are comparable to the reference drug Trolox C in antioxidant activity, and four compounds (23.5%) exceed its activity.

For the $\Delta(10^{-6})$ index, the training set contains 45.2% compounds with an expressed activity, including 24.0% compounds with a high activity. The primary screening did not employ *in silico* methods, so the percentage of the obtained active substances can be regarded as an indication of the accuracy of the intuitive human prediction performed

Table 12.12 Indices of the prediction accuracy for the antioxidant-activity level of condensed azole derivatives

Activity level	ST, %				LOOCV, %				SHCV, %			
	F_0	F_a	F_n	F_u	F_0	F_a	F_n	F_u	F_0	F_a	F_n	F_u
EC_{50}[a]												
CS[b]												
High	96	91	98	5	91	74	97	8	90	75	95	5
Moderate	96	86	99	18	74	35	84	23	79	28	90	18
High or moderate	97	94	99	10	80	73	86	11	82	76	87	11
NS[c,d]												
High, BA	94	87	96	–	90	83	93	–	89	79	92	–
Moderate, DM	72	75	71	–	66	56	68	–	66	39	73	–
High or moderate, NNM	99	100	99	–	74	73	75	–	79	75	83	–
RS[e]												
High, LDM10	94	88	96	–	90	79	94	–	89	77	93	–
Moderate, BA11	73	83	70	–	60	58	61	–	66	60	68	–
High or moderate, DM10	78	78	79	–	72	72	72	–	74	74	74	–
Ind_{10}[a]												
CS[b]												
High	97	95	98	4	91	77	96	8	92	83	96	10
Moderate	97	90	99	17	79	30	89	24	82	41	90	18
High or moderate	97	94	99	10	80	73	86	11	82	76	87	11
NS[c,d]												
High, NNM	99	100	98	–	86	80	88	–	85	78	88	–
Moderate, DM	75	75	75	–	68	54	71	–	65	44	70	–
High or moderate, NNM	99	100	99	–	74	73	75	–	79	75	83	–
RS[e]												
High, DM10	88	88	88	–	85	82	85	–	86	86	85	–
Moderate, BA11	73	90	70	–	62	59	62	–	66	63	67	–
High or moderate, DM10	78	78	79	–	72	72	72	–	74	74	74	–

[a] Criteria for distinguishing an activity class.
[b] The refusal of prediction was only defined in the conservative strategy.
[c] The best prediction method is indicated.
[d] The DLOOCV and LOOCV results are identical.
[e] The best method of prediction and the type of descriptor are indicated.

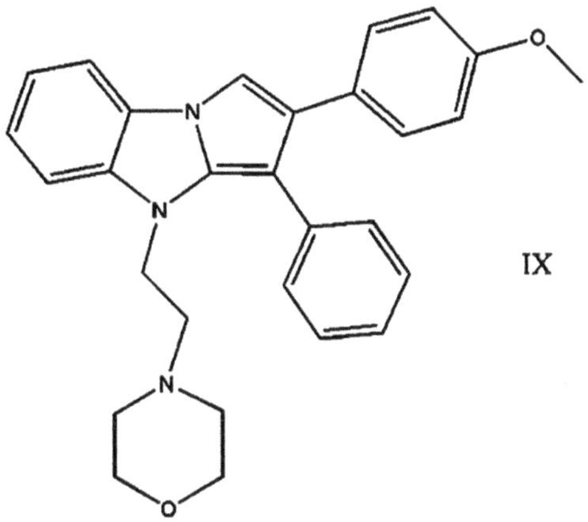

Fig. 12.3 Leading compound with a high antioxidant activity

by an experienced experimental pharmacologist based on a conventional comparative analysis of literature data. Thus, the effectiveness of IT Microcosm in the search for compounds with an expressed antioxidant activity is 1.51 times greater than the accuracy of intuitive human prediction; its effectiveness in a search for compounds with a high antioxidant activity is 1.73 times greater.

Of the four most active compounds, compound (IX) (Fig. 12.3) was recognized as the most promising for a further, in-depth study (according to accessory *in vivo* tests); this compound showed antioxidant-activity indices $\Delta\,(10^{-6})=94.4\,\%$, $EC_{50}=3.2\cdot10^{-7}$ M and $Ind_{10}=7$. The compound is now covered by a patent [6].

12.4.2 Antiarrhythmic Activity

The training set was also constructed on the basis of an experimental study of 305 condensed azole derivatives of four chemical classes (Fig. 12.2): 223 imidazo[1,2-a]benzimidazoles (IV), 31 benzimidazoles (II), 27 pyrrolo[1,2-a]benzimidazoles (V) and 24 pyrazolo[1,5-a]benzimidazoles (VI) [3, 37, 38, 69, 105].

The antiarrhythmic activity was evaluated *in vitro* according to the extent to which the compounds affected the prolongation of the myocardial absolute refractory period in an isolated rat atrium of the heart stimulated by electric impulses [3, 138]. This method does not allow for the identification of all of the antiarrhythmic effects of new substances, but the technique is valid at the stage of primary screening. MEC, the minimum effective concentration of a substance (mol/L) preventing the atrium from adopting the rhythm forced on it, served as a measure of the antiarrhythmic activity. Moricizine (CAS 31883-05-3) was studied as the drug for comparison; its MEC was found to be $5.10\cdot10^{-5}$ M.

A cluster analysis helped to single out the following classes of activity:

- high—$MEC<7.3\cdot10^{-5}$ M (71 compounds);
- moderate—$7.3\cdot10^{-5}\leq MEC<2.8\cdot10^{-4}$ M (76 compounds);
- high or moderate—$MEC<2.8\cdot10^{-4}$ M (147 compounds); and
- low— $MEC\geq2.8\cdot10^{-4}$ M (158 compounds).

Table 12.13 Indices of the prediction accuracy for the antiarrhythmic activity level of condensed azole derivatives

Activity level	ST, %				LOOCV, %				SHCV, %			
	F_0	F_a	F_n	F_u	F_0	F_a	F_n	F_u	F_0	F_a	F_n	F_u
CS												
High	98	97	98	12	87	72	91	14	89	78	92	14
Moderate	98	97	99	18	72	40	81	21	74	49	81	18
High or moderate	94	94	94	13	75	77	73	15	78	81	75	16
NS												
High, BA	81	82	80	–	78	79	78	–	79	82	78	–
Moderate, BA	66	91	58	–	56	67	52	–	59	61	58	–
High or moderate, LDM	90	90	91	–	74	74	75	–	75	76	75	–
RS												
High, DM11	83	82	84	–	81	80	81	–	80	80	80	–
Moderate, BA11	73	91	68	–	59	59	59	–	66	66	66	–
High or moderate, DM11	79	78	81	–	72	71	73	–	72	72	72	–

The designations are the same as in Table 12.12

The results of the computational-accuracy testing for the prediction of the extent of the antiarrhythmic activity of condensed azole derivatives are shown in Table 12.13. Not a single adequate decision rule was obtained for the moderate activity class.

According to the predicted results, 25 substances were selected out of 752 novel, untested condensed azole derivatives for an experimental study. The selected substances showed 3A for a high activity level in all three strategies, a conformity coefficient of prediction estimates spectrum (21) $K_{\text{High}} = 1.0$ and a membership function in the conservative strategy (17) $Fb_{\text{High}} > 0.65$.

Experimentally, all of the 25 compounds (100%) showed an expressed antiarrhythmic activity MEC $< 2.8 \cdot 10^{-4}$ M; 20 of them (80.0%) showed a high activity level MEC $< 7.3 \cdot 10^{-5}$ M. Of 20 highly active substances, 11 (55.0%) are comparable with Moricizine in the extent of the antiarrhythmic activity; nine substances (45.0%) showed a higher extent of activity.

Previously, the primary screening of condensed azole derivatives with a high antiarrhythmic activity employed, apart from intuitive prediction, the Hansch meth-

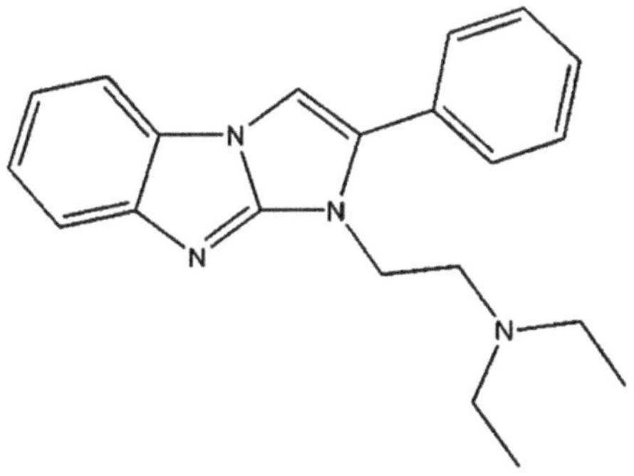

Fig. 12.4 Leading compound with a high antiarrhythmic activity

od, discriminant analysis and substructure analysis. Thus, according to the training set structure, the accuracy of combined primary machine-human prediction of an expressed antiarrhythmic activity was 48.2% and was 23.3% for high activity. Consequently, the effectiveness of IT Microcosm in a search for compounds with an expressed antiarrhythmic activity among condensed azole derivatives was 2.07 times greater, and the search for compounds with a high antiarrhythmic activity was 3.43 times more accurate than the combined machine-human prediction performed with the help of three other QSAR methods.

Out of the nine most active compounds, compound (X) (Fig. 12.4) with a $MEC = 2.0 \cdot 10^{-5}$ M was selected for a further, in-depth study. This compound is now covered by a patent [83].

12.4.3 Antiplatelet Activity

The training set was constructed on the basis of the experimental results for 312 condensed azole derivatives of eight chemical classes (Fig. 12.2): 192 imidazo[1,2-a]benzimidazoles (IV), 40 benzimidazoles (II), 28 pyrimido[1,2-a]benzimidazoles (VIII), 23 purines (III), 10 pyrrolo[1,2-a]benzimidazoles (V), eight 1,2,4-triazoles (I), 8 pyrazolo[1,5-a]benzimidazoles (VI) and three 1,2,4-triazolo[1,5-a]benzimidazoles (VII) [4, 49, 105, 125].

The antiplatelet activity of the substances was studied *in vitro* on a model of ADP-induced rabbit platelet aggregation [12, 26]. The indices of the extent of the antiplatelet activity were as follows: EC_{50} was the compound concentration (mol/L) decreasing platelet aggregation by 50%; and $\Delta\,(10^{-4})$ was the percentage of platelet-aggregation decrease at a concentration of the studied substance of $1 \cdot 10^{-4}$ M. Aspirin (CAS 50-78-2) served as the drug for comparison; the indices of its antiplatelet activity were found to be $EC_{50} = 7.10 \cdot 10^{-4}$ M and $\Delta\,(10^{-4}) = 29.3\%$.

The following activity classes were singled out on the basis of cluster analysis in combination with an expert evaluation:

- high—$EC_{50} < 3.3 \cdot 10^{-4}$ M or $\Delta\,(10^{-4}) > 25.0\%$ (71 compounds);

Table 12.14 Indices of the prediction accuracy for the antiplatelet-activity level of condensed azole derivatives

Activity level	ST, %				LOOCV, %				SHCV, %			
	F_0	F_a	F_n	F_u	F_0	F_a	F_n	F_u	F_0	F_a	F_n	F_u
CS												
High	97	96	98	17	78	47	87	18	80	53	88	13
Moderate	96	97	95	19	63	41	70	24	71	51	77	23
High or moderate	96	97	96	15	73	78	67	19	78	81	74	17
NS												
High, NNM	100	100	100	–	73	55	79	–	72	61	78	–
Moderate, BA	71	91	63	–	56	60	54	–	53	43	56	–
High or moderate, NNM	100	100	100	–	72	73	72	–	76	73	79	–
RS												
High, NNM11	100	100	100	–	74	51	81	–	76	52	83	–
Moderate, BA11	75	90	69	–	60	62	59	–	64	62	64	–
High or moderate, NNM8	100	100	100	–	71	73	71	–	75	72	79	–

The designations are the same as in Table 12.12

- moderate—$3.3 \cdot 10^{-4} \leq EC_{50} < 1.0 \cdot 10^{-3}$ M or $10.0 < \Delta\ (10^{-4}) \leq 25.0\%$ (82 compounds);
- high or moderate—$EC_{50} < 1.0 \cdot 10^{-3}$ M or $\Delta\ (10^{-4}) > 10.0\%$ (153 compounds); and
- low—$EC_{50} \geq 1.0 \cdot 10^{-3}$ M or $\Delta\ (10^{-4}) \leq 10.0\%$ (159 compounds).

The results of the computational-accuracy testing for the prediction of the extent of antiplatelet activity of the condensed azole derivatives are shown in Table 12.14.

It can be concluded from the data that no adequate decision rules were developed for the moderate- and high-activity gradations.

As discussed above (see Sect. 2), if there is a complete consensus on the positive estimates for a high-level activity in three prediction strategies, and a high value of the conformity coefficient of the prediction-estimates spectrum (21), the accuracy of a search for highly active compounds is considerably higher. This property of integral-consensus decision rules was successfully exploited in a search for substances with a high antiplatelet activity.

Fig. 12.5 Leading compound with a high antiplatelet activity

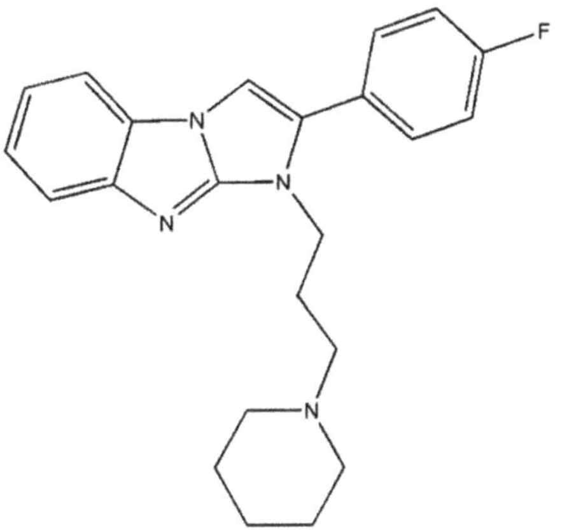

XI

On the basis of the predicted results, 49 substances out of 745 novel, untested condensed azole derivatives were selected for an experimental study. The substances had 3A for a high activity in all three strategies and a conformity coefficient of the prediction-estimates spectrum $K_{High} > 0.9$.

According to the experimental data, of 49 compounds, 48 (98.0%) showed an expressed antiplatelet activity $\Delta\ (10^{-4}) > 10.0\%$; of these, 39 compounds (79.6%) showed a high activity $\Delta\ (10^{-4}) > 25.0\%$. Of 39 highly active compounds, 19 (48.7%) showed an activity comparable to that of Aspirin, the drug for comparison, and 17 compounds (43.6%) exceeded the antiplatelet activity of Aspirin.

When 312 training-set compounds were tested for their antiplatelet activity, no computational methods of prediction were used. The primary screening detected 153 and 71 substances with expressed and high activity, respectively; that is, the accuracy of the intuitive prediction amounts to 49.0% and 22.8%, respectively. Thus, the accuracy of the search for an expressed antiplatelet activity with IT Microcosm was twice as high, and of those with a high antiplatelet activity, it was 3.49 times higher than the accuracy of intuitive prediction.

Of the 17 most active compounds, compound (XI) (Fig. 12.5) was selected for a further, in-depth study (according to accessory *in vivo* tests); this compound showed antiplatelet activity $\Delta\ (10^{-4}) = 46.8\%$. The compound is now covered by a patent [81].

The consensus search with IT Microcosm for condensed azole derivatives with high antioxidant, antiarrhythmic and antiplatelet activities can be summarized as follows:

1. 115 compounds were studied experimentally; according to the predicted results, they were expected to show high-level activity;
2. 101 compounds (87.8%) with expressed activity were detected; among them, 76 compounds (66.1%) showed high-level activity;

3. 68 compounds (67.3% of those with high-level activity) are comparable to or exceed the activity of comparison drugs; 30 of them (44.1%) are more active than the drugs for comparison;
4. The effectiveness of the search for compounds with an expressed activity was 2.07 times higher, and for those with a high activity, it was 3.49 times higher, in comparison to a prediction without IT Microcosm.

Therefore, the complex methodology of predicting the pharmacological activity employed in IT Microcosm and comprising a consensus approach to prediction as one of its constituent parts proved to be effective in a search for novel drugs among condensed azole derivatives of 15 different chemical classes.

12.5 IT Microcosm in Prediction of Pharmacological Activity of Complex Molecular Systems and of Their Component Synergism

Prediction of the biological activity of unconventional chemical structures is one of the most challenging problems faced by QSAR. In particular, such unconventional systems encompass organic salts, including those with organic acids and bases, supramolecular complexes formed by noncovalent intermolecular interactions between certain compounds and mixtures containing several individual substances. To make a successful *in silico* assessment of the pharmacological activity of such complex chemical constructs, one should consider both their qualitative and quantitative composition as well as the mutual effect of the components constituting these systems, in particular, the synergistic effects. The QL language in IT Microcosm provides an option of taking into account the noncovalent interactions [110] and making a successful prediction of the biological activity of complex molecular systems [35, 74, 75, 97, 105, 111, 112, 114, 119, 120, 123, 128, 129].

12.5.1 Organic Salts

Varying the salt-forming residue is a common method in drug design. Several examples are the antitussive drug Codeine (manufactured as a hydrochloride or a phosphate), the spasmolytic drug Prenoverine (in the form of a citrate) and the antibiotic Fumagillin (dicyclohexylammonium salt). A salt of a complex organic compound can be likened to a complete supramolecular system because the stability in both cases is achieved through noncovalent interactions.

The results of using IT Microcosm for predicting the presence/absence or the level of various types of pharmacological activity among the structurally similar and structurally diverse compounds discussed in Sect. 3 were obtained for the salts of those compounds; the salt-forming residue associated with the main chemical structure was used in the computations [105, 109, 123]. In the sets, there were salts of the main inorganic acids (HCl, HBr, HNO_3, H_2SO_4, H_3PO_4, $HClO_4$), of various

Table 12.15 Comparing the accuracy of predicting the level of the pharmacological activity of salts and bases of condensed azole derivatives

Activity	F_0 for salts, %[a]	F_0 for bases, %[a]	Δ, %[b]
CS			
Antioxidant[c]	87.5	83.8	8.6
Hemorheologic[d]	72.8	72.4	2.6
5-HT_3 antagonist[e]	55.7	56.8	10.9
Average for activities	72.0	71.0	7.4
NS			
Antioxidant	88.6	87.4	4.0
Hemorheologic	74.5	74.1	8.2
5-HT_3 antagonist	81.9	78.8	12.5
Average for activities	81.7	80.1	8.2
RS			
Antioxidant	91.1	89.6	3.7
Hemorheologic	71.2	71.7	2.1
5-HT_3 antagonist	72.5	73.8	6.3
Average for activity	78.3	78.4	4.0
Average for strategy	77.3	76.5	6.5

[a] The average value in all gradations in leave-one-out cross validation
[b] The maximum value in all gradations for F_0 surplus for salts over the F_0 for bases
[c] The 63 compounds and five activity gradations in the model set
[d] The 146 compounds and six activity gradations in the model set
[e] The 94 compounds and five activity gradations in the model set

organic acids (e.g., propionic, oxalic, citric, cyclamic, adamantane-1-carboxylic, and p-toluenesulfonic), of various metals (e.g., Na, K, Ca, Mg, Al, Bi, and Cu) and of organic bases (e.g., ethylenediamine, benzylamine, tris, and piperazine). A directed search for novel drugs among condensed azole derivatives (see Sect. 4) was also performed with consideration of the salt component associated with the main structure.

The following test was performed with the express purpose of understanding how taking into consideration the presence of an acid residue affects the accuracy of a prediction [105, 123]. Six model sets were constructed with three types of activity studied for condensed azole derivatives (I-VIII): the antioxidant, hemorheological and 5HT_3-antiserotonin activities. Three of them included the compound structures with acid residues added to them; the other three showed only the main structure of the same substances. The F_0 indices of the total accuracy of predicting various levels of these three activity types were evaluated in three strategies by a method of leave-one-out cross validation; the results of the evaluation are summed up in Table 12.15. One can see that taking the acid residue in consideration increased the accuracy of predicting the level of the pharmacological activity in IT Microcosm: by 12.5 % as a maximum, and by 6.5 % on the average.

The obtained results indicate that the prediction made with IT Microcosm shows an adequate method of taking account of the salt-forming residue associated with the main structure, which permits great accuracy and effectiveness in predicting the

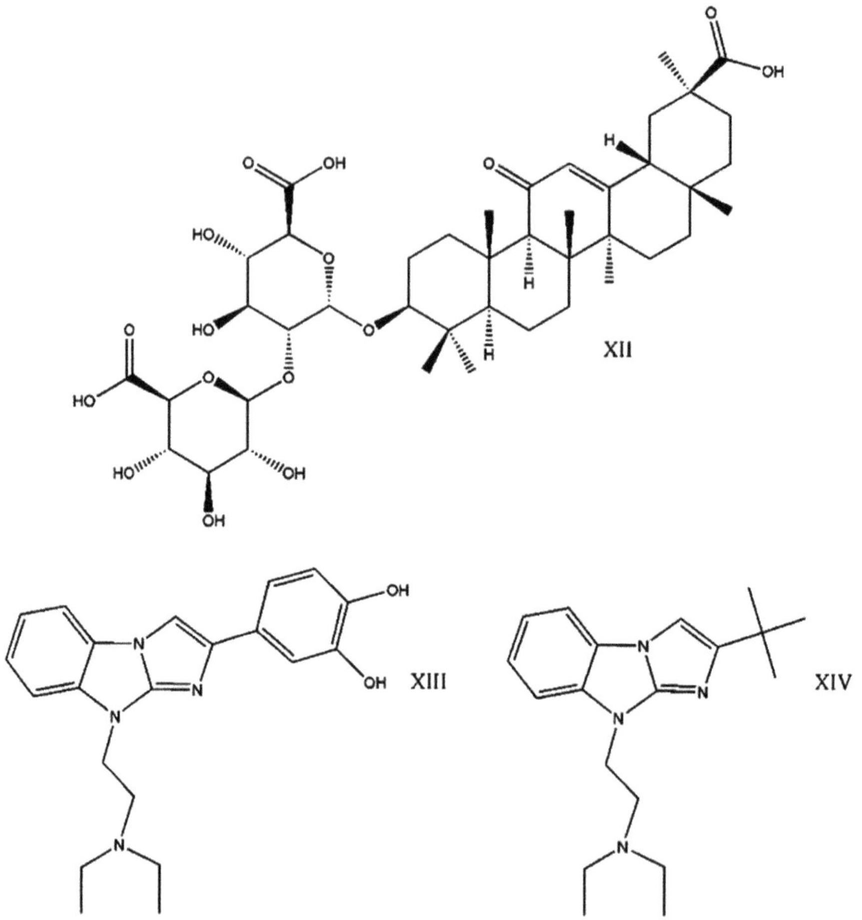

Fig. 12.6 Glycyrrhizinic acid (XII) and condensed azole derivatives with antioxidant (XIII) and antiarrhythmic (XIV) activity

presence and extent of the various types of pharmacological activity of the various salts of organic compounds in the most diverse chemical classes.

12.5.2 Supramolecular Compounds

The synthesis of supramolecular complexes of known drugs offers a promising approach to novel drug search. For example, let us look at the development of clathrates of nifedipine, allapinin, fluoxetine and phenibut with glycyrrhizinic acid and stevioside, which showed effects exceeding those of the nonclathrated compounds by 10–290 times [88].

Table 12.16 Prediction of the antioxidant activity of the clathrate complexes of compound (XIII) with glycyrrhizinic acid

Composition	Fb_{Norm} [a]	Fb_{Risk} [b]	$Rank_{Norm}$ [c]	$Rank_{Risk}$ [d]	$\sum Rank$ [e]	
Presence of the activity						
XIII	0.9999	0.5906	3	1	4	
XIII–GA 1:1	0.9999	0.5932	3	2	5	
XIII–GA 1:2	0.9999	0.5945	3	3	6	
XIII–GA 1:3	0.9999	0.5950	3	4	7	
XIII–GA 1:4	0.9999	0.5953	3	5	8	
High activity						$\sum\sum Rank$ [f]
XIII	0.0001	0.4349	1	1	2	6
XIII–GA 1:1	0.6116	0.4887	5	2	7	12
XIII–GA 1:2	0.5908	0.4929	3	3	6	12
XIII–GA 1:3	0.5908	0.4949	3	4	7	14
XIII–GA 1:4	0.5908	0.4974	3	5	8	16

[a] Membership function for the normal strategy of prediction
[b] Membership function for the risk strategy of prediction
[c] The Fb_{Norm} values ranked in ascending order
[d] The Fb_{Risk} values ranked in ascending order
[e] The sum of the ranks in the strategies
[f] The sum of the ranks in the strategies and the activity gradations

IT Microcosm performed an optimization of the composition of supramolecular clathrate complexes of glycyrrhizinic acid (XII) with two pharmacons: an antioxidant agent (XIII) [5] and an antiarrhythmic agent (XIV) [78] (Fig. 12.6) with the purpose of achieving the maximal pharmacologic effect.

A prediction of the activity of pure pharmacons and pharmacon clathrates with glycyrrhizinic acid with molar compositions of 1:1, 1:2, 1:3 and 1:4 was performed in the normal and risk strategies using training sets for the antioxidant and antiarrhythmic activity of condensed azole derivatives (see Sect. 3 and 4). The values of the membership functions for the classes of active and highly active compounds, as well the ranks of these values in ascending order, served as a metric of the expected activity. The conservative strategy was not employed due to the nonstandard structures of the predicted compounds.

The predicted results for the antioxidant activities of the compound (XIII) clathrates with glycyrrhizinic acid are shown in Table 12.16.

According to the sum of the rank estimates $\sum\sum Rank$, the antioxidant activity of all compound (XIII) clathrates should exceed the activity of the pure nonclathrated compound (XIII). In this case, the comparable value of the antioxidant activity should decrease in the series: (XIII–GA 1:4)>(XIII–GA 1:3)>(XIII–GA 1:2)≈(XIII–GA 1:1)>(XIII). The clathrate (XIII–GA 1:4) should show the highest activity. The (XIII–GA 1:2) and (XIII–GA 1:1) complexes should show the lowest activities.

Experimentally, the substances were studied *in vivo* on rats at a dose equivalent to 10 mg/kg of the pharmacon *per os*. The total antioxidant activity of the blood plasma was assessed by a method based on the ability of the biological agents to inhibit the accumulation of lipid peroxidation products in a suspension of egg-yolk lipoproteins [44]. Δ_{max}, the maximum percentage of the lipid-peroxidation inhibition (in relation to the outcome) observed for nine hours of the experiment, served as an index of the antioxidant activity.

Table 12.17 Prediction of the antiarrhythmic activity of the clathrate complexes of compound (XIV) with glycyrrhizinic acid

Composition	Fb_{Norm}	Fb_{Risk}	$Rank_{Norm}$	$Rank_{Risk}$	$\sum Rank$	
Presence of the activity						
XIV	0.9431	0.5028	5	5	10	
XIV–GA 1:1	0.4287	0.5008	1	4	5	
XIV–GA 1:2	0.6318	0.5004	3.5	3	6.5	
XIV–GA 1:3	0.5334	0.5003	2	2	4	
XIV–GA 1:4	0.6318	0.5002	3.5	1	4.5	
High activity						$\sum\sum$Rank
XIV	0.4852	0.4986	1	1	2	12
XIV–GA 1:1	0.5491	0.5004	3.5	5	8.5	13.5
XIV–GA 1:2	0.5491	0.5002	3.5	3.5	7	13.5
XIV–GA 1:3	0.5491	0.5002	3.5	3.5	7	11
XIV–GA 1:4	0.5491	0.5001	3.5	2	5.5	10

The designations are the same as in Table 12.16

The following Δ_{max} values were obtained: (XIII)—36.9%, (XIII–GA 1:1)—55.8%, (XIII–GA 1:2)—55.9%, (XIII–GA 1:3)—54.7% and (XIII–GA 1:4)—65.5%. The clathrates of the compositions (XIII–GA 1:1), (XIII–GA 1:2) and (XIII–GA 1:3) showed the same activity, which was approximately 1.5 times higher than the activity of the pure compound (XIII), whereas the clathrate of compound (XIII–GA 1:4) was 1.77 times more active than substance (XIII). The correlation coefficient between the calculated estimate $\sum\sum Rank$ and the experimental value Δ_{max} amounted to $R=0.965$.

Thus, the optimal composition of the (XIII–GA 1:4) clathrate with the highest level of antioxidant activity was established computationally and confirmed experimentally.

The results of predicting the antiarrhythmic activity of the compound (XIV) clathrates with glycyrrhizinic acid are shown in Table 12.17. According to the sum of the rank estimates $\sum\sum Rank$, the expected antiarrhythmic activity of compound (XIV) clathrates did not differ significantly from the activity of the pure nonclathrated compound (XIV); the maximum difference was only two units, whereas the minimum difference was 10 in the case of the antioxidant activity. Thus, the antiarrhythmic activity of all clathrates of compound (XIV) should be comparable with the activity of the pure nonclathrated compound (XIV).

The substances were studied *in vivo* in rats at a dose equivalent to 30 mg/kg of the pharmacon *per os*, in an experimental model of heart-rhythm disturbance induced by the intravenous administration of aconitine [28]. The relative time (t) (compared with the control) until the onset of arrhythmia after aconitine administration served as an index of the antiarrhythmic activity.

The following t values were obtained: (XIV)—1.66, (XIV–GA 1:1)—1.50, (XIV–GA 1:2)—1.68, (XIV–GA 1:3)—1.31 and (XIV–GA 1:4)—1.7. The indices of the activities of the four clathrates did not differ statistically from the activities of the pure compound.

Thus, it was established computationally and confirmed experimentally that the antiarrhythmic activity of the clathrates of compound (XIV) with glycyrrhizinic acid equaled the activity of the pure compound (XIV).

The results obtained show that IT Microcosm makes it possible to predict the pharmacological activity of supramolecular complexes and to optimize their composition.

12.5.3 Mixtures of Natural and Synthetic Organic Compounds

Combination drugs that include several active substances are used widely in clinical practice. Such drugs show effects of interaction between pharmacologically active compounds, including the synergistic effect that is of the greatest interest for practical purposes. The difference between mixtures of compounds and molecular complexes is more or less relative; it is determined by the energy of the intermolecular interactions; it is believed that mixtures show a prevalence of weak-dispersion interactions, whereas more energetically stable bonds, such as ion-dipole bonds and hydrogen bonds that are due to charge transfer, are more typical of molecular complexes [85]. We defined a substance as a mixture if it consisted of two or more individual substances and if it is not possible make any reasonable assumptions about the type of the molecular interactions between them. The prediction of the biological activity of mixtures has been successfully performed using the software HIT QSAR [50], PASS [52], ISIDA [66] and CORAL [90].

Several examples are given below of using IT Microcosm successfully in the prediction of the pharmacological activity of mixtures; these examples include predictions accounting for the synergism of the components [35, 74, 75, 97, 105, 111, 112, 114, 119, 120].

12.5.3.1 Spectrum of the Pharmacological Activity of *Juglans Regia* Extract

Juglans regia extracted from fruits of milky-wax ripeness is a veterinary drug [35]. It consists of 23 main active substances: juglone (CAS 481-39-0), leucoanthocyan (2-phenylchromen-2-ol), inosite (CAS 6917-35-7), β-sitosterol (CAS 83-46-5) and chlorophyll A (CAS 479-61-8); gallic (CAS 149-91-7), ellagic (CAS 476-66-4), palmitic (CAS 57-10-3), stearic (CAS 57-11-4), lauric (CAS 143-07-7), myristic (CAS 544-63-8), arachidic (CAS 506-30-9), linolenic (CAS 463-40-1), linoleic (CAS 60-33-3), oleic (CAS 112-80-1) and palmitoleic (CAS 2091-29-4) acids; and vitamins A (CAS 68-26-8), B_{12} (CAS 68-19-9), C (CAS 50-81-7), D (CAS 67-97-0), E (CAS 10191-41-0), P (CAS 153-18-4) and PP (CAS 98-92-0).

For each one of these 23 compounds, estimates of the presence of 13 types of pharmacological activity were calculated using decision rules developed by IT Microcosm on training sets with known drug substances (see Sect. 3).

Table 12.18 Pharmacological activity spectrum of *Juglans regia* extract according to *in silico* prediction results

Activity	Number of estimates			Total
	Active	Inactive	Unknown	for the extract
Central effects				
Neuroleptic	0	18	5	Inactive
Tranquilizer	0	17	6	Inactive
Antidepressant	0	18	5	Inactive
Enhancing CNS energy metabolism	2	4	17	Low
Analgesic narcotic	3	4	16	Low
Peripheral effects				
Cardiotonic	9	1	13	Moderate
Local anesthetic	0	10	13	Inactive
System-wide effects				
Immunostimulatory	14	1	8	High
Enhancing protein synthesis	15	0	8	High
Effects on the Protozoa				
Tuberculostatic	1	9	13	Inactive
Antiseptic	0	8	15	Inactive
Antifungal	7	2	14	Moderate
Antiviral	6	9	8	Moderate

The estimates of the existence of pharmacological activity were made on 11 levels of QL description by three methods: the Bayesian approach, distance method and chance method [97, 105, 119, 120]. The conclusion on whether the compound showed a particular activity was made using the conservative strategy and a qualified 2/3 majority. The compound was deemed active/inactive within the framework of one method if at least 8 out of 11 prediction estimates coincided. The final judgment about the activity of the compound was made if the evaluations of all three methods coincided; otherwise, the result was regarded as indefinite. An integral conclusion about the activity of an extract as a whole was made according to the ratio of the positive and negative final estimates of the 23 compounds.

According to the prediction (Table 12.18), *Juglans regia* extract should show the following activity types: an expressed immunostimulatory effect, the ability to enhance protein synthesis and energy metabolism of organs and tissues; moderate fungicidal, antiviral, cardiotonic effects; and a weak analgesic effect, the ability to enhance the CNS energy metabolism. There is no effect on the behavioral reactions of animals nor a direct effect on protozoa nor a local anaesthetizing effect. An adaptogenic effect due to combined immunostimulatory and metabolic activities is possible, as well as the ability to promote wound healing, which is due to the immunostimulatory activity and the ability to enhance protein synthesis.

On the whole, the pharmacologic effects of the extract become manifest as a result of the complex effect of all 23 components; they are due to its combined immunostimulatory and metabolic action, which is predominantly peripheral.

An experimental verification of the prediction results was performed on 60 newborn Simmental calves on three bred livestock farms [35]. In comparison with the controls, the groups of calves receiving the agent showed almost no diarrhea morbidity; there were no mortality cases; the rate of liveweight gain was considerably higher; and the protective properties of the blood were considerably better. All of the detected differences were statistically significant. The findings obtained indicate an enhanced natural resistance of the newborn calves' organisms, which confirms the results of the *in silico* prediction that the extract has an ability to enhance protein synthesis and adaptogenic, immunostimulatory and metabolic effects. A powerful wound-healing effect of the extract was shown in experiments on dogs. All of the animals showed a good tolerance for the drug; no local irritation, allergy or deviation in behavioural reactions was noted.

Consequently, the results of the experiments confirmed the estimates of the *in silico* prediction of a series of beneficial pharmacological effects of *Juglans regia* extracted from the fruits of milky-wax ripeness [35, 97, 105, 119, 120].

12.5.3.2 Synergism of the Active Compounds in *Gymnema Sylvestre* Extract

The extract of *Gymnema sylvestre* leaves is an effective hypoglycemic nutritional supplement; it is widely used as an adjuvant agent in antidiabetic therapy [55]. Our task was to determine with the help of IT Microcosm how the ratio of the main substances in the extract affects the extent of its hypoglycemic activity [74, 75, 105, 112]. The normal strategy was chosen for prediction because it makes it possible to account for nonstandard QSAR regularities, including those associated with the pharmacologic interaction of several substances, and it achieves this with fewer errors than with the risk strategy.

Gymnema sylvestre extract is composed of seven main components (Fig. 12.7): gymnemic acids I, II, III, IV (XV-XVIII), gymnemosides A and B (XIX, XX), and conduritol B (XXI). The structural formulas of six mixtures of these substances with varying compositions were constructed. A prediction of the presence of hypoglycemic activity in the individual components of *Gymnema sylvestre* extract and their mixtures was made with the help of decision rules calculated for a set of known hypoglycemic substances (see Sect. 3). The expected extent of hypoglycemic activity was estimated according to the membership-function value. The results are shown in Table 12.19.

According to the results of the predicted hypoglycemic activity of the individual components of the extract, it was established that gymnemic acid III (XVII) should be the most active among all of the components; gymnemic acid II (XVI) should show a high or medium activity; gymnemic acid IV (XVIII) and gymnemoside A (XIX) should show a moderate activity; and gymnemoside B (XX) and conduritol B (XXI) should be inactive.

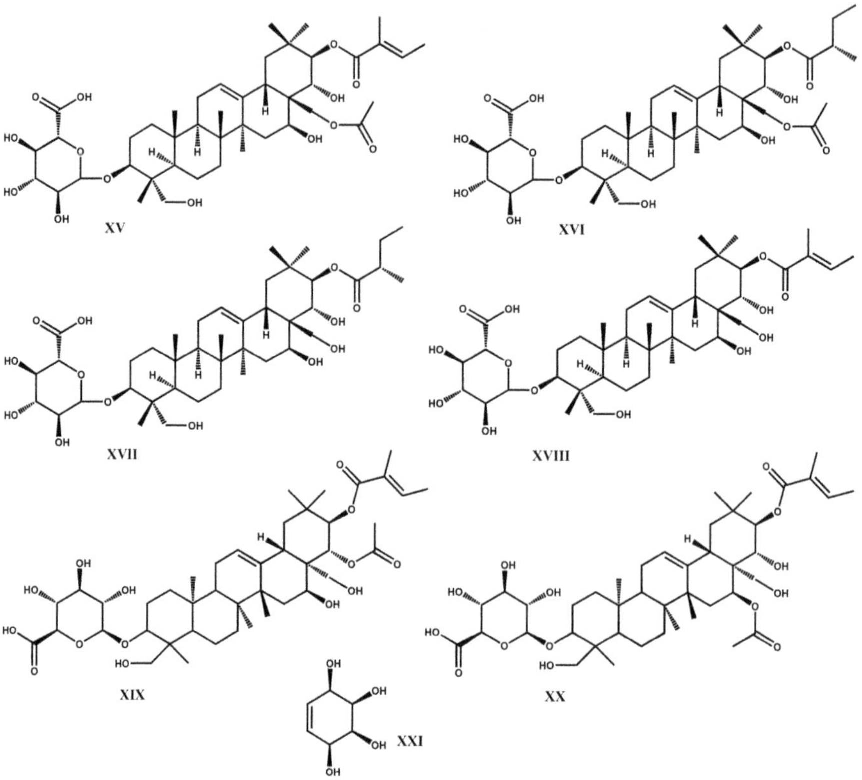

Fig. 12.7 The main components of *Gymnema sylvestre* extract

Table 12.19 Results of predicting the hypoglycemic activity of the main components of *Gymnema sylvestre* extract and of their mixtures

Substance, mixture	Estimate[a]	*Fb*[b]
Gymnemic acid I (XV)	A	0.60085
Gymnemic acid II (XVI)	A	0.68467
Gymnemic acid III (XVII)	A	0.83221
Gymnemic acid IV (XVIII)	A	0.60085
Gymnemoside A (XIX)	A	0.60085
Gymnemoside B (XX)	N	0.50802
Conduritol B (XXI)	N	0.99996
Mixture 1– (XV-XXI) 1:1:1:1:1:1:1	A	0.91614
Mixture 2– (XV-XIX) 1:1:1:1:1	A	0.91614
Mixture 3– (XV-XIX) 1:1:2:1:1	A	0.83010
Mixture 4– (XV-XIX) 1:1:3:1:1	A	0.83010
Mixture 5– (XVI, XVII) 1:2	A	0.81400
Mixture 5– (XVI, XVII, XXI) 1:2:3	A	0.81400

[a]Prediction with the normal strategy, nearest neighbor method
[b]Membership function

As a result of predicting the hypoglycemic activity of mixtures, it was established that the mixture of all of the main components of the extract in equal molar fractions should show a higher activity than any individual component; excluding two inactive components (XX, XXI) from the complete mixture should not result in a higher activity; increasing the ratio of the two most active components (XVI, XVII) and excluding two inactive compounds (XX, XXI) from the complete mixture should result in a lower activity; combinations of the two most active components (XVI, XVII) should show a considerably lower activity compared with the complete mixture; and adding a considerable amount of an inactive compound (XXI) to a combination of the two most active components (XVI, XVII) should not affect the level of activity [74, 105].

It follows from the data obtained from the prediction that a complex of active substances in *Gymnema sylvestre* extract should show a more powerful and, most likely, a more stable and prolonged hypoglycemic effect than any component of the extract, either individually or in limited combinations, due to a mutual potentiating synergistic effect of all of the components [74, 105, 112].

An experimental study of the hypoglycemic activity of hydroalcoholic *Gymnema sylvestre* extracts at gravimetric concentrations of 25, 50 and 75 % was made on outbred rats at a dose of 280 mg/kg *per os*; the glucose content in the blood was determined by the glucosidase method prior to extract administration and two hours after the administration. According to the experimental data, a medium reduction in the glucose concentration was 41, 29 and 26 % for the extracts with concentrations of 25, 50 and 75 %, respectively [75].

The method of obtaining *Gymnema sylvestre* extract foresees that, as its gravimetric concentration grows, the portion of the active gymnemic acids and gymnemosides increases, whereas the portion of the inactive components diminishes [76].

Thus, the experimental results confirm the presence of the synergistic effects predicted with IT Microcosm, which enhance the level of the hypoglycemic activity when administering *Gymnema sylvestre* extract with a lower content of active substances but with a more varied composition.

12.5.3.3 Synergism of Antidiabetic Drug Combinations

In this study, we used IT Microcosm for a prediction of the synergism of the hypoglycemic activity shown by Metformin combinations with five other antidiabetic drugs used in clinical practice for antidiabetic therapy (Fig. 12.8) [114].

Nine Metformin combinations with these drugs were used in protracted antidiabetic therapy in multicenter clinical studies with durations of 16 to 26 weeks and 328 to 1250 patients.

For the purposes of the *in silico* prediction, the administered doses were translated into molar ratios at a rate of 1 mol of the second agent per corresponding Metformin moles. In all nine cases, we observed a synergism manifested as a considerable enhancement of the hypoglycemic activity of the mixture, which far exceeded the effect produced by the portion of the added agent compared with Metformin *per se* (Table 12.20).

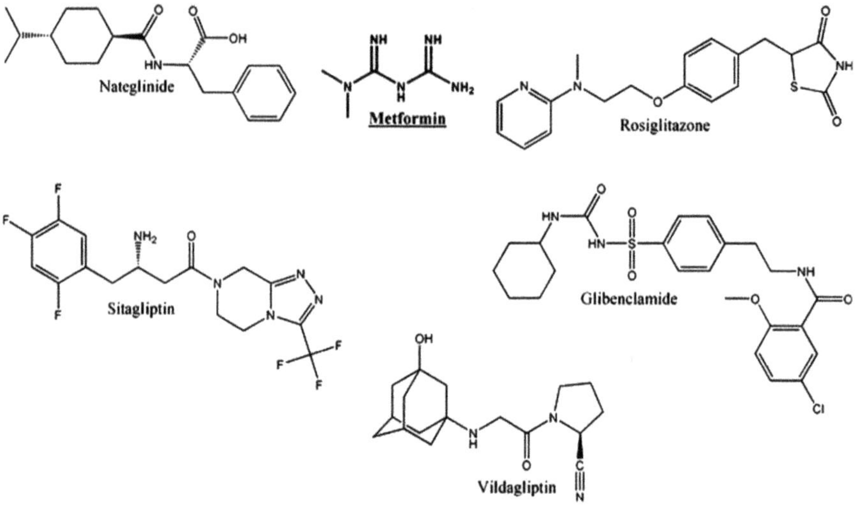

Fig. 12.8 Antidiabetic drugs

Table 12.20 Experimental clinical indices and predictive estimates of the hypoglycemic activity synergism of antidiabetic drug combinations

Drug (1 mol)	Metformin, mol	Number of patients	Duration, weeks	Synergism, %[a]	Fb_{Dist} [b]
Nateglinide	10	701	24	55	0.8244
Rosiglitazone	865	328	26	75	0.9974
Glibenclamide	765	806	20	39	0.9132
Glibenclamide	383	411	16	14	0.9131
Rosiglitazone	346	468	32	109	0.9975
Vildagliptin	23	1179	24	64	0.9132
Vildagliptin	47	1179	24	14	0.9131
Sitagliptin	32	1250	18	218	0.9999
Sitagliptin	63	1250	18	129	0.9999

[a] An average increase in the hypoglycemic activity of a combination of two drugs compared with Metformin *per*per *se*, according to clinical study data

[b] The membership function (20) in the class of hypoglycemic compounds in predicting with the normal strategy by the distance method

As in the case with *Gymnema sylvestre* extract, the IT Microcosm prediction of the hypoglycemic activity of these mixtures was made with the normal strategy on a training set of known hypoglycemic substances. The membership function in the class of hypoglycemic compounds served as the evaluation metric of the extent of activity.

Spearman's rank correlation coefficient R_S between the synergistic-effect value and the membership-function values was calculated, and the value $R_S = 0.8169$ was achieved, which corresponds to statistical significance $p = 7.192 \cdot 10^{-3}$. Consequently, there is a statistically highly significant dependence between the synergistic effect of antidiabetic drug combinations and the metric of membership of these compounds in the hypoglycemic compound class calculated by IT Microcosm.

Thus, IT Microcosm permits pinpoint accuracy for predicting the presence and levels of various pharmacological activities of complex organic compound mixtures with inclusion of the component synergism. This allows an optimisation of the qualitative and quantitative composition of the mixtures, which can lead to the design of novel, powerful drugs based on several gently acting, nontoxic compounds and several synergistic admixtures potentiating their effects.

Particular emphasis should be given to the fact that the successful prediction of mixture activity, including prediction with the inclusion of the synergistic effects, was performed on training sets comprising individual compounds only. For this purpose, a special toolkit for processing complex mixtures of up to 9999 components in ratios of 1 to 9999 mol fraction of each substance was developed in IT Microcosm.

The results obtained indicate that IT Microcosm permits a prediction of the presence and level of the various pharmacological activity types of complex molecular systems with consideration paid to the mutual effect of their components and optimisation of their quantitative and qualitative composition. This makes the technology helpful in designing novel, powerful drugs with minimal side effects.

12.6 Conclusions

The results of the computational and experimental studies described above indicate that the complex methodology of IT Microcosm for predicting the pharmacological activity of organic compounds, which includes consensus methods of constructing integral prediction regularities, is a universal, highly effective tool that solves most of the diverse problems associated with the search for novel drugs. The technology makes it possible to predict the presence and level of various types of pharmacological activity of structurally diverse and structurally similar compounds in various chemical classes (including chiral compounds) and to perform a directed search for highly active substances with predetermined properties. The possibility to consider noncovalent interactions provides a successful prediction of the spectrum and level of pharmacological activity. It also enables optimisation of the qualitative and quantitative composition of complex molecular systems, such as organic salts, supramolecular complexes and mixtures of individual compounds, with account of the mutual effect of the constituent components and of their synergism. This enables the design of novel, multicomponent drugs showing powerful beneficial effects and minimal side effects, on the basis of nontoxic, gently acting compounds and synergistic admixtures that potentiate their action.

The precision of search for highly active drug substances in IT Microcosm exceeds the precision of noncomputerized "intuitive" prediction performed by qualified pharmacologists. A considerable number of novel pharmacologic compounds were discovered with the help of IT Microcosm; these compounds have subsequently been patented. The technology was implemented in the form of the IT Microcosm software package that consists of 20 basic computer programs and a number of auxiliary utilities. The free computer program Microcosm White was developed separately. The software is available upon request to any interested persons.

References

1. Aivazyan SA, Buchstaber VM, Yenyukov IS et al (1989) Applied statistic: classification and reduction of dimensionality. Finansy i statistika, Moscow
2. Altunin AE, Semukhin MV (2000) Models and algorithms of decision making in fuzzy conditions. Izd-vo TyumenGU, Tyumen
3. Anisimova VA, Kuz'menko TA, Spasov AA et al (1999) Synthesis and study of the hypotensive and antiarrhythmic activity of 2,9-disubstitued 3-alkoxycarbonyl-imidazo[1,2-a]benzimidazoles. Pharm Chem J 33(7):361–365
4. Anisimova VA, Spasov AA, Kucheryavenko AF et al (2002) Synthesis and pharmacological activity of 2-(hetaryl)-imidazo[1,2-a]benzimidazoles. Pharm Chem J 36(10):528–534
5. Anisimova VA, Kosolapov VA, Minkin VI et al (2008) Dihydrobromide of 2-(3,4-dihydroxyphenyl)-9-diethylaminoethyl-imidazo[1,2-a]benzimidazole and a pharmaceutical composition based on it. Patent RU 2391979, 12 May 2008
6. Anisimova VA, Spasov AA, Kosolapov VA et al (2010) Sulfates of 2-aryl-4-dialkylaminoethyl-3-phenyl-pyrrolo[1,2-a]benzimidazoles with antioxidant and antiradical properties. Patent RU 2443704, 29 Oct 2010
7. Ayrton A, Morgan P (2008) Role of transport proteins in drug discovery and development: a pharmaceutical perspective. Xenobiotica 38(7–8):676–708
8. Baskin II, Varnek A (2008) Fragment descriptors in SAR/QSAR/QSPR studies. In:Varnek A, Tropsha A (eds) Chemoinformatics approaches to virtual screening. Royal Society of Chemistry, Cambridge, pp 1–43
9. Baskin II, Palyulin VA, Zefirov NS (2008) Neural networks in building QSAR models. Methods Mol Biol 458:137–158
10. Baurin N, Mozziconacci JC, Arnoult E et al (2004) 2D QSAR consensus prediction for high-throughput virtual screening. An application to COX-2 inhibition modeling and screening of the NCI database. J Chem Inf Comput Sci 44(1):276–285
11. Bemis GW, Murcko MM (1996) The properties of known drugs. 1. Molecular frameworks. J Med Chem 39(15):2887–2893
12. Born GV (1962) Aggregation of blood platelets by adenosine diphosphate and its reversal. Nature 194:927–929
13. Boser BE, Guyon IM, Vapnik VN (1992) A training algorithm for optimal margin classifiers. In: Proceedings of the fifth Annual ACM workshop on computational learning theory. ACM Press, New York, pp 144–152
14. Buccafusco JI (ed) (2001) Methods of behavior analysis in neuroscience. CRC Press, New York
15. Butov GM, Vassiliev PM, Parshin GYu et al (2006) Computer prediction of biologic activity of novel adamantane derivatives using information technology Microcosm. Bull Volgogr Res Cent RAMS 2:5–6
16. Butov GM, Vassiliev PM, Parshin GYu et al (2007) Synthesis, computer prediction and experimental testing of biological activity of new adamantane derivatives. In: Proceeding of the III Congress of Pharmacologists of Russia "Pharmacology—practical health", Saint Petersburg, 23–27 Sept 2007. Psycopharmacol biol narcol 7:1627

17. Butov GM, Vassiliev PM, Parshin GYu et al (2008) Synthesis and virtual screening for biological activity of adamantyl-containing trimethylbicyclo[2.2.1]heptan-2-one derivatives. Bull Volgogr Res Cent RAMS 3:67
18. Cammarata A (1972) Interrelationship of the regression models used for structure activity analisis. J Med Chem 15(6):573–577
19. Chernikov MV, Vassiliev PM (2006) Computer screening for novel benzimidazole derivatives with a high 5-HT_3-antiserotonin activity using information technology Microcosm. Bull Volgogr Res Cent RAMS 2:7–8
20. Daintith J (ed) (2004) Dictionary of computing: Oxford paperback reference. Oxford University Press, New-York
21. DeSimone RW, Currie KS, Mitchell SA et al (2004) Privileged structures: applications in drug discovery. Comb Chem High Throughput Screen 7(5):473–494
22. Draper NR, Smith H (1998) Applied regression analysis. Wiley series in probability and statistics. Wiley, New York
23. Duch W, Swaminathan K, Meller J (2007) Artificial intelligence approaches for rational drug design and discovery. Curr Pharm Des 13(14):1497–1508
24. Esbensen KH (2010) Multivariate data analysis—in practice. An introduction to multivariate data analysis and experimental design. CAMO AS, Oslo
25. Freson K, Thys C, Wittevrongel C et al (2006) Mechanisms of action and targets for actual and future antiplatelet drugs. Mini Rev Med Chem 6(6):719–726
26. Gabbasov ZA, Popov EG, Gavrilov IYu et al (1989) New highly sensitive method for analyzing platelet aggregation. Labor delo 10:15–18
27. Gasteiger J, Engel T (eds) (2003) Chemoinformatics: a textbook. Wiley-VCH Verlag GmbH & Co KGaA, Weinheim
28. Gendenshtein EI, Khadzsay YaI (1961) On pharmacologic properties of ajmalin, a new antiarrhythmic agent. Farmacol i toxicol, 24(1):49–57
29. Germashev IV, Derbisher VE, Vasil'ev PM (1998) Prediction of the activity of low-molecular organics in polymer compounds using probabilistic methods. Theor Found Chem Eng 32 (5):514–517
30. Germashev IV, Derbisher VE, Zotov YuL et al (2001) Computer-assisted design of active additives for polyvinyl chloride. Int Polym Sci Technol 28(7):36–38
31. Gertrudes JC, Maltarollo VG, Silva RA et al (2012) Machine learning techniques and drug design. Curr Med Chem 19(25):4289–4297
32. Golender VE, Rosenblit AB (1978) Computer methods for drug design. Zinatne, Riga
33. González MP, Terán C, Saíz-Urra L et al (2008) Variable selection methods in QSAR: an overview. Curr Top Med Chem 8(18):1606–1627
34. Gorelik AL, Skripkin VA (1984) Recognition methods. Vysshaya shkola, Moscow
35. Gorlov IF, Yurina OS, Vasiliev PM (2002) An experimental testing of results of the computer-aided prediction of a pharmacological activity spectrum of walnut's extract. Russ Agric Sci 5:45–47
36. Gupta SP (ed) (2006) QSAR and molecular modeling studies in heterocyclic drugs I. Topics in Heterocyclic Chemistry, vol 3. Springer-Verlag, Heidelberg
37. Gurova NA, Vassiliev PM (2009) Testing the prediction of antiarrhythmic activity of imidazonezimidazole derivatives. In: Proceedings of the XVI Russian National Congress "Man and drug", Moscow, 6–10 Apr 2009, p 79
38. Gurova NA, Vassiliev PM, Anisimova VA (2007) Computer prediction and experimental testing of antiarrhythmic activity of derivatives of nitrogen-containing heterocycles. In: Proceeding of the III Congress of Pharmacologists of Russia "Pharmacology—practical health", Saint Petersburg, 23–27 Sept 2007. Psycopharmacol biol narcol 7:1671
39. Hawkins DM (2004) The problem of overfitting. J Chem Inf Comput Sci 44(1):1–12
40. Haykin S (2009) Neural networks and learning machines. Pearson Education, New Jersey
41. Hollander M, Wolfe DA (1999) Nonparametric statistical methods. Wiley Series in Probability and Statistics. Wiley, New York

42. Izenman AJ (2008) Modern multivariate statistical techniques: regression, classification, and manifold learning. Springer texts in statistics. Springer Science+Business Media LLC, New-York
43. Jahnke W, Erlanson DA (eds) (2006) Fragment-based approaches in drug discovery. Methods and principles in medicinal chemistry, vol 34. Wiley-VCH Verlag GmbH & Co KGaA, Weinheim
44. Klebanov GI, Babenkova IV, Teselkin YuO et al (1988) Estimation of antioxidant activity of blood plasma using yolk lipoproteins. Labor delo 5:59–62
45. Kochetkov AN, Vassiliev PM, Breslaukhov AG (1991) Micro-COSM as a system of computer aided design of novel chemical compounds with desired biological activity. In: WATOC. Proceedings of the first All-Union conference on theoretical organic chemistry, Volgograd, 29 Sept–5 Oct 1991, p 500
46. Kosolapov VA (2005) Antioxidant agents: strategy of choice, perspectives of administration. Dissertation, Volgograd State Medical University
47. Kosolapov VA, Vassiliev PM, Tibir'kova EV et al (2007) Experimental testing the accuracy of computer prediction of antioxidant activity of novel heterocyclic compounds. In: Proceeding of the III Congress of Pharmacologists of Russia "Pharmacology—practical health", Saint Petersburg, 23–27 Sept 2007. Psycopharmacol biol narcol 7:1743
48. Kosolapov VA, Spasov AA, Vassiliev PM et al (2009) Directed search for and study of antioxidant substances. In: Proceedings of the VII Russian scientific conference "Chemistry and medicine (Orkhimed-2009)", Ufa, 1–5 July 2009, p 193
49. Kucheryavenko AF, Vassiliev PM, Salaznikova OA et al (2007) Computer search for heterocyclic compounds with a high antiplatelet activity. In: Proceeding of the III Congress of Pharmacologists of Russia "Pharmacology—practical health", Saint Petersburg, 23–27 Sept 2007. Psycopharmacol biol narcol 7:1760
50. Kuz'min VE, Muratov EN, Artemenko AG et al (2009) Consensus QSAR modeling of phosphor-containing chiral AChE inhibitors. QSAR Comb Sci 28(6–7):664–677
51. Kuz'min VE, Artemenko AG, Muratov EN et al (2010) Virtual screening and molecular design based on hierarchic QSAR technology. In: Puzyn T, Leszczynski J, Cronin MTD (eds) Recent advances in QSAR studies: methods and applications. Challenges and advances in computational chemistry and physics, vol 8. Springer Science+Business Media BV, Dordrecht, pp 127–176
52. Lagunin AA, Filimonov DA, Poroikov VV (2010) Multi-targeted natural products evaluation based on biological activity prediction with PASS. Curr Pharm Des, 16(15):1703–1717
53. Lankin VZ, Gurevich SM, Burlakova EB (1975) Study of ascorbate-dependent lipid peroxidation of tissues using 2-thiobarbituric acid test. Trudy moskovskogo obstchestva ispitateley prirody 52:73–78
54. Larichev OI (2006) Theory and methods of decision making. The new university library. Logos, Moscow
55. Leach MJ (2007) *Gymnema sylvestre* for diabetes mellitus: a systematic review. J Altern Complement Med 13(9):977–983
56. Lipinski CA, Lombardo F, Dominy BW et al (1997) Experimental and computational approaches to estimate solubility and permeability in drug discovery and development settings. Adv Drug Deliv Rev 23(1):3–25
57. Livingstone DJ (1991) Pattern recognition methods in rational drug design. Methods Enzymol 203:613–638
58. Mahe P, Ralaivola L, Stoven V et al (2006) The pharmacophore kernel for virtual screening with support vector machines. J Chem Inf Model 46(5):2003–2014
59. Mandel ID (1988) Cluster analysis. Finansy i statistika, Moscow
60. Mehvar R, Brocks DR (2001) Stereospecific pharmacokinetics and pharmacodynamics of beta-adrenergic blockers in humans. J Pharm Pharmaceut Sci 4(2):185–200
61. Merz KM, Ringe D, Reynolssds CH (eds) (2010) Drug design: structure- and ligand-based approaches. Cambridge University Press, New York

62. Morphy R, Kay C, Rankovic Z (2004) From magic bullets to designed multiple ligands. Drug Discov Today 9(15):641–651
63. No BI, Zotov YuL, Shishkin EV et al (2001) Adamantyl-containing derivatives of imidic acids with predicted high psychotropic, antiviral and antifungal activity. In: Proceedings of the IX International scientific conference "Chemistry and technology of carcass compounds", VolgGTU, Volgograd, 5–7 June 2001, pp 194–195
64. No BI, Vassiliev PM, Zotov YuL et al (2003) Computer design and directed synthesis of adamantyl-containing compounds, highly effective additives to polymer compositions. Int Polym Sci Technol 30(4):27–32
65. Opitz D, Maclin R (1999) Popular ensemble methods: an empirical study. J Artif Intell Res 11:169–198
66. Oprisiu I, Varlamova E, Muratov E et al (2012) QSPR approach to predict nonadditive properties of mixtures. Application to bubble point temperatures of binary mixtures of liquids. Mol Inf, 31(6–7):491–502
67. Orlov VV, Vassiliev PM, Derbisher VE (1999) Computer system for estimation of carcinogenic risk of chemical production. In: Proceeding of the 4th Russian scientific and practical conference with international participation "Novelties in Envionmentalism and Safety of Living", Saint Petersburg, 16–18 June 1999, p 212
68. Orlov VV, Derbisher VE, Zotov YuL et al (2003) Diagnostics of possible activity of adamantane derivatives in polymer compositions by molecular design. Khimicheskaya promishlennost 80(2):46–55
69. Petrov VI, Spasov AA, Anisimova VA et al (2003) The development and clinical testing of antiarrhythmia drugs of a new chemical class. Vestn Ross Akad Med Nauk 12:15–20
70. Poroikov V, Lagunin A, Filimonov D (2005) PASS: prediction of biological activity spectra for substances. In: Helma C (ed) Predictive toxicology. Taylor & Francis, New-York, pp 459–478
71. Preston Mason R (2012) Pleiotropic effects of calcium channel blockers. Curr Hypertens Rep 14(4):293–303
72. Puzyn T, Leszczynski J, Cronin MTD (eds) (2010) Recent advances in QSAR studies: methods and applications. Challenges and advances in computational chemistry and physics, vol 8. Springer Science+Business Media BV, Dordrecht
73. Radchenko EV, Palyulin VA, Zefirov NS (2008) Molecular field topology analysis in drug design and virtual screening. In:Varnek A, Tropsha A (eds) Chemoinformatics approaches to virtual screening. Royal Society of Chemistry, Cambridge, pp 150–181
74. Samokhina MP, Vassiliev PM, Chepljaeva NI (2007) QSAR-modeling of synergism between active compounds of *Gymnema sylvestris* extract. In: Proceeding of the III Congress of Pharmacologists of Russia "Pharmacology—practical health", Saint Petersburg, 23–27 Sept 2007. Psycopharmacol biol narcol 7:1936
75. Samokhina MP, Vassiliev PM, Chepljaeva NI (2008) Experimental study of hypoglycemic activity of *Gymnema sylvestris* extract and computer modeling of its components synergism. In: Collection of scientific papers "Development, research and marketing of new pharmaceutical products", PyatGFA, Pyatigorsk, p 490
76. Shanmugasundaram ER, Gopinath KL, Radha Shanmugasundaram K et al (1990) Possible regeneration of the islets of Langerhans in streptozotocin-diabetic rats given *Gymnema sylvestre* leaf extracts. J Ethnopharmacol 30(3):265–279
77. Siebert CD, HaÑ'nsicke A, Nagel T (2008) Stereochemical comparison of nebivolol with other b-blockers. Chirality 20(2):103–109
78. Simonov AM, Kovalev GV, Anisimova VA et al (1996) Antiarrhythmic agent. Patent RU 2068261, 27 Oct 1996
79. Spasov AA, Chernikov MV, Vassiliev PM et al (2007) Histamine receptors (molecular biological and pharmacological aspects). Izd-vo VolgGMU, Volgograd
80. Spasov AA, Vassiliev PM, Grechko OY et al (2010) *In silico* screening of condensed azoles derivatives with high pharmacological activity. In: Proceedings of the 2nd International congress EurasiaBio-2010, Moscow, 13–15 Apr 2010, pp 372–374

81. Spasov AA, Anisimova VA, Kucheryavenko AF et al (2010) Agent with antithrombogenic activity. Patent RU 2440814, 10 Nov 2010
82. Spasov AA, Iezsitsa IN, Vassiliev PM et al (2011) Pharmacology of drug stereoisomers. Izd-vo VolgGMU, Volgograd
83. Spasov AA, Anisimova VA, Gurova NA et al (2011) Agent with antiarrhythmic, antifibrillatory, antiischemic effects and a pharmaceutical composition based on it. Patent RU 2477130, 12 July 2011
84. Starovoitov MK, Vassiliev PM, Rudakova TV et al (2002) The computer aided prediction of carcinogenic risks of sulfenamide vulcanization accelerators. Kauchuk i rezina 1:28–31
85. Steed JW, Atwood JL (2009) Supramolecular chemistry, 2nd edn. Wiley, Chichester
86. Tallarida RJ (2001) Drug synergism: its detection and applications. J Pharmacol Exp Ther 298(3):865–872
87. The Cheminformatics and QSAR Society (2009) Martin Data Set II (Dopamine D-2 agonists). http://www.qsar.org/resource/datasets.htm. Accessed 21 Apr 2009.
88. Tolstikova TG, Bryzgalov AO, Sorokina IV et al (2007) Increase in pharmacological activity of drugs in their clathrates with plant glycosides. Lett Drug Des Discov 4(3):168–170
89. Toropova AP, Toropov AA, Benfenati E et al (2011) CORAL: Quantitative structure–activity relationship models for estimating toxicity of organic compounds in rats. J Comput Chem, 32(12):2727–2733
90. Toropova AP, Toropov AA, Benfenati E et al (2012) CORAL: models of toxicity of binary mixtures. Chemom Intell Lab Syst, 119:39–43
91. Vapnik VN (1998) Statistical learning theory. Wiley, New York
92. Varnek A, Fourches D, Horvath D et al (2008) ISIDA—platform for virtual screening based on fragment and pharmacophoric descriptors. Curr Comput Aided Drug Des, 4(3):191–198
93. Vassiliev PM (1989) A complex statistical approach to computer system creation for designing pharmacologically active compounds. In: Proceedings of All-Union scientific conference "Estimating Pharmacological activity of Chemical Compounds: Principles and Approaches", Moscow, 15–19 Nov 1989, part 1, p 56
94. Vassiliev PM (1991) Generalized pattern of biologically active compound class as an alternative to pharmacophores. In: WATOC. Proceedings of the first All-Union conference on theoretical organic chemistry, Volgograd, 29 Sept–5 Oct 1991, pp 77, 497
95. Vassiliev PM (2000) High-complexity dynamic chemical systems. In: Proceedings of the first Russian electronic conference on bioinformatics (RECOB–2000), Moscow, 15 Mar–21 Apr 2000, p G05
96. Vassiliev PM (2000) Strategies of computer prediction of organic compound properties. In: Proceedings of International scientific and technical conference "Modern information technology. Section "Information technologies in scientific experiments", PenzTI, Penza, p 7
97. Vassiliev PM (2001) Computer prediction of a spectrum of pharmacological activity of multicomponent drugs using conservative strategy in Microcosm system. In: Proceedings of International scientific and practical conference "Modern techniques and technologies in medicine and biology", Novocherkassk, 25 Dec 2000, part 2, pp 48–49
98. Vassiliev PM (2001) Microcosm, a software system for prediction of organic compound properties. In: Proceedings of the 2nd Russian conference "Molecular modeling", Moscow, 24–26 Apr 2001, p 21
99. Vassiliev PM (2002) Paradigm of property prediction of organic compounds in software complex Microcosm. In: Proceedings of the first National conference "Information Technology in Solving Fundamental Scientific Problems and Applied Problems in Chemistry, Biology, Pharmaceutics, Medicine (IVTN-2002)", Moscow, 9 Apr–13 June 2002, p 54
100. Vassiliev PM (2002) Virtual screening for anti-HIV active compounds using risk strategy in software system Microcosm. In: Proceedings of the IX Russian National Congress "Man and drug", Moscow, 8–12 Apr 2002, p 592

101. Vassiliev PM (2002) Generalized pattern as a meta-model of biologically active compound class. In: Proceedings of the I Russian school-conference "Molecular modeling in chemistry, biology and medicine", Saratov, 18–20 Sept 2002, pp 19–20
102. Vassiliev PM (2007) Prediction of biological activity in IT Microcosm with check for non-contradiction of prediction estimates spectrum. In: Proceedings of the 5th Russian conference "Molecular modeling", Moscow, 18–20 Apr 2007, p 42
103. Vassiliev PM (2007) Information technology of computer search for novel drugs. In: Proceeding of the III Congress of Pharmacologists of Russia "Pharmacology—practical health", Saint Petersburg, 23–27 Sept 2007. Psycopharmacol biol narcol 7:1631–1632
104. Vassiliev PM (2008) Mirages of 3D molecular modelling. Bull Volgogr Res Cent RAMS 3:69–71
105. Vassiliev PM (2009) Information technology for prediction of pharmacological activity of chemical compounds. Dissertation, Volgograd State Medical University
106. Vassiliev PM (2012) IT Microcosm for property prediction of organic compounds. In: Proceedings of the 2nd French-Russian workshop in chemoinformatics and bioinformatics, KFU, Kazan, 17 Sept 2012, pp 8–9
107. Vassiliev PM, Breslaukhov AG (1990) Hierarchical language for description of structures of bioactive compounds. In: Proceedings of the second World congress of theoretical organic chemists, University of Toronto, Toronto, p AA–38 (8–14 July 1990)
108. Vassiliev PM, Derbisher VE (1998) Principles of computer appraisal of environmental hazard of chemical production. In: Proceedings of the IV Traditional scientific and technical conference of CIS countries, VolgGTU, Volgograd, 15–16 Sept 1998, pp 211–213
109. Vassiliev PM, Spasov AA (2005) Computerized information technology in prognosis of pharmacological activity of structurally heterogeneous compounds. Vestn Volgogr State Med Univ 1:23–30
110. Vassiliev PM, Spasov AA (2006) Fragmentary encoding languages of compound structure for computer prediction of biological activity. Ross Khim Zh 50(2):108–127(Zhurn Ross Khim ob-va im DI Mendeleeva)
111. Vassiliev PM, Spasov AA (2009) Computer prediction of synergism of hypoglycemic and antioxidant compounds using IT Microcosm. In: Proceedings of the XVI Russian National Congress "Man and drug", Moscow, 6–10 Apr 2009, pp 528–529
112. Vassiliev PM, Spasov AA (2009) QSAR-modeling of component synergism in prediction of pharmacological activity of mixtures of naturally occurring and synthetic organic compounds. In: Proceedings of the 6th Russian conference "Molecular modeling", Moscow, 8–10 Apr 2009, p 58
113. Vassiliev PM, Kochetkov AN (2011) IT Microcosm. State Registration Certificate for software program 2011618547 (Russian), 31 Oct 2011
114. Vassiliev PM, Spasov AA (2012) Computer prediction of hypoglycemic activity of antidiabeic drugs using IT Microcosm. In: Proceedings of the XIX Russian National Congress "Man and drug", Moscow, 23–27 Apr 2012, p 361
115. Vassiliev PM, Breslaukhov AG, Kochetkov AN (1991) SOS, a system of computerized evaluation of environmental hazard of chemical compounds and production. In: WATOC. Proceedings of the first All-Union conference on theoretical organic chemistry, Volgograd, 29 Sept–5 Oct 1991, p 507
116. Vassiliev PM, Orlov VV, Khortik KV et al (2000) Principles of computer system creation for prediction of environmentally hazardous properties of chemical compounds. In: Proceedings of International ecological congress "Novelties of environmentalism and safety of living", BaltSTU, Saint Petersburg, 14–16 June 2000, p 202
117. Vassiliev PM, Orlov VV, Derbisher VE (2000) Prediction of carcinogenic risks of organic compounds by method of chances. Pharmac Chem J 34(7):19–22
118. Vassiliev PM, Kablov VF, Khortik KV et al (2001) The computer prediction method of the vulcanization accelerator properties. Kauchuk i rezina 3:22–25
119. Vasiliev PM, Gorlov IF, Yurina OS (2002) A computer-aided prediction of a pharmacological activity spectrum of waln's extract. Russ Agric Sci 2:55–58

120. Vassiliev PM, Gorlov IF, Yurina OS (2003) Prediction of pharmacological activity of multicomponent mixtures of organic compounds in information technology Microcosm. In: Proceedings of the 3rd Russian conference "Molecular modeling", Moscow, 15–17 Apr 2003, p 54
121. Vassiliev PM, Rudakova TV, Beloussov EK et al (2004) Computer forecast of carcinogenic hazard of rubber sulphonamide accelerators. In: Proceedings of the 16th international congress of chemical and process engineering (CHISA 2004), Praha, 22–26 Aug 2004, vol 5 "Systems and Technology", pp 1944–1945
122. Vassiliev PM, Spasov AA, Kosolapov VA et al (2004) Information technology Microcosm in prediction of pharmacological activity of novel heterocyclic compounds. In: Proceedings of international conference "Information technology in education, technology and medicine", Volgograd, 18–22 Oct 2004, vol 3, pp 180–186
123. Vassiliev PM, Spasov AA, Kosolapov VA et al (2005) Information technology Microcosm in prediction of pharmacological activity of organic compound salts. In: Proceedings of the 4th Russian conference "Molecular modeling", Moscow, 12–15 Apr 2005, p 53
124. Vassiliev PM, Spasov AA, Kosolapov VA et al (2005) A computer prognosis and experimental testing of antioxidant activity of new chemical compounds using QSAR-dependences. Vestn Volgogr State Med Univ 2:16–19
125. Vassiliev PM, Kucheryavenko AF, Salaznikova OA et al (2007) Computer prediction and experimental testing of antiplatelet activity of benzimidazole derivatives. In: Proceedings of the XIV Russian National Congress "Man and drug", Moscow, 16–20 Apr 2007, p 272
126. Vassiliev PM, Yakovlev DS, Poroikov VV et al (2007) The 3D-modeling, computer prediction and experimental testing of 5-HT_3-antiserotonin activity of new chemical compounds. In: Proceedings of the forth International symposium on computational methods in toxicology and pharmacology integrating internet resources (CMTPI-2007), Moscow, 1–5 Sept 2007, p 169
127. Vassiliev PM, Naumenko LV, Spasov AA (2007) The Microcosm information technology system for biological activity prediction of organic compounds: the prediction of hemorheological activity. In: Proceedings of the forth International symposium on computational methods in toxicology and pharmacology integrating internet resources (CMTPI-2007), Moscow, 1–5 Sept 2007, p 126
128. Vassiliev PM, Perfilova VN, Tyurenkov IN (2008) Computer prediction and experimental testing of antianginal activity of molecular complexes of phenibut derivatives. In: Proceedings of the XV Russian National Congress "Man and drug", Moscow, 14–18 Apr 2008, p 599
129. Vassiliev PM, Perfilova VN, Tyurenkov IN (2008) Comparative pharmacophore analysis of antiischemic activity of known drugs and molecular complexes of GABA derivatives. Bull Volgogr Res Cent RAMS 3:73–75
130. Vassiliev PM, Stukovina AYu, Spasov AA et al (2008) Directed search for new derivatives of benzimidazole and indole with a high $P2Y_1$—activity using IT Microcosm. Bull Volgogr Res Cent RAMS 3:71–72
131. Vassiliev PM, Spasov AA, Yakovlev DS et al (2009) Directed *in silico* search for derivatives of condensed azoles with a high 5-HT3-antiserotonin activity. In: Proceedings of the VII Russian scientific conference "Chemistry and medicine (Orkhimed-2009)", Ufa, 1–5 July 2009, pp 144–145
132. Vassiliev PM, Kruglikov ME, Kochetkov AN (2011) *In silico* search for condensed azol derivatives with a high pharmacological activity. In: Materials of Indo-Russian seminars "From generics to innovative pharmacological agents", VSMU, Volgograd, 8–10 Sept 2011. Vestnik of the Volgograd State Medical University, Addendum, pp 21–22
133. Vassiliev PM, Spasov AA, Jakovlev DS et al (2013) *In silico* bioinformation technology in search for selective ligands. In: Proceedings of the VII Moscow international congress "Biotechnology: state of the art and prospects of development", Moscow, 19–22 Mar 2013, part 1, pp 259–260

134. Vassiliev PM, Spasov AA, Maltsev DV et al (2013) *In silico* consensus search for condensed azole derivatives with a high 5-HT2A-antiserotonin activity. In: Proceedings of the XX Jubilee Russian national congress "Man and drug", Moscow, 15–19 Apr 2013, p 307
135. Verma RP, Hansch C (2005) An approach toward the problem of outliers in QSAR. Bioorg Med Chem 13(15):4597–4621
136. Vol'sky VI, Lezina ZM (1991) Small group voting: procedures and methods of comparative analysis. Nauka, Moscow
137. Yap CW, Li H, Ji ZL et al (2007) Regression methods for developing QSAR and QSPR models to predict compounds of specific pharmacodynamic, pharmacokinetic and toxicological properties. Mini Rev Med Chem 7(11):1097–1107
138. Zaidler YaI (1967) Search for antiarrhythmic compounds on an isolated rat atrium preparation. In: Modeling, methods of study and experimental therapy of pathological conditions, part 3. Meditsina, Moscow, p 191
139. Zefirov NS, Palyulin VA (2002) Fragmental approach in QSPR. J Chem Inf Comput Sci 42(5):1112–1122
140. Zefirova ON, Baranova TYu, Lyssenko KA et al (2012) Synthesis and biological testing of conformationally restricted serotonin analogues with bridgehead moieties. Mendeleev Commun 22(2):75–77

Chapter 13
Continuous Molecular Fields Approach Applied to Structure-Activity Modeling

Igor I. Baskin and Nelly I. Zhokhova

Abstract The Method of Continuous Molecular Fields is a universal approach to predict various properties of chemical compounds, in which molecules are represented by means of continuous fields (such as electrostatic, steric, electron density functions, etc.). The essence of the proposed approach consists in performing statistical analysis of functional molecular data by means of joint application of kernel machine learning methods and special kernels which compare molecules by computing overlap integrals of their molecular fields. This approach is an alternative to traditional methods of building 3D "structure-activity" and "structure-property" models based on the use of fixed sets of molecular descriptors. The methodology of the approach is described in this chapter, followed by its application to building regression 3D-QSAR models and conducting virtual screening based on one-class classification models. The main directions of the further development of this approach are outlined at the end of the chapter.

13.1 Introduction

Currently, the leading role in predicting biological activity and physicochemical properties of chemical compounds belongs to methods of chemoinformatics [1–3], which are based on revealing "structure-activity" and "structure-property" relationships using modern statistical, data mining, machine learning and artificial intelligence approaches. Rapid progress of these techniques requires the development of new tools of machine learning and data mining specially adapted to work with molecular structures of chemical compounds [4].

In this section, we consider a new approach to building "structure-activity" and "structure-property" models based on the use of continuous functions on space coordinates (called hereinafter *continuous molecular fields*) to represent molecular

---

I. I. Baskin (✉) · N. I. Zhokhova
Faculty of Physics, M. V. Lomonosov Moscow State University, Moscow, Russia 119991
e-mail: igbaskin@gmail.com

N. I. Zhokhova
e-mail: zhokhovann@gmail.com

L. Gorb et al. (eds.), *Application of Computational Techniques in Pharmacy and Medicine,*
Challenges and Advances in Computational Chemistry and Physics 17,
DOI 10.1007/978-94-017-9257-8_13

structures. Different types of molecular fields can be used for this purpose, including electrostatic, steric, hydrophobic, hydrogen bond donor and acceptor fields, etc. This way to describe chemical structures corresponds well to the physical nature of the molecules, which interact with the environment through molecular fields. The suggested approach is an alternative to traditional methods of representing chemical structures in SAR/QSAR/QSPR (Structure-Activity Relationships/Quantitative Structure-Activity Relationships/Quantitative Structure-Property Relationships) studies by means of fixed-sized vectors of descriptors derived from topological molecular graphs, as well as interaction energies with certain probes calculated at specific points in space (e.g., at the nodes of a hypothetic lattice).

So far, the direct use of continuous molecular fields in their functional form in statistical analysis was not possible because standard data analysis procedures can only work with finite and fixed number of features (molecular descriptors). Only recently, thanks to the development of the statistical learning theory [5] and the methodology of using kernels [6] in machine learning instead of fixed-sized feature vectors, it has become possible to process data of any form and complexity.

> The essence of the Continuous Molecular Fields (CMF) approach consists in performing statistical analysis of functional molecular data by means of joint application of kernel machine learning methods and special kernels which compare molecules by computing overlap integrals of their molecular fields. [7, 8]

The principal novelty of our approach is the ability to conduct a statistical analysis of chemical data represented in the form of continuous molecular fields, i.e. an infinite number of attributes organized in a functional form. In this case, statistical model is not a function relating the values of some properties of chemical compounds with the values of several molecular descriptors, like in the case of traditional SAR/QSAR/QSPR models, but a functional relating the properties of chemical compounds with functions describing spatial distribution of molecular fields. As a result, the resulting models are characterized by continuous fields of model coefficients, which can be visualized in the same manner as molecular fields themselves to provide intuitive interpretation of the models and a deep insight into the nature of the corresponding chemical phenomenon. The principal advantage of this approach follows from a natural, accurate and comprehensive representation of molecules by means of continuous molecular fields.

In combination with kernel based machine learning methods, such as the Support Vector Machines (SVM) [9], the Kernel Ridge Regression (KRR) [10], the Gaussian Processes (GP) [11], etc., continuous molecular fields can be used to provide the quantitative prediction of various properties of chemical compounds. The feasibility of this was demonstrated by us earlier [7, 8]. In conjunction with kernel-based methods of one-class classification (novelty detection), such as the One-Class Support Vector Machines (1-SVM) [12], the use of continuous molecular fields allows one to set up a ligand-based virtual screening procedure for searching huge databases of available chemical compounds and retrieving from them potentially biological active compounds (hits), molecular fields of which are similar to the “idealized configuration” (which reflects the structure of the corresponding binding site in biological macromolecule and can be viewed as a "negative image" of its molecular fields) learned by the corresponding one-class model. The feasibility of this approach in practice has also been demonstrated by us [13, 14].

13.2 Method of Continuous Molecular Fields

3D-QSAR approaches, in which information concerning the spatial structure of molecules is explicitly taken into account, play very important role in medicinal chemistry and drug design [15–17]. Most of them are based on the use of the molecular fields reflecting different types of intermolecular interactions in which molecules under study can be involved. Historically the first and still the most popular 3D-QSAR method is CoMFA (Comparative Molecular Field Analysis) [18], in which electrostatic interactions are approximated by means of the Coulomb law with point partial charges computed for each atom, whereas steric interactions are expressed using the Lennard-Jones potentials with standard force field parameters. Several other types of molecular fields, such as the hydrophobic [19], hydrogen bond donor and acceptor fields [20], molecular orbital fields [21], E-state fields [22], fields of atom-based indicator variables [23] are also used in the framework of the CoMFA method. In the CoMSIA (Comparative Molecular Similarity Indices Analysis) approach [24] the same types of molecular fields are approximated using the Gaussian radial basis functions. In the GRID method [25] the values of various types of molecular fields are computed as interaction energies with certain probe atom or group of atoms placed at grid nodes.

In all of the above approaches molecular fields are evaluated at node points of some imaginary grid surrounding the set of aligned molecules. The advantage of using such lattices lies in the possibility to use the values of molecular field potentials calculated at grid nodes as a vector of descriptors, which can further be fed to some standard statistical analysis procedure, usually PLS (Partial Least Squares) [26], in order to build regression QSAR models. Another appealing feature of this approach is the possibility to visualize the resulting regression coefficients using easily interpretable isosurfaces (usually colored according to the sign of coefficients and the type of the corresponding molecular field) surrounding the molecules. Nonetheless, this approach has certain considerable drawbacks. Indeed, it is necessary to: (1) choose biologically active conformation for each molecule; (2) align in space the training set of molecules; (3) build a lattice around such set of molecules; (4) choose molecular fields and compute their potentials at grid points; (5) build regression QSAR models. The problems associated with each of these stages are well known and present a challenge for the current stage of the development of the 3D-QSAR methodology.

The CMF approach addresses the problems caused by the necessity to choose a grid of points around molecules. It is known that 3D-QSAR models sharply depend on the spatial orientation and extent of such grid, as well as on the step size (i.e. the distance between the closest points) in it [16]. Another problem caused by the use of grids is very high dimensionality of the regression task caused by the big number of grid points, which precludes the use of many efficient statistical methods. Unfortunately, decrease of the number of grid points through increase of the step size (i.e. the use of coarser grid) or decrease of its extent causes the loss of important information. In order to tackle the problem, we suggest not using grids or fixed probe positions in 3D-QSAR studies. Instead of computing descriptor values at a discrete set of points, we propose to work directly with descriptions of molecular fields in the form of continuous functions on radius vector **r**, by means of specially constructed kernels. The use of such continuous molecular fields seems to be much more natural and corresponding to the real physical picture of the world than the application of certain speculative grids arbitrary chosen to approximate such fields.

It should be mentioned that continuous molecular fields have already been used in QSAR studies. So, indexes of R. Carbó-Dorca, which were used in certain QSAR studies [27], can also be considered as a particular case of continuous molecular fields.

The method of Continuous Molecular Fields (CMF) performs statistical analysis of functional molecular data by means of joint application of kernel machine learning methods and special kernels which compare molecules by computing overlap integrals of their molecular fields [7, 8].

13.2.1 Procedure of Kernel Calculation

The principal element of the CMF approach is the procedure of calculating molecular field kernels. The joint molecular field kernel $K(M_i, M_j)$ that describes the similarity between all molecular fields of molecules M_i and M_j can be calculated as a linear combination of kernels corresponding to each of N_f types of molecular fields:

$$K(M_i, M_j) = \sum_{f=1}^{N_f} h_f K_f(M_i, M_j), \tag{13.1}$$

where h_f is the mixing coefficient of molecular fields; $K_f(\mathrm{M}_i, \mathrm{M}_j)$ is the kernel describing the similarity between the molecular fields of the *f*th type for the *i*th and *j*th molecules. Function $K(M_i, M_j)$ represents correctly constructed kernel, because a linear combination of kernels is a kernel.

In some cases, we also use the normalized version of the kernels:

$$K_f^{'}(M_i, M_j) = \frac{K_f(M_i, M_j)}{\sqrt{K_f(M_i, M_i) \cdot K_f(M_j, M_j)}} \tag{13.2}$$

The molecular field kernel for each the *f*th type of molecular field is calculated in the framework of the CMF approach by summation of the kernels for each pair of atoms for the *i*th and *j*th molecules:

$$K_f(M_i, M_j) = \sum_{l=1}^{N_i} \sum_{m=1}^{N_j} k_f(A_{il}, A_{jm}), \tag{13.3}$$

where $k_f(A_{il}, A_{jm})$ is the kernel that describes the similarity between the field of the fth type of the lth atom in the ith molecule and mth atom in the jth molecule; N_i is the number of atoms in the ith molecule; N_j is the number of atoms in the jth molecule. $K_f(M_i, M_j)$ can be considered as a 1-tuple convolution kernel corresponding to decomposition of a molecule into atoms. The value of kernel $k_f(A_{il}, A_{jm})$ can be calculated by integration of the product of the fields for a pair of atoms over the entire physical space:

$$k_f(A_{il}, A_{jm}) = \iiint \rho_{fil}(\mathbf{r}) \rho_{fjm}(\mathbf{r}) d^3\mathbf{r} \tag{13.4}$$

where $\rho_{fil}(\mathbf{r})$ is the value of molecular field of the fth type induced by the lth atom of the ith molecule at the point $\mathbf{r}$ of the physical space; the $\rho_{fjm}(\mathbf{r})$ is the same magnitude for the mth atom of the jth molecule. To simplify the integration, one can approximate any molecular field as a weighted sum of Gaussian basis functions. We have found empirically that in most of cases it is sufficient to use a single Gaussian function to represent any kind of fields produced by a single atom, exactly like in the CoMSIA method [24]:

$$\rho_{fil}(\mathbf{r}) = w_{fil} \exp\left(-\frac{1}{2} \alpha_f \left\| \mathbf{r} - \mathbf{r}_{il} \right\|^2 \right), \tag{13.5}$$

where $\mathbf{r}_{il}$ is the location of the lth atom of the ith molecule in the physical space; α_f is the fitting parameter for molecular field of the fth type; w_{fil} is the weight of the contribution of lth atom of the ith molecule to the molecular field of the fth type. For example, in the case of electrostatic field the w_{fil} is the partial charge on the lth atom of the ith molecule, for the steric field—the Lennard-Jones potential parameters, for the hydrophobic field—the contribution of a given atom to the total hydrophobicity. Evidently different sets of the values w_f constitute different parameterizations of the CMF approach.

Due to the afore-mentioned approximation, the foregoing integral can be calculated analytically:

$$k_f(A_{il}, A_{jm}) = \iiint \rho_{fil}(\mathbf{r}) \rho_{fjm}(\mathbf{r}) d^3\mathbf{r} = w_{fil} w_{fjm} \sqrt{\frac{\pi^3}{(\alpha_f)^3}} \exp\left(-\frac{\alpha_f}{4} \left\| \mathbf{r}_{il} - \mathbf{r}_{jm} \right\|^2 \right) \tag{13.6}$$

It should be pointed out that the CMF approach is not confined to the simplest approximations introduced by Eq. (13.5). Any number of Gaussian functions as well as any other sets of basic functions (such as splines, wavelets, etc.) can be used for approximating continuous molecular fields. This provides the ability to use complex types of molecular fields, including those derived from quantum chemistry, such as electron density functions.

13.2.2 The Use of Continuous Molecular Fields in Conjunction with Regression Kernel-based Machine Learning Methods

Kernel $K(M_i, M_j)$, or its normalized version $K'(M_i, M_j)$, can be plugged in any kernel-based machine learning method (such as Support Vector Machine, Kernel Ridge Regression [10], Kernel Partial Least Squares [28], Gaussian Processes, etc.) in order to build regression, classification, novelty detection models, etc.

In the case of 3D-QSAR kernel-based regression models, the value of the predicted property y_t for a new molecule M_t can be calculated using the following expression:

$$y_t = \sum_{j=1}^{N_m} a_j K(M_t, M_j) + b, \qquad (13.7)$$

where N_m is the number of molecules in the training set. If Support Vector Regression (SVR) is used for deriving the values of a_j and b, the vector a_j appears to be sparse, with non-zero values corresponding to a certain subset of compounds from the training set. In contrast, contributions of all molecules from the training set are needed to make predictions based on regression models built using the Kernel Ridge Regression (KRR) [10], the Kernel Partial Least Squares [28] or the Gaussian Processes machine learning methods.

In addition to the set of adjustable coefficients a_j and b contained in Eq. (13.7), the method CMF also requires calculation of a certain number of adjustable parameters. Among them are the parameter ν for the support vector regression method ν-SVR and the ridge parameter γ for KRR. Their values should be optimized with the aim to improve the predictive capability of the model constructed. In addition, for each molecular field one can adjust the values of up to two parameters: α_f (attenuation factor, which is related to the width of the Gaussian function) and h_f (mixing coefficient, which has the meaning of the relative contribution of molecular field of the fth type).

13.2.3 Fields of Model Coefficients

Equation (13.7) represents the dual form of the regression model, since in it the activity y_t is predicted by considering similarity measures of a test compound M_t in relation to the training set compounds M_j. In order to obtain the traditional primal form of the 3D-QSAR model, which involves an explicit consideration of molecular descriptors and regression coefficients, one can make the substitution of Eqs. (13.1) and (13.3–13.6) to Eq. (13.7) to obtain:

$$y_t = \sum_{f=1}^{N_f} h_f \iiint C_f(\mathbf{r}) X_f(\mathbf{r}) d^3\mathbf{r} + b, \tag{13.8}$$

$$X_f(\mathbf{r}) = \sum_{l=1}^{N_t} \rho_{ftl}(\mathbf{r}) = \sum_{l=1}^{N_t} w_{ftl} \exp\left(-\frac{\alpha_f}{2} \left\| \mathbf{r} - \mathbf{r}_{tl} \right\|^2 \right), \tag{13.9}$$

$$C_f(\mathbf{r}) = \sum_{j=1}^{N_m} a_j \sum_{m=1}^{N_j} \rho_{fjm}(\mathbf{r}) = \sum_{j=1}^{N_m} a_j \sum_{m=1}^{N_j} w_{fjm} \exp\left(-\frac{\alpha_f}{2} \left\| \mathbf{r} - \mathbf{r}_{jm} \right\|^2 \right) \tag{13.10}$$

The primal form of the regression model is expressed by Eq. (13.8), in which molecular field $X(\mathbf{r})$ represents the continuous field of molecular descriptors for the test molecule t, whereas $C_f(\mathbf{r})$ is a continuous field of the corresponding regression coefficients. The principal distinction of the CMF Eq. (13.8) from that of an ordinary 3D-QSAR linear model lies in the infinite number of point descriptors in the CMF model. As a result, continuous field of molecular field descriptors is used in it instead of several thousands of descriptors computed in CoMFA or CoMSIA at lattice points, continuous field of regression coefficients is used instead of several thousands of regression coefficients obtained by means of the PLS regression, and integration over the entire physical space substitutes summation over the grid points. It also follows from this analysis that iso-surfaces of the fields of regression coefficients $C_f(\mathbf{r})$ could be used in the same manner and for the same purposes as CoMFA and CoMSIA contour maps. Table 13.1 lists three types of molecular fields, shows isosurfaces of KRR regression coefficients for the case of thrombin inhibitors (2-amidinophenylalanines), and also provides chemical interpretation.

The benefit of such visualization for drug design is evident. Note, however, an important difference between isosurfaces in CMF models, from one side, and CoMFA, CoMSIA and GRID contour maps, from the other: the former are centered on atoms (as follows from Eq. 13.10), whereas the latter are situated around molecules (as a consequence of the impossibility to place probe atoms and groups inside atoms). Although such location of isosurfaces might seem unusual from the first glace, but it offers more direct answer to the question as to what changes should be introduced in order to increase biological activity of chemical compound.

Replacement of binding measures (such as $\log(1/IC_{50})$) in relation to individual targets for the difference in binding to two targets as an output of regression equation leads to the concept of selectivity fields, formulated by us earlier for CoMFA analysis [29]. In the framework of the continuous molecular field approach analogous fields of regression coefficients can also be built. Graphical visualization of their isosurfaces clearly indicates the changes that should be introduced into chemical structure in order to tune their selectivity towards different biological targets.

Note also that, due to the lack of discrete grids, molecular fields are treated in CMF as continuous functions with respect to spatial coordinates and therefore can be differentiated and integrated (even analytically, due to the use of Gaussian

Table 13.1 Types of molecular fields, isosurfaces of KRR regression coefficients and chemical interpretation for the case of thrombin inhibitors (2-amidinophenylalanines). In all cases the isosurfaces are superimposed over the structure of one of inhibitors

Type of molecular field	Isosurfaces of the fields of regression coefficients	Chemical interpretation
Electrostatic	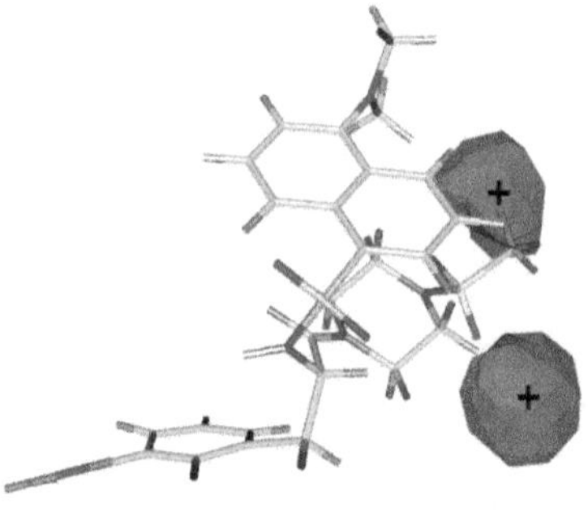	Red color (marked with sign "+") means increase of biological activity upon increase of partial electric positive charge on the nearest atoms
Steric	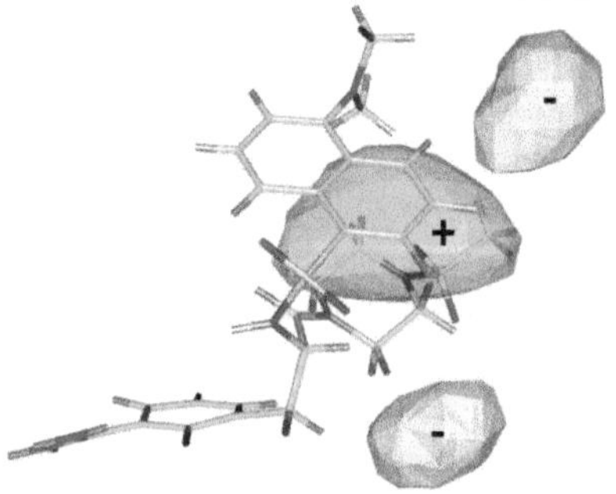	Green color (marked with sign "+") means increase, yellow color (marked with sign "-") means decrease of biological activity upon increase of bulkiness of the nearest atoms.
Hydrophobic	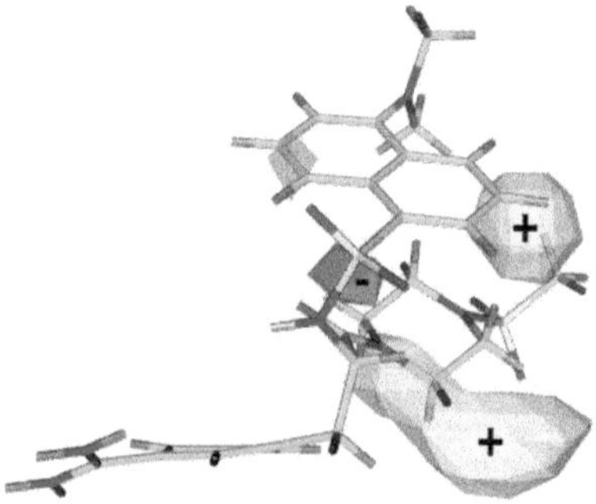	Yellow color (marked with sign "+") means increase, violet color (marked with sign "-") means decrease of biological activity upon increase of lipophilicity of the nearest atoms.

basis function). Easy differentiability of molecular field functions allows for applying very powerful although still unexplored in chemistry apparatus of data analysis and visualization offered by a newly emerged branch of mathematical statistics, functional data analysis [30], to visualize both molecular fields and fields of model coefficients. One should also add that, by applying the above-discussed methodology, in the framework of CMF, due to modularity of kernel-based approaches, it is possible to deduce fields of model coefficients not only for 3D-QSAR regression, but also for: (a) classification (in the case of Support Vector Machines they describe angular orientation of the hyperplane separating active from inactive compounds in the infinite-dimensional feature space); (b) novelty detection (or one-class classification [31], which can be used for virtual screening); (c) clustering; (d) dimensionality reduction; etc. We hope that numerous methods of analyzing and visualizing chemical databases and SAR/QSAR/QSPR models offered by the use of continuous molecular fields would deepen insights into the nature of structure–activity relationships and facilitate drug design. This is one of the main directions of our current studies in this direction and a topic of future publications.

13.3 Modelling Biological Activity

We validated the CMF approach in two case studies and obtained preliminary results, which have been published as short communications [7, 13]. The first one dealt with the use of the CMF to build 3D-QSAR regression models [7]. In the second case study [13], the performance of a new method for virtual screening of organic compounds based on the combination of the CMF methodology with the one-class SVM method (1-SVM) has been assessed. In both cases the CMF has not only proven its efficiency, but has also demonstrated some advantages compared to state-of-the-art approaches in chemoinformatics.

13.3.1 Building 3D-QSAR Regression Models

We have tested the performance of the CMF approach in building 3D-QSAR regression models by using eight data sets. These sets were selected as expanded benchmark for 3D-QSAR methods as well as examples of various types of biologycal activity of organic ligands. We have studied 114 angiotensin converting enzyme (ACE) inhibitors [32], 111 acetylcholinesterase (AChE) inhibitors [33], 163 ligands for benzodiazepine receptors (BZR) [33], 322 cyclooxygenase-2 (COX-2) inhibitors [33], 397 dihydrofolatereductase (DHFR) inhibitors [33], 66 glycogen phosphorylase b (GPB) inhibitors [34], 76 thermolysin (THER) inhibitors [34], and 88 thrombine (THR) inhibitors [35].

All data on these sets were taken from the supplementary materials to Sutherland's paper [33]. They included chemical structures, activity values, splitting into the training and test sets, ionization states and conformations for all molecules, their spatial alignment and partial charges on atoms. As indicated in [25], ionization states of molecules had been prepared by deprotonating carbocyclic acids and phosphates and protonating non-aryl basic amines (except NH_2 groups that coordinate Zn in the ACE set), energy-minimizing the aligned molecules with MMFF94S force field in Sybyl was used for determination of atomic coordinates, scaled MNDO ESP-fit

Table 13.2 QSAR DataSets

Ligand data set	Training set	Test set	Activity ranging
Angiotensin converting enzyme (ACE) inhibitors	76	38	pIC_{50} 2.1–9.9
Acetylcholinesterase (AchE) inhibitors	74	37	pIC_{50} 4.3–9.5
Ligands for benzodiazepine receptors (BZR)[a]	98	49	pIC_{50} 5.5–8.9
Cyclooxygenase-2 (COX-2) inhibitors[a]	188	94	pIC_{50} 4.0–9.0
Dihydrofolatereductase (DHFR) inhibitors[a]	237	124	pIC_{50} 3.3–9.8
Glycogen phosphorylase b (GPB) inhibitors	44	22	pKi 1.3–6.8
Thermolysin (THER) inhibitors	51	25	pKi 0.5–10.2
Thrombine (THR) inhibitors	59	29	pKi 4.4–8.5

[a] In the BZR, COX-2, and DHFR data sets several compounds (16, 40, 36, respectively) were excluded because they did not have the exact experimental values [33]

partial charges [36] had been calculated for all atoms with MOPAC 6.0, except that for the THER set all partial charges on atoms had been computed using the Gastei-ger-Marsili method [37] as also implemented in Sybyl. Several characteristics of benchmarking data sets are presented in Table 13.2.

In the course of building CMF models, the KRR regression method was applied to obtain the values of coefficients a_j and b, which are necessary for making predictions according to Eq. (13.7) as well as for visualizing the fields of regression coefficients using formula (13.10). The optimal values of attenuation factor α_f (which was kept the same for all types of molecular fields) and regularization coefficient γ were determined by minimizing the root-mean-square error in internal 10-fold cross-validation performed inside the training set. The same fixed value for all mixing coefficients, $h_f = 1$, was used.

Statistical characteristics of CMF models obtained for these data sets were compared with the same characteristics built for corresponding data sets using the common 3D-QSAR methods, CoMFA (Comparative Molecular Fields Analysis) [18] and CoMSIA (Comparative Molecular Similarity Index Analysis) [24], based on the use of molecular fields. Data on CoMFA and CoMSIA models were taken from Ref. [25].

Statistical parameters of CMF, CoMFA and CoMSIA models are shown in Table 13.3. They include the values of 4 statistical parameters: q^2 and $RMSE_{cv}$ characterizing internal predictive performance, R^2_p and $RMSE_p$—external predictive performance estimated using a single external validation set.

CoMSIA (in Ref. [25]—CoMSIA2) models are based on electrostatic and steric fields molecular fields and also involve contributions from the hydrophobic and two hydrogen-bonding molecular fields. All CMF models are based on the use of all afore-mentioned five types of molecular fields. All CoMFA and CoMSIA models were obtained by using a lattice with 2 Å spacing expanding at least 4 Å in each direction beyond aligned molecules. Only the most predictive CoMFA and CoMSIA models are included in the Table 13.3.

For comparing predictive ability of CMF models with those published in literature, all the data sets were split into the training and the test sets as specified

Table 13.3 Statistical parameters of 3D-QSAR CMF, CoMFA and CoMSIA (in Ref. [33]—CoMSIA2) models

	CMF			CoMFA[a]			CoMSIA [b]		
	q^2	R^2_p	$RMSE_p$	q^2	R^2_p	$RMSE_p$	q^2	R^2_p	$RMSE_p$
ACE	0.72	0.65	1.24	0.68	0.49	1.54	0.66	0.49	1.53
AChE	0.58	0.64	0.77	0.52	0.47	0.95	0.49	0.44	0.98
BZR	0.40	0.51	0.79	0.65	0.00	0.97	0.45	0.12	0.91
COX-2	0.57	0.14	1.23	0.49	0.29	1.24	0.57	0.37	1.17
DHFR	0.67	0.65	0.80	0.49	0.59	0.89	0.57	0.53	0.95
GPB	0.69	0.51	0.84	0.42	0.42	0.94	0.61	0.59	0.79
THER	0.60	0.31	1.86	0.52	0.54	1.59	0.51	0.53	1.60
THR	0.73	0.63	0.66	0.59	0.63	0.70	0.72	0.63	0.69

[a] PLS components: 3 (Ace, BZR), 4 (GPB, THER, THR), 5 (AChE, COX-2, DHFR)

[b] PLS components/*Additional fields:* 2/hydro (Ace), 3/hydro (BZR, THER), 4/hydro (GPB), 4/hydro + H-bonding (AChE, COX-2, DHFR, THR)

in Table 13.2. The training sets were used for building 3D-QSAR models and for assessing their internal predictive performance using the 10-fold cross-validation procedure. The test sets were used for assessing the external predictive performance of the models.

As it is clear from Table 13.3, models built for 7 data sets by using the CMF approach almost in all cases show better internal (cross-validation) predictive performance (i.e., higher q^2) then the corresponding models obtained by the CoMFA and CoMSIA methods. The q^2 values of the CMF models obtained for the COX-2 and BZR data sets are, respectively, equal and lower as compared to the corresponding CoMSIA models.

There is also a moderate advantage in external predictive performance (estimated on external test sets using the parameters R^2_p and $RMSE_p$) of the CMF models over the CoMFA, models for 5 data sets (ACE, AChE, BZR, DHFR, and GPB), and CoMSIA models for 4 data sets (ACE, AChE, BZR and DHFR).

One can notice that the performance of CMF is closer to that of the CoMSIA approach in comparison with CoMFA. This could be attributed to the fact that the mathematical form of Eq. (13.5) resembles expressions for similarity indices in CoMSIA. So, in spite of absolutely different underlying ideas, CoMSIA can formally be regarded as a discretized approximation of the current version of CMF, or, *vice versa*, CMF—as a continuous functional extension of CoMSIA. Therefore, the difference between the models produced by these methods might result from the effect of field discretization, different statistical procedure and parameterization of molecular fields.

Thus, the 3D-QSAR models obtained by CMF are comparable by the predictive ability with models built by means of such popular state-of-the-art approaches as CoMFA and CoMSIA. Moreover, in some cases, e.g. for data sets ACE, AChE, BZR and DHFR, the CMF approach is clearly advantageous.

Table 13.4 Cross-validated external predictive performance of 3D-QSAR CMF models

Parameters/Datasets	ACE	AChE	BZR	COX-2	DHFR	GPB	THER	THR
q^2_{ex}	0.67	0.54	0.27	0.41	0.67	0.59	0.40	0.71
$RMSE_{cvex}$	1.31	0.84	0.65	0.89	0.75	0.71	1.56	0.54

Parameters of external predictive performance of 3D-QSAR CMF models (q^2_{ex} and $RMSE_{cvex}$) estimated using external 5-fold cross-validation procedure are shown in the Table 13.4.

As follows from Table 13.4, in all cases the value q^2_{ex}, which characterizes the external predictive performance, is lower than the value q^2 computed using the internal cross-validation. This means that the use of only two adjustable hyper-parameters may cause the "model selection bias". Almost in all cases the value q^2_{ex} lies between R^2_p and q^2. It is interesting to note that q^2_{ex} for CMF models are usually higher than R^2_p for CoMFA and CoMSIA models. The predictive performance assessed using the external 5-fold cross-validation procedure is especially high for ACE, DHFR and THR.

Despite the important role played by the concept of applicability domain (AD) in construction and application of QSAR models [38–40], this issue is usually not considered for 3D-QSAR models. Possible reason for this is that the process of constructing such models is rather complex and involves several important stages: selection of a congeneric set of chemical compounds with the same putative mechanism of action, selection of "biologically active" conformation and alignment rule, etc. At some of these steps a separate rule for defining AD can be introduced. For example, specifications for the congeneric set of chemical compounds, the occurrence of a common template substructure or of a pharmacophore needed for aligning molecules and choosing their conformations, etc. The question arises, however, as to does it make sense to apply additional AD criteria to compounds that have already passed all the above filters. The following analysis suggests the affirmative answer.

We made the following computational experiment. The external 5-fold cross-validation procedure [39] was applied 20 times to the aforementioned thrombin dataset, each time after a random reshuffle of compounds in it. This produced 20 predicted values for each compound. The variance of these values was considered as a "distance to model" (DM) for it, as suggested in paper [39]. All compounds in the dataset were sorted in accordance with their DM. All compounds with DM below a certain threshold were considered to belong to AD. For each threshold value the mean absolute cross-validation error was computed for compounds within AD. The ratio of compounds within AD to the total number of compounds defines its coverage. A plot of the mean absolute cross-validation error *vs* coverage is given in Fig. 13.1, along with a smoothing line. Each point in it corresponds to a chemical compound from the dataset, while its coordinates correspond to the coverage and the mean absolute error, respectively, calculated for a subset of compounds with DM not exceeding the DM of this compound. Such subset belongs to the AD defined by the threshold value of DM equal to the DM of this compound. So, compounds

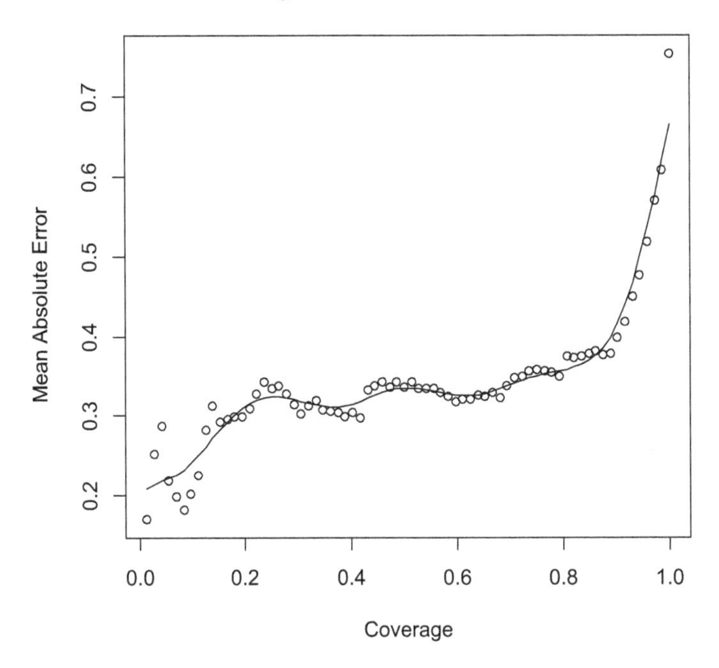

Fig. 13.1 Plot of the mean absolute cross-validation error (calculated for compounds outside AD) *vs* coverage for the thrombin dataset

with low coverage value lie in the regions of chemical space densely populated by compounds from the dataset, whereas those with high coverage values lie in non-populated areas rather far from other compounds. One can see that starting from the coverage 0.9 the mean absolute cross-validation error of compounds outside AD is considerably higher than that of compounds within AD. This means that in this case the DM corresponding to coverage 0.9 can be used as a threshold for defining the AD of CMF 3D-QSAR model.

13.3.2 Virtual Screening via Combination CMF with 1-SVM Technique

As it follows from our earlier publications, one-class classification (novelty detection) machine learning methods is a mathematical base of a new general approach to conducting ligand-based virtual screening of chemical compounds [41, 42]. Although several dozens of different algorithms for building one-class classification (novelty detection) models are known [43, 44], only those of them which are based on using kernels are suited for working with continuous molecular fields. The One-Class Support Vector Machines (1-SVM) [12] machine learning method is one of

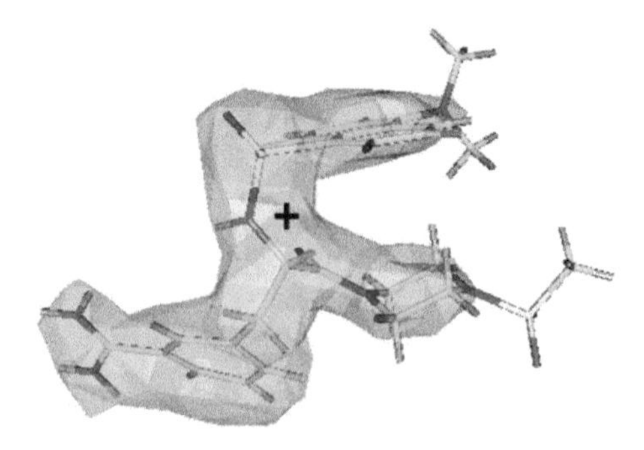

Fig. 13.2 The "ideal" steric molecular field corresponding to the vector perpendicular to separating hyperplane in the 1-SVM model built for thrombin inhibitors (2-amidinophenylalanines). Its isosurface can be viewed as a "negative image" of the binding site in biological target

them. It builds one-class classification models by seeking for hyperplane in infinite-dimensional functional Hilbert space (feature space) with maximum distance from the coordinate origin and separating a given proportion of training examples from it. In this case, the field of model coefficients is formed by leading cosines of the normal to this hyperplane. In the physical space they form description of "ideal" molecular fields (shapes), which are compared with molecular fields of test molecules. This "ideal" combination of fields reflects the structure of the corresponding binding site in biological macromolecule and can be viewed as a "negative image" of its molecular fields (see Fig. 13.2).

So, 1-SVM models in conjunction with molecular field kernels perform ligand-based virtual screening based on similarity of molecular fields (shapes). In comparison with other shape-based similarity search methods, they are more flexible, because one can find the optimal degree of generalization (simplification). See Fig. 13.3, in which the level of generalization of model coefficient field description increases from the lower to the upper row. This leads to high performance in ligand-based virtual screening [13].

In the following case study [13], we have assessed the performance of a new promising method for virtual screening of organic compounds based on a combination of the CMF methodology with the one-class SVM method (1-SVM). The first step of constructing the model was spatial alignment of the structures of organic ligands. In this work, the alignment was performed with the SEAL algorithm [45] implemented by us in the framework of the software for CMF modeling. Then, kernel values were calculated, and the model was constructed using the LibSVM program [46].

In the one-class classification method, only active structures are used. Sequentially excluding one structure at a time and constructing the model based on the remaining structures, one can predict the activity of all active compounds. However, for assessing the statistical characteristics of classification models, it is necessary to determine not only the number of active compounds predicted to be active (true positive, TP) and the number of active compounds predicted to be inactive (false

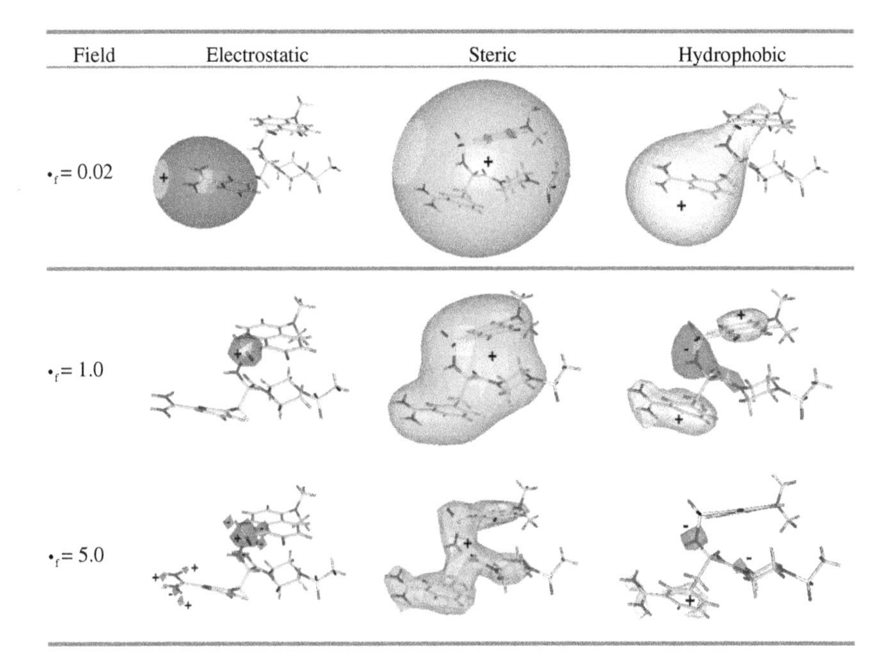

Fig. 13.3 Isosurfaces for the fields of 1-SVM model coefficients for thrombin inhibitors (2-amidinophenylalanines)

negative, FN) but also the number of inactive compounds predicted to be inactive (true negative, TN) and the number of inactive compounds predicted to be active (false positive, FP). To determine the last two characteristics, we used structures resembling the structures of active ligands in their physicochemical properties but presumed to be inactive (so-called decoys).

We have built one-class models for data taken from the DUD database [47], which contains the structures of active ligands for different biological targets, as well as the structures of corresponding decoys. It is worth noting that the latter were used only for assessing the statistical characteristics of classification models and were not involved in their construction. In particular, decoys were used for determining the TN and FP models constructed with the use of active compounds.

The suggested one-class classifier calculates a continuous quantity (a classifier function), for which the threshold value is determined. If the classifier function calculated for a certain ligand exceeds the threshold value, the compound is considered active; otherwise, the structure is discarded from further consideration.

Dependence of the FN, FP, TN, and TP on the threshold value is clearly reflected by a Receiver Operator Characteristic (ROC) curve [48] in the TPR–FPR coordinates (true positive rate versus false positive rate), where TRP=TP/(TP+TN) and FPR=FP/(FP+TN). The larger the area under the curve (AUC), the higher is the classifier efficiency.

Table 13.5 Parameters and AUC of the models for HIV reverse transcriptase inhibitors obtained by the 1-SVM method with the use of kernels in the framework of the CMF

Molecular field	ν	α_{el}	α_{st}	α_{hyd}	AUC
Electrostatic	0.082	0.031	–	–	0.60
Steric	0.002	–	0.010	–	0.75
Hydrophobic	0.466	–	–	0.009	0.65

Table 13.6 Parameters and AUC of the models for trypsin inhibitors obtained by the 1-SVM method with the use of kernels in the framework of the CMF

Molecular Field	ν	h_{el}	α_{el}	h_{st}	α_{st}	h_{hyd}	α_{hyd}	AUC
Electrostatic	0.53	–	0.30	–	–	–	–	0.91
Steric	0.47	–	–	–	0.00	–	–	0.87
Hydrophobic	0.66	–	–	–	–	–	0.30	0.86
Linear combination of the fields	0.45	0.32	0.30	0.58	0.04	0.11	0.15	0.94

Constructing the classification model necessitates maximizing the AUC value by optimizing the 1-SVM parameter ν and parameter α_f of the CMF kernel from Eq. (13.5). We have studied both the individual electrostatic, steric, and hydrophobic kernels and their linear combinations. For individual kernels, two parameters have been optimized: ν and the α_f parameter corresponding to a given type of molecular field. At the first step, the optimization algorithm was launched ten times, each time starting from a set of random parameter values in the ranges $\nu \in [0.01; 0.80]$, $h_f \in [0.0001;0.3000]$, and $\alpha_f \in [0.001; 1.000]$. At the second step, the Nelder–Mead algorithm was used for refining the optimal parameters; the set of the best-fit parameters obtained at the first step of optimization were used as the initial approximation.

Tables 13.5 and 13.6 summarize the results of building 1-SVM models on the basis of continuous molecular fields for HIV reverse transcriptase (HIVRT) and trypsin inhibitors. As follows from Table 13.5, the best performance for HIVRT is obtained by the model constructed using the steric kernel and resulting in an AUC value of 0.75. For this target, the use of a linear combination of several kernels does not improve the AUC value. At the same time, for trypsin inhibitors, rather high AUC values (0.86–0.91) were obtained on the basis of individual models constructed with the use of all three kernels, which is likely due to their mutual correlation. However, for this target, the use of a linear combination of all kernels increases the AUC value up to 0.94.

Our results demonstrate the effectiveness of the suggested methodology. Application of 1-SVM models to virtual screening can be viewed as a special kind of similarity search, in which similarity is considered with respect not to individual actives but to the whole dataset of active compounds. Being an alternative to the traditional similarity search for active compounds using of the Tanimoto coefficient and to the binary classification methods, this approach possesses unique properties. In contrast to the binary classification methods, the described method is not sensitive to the choice of counterexamples. Moreover, with default values of parameters, this method does not require counterexamples at all and therefore, in distinction from traditional SVM, can be used in the cases (typical for drug discovery) when counterexamples are not known or the dataset is highly imbalanced. As distinct from the traditional similarity search, the suggested method can adapt to complex structure-activity landscapes and, thus, makes it possible to avoid activity cliffs [49]. In addition, as compared to similarity search methods based on the use of fragmental descriptors, the suggested approach implies using the same model to the sets of compounds belonging to different structural classes.

13.4 The Main Directions of Further Development of the CMF Approach

In this article, we have considered a particular implementation of the CMF approach aimed at building 3D-QSAR models. This implementation is being actively developed, and its future version will surely be better than the current one. This section, however, concerns more global, strategic directions of further development of the whole CMF approach.

13.4.1 Introduction of Additional Types of Molecular Fields, Integration with Quantum Chemistry

The CMF approach is not confined to the simplest approximation scheme introduced by Eq. (13.5). Any number of Gaussian functions, (both isotropic, i.e. spherically symmetrical, and non-isotropic) as well as any other set of basic functions (such as splines, wavelets, etc.) can be used for approximating continuous molecular fields. This provides the ability to work with complex types of molecular fields, including those derived from the electron density function.

Close integration with quantum chemistry is a promising direction of further development of the CMF approach. Indeed, wave and electron density functions are the most natural, comprehensive and accurate way to describe molecules. Moreover, all possible continuous molecular fields can be generated from the electron density functions with the help of integral transforms [50]. Especially promising is the use of conceptual DFT molecular fields [51] based on conceptual DFT [52]. Molecular fields (steric, electrostatic, local softness and LUMO) derived from the electron density function have already been successfully used for building 3D-QSAR models in the field of metal complex catalysis using the traditional grid-based CoMFA approach [53, 54]. We can expect that due to the ability to approximate all kinds of molecular fields with any degree of accuracy the CMF approach will be very useful in the development of new catalysts and supramolecular complexes, which require high accuracy in representing the electronic structure of molecules.

13.4.2 The Issue of Molecular Alignment

All methods of molecular alignment useful for building traditional lattice-based 3D-QSAR models can also be applied in the framework of the CMF approach. Meanwhile, thanks to the integrability of continuous functions describing molecular fields, the latter approach offers additional possibilities.

Molecular alignment exactly corresponds to the "curve registration" procedure, which is the first necessary step in any functional data analysis [30]. In the case of biological activity caused by protein–ligand interactions, mutual orientation of different ligands inside binding sites defines their "natural alignment". Such alignment can be checked by means of independent physical experiments (X-ray diffraction, NMR, etc.), and hence its explicit consideration is equivalent to the inclusion of "external domain knowledge", which is always preferred in machine learning [4]. Therefore, the choice of the strategy for molecular alignment should be governed by the necessity to mimic mutual orientation of molecules in the underlying physical processes. It should however be pointed out that this does not mean that the "natural alignment" should always provide the strongest 3D-QSAR models.

In the case when the exact information concerning the structure of binding pockets of biological macromolecules is not available, the CMF approach provides a consistent criterion for the pairwise alignment of molecules i and j: maximization of the kernel $K(M_i, M_j)$. Indeed, its form appears to be closely related to the function used in the SEAL program for aligning molecules [45]. So, the CMF approach offers the possibility to use the same function both for aligning molecules and building structure–activity models. This could lead to more close integration of molecular alignment into the process of chemical data analysis. Moreover, the CMF approach provides an additional criterion for the multiple alignment of molecules—the "compressibility" of molecular fields, which can be assessed using unsupervised dimensionality reduction approaches, such as the kernel (functional) principal component analysis. Both criteria can also be applied to choose molecular conformations for tackling the problem of molecular flexibility.

The issue of alignment-free approaches deserves special attention. Sometimes the necessity to perform alignment of molecules in several grid-based 3D-QSAR methods, such as CoMFA and CoMSIA, is considered as limitation of such approaches, which should be avoided [55]. This has led to the development of alignment-free approaches, such as those based on autocorrelation vectors [56], molecular moments (CoMMA) [57], 3D WHIM [58], EVA [55], GRIND [59], FLAP [60], VolSurf [61] descriptors, etc. The CMF approach provides a new kind of solutions based on the use of 3D-rotation invariant kernels [62, 63] in the frame of the concept of invariant pattern recognition [64]. The feasibility of this approach results from the ability to apply the required integral transforms to continuous functions describing molecular fields.

13.4.3 Taking into Account Molecular Flexibility

A universal way of tackling the problem of molecular flexibility was suggested in paper [65] for kernel-based methods. It consists in averaging kernels over all conformations for each molecule. This approach is however computationally feasible only for very small number of conformations per molecule. Indeed, consideration of only 20 conformations for each molecule results in the necessity to consider 400 pairs of conformations in order to fill each cell in kernel matrix. In addition to huge computational burden, this approach does not solve the problem of alignment and therefore applicable only for alignment-free approaches.

The CMF approach can offer an alternative solution to this problem. Instead of using discrete sets of "representative" conformations, one can consider for each molecule an infinite number of conformations organized into a continuous manifold, so-called "conformational space". This provides the ability to apply functional data analysis not only to molecular fields but also to molecular geometry in a consistent way. Such "conformational space" can be described by means of some probability density function (*pdf*) in 3N-dimensional Euclidean space, where N is the number of atoms in the molecule under study. Having applied several approximations from the arsenal of statistical physics, one can obtain the following expression for calculating atomic kernels instead of Eq. (13.6):

$$k_f(A_{il}, A_{jm}) = w_{fil} w_{fjm} \sqrt{\frac{\pi^3}{(\alpha_f)^3}} \int_{\Re^3} \int_{\Re^3} \exp\left(-\frac{\alpha_f}{4} \left\| \mathbf{r}_{il} - \mathbf{r}_{jm} \right\|^2 \right) p(\mathbf{r}_{il}) p(\mathbf{r}_{jm}) d\mathbf{r}_{il} d\mathbf{r}_{jm} \tag{13.11}$$

where $p(\mathbf{r}_{il})$ is the one-particle *pdf* for atom l in molecule i. In order to compute all necessary *pdf*, one should perform molecular dynamics or Monte-Carlo studies for all molecules, collect conformations along trajectories, align them (e.g., using the common template), and apply the GMM algorithm [66] to approximate *pdf* for each atom as a mixture of several Gaussian functions. In this case the integral in Eq. (13.11) contains the product of Gaussian functions and hence can be computed analytically. Therefore, due to easy integrability of continuous molecular field functions, it is possible to build models taking into account the whole "conformational spaces". For modeling receptor–ligand interactions, it is important to perform molecular dynamics simulations of ligands either inside the binding pockets or using their simplified surrogates. In the simplest case this amounts to choosing a single "biologically active" conformation from the results of molecular docking.

13.4.4 Prediction of Physico-Chemical Properties

It is expected that the CMF approach will be extended to predict physico-chemical properties of chemical compounds and their supramolecular complexes. The theoretical possibility of this follows from the following analysis.

Due to the ability to apply methods of functional analysis, continuous molecular fields can be tailored for solving many different tasks in chemoinformatics. Consider, for example, prediction of physico-chemical properties in diverse datasets. In this case, "natural alignment" corresponds to the uniform probability of molecules to adopt any possible mutual orientation. Therefore, kernel $K_f(M_i, M_j)$ describing the similarity between the molecular fields of the *f*th type for the *i*th and *j*th molecules can be computed by averaging over all possible mutual orientations:

$$K_f(M_i, M_j) = \sum_l \sum_m \iiint k_f(A_{il}, A_{jm}) d^3\mathbf{r}_{jm} = \frac{8\pi^3}{(\alpha_f)^3} \sum_l \sum_m w_{fil} w_{fjm} \quad (13.12)$$

The resulting kernel does not depend on molecular geometry, although it includes all constants describing continuous molecular fields. This kernel can be considered as a particular case of the molecular convolution kernels for which the ability to model additive physicochemical properties was shown by us earlier [67]. Extension to the case of modeling metal complexation would require integration over only two angles θ and φ of spherical coordinate system, whereas for modeling cyclodextrine complexation it is sufficient to integrate over the single angle φ of the cylindric coordinate system, as follows from consideration of the "natural alignment" in each of these systems. One can show that in these cases all necessary integrals can be taken in analytical form using Bessel functions. Such universality and flexibility results from integrability of continuous molecular fields. This opens a direct way to extending the CMF approach to predicting physico-chemical properties of chemical compounds.

13.4.5 Taking into Account Different Ionization States, Tautomers and Conformers

The CMF approach can be extended to the case of the existence of several tautomers, ionization (protonation) states and conformers by replacing the Eq. (13.5) with its more general form:

$$\rho_{fil}(\mathbf{r}) = \sum_s v_s \sum_t v_{st} \sum_c v_{stc} w_{fil} \exp\left(-\frac{1}{2} \alpha_f \left\| \mathbf{r} - \mathbf{r}_{il} \right\|^2 \right), \quad (13.13)$$

where index s counts different ionization (protonation) states, index t counts tautomers, index c counts conformers, v_s is the population of the ionization state s, v_{st} is the relative population of tautomer t in ionization state s, while v_{stc} is the relative population of the conformer c of the molecule in ionization state s and in tautomeric state t. Populations v_s, v_{st} and v_{stc} can be assessed using molecular modeling simulations.

13.4.6 Further Extension of the Approach to Encompass Biological Macromolecules and Their Interactions with Both Macro- and Small Molecules

One of the most promising ways to develop further the CMF approach is its further extension to the description of the binding sites of biomolecules and their interactions with ligands. This seems to be feasible, because molecular fields of biological targets are identical by their nature to those of small ligands. Because of this, many of the approaches and methods originally developed for working with small molecules can be transferred to biological macromolecules.

One can suggest several ways of conducting research in this direction. First, kernels for comparing molecular fields of biological targets (or their binding sites) could be constructed in exactly the same way as it has been done for small molecules and described in this paper. By combining them with various kernel machine learning methods, various regression, classification, novelty detection, ranking and dimensionality reductions tasks could be formulated and solved for biological macromolecules and their binding sites. This could lead to promising applications in biology-related sciences.

Second, kernels for protein-ligand pairs can be constructed by combining kernels for small molecules (ligands) and kernels for macromolecules (proteins), as it was done by Erhan et al. [68], Faulon et al. [69], Jacob and Vert [70], and Bajorath et al. [71]. The main advantage of using continuous molecular fields in this case is that this approach can be applied consistently to construct kernels for both small organic and big biological macromolecules using the same typed of molecular fields. Thanks to this, the same data analysis methods could be used to describe also protein–peptide, protein-protein, protein-DNA, protein-RNA interactions, as well as properties of peptides and proteins with non-standard residues.

Third, kernels for protein-ligand interactions can be constructed in the frame of the CMF approach by encapsulating products of molecular fields of ligand and protein into kernels. Such combined kernels could easily be used in conjunction with various kernel machine learning methods to solve different tasks relating to protein-ligand interactions.

It is expected that the results obtained in this direction might be useful in the field of chemogenomics for target profiling, in bioinformatics and proteomics for classifying and annotating protein macromolecules and their binding sites by considering and comparing their molecular fields, and in system biology for predicting interaction graphs for biomolecules, and in drug design.

13.4.7 Integration of Approaches to Prevent the Model Selection Bias

As it has already been discussed, one of the drawbacks of the CMF approach in its present state is the danger of over-fitting because of the model selection bias [72].

Table 13.7 The use of CMF in conjunction with different kernel-based machine learning methods

Machine Learning Task	Machine Learning Methods	Role in Chemoinformatics
Regression	Support Vector Regression (SVR) [5], Kernel Ridge Regression (KRR) [10], Kernel Partial Least Squares (KPLS) [28], Gaussian Processes for Regression (GP-R) [11]	QSAR/QSPR
One-class classification (novelty detection)	One-Class Support Vector Machine (1-SVM) [12], Support Vector Data Description (SVDD) [76]	Virtual screening based of similarity of molecular fields
Binary and multi-class classification	Support Vector Machines (SVM) [9], Gaussian Processes for Classification (GP-C) [11]	Classification of chemical compounds (active/inactive), predicting profiles of biological activity for chemical compounds
Dimensionality reduction and data visualization	Kernel Principal Component Analysis (KPCA) [6], Kernel Feature Analysis (KFA) [77]	Drawing maps of chemical space
Cluster analysis	Kernel k-means [78]	Classification of chemical compounds by mechanism of action (including binding mode)
Canonical correlation	Kernel Canonical Correlation Analysis (KCCA) [79]	Relationships between molecular fields of ligands and molecular fields of their binding sites

Several ways to prevent this phenomenon have been suggested in literature, including Bayesian regularization of hyper-parameters [73] and hyper-parameter averaging [74]. Algorithms of multiple kernel learning [75] might also be useful in this case. For small datasets the preferred solution would be to apply the full Bayesian approach [66], where the hyper-parameters are integrated out rather than optimized. Integration of approaches to prevent the model selection bias is one of the most important tasks for further development of the CMF approach.

13.4.8 Combining with Different Machine Learning Methods

The CMF approach is easily extensible thanks to its modularity. By combining different types of molecular fields, different types of kernels with different types of kernel-based machine learning methods, one can obtain various methods for building SAR/QSAR/QSPR models and conducting virtual screening. Although some of such methods may be similar to the existing ones, nonetheless it is likely that some of them will be fundamentally novel approaches. Table 13.7 lists different tasks being solved by kernel-based machine learning methods, the names of such methods, and the roles that the CMF approach could play in conjunction with them in chemoinformatics. The first row in this table deals with the regression task considered in this paper. The second row concerns the use of molecular kernels in combination with one-class classification kernel-based methods for conducting virtual screening based on the similarity of molecular fields. The feasibility of this approach has already been proved by us, see preliminary communication [13]. The rest of the table shows the promising directions for further development of the CMF approach.

13.5 Conclusion

The CMF approach describes molecules by ensemble of continuous functions (molecular fields), instead of finite sets of molecular descriptors (such as interaction energies computed at grid nodes). The potential advantages of this approach results from the ability to approximate electronic molecular structures with any desirable accuracy level, the ability to leverage the valuable information contained in partial derivatives of molecular fields (otherwise lost upon discretization) to analyze models and enhance their predictive performance, the ability to apply integral transforms to molecular fields and models, etc.

The most attractive features of the CMF approach are its versatility and universality. By combining different types of molecular fields and methods of their approximation, different types of kernels with different types of kernel-based machine learning methods, it is possible to present lots of existing methods in chemoinformatics and medicinal chemistry as particular cases within a universal methodology. The CMF methodology can easily be extended to building classification and novelty detection models, visualizing them, performing virtual screening, processing diverse datasets.

We see the following main directions for further development of the CMF approach: introduction of additional types of molecular fields, including conceptual DFT molecular fields; tackling the issue of molecular alignment and flexibility; taking into account molecular flexibility; prediction of physico-chemical properties of chemical compounds; taking into account different ionization states of molecules, their tautomers and conformers; extension of the approach to work with biological macromolecules and supramolecular complexes; integration of special approaches for preventing model selection bias; combining with different machine learning methods aimed at solving various tasks.

The CMF approach is implemented as a set of scripts operating under the R environment for statistical computing and graphics [80]. A version of the software is available at http://sites.google.com/site/conmolfields/.

Acknowledgments The authors thank Prof. Yu.A.Ustynyuk for stimulating discussion and advice. The authors also thank Prof. A.Varnek and Dr. G.Marcou for valuable comments regarding the developed approach. This work was supported by Russian Foundation for Basic Research (Grant 13-07-00511).

References

1. Varnek A, Baskin II (2011) Chemoinformatics as a theoretical chemistry discipline. Mol Inf 30(1):20–32. doi:10.1002/minf.201000100
2. Gasteiger J, Engel T (2003) Chemoinformatics: a textbook. Wiley-VCH, Weinheim
3. Gasteiger J (2003) Handbook of chemoinformatics: from data to knowledge. Wiley-VCH, Weinheim
4. Varnek A, Baskin I (2012) Machine learning methods for property prediction in chemoinformatics: quo vadis? J Chem Inf Mod 52(6):1413–1437. doi:10.1021/ci200409x
5. Vapnik V (1998) Statistical learning theory. Wiley-Interscience, New York
6. Schölkopf B, Smola AJ (2002) Learning with kernels: support vector machines, regularization, optimization, and beyond. MIT, Cambridge
7. Zhokhova NI, Baskin II, Bakhronov DK, Palyulin VA, Zefirov NS (2009) Method of continuous molecular fields in the search for quantitative structure-activity relationships. Dokl Chem 429(1):273–276
8. Baskin II, Zhokhova NI (2013) The continuous molecular fields approach to building 3D-QSAR models. J Comput-Aided Mol Des 27(5):427–442. doi:10.1007/s10822-013-9656-4
9. Cortes C, Vapnik V (1995) Support-vector networks. Mach Learn 20(3):273–297. doi:10.1007/bf00994018
10. Saunders C, Gammerman A, Vovk V (1998) Ridge regression learning algorithm in dual variables. In: proceedings of the Fifteenth International Conference on Machine Learning (ICML-98). Morgan Kaufmann, Burlington, pp 515–521
11. Rasmussen CE, Williams CKI (2006) Gaussian processes in machine learning. Adaptive computation and machine learning. MIT, Cambridge
12. Schölkopf B, Platt JC, Shawe-Taylor J, Smola AJ, Williamson RC (2001) Estimating the support of a high-dimensional distribution. Neural Comput 13(7):1443–1471
13. Karpov PV, Baskin II, Zhokhova NI, Zefirov NS (2011) Method of continuous molecular fields in the one-class classification task. Dokl Chem 440(2):263–265
14. Karpov PV, Baskin II, Zhokhova NI, Nawrozkij MB, Zefirov AN, Yablokov AS, Novakov IA, Zefirov NS (2011) One-class approach: models for virtual screening of non-nucleoside HIV-1 reverse transcriptase inhibitors based on the concept of continuous molecular fields. Russ Chem Bull 60(11):2418–2424. doi:10.1007/s11172-011-0372-8
15. Kubinyi H (ed) (2000) 3D QSAR in drug design. Volume 1: theory methods and applications (Three-dimensional quantitative structure activity relationships). Kluwer/Escom, Dordrecht
16. Kubinyi H, Folkers G, Martin YC (eds) (2002a) 3D QSAR in drug design. Volume 2: ligand-protein Interactions and Molecular Similarity. Kluwer Academic Publishers, Dordrecht
17. Kubinyi H, Folkers G, Martin YC (eds) (2002b) 3D QSAR in drug design. Volume 3: Recent advances. Kluwer Academic Publishers, Dordrecht
18. Cramer RD, Patterson DE, Bunce JD (1988) Comparative molecular field analysis (CoMFA). 1. Effect of shape on binding of steroids to carrier proteins. J Am Chem Soc 110(18):5959–5967. doi:10.1021/ja00226a005
19. Testa B, Carrupt PA, Gaillard P, Billois F, Weber P (1996) Lipophilicity in molecular modeling. Pharm Res 13(3):335–343. doi:10.1023/a:1016024005429
20. Kim KH, Greco G, Novellino E, Silipo C, Vittoria A (1993) Use of the hydrogen bond potential function in a comparative molecular field analysis (CoMFA) on a set of benzodiazepines. J Comput-Aided Mol Des 7(3):263–280
21. Waller CL, Marshall GR (1993) Three-dimensional quantitative structure-activity relationship of angiotesin-converting enzyme and thermolysin inhibitors. II. A comparison of CoMFA models incorporating molecular orbital fields and desolvation free energies based on active-analog and complementary-receptor-field alignment rules. J Med Chem 36(16):2390–2403
22. Kellogg GE (1996) E-state fields: applications to 3D QSAR. J Comput-Aided Mol Des 10(6):513–520

23. Kroemer RT, Hecht P (1995) Replacement of steric 6-12 potential–derived interaction energies by atom-based indicator variables in CoMFA leads to models of higher consistency. J Comput-Aided Mol Des 9(3):205–212
24. Klebe G, Abraham U (1999) Comparative molecular similarity index analysis (CoMSIA) to study hydrogen-bonding properties and to score combinatorial libraries. J Comput-Aided Mol Des 13(1):1–10
25. Goodford P (2006) The basic principles of GRID. In: Cruciani G (ed) Molecular interaction fields. Applications in drug discovery and ADME prediction. Methods and principles in medicinal chemistry, vol 27. Wiley-VCH, Weinheim, pp 3–26
26. Höskuldsson A (1988) PLS regression methods. J Chemom 2(3):211–228
27. Fradera X, Amat L, Besalu E, Carbo-Dorca R (1997) Application of molecular quantum similarity to QSAR. Quant Struct-Act Rel 16(1):25–32
28. Rosipal R, Trejo LJ (2002) Kernel partial least squares regression in reproducing Kernel Hilbert Space. J Mach Learn Res 2(2):97–123. doi:10.1162/15324430260185556
29. Baskin II, Tikhonova IG, Palyulin VA, Zefirov NS (2003) Selectivity fields: comparative molecular field analysis (CoMFA) of the glycine/NMDA and AMPA receptors. J Med Chem 46(19):4063–4069
30. Ramsay JO, Silverman BW (2005) Functional data analysis. Springer series in statistics, 2nd edn. Springer, New York
31. Baskin II, Kireeva N, Varnek A (2010) The One-class classification approach to data description and to models applicability domain. Mol Inf 29(8–9):581–587. doi:10.1002/minf.201000063
32. DePriest SA, Mayer D, Naylor CB, Marshall GR (1993) 3D-QSAR of angiotensin-converting enzyme and thermolysin inhibitors: a comparison of CoMFA models based on deduced and experimentally determined active site geometries. J Am Chem Soc 115(13):5372–5384. doi:10.1021/ja00066a004
33. Sutherland JJ, O'Brien LA, Weaver DF (2004) A comparison of methods for modeling quantitative structure-activity relationships. J Med Chem 47(22):5541–5554
34. Gohlke H, Klebe G (2002) DrugScore meets CoMFA: adaptation of fields for molecular comparison (AFMoC) or how to tailor knowledge-based pair-potentials to a particular protein. J Med Chem 45(19):4153–4170. doi:10.1021/jm020808p
35. Böhm M, StüÑrzebecher J, Klebe G (1999) Three-Dimensional quantitative structure-activity relationship analyses using comparative molecular field analysis and comparative molecular similarity indices analysis to elucidate selectivity differences of inhibitors binding to trypsin, thrombin, and factor Xa. J Med Chem 42(3):458–477. doi:10.1021/jm981062r
36. Besler BH, Merz KM, Kollman PA (1990) Atomic charges derived from semiempirical methods. J Comp Chem 11(4):431–439. doi:10.1002/jcc.540110404
37. Gasteiger J, Marsili M (1980) Iterative partial equalization of orbital electronegativity-a rapid access to atomic charges. Tetrahedron 36(22):3219–3228
38. Jaworska J, Nikolova-Jeliazkova N, Aldenberg T (2005) QSAR applicability domain estimation by projection of the training set in descriptor space: a review. Altern Lab Anim 33(5):445–459
39. Tetko IV, Sushko I, Pandey AK, Zhu H, Tropsha A, Papa E, Oberg T, Todeschini R, Fourches D, Varnek A (2008) Critical assessment of QSAR models of environmental toxicity against tetrahymena pyriformis: focusing on applicability domain and overfitting by variable selection. J Chem Inf Model 48(9):1733–1746. doi:10.1021/ci800151m
40. Sushko I, Novotarskyi S, Korner R, Pandey AK, Cherkasov A, Li J, Gramatica P, Hansen K, Schroeter T, Muller KR, Xi L, Liu H, Yao X, Oberg T, Hormozdiari F, Dao P, Sahinalp C, Todeschini R, Polishchuk P, Artemenko A, Kuz'min V, Martin TM, Young DM, Fourches D, Muratov E, Tropsha A, Baskin I, Horvath D, Marcou G, Muller C, Varnek A, Prokopenko VV, Tetko IV (2010) Applicability domains for classification problems: Benchmarking of distance to models for Ames mutagenicity set. J Chem Inf Model 50(12):2094–2111. doi:10.1021/ci100253r

41. Karpov PV, Baskin II, Palyulin VA, Zefirov NS (2011a) Virtual screening based on one-class classification. Dokl Chem 437(2):107–111
42. Karpov PV, Osolodkin DI, Baskin II, Palyulin VA, Zefirov NS (2011b) One-class classification as a novel method of ligand-based virtual screening: the case of glycogen synthase kinase 3ĐĐ inhibitors. Bioorg Med Chem Lett 21(22):6728–6731
43. Markou M, Singh S (2003a) Novelty detection: a review—part 1: statistical approaches. Signal Process 83(12):2481–2497
44. Markou M, Singh S (2003b) Novelty detection: A review—part 2: neural network based approaches. Signal Process 83(12):2499–2521
45. Kearsley SK, Smith GM (1990) An alternative method for the alignment of molecular structures: maximizing electrostatic and steric overlap. Tetrahedron Comput Methodol 3(6 PART C):615–633
46. Chang C-C, Lin C-J (2001) LIBSVM: a library for support vector machines. ACM Trans Intel Syst Technol 2(3):27:21–27:27
47. Huang N, Shoichet BK, Irwin JJ (2006) Benchmarking sets for molecular docking. J Med Chem 49(23):6789–6801
48. Fawcett T (2006) An introduction to ROC analysis. Pattern Recogn Lett 27(8):861–874
49. Maggiora GM (2006) On outliers and activity cliffs why QSAR often disappoints. J Chem Inf Mod 46(4):1535–1535. doi:10.1021/ci060117s
50. Carbo-Dorca R, Besalu E (2006) Generation of molecular fields, quantum similarity measures and related questions. J Math Chem 39(3–4):495–510. doi:10.1007/s10910-005-9046-9
51. Van Damme S, Bultinck P (2009) 3D QSAR based on conceptual DFT molecular fields: antituberculotic activity. J Mol Struct—THEOCHEM 943 (1–3):83–89. doi:10.1016/j.theochem.2009.10.031
52. Geerlings P, De Proft F, Langenaeker W (2003) Conceptual density functional theory. Chem Rev 103(5):1793–1874. doi:10.1021/cr990029p
53. Cruz V, Ramos J, Munoz-Escalona A, Lafuente P, Pena B, Martinez-Salazar J (2004) 3D-QSAR analysis of metallocene-based catalysts used in ethylene polymerisation. Polymer 45(6):2061–2072. doi:10.1016/j.polymer.2003.12.059
54. Cruz VL, Ramos J, Martinez S, Munoz-Escalona A, Martinez-Salazar J (2005) Structure–activity relationship study of the metallocene catalyst activity in ethylene polymerization. Organometallics 24(21):5095–5102. doi:10.1021/om050458f
55. Heritage TW, Ferguson AM, Turner DB, Willett P (1998) EVA: a novel theoretical descriptor for QSAR studies. In: Kubinyi H, Folkers G, Martin YC (eds) 3D QSAR in drug design. Ligand-protein complexes and molecular similarity, vol 2. Kluwer Academic Publishers, London, pp 381–398
56. Wagener M, Sadowski J, Gasteiger J (1995) Autocorrelation of molecular surface properties for modeling corticosteroid binding globulin and cytosolic Ah receptor activity by neural networks. J Am Chem Soc 117(29):7769–7775. doi:10.1021/ja00134a023
57. Silverman BD, Platt DE (1996) Comparative molecular moment analysis (CoMMA): 3D-QSAR without molecular superposition. J Med Chem 39(11):2129–2140. doi:10.1021/jm950589q
58. Todeschini R, Gramatica P (1998) New 3D molecular descriptors: the WHIM theory and QSAR applications. In: Kubinyi H, Folkers G, Martin YC (eds) 3D QSAR in drug design. Ligand–protein complexes and molecular similarity, vol 2. Kluwer Academic Publishers, London, pp 355–380
59. Pastor M, Cruciani G, McLay I, Pickett S, Clementi S (2000) GRid-INdependent descriptors (GRIND): a novel class of alignment-independent three-dimensional molecular descriptors. J Med Chem 43(17):3233–3243. doi:jm000941m
60. Baroni M, Cruciani G, Sciabola S, Perruccio F, Mason JS (2007) A common reference framework for analyzing/comparing proteins and ligands. Fingerprints for Ligands and Proteins (FLAP): theory and application. J Chem Inf Mod 47(2):279–294
61. Cruciani G, Pastor M, Guba W (2000) VolSurf: a new tool for the pharmacokinetic optimization of lead compounds. Eur J Pharm Sci 11(Suppl. 2):S29–S39. doi:S0928098700001627

62. Hamsici OC, Martinez AM (2009) Rotation invariant kernels and their application to shape analysis. IEEE Trans Pattern Anal 31(11):1985–1999. doi:10.1109/tpami.2008.234
63. Haasdonk B, Burkhardt H (2007) Invariant kernel functions for pattern analysis and machine learning. Mach Learn 68(1):35–61. doi:10.1007/s10994-007-5009-7
64. Wood J (1996) Invariant pattern recognition: A review. Pattern Recogn 29(1):1–17. doi:10.1016/0031-3203(95)00069-0
65. Azencott CA, Ksikes A, Swamidass SJ, Chen JH, Ralaivola L, Baldi P (2007) One- to four-dimensional kernels for virtual screening and the prediction of physical, chemical, and biological properties. J Chem Inf Mod 47(3):965–974
66. Bishop CM (2006) Pattern ecognition and machine learning. Information science and statistics. Springer, New York
67. Baskin II, Zhokhova NI, Palyulin VA, Zefirov NS (2008) Additive inductive learning in QSAR/QSPR studies and molecular modeling. In: 4th German conference on chemoinformatics, November 9–11, 2008, Goslar, Germany, p 78
68. Erhan D, L'Heureux P-J, Yue SY, Bengio Y (2006) Collaborative filtering on a family of biological targets. J Chem Inf Model 46(2):626–635
69. Faulon J-L, Misra M, Martin S, Sale K, Sapra R (2008) Genome scale enzyme-metabolite and drug-target interaction predictions using the signature molecular descriptor. Bioinformatics 24(2):225–233. doi:10.1093/bioinformatics/btm580
70. Jacob L, Vert JP (2008) Protein-ligand interaction prediction: an improved chemogenomics approach. Bioinformatics 24(19):2149–2156
71. Geppert H, Humrich J, Stumpfe D, Gaertner T, Bajorath J (2009) Ligand prediction from protein sequence and small molecule information using support vector machines and fingerprint descriptors. J Chem Inf Mod 49(4):767–779. doi:10.1021/ci900004a
72. Cawley GC, Talbot NLC (2010) On over-fitting in model selection and subsequent selection bias in performance evaluation. J Mach Learn Res 11:2079–2107
73. Cawley GC, Talbot NLC (2007) Preventing over-fitting during model selection via bayesian regularisation of the hyper-parameters. J Mach Learn Res 8:841–861
74. Hall P, Robinson AP (2009) Reducing variability of crossvalidation for smoothing-parameter choice. Biometrika 96(1):175–186. doi:10.1093/biomet/asn068
75. Gönen M, Alpaydin E (2011) Multiple kernel learning algorithms. J Mach Learn Res 12:2211–2268
76. Tax DMJ, Duin RPW (2004) Support vector data description. Mach Learn 54(1):45–66
77. Smola AJ, Mangasarian OL, Scholkopf B (2002) Sparse kernel feature analysis. In: classification, automation, and new media. Studies in classification, data analysis, and knowledge organization, pp 167–178
78. Filippone M, Camastra F, Masulli F, Rovetta S (2008) A survey of kernel and spectral methods for clustering. Pattern Recogn 41(1):176–190. doi:10.1016/j.patcog.2007.05.018
79. Hardoon DR, Szedmak S, Shawe-Taylor J (2004) Canonical correlation analysis: an overview with application to learning methods. Neural Comput 16(12):2639–2664
80. R: a language and environment for statistical computing. (2012). http://www.R-project.org/. Accessed 11 August 2014.

Chapter 14
Quantitative Structure-Property Relationship Analysis of Drugs' Pharmacokinetics Within the Framework of Biopharmaceutics Classification System Using Simplex Representation of Molecular Structure

N. Ya. Golovenko, I. Yu. Borisyuk, M. A. Kulinskiy, P. G. Polishchuk, E. N. Muratov and V. E. Kuz'min

Abstract The Biopharmaceutics Classification System (BCS) categorizes drugs into one of four biopharmaceutical classes according to their water solubility and membrane permeability characteristics and broadly allows the prediction of the rate-limiting step in the intestinal absorption process following oral administration. When combined with the *in vitro* dissolution characteristics of the drug product, the BCS takes into account three major factors: solubility, intestinal permeability, and dissolution rate, all of which govern the rate and extent of oral drug absorption from immediate-release (IR) solid oral-dosage forms. The concept of BCS provides a better understanding of the relationship between drug release from the product and the absorption process. This report reviews the current status of computational tools in predicting the base properties (aqueous solubility, and passive absorption) of the BCS and explores the application of the Simplex representation of molecular structure (SiRMS) QSAR approach in absorption (bioavailability) research. The main advantages of SiRMS are consideration of the different physico–chemical properties of atoms, high robustness, predictivity, and interpretability of developed models that creates good opportunities for molecular design. The reliability of developed QSAR models as predictive virtual screening tools and their utility for targeted drug design were validated by subsequent synthetic and biological experiments. The SiRMS approach was implemented as "HiT QSAR" software. In addition, we provide our perspective on the progress of research into an *in silico* equivalent to the BCS.

N. Ya. Golovenko (✉) · I. Yu. Borisyuk · M. A. Kulinskiy · P. G. Polishchuk ·
E. N. Muratov · V. E. Kuz'min
A.V. Bogatsky Physico-Chemical Institute NAS of Ukraine,
86 Lustdorfskaya Doroga, Odessa 65080, Ukraine
e-mail: n.golovenko@gmail.com

L. Gorb et al. (eds.), *Application of Computational Techniques in Pharmacy and Medicine*,
Challenges and Advances in Computational Chemistry and Physics 17,
DOI 10.1007/978-94-017-9257-8_14

14.1 Introduction

During the last few years, the role of pharmacokinetic (absorption, distribution, metabolism, and elimination, i.e., ADME) properties in the drug research and development increased dramatically [1–3]. Oral bioavailability is one of these ADME components that have been widely studied. Screening of absorption ability is an important part of assessing oral bioavailability and attracts efforts from industry and academia. Although plethora of *in vitro* and *in vivo* ADME screening methods have been applied to boost drug discovery process in pharmaceutical industry, this process is still resource-intensive and time-consuming. The prediction of oral bioavailability is very challenging because bioavailability is a complex function of many biological and physico-chemical factors, such as dissolution in the gastrointestinal tract, intestinal membrane permeation, intestinal and hepatic "first-pass" metabolism, etc.

Although ADME assays have been the gold standards in pharmacokinetics, there are additional tests that should be incorporated, since they play a key role in drug discovery and further development. Liberation and dissolution of the drug from the pharmaceutical form is a key parameter in bioequivalence studies The solubility and permeability of a drug are considered to be the most important properties that determine absorption and the influence of these two properties on the extent of absorption from the intestinal tract has received considerable attention [4–6]. An experimental system for classification of drugs based on their aqueous solubility and membrane permeability was implemented recently by the Food and Drug Administration (FDA). It was named as the biopharmaceutics classification system (BCS). BCS was originally implemented to waive clinical studies of generic high-permeability/high-solubility drugs. The original BCS categorizes drugs into four different classes based on combinations of high/low solubility and high/low permeability [7]. As an alternative to experimental measurements, the *in silico* prediction of ADME properties is very attractive, because it provides an inexpensive and highthroughput way to assess the ADME properties of a molecule prior to synthesis and biological testing.

The purpose of this monograph is to discuss basic principles associated with the process of drug absorption. Special attention will be given to the use of the BCS as a predictive tool for identifying compounds whose absorption characteristics may be sensitive to physiological and formulation variables. Investigation of drug absorption requires a fundamental understanding of the molecular properties that determine passive membrane transport. These properties will be briefly discussed in the following section.

14.2 Methodology of the Biopharmaceutics Classification System (BCS)

The rate and extent of drug absorption from the gastrointestinal (GI) tract are very complex and affected by many factors. These include physico-chemical factors (solubility, lipophilicity, stability, pK_a, polar surface area, presence of hydrogen bonding functionalities, particle size, and crystal form), physiological (GI blood flow, GI pH, gastric emptying, GI transit time, and absorption mechanisms), and factors related to the dosage form (tablet, capsule, solution, suspension, emulsion, or gel) [1–4]. Despite this complexity, the work of the various authors [8–10] revealed that the fundamental events controlling oral drug absorption are the permeability of the drug through the GI membrane and the solubility/dissolution of the drug dose in the GI milieu. In the solid oral dosage form active pharmaceutical ingredient (API) is characterized by the following interrelated processes occurring in GI: (a) tablet degradation and release of solid particles of API; (b) solubilization of solid particles in a liquid medium (C_s) of GI; (c) penetration of API molecules from an intestinal liquid through an immobile layer (coated mucosal surface) to mucosal surface; (d) transfer of API molecules from the liquid medium into the mucosal layer; (e) penetration (P_{eff} or P_{app}) of API from the mucosal layer into systemic blood circulation.

The P_{eff} and P_{app} values are alternative parameters of F (fraction absorption). P_{eff} is an effective permeability of organotypic models, P_{app} is apparent permeability determined *in vitro* [11, 12].

Key parameters are characterized in the BCS by three dimensionless numbers: absorption number (A_n), dissolution number (D_n), and dose number (D_0). These numbers take into account both physico-chemical and physiological parameters and are fundamental to the oral absorption process based on these properties. Amidon et al. [1, 13] proposed biopharmaceutic classification system (BCS), which in present times is serving as a guide for regulatory and industrial purposes. This concept exploring dose number, dissolution number, and absorption number of an orally administered drug clearly dictates its systemic availability. These three numbers are associated with a number of multifaceted hurdles, which include: (a) physico-chemical properties of the molecule (solubility/dissolution); (b) stability of drug in GI environment (acid degradation); (c) enzymatic stability in GI lumen, epithelium and liver; (d) permeability (molecular weight, log P, H-bonding efficiency); and (e) substrates specificity to various biological transporters and efflux systems of intestinal epithelium.

Dose number (D_0) is characterized by the volume required for solubilising the maximum dose strength of the drug.

$$D_0 = \frac{M / V_0}{C_S}$$

Where C_S is the solubility, M is the dose and V_0 is the volume of water taken with the dose, which is generally set to 250 ml.

Dissolution number (D_n) is characterized by the time required for drug dissolution which is the ratio of the intestinal residence time $< T_{sit} >$ and the dissolution time $< T_{diss} >$

$$D_n = \frac{\langle T_{sit} \rangle}{\langle T_{diss} \rangle} = \frac{3DC_S \langle T_{sit} \rangle}{r^2 \rho}$$

Where D is the diffusivity of the dissolved drug, ρ is the density of the dissolved drug, C_S is the drug solubility, and r is the initial radius of the drug particle.

Absorption number (A_n) is characterized by the time required for absorption of the dose administered which is a ratio of residence time and absorptive time $<T_{abs}>$

$$A_n = \frac{\langle T_{sit} \rangle}{\langle T_{abs} \rangle} = \frac{P_{eff} \langle T_{sit} \rangle}{R}$$

where P_{eff}-effective permeability, R-the gut radius, and $<T_{sit}>$ the residence time of the drug within the intestine.

However, all these numbers are related to two important parameters controlling drug absorption, i.e., solubility and permeability. Drug with complete absorption show $D_0 < 1$, while D_n and $A_n > 1$. If the P_{eff} of a drug is less than $2 \cdot 10^{-4}$ cm/s, then drug absorption will be incomplete, whereas complete absorption can be expected for substances whose P_{eff} exceeds this value. For poorly soluble drugs, critical variables include the volume of the intestinal fluids, GI pH, and GI transit time (where adequate time is needed to dissolve poorly soluble materials). For these lipophilic compounds, food and bile salts may increase drug solubility.

Based on these two parameters, drug API have been classified into one of four categories according to the BCS.

Class I API is characterized by the high A_n and low D_n and D_0, indicating that they are in solution form throughout the intestine and is available for permeation. In this case, F can be expressed as follows [10]:

$$F = 1 - \exp(-2A_n)$$

For these agents, as "A_n" increases, the fraction of drug absorbed increases, with 90% absorption (highly permeable compounds) occurring when $A_n = 1.15$. Using the equation for A_n, we conclude that F can be affected by a change in the compound's membrane permeability, the gut radius of the host, or the intestinal transit time. Based on these factors alone, it is evident that differences in GI physiology caused by such factors as disease, age, or animal species, can alter the value of A_n and, therefore, the fraction of drug absorbed. Drugs are highly soluble and highly permeable and are ideal candidates for oral delivery.

Class 2 drugs are highly permeable across the GI membrane, primarily by passive transport, because of their high lipophilicity. These drugs are characterized by mean absorption time more than mean dissolution time, and thus gastric emptying and GI transit are important determinants of drug absorption. High correlation between *in vitro* dissolution and *in vivo* rate and extent of absorption is expected; however, since the rate of drug getting into solution is rate-limiting, the *in vitro* permeability may not predict the *in vivo* absorption.

Class 3 drugs are either having less intrinsic permeability due to their unfavorable physico-chemical properties. In order for these molecules to permeate the lipophilic epithelial cell membranes lining the gastrointestinal tract, they must possess optimum lipophilicity. Thus for highly polar compounds, administration of less polar and more lipophilic prodrugs may improve absorption.

Low and variable absorption for class 4 drugs is anticipated because of the combined limitation of solubility and permeability.

Experience gained through development of traditional *in vitro–in vivo* correlations, *IVIVC* (level A, B, or C correlations) for IR products containing poorly soluble drugs and for extended release products suggests a significant degree of formulation dependency or specificity associated with such correlations. Therefore, for products that are likely to exhibit slow *in vivo* dissolution, *IVIVC* need to be established and their predictive performance verified through experimentation. Future research in this area should address how to *a priori* identify dissolution test conditions that yield robust *IVIVC* that are applicable to a wide range of formulations [14].

In recent years; the validity and broad applicability of the BCS have been the subject of extensive research and discussion [4–6, 15]. It has been adopted by the US Food and Drug Administration (FDA), the European Medicines Agency (EMEA), and the World Health Organization (WHO) for setting bioavailability/bioequivalence standards for IR oral drug product approval; and the BCS principles are extensively used by the pharmaceutical industry throughout drug discovery and development [7, 16–18].

The introduction of the simplification of the BCS in FDA guidelines represents a major step forward in the regulation of oral drug products. The FDA guideline suggests internal standards and marker substances to characterize the permeability of drug substances *in vitro* and *in vivo*. The BCS is used to set drug product dissolution standards to reduce the *in vivo* bioequivalence requirements [1, 19]. Knowledge of the BCS can also help the formulation scientist to develop a dosage form based on mechanistic, rather than empirical, approaches [20]. This allows one to determine the potential for *IVIVC*, and can significantly reduce *in vivo* studies.

According to the current FDA guidance [17, 18], drug API is considered highly permeable when the extent of absorption in humans is determined to be 90 % or more of an administered dose based on a mass balance determination or in comparison to an intravenous reference dose. The solubility classification of a given drug is based on the highest dose strength in an IR product. In this guidance, an IR drug product is considered rapidly dissolving when no less than 85 % of the labeled amount of the drug substance dissolves within 30 min, using U.S. Pharmacopeia (USP) apparatus I at 100 rpm (or apparatus II at 50 rpm) in a volume of 900 ml or less in each of the following media: (1) 0.1 N HCl or simulated gastric fluid USP without enzymes; (2) a pH 4.5 buffer; and (3) a pH 6.8 buffer or simulated intestinal fluid USP without enzymes.

14.3 Pharmacokinetic Characteristics of Absorption and Bioavailability

Bioavailability is based on the physiological process of absorption, which include three stages: (1) transfer an API through apical plasma membrane inside cells; (2) intracellular transport of substances followed by their possible metabolism; (3) transfer of the transported and transformed API from cells into blood or lymph.

Bioavailability is the degree to which or the rate at which an API from corresponding drug dosage form and at the targeted place of administration in systemic circulation becomes available in the biophase. Such definition reflects relative character of the notion of biological availability of drugs. In experiments, intestinal absorption is usually measured by fraction absorption, $\%F$, which is defined by the fraction of total mass absorbed to the given dose of the drug.

Absolute bioavailability, F, is the fraction of an administered dose which actually reaches the systemic circulation:

$$F = \frac{AUC_{ev}}{AUC_{iv}} \tag{2.1}$$

where AUC_{ev} and AUC_{iv} are, respectively, the area under the plasma concentration-time curve following the extravascular and intravenous administration of a given dose of drug.

Relative or comparative bioavailability refers to the availability of a drug product as compared to another dosage form or product of the same drug given in the same dose. These measurements determine the effects of formulation differences on drug absorption. The relative bioavailability of product A compared to product B, if both products containing the same dose of the same drug, is obtained by comparing their respective AUC_S. When the bioavailability of a generic product is considered, it is usually the relative bioavailability that is referred to.

Bioavailability mainly depends on intestinal and hepatic clearance of API. In the case when clearance rate depends on blood API concentration absorption and bioavailability are identical. However, if clearance process employs active secretion or metabolic pathways and it becomes saturable the pharmacokinetic dependence becomes nonlinear. In this case changes in absorption is not accompanied by proportional change in bioavailability.

14.4 Prognosis of Bioavailability

Study of drug bioavailability still remains the most complex and expensive test. It is based on elucidation of API concentration in certain biological fluids (blood, urine, et al.). Unfortunately, there is no common method, which would meet all requirements during evaluation of various drugs. In each case, it is a unique method, which should provide selective, accurate, and reproducible monitoring of drug concentration under chosen conditions of a pharmacokinetic study, particularly, its duration. In the case of evaluation of bioequivalence the kinetic studies are carried out under conditions of certified bioanalytical laboratories independently from drug firms. Given that in most cases experimental bioavailability on animals employs radioactively labeled API and clinical evaluation of bioequivalence employs volunteers, the difficulties for researchers become clearly evident. Consequently, simplification of such procedures would be achieved by means of reliable rather simple method predicting API bioavailability. Realization of such project requires information on physiological nature of bioavailability and physico-chemical properties of API.

Now the market of software for prediction of biological properties of chemical substances is saturated with various products. Analysis of this problem has been reviewed in [21–23].

For significant prediction of bioavailability a training set of drugs is subdivided into either two (e.g. Yes/No, Positive/Negative), three [24] (high with $F > 80\,\%$; moderate with $F = 21–79\,\%$, and low with $F < 20\,\%$), or four groups [25] (1: <20 %, 2: 20–49 %; 3: 50–79 %; 4: >80 %).

While experimental methods always require sufficient amount of chemicals for the estimation of drug absorption, computational (*in silico*) methods can lead to the prediction of intestinal absorption based on chemical structure, and can thus be used before synthesis of compounds. *In silico* predictions could be based both on relatively simple quantitative structure-activity relationships (QSAR) analysis and more complex physiologically based pharmacokinetic and/or pharmacodynamic models. Whichever the approach used for model building, computational methods should be based on experimental data that were obtained for a wide range of structurally diverse compounds (training set). It should be noted, however, that current *in silico* methods, are not as reliable as experimental models.

The models can be divided into two categories: regression and classification ones. Most of the published studies are based on small or large sets of permeability data collected from literature sources. Since the validity of any model primarily depends on the data, a large and accurate dataset is required for the development of global models with general applicability. Model validation is another critically important step in building robust QSPR models. Ideally, one should report results based on the training set, the cross-validation set, and an external dataset to increase the user's confidence level. Consistent reporting of model statistics is highly desirable so that readers can objectively evaluate the model quality and applicability in a real-life drug discovery setting [22].

Rather than trying to predict specific absorption-related quantities, researchers have tried to find general principles to distinguish drug-like from non-drug-like molecules by analyzing databases of drugs and non-drugs. Generally, these rules obtained from database analysis can be used to distinguish well-absorbed molecules from poorly-absorbed molecules. Among numerous attempts to find relevant correlation between physico-chemical properties of API and their bioavailability the "rule of five" is the most popular ADME-concerned filters, and most widely used [24]. The "rule of five" defined several rules for identifying compounds with possible poor absorption and permeability: (1) molecular weight > 500, (2) calculated $\log P > 5$ (CLOGP) or > 4.15 (MLOGP), (3) number of hydrogen bond donors (OH and NH groups) > 5, and (4) number of hydrogen-bond acceptors (N and O atoms) > 10. A disadvantage of the "rule of five" is that it can only give a quite rough classification of molecules, allowing the elimination of only a very limited set of molecules. Later Lipinski [26] indicated that this rule does not belong to ideal filters, because predictions based on these rules are basically related to one parameter Mlog P or Clog P, where errors are possible during additive evaluation (CLOGP program). Moreover, "the rule of five" was developed when information about drug transporters was very limited and so it described links between physico–chemical properties of drugs and their bioavailability based on API absorption determined by simple diffusion.

In addition to the molecular properties discussed by Lipinski, other properties have been discussed in regard to oral bioavailability. Since then, numerous classification and regression models for the predictions of absorption were reported by applying a variety of statistical and machine-learning approaches, which include multiple linear regression (MLR), nonlinear regression, partial least squares (PLS) regression, linear discriminant analysis, classification and regression trees, artificial neural networks (ANN), support vector machines (SVM), and so forth. Because of the fact that many factors are related to intestinal absorption, many physicochemical descriptors were introduced into the prediction of absorption and bioavailability, such as polar surface area (PSA), partition coefficients, molecular size, hydrogen-bonding descriptors, topological descriptors, and even quantum chemical descriptors. Predictivity of models bioavailability is by prediction of estimated independent compounds. To be useful in drug development, models should ultimately be developed to predict unknown compounds.

In subsequent work, other researchers have introduced rules-of-thumb which can increase the chances of drug compounds being well absorbed. In 2002, Veber and colleagues reported studies on rat bioavailability data for 1100 drug candidates [27]. They proposed two other descriptors and suggested that compounds which meet only two criteria of (a) 10 or fewer rotatable bonds and (b) polar surface area equal to or less 140 $Å^2$ (or 12 or fewer H-bond donors and acceptors) will have a high probability of good oral bioavailability in rat. In 2004, Lu and colleagues investigated the relationship between the number of rotatable bonds and PSA for rat oral bioavailability using 434 molecules [28]. Compared to Veber's work, authors reported that the prediction results were dependent on the calculation methods. In 2007, Hou and colleagues [29] collected a dataset of 773 compounds with experimental human oral bioavailability values. They showed that the percentages of compounds meeting the criteria based on molecular properties does not distinguish compounds with poor oral bioavailability from those with acceptable values. A dataset of intestinal absorption was also examined and compared with that of oral bioavailability. The performance of these rules based on molecular properties in the prediction of intestinal absorption is obviously much better than that of oral bioavailability in term of false positive rate, and, therefore, the applications of the "rule-based" approaches on the prediction of human bioavailability should be very cautious. Then, Veber's rules were used for the entire dataset to see if these rules could be applied for the prediction of human oral bioavailability. Afterwards, the correlations between several important molecular descriptors and human oral bioavailability were examined. They conjectured that there are no simple rules based on molecular descriptors that can be used to predict human oral bioavailability truly well compared to the rules based on analyzing rat oral bioavailability data [29]. However, the value of developed models and all the conclusions mentioned above is significantly decreased by the very low quality of collected dataset that contained plenty of duplicates and high inconsistencies in bioavailability values. In the same time, it is still clear that powerful descriptors related to carrier-mediated transport and first-pass metabolism are needed for building a useful prediction model for human oral bioavailability.

Recently, Hirono et al. published a study of the quantitative physico-chemical propertybioavailability relationships for 188 noncongeneric diverse organic API using the "fuzzy adaptive least squares" method [30]. The model was validated by "leave one out" cross-validation, and not an independent test set. The compounds were divided into three groups, non-aromatics, aromatics, and heteroaromatics, and separate equations were formulated for each group which were statistically reliable and satisfactory. However, in addition to the need for prior classification of the compounds into one of the three groups, the lipophilicity of the compounds was not separately identified as a factor, although many studies have reported that this is one of the most important properties which determines absorption and metabolism. Finally, Bains et al. proposed evolutionary and adaptive methods for classifying drug bioavailability into "high" and "low" classes [31], and showed that obtaining predictive models on the basis of the molecular structure alone is possible. Innovative concepts for correlating molecular structures with biological activities are represented by fuzzy logic (FL) [32]. In fact, FL methods based on the possibility to handle the "concept of partial truth", provide interesting solutions to classification problems within the context of imprecise categories, in which ADME properties can be included. Fuzzy classification represents the boundaries between neighbouring classes as continuous, assigning to the compounds a degree of membership of each class. FL has been widely used in the field of process control, where the idea is to convert human expert knowledge into fuzzy rules, and it is able to extract relevant structure-activity relationships (SAR) from a database, without a priori knowledge. Wessel *et al.* have reported a model based on 76 compounds with reported human intestinal absorption data, using GA with an ANN scoring function [33]. A standard error of 16 % was obtained for the test set of 10 molecules. Clark reports the use of polar surface area (PSA) to create a classification model to separate poor (< 10 %) from well absorbed (≥ 10 %) compounds [5], using the same dataset as reported by [33]. Egan et al. published a model for intestinal absorbtion HIA based on PSA and Alog*P* descriptors alone [34]. This classification model had accuracy of 74–92 % for different validation sets. Abraham and collaborators have recently reported a model based on a comprehensive intestinal absorption dataset [35]. The absorption of 111 drug and drug-like compounds was evaluated from 111 references based on the ratio of urinary excretion of drugs following oral and intravenous administration to intact rats and biliary excretion of bile duct-cannulated rats. Ninety-eight drug compounds with known both human and rat absorption data were selected for correlation analysis between the human and rat absorption. The results showed that the extent of absorption in these two species is similar. For 94 % of the drugs the absorption difference between humans and rats is less than 20 % and for 98 % of drugs the difference is less than 30 %. There is only one drug for which human absorption is significantly different from rat absorption. The standard deviation is 11 %

between human and rat absorption. The linear relationship between human and rat absorption forced through the origin, as determined by least squares regression. It is suggested that the absorption in rats could be used as an alternative method to human absorption in preclinical oral absorption studies.

Using Abraham descriptors, described in the same paper, the authors report $R^2=0.74$ when trained on the whole set. However, the authors also highlight the fact that the training set is heavily biased towards well-absorbed compounds (over 30 % absorbed). Yoshida and Topliss [36] proposed a classification model to predict bioavailability by using a dataset of 272 drugs, fingerprints and pharmacokinetics descriptors, with help of a method named ORMUCS (ordered multicategorical classification method using the simplex technique) This approach, after dividing the bioavailability data in four classes, allowed the authors to get a correct classification rate of 71 % for the training set and 60 % for the 40 compounds included in the test set. A quantitative structure-bioavailability relationship model was developed by Andrews et al. on a dataset including 591 compounds [37]. A stepwise regression procedure was used to relate oral bioavailability in humans and structural fragments in drugs. Compared to the Lipinski's "rule of five", this model allowed to reduce the amount of false negatives and positives. Poor bioavailability predictions were false in only 3 % of cases, but 53 % of predictions of high bioavailability were incorrect.

In addition to the molecular properties discussed by Lipinski, the relation of other properties to oral bioavailability was also extensively investigated. Navia et al. has postulated the desirability of molecular flexibility for membrane permeation [38]. Hirschmann et al. has focused on the undesirable property of water complexation by amide bonds as a negative factor for oral bioavailability [39].The negative impact of a high polar surface area on intestinal absorption is recognized [40, 41]. Membrane permeation is recognized as a common requirement for oral bioavailability in the absence of active transport, and failure to achieve this usually results in poor oral bioavailability.

Similar observations were made by others in an effort to define descriptors that can provide a rationale for establishing qualitative, semiquantitative, and quantitative structure-absorption relationship models [28, 31, 32]. The dependence of human intestinal absorption on the readily accessible physico-chemical properties like lipophilicity, molecular size, hydrogen bonding capacity, PSA, and number of free rotatable bonds has been demonstrated. Identification of these basic physicochemical properties as determinants is consistent with notions regarding the ability of small organic molecules to pass through lipid bilayer membranes.

14.4.1 QSAR Analysis Using Simplex Representation of Molecular Structure

Historically, the Simplex approach was developed as a method for the characterization of chirality [42] and only later it was used in QSAR analysis [43]. In the frameworks of Simplex representation of molecular structure (SiRMS) any molecule can be represented as a system of different simplexes (tetratomic fragments of fixed composition, structure, chirality and symmetry) (Fig. 14.1). Atoms in a simplex can be differentiated on the base of different characteristics:

1. Atoms individuality expressed through the nature (e.g., nitrogen or carbon) or more detailed type of atoms (C-sp^3 or C-sp^2);
2. Partial atom charge that reflect their electrostatic properties;
3. Lipophilicity of atoms that reflects its hydrophobic/hydrophilic properties;
4. Atomic refraction that partially reflects the ability of atoms to dispersion interactions;
5. A mark that characterizes the atom as a possible hydrogen donor or acceptor during H-bond formation (A: hydrogen acceptor in H-bond; D: hydrogen donor in H-bond, I: indifferent atom).

For atom characteristics, that have real values (for example, charge, lipophilicity and refraction) the division of continuous values into definite discrete groups is carried out at the preliminary stage. The number of groups (G) is a tuning parameter and can be varied (usually G=3–7).

The use of sundry variants of simplex vertexes (atoms) differentiation represents an important part of SiRMS we consider that specification of atoms by their nature alone (this actually reflects atom identity, for example, C, N or O), which is realized in many QSAR methods limits the possibilities of active fragments selection. For example, if the –NH– group has been selected as the determining activity fragment and the ability of H-bond formation is a factor determining its properties, then we will miss donors of H-bonds, such as OH-groups. The use of atom differentiation by donor/acceptor of H-bond allows one to avoid the situation illustrated above. One can make analogical examples for other atom properties (lipophilicity, partial charge and refraction, for example). Different types of simplexes are generated depending on the level of detail of molecular structure.

1D models. 1D simplex is a combination of any four atoms contained in the molecule (Fig. 14.1). The simplex descriptor (SD) at this level is a number of quadruples of atoms of the definite composition. 1D simplexes were not used in our studies.

2D models. The connectivity of atoms in simplex, atom type and bond nature (single, double, triple, or aromatic) have been considered. Thus, the SD at the 2D level is a number of simplexes of fixed composition and topology (Fig. 14.1). Other structural parameters corresponding to molecular fragments of different size, can be used on 1 and 2D levels. The use of 1–4 atomic fragments is preferable, because further extension of the fragment length could increase the probability of model over-fitting and decrease its predictivity and DA.

3D models. Not only topology, but also the stereochemistry of the molecule is taken into account at the 3D level. It is possible to differentiate all the simplexes as right (R), left (L), symmetrical (S) and plane (P) achiral (Fig. 14.2). Stereochemical configuration of simplexes is defined by a modified Kahn–Ingold–Prelog rule. An SD at this level is a number of simplexes of fixed composition, topology, chirality and symmetry (Fig. 14.1).

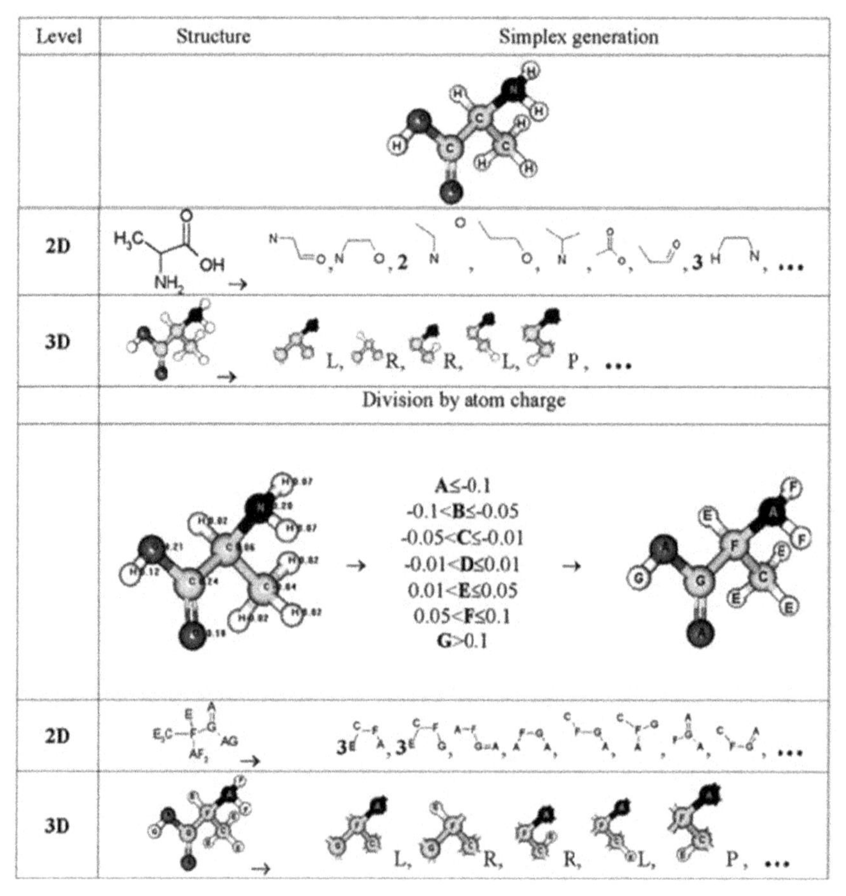

Fig. 14.1 Example of simplex descriptors generation at 2D and 3D levels (*L* Left, *P* Achiral, *R* Right)

Fig. 14.2 Four possible stereochemical types of simplexes

Right Left Symmetrical Achiral

4D models. Each SD is calculated by the summation of the products of descriptor value for each conformer (SD_k) and the probability of realizing the corresponding conformer (P_k).

$$SD = \sum_{\kappa=1}^{N} (SD_{\kappa} \cdot P_{\kappa}) \tag{14.1}$$

where N is the number of conformers being considered.

As is well known, the probability of conformation Pk is defined by its energy:

$$P_K = \left\{1 + \sum_{i \neq K} \exp\left(\frac{-(E_i - E_K)}{RT}\right)\right\}^{-1}, \sum_K P_K = 1 \tag{14.2}$$

This formula corresponds to the Boltzmann distribution of conformers, where E_i and E_k are the energies of conformations i and k, respectively. The conformers are analyzed within the energy band of 5–7 kcal/mol. Thus, the molecular SD at 4D level takes into account the probability of realization of 3D level SD in the set of conformers. On a 4D level the other 3D whole-molecule parameters efficient for the description of the spatial form of conformer (e.g., characteristics of inertia ellipsoid or dipole moment) can be used along with SD.

Plethora of simplex descriptors is usually generated in SiRMS. The PLS method proved efficient at the work with a large number of variables and was described well elsewhere [44, 45]. Briefly, a PLS regression model could be represented as Eq. 14.3 [45]:

$$Y = b_0 + \sum_{i=1}^{N} b_i x_i \tag{14.3}$$

where Y is an appropriate activity, b_i is PLS regression coefficients, x_i is an i-th descriptor value, N is a total number of simplexes. In PLS one assumes the x-variables to be colinear and PLS estimates the covariance structure in terms of a limited number of weights and loadings. In this way PLS can analyze any number of x-variables regarding to the number of objects [45].

The removal of constant and highly correlated ($r > 0.9$) descriptors, genetic algorithm (GA) [46], trend-vector approach (TV) [47] and automatic variable selection (AVS) [48] strategy have been used for the selection of descriptors in PLS. The removal of highly correlated descriptors is not necessary for PLS analysis, since descriptors are reduced to the series of uncorrelated latent variables. However, this procedure frequently helps in obtaining more adequate models and reducing the number of used variables by up to five times [48]. Usage of methods of exhaustive or partial search (depending on the number of selected descriptors) after AVS or GA very often allow one to increase the quality of the models obtained (PLS, MLR and TV). After the mentioned statistic-processing model or models with the best combinations of statistic characteristics (R^2 and Q^2) have been selected from the obtained resulting list for subsequent validation using an external test set. The general scheme of the PLS models generation and selection applied in Hierarchical QSAR technology (HiT QSAR) has been presented elsewhere [49]. This procedure can be repeated several times using as an input (initial set) the SD of different levels of the molecular structure representation (usually 2–4D) and/or with various kinds of atom differentiation with the purpose of developing several resulting QSAR models for consensus modeling. This approach is believed to yield more accurate predictions.

Cross validation is the statistical practice of partitioning a sample of data into subsets, such that the analysis is initially performed on a single subset, while the other subset(s) are retained for subsequent use in confirming and validating the initial analysis. The initial subset of data is known as the training set, while the other subset(s) are known as validation sets. Two types of cross validation can be used in QSAR analysis: leave-one-out and leave-group-out cross validation. The latter is a more severe method for model robustness estimation [50].

Determination coefficient Q^2 calculated in cross-validation terms is the main characteristic of model robustness. Q^2 is calculated by the following formula:

$$Q^2 = 1 - \frac{\sum_{\gamma} (Y_{pred} - Y_{actual})^2}{\sum_{\gamma} (Y_{actual} - Y_{mean})^2} \qquad (14.4)$$

where Y_{pred} is a predicted value of activity, Y_{actual} is an actual or experimental value of activity and Y_{mean} is the mean activity value. However, goodness-of-fit or robustness should not be confused with the ability of a model to make predictions [51]. Usage of external validation (test) set is the only way for the estimation of model predictivity. Thus, a certain fraction of the dataset molecules is removed into a test set before the modeling process begins (remaining compounds form the training set). Once a model has been developed, predictions can be made for the test set. However, even in the case of a beneficial effect, one should be aware that this may only represent the model's ability to predict a certain test set. It is important, therefore, that both training and test sets cover the structural space of the complete dataset as large as possible.

The most similar or dissimilar compounds as well as those randomly chosen, taking into account activity variation, are selected for the external set in our studies [43]. Structural dissimilarity/similarity is obtained for all initial training set molecules on the basis of relevant structural descriptors. In our opinion, use of the whole set of descriptors generated in the very beginning is not completely correct, because during QSAR research we are interested not only in structural similarity by itself, but from the point of view of the investigated properties. Thus, mentioned descriptors selection will help in avoiding some distortions caused by the insignificance of structural parameters from the initial set for this concrete task.

The targets of the first level are activity prediction or virtual screening. Any descriptors could be used here, even those hardly interpretable or noninterpretable, such as different topological indices, informational-topological indices or eigenvalues of various structural matrices. The aims of the second level must include the interpretability of QSAR models obtained. Only descriptors that have clear physico–chemical sense (e.g., reflecting parameters of the molecule such as dipole moment, lipophilicity, polarizability and van der Waals volume) can be used at this level. Analysis of QSAR models corresponding to this level allows one to reveal structural factors promoting or interfering the property investigated.

Finally, the presence of information useful for molecular design is expected from QSAR models corresponding to the third level of purposes. Fragmentary descriptors are usually used in such models. In this case, the analysis of the degree and direction of influence of such descriptors on activity can give immediate information for the optimization of known structures and the design of novel substances with desired properties.

However, one should be aware that, first of all, any selected model must be predictive and only after that interpretable and informative, and so forth. It is a necessary condition of its subsequent usage that mechanistic-based or interpretability-oriented models, which are not predictive, are not acceptable and senseless.

RF is an effective nonparametric statistical technique for large databases analysis [52]. The main features of RF are listed below: (1) it is possible to analyze compounds with different mechanism of action within one dataset; (2) there is no need to pre-select descriptors; (3) the method has its own reliable procedure for the estimation of model quality and its internal predictive ability; (4) models obtained are tolerant to "noise" in source experimental data.

RF model construction is based on an original algorithm [53]. RF is an ensemble of single decision trees built by a classification and regression trees algorithm (CART) [54]. Decision trees are the ensemble of hierarchically structured rules. Every rule is a logical construction that can be represented as "if … then …" criterion. An RF algorithm recursively tries to find common criteria for objects from the same class, using some randomly selected descriptors.

Each tree grows according to the following algorithm: (1) Bootstrap sample which will be a training set for current tree is produced from the whole training set of N compounds. About one-third of the compounds which aren't in the current training set are placed in out-of-bag (OOB) set. It is used to get a running unbiased estimate of the model error and variable importance; (2) The best split among the m randomly selected descriptors taken from the whole set of M ones in each node is chosen. The value of m is the only tuning parameter for which RF models are sensitive; (3) Each tree is grown to the largest possible extent. There is no pruning. RF possesses its own reliable statistical characteristics, which could be used for validation and model selection. Determination coefficients for training set (R^2) and out-of bag set (R^2_{oob}) are two main characteristics of the model. The major criterion for estimation of internal predictive ability of the RF models and model selection is the value of R^2_{oob} [55, 56].

Determination of optimal values of T (trees count) and m (descriptors count) are the traditional starting points for every RF investigation. The following procedure is recommended for determination of an optimal T value: starting from a small constant number of descriptors, increase the number of trees stepwise until R^2_{oob} does not change significantly. Once an optimal T value has been identified, build models with the optimal number of trees by increasing the variable number in a stepwise manner. The final model is determined by the highest value of OOB set prediction. In this study, T and m values equal to 100 and 200, respectively, were found as optimal and were chosen for subsequent models.

The approach outlined above has already been shown to be highly efficient for solving various "structure-activity" tasks [23, 27, 30, 57–59]. SiRMS methodology does not have the restrictions of such well-known and widely used approaches as CoMFA, CoMSIA, and HASL, in which the application is limited to a structurally homogeneous set of molecules only. SiRMS approaches also do not have the disadvantages of the HQSAR approach [36, 52] that are related to the ambiguity of the descriptors system formation.

14.4.1.1 QSPR Prediction of the Drugs Bioavailability on the Base of the SiRMS Approach

Influence of the molecular structure of drug API on absorption was investigated using SiRMS QSAR approach [26, 60]. Data for human oral bioavailability were obtained from the literature and an internal database. SMILES strings were retrieved from the World Drug Index (WDI, Derwent Publishers, London) or created manually. Finally, 628 structures with SMILES, generic name, and bioavailability value were obtained. Any compounds whose bioavailability is strongly affected by the dose and formulation was excluded from the dataset. Random forest method [61, 62] was used for the development of QSPR classification models. All compounds were divided into three (high, medium and low bioavailability, Table 14.1) or two (high and acceptable, Table 14.2) classes. To determine whether the classes limits shift affects the quality of predictive models and how it affects, were considered different limits of the classes, as shown in Table 14.1.

As seen from the results presented above, misclassification error is high and hence predictive ability of models is rather low. Model II has the best prediction performance, especially for classes 1 and 2. Regression model with low predictive ability ($R^2_{oob}=0.29$) was also obtained using this dataset of 628 compounds.

Binary classification model has better predictive ability (Table 14.2).

Table 14.2 shows that the shift of models boundaries reduces accuracy of their prediction. Consequently, in order to identify the causes of this phenomenon, we conducted a thorough analysis of medicinal agents by all relevant models (IV—VI). The analysis of the results showed the good concordance between models IV-VI. It have been proved that prediction error for the class 1 (bioavailability of 90–100, or 80–100, or 70–100) is associated with the active participation of specific transporters during absorption. For class 2, insignificant participation of transporters in the absorption of medicinal agents has been observed, but there is an intensive metabolism of compounds and their binding to targets (furthermore, the most of medicinal agents have multiple targets, binding with which is strong and durable).

Taking into account intersection of models and variation of bioavailability in some range of values, we have introduced to the obtained model a confidence interval within 5 % on either side of the chosen threshold value (Fig. 14.3).

Thus, the boundary between the classes has a certain thickness. Errors of classification shall be specified in the case when the molecules are predicted to be in the opposite class beyond the boundary limits.

Table 14.1 Classification models for three classes of bioavailability

ID of the model	Classes boundaries	Number of molecules in the class	Total error, %
I	1) 0–20 2) 20–80 3) 80–100	114 304 210	36
II	1) 0–10 2) 10–90 3) 90–100	68 432 128	27
III	1) 0–20 2) 20–70 3) 70–100	114 250 264	36

Table 14.2 Classification models by two classes bioavailability

Number of the model	Classes boundaries	Number of molecules in the class	Error of classification, %
IV	1) 90–100 2) 0–90	156 472	20
V	1) 80–100 2) 0–80	222 406	24
VI	1) 70–100 2) 0–70	280 348	27

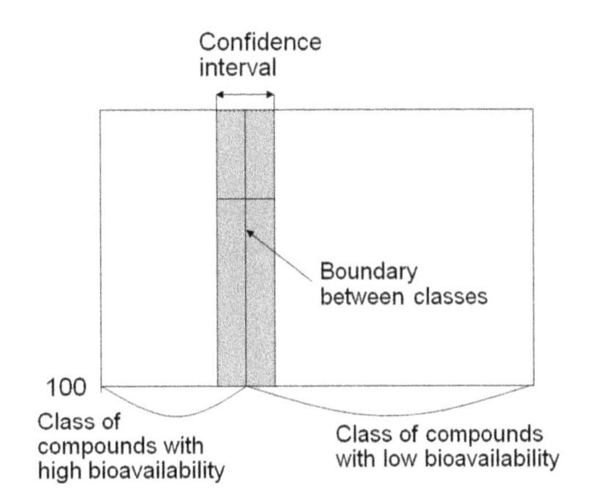

Fig. 14.3 The confidence interval of classification

As a result, classification models considering the confidence interval were obtained (Table 14.3). The presented models (IV-1 – VI-1) significantly increase the predictive ability. For the model IV-1 with the lowest classification error, we have cited the medicinal agents for which a detailed analysis, considering various physiological factors that reduce the bioavailability of drugs prior to their entering the systemic blood circulation, was carried out. These factors include: physical properties of medicinal agents, such as hydrophobicity, dissociation rate, solubility, drug formulation (immediate, delayed, extended or prolonged release, the use of supplementary substances, methods of production); the fact, whether a medicinal agent was entered on an empty stomach or after a meal; differences during the day;

Table 14.3 Classification models for two classes with a confidence interval

Number of the model	Class boundaries	Number of molecules in the class	Error of classification, %
IV-1	1) 90–100 2) 0–90	156 472	14
V-1	1) 80–100 2) 0–80	222 406	16
VI-1	1) 70–100 2) 0–70	280 348	18

speed of gastric emptying; induction/inhibition of other drugs or food; interactions with other drugs (antacids, alcohol, nicotine), interaction with some food products (grapefruit juice, pomelo, cranberry juice); transport proteins, substrates for transporters (e.g., P-glycoprotein); the state of the gastrointestinal tract, its function and morphology.

Thus, drugs that give the error in predicting can be divided into the following groups:

a. during their absorption transporters play the active role and they have certain targets (class 1-Bumetanide, Cefradine, Diflunisal, Folic acid, Levetiracetam, Loracarbef, Pramipexole)
b. are intensively subjected to metabolism and have the appropriate targets (class 1-Amobarbital, Anastrozole, Clofibratum, Clonazepam, Cyproterone, Dapsone, Dofetilide, Dolasetron, Donepezil, Ethosuximide, Felbamate, Galantamine, Glimepiride, Hexobarbitalum, Linezolid, Methimazole, Midodrine, Nevirapine, Oxaprozin, Pentobarbital, Phenprocoumon, Phenylpropanolamine, Primakvin, Pseudoephedrine, Tamsulosin, Tiagabine, Tocainide, Zonisamide; class 2-Busulfan, Etofyllinum)
c. have targets (class 1-Diazoxide, Indapamide, Minoxidil, Penbutolol, Pheniramine, Practolol, Rimantadine, Roxatidine, Sotalol, Trihexyphenidyl; class 2-Butabarbital)
d. all the factors have a place (class 1-Acetazolamide, Bezafibrate, Clonidine, Corticosterone, Cyclopenlhiazide, Doxycycline, Etodolac, Gemfibrozil, Hydrocortisone, Imatinib, Lamotrigine, Liothyronine, Phenylbutazone, Probenecid, Reboxetine, Rosiglitazone, Sertraline)
e. have other factors that affect the bioavailability (class 1-Acipimoks, Amosulalol, Antipyrinum, Betaxolol, Cicaplast, Cycloprolol, Chloroquine, Fenspiride, Flupirtine, Gestrinone, Digoxin, Isosorbide mononitrate, Letrozole, Lorazepam, Nicorandil, Pirprofen, Rilmenidine, Sulfadimezin, Tianeptine, Trapidil, Treosulfan; class 2-Renicin, Rizatriptan, Sulfadimidine)

Thus, the method of Random Forest is a quite promising tool for preliminary analysis for activity (bioavailability) of potential medicinal agents. However, there is a significant effect of various physiological factors that reduce the bioavailability of drugs prior to their entering the systemic blood circulation, and these factors are difficult to determine by modeling, as it requires additional experimental studies. In our view, this method is well for predicting the bioavailability of low molecular weight compounds, absorption of which occurs by simple diffusion.

14.4.1.2 Investigation of Permeability and Solubility of Drugs on the WHO List Using SiRMS Approach

Creation of *in silico* BCS to study the relationship between the structure of APIs and their bioavailability is highly important. Similarly to permeability classification, this would be based on experimental human intestine permeability data, or well-defined mass balance studies and/or comparison to an intravenous reference dose. However, since such data are available only for a small number of drugs, the provisional permeability classification was based on correlation of the estimated n-octanol/water partition coefficient using both Mlog*P* and Clog*P* of the uncharged form of the drug molecule [63, 64]. In order to determine the broad applicability and significance of the BCS, we developed a provisional classification of the WHO Essential Medicines List [65] and then extended this analysis to the top 200 drugs on the United States and others country lists [66]. Values for drug solubility were obtained from standard references (e.g., Merck Index, USP, etc.). The BCS classification of the WHO medicines was conducted using two criteria. The first, a solubility classification, was based on the calculated dose number. Drugs were categorized as "soluble" if they had a dose number of 1. The finding that 67 % of the drugs on the WHO list and 68 % on the top 200 U.S. list were classified as "high solubility" drugs suggests that major differences in drug BCS classification of the two lists are unlikely. A total of 43 drugs on the WHO list and 49 drugs on the U.S. list exhibited a solubility of < 0.1 mg/ml. However, a few of these drugs were classified as "soluble" drugs on the basis of dose numbers and may reflect recent trends toward development of highly lipophilic, low-solubility drugs that are quite potent. The percentages of the drugs in IR dosage forms on the WHO list that were classified as class 1 drugs based on Mlog *P* or Clog *P* were 23.6 and 28.5 %, respectively. The corresponding percentage of drugs classified as class 3 drugs were 31.7 and 35.0 %, respectively, and regulatory approval of biowaiver for this class of drugs is scientifically justified and recommended by WHO [67]. Hence, the majority of IR oral drug products on the WHO List of Essential Drugs are candidates for waiver of *in vivo* BE testing based on an *in vitro* dissolution test. The impact of waiving an expensive *in vivo* BE testing and its replacement by rapid and affordable in vitro dissolution standards in developing countries is expected to be profoundly significant. Similar results were obtained in a subsequent classification of the WHO list of Essential Medicines that was based primarily on human fraction absorbed (F_{abs}) literature data for the permeability assignment [49]. Out of 61 drugs that could be reliably classified, 34 % were classified as class I, 17 % as class II, 39 % as class III, and 10 % as class IV. In this analysis, hence, more than 70 % of the classified drugs proved to be candidates for waiver of *in vivo* BE testing based on *in vitro* dissolution test. Of course, other drug product characteristics, such as the therapeutic index and the potential influence of the excipients on the rate and extent of absorption, should also be considered.

In our studies, we have used the following drugs relating to different classes of BCS. SiRMS approach was used to develop the models that can classify compounds within the framework of BCS. To the set of studied compounds, 95 representatives of listed drugs relating to 4 classes of BCS, were included.

Class 1 (High Permeability, High Solubility): Albuterol, Allopurinol, Amlodipine (Amlo), Amoxicillin, Antipyrine, Dexamethasone, Diltiazem, Zidovudine, Isosorbide mononitrate, Ketoprofen, Lamivudine, Levonorgestrel, Levofloxacin, Metronidazole, Midazolam, Minocycline, Morphine, Nifedipine, Ofloxacin, Prednisolone, Propylthiouracil, Stavudine, Phenobarbital, Fluconazole, Chinin, Enalapril, Acetaminophen*, Diazepam*, Isoniazid*, Levodopa*, Metoprolol*, Paracetamol*, Pyrazinamide*, Salicylic acid*, Ethinylestradiol*.

Class 2 (High Permeability, Low Solubility): Azathioprine, Azithromycin, Alprazolam, Warfarin, Haloperidol, Glipizide, Griseofulvinum, Danazol, Dapsone, Diclofenac, Indometacin, Itraconzol, Carbamazepine, Carvedilol, Ketoconazole, Lansoprazol, Mebendazole, Mefloquine, Nalidixic acid, Nevirapine, Piroxicam, Praziquantel, Ritonavir, Rifampicinum, Spironolactone, Tamoxifen, Terfenadine, Trimethoprim, Ibuprofen*, Iopanoic acid*, Lovastatin*, Naproxen*, Oxaprozin*, Flubiprofen*, Cisapride*.

Class 3 (Low Permeability, High Solubility): Atropine, Aciclovir, Valsartan, Gancyclovir, Didanosine, Dicloxacillin, Zalcitabine, Lomefloxacin, Methyldopa, Methotrexatum, Nadolol, Pravastatin, Ranitidine, Tetracycline, Famotidine, Cefazolin, Ciprofloxacin, Erythromycin, Atenolol*, Hydrochlorothiazide*, Metformin*.

Class 4 (Low Permeability, Low Solubility): Sulfasalazine, Talinololum, Furosemide, Chlorothiazide. (drugs marked with asterisks * are included to the test set). They were selected randomly and not used for constructing of models, they were only used to examine the predictive ability of models.

In order to determinate both parameters simultaneously, at first, for all molecules, simplex (local) and some integral (describing the entire molecule) descriptors (just over 10 thousand) were calculated by us. Differentiation of atoms in simplex was conducted based on the following characteristics: (1) the type of atom; (2) partial charge; (3) lipophilicity of atom, (4) atomic refraction; (5) possibility of atom to act as a donor/acceptor of hydrogen in the formation of hydrogen bonds.

Total solubility (C_S) of the active pharmaceutical ingredient in an aqueous medium and the coefficient of penetration (P) of an active pharmaceutical ingredient through the lipophilic part of biomembranes have been represented as binary scale (0—low property value, 1—high property value).

Two models that adequately describe the solubility and penetration of compounds were obtained using classification and regression trees approach (Fig. 14.4). Model 1 (Fig. 14.4a) can be described as follows: compounds characterized by a high level of penetration, if: (a) molecule have no N–H groups; (b) in the presence of N–H groups, number of groups H H is more than 2. If the molecule has at least one N–H group and two or fewer H H groups, the compounds characterized by a low level of penetration. Model 2 (Fig. 14.4b) can be described as follows: a

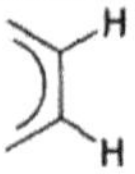

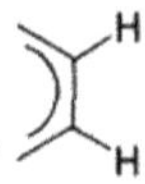

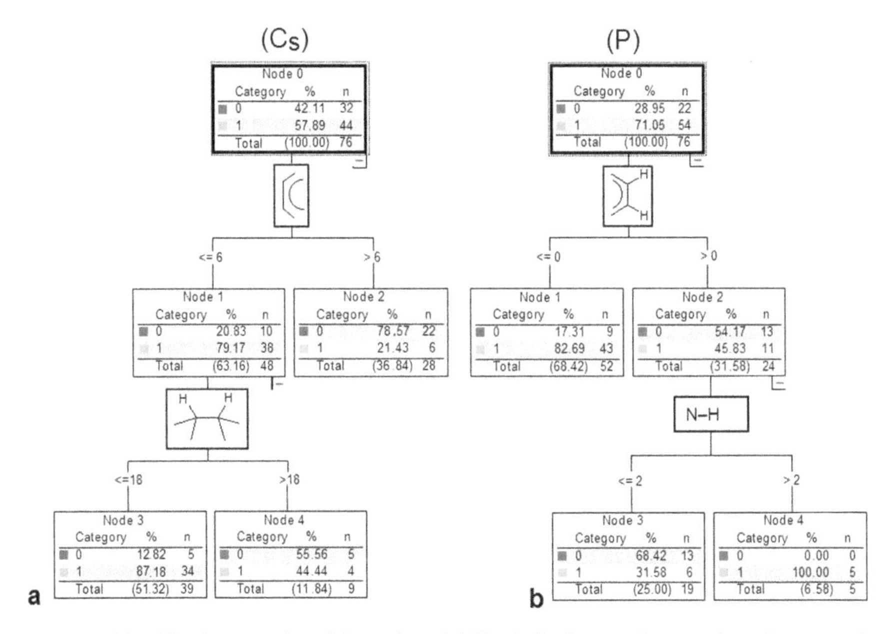

Fig. 14.4 Classification trees describing **a** the solubility, **b** the degree of penetration of compounds

compound is characterized by a low solubility if the molecule has more than 6 fragments. Compound is characterized by a high solubility if a molecule has six or fewer fragments and at the same time no more than 18 H–C–C–H fragments.

Statistical models 3 and 4 that quite fairly describe both studied properties (Table 14.4) were obtained using PLS method [10].

It is important to note that these models are characterized by lesser prediction errors for the training and test sets than models 1 and 2. Ouite low Q^2 values can be explained by the fact that the PLS method is more suited to the analysis of continuous (but not binary) data. As a result of the interpretation of models obtained by PLS method, molecular fragments that promote or prevent penetration and solubility of drugs were identified (Table 14.5).

The following rules were established: (1) The degree of penetration of drugs remains practically unchanged, and the solubility decreases with an increase in the length of alkyl chains (-C_4H_8- and above); (2) Amides of carboxylic acids are dissolved better than esters of acids; (3) The degree of penetration increases in the acid-amide-ester row; (4) To a large extent, the penetration is prevented due to the

Table 14.4 Statistical characteristics of the of the obtained models

№	R^2	Q^2	S_{OB}	S_{TB}	A	N	M
1	–	–	20%	21%	–	2	76
2	–	–	20%	26%	–	2	76
3	0.957	0.62	5%	16%	2	35	76
4	0.982	0.66	3%	0%	3	16	75

R^2 coefficient of determination (sign), Q^2 coefficient of determination in the conditions of sliding control, S_{OB} percentage of classification errors for the training set, S_{TB} percentage of classification errors for the test set, *A* number of latent variables, *M* number of molecules in the training set, *N* number of structure parameters in a model

presence tert-butylamino and isopropylamino groups; (5) In the series of saturated cycles, heterocycles reduce the degree of penetration compared to carbocycles.

14.4.1.3 Structure-Pharmacokinetic Relationships of Selected 1, 4-Benzodiazepines Derivatives

Today the number of 1,4-benzodiazepines derivatives synthesized in various laboratories of the world is over 3000 and 30 of them are drugs. This allows to choose the compond or drug, which is the most suitable for a given purpose. However, the physician must have the appropriate criteria (pharmacodynamics and pharmacokinetics) in order to declare that the drug has a certain advantage. The pharmacodynamic criteria include the duration of action of the benzodiazepines, which fall into three groups: (a) short-action of 2–10 h (oxazepazepam, temazepam, triazolam); (b) medium-action of about 10–15 h (alprozalam, bromazepam, lorazepam); and c) long-action of 15–30 h (clobazam, clonazepam, diazepam, nitrazepam) [68].

The half-life of elimination ($t_{1/2}$) of benzodiazepines is a special pharmacokinetic parameter. It clearly divides them into three groups: (a) long half-life 48 h and above; (b) medium, 24–48 h; and (c) short, less than 24 h [69]. We note also that clinical pharmacologists have recently linked the development of dependence on benzodiazepines to $t_{1/2}$ [70]. The half-life of elimination is closely related through the clearance (*Cl*) to other pharmacokinetic parameters such as the distribution volume (V_d) and the bioavailability (*F*).

Drugs with a short $t_{1/2}$ (alprozalam) are prescribed to patients in doses of 0.2–0.5 mg peroral 2–3 times per day; triazolam, 0.125–0.5 mg. For drugs with a medium $t_{1/2}$ (diazepam), the dose is 2–10 mg 2–4 times per day; long (flunitrazepam, 15–30 mg; gidazepam, 20–50 mg), 3 times per day.

These and other properties of benzodiazepine drugs make it critical to establish the quantitative relationship between their structure and pharmacokinetic properties in order to optimize their action and predict the properties of innovative drugs using QSPR models developed by us.

Herein the effects of structure and physico-chemical properties of substituted benzodiazepines (27 drugs, Table 14.6) on the change of their pharmacokinetic parameters ($t_{1/2}$, *Cl*, V_d, *F*, t_{max}) in the human body are studied.

Table 14.5 Relative effect of some fragments on the change of the penetration coefficient (ΔP) and solubility (ΔC_S) for active pharmaceutical ingredient in an aqueous medium

Fragment	ΔP	ΔC_S
1	2	3
Linker		
$-C_2H_4-$,	−0.003	−0.016
$-C_3H_6-$	0.003	−0.033
$-C_4H_8-$	−0.014	−0.081
$-C_5H_{10}-$	−0.006	−0.197
–CH=CH–	−0.005	−0.109
Functional group		
–COOH	−0.126	−0.011
$-CONH_2$	−0.181	0.252
$-NH_2$	−0.103	0.059
–CONH–	−0.044	−0.009
$-N(CH_3)_2$	−0.024	−0.016
–OH	−0.032	0.03
$-OCH_3$	−0.018	0.006
$-NO_2$	0.000	−0.017
$-COOC_2H_5$	−0.022	0.053
Terminal fragment		
–F	−0.008	−0.005
–Cl	0.003	−0.002
–I	0.0	−0.115
$-CF_3$	−0.046	−0.048
$-CH_3$	−0.011	−0.004
$-C_2H_5$	0.0	−0.032
$-C_3H_7$	−0.004	0.0
–iPr	−0.05	0.005
–NH–iPr	−0.282	0.049
–NH–tBu	−0.737	0.049
Saturated cycle		
Cyclohexane	−0.022	−0.074
Cyclopentane	0.0	−0.063
Piperidine	−0.083	−0.001
Tetrahydrofuran	−0.187	−0.016
Piperazine	−0.035	−0.027
Pyrrolidine	−0.156	0.0
Unsaturated cycle		
Phenyl	−0.007	−0.089
Pyridine	−0.038	−0.152
Pyrazine	−0.134	−0.048
Pyrimidine	−0.15	−0.148
Cyclohexene	−0.023	−0.125
1H-imidazole	−0.102	−0.017

Table 14.6 Investigated compounds

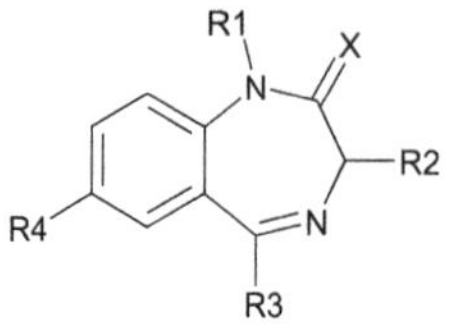

Preparation name	Skeleton	X	R^1	R^2	R^3	R^4
1	2	3	4	5	6	7
Alprazolam	B	N	CH_3	H	Ph	Cl
Bromazepam	A	O	H	H	2–pyridyl	Br
Halazepam	A	O	CH_2CF_3	H	Ph	Cl
Gidazepam	A	O	CH_2–CO–$NHNH_2$	H	Ph	Br
Diazepam	A	O	CH_3	H	Ph	Cl
Estazolam	B	N	H	H	Ph	Cl
Camazepam	A	O	CH_3	O–CO–$N(CH_3)_2$	Ph	Cl
Quazepam	A	S	CH_3	H	*o*–F–Ph	Cl
Clobazam	C		CH_3		Ph	Cl
Clonazepam	A	O	H	H	*o*–Cl–Ph	NO_2
Clorazepate	A	O	H	COOK	Ph	Cl
Lorazepam	A	O	H	OH	*o*–Cl–Ph	Cl
Medazepam	A	2H	CH_3	H	Ph	Cl
Mendon	A	O	H	COOH	Ph	Cl
Midazolam	B	C	CH_3	H	*o*–F–Ph	Cl
Nitrazepam	A	O	H	H	Ph	NO_2
Nordazepam	A	O	H	H	Ph	Cl
Oxazepam	A	O	H	OH	Ph	Cl
Pinazepam	A	O	CH_2CCH	H	Ph	Cl
Prazepam	A	O	CH_2–cyclopropyl	H	1–cyclohex-enyl	Cl
Temazepam	A	O	CH_3	OH	Ph	Cl
Tofisopam	D	–	C_2H_5	–	3,4–dime-thoxyphe-nyl	OCH_3

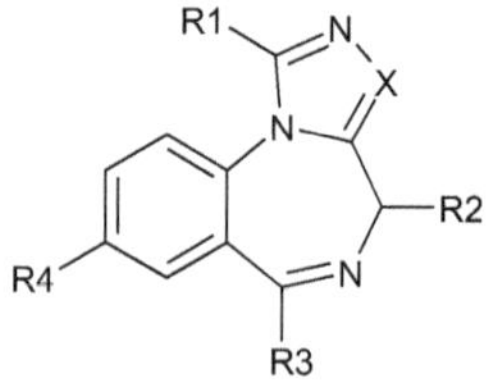

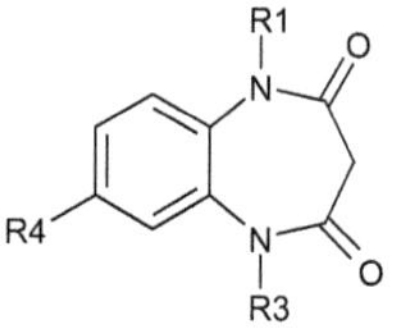

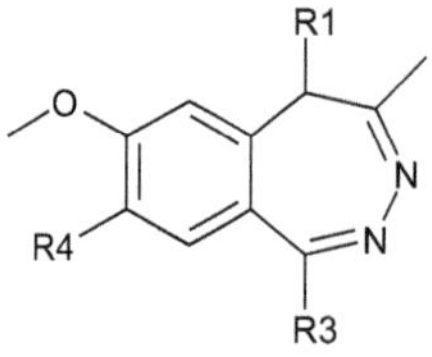

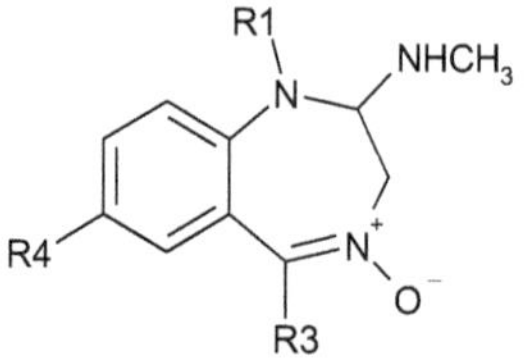

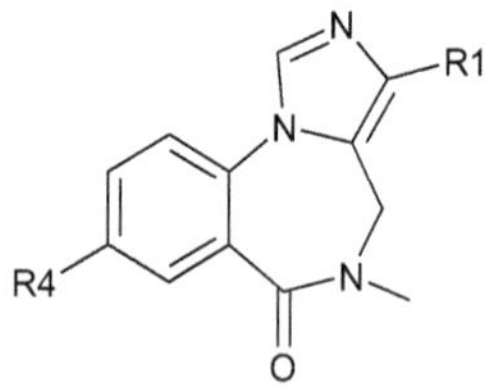

Table 14.6 (continued)

Preparation name	Skeleton	X	R[1]	R[2]	R[3]	R[4]
1	2	3	4	5	6	7
Triazolam	B	N	CH_3	H	*o*–Cl–Ph	Cl
Chlordiaz-epoxide	E		H		Ph	Cl
Phenazepm	A	O	H	H	*o*-Cl–Ph	Br
Flumazenil	F	–	$CO–OC_2H_5$	–	–	F
Flurazepam	A	O	$C_2H_5N(C_2H_5)_2$	H	*o*–F–Ph	Cl
Compound 1	A	O	H	OH	*o*–Cl–Ph	Br
Compound 2	A	O	H	$OCOCH_3$	*o*–Cl–Ph	Br
Compound 3	A	O	H	$OCO–CH(C_3H_7)_2$	*o*-Cl–Ph	Br
Compound 4	A	O	H	$OCO–(CH_2)_2COOH$	*o*–Cl–Ph	Br
Compound 5	A	O	H	$OCO–(CH_2)_3COOH$	*o*–Cl–Ph	Br
Compound 6	A	O	H	$OCO–C_6H_5$	*o*–Cl–Ph	Br
Compound 7	A	O	H	OCO–4-pyridyl	*o*–Cl–Ph	Br

Several integral parameters (describing the properties of the molecule in general) that were generated by the Dragon program were also used to construct statistical equations [71].

Thus, simplex and integral descriptors (over 10,000 total) were calculated in the initial stage for all molecules. Atoms in simplexes were differentiated based on the following characteristics: (1) atom type, (2) partial charge, (3) rigidity, (4) nucleophilicity, (5) electrophilicity [72], (6) lipophilicity of the atom [73], (7) atomic refraction, (8) ability of the atom to act as a H donor or acceptor in forming H-bonds, and (9) ability of the atom for Van-der-Waals attraction and (10) repulsion [74]. MLR, PLS [75] and classification trees [76] were used for building the models. Descriptors were selected using genetic algorithm [77] and trend-vector method [78]. Considering the small number of studied compounds, the model was not validated for a test set to avoid a loss of the required structural information. The model was checked for consistency using an iterative procedure [41].

Table 14.7 lists the observed values of the studied pharmacokinetic properties. It should be noted that this series is not sufficiently representative at high distribution volumes (V_d) and small bioavailability. Therefore, developed QSPR models have limited applicability domain.

Analysis of the data in Table 14.6 showed that the drugs could be divided into three groups according to values of one of the most important pharmacokinetic parameters, $t_{1/2}$ (elimination half-life) (Table 14.7).

Benzodiazepines in most instances are highly lipophilic compounds ($\log P = 2-4$). They are eliminated through excretion and metabolism. Benzodiazepine compounds with high $t_{1/2}$ values undergo during metabolism N_1-dealkylation (CYP3A4) and C_3-hydroxylation (CYP2C19). Intrinsic clearance [79] due to intestinal and hepatic CYP450 has been reported for this group. All metabolites of this group have pharmacometabolic profiles [80] with an analogous spectrum of psychotropic action.

Table 14.7 Observed values of studied pharmacokinetic characteristics

No.	Preparation	F^a	Cl^a	t_{max}	$t_{1/2}{}^a$		$V_d{}^a$
					Value	Class[b]	
1	2	3	4	5	6	7	8
1.	Alprazolam	0.88	0.74	1–2	10–12	3	0.72
2.	Bromazepam	0.84	0.5–1.5	8–20		3	
3.	Halazepam			1–3			
4.	Gidazepam		3.03		86	1	
5.	Diazepam	0.99	0.38	0.5–1.5	43	2	1.1
6.	Estazolam	1.00		1–1.5	10–24	3	
7.	Camazepam				12–24	3	
8.	Quazepam			0.4–1	36–120	1	
9.	Clobazam			2–4	10–30	3	
10.	Clonazepam	0.90	92.0	1–4	20–60	2	3.2
11.	Clorazepate	0.98		0.75–1	48	2	
12.	Lorazepam	0.93	1.10	1–2	10–12	3	1.3
13.	Medazepam	0.60		1–2	48–60	1	
14.	Mendon			1	12–24	3	
15.	Midazolam	0.40	6.6	0.5–1	1.5–3.5	3	1.1
16.	Nitrazepam	0.78	0.86	1.5–2	18–25	2	1.9
17.	Nordazepam				24	2	
18.	Oxazepam	0.97	1.05	1–2	8–10	3	0.6
19.	Pinazepam				48–60	1	
20.	Prazepam	0.25	1400	0.5	1.3	3	14.4
21.	Temazepam	0.90	1.00	0.3–0.7	5–15	3	0.95
22.	Tofisopam			2	6–8	3	
23.	Triazolam	0.44	5.60	0.5–1	1.5–5.5	3	1.1
24.	Chlordiazepoxide	1.00	0.54	2–4	5–30	3	0.3
25.	Phenazepam				10–18	3	
26.	Flumazenil	0.20	17.0	0.25	0.9	3	1
27.	Flurazepam	0.93	4.5	0.5–1	36–120	1	22

[a] *F* absolute bioavailability, *Cl* total preparation clearance, t_{max} time to reach maximum preparation concentration in blood, $t_{1/2}$ preparation elimination half-life, V_d preparation distribution volume
[b] Classification by elimination half-life: 1-long period, $t_{1/2}$ 48 h; 2-average period, $t_{1/2}$=24–48 h; 3-short period, $t_{1/2}$ h

Drugs of the second group eliminate the nitro group during metabolism to amino derivatives (NAT1), which do not possess psychotropic activity.

Some of the drugs of the third group (3-hydroxy derivatives) form in the human organism the corresponding inactive glucuronides (UGT 1). The others form triazolobenzodiazepines and imidazolobenzodiazepines, which are oxidized in the human organism to 3-hydroxy derivatives (CYP3A4).

We attempted to integrate the proposed classification system based on structural parameters (molecular simplexes). A classification tree was constructed (Fig. 14.5) and could be used to correctly classify the whole set of benzodiazepines using only four structural parameters (S2, S5, S7, S76).

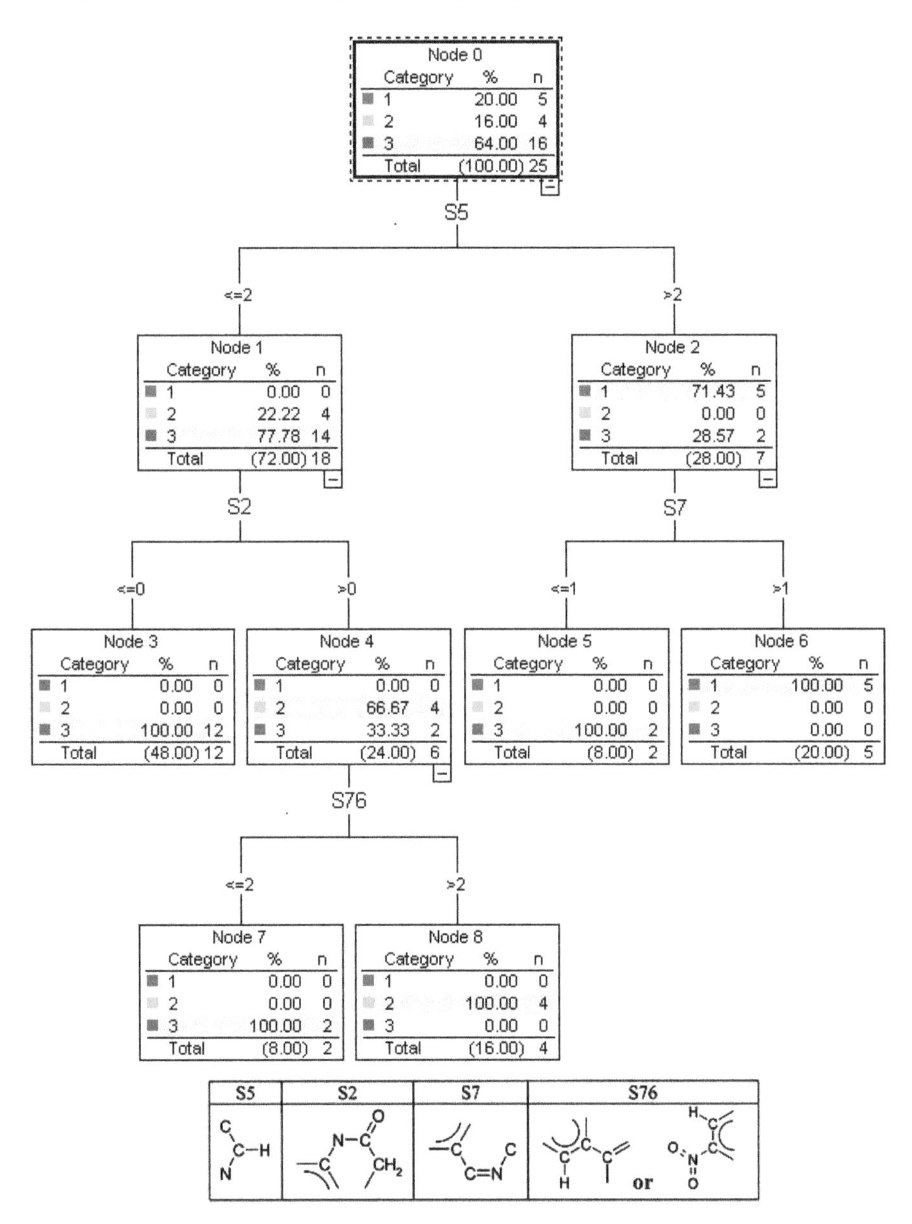

Fig. 14.5 Classification tree for deetermining elimination half-life of substituted benzodiazepines

The results showed that the third class of benzodiazepines usually contained two or less S5 fragments and no S2 fragments. Compounds of the second class also contained two or less S5 fragments, S2 fragments, and more than two S76 fragments. The first class of benzodiazepines had more than two S5 fragments and more than one S7 fragment. Thus, it seemed to us that the constructed classification tree was

Table 14.8 Statistical characteristics of developed PLS QSPR models

Statistical characteristics	Pharmacokinetic characteristics				
	F^a	$\log(Cl)^a$	$\log(t_{1/2})^a$	$\log(V_d)^b$	t_{max}^a
M, number of molecules in learning set	17	13	26	13	22
N, number of structural parameters in model	9	7	6	3	7
A, number of latent variables	1	1	1	1	1
R^2, determination coefficient	0.95	0.91	0.91	0.82	0.93
Q^2, determination coefficient from iterative procedure	0.94	0.81	0.87	0.81	0.85
SE, standard error of prediction for learning set	0.06	0.25	0.16	0.22	0.20

[a] Model constructed based on Simplex structure parameters
[b] Model constructed based on integral parameters of Dragon program

entirely suitable for qualitative rapid evaluations of the elimination half-life of benzodiazepine drugs.

Statistical characteristics of the QSPR models obtained by PLS method are shown in Table 14.8. Rather high values of R^2 and Q^2 indicated that the resulting models could be used to predict the studied pharmacokinetic properties.

The most influential structural fragments were determined based on the resulting QSPR models (Table 14.9, Fig. 14.6). Although a correlation between F and the lipophilicity was not observed ($R \approx 0$), the trend toward an increased contribution of the molecular fragment to the total bioavailability upon increasing its lipophilicity was clearly visible. This was especially evident for aromatic fragments A01—A03 and D01—D04 (Fig. 14.6). This trend did not hold for non-aromatic fragments (e.g., for fragments B11, B14, and D05).

The fragments 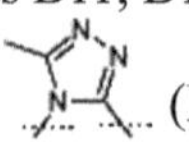 (B01) and 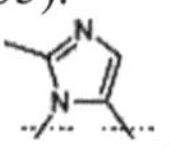 (B09) had the highest negative effect on the bioavailability. Similar fragment 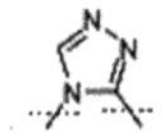 (B06), which does not contain a methyl group, had the highest bioavailability. All molecules containing a group with a double bond and without the negative-effect B01 and B09 groups (Table 14.9) had high levels of bioavailability. These observations were confirmed by the model obtained using integral molecular parameters (Dragon program) and multiple linear regression:

$$F(\%) = 0.44 \cdot C_{024} + 1.63Me - 1.05n_{CaR} - 1.53n_O$$
$$R^2 = 0.886;\ F = 23.3,\ SE = 0.11,\ Q^2 = 0.813$$

where C_{024} is the number of benzene rings, Me, the average Sanderson atomic electronegativity, n_{CaR}, the number of substituted aromatic carbons, and n_O, the number of O atoms.

Thus, the presence of benzene rings increased the bioavailability of the substituted benzodiazepines. Substitution in the aromatic rings reduced the bioavailability. It

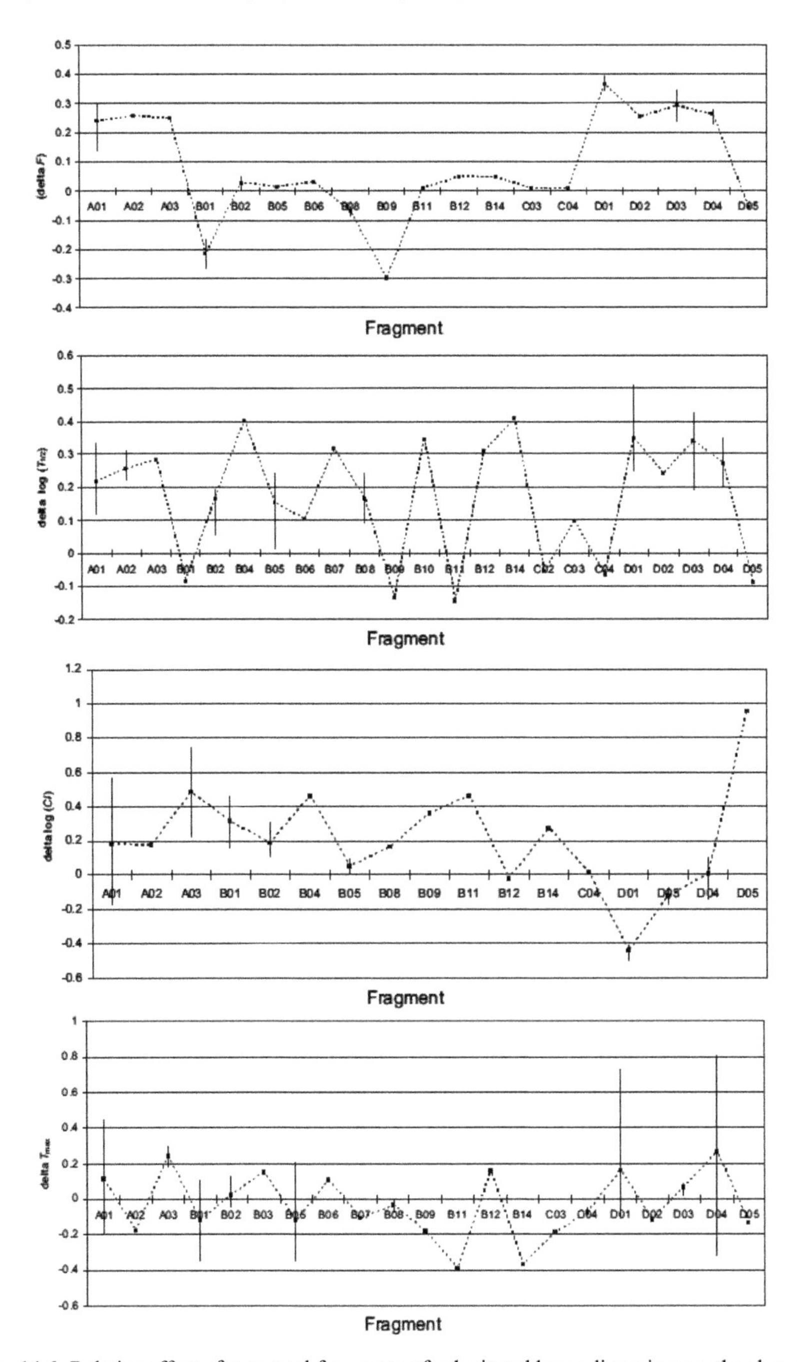

Fig. 14.6 Relative effect of structural fragments of substituted benzodiazepines on the change of their pharmacokinetic characteristics. Points denote average contributions of fragments. The range of their change due to the environment is shown by vertical lines

Table 14.9 Relative effect of structural fragments of substituted benzodiazepines on the change of their pharmacokinetic characteristics (ΔX) calculated from the corresponding PLS models

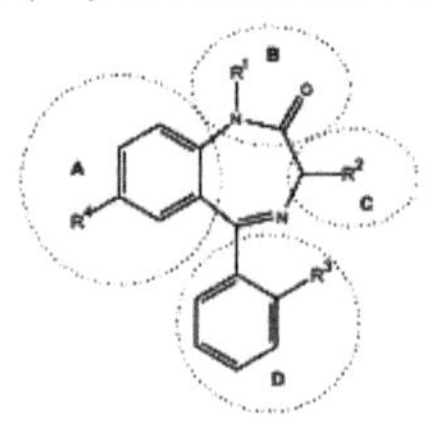

Fragment		ΔX				
		$\Delta \log P$	ΔF	$\Delta \log(Cl)$	$\Delta \log(T_{1/2})$	Δt_{max}
1	2	3	4	5	6	7
	A01	2.19	0.24	0.18	0.22	0.11
	A02	2.36	0.26	0.18	0.26	– 0.18
	A03	0.61	0.25	0.48	0.28	0.24
	B01	– 0.04	– 0.22	0.31	– 0.08	– 0.12
	B02	– 0.37	0.03	0.19	0.17	0.02
	B03	0.55	–	–	–	0.15
	B04	– 1.64	–	0.46	0.40	–
	B05	0.03	0.01	0.05	0.15	– 0.13
	B06	– 0.46	0.03	–	0.10	0.11
	B07	0.79	–	–	0.32	– 0.11
	B08	0.17	– 0.07	0.16	0.17	– 0.04
	B09	0.06	– 0.30	0.36	– 0.14	– 0.18
	B10	0.23	–	–	0.34	–
	B11	0.80	0.01	0.46	– 0.15	– 0.40
	B12	– 0.25	0.05	– 0.02	0.31	0.16
	B14	0.76	0.05	0.27	0.41	– 0.37

Table 14.9 (continued)

1	2	3	4	5	6	7
	C02	0.12	–	–	– 0.06	–
	C03	– 0.77	0.01	–	0.10	– 0.19
	C04	– 0.75	0.01	0.01	– 0.07	– 0.08
	D01	1.93	0.36	– 0.45	0.35	0.16
	D02	0.56	0.25	–	0.24	– 0.12
	D03	2.05	0.29	– 0.12	0.34	0.07
	D04	2.57	0.26	0.01	0.27	0.26
	D05	1.53	– 0.06	0.95	– 0.09	– 0.14

should also be noted that more O atoms in the molecule led to lower bioavailability. In all probability, this was caused to a large extent by the ability of the O atoms to act as acceptors in forming H-bonds. This parameter (n_O) correlated to a large extent with the number of acceptor centers for H-bonds ($R = 0.9$). This agreed completely with the classical "rule of 5" [81]. Tendencies opposite to bioavailability (see above) were observed in general for the total clearance (Cl) of the drug. Thus, the presence of unsubstituted benzene rings (D01) and π-donor substituents (–F, –Cl, –Br) in the aromatic ring decreased the clearance. The presence of π-acceptors (–NO_2, fragment A03) increased it substantially. Replacement of an aromatic ring in the D position by a cyclohexene (D05) and the presence of B04 fragments 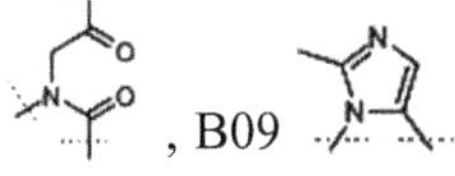 , B09 B11 (Table 14.9) increased the clearance. The clearance (Cl) was inversely related to the lipophilicity only for aromatic fragments.

The following regression model was obtained using integral parameters of the molecules (Dragon program):

$$\log(Cl) = -1.88Ui + 0.99n_{HDon} + 1.70n_{CaR}$$
$$R^2 = 0.830, F = 14.68, SE = 0.39, Q^2 = 0.62$$

where Ui is the unsaturation index [82], n_{HDon}, the number of H-donors; n_{CaR}, the number of substituted aromatic carbons . Thus, the presence of H-donors in the molecule and substitution in the aromatic rings increased the clearance. It was also noticed that the clearance was higher for the more substituted molecules.

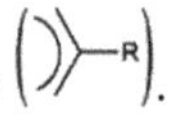

The effect of structural fragments on the change of elimination half-life ($t_{1/2}$) was somewhat similar to that described above for the bioavailability. Thus, all lipophilic aromatic fragments had high $t_{1/2}$ values (Fig. 14.6). The following model was obtained with Dragon descriptors and MLR method:

$$\log(t_{1/2}) = 0.77C_{-024} - 0.28n_{CIC} + 0.35n_{C-N} + 0.35C_{-006}$$
$$R^2 = 0.860, F = 31.1, SE = 0.21, Q^2 = 0.814$$

where C_{-024} is the number of benzene rings, n_{CIC}, the number of rings; n_{C-N}, the number of aliphatic imino groups; C_{-006}, the number of $-CH_2RX$ fragments, where X=halogen.

A study of the effect of the structure of substituted benzodiazepine derivatives on the volume of distribution (V_d) revealed common tendencies in the effect of the substituents that were similar to those for the clearance. Thus, V_d increased for B11 fragments 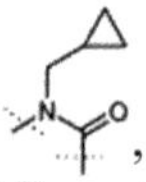 , B14 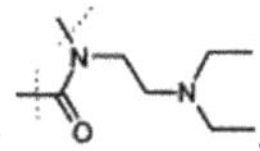 , and decreased for B12 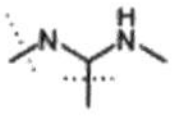 (Table 14.7).

The MLR model that was obtained using integral parameters (Table 14.7) had the form:

$$\log(V_d) = 0.480MR - 0.422ARR - 0.375Hy$$

where MR is the molar refractivity; ARR, the aromaticity index and Hy, the hydrophilic factor [83]. Thus, it can be stated that refractivity (electronic polarizability) is capable of increasing the distribution volume whereas high aromaticity and hydrophilicity decrease it.

Consistent tendencies for the effect of various fragments could not be found for the resulting PLS model that described the relationship of the structure of substituted benzodiazepines and the time to reach the maximum concentration (t_{max}) (Tables 14.8 and 14.9). Adequate statistical equations for this property could not in general be obtained using the parameters of the Dragon program.

The resulting PLS models were used to predict the pharmacokinetic properties of several benzodiazepine compounds (Table 14.10).

All predicted compounds will presumably possess high bioavailability and short elimination half-lives. Most likely high bioavailability values are caused by the presence of two aryl fragments and small number of H-bond donors and acceptors whereas short elimination half-lives are caused by the presence in these compounds of less than two S5 fragments and the absence of S2 fragments (Fig. 14.6, Node 3).

Table 14.10 Predicted values of pharmacokinetic characteristics for certain benzodiazepines

Compound	F	log (Cl)	t_{abs}	log ($t_{1/2}$)	Class by $t_{1/2}$	log (V_d)
Phenazepam	0.90	0.34	1.75	1.41 (1.15)[a]	3 (3)	0.21
Compound 1	0.90	0.34	1.40	0.92	3	0.00
Compound 2	≈1.0	0.61	0.96	0.79	3	0.66
Compound 3	≈1.0	0.84	1.01	0.79	3	1.68[b]
Compound 4	≈1.0	1.18	2.07	0.66	3	1.08
Compound 5	≈1.0	1.13	2.01	0.66	3	1.25[b]
Compound 6	≈1.0	−0.52	1.52	1.47	3	1.11
Compound 7	≈1.0	−0.12	1.35	1.26	3	1.02

[a] Observed values given in parentheses

[b] Unreliable predictions because of significant structural differences of these molecules from the learning set according to model

14.4.2 Prediction of Absorption and Bioavailability Using Physiologically Based Modeling

Oral drug absorption is a complex process. It consists of multiple steps that may include drug disintegration and dissolution, degradation, gastric emptying, intestinal transit, intestinal permeation and transport, intestinal and hepatic metabolism. The factors that may have impact on the rate and extent of drug absorption are dosage form, physico-chemical and biopharmaceutical properties of the API, and physiology of the GI. Knowledge of how these steps and factors influence absorption has fostered the development of predictive models for oral drug absorption. Pharmacokinetic models are used to describe the time-dependent distribution and disposition of an API in a living system. In contrast to the QSAR, gastrointestinal absorption model (Physiologically Based Absorption Modeling, PBPK) models vary in complexity and capability according to their intended purpose. PBPK models differ from one or two compartment models by representing physiological, physico-chemical and biochemical processes in the species of interest. In most PBPK models, tissues are represented by specific compartments, each with a unique set of physiological (blood flows, drug fate in the GI tract), physico-chemical (partition coefficients) and biochemical (metabolic rates) parameter values. Target tissues are generally represented individually. The absorption model connected to a distribution model predicts the *in vivo* pharmacokinetic profiles.

PBPK approaches are classified into three categories [82]: quasiequilibrium models, steady-state models, and dynamic models. The classification of these models is based upon their dependence on spatial and temporal variables. The quasiequilibrium models, which are independent of spatial and temporal variables, include the *pH*-partition hypothesis and absorption potential concept. The steady-state models are limited to prediction of the extent but not the rate of oral drug absorption. The dynamic models consider spatial and temporal variables and can predict both the rate and extent of oral drug absorption. The dynamic models include dispersion models and compartmental models. Both models can be linked to pharmacokinetic models to predict plasma concentration-time profiles of drugs. The dispersion models defines the GI tract as a single tube with spatially varying properties (*pH*, surface area, etc.). Instead of treating the small intestine as one long cylindrical tube in a dispersion model, compartmental models assume the GI tract as one compartment or a series of compartments with linear transfer kinetics, and each compartment is well mixed with a uniform concentration. There have been several reports on physiologically based mathematical compartmental models (CAT, ACAT, Grass, GITA, ADAM) that are capable of producing such predictions, and there are a few commercially available software packages (GastroPlusTM, IDEA™, INTELLIPHARM PK, PK-SIM, etc.) that have been shown to predict the human absorption properties with a fairly high degree of accuracy. As these models use the interplay between the drug characteristics and the human physiology to simulate the processes involved in the oral absorption of drugs, they also give information about the underlying mechanisms for absorption limitations and as such they are well suited to be used in progressing new chemistry and to support formulation development.

Study [84] has attempted to apply gastrointestinal simulation technology and integration of physiological parameters to predict biopharmaceutical drug classification. GastroPlus® was used with experimentally determined physico-chemical and pharmacokinetic drug properties to simulate the absorption of several weak acid and weak base BCS class 2 compounds. Simulation of oral drug absorption given physico-chemical drug properties and physico-chemical parameters will aid justification of biowaivers for selected BCS class 2 compounds. In silico models are useful to identify BCS class 2 biowaiver candidate drugs. The risk of bioinequivalence in terms of C_{max} has shown to be higher than for *AUC*. Class 2 weak acids and bases in immediate release dosage forms may be eligible for biowaivers provided that the dose dissolves completely before reaching middle jejunum. Biowaivers for some class 2 drugs also necessitate the availability of a discriminative and *in vivo* predictive *in vitro* dissolution method. Thus they should be complemented by prospective *IVIVC* studies to validate the proper selection of biowaiver candidate drugs.

The *in vivo* absorbability of drugs categorized into the BCS class 2 is very difficult to predict because of the large variability in the absorption and/or dissolution kinetics and the lack of an adequate *in vitro* system for evaluating the dissolution behavior. Fujioka et al. [85] tried to predict the plasma concentration-time profile of griseofulvin after oral administration into rats, based on GI-Transit-Absorption model (GITA model). Griseofulvin, with Clog$P$=2.88, was taken as a model drug of BCS class 2, and the *in vitro* dissolution study was performed to evaluate the dissolution rate constant by employing several different dissolution media including FaSSIF, FeSSIF and a couple of novel media. Using the dissolution rate constants (k_{dis}) of griseofulvin obtained with JP 1st solution, JP 2nd solution, FaSSIF, FeSSIF, and modified SIBLM as a medium, simulation lines were not able to describe the observed mean plasma profile at all. On the other hand, a calculated line provided by employing k_{dis} obtained with MREVID 2 (medium reflecting *in vivo* dissolution 2), a new medium, was in better agreement with the observed mean plasma profile than existing media.

Grass Model describes fluid movement (emptying and transit) in the GI tract and calculates the drug absorption in each compartment (stomach, duodenum, jejunum, ileum, and colon) over time based primarily on three parameters-solubility, permeability, and tissue surface area. This model was shown to predict plasma concentration time profiles including the *AUC*, C_{max}, and t_{max} for ketorolac (BCS 1) and gancyclovir (BCS 3) well. The software IDEA™ was developed based upon compartmental models (IDEA™ is not currently available). By using IDEA™, the fractions of dose absorbed for atenolol (BCS 3), naproxen (BCS 2), and gancyclovir (BCS 2) in animals and humans were successfully predicted [86]. Similar to GastroPlus™, factors including dose, solubility, and permeability are considered. However, IDEA™ does not take first pass metabolism and drug transport into account.

14.5 Conclusions

Sufficient intestinal absorption of orally administered drugs from the gastrointestinal tract is one of the prerequisites for successful oral drug therapy. It is generally accepted that the main barrier interrupting drug absorption is formed by the intestinal epithelium and several routes can be followed to pass it. Passive transport often occurs through the cell membrane of the enterocytes (transcellular transport), this is the predominant transport route for hydrophobic drugs. Another passive route is transport via the tight junctions between the enterocytes, i.e., paracellular transport. Hydrophilic compounds are mostly transported via the paracellular route. Finally, carrier-mediated transport (influx or efflux) can be observed. Intestinal drug absorption is controlled by dissolution rate and solubility, determining how fast a drug reaches maximum concentration in the lumenal intestinal fluid, and permeability coefficient, which relates to the rate at which dissolved drug will cross the intestinal wall to reach the portal blood circulation. Determination of the dissolution, solubility (at different pH values), and permeability properties of drug candidates can thus provide information about their absorption potential, and thus allows evaluation of the compounds according to BCS. The aim of the BCS is to provide a regulatory tool for replacing certain bioequivalence studies by accurate *in vitro* dissolution tests. This will certainly reduce costs and time in the drug development process, both directly and indirectly, and reduce unnecessary drug exposure in healthy subjects, which is normally the study population in bioequivalence studies. The BCS is today only intended for oral immediate-release products that are absorbed throughout the intestinal tract. It has been reported that an application of a BCS strategy in drug development will lead to significant direct and indirect savings for pharmaceutical companies.

Usage of *in silico* tools for modeling of the properties underlying the BCS can be used for fast preliminary screening of new compounds to determine their positions in the BCS. Nowadays a lot of different models and systems for prediction of bioavailability, permeability, solubility, etc., are available. But all of them lack the prediction accuracy. There are two ways to overcome this problem: (1) improve algorithms of model development, (2) expand the training sets which are used for model development by novel experimental data. We believe that the usage of *in silico* modeling for prediction of pharmacokinetic properties will be boosted by the recent explosion of chemogenomics and other types of data.

References

1. Amidon GL, Lennernas H, Shah VP et al (1995) A theoretical basis for a biopharmaceutics drug classification: the correlation of in vitro drug product dissolution and in vivo bioavailability. Pharm Res 12:413–420
2. Golovenko NYa, Baula OP, Borisyuk IYu (2010) Biofarmatsevticheskaya klassifikatsionnaya sistema (The biopharmaceutics classification system). Kiev
3. Sachan NK (2009) Biopharmaceutical classification system: a strategic tool for oral drug delivery technology. Asian J Pharm 3:76–81
4. Dressman J, Butler J, Hempenstall J et al (2001) The BCS: where do we go from here? Pharm Tech 25:68–76
5. Thiel-Demby VE, Humphreys JE, Williams LA et al (2009) Biopharmaceutics classification system: validation and learnings of an in vitro permeability assay. Mol Pharm 6:11–18
6. Kovacevic I, Parojcic J, Homsek I et al (2009) Justification of biowaiver for carbamazepine, a low soluble high permeable compound, in solid dosage forms based on IVIVC and gastrointestinal simulation. Mol Pharm 6:40–47
7. FDA, Draft guidance for industry: bioavailability and bioequivalence studies for orally administered drug products-general considerations, US Department of Health, Food and Drug Administration, Center for Drug Evaluation and Research BP, August 1999
8. Golovenko NYa (2004) Fiziko-khimicheskaya farmakologiya (Physico-Chemical Pharmacology). Astroprint, Odessa
9. van de Waterbeemd H (2003) Physico-chemical approaches to drug absorption. In: van de Waterbeemd H, Lennernas H, Artursson P (eds) Drug bioavailability: estimation of solubility, permeability, absorption and bioavailability. Wiley-VCH, pp 3–20
10. Martinez M, Amidon GA (2002) Mechanistic approach to understanding the factors affecting drug absorption: a review of fundamentals. J Clin Pharmacol 42:620–643
11. Barthe L, Woodley J, Houin G (1999) Gastrointestinal absorption of drugs: methods and studies. Fundam Clin Pharmacol 13:154–168
12. Hidalgo IJ, Li J (1996) Carrier-mediated transport and efflux mechanisms in Caco-2 cells. Adv Drug Deliv Rev 22:53–66
13. Lobenberg R, Amidon GL (2000) Modern bioavailability, bioequivalence and biopharmaceutics classification system: new scientific approaches to international regulatory standards. Eur J Pharm Biopharm 50:3–12
14. Emami J (2006) In vitro-in vivo correlation: from theory to applications. J Pharm Pharm Sci 9(2):31–51
15. Golovenko NYa, Borisyuk IYu (2008) The biopharmaceutical classification system-experimental model of prediction of drug bioavailability. Biochem Suppl Series B: Biomed Chem 2(3):235–244
16. EMEA (1998) Note for guidance on the investigation of bioavailability and bioequivalence, (CPMP/EWP/QWP/1401/98), Committee for proprietary medicinal product
17. FDA, Draft guidance for industry: in vivo drug metabolism/drug interaction studies study design, data analysis, and recommendations for dosing and labeling, US Department of Health, Food and Drug Administration, Center for Drug Evaluation and Research Clin Pharm, November 1998

18. FDA, Draft guidance for industry: waiver of in vivo bioavailability and bioequivalence studies for immediate release solid oral dosage forms containing certain active moieties/active ingredient based on a biopharmaceutics classification system, US Department of Health, Food and Drug Administration, Center for Drug Evaluation and Research BP2, January 1999
19. Dressman JB, Amidon GL, Reppas C et al (1998) Dissolution testing as a prognostic tool for oral drug absorption: immediate release dosage forms. Pharm Res 15:11–22
20. Meyer MC, Straughn AB, Jarvi EJ et al (1992) The bioinequivalence of carbamazepine tablets with a history of clinical failures. Pharm Res 9:1612–1616
21. Stephen RJ, Zheng W (2006) Recent progress in the computational prediction of aqueous solubility and absorption. AAPS J 8(1):E27–E40
22. Houa T, Wangb J, Zhangc W et al (2006) Recent advances in computational prediction of drug absorption and permeability in drug discovery. Curr Med Chem 13:2653–2667
23. Golovenko NYa, Borisyuk IYu, Kuz'min VE (2007) The dependence of "structure-property" in the models, predictive bioavailability of drugs. Farmacom 3:27–36
24. Raevskii OA, Kazachenko IV, Raevskaya OE (2004) The calculation of the bioavailability of drugs based on the similarity of the molecular structures. Chim-Farm Zhurn 38(10):3–8
25. Norinder U, Haeberlein M (2003) Calculated molecular properties and multivariate statistical analysis in absorption prediction. In: Drug bioavailability, methods and principles in medicinal chemstry, Van de Waterbeemd H, Lennern SH, Artursson P. (eds.), WILEY-VCH, Weinheim, 18:358–405
26. Lipinski CA (2000) Drug-like properties and the causes of poor solubility and poor permeability. J Pharmacol Toxicol Methods 44:235–249
27. Veber DF, Johnson SR, Cheng H-Y et al (2002) Molecular properties that influence the oral bioavailability of drug candidates. J Med Chem 45:2615–2623
28. Lu JJ, Crimin K, Goodwin JT et al (2004) Influence of molecular flexibility and polar surface area metrics on oral bioavailability in rat. J Med Chem 47:6104–6107
29. Hou T, Wang J, Zhang W et al (2007) ADME evaluation in drug discovery. 6. Can oral bioavailability in humans be effectively predicted by simple molecular property-based rules? J Chem Inf Model 47:460–463
30. Hirono S, Nakagome I, Hirano H et al (1994) Non-congeneric structure–pharmacokinetic property correlation studies using fuzzy adaptive least-squares: oral bioavailability. Biol Pharm Bull 17:306–309
31. Bains W, Gilbert R, Sviridenko L et al (2002) Evolutionary computational methods to predict oral bioavailability QSPRs. Curr Opin Drug Discov Devel 5(1):44–51
32. Zadeh LA (1977) Fuzzy sets and their applications to classification and clustering. In: Ryzin JV (ed) Classification and clustering. Academic, NY, pp 251–299
33. Wessel MD, Jurs PC, Tolan JW et al (1998) Prediction of human intestinal absorption of drug compounds from molecular structure. J Chem Inf Comput Sci 38:726–735
34. Egan WJ, Merz JKM, Baldwin JJ (2000) Prediction of drug absorption using multivariate statistics. J Med Chem 43:3867–3877
35. Zhao YH, Abraham MH, Le J et al (2003) Evaluation of rat intestinal absorption data and correlation with human intestinal absorption. Eur J Med Chem 38(3):233–243
36. Yoshida F, Topliss JG (2000) QSAR model for drug human oral bioavailability. J Med Chem 43:2575–2585
37. Andrews CW, Bennett L, Yu LX et al (2000) Predicting human oral bioavailability of a compound: development of a novel quantitative structure–bioavailability relationship. Pharm Res 17:639–644
38. Navia MA, Chaturvedi PR (1996) Design principles for orally bioavailable drugs. Drug Discov Today 1:179–189
39. Smith AB, Hirschmann R, Pasternak A et al (1997) An artificial antiparallel β-sheet containing a new peptidomimetic template. J Med Chem 40:2440–2444
40. Palm K, Stenberg P, Luthman K et al (1997) Polar molecular surface properties predict the intestinal absorption of drugs in humans. Pharm Res 14:568–571

41. Clark DE (1999) Rapid calculation of polar molecular surface area and its application to the prediction of transport phenomena. 1. Prediction of intestinal absorption. J Pharm Sci 88:807–814
42. Kuz'min VE, Artemenko AG, Polischuk PG et al (2005) Hierarchic system of QSAR models (1D–4D) on the base of simplex representation of molecular structure. J Mol Model 11:457–467
43. Kuz'min VE, Artemenko AG, Lozitsky VP et al (2002) The analysis of structure–anticancer and antiviral activity relationships for macrocyclic pyridinophanes and their analogues on the basis of 4D QSAR models (simplex representation of molecular structure). Acta Biochim Pol 49:157–168
44. Lindgren F, Geladi P, Rannar S et al (1994) Interactive variable selection (IVS) for PLS. Part 1: theory and algorithms. J Chemom 8:349–363
45. Rannar S, Lindgren F, Geladi P et al (1994) A PLS kernel algorithm for data sets with many variables and fewer objects. Part 1: theory and algorithm. J Chemom 8:111–125
46. Rogers D, Hopfinger AJ (1994) Application of genetic function approximation to quantitative structure–activity relationships and quantitative structure-property relationships. J Chem Inf Comput Sci 34(4):854–866
47. Carhart RE, Smith DH, Venkataraghavan R (1985) Atom pairs as molecular features in structure–activity studies. Definition and application. J Chem Inf Comput Sci 25:64–73
48. Kuz'min VE, Artemenko AG, Muratov EN (2008) Hierarchical QSAR technology on the base of Simplex representation of molecular structure. J Comput Aided Mol Des 22:403–421
49. Lindenberg M, Kopp S, Dressman JB (2004) Classification of orally administered drugs on the World Health Organization Model list of essential medicines according to the biopharmaceutics classification system. Eur J Pharm Biopharm 58:265–278
50. Kuz'min VE, Artemenko AG, Muratov EN et al (2010) Virtual screening and molecular design based on hierarchical QSAR technology. In: Puzyn T, Cronin M, Leszczynski J (eds) Recent advances in QSAR studies. Springer, London, pp. 127–176
51. Golbraikh A, Tropsha A (2002) Beware of Q2. J Mol Graph Model 20:269–276
52. Seel M, Turner DB, Willett P (1999) Effect of parameter variations on the effectiveness of HQSAR analyses. Quant Struct-Activ Relat 18:245–252
53. Breiman L (2001) Random forests. Mach Learn 45:5–32
54. Breiman L, Friedman JH, Olshen RA et al (1984) Classification and regression trees. Wadsworth, Belmont, p 368
55. Svetnik V, Liaw A, Tong C, Culberson JC, Sheridan RP, Feuston BP (2003) Random Forest: a classification and regression tool for compound classification and qsar modeling. J Chem Inf Comput Sci 43(6):1947–1958
56. program CF, Polishchuk PG (2010–2013) http://qsar4u.com. Accessed 10 March 20112
57. Kuz'min VE, Artemenko AG, Lozitska RN et al (2005) Investigation of anticancer activity of macrocyclic Schiff bases by means of 4D-QSAR based on simplex representation of molecular structure. SAR QSAR Environ Res 16(3):219–230
58. Kuz'min VE, Artemenko AG, Muratov EN et al (2007) Quantitative structure-activity relationship studies of [(biphenyloxy)propyl]isoxazole derivatives-human rhinovirus 2 replication inhibitors. J Med Chem 50:4205–4213
59. Muratov EN, Varlamova EV, Artemenko AG et al (2011) QSAR analysis of[(biphenyloxy) propyl] isoxazoles: agents against coxsackievirus B3. Future Med Chem 3(1):15–27
60. Muratov EN, Artemenko AG, Varlamova EV et al (2010) Per aspera ad astra: application of Simplex QSAR approach in antiviral research. Future Med Chem 2:1205–1226
61. Kovdienko NA, Polischuk PG, Muratov EN et al (2010) Application of random forest and multiple linear regression technologues to QSPR prediction of an aqueous solubility for military compounds. Mol informatics 29:394–406
62. Golovenko NYa, Kuz'min, Artemenko AG et al (2011) Prediction of bioavailability of drugs by the method of classification models. Clin Inform Telemed 8(7):88–92
63. Oh DM, Curl RL, Amidon GL (1993) Estimating the fraction dose absorbed from suspensions of poorly soluble compounds in humans: a mathematical model. Pharm Res 10(2):264–270

64. Winiwarter S, Ax F, Lennernas H et al (2003) Hydrogen bonding descriptors in the prediction of human in vivo intestinal permeability. J Mol Graph Model 21:273–287
65. Kasim NA, Whitehouse M, Ramachandran C et al (2004) Molecular properties of WHO essential drugs and provisional biopharmaceutical classification. Mol Pharm 1:85–96
66. Takagi T, Ramachandran C, Bermejo M et al (2006) A provisional biopharmaceutical classification of the top 200 oral drug products in the United States, Great Britain, Spain, and Japan. Mol Pharm 3:631–643
67. Proposal to waive in vivo bioequivalence requirements for the WHO model list of essential medicines immediate release, solid oral dosage forms. http://www.who.int/medicines/services/expertcommittees/pharmprep/QAS04_109Rev1_Waive_invivo_bioequiv.pdf.
68. Uchimura N, Takeuchi N, Kuwahara U et al. (2002) Situation and problem of administration methods and the intermission of hypnotics. Psych Clin Neurosci 56(3):295–296
69. Caccia S, Yaratini S (1985) Antiepileptic drug. Springer-Verlag, Berlin, pp 575–593
70. Hallfors DD, Saxe L (1993) The dependence potential of short half-life benzodiazepines: a meta-analysis. Am J Public Health 83(9):1300–1304
71. Dragon Software v. 3.0. Milano Chemometrics and QSAR Research Group. www.disat.unimib.it _chm. Accessed 5 May 20112
72. Jolly WL, Perry WB (1973) Estimation of atomic charges by an electronegativity equalization procedure calibrated with core binding energies. J Am Chem Soc 95:5442–5450
73. Wang R, Fu Y, Lai L (1997) A new atom-additive method for calculating partition coefficients J Chem Inf Comput Sci 37:615–621
74. Rappe AK, Casewit CJ, Colwell KS et al (1992) UFF, a full periodic-table force-field for molecular mechanics and molecular-dynamics simulations. J Am Chem Soc 114:10024–10035
75. Hoskuldsson A (1988) PLS regression methods. J Chemometrics 2(3):211–228
76. Breiman L, Friedman JH, Olshen RA et al (1984) Classification and regression trees, 1st edn. Wadsworth International Group, Belmont, pp 102–116
77. Hasegawa K, Miyashita Y, Funatsu K (1997) GA strategy for variable selection in QSAR studies: GA-based PLS analysis of calcium channel antagonists J Chem Inf Comput Sci 37:306–310
78. Carhart RE, Smith DH, Venkataraghavan R (1985) Atom pairs as molecular features in structure activity studies definition and applications. J Chem Inf Comput Sci 25:64–73
79. Perloff M, Von Moltke L, Court M et al (2000) Midazolam and triazolam biotransformation in mouse and human liver microsomes: relative contribution of CYP3A and CYP2C isoforms. Pharmacol Exper Ther 292(2):618–628
80. Golovenko NYa, Kravchenko IA (2007) Biochemicheskaya Pharmacologiya Prolekarstv (Biochemical Pharmacology of Prodrugs). Ekologiya, Odessa
81. Lipinski CA, Lombardo F, Dominy BW et al (1997) Experimental and computational approaches to estimate solubility and permeability in drug discovery and development settings. Adv Drug Deliv Rev 23:3–25
82. Zhang X, Lionberger RA, Davit BM et al (2011) Utility of physiologically based absorption modeling in implementing quality by design in drug development. AAPS J 13(1):59–71
83. Mannhold R, Kubinyi H, Timmerman H (eds) (2000) Handbook of molecular descriptors In: Todeschini R, Consonni V (eds) Methods and principles in medicinal chemistry, vol. 11. Wiley-VCH, Weinheim
84. Tubic-Grozdanis M, Bolger MB, Langguth P (2008) Application of gastrointestinal simulation for extensions for biowaivers of highly permeable compounds. AAPS J 10(1): 213–226
85. Fujioka Y, Kadono K, Fujie Y et al (2007) Prediction of oral absorption of griseofulvin, a BCS class II drug, ased on GITA model: utilization of a more suitable edium for in vitro dissolution study. J Control Release 119:222–228
86. Norris DA, Leesman GD, Sinko PJ et al (2000) Development of predictive pharmacokinetic simulation models for drug discovery. J Control Release 65:55–62

Chapter 15
(How to) Profit from Molecular Dynamics-based Ensemble Docking

Susanne von Grafenstein, Julian E. Fuchs and Klaus R. Liedl

Abstract Computational techniques have provided the field of drug discovery with enormous advances over the last decades. The development of methods covering dynamical aspects in protein–ligand binding is currently leading computer-aided drug design to new levels of complexity as well as accuracy. In this book chapter we focus on molecular docking to structural ensembles generated by molecular dynamics (MD) simulations. Does the incorporation of multiple receptor conformations allow pushing the borders for molecular docking or does it just lead to an artificial increase in false positive hit rates due to a broader conformational space of the receptor? We aim to identify guidelines for the best practice of molecular dynamics simulation-based ensemble docking from recent studies in the literature. Hence, we split the computational workflow for MD-based ensemble docking into the respective steps starting from protein structure and compound database to *in silico* hit lists. Thereby, we focus on the identification of successful strategies for virtual screening.

15.1 Introduction

15.1.1 Structural Ensembles of Proteins

The view on proteins has evolved rapidly over past decades. Starting out from a static view on protein structures, it has been recognized that proteins are intrinsically dynamic at room temperature. X-ray structures of proteins represent at best (assuming no structural perturbation due to crystallization) one conformational state trapped in

---

Susanne von Grafenstein and Julian E. Fuchs contributed equally.

K. R. Liedl (✉) · S. von Grafenstein · J. E. Fuchs
Institute of General, Inorganic and Theoretical Chemistry,
University of Innsbruck, 6020 Innsbruck, Austria
e-mail: klaus.liedl@uibk.ac.at

L. Gorb et al. (eds.), *Application of Computational Techniques in Pharmacy and Medicine*,
Challenges and Advances in Computational Chemistry and Physics 17,
DOI 10.1007/978-94-017-9257-8_15

a local energy minimum. As proteins may undergo major conformational rearrangements in solution even at room temperature, information accessible from static protein structures is inherently limited [1].

Hence, also the paradigm of protein–ligand recognition via a lock-and-key model [2] had to be adapted several times over the last years. Koshland proposed an induced fit of a single protein conformation to adapt to the ligand or substrate upon complex formation [3]. This idea has been expanded by Tsai et al. to include the whole conformational ensemble thermally accessible to a protein [4, 5]. They propose a mechanism of conformational selection picking a protein conformation favorable to bind the ligand from a pre-existing conformational thermodynamic equilibrium, therefore covering the whole free energy landscape. Consequently, higher energy conformations may contribute to ligand binding although not available from a static view of a protein [6].

These novel findings of protein chemistry are currently being incorporated into structure-based drug design [7]. Structural ensembles available from various sources are included in the computer-aided design process of novel therapeutics. Protein–ligand docking may be extended from static docking to a single X-ray structure to multiple X-ray structures, an NMR ensemble or an *in silico* generated conformational ensemble [8]. Besides covering a more realistic view on protein–ligand recognition, these novel approaches including conformational plasticity of the receptor have been recognized as important step to treat these macromolecular binding events with higher accuracy [9, 10]. The acceptance of flexibility as an important issue in molecular recognition is also reflected in attempts to include backbone-flexibility in protein-protein docking [11].

15.1.2 Molecular Dynamics Simulations

Molecular dynamics (MD) simulations are a computational technique giving access to a conformational ensemble for, e.g., a receptor protein [12]. The evolution of an atomic system is simulated via a numerical solution of Newton's law of motion over time. The potential energy is evaluated at every time step by a molecular mechanics force field (e.g., AMBER force field FF99SB-ILDN [13]). The starting configuration of the system is generated from an X-ray or NMR structure of the macromolecule and a surrounding water shell is added. A time interval according to the fastest movements in the system (hydrogen positions)—typically in sub-femtosecond scale—is chosen allowing for a numerical integration over the differential equation for the particle movements. Hence, after every time step new positions, new forces and new velocities are calculated for each atom of the system. After a plethora of repeats a so-called trajectory of the system is successfully generated containing an *in silico* image of biomolecular movements in solution. Nowadays, typical trajectories realistically capture protein dynamics at atomistic level of detail at nano to microsecond time scale [14]. The latter has been facilitated by the broad application of graphics processing units (GPUs) speeding up molecular dynamics simulations and hence allowing for routine microsecond simulations [15].

As all atomic coordinates as well as the system energy are stored, a conformational ensemble capturing thermally accessible states for subsequent protein–ligand docking can be extracted from the trajectory. In addition to capturing enthalpic contributions, these thermodynamic ensembles also include entropic effects rendering a valid free energy for the set of conformations. This fact is especially important, as flexible regions both close to the binding site as well as at distinct allosteric sites were shown to influence protein–ligand binding sites [16]. Thus, inherently flexible regions are of considerable interest for both lead identification and lead optimization.

15.1.3 Molecular Docking

Molecular docking aims at predicting valid protein–ligand geometry (pose) from a receptor conformation and a set of ligand conformations. To identify a promising binding mode, the poses are scored according to molecular interactions and steric characteristics aiming at the identification of the optimal binding pose of a ligand. More important than the identification of a binding mode is the application of molecular docking on large compound databases. The score is used to rank potential ligands to obtain an enrichment of potential high affinity binders amongst highest scoring hits. Molecular docking is applied as standard structure-based virtual screening technique for the identification of early hits in drug discovery as well as idea generator for *de novo* design [17]. Another application of docking protocols is the elucidation of a binding mode for an experimentally verified ligand at atomic level.

The described rigid docking approach does not account for receptor flexibility, solely aiming at the identification of the most suitable three–dimensional arrangement of two rigid bodies. The inclusion of side chain flexibility to the receptor poses additional challenges in form of additional degrees of freedom to be sampled within the docking workflow [18]. Soft docking applications were described including side chain flexibility inherently by allowing some degree of overlap between ligand and receptor atoms [19].

Degrees of freedom are further increased if also the protein backbone is allowed to show conformational heterogeneity leading to so-called ensemble docking. This can either be achieved by a united description of several receptor conformations by a grid averaging approach [20], the remodeling of loop regions after docking [21], or the explicit treatment of different conformations during docking. This last technique will be the focus of our review, as it largely resembles a mechanism of conformational selection.

15.1.4 *The Ensemble Docking Workflow*

Docking to multiple receptor conformations (sometimes abbreviated MRC [8]) can be applied on a set of conformations from any source. Experimentally different X-ray conformations or NMR solution structures can represent the same biomolecular system. NMR solution structures and conformations sampled during an MD simulation represent a valid thermodynamic ensemble, e.g., a canonical ensemble, for simulations with constant number of particles, volume, and temperature. Although, a collection of X-ray structures is not a valid thermodynamic ensemble, docking to multiple X-ray structures is sometimes also termed ensemble docking.

The idea to combine molecular dynamics simulations and molecular docking is an obvious way to include protein flexibility within docking if flexibility is not covered within experimental structures [22]. Here, we describe the concept of ensemble docking to an MD-based ensemble by following the workflow and present recent interesting examples applying this technique. The principle workflow is intuitive: First, a structural ensemble of the receptor is generated via an MD simulation. Second, molecular docking into one or multiple representations of the ensemble is performed. Finally, the docking results for multiple representations are pooled into a single score for each compound.

Several practical questions arise along this ensemble docking workflow where different strategies can be followed. Moreover, we will discuss that they are not only practical questions but are directly related to the underlying theory and should be considered regarding the physical interpretation of final results.

- How to setup the MD simulation?
- How to select the right representation of the ensemble for docking?
- How to score given multiple hit lists?

At each step, we tried to systematically cover alternative approaches followed in literature. Multiple branching points introduce a lot of variables in the workflow making it hard to compare test cases from literature directly with each other. Therefore, we set up a test case where we can consistently address individual questions and systematically show the influence of distinct strategies.

We will focus on influenza neuraminidase, an established drug target, where computational techniques were already successfully applied in the development of established inhibitors [23]. Influenza neuraminidase is a classical target to investigate and visualize the adaption of side chain orientation upon binding of alternative ligands [24]. Additionally, backbone flexibility is an issue for ligand recognition [25] and several studies applied molecular dynamic simulations in combination with molecular docking for the investigation of novel lead structures [26, 27].

Data presented here are extracted from all-atom simulations of influenza neuraminidase with an explicit water model and represent a sampling time of 30 ns (Fig. 15.1). The starting structures are three alternative X-ray structures of neuraminidase of recent forms of influenza A H1N1 without an inhibitor (Fig. 15.1a), bound to the inhibitor zanamivir (compound 1; Fig. 15.1b) and an active site

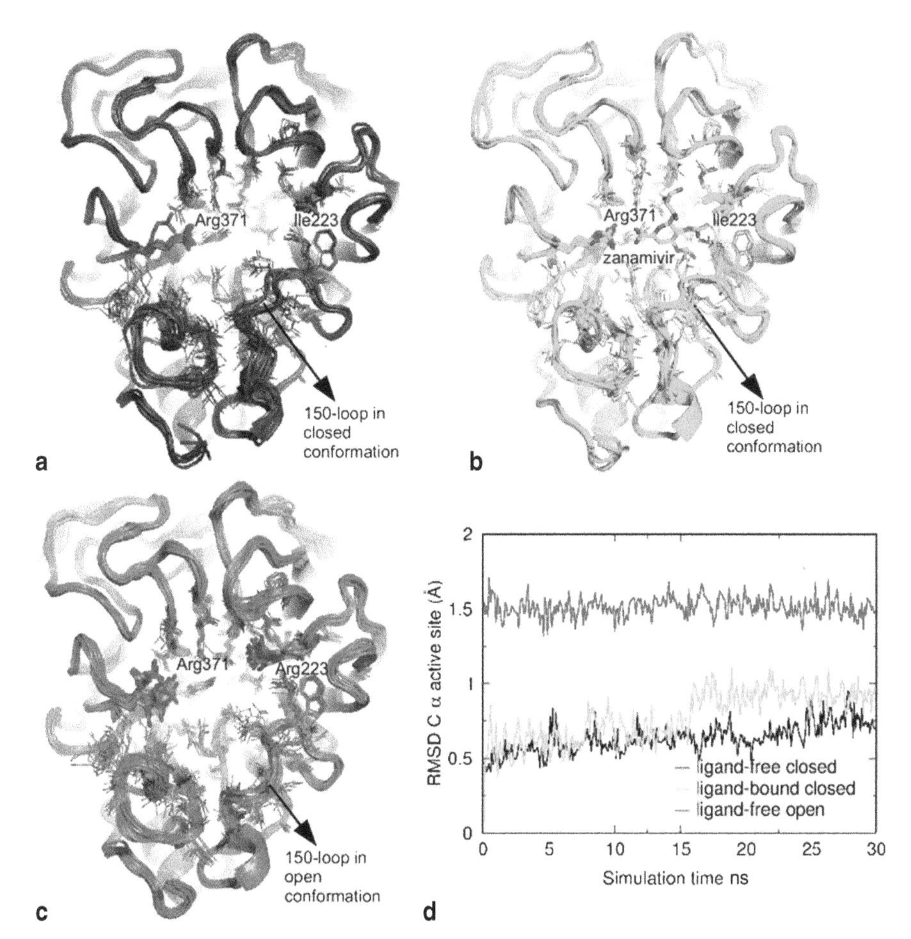

Fig. 15.1 Structural comparison of three influenza neuraminidase trajectories: Representative structures from the trajectories of the ligand-free closed simulation (**a**), the ligand-bound closed simulation (**b**), and the ligand-free open simulation (**c**). Root mean square deviations (RMSDs) of active site C-α atoms in respect to the first snapshot of the ligand-bound trajectory indicate non-overlapping conformational spaces between open and closed state simulations (**d**)

mutation I223R (PDB codes 3NSS, 3TI5 and 4B7M [28–30]). The starting structure of the mutant I223R has an alternative backbone conformation at a loop region next to the binding site which enlarges the binding site cavity (Fig. 15.1c) [25]. This so-called "open" state of the binding site is observed for several ligand-free X-ray structures of influenza neuraminidase and was shown to allow the binding of larger inhibitors in comparison to the classical neuraminidase inhibitors [27, 31]. We included 3-(p-tolyl)allyl-2-deoxy-2,3-didehydro-D-N-acetylneuraminic acid [31] (compound 2) in our docking study to examine which protocol can reproduce this alternative binding mode. Transitions between the closed state, resembling the zanamivir-bound state and the alternative, open state were sampled in simulations extending over several dozens of nanoseconds in the absence of a ligand [32]. Herein presented trajectories do not overlap in the sampling of alternative conformations of the critical loop region. This gives a further hint that the prediction of a specific ligand-bound conformation needs extensive sampling (in this case exceeding 30 ns) or enhanced sampling methods. We decided to start sampling from three alternative starting structures which allows covering a broader conformational space. However, this strategy is only feasible when information about alternative conformations has already been explored experimentally. Equilibration and sampling simulation followed an established protocol [33] using Amber12 [34] and Ambertools13 for analysis [35]. Influenza neuraminidases were sampled in their native tetrameric state [36].

15.2 Molecular Dynamics-based Ensemble Docking

15.2.1 Sampling the Free Energy Landscape

15.2.1.1 Principles of Molecular Dynamics Simulations

MD simulations provide a complete atomistic view of (bio-)molecular motions on the femtosecond to microsecond scale. The free energy landscape determines which states will contribute to an ensemble of structures at a given temperature. Also kinetic aspects, e.g., transition frequencies between different states are determined by the energy barriers within the free energy landscape of the system [12]. Molecular dynamics simulations explore the landscape given an energy distribution determined by the system's temperature.

First computer simulations of a protein system were described in 1977 [37] with a trajectory length of 9 ps *in vacuo*. Within this time scale side chain movements could be observed, whereas the backbone geometry remained virtually unaffected. Since then continuous increase in computing power allowed extension of sampling time. Time scale of most publications we will refer to is in the nanosecond range. Loop movements including fast domain motions can be observed within this time scale [12]. There is evidence that increasing sampling time positively affects the results of subsequent docking to the structural ensemble: the fraction of ligand poses similar to the native pose increased for the test cases thrombin and acetylcholinesterase, when comparing a 50 ps simulation with a 30 ns simulation [38]. Also, single observations of transitions between ligand-bound and free state were observed in MD simulations of this range, e.g., for influenza neuraminidase in a 100 ns simulation [32]. However, statistical sampling of transition rates between these states is mostly not reached and we must expect that the simulations discussed within this overview do not cover larger loop movements.

Current simulation techniques are capable of routinely capturing molecular motions in the microsecond time scale covering water molecules as explicit solvent by usage of GPUs [15]. It will show up in the near future how the more thorough sampling of the receptor free energy landscape will influence results of ensemble docking. This technical advance and the use of enhanced sampling methods (see Sect. 15.2.1.3) facilitate coverage of dynamical events such as domain rearrangements associated with slower kinetics [12].

15.2.1.2 Setting up an MD Simulation to Generate an Ensemble for Docking

Molecular dynamics simulate the evolution of an atomic system via a numerical solution of Newton's law of motion. The potential energy is evaluated at every time step by a molecular mechanics force field (e.g., AMBER force field FF99SB-ILDN [13]). Small organic molecules can be incorporated into simulations after fitting charges from quantum mechanical calculations and atom typing via the Generalized Amber Force Field GAFF [39].

The starting configuration of the system is generated from an X-ray or NMR structure of the (bio)-molecule and a surrounding water shell is added. A time interval according to the fastest movements in the system (hydrogen positions)—typically in sub-femtosecond scale—is chosen allowing for a numerical integration over the differential equation for the particle movements. Hence, after every time step new positions, new forces and new velocities are calculated for each atom of the system. After a plethora of repeats a so-called trajectory of the system was successfully generated containing an *in silico* image of biomolecular movements in solution.

In a perspective article from 2008, molecular dynamic simulations are pointed out to be an "expert system" [22]. The authors highlighted that this technique is challenging with a lot of practical caveats to be considered in system preparation and simulation protocol. Moreover, the simulation data need to be followed and checked for reasonable credibility of the results. As all atomic coordinates as well as the system energy are stored, several analysis techniques can be applied after performing molecular dynamics simulations. Standard analyses include the calculation of one-dimensional root mean square deviation (RMSD) to a reference structure yielding measures for stability of the simulation (see Fig. 15.1d). Additionally, two-dimensional RMSD analysis reflects convergence of the simulations (compare [33, 36, 40]).

Some of the questions when setting up the simulation are: Should a ligand-bound or a ligand-free simulation serve as starting structure? Should the ligand be included in the simulation? Which part of the protein should be treated as flexible? How to handle protonation and solvation? Considering the high computational costs to run the MD simulation an appropriate selection and preparation of the starting coordinates is essential.

Ideally, the impact of alternative starting structures will compensate during the simulation when the conformational space is sampled exhaustively. Following the conformational selection paradigm the ensemble of a ligand-free simulation of a receptor should cover a conformation prone to ligand binding. In agreement with this concept, the McCammon group selects ligand-free starting structures to generate trajectories for subsequent ensemble-based docking studies [41]. Systematic analysis indicate that conformations similar to the ligand-bound state can be sampled in trajectories where no ligand was present during sampling. For our test case, the active site C-α atoms of the ligand-free simulation remain close to the starting conformation of the ligand-bound simulations reflected in RMSD values below 1 Å in Fig. 15.1d.

However, the conformational space of the protein–ligand complex is different from the ligand-free state and the overlap is not necessarily large. Upon ligand binding a shift in population and accessible conformations is expected [24]. Examples where the crystal structures of bound and ligand-free state differ in several Å in RMSD are known, e.g., the backbone between the L-leucine binding protein differs in 7.1 Å RMSD in free state versus phenylalanine-bound state [42]. In such cases sophisticated protocols are necessary to generate a conformation resembling a ligand-bound state [42]. However, there are cases where a ligand-bound receptor state is never sampled in an unbiased ligand-free simulation.

Conformations sampled in the ligand-bound trajectory are adapted to the ligand. This is reflected when docking performance is compared for ligand-bound simulation and ligand-free simulation. Enrichment of known ligands within sets of decoys is higher for snapshots extracted from trajectories where the ligand is present [43]. Ligand-bound X-ray structures representing the holo state are prepared to accommodate a ligand. The presence of a ligand in the MD simulations also influences conformational sampling. In our test case influenza neuraminidase, the side chain of Arg371 stays in an stretched conformation forming a salt bridge with the carboxylate of zanamivir (see Fig. 15.1b), whereas it explores alternative states in ligand-free simulations (see Fig. 15.1a, c). However, the presence of a specific ligand biases the sampled conformations of the receptor towards a distinct chemotype. This leads to restricted recovery of the correct binding mode of alternative scaffolds [44]. For X-ray structures it was shown that ligand-bound structures might be considered as overspecialized [44]. Especially, in virtual screening, when the search for novel inhibitor scaffolds motivates a docking project, this overspecialization needs to be avoided to allow an enrichment of different chemotypes [45]. Therefore, a simulation starting from a ligand-free simulation might be more appropriate. In order to allow the binding site to be adapted for ligand binding without bias toward a specific ligand, Xu et al. developed an implicit ligand-model [38]. The ligand–model concept (LIMOC) uses small molecular probes covering the accessible surface of the binding site dependent on the local environment. Subsequent sampling follows an algorithm which allows the probes to explore different sites as well as an adaption of the binding site to the fragment-like structures. Application of the concept on thrombin and acetylcholinesterase shows that the recovery of the native binding pose is between the one for the simulation with the native ligand and a ligand-free simulation [38].

With focus on subsequent investigation of the protein–ligand interactions in a defined binding site within docking the flexibility of regions away from this binding site might be negligible. Therefore, flexibility is frequently limited on specific regions during the MD simulations. The influence of restricted flexibility was analyzed by Bolstad et al. for dihydrofolate reductase of different organisms and by Armen et al. for the case of p38α mitogen-activated protein kinase [46, 47]. Bolstad

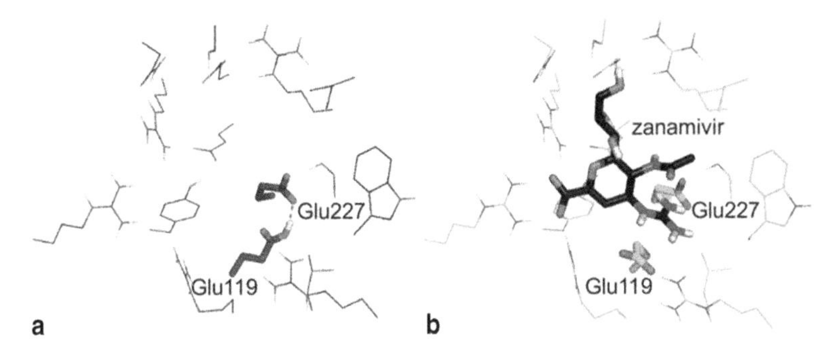

Fig. 15.2 Active site of influenza neuraminidase protonation can dependent on side chain geometry and ligand binding. **a** In a ligand-free structure Glu119 orients towards Glu227. A protonated state of Glu119 explains this conformation via a hydrogen bond. **b** Glu119 shows an alternative side chain conformation in a ligand-bound structure. Therein, Glu119 faces the positively charged guanidinium group of the ligand zanamivir and is deprotonated. Protonation states were predicted with the Protonate3D protocol [54]. (PDB accession codes 3BEQ and 3B7E)

et al. limited the flexibility on residues within 3.5 or 6 Å around the ligand while keeping the ligand and the co-factor restrained. They compared those simulations regarding the subsequent docking performance which showed an increase in the rate of improper binding poses with an increased flexible region. These results were not completely consistent and seem to depend on the quality of the starting structure. When homology models or NMR structures were used as starting structures, docking results improved with the inclusion of more residues within the flexible part [46]. Similarly, Armen et al. observed a reduced fraction of poses close to the native structure when they allowed the complete protein to be flexible in comparison to models with restricted flexibility [47]. Both studies used simulation conditions without specifying a water model, and we agree with the authors of the second study that especially the fully flexible model will benefit from an alternative simulation protocol, e.g., inclusion of solvent. We recommend the use of explicit solvent environment or at least an implicit solvent model. Otherwise, an all-atom simulation with an empty active site will in general tend to collapse over the active site. Explicit water models increase the computing time. Therefore, a water cap around the active site is applied in some studies to reduce the water fraction in comparison to a simulation in a water-filled periodic simulation box [38, 48]. However, this restriction in contrast to a fully solvated state was not yet evaluated in respect to docking results. The crucial role of water in docking experiments is well accepted [49] and therefore will also be discussed for preparation of structures for molecular docking (see Sect. 15.2.3.2).

Similar to the question of solvent treatment, protonation of amino acid side chains is an essential step in protein–ligand docking [8, 50]. Therefore, the protonation state should also be critically considered during setup of the simulation. In the case of neuraminidase, we show an example where the ligand-free and ligand-bound state do not only differ in starting conformation but also in protonation states (Fig. 15.2). For the test case we present in other parts of this overview, we used structures having consistent protonation states for both ligand-bound and ligand-free structure. All glutamic acid residues in the binding site are deprotonated. Although there are attempts to include reprotonation events in MD simulations with constant pH simulations [51], classical force fields do not allow an adaption of the protonation state. With the assignment of atom types in simulation setup, protonation and tautomer states of amino acids are determined for the simulation. The effect of protonation on docking is essential as the interaction potentials, especially electrostatic interactions, will differ for a deprotonated or protonated glutamate. Recently, two studies addressed the aspect of binding site protonation for rigid molecular docking [52, 53]. Especially docking results for binding sites containing a histidine residue were shown to be dependent on protonation and tautomer states, as for this amino acid various alternative states can occur [53]. The altered interactions of a protonated versus deprotonated amino acid will also influence the conformational sampling during the MD simulation.

15.2.1.3 Alternative Sampling Methods

The conformational transitions observed between X-ray structures in the ligand–free and ligand-bound state can include loop shifts of several Ångströms or the reorientation of whole protein domains [42, 55]. The time scale of such events is typically not covered statistically by all-atom simulations ranging in the area of several dozens of nanoseconds [12]. Focusing on these transitions several groups established different procedures to improve the sampling in comparison to classical unbiased MD simulation. We refer the interested readers to a recent review on enhanced sampling techniques in the context of drug design and the references therein [56]. Here we will only focus on methods used in connection with molecular docking.

We group methods which rely on the same principles as MD simulation, e.g., torsional angle MD [47], replica exchange MD (REMD) [57, 58] and targeted MD [59]. Also other **enhanced MD sampling techniques**, such as accelerated MD could be applied [56]. Torsional angle MD is not an enhanced sampling technique in the classical sense as no bias is applied to overcome energy barriers more easily. Sampling using torsional angle MD is more efficient than standard MD as it takes advantage of an internal coordinate system based on the relative orientations along the dihedral angles in a protein [47]. However, torsional angle MD cannot be applied to systems with multiple molecular entities such as solute plus solvent molecules hindering its broad application.

In REMD techniques higher temperatures allow an enhanced sampling [60]. Multiple copies of a system are simulated at different temperatures and an exchange between the replicas is regulated by a Metropolis criterion [61]. The technique allows to overcome energy barriers more easily and was shown to be suitable to reproduce a ligand-bound-like conformation of a flexible RNA target [58]. Osgurthorpe et al. compared the performance of docking to REMD-derived versus classical MD-derived ensembles. However, they could not observe an improvement on the discrimination between active and inactive ligands for the subsequent docking [57].

Targeted MD can be used when two states of a given protein are known, e.g., the open and closed states of the binding pocket of serine racemase in a study of Bruno et al [59]. An artificial force is applied during the MD to bias the transition between the two states. The need of a reaction coordinate limits this application to systems where a target structure is known beforehand. The trajectory of a targeted MD simulation can be used in the same way as ensembles derived from unbiased MD simulations [59].

Alternatively, biased simulations allow the reconstruction of the potential of mean force or a free energy landscape [56]. If the reaction coordinates are selected in a way that they cover the ligand binding or unbinding process, enhanced sampling methods allow an estimation of the free energy of binding. This was shown for metadynamics simulations using an artificial potential which promotes the system to leave conformational spaces already visited [62]. Such a "docking" strategy is inherently considering the flexibility of both the ligand and the target, but is not suitable for high-throughput screening due to the necessity of a defined reaction coordinate and the exhaustive sampling for individual ligands.

Other sampling methods do not rely on the time-dependent evolution of a structure, but try to cover protein flexibility in **collective variables**. Collective variables describe the global motions within a system and can be derived from an MD simulation or from network models. From an MD simulation a principal component analysis of the covariance matrix of the N atom positions yields a set of N eigenvectors. The first ten to twenty eigenvectors contribute significantly to the overall atomic fluctuations and are therefore called essential modes or soft modes [63].

Alternatively, collective variables can be obtained by normal mode analysis (NMA) when diagonalizing the second derivative of the energy function with respect to the coordinates (the Hessian matrix) [24]. NMA is frequently applied on network models of biomolecular systems where heavy atoms, C-α atoms or residue representing beads (N) are connected with springs of an adapted force constant. The various approaches suggested for subsequent docking have limited comparability as they differ by the form of the network model, the underlying energy function (origin of spring forces), the extraction of relevant normal modes and the weighting of deformation along a normal mode. One of the models applied in the community is the elastic network model (ENM) by Hinsen [64] on C-α atoms (N) with distance-dependent force constants. NMA yields 3N-6 eigenvectors, which can be used as coordinates for deformation of the protein and generation of an ensemble of alternative structures [24]. To limit the number of modes for subsequent structure generation, one can decide to focus on the first (up to 10) softest modes with the highest eigenvalues [65, 66]. Alternative approaches to reduce dimensions are the identification of the relevant modes for loops in the binding site after NMA of an ENM [67] or the limitation of the ENM on heavy atoms in the binding site region and analysis of the first 100 modes [68]. Subsequently, normal modes can be used to generate conformations, which are then used similarly as conformations generated during an MD trajectory [66, 67, 68, 69]. Rueda et al. used a larger test set of targets and could show that their approach could successfully reproduce a near-native ligand pose in cross-docking for flexible targets [68]. Alternatively, collective variables can be integrated in the docking algorithms as additional degree of freedom [63, 65, 66]. Zaccharias integrated the MD-derived collective variables [63] in a new docking algorithm and later NMA-derived collective variables and side chain flexibility were combined [65].

Alternative sampling strategies use **Monte Carlo (MC) algorithms** to sample alternative protein structures. Here, we will present two cases where active site backbone flexibility is integrated in presence of a ligand. The IREDA algorithm generates multiple pocket conformations by an algorithm reorienting the co-crystallized ligand in the pocket (seeding), minimizing the resulting complexes and finally optimizing the energy with a global MC protocol [70]. During the optimization sampling of protein flexibility is enhanced with random perturbation of torsion angles around the binding site and small displacement and rotation of the ligand. The procedure uses the internal coordinate system of the ICM docking protocol and the generated structures are evaluated using a Poisson solvation energy term [70]. For a set of kinases the novel generated structures perform equally well as multiple X-ray structures in cross-docking experiments and for the discrimination of active from inactive ligands. Moreover, novel binding pockets were not overoptimized for the ligand used for structure generation and allowed near-native poses for alternative ligands.

In contrast, ROSETTALIGAND docking uses MC sampling in presence of the ligand to be docked [71]. The algorithm presented in this study is improved over the first published ROSETTALIGAND algorithm [72] by integrating backbone receptor flexibility and additionally, a more exhaustive sampling of ligand flexibility. The presented study used about 5000 trajectories for each ligand conformer limiting its large scale application. A potential reduction of this extensive sampling is discussed by the authors. Still, the protocol is highly parallel and therefore scales with the number of used CPUs. Each trajectory consists of three phases: first, a coarse-grained placement of the ligand is performed and evaluated for shape complementarity focusing only on receptor backbone and C-β atoms; second, a MC-based minimization scheme is repeated for six cycles exploring side chain rotamer libraries, ligand positions and torsions; third, a final minimization includes backbone flexibility. The energy is evaluated with the all-atom ROSETTA scoring function for the second and third step with soft van-der-Waals repulsion in the second step. The bench–marking results show a clear improvement in comparison to the previous ROSETTALIGAND implementation indicating the beneficial effect of integrating backbone flexibility in terms of identification of the near-native ligand poses for the ROSETTALIGAND test cases [72] and the Astex test set. However, for a set of urokinase structures inclusion of backbone flexibility lead to generation of highly ranked but non-native poses indicating an increased rate of false positive hits by the additional degrees of freedom [71].

15.2.1.4 Representation of the Structural Ensemble

Choosing the optimal representation of a collection of conformers (ensemble) is a crucial step independent of the origin of alternative conformations. Therefore, most

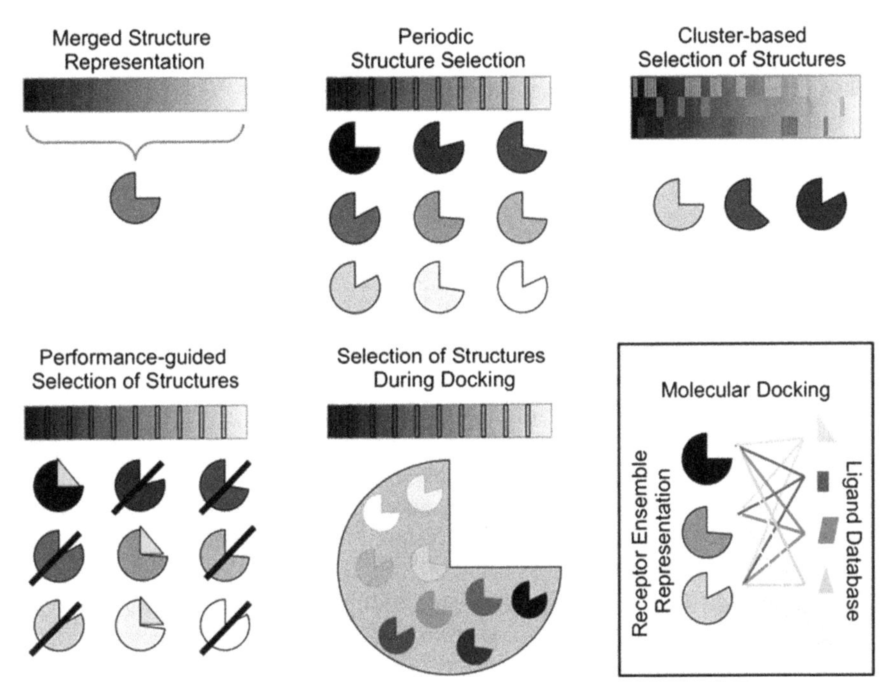

Fig. 15.3 We classify the approaches to represent an MD trajectory for the subsequent molecular docking into five alternative strategies: merged structure representation, periodic structure selection, selection of structures based on a clustering, guided by the performance of individual structures, or selection of structures during docking. Subsequent molecular docking relies on chosen the receptor ensemble representation

reviews dealing with the integration of receptor flexibility in molecular docking discuss this aspect extensively [8, 24, 73]. Here, we try to systematically classify different strategies and discuss their application in respect to conformations generated by MD simulations. However, some of the concepts are applied mainly to collections of structures derived by experimental sources (X-ray or NMR structures) rather than to ensembles generated by MD simulations. In this chapter we will present the diverse strategies according to Fig. 15.3 and discuss their individual potential.

The easiest strategy is a systematic **merging** of conformational plasticity sampled during the MD trajectory into one single representation. Prominent examples are the unified representation of multiple conformations in FlexE [20] or ensemble-based grid representation [74].

FlexE is an extension of the docking program FlexX [75]. FlexX implements side chain flexibility via sampling of alternative rotamers during docking. Similarly, alternating backbone conformations are sampled in FlexE, whereas equivalent structures are unified in one representation based on superposition [20]. The concept relies on a main preserved geometry of the binding site for the different conformations; for larger conformational differences FlexE fails to unify the alternative conformations, e.g., for the two test cases: X-ray structures of c-jun N-terminal kinase 3 and β-secretase [76]. This might be one of the reasons why FlexE was more frequently applied on ensembles of alternative X-ray structures and not on MD-derived ensembles where larger transitions are expected.

Likewise, ensemble-based grid representation of binding sites is rarely used for MD-derived ensembles. Ensemble-based grid representations, also called composite-grids, are one of the oldest concepts to include protein flexibility within docking [74]. Technically, the derivation of one grid from an MD ensemble follows the same approach as for multiple X-ray structures or NMR ensembles. Different averaging strategies were evaluated to calculate interaction potentials of the grid points for experimental ensembles [74, 77]. Broughton averaged the grid representation of conformations extracted from short MD simulations (< 100 ps) of cyclooxygenase 2 and observed an improvement over docking to a rigid protein in terms of ranking using the docking program FLOG [78].

Obviously, merging strategies are especially time efficient [79]. A single docking run covers all information from an ensemble, so the time per docked compound is minimized compared to methods explicitly relying on multiple independent receptor conformations. However, taking into account the computational need for the generation of an ensemble using all-atom MD simulations, the argument of computational demand while docking loses weight. Additionally, atomistic information of specific conformational states sampled during MD simulation is lost by the unification into one representation. With these aspects in mind it is not surprising that alternative strategies of ensemble representation are preferentially used in combination with MD-generated ensembles.

Parallel docking to multiple conformations selected from the trajectory is frequently used in combination with MD simulations. The **periodic extraction** of snapshots from an MD simulation is a straightforward way to select multiple conformations. Subsequently, molecular docking is performed on the individual snapshots independently from each other. The continuous and unbiased selection should ideally represent the state space sampled in the simulation. In consequence, the resulting collection of snapshots represents a valid thermodynamic ensemble and the subsequent ensemble of docking results are interpreted via statistical thermodynamics (see Sect. 15.2.2.3).

A protocol using time-dependent extraction of conformers from MD simulations and subsequent docking is developed by the McCammon group and known as "relaxed-complex-scheme" (RCS) [41, 80, 81, 82]. Classically, RCS applications use AutoDock, an established docking program for flexible ligand docking based on a genetic algorithm. Recently, AutoDock Vina a new implementation of the algorithm is used as it is more suitable for parallel application [83]. A major success of this protocol was the identification of a cavity next to the binding site in HIV integrase [81] allowing the development of inhibitors specifically addressing this cavity. The approved drug raltegravir was developed based on the binding mode suggested by Schames et al. Besides this one significant contribution to drug discovery, several more applications of the RCS are described in literature and represent MD-derived ensemble docking studies using periodic snapshot extraction.

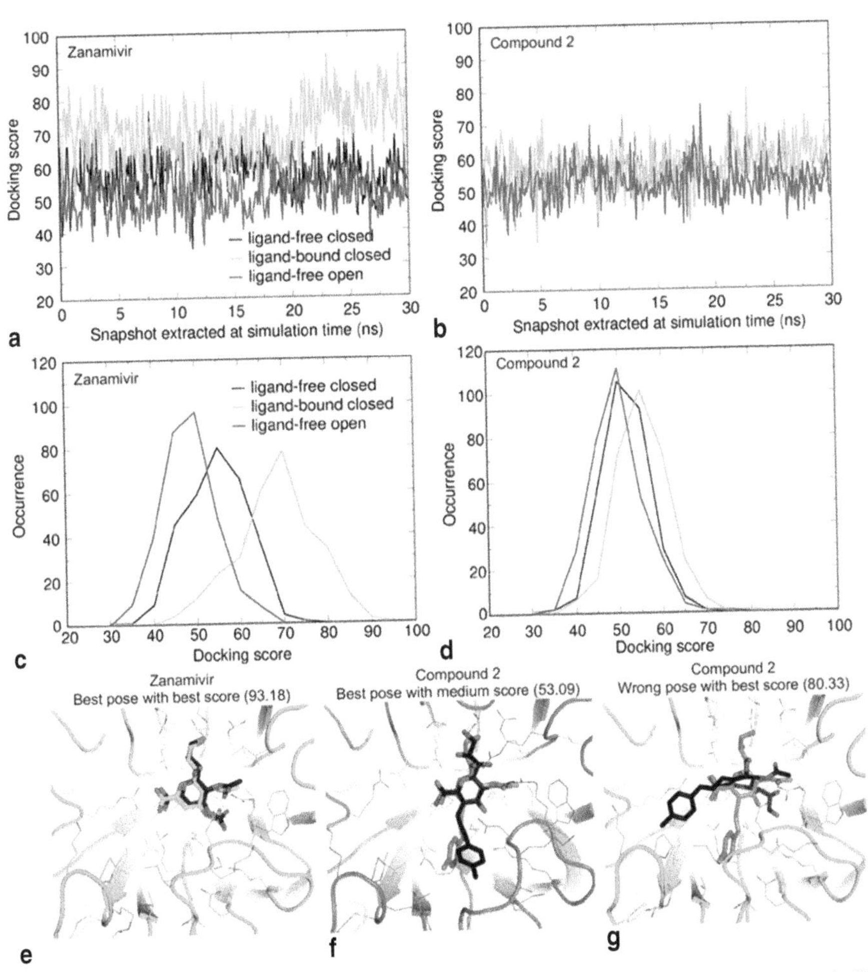

Fig. 15.4 Docking results: Typical score profile predicted by docking to 300 snapshots periodically extracted from several simulations of influenza neuraminidase over 30 ns sampling time for **a** zanamivir and **b** compound 2, known to bind the open conformation of influenza neuraminidase. The histogram visualizations (**c**, **d**) show that best scoring values are obtained with conformations from the ligand-bound simulation. Docking poses (*black*) discussed in the text are shown in comparison with native poses of zanamivir (*green*) and compound 2 (*blue*) (**e–g**)

In early applications, snapshots are extracted in a time interval of 10 ps for simulation lengths of 2 ns [80] or 22 ns [82], whereas in later applications every 20 ps for simulations of lengths up to 50 ns yielding in hundreds to thousands of snapshots for docking [43]. Consequently, docking runs for individual snapshots produce a range of scoring values for each compound instead of single scores [41, 82, 84, 85]. Visualization of such scoring profiles over the simulation time shows the fluctuation of the docking scores and typically a histogram is generated for scoring values (Fig. 15.4). We docked the approved drug zanamivir and an experimental inhibitor, compound 2, known to bind to the open 150-loop conformation of neuraminidase (see Fig. 15.1c). Each of the three different simulations were represented by 300 periodically extracted snapshots and docking to in total 900 structures was performed with the program GOLD using the knowledge-based fitness function GOLDSCORE. As discussed before (Sect. 15.2.1.2) conformations from the ligand-bound simulations are optimized for zanamivir and result in the most favorable poses, reflected by high docking scores (Fig. 15.4a, c). The pose with the best score represents a near-native structure (Fig. 15.4e). As expected a near-native pose of compound 2 is generated with a conformation from the open simulation. The low scoring value for this pose might be related to the conformation of the Arg371 side chain in the receptor conformation which is unfavorable to establish the close contact to the ligand's carboxylate group (Fig. 15.4g). Consequently, a higher score is generated with a snapshot of the zanamivir-bound simulation showing a flipped ligand orientation in comparison to the native-pose from the X-ray structure but an optimal contact between Arg371 and the ligand (PDB code 3O9K) (Fig. 15.4g).

It was shown that the best scoring value does not necessarily represent a binding pose close to the crystal structure [80]. Single values showed artificially high non-physical binding energies generated for docking poses with clashes between ligand and protein. Subsequent minimization or post-processing using MM/PBSA as alternative free energy estimation techniques yielded top ranking for the pose near to the native X-ray structure [82]. Without further post-processing ranking of different compounds by the arithmetic mean over the scoring profile is suggested and the mean of scoring values was successfully applied in selecting a lead modification for HIV protease inhibitors [82].

Concepts similar to the RCS are applied by other groups using alternative simulation protocols or docking algorithms [24, 86]. Most of them do not present applications but rather evaluate the technique on multiple test cases in comparison to other virtual screening techniques or aim at the identification of useful parameters for the workflow. For example, Paulsen et al. investigate how the pooling of docking scores impacted the ranking accuracy for two test cases: *Candida albicans* dihydrofolate reductase and influenza neuraminidase [86]. Handling of multiple scoring values is a major question in ensemble docking, which we will address later (see Sect. 15.2.2.3). An exemplary application of a similar method is the virtual screening study on the estrogen receptor predicting novel leads *in silico,* unfortunately lacking experimental verification [85].

Successful virtual screening projects using the RCS are summarized in a paper by Amaro et al [84]. However, for these studies an alternative method to select the conformations from the ensemble was chosen as multiple receptor conformations increase the docking time per compound linearly with the number of conformations considered. Therefore, especially focusing on virtual screening application, where the compounds to be docked can easily exceed 105, the number of considered snapshots has to be limited. In addition to the increase in docking time per compound, more conformations demand a higher overhead time as each protein conformation must be prepared separately for the docking run depending on the docking algorithm (e.g., grid generation in GLIDE).

Reducing the number of snapshots resembles the stochastic experiment of picking individuals out of a larger collection: The possible combinations to select r different snapshots out of N different snapshots registered in a trajectory scales with an N over r relation leading to an explosion of possibilities with N [87, 88]. For ensembles of X-ray structures an extensive analysis over potential combinations was performed. Indeed, the performance of the individual combinations is not equal [87]. In our test case we observed that reducing 300 snapshots to 100 results in a similar score profile if the snapshots are selected continuously. The further reduction by increasing the incremental interval of extraction to 3 ns yielding 10 snapshots, which would be a rational number of structures for virtual screening, is not sufficient to reproduce the original scoring profile (data not shown). This shows that this large extraction interval is not sufficient as one might lose crucial structural diversity.

More appropriate is the selection of structures representing specific regions of conformational space to ensure coverage of the structural variation in the trajectory. Such a **clustering approach** needs a metric to compare the binding site characteristics of the snapshots and groups them by similarity or distance. Subsequently, single structures are selected to represent the individual cluster groups representing a specific fraction of a trajectory (population). A plethora of clustering algorithms [89] is available for conformations from MD simulations, e.g., as implemented in the analysis tools of widely used simulation packages such as AmberTools PTRAJ [34]. For the case of the mutant W191G of cytochrome c peroxidase [84], Amaro et al. discuss the application of root mean square deviation (RMSD) of atom positions between pairs of conformations as distance criterion and subsequent clustering. The applied exclusion sphere clustering based on a similarity criterion of 1 Å RMSD between backbone atoms was especially developed to analyze the conformational space sampled during an MD simulation [90]. As most clustering algorithms are based on atom positions, a proper alignment of the overall structures is needed. Similarly, the tool "QR factorization" uses an RMSD criterion between the snapshots to perform a hierarchical clustering. This procedure allows to select a distance threshold after clustering, and hence, selection of snapshots by cluster level. It was shown that similar results regarding the scoring profile could be produced for the RNA editing ligase when docking to an ensemble of 400 snapshots extracted periodically or to reduced set of 33 representative structures selected by a certain RMSD threshold [84, 91].

Clustering based on active site atoms RMSD was selected as method of ensemble representation in a study to rationalize the binding mode of the natural product katsumadain A to influenza neuraminidase [27]. A similar strategy was followed for the same target in a lead discovery study [26]. For virtual screening of the database the authors first used representative structures of the three most populated clusters for two simulations with and without ligand and the two crystal structures resulting in 8 parallel docking runs. Subsequently, top ranked compounds were rescored using the mean of scores from docking to all 27 cluster representatives of the holo simulation following the RCS protocol [84]. Discrimination of active and inactive compounds by docking could be improved for serine racemase with cluster

Table 15.1 Scoring results (GOLDSCORE) for periodical extracted structures versus ensemble representation by cluster representatives generated by RMSD-based clustering on the active site atoms shown in Fig. 15.1a–c. In the weighted average score, the scoring values for cluster representative structures are multiplied with the population of the respective cluster in the MD simulation

Ligand	Simulation	Periodic structure selection			Cluster-based selection		
		Docking runs	Best score	Average score	Docking runs	Best score	Weighted average score
Zanamivir	Ligand-free closed	300	76.87	57.01	8	71.29	60.13
	Ligand-bound closed	300	93.18	70.96	5	85.29	71.16
	Ligand-free open	300	67.47	51.14	14	64.12	52.60
Compound 2	Ligand-free closed	300	72.66	54.24	8	67.75	55.45
	Ligand-bound closed	300	80.33	57.85	5	69.60	62.81
	Ligand-free open	300	75.61	52.25	14	75.61	56.20

representatives versus two X-ray structures representing the start and target state of the targeted MD simulation [59]. Six duster representatives were extracted from a targeted MD trajectory by RMSD-based clustering. Xu et al. applied an RMSD-based clustering to reduce the number of snapshots from different trajectories of 10 ns or 30 ns simulation time to proceed with a fairly equal number of 200–250 representatives for the subsequent validation [38].

We performed an RMSD-based clustering on the active site atoms revealing the representative structures shown in Fig. 15.1a–d. The clustering was performed with CPPTRAJ from AmberTools13 [35] and followed an agglomerative clustering algorithm using complete linkage with an RMSD of 1.5 Å for the formation of a new cluster. For each cluster one central conformation is defined as representative. Using this representation for the subsequent docking the number of necessary docking runs is reduced by a factor of around 10 in comparison to the periodical extraction using one snapshot every 100 ps (Table 15.1). Still, docking results show that the ligand-bound simulations result in the most favorable scoring values. For both ligands the maximal scoring values are observed for a structure representing a cluster populated only by 11.7% of the ligand-bound simulation. The respective docking poses resemble the near-native pose of zanamivir and the flipped pose of compound 2 as shown in Fig. 15.4e, g.

The RMSD of C-α or active site atoms is the most popular measure for distance between conformations. Further criteria suggested are for example relative distances measured between binding site atoms forming a conserved ligand binding framework in dihydrofolate reductase [46]. The authors combine these distances into a one-dimensional measure, compare it to the starting structure and apply it to reduce the number of ensemble representatives. Furthermore, the shape of the active site can be considered as distance criterion to cluster conformations as applied on MD snapshots from a simulation of HIV protease, cyclin-dependent kinase 2 and androgen receptor [57]. In principle, any other metric allowing to differentiate binding site geometries from each other could be used for clustering.

Clustering methods are commonly used to reduce the number of representative structures for docking to ensembles derived from MD simulations. Clearly, increase in efficiency by the reduction of time per docked compound makes clustering attractive compared to the periodic selection of snapshots. The major assumption with this strategy is that a suitable selection of representatives should allow to cover the important structural variability. By now, a solid statistical evaluation of a standard recipe is out of reach considering the variety of clustering criteria and algorithms.

A systematic reduction of structures from the MD simulation can also be achieved via a **performance assessment** of individual snapshots, e.g., selected by periodic extraction. Subsequent applications such as virtual screening of large databases would only proceed with the snapshot(s) performing best. Docking performance can be evaluated in, e.g., three categories: reproduction of native binding poses, ranking of active compounds, and discrimination of active and inactive compounds [88, 92]. Dependent on the purpose of docking one might choose between different performance measures. Various binding site properties such as volume, hydrophobicity and opening state were evaluated for ensembles of X-ray structures in respect to their performance in discriminating active from inactive compounds [93]. The authors could develop some rules depending on the inherent characteristics of the protein binding pocket which might be useful to select a specific X-ray structure prone to outperform alternative ones prior of extensive evaluation. By now, similar rules are not available for MD-derived structural ensembles [43]. Therefore, evaluation strategies are necessary and their dependence on experimental data might be a general limit of this structure selection strategy. When evaluating the performance of periodically extracted snapshots from MD simulations, Nichols et al. could identify individual conformations that outperformed even the best available X-ray structure in terms of discrimination between known ligands and decoys for all investigated cases [43]. Xu et al. suggest a similar strategy to identify suitable snapshots generated via implicit ligand simulations (LIMOC) according to their ability to score ligands with more favorable scores than decoys [38]. Besides lack of experimental data, the limited chemical diversity of experimentally verified ligands hinders performance-based selection of structures. Performance can only be evaluated within the chemical space covered by active and inactive compounds. However, a conformation recovering known active ligands is not necessarily the best conformation to identify novel ligand scaffolds. Searching for novel chemical scaffolds by virtual screening one intentionally leaves the applicability area of the underlying performance metric.

The restriction on specific conformations selected either by clustering or by performance reduces the complexity of ensemble docking by neglecting that individual ligands might have preference for individual snapshots sampled during the MD. To overcome this limitation a last strategy attempts to allow a selection of the conformation for each ligand individually. The challenge is the implementation of this additional degree of freedom without exhaustively sampling every ligand versus snapshot combination. Some ensemble docking strategies integrate the selection of the conformation from the ensemble within the docking procedure. The **selection of the optimal conformation** is also subject of the docking algorithm in analogy to the translational and rotational degrees of freedom of the ligand optimized during posing. In this context the ensemble is not be pre-generated. Zaccharias proposed the use of collective variables sampling the plasticity of the protein in presence of a ligand to allow ligand-specific adaption of the protein conformation [63]. Selection of the ligand-specific conformations from a set of pre-generated conformers corresponds to the challenge of cross docking into alternative X-ray structures [94]. Applications focusing on MD structures used a consensus representation for all conformations as starting point and then included the sampling of individual conformations within the simplex optimization of the docking pose using the program DOCK [95]. In a review, Totrov and Abagyan suggest that other optimization algorithms might be more suitable to address the optimization of this parameter [8]. The concept of four dimensional (4D) docking follows a similar concept using an optimization based on MC sampling established by Bottegoni et al. for the docking program ICM [96]. For a large set of X-ray derived ensembles, they showed a similar performance in reproducing the native binding pose in comparison to the parallel use of the X-ray ensembles without linear increase of computational time with number of structures. However, the method was neither evaluated for the performance to discriminate ligands from decoys nor was it applied on MD-derived ensembles. Using structures generated with a normal mode approach Leis et al. included the selection of the structure in the genetic algorithm implemented in AutoDock [66]. Studying binding to protein kinase A, they showed that this approach worked best to reproduce the binding mode for cases where the additional degrees of freedom, e.g., flexible torsion angles in the ligand are limited. The presented cases did not use conformations from an MD simulation. However, apart from sampling issues no limitations for the application on MD-derived ensembles are known by now.

The ongoing research in the field indicates that the optimal way to tackle the selection of conformers is not yet identified within the variety of strategies followed. Moreover, this issue is not an isolated challenge within the workflow of ensemble docking, but is directly related to the evaluation of docking results. Depending on the strategy followed for structure selection docking results will either be a single hit list or a collection of several hit lists. Subsequent pooling strategies for multiple hit lists depend on the principal approach employed for selection of receptor representation.

15.2.2 Molecular Docking

15.2.2.1 Docking Algorithms and Scoring Functions

Molecular docking can be considered as one of the major pillars of structure-based drug design. Docking flexible ligands, mostly small organic molecules, to a specific site of a receptor representing a rigid bio-molecular structure is an established procedure [97, 98, 99, 100, 101]. Therefore, there is a critical awareness for the potential as well as the limitations of this technique [92, 102]. Well accepted but still open challenges are the inclusion of backbone flexibility (which is subject of this overview), treatment of solvent and solvent-related effects [49], and quantitative estimation of ligand binding [92, 103]. Prediction of accurate binding affinities is out of the scope of the scoring functions implemented in established docking applications [103]. In contrast to rigorous physics-based approaches aiming at the prediction of free energy of binding, scoring functions are fitting functions designed to be fast and transferable, thereby allowing high-throughput screening of databases.

Three types of scoring functions can be identified in general: force field-based, knowledge-based and empirical scoring functions. Classically, they all treat molecular interactions as pair-wise additive terms on atom level aiming to score a ligand pose at high-throughput speed. In force field-based methods, the energy of the protein–ligand interaction is covered by van-der-Waals terms as well as electrostatic contributions from molecular mechanics. The ligand's force field energy is also included in some scoring functions to consider the conformational strain of the pose [104]. It has been shown that empirical scoring functions including entropy penalties for the ligand fixation and desolvation terms are more accurate to estimate binding affinities [105]. Knowledge-based scoring functions derive the terms for interaction potential based on experimental geometry distributions. Coefficients are directly fitted to experimental data to reproduce binding affinities of protein–ligand complexes with known X-ray structures [106]. For a review on scoring functions and docking algorithms we recommend the review by Kitchen et al. as an overview article [98]. Recent advances in scoring functions are the inclusion of non-additive, cooperative effects, e.g., for hydrogen bond networks established upon ligand binding [107]. Also in the field of solvent effects some progress can be expected by inclusion of entropic and enthalpic contributions depending on interaction with explicitly investigated water molecules by WaterMap [108, 109].

However, the universal scoring function has not yet been developed and the optimal choice of the best function is project-dependent [103]. We propose to consider results of unbiased docking challenges which are periodically repeated to reflect the current state of the technique. One example is the challenge organized by the Community Structure-Activity Resource (CSAR) around Heather Carlson [92]. Focusing on the application to multiple structures, the speed of the docking and scoring algorithm as well as the capability of the program to handle multiple structures in parallel might be an issue. When docking to multiple structures derived from an MD simulation, the major steps remain the same as for docking to rigid structures. However, several docking programs offer a workflow especially tailored for parallel docking to optimize computing time. For example, in GOLD up to 20 protein conformations can be processed by the "ensemble docking" feature and allow the user to weight the scores for individual protein conformations.

15.2.2.2 Preparation of Structures for Molecular Docking

As important as the choice of the program and scoring function is the preparation of the ligand structures or the database of potential ligands used for a virtual screening [110]. The protein structure must be prepared appropriately as for rigid docking [50]. The question of proper protonation of protein residues should already have been focused during the set-up of the MD simulation topology (see Sect. 15.2.1.2). Nevertheless, structure preparation before docking could allow an adaptation of the protonation state depending on the conformation; an aspect which was not yet investigated to our knowledge. A greater issue might be the treatment of solvent molecules [49]. Valuable information on the fluctuation, positional and rotational degrees of freedom of water molecules in the binding site is generated when one uses an explicit solvent model in the MD simulation. This information is available from the trajectory stored for conformer generation. It might be worth to perform an in-depth analysis to extract this information as it will guide the decision on how to proceed with water molecules in receptor preparation. Fixed tightly bound water molecules are unlikely to be displaced by ligand molecules and can be treated as part of the receptor [49]. If a water position from the X-ray structure is not characterized by high density during the MD simulation, this would support to remove the water molecule before docking. Displacement of "unhappy" water molecules is expected to contribute favorable to the binding of a ligand [111]. Attempts to integrate MD-derived information on water behavior in scoring are made for example in the scoring function Wscore. This scoring function is currently under development by the group of Richard Friesner using the WaterMap concept and is likely to be implemented in the docking program Glide [108, 109].

Depending on the scoring function one might also consider minimization of snapshots or representative structures [48, 86, 112]. For several targets Marelius et al. applied an empirical scoring function to evaluate poses sampled during an MD simulation. Paulsen et al. used minimized structures from an MD sampling of several picoseconds of *Candida albicans* dihydrofolate reductase and influenza neuraminidase to dock and rank active ligands. Ranking accuracy increased for the empirical scoring function Surflex-Dock when minimizing the conformations sequentially taken from an MD simulation. Additionally, they discuss that an adaption of the subsequent pooling strategy of snapshots should be considered using minimized structures [86]. Especially the hydrogen bond term showed high sensitivity towards structural variations sampled during the trajectory. Energy minimization of structures selected from the MD trajectory consistently led to binding free energy predictions which were lower and closer to the experimental value [48]. For scoring functions fitted on minimized protein structures this step might improve scoring, as observed for an empirical function [113]. Fitting to crystal structures, which per se should represent a local energy minimum conformation, might in general limit the applicability of empirical scoring functions on MD snapshots. During an MD simulation performed at a given temperature, e.g., at 300 K, the conformations are explicitly allowed to sample non-minimum configurations. A non-optimal hydrogen position at an amino acid will result in an unfavorable scoring of a hydrogen bond, when the scoring function judges hydrogen bonds via geometry. In contrast to these observations in literature, we observe that minimization does not affect docking scores or they show even a decrease indicating less favorable poses in GOLD (see Fig. 15.5). The minor effect of minimization of hydrogen positions is not unexpected as GOLD optimizes hydrogen orientations during docking (Fig. 15.5a). Minimization of side chain atoms with the Amber force field FF99SB-ILDN especially

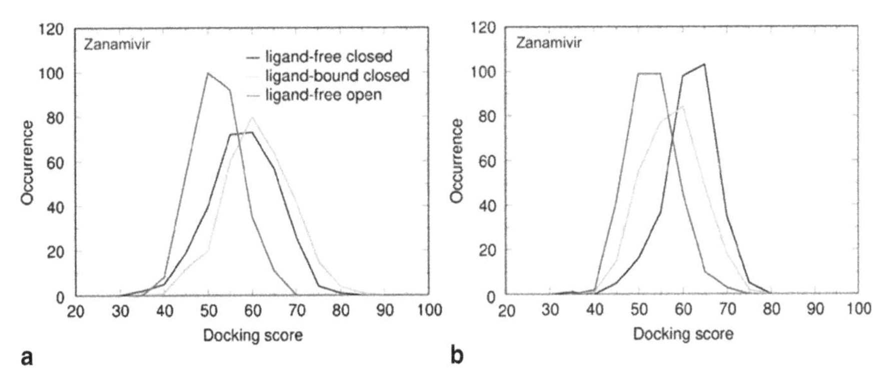

Fig. 15.5 Effect of energy minimization before docking on the docking results of zanamivir. Minimization of the hydrogen positions (**a**) or the side chain atom positions (**b**) was performed with the force field used for simulation in absence of water and ligand using 500 cycles steepest descent and 500 cycles of conjugate gradient minimization. In comparison to the corresponding results without minimization (Fig. 15.4c) especially the scoring distributions for the ligand-bound simulation is shifted towards more unfavorable poses

decreases scores obtained with receptor conformers from the ligand-bound simulation (Fig. 15.5b). This might be related to the fact that we minimized the structures in absence of the ligand and water molecules. Thereby, minimization favors a collapse of the binding site, resulting in structure less suitable for ligand docking. In summary, an appropriate preparation of selected structures might be as crucial as the setup of the simulation and for this aspect a lot of optimization potential remained unexploited.

15.2.2.3 Scoring of Multiple Protein–Ligand Poses

Handling of multiple scores for one ligand is an issue, when parallel dockingto multiple conformations representing the MD ensemble is performed. In contrast to classical molecular docking, ensemble-based docking to multiple structures in parallel does not result in a single scoring value but rather a scoring profile (see Fig. 15.4). We will present two alternative strategies how these scoring profiles are handled: considering the best score versus averaging over multiple scores. Subsequently, we discuss both strategies in respect to the thermodynamic interpretations.

Multiple docking results are also generated when a docking algorithm is applied several times for one protein conformation. This option is selected for docking algorithms with random elements in sampling of the ligand pose such as genetic algorithms. With increasing number of trials the probability to identify the optimal pose is higher and therefore, the pose with the most favorable scoring represents the docking result. Similarly, using multiple receptor conformations **the pose scoring best** can be considered as result of the docking run. The resulting docking poses and scores are expected to be the same as if the selection of the optimal conformer is a degree of freedom in the docking algorithm (see Sect. 15.2.1.4). In both cases the best suitable receptor conformation for each ligand is selected individually. Selecting the best pose from multiple hit lists is for example used in a study using ensembles of X-ray structures and is also referred to as "merging and shrinking procedure" [70]. For various targets it was shown that the enrichment of active ligands is improved in comparison to the use of a single X-ray structure [114]. The merging strategy was used to compare the performance of representatives derived from standard MD simulation and replica exchange MD simulations to X-ray structures of three systems [57]. Applying the hit list ranking by best pose, MD-derived ensemble representatives perform at least similar to X-ray structures. This indicates, that MD-derived ensemble docking will be useful in cases where alternative X-ray structures are not available [57]. Xu et al. compare the effect of selecting the most favorable score over an averaging of multiple scores which will be discussed below [115].

No difference is made if the best score is a single event calculated on a receptor conformation occurring only once in the whole trajectory. Taking experimentally derived structural ensembles the composition of the available structures has no physical meaning. In some cases, there is experimental evidence that a structure representing a specific conformation should be favored or penalized. For example, the formation of a subpocket might be an endothermic process and the energy of pocket formation is needed to be compensated by ligand binding [116]. Attempts to include the energy of individual conformations in scoring similar to the conformational energy for a ligand pose are made for ensemble docking with X-ray structures. Wei et al. show that the integration of a conformational strain for the receptor conformation improved the enrichment of the ensemble docking [116]. They used a docking algorithm where flexible parts are sampled from alternative conformations during docking for a study on T4 lysozyme and thymidylate synthase. The energy for deformation of the binding pocket in reference to the ligand-free conformation was estimated by a special function for cavity formation within T4 lysozyme [117] or an approximation via the change in total electrostatics and the change in accessible non-polar surface area for thymidylate synthase. Also Barril and Morley showed that a penalty term for the formation of an alternative conformation elevated the number of ligands docked with RMSD lower 2 Å compared to the native binding pose using a set of 149 heat shock protein 90 structures [118].

Inclusion of a conformational strain for the receptor conformation is not common for MD-derived ensembles. Assuming converged sampling, the distribution of the conformations within the MD ensemble will reflect not only the strain enthalpy of individual conformers but also the free energy. Conformations with unfavorable energy will be visited less often in comparison to conformations of lower free energy. This might be the physical background why averaging over multiple scoring values is frequently used to combine scoring values from docking to MD ensembles.

The relaxed-complex-scheme implements sampling information in a straightforward way by taking the **average of scoring values** of a ligand when docked to the individual periodically extracted conformations. The developers of RCS observed in a first application that the most favorable scoring values resulted from non-native poses [41]. Averaging the scoring values, these rarely sampled conformations and the respective scoring values contribute less to the final score [41].

In a virtual screening study on influenza neuraminidase, Cheng et al. used 27 cluster representatives of the holo ensemble and the score of binding to each representative was weighted by cluster size [26]. Known inhibitors, zanamivir and oseltamivir, were ranked within the top compounds, and superior to the less active inhibitor DANA and the natural substrate sialic acid. However, peramivir was scored more unfavorable compared to experimental expectations independent from the weighting method. In this publication also other ranking metrics were evaluated. Taking the arithmetical mean assumes a normal distribution of the energies of the conformations underlying the docking experiment. Alternatively, an averaging assuming a harmonic distribution for the K_i values [26] or a Boltzmann distribution for the predicted binding free energy [86, 115] can be applied. Xu et al. discuss the effects of averaging regarding enrichment of active versus inactive compounds [115]. They used two ways of merging the hit list: either selection of the best score from multiple hit list or averaging over multiple scores clustered by docking poses. This averaging strategy performed slightly better for two cases (thrombin, *Pneumocystis carinii* dihydrofolate reductase) and both methods performed equally on the other three out of five cases (acetylcholinesterase, cyclin-dependent kinase 2, estrogen receptor α) [115].

Average scores for our test scenarios and most favorable scores are shown in Table 15.1. Regarding the bell-shaped symmetric distribution of scoring values for periodically extracted structures (Fig. 15.4c, d) we calculated the arithmetic mean of these values. Zanamivir preserves a higher average score than compound 2 in docking to conformations from the two closed simulations, which is in agreement with experimental data. The average scores for compound 2 are in the range of the score calculated for the near-native pose (Fig. 15.4f), as docking scores between 50 and 60 dominate the distribution of scoring values (Fig. 15.4d). To average the results for cluster representatives, the scores for the representatives were weighted by the population of the respective cluster. The ranking trends between the different simulations are preserved independently from the strategy of representation or score extraction.

Choosing the best scoring pose to determine the docking result corresponds to the dominant state approximation [119]. In the picture of conformational selection the optimal conformation for formation of the complex imposes a shift in state populations. Moreover, the paradigm of conformational selection states that this optimal conformation will be part of the ligand-free as well as the ligand-bound ensemble. All other states with less favorable score for the docking pose sampled during the MD are ignored. This assumes that a ligand selects one single ideal conformation from the MD ensemble. Classical scoring functions are trained to reflect the free energy of the bound state based on a single complex structure. The selection procedure would allow integrating force field energies for the protein conformation in analogy to the ligand conformational strain term in scoring functions.

The free energy of a conformation is reflected in its state probability within the MD ensemble. Going with the recently formulated implicit ligand theory, an averaging over the binding energies of the individual states would be physically rigorous as the free energy of binding is expected to be a function of the whole thermodynamic ensemble [119]. Still, classical docking algorithms and scoring functions do not fulfill the needs for this approach. This theory relies on a thermodynamic ensemble of docking poses whereas most docking algorithms are optimized to generate a single pose with optimal score [119]. Moreover, scoring functions are fitted to experimentally-derived thermodynamic data as free energy of binding or related binding constants (K_i) intrinsically representing ensemble properties. Therefore, scoring functions inherently include ensemble properties and are trained to mirror these effects in the score of a single pose. In contrast, this theory needs alternative scoring functions which estimate the free energy of binding directly via averaging over an ensemble. Such scoring functions might rather focus on an estimate of the enthalpy of binding for each individual receptor conformation in contrast to a global free energy of binding. In addition, these scoring functions would need a training based on ensemble data rather than single X-ray structures.

15.2.2.4 Validation of Molecular Docking Results

Having decided for a merging strategy of the multiple hit list, the final scoring value used for ranking can be evaluated with the same criteria and measures as a classical docking run. Performance metrics of molecular docking rely on experimental data and can be categorized in: (1) evaluation of the binding pose for ligands with known native pose, (2) ranking of known ligands by the experimental activity, (3) discrimination of active ligands and non-active ligands or decoy molecules.

First, identification of a native binding pose similar to the ligand's structure in an X-ray structures is in the simplest case a redocking of a ligand into the empty binding pocket. Similarity is classically measured in terms of root mean square deviation (RMSD) of (heavy) atom positions of the ligand. The recovery of a near-native pose is a confirmation of a successful sampling of the pose. Moreover, successful pose identification alone is not sufficient: a scoring function must guide most docking algorithms and it should be capable to identify the best pose as the top solution or within the top solutions. Therefore, some studies give the RMSD of the highest ranked as well as the first occurrence of near-native pose within the solutions. The interdependence of posing and scoring is a problem. The posing algorithm SKATE tries to provide exhaustively enumerated poses which can be used for evaluation and optimization of scoring without interference on pose sampling and show an attempt to decouple posing and scoring [120].

In rigid docking, values below 2 Å are classically considered near-native structures and a successful docking pose. The tolerance is related to the resolution of the X-ray experiments. Considering docking algorithms covering flexibility, the generation of a ligand pose from an alternative crystal structure is a more appropriate measure for posing. Especially when dealing with X-ray ensembles the exclusion of the specific ligand complex within the ensemble should be considered. It reduces the challenge to a classical redocking challenge if the right combination of ligand and

X-ray structure is found [95]. Korb et al. formulated the ensemble pose prediction index—a metric to evaluate the ranking of poses for an individual ligand for a set of ensemble members [121]. A value over 0.5 indicates that the ranking according to the scoring function identifies the pose with the lowest RMSD to the reference on the top position of the ensemble results. For the recovery of a near-native pose from an ensemble without the corresponding protein structure, a higher RMSD tolerance up to 2.5 Å can sometimes be allowed. This is due to the necessary superposition of the protein complexes, which might not give a unique solution regarding alternative protein conformations [38]. Another aspect in docking considering protein flexibility is the recovery of a native complex geometry. Not only could the position of the ligand but also the conformation of the protein be evaluated in respect to the experimental structures of alternative conformations. This recovery of the proper protein geometry of the complex is sometimes performed visually (e.g., see Fig. 4 in [70], Fig. 10 in [69]). However this part was not yet part of systematic evaluations.

Second, the ranking of different ligands according to their activity is another criterion. Direct correlation of docking scores with experimental data, not being part of the training, is currently out of reach for docking, as discussed before. Paulsen et al. classified a set of active ligands into bins of activity classes and evaluated if the compounds are scored within the bins or neighboring categories. The ranking accuracy by binning is then expressed as success fraction [86]. Categorization of activity data is a way to deal with the inaccuracies of the experimental data. The lack of consistent activity data might be a reason why this kind of metric is not that frequently applied. Moreover, the ranking of a compound series is a question of lead optimization where other methods than docking play a major role (see Sect. 15.2.2.5).

In contrast, molecular docking is a classical high-throughput technique with its major application in virtual screening. Therefore, the discrimination between active and inactive compounds is a suitable criterion. The enrichment factor is the rate of recovered active compounds for a given fraction of a database. Its absolute value depends on the data set composition and this limits the comparison of this measure for different applications [122]. Displaying the true positive rate (selectivity) versus the false positive rate (1 − specificity) according to the ranking by docking corresponds to a receiver-operator characteristic curve (ROC) of a docking run [123]. The area under this curve (AUC) is an established measure in the field of virtual screening which gives 1 for a perfect model and 0.5 for a model with random performance [122]. This metric is sometimes also called the "predictive power" [43]. To focus on early enrichment the logarithmic scaling of the fraction of inactive compounds or decoys can be chosen [115] or some weighting can be applied, e.g., the robust initial enrichment (RIE) and the Boltzmann enhanced discrimination of ROC (BedROC) [122]. If an ensemble of protein conformations is evaluated, each conformation has its own receiver operator characteristic and the distribution of the AUC values seems to be normally distributed [43]. Individual conformations outperform the X-ray structures whereas the majority results in AUC values below the corresponding X-ray structure. Such a comparison can be useful for the selection of conformations representing the ensemble for a subsequent docking application (see Sect. 15.2.1.4) and the validation of an MD-derived ensemble over experimental structures. Comparison to the experimental structure in the validation is a necessary step in order to judge the value of the MD-derived ensemble. Moreover, different performance of two methods should be tested for significance [122]. Okimoto et al. evaluated their screening method using a combination of docking and subsequent MM/PBSA calculation against classical docking and showed that the new method performed better for 3 out of 4 test scenarios with a confidence of 95 % [124].

One major motivation for ensemble docking is the assumption that ligands with an alternative scaffold might bind to another protein conformation. To include this aspect the ligands for evaluation should ideally represent a collection of chemically diverse ligands. A metric covering this aspect would be the chemotype enrichment [45, 114, 121]. The definition of a chemotype is not trivial in this context. Considering for example the two molecules used in this test case, they might fall into the same chemotype based on the sialic acid scaffold. However, they bind to different conformations of neuraminidase. Ensembles of X-ray structures were able to score more diverse ligands compared to single structures [114, 121].

The plethora of measures for the evaluation makes an objective comparison between different studies nearly unfeasible. The evaluation in respect to experimental data is always limited by existing knowledge. For some targets the chemical space of known inhibitors does not allow to include a measure of diversity enrichment. Additionally, the data we refer to during the validation is subject of change with increasing number of available X-ray structures and activity data.

Ultimately, to prove that a method is able to identify new binders the experimental verification upon application to virtual screening is the ideal proof of concept. Successful identification of novel inhibitors for the potential *Trypanosoma brucei* drug target RNA editing ligase 1 by the relaxed-complex-scheme represents such a verification [84, 91]. Additionally, for the alternative target UDB-galactose 4′-epimerase of this pathogen a hit compound was experimentally confirmed and led to a series of inhibitors with low μM inhibition. Similar promising results were obtained for neuraminidase and were patented based on the μM activity in a flourescence-based enzyme inhibition assay [91, 125].

15.2.2.5 Rescoring of Highly Ranked Docking Poses

Apart from rescoring with alternative scoring functions post-processing of docking results can include subsequent MD simulations based on the docking pose(s). As post-processing requires additional calculation time these calculations are normally restricted to highly ranked poses. Besides of the classical techniques for free energy estimation with the help of MD simulations, we will first discuss an alternative MD-based metric to evaluate the poses. Classically, the starting point of the protocols is an X-ray structure of a ligand-protein complex or a docking pose from a rigid docking protocol.

Proctor et al. evaluated multiple docking poses by the stability of the pose during a subsequent simulation [126]. They used non-classical discrete molecular dynamics simulations which provide a fast sampling based on multiple independent repeats. As a criterion for the stability they calculate the residence time of the ligand. They showed that on average docking poses close to the native structure have a higher residence time and non-native poses can be eliminated based on a low residence time. Consequently, they suggest using the residence time as a criterion for scoring of ligand poses [126].

Molecular dynamics allow estimating binding free energies of small molecules to proteins in a physics-based framework. These methods predict Gibbs free energies of protein–ligand binding. In post-processing they can complement the estimates from docking and be used for rescoring. Depending on the level of accuracy necessary, several different methods for free energy calculation can be applied [127]. Free energy calculations are especially suited to provide relative binding free energies of congeneric ligand series as errors introduced, e.g., with the force field are canceling. So, some of these techniques are applied in lead optimization where series of similar compounds are evaluated (thermodynamic integration, free energy perturbation). In virtual screening diverse chemical entities need to be evaluated. This task can be performed with MM/PBSA, MM/GBSA and LIE approach.

Post-processing of molecular dynamics trajectories allows to perform molecular mechanics/Poisson-Boltzmann Surface Area (MM/PBSA) [128] or faster and more approximate molecular mechanics/Generalized Born Surface Area (MM/GBSA) [129] calculations. These statistical methods estimate binding energies by summing up *in vacuo* force field energies, polar and non-polar solvation energies as well as entropic contributions from normal mode analysis. The difference between both approaches lies in the calculation of the polar solvation energy, with GBSA using more approximations to yield a numerical estimate of the electrostatic solvation energy than PBSA. The non-polar solvation term captures the hydrophobic effect, forcing apolar groups to bury inside of a protein. A final free energy estimate is generated by subtracting respective energies of free receptor and free ligand from the complex. MM/PBSA was used in first applications of the relaxed-complex-scheme to rescore the best poses for each ligand. Artificially high binding energies predicted by the force field-based scoring function of AutoDock could be avoided [80]. Negri et al. applied docking and subsequent MD simulations evaluated by MM/PBSA to investigate the binding mode of bis(hydroxyl)phenylarenes to 17β-hydroxysteroid dehydrogenase type 1 [130]. On a larger scale Okimoto et al. suggested to use docking and subsequent scoring with MM/PBSA in virtual screening application. The authors showed that MM/PBSA-based scoring enables to identify the near-native pose for a trypsin inhibitor whereas docking scores failed to rank this pose accurately. ROC values improved for 3 out of 4 test cases using MM/PBSA over a docking protocol using the program GOLD [124]. Performing MD-based free energy estimation for multiple ligands and even for multiple poses per ligand (as done in this study) leads to an explosion of computational cost. The application of this approach in high-throughput scale is only feasible for facilities where tailored computational infrastructure is accessible [124].

An alternative approach is the linear interaction energy (LIE) [131]. Herein, the binding energy is estimated via summation of weighted contributions of ensemble-averaged electrostatics and van-der-Waals interactions. The difference between bound and free state gives an estimate of protein-ligand interactions and hence free energy of binding. Targets with a wide and flexible binding pocket, like the promiscuous monooxygenase cytochrome P450 isoform 2D6, potentially allow alternative binding modes. To include this flexibility Stjernschantz et al. developed a LIE model on different initial poses generated by docking for a series of thiourea compounds [132]. Marelius et al. included the free energy of the LIE in comparison to the results by scoring to illustrate the fluctuation over the simulation time for the arabinose binding protein as for this target a LIE model had been already adapted [48]. A problem of LIE is the choice of weighting parameters for the energy contributions. Depending on the system these coefficients need readjustments, hence hindering the establishment of global models.

15.3 Potential of Ensemble Docking

We gave an overview on recent applications of ensemble docking with the focus on MD-derived ensembles. Following the practical workflow it becomes clear that no common recipe for a best practice has been established to date. At multiple stages variability occurs and at each branching point the possible number of workflows increases. This heterogeneity of the different applications limits the comparability between them. Moreover, lack of consistent performance metrics and performance comparison in respect to docking to single or multiple X-ray structures also limits to quantify the benefit from ensemble docking to MD-derived ensembles. A systematic evaluation on a tailored benchmark set could allow to characterize the impact of variability within the field and also to identify an optimal workflow. However, considering the complexity of the ligand recognition process individual contributions depend on the system investigated. Therefore, we also suspect the "optimal" workflow to be dependent on the system of interest. One of the most relevant points of opposing concepts is the scoring of multiple poses. We expect that consideration of thermodynamic concepts will aid to adapt molecular docking algorithms and scoring functions on ensembles. The application of scoring functions and especially the averaging of scoring values need a conscious handling and interpretation of the resulting scores. These values might be valid to rank hit lists, which is the primary aim of scoring. However, interpretation in terms of thermodynamic properties should be reflected carefully.

Despite of current absence of a clear answer to the question "How to?", single success stories and also attempts to benchmark the method provide a growing number of pieces of evidence to establish this method within virtual screening. MD simulations or alternative computational sampling methods are an essential way to investigate the target's flexibility, especially if the flexibility is not covered by experimental ensembles. The gained knowledge on the target gathered while considering a proper simulation setup or gained directly from the simulation data will be of benefit for a structure-based drug discovery project anyway.

Besides the expected methodological improvements the high-throughput application of the ensemble docking becomes more and more realistic with continuously increasing computational resources. With the current computational power and the efficient use of GPU facilities for simulation, the time for the generation of an MD ensemble will no longer be a limit for the its application for subsequent docking. These circumstances will also enforce the use of computationally more demanding methods which are more accurate than docking scores and ensemble docking might be facing a competition with free energy estimation techniques inherently including flexibility aspects. Nevertheless, the potential benefit of a defined workflow for virtual screening based on an MD-derived ensemble as strategy including target flexibility is evident. MD-based ensemble docking has the potential to become a standard procedure particularly for detection of chemically different ligands within next decades.

Acknowledgments Presented work was supported by funding of the Austrian Science Fund FWF: project "Targeting Influenza Neuraminidase" (P23051). Julian E. Fuchs is a recipient of a DOC-fellowship of the Austrian Academy of Sciences at the Institute of General, Inorganic and Theoretical Chemistry at University of Innsbruck.

References

1. Boehr DD, McElheny D, Dyson HJ, Wright PE (2006) The dynamic energy landscape of dihydrofolate reductase catalysis. Science 313(5793):1638–1642. doi:10.1126/science.1130258
2. Fischer E (1894) Einfluss der Configuration auf die Wirkung der Enzyme. Ber Dtsch Chem Ges 27
3. Koshland DE (1958) Application of a theory of enzyme specificity to protein synthesis. Proc Natl Acad Sci U S A 44 98–104
4. Tsai CJ, Ma BY, Nussinov R (1999) Folding and binding cascades: shifts in energy landscapes. Proc Natl Acad Sci U S A 96(18):9970–9972. doi:10.1073/pnas.96.18.9970
5. Tsai CJ, Kumar S, Ma BY, Nussinov R (1999) Folding funnels, binding funnels, and protein function. Protein Sci 8(6):1181–1190
6. Boehr DD, Nussinov R, Wright PE (2009) The role of dynamic conformational ensembles in biomolecular recognition. Nat Chem Biol 5(11):789–796. doi:10.1038/nchembio.232
7. Durrant JD, McCammon JA (2010) Computer-aided drug-discovery techniques that account for receptor flexibility. Curr Opin Pharmacol 10(6):770–774. doi:10.1016/j.coph.2010.09.001
8. Totrov M, Abagyan R (2008) Flexible ligand docking to multiple receptor conformations: a practical alternative. Curr Opin Struct Biol 18(2):178–184. doi:10.1016/j.sbi.2008.01.004
9. Carlson HA, McCammon JA (2000) Accommodating protein flexibility in computational drug design. Mol Pharmacol 57(2):213–218
10. Carlson HA (2002) Protein flexibility is an important component of structure-based drug discovery. Curr Pharm Des 8(17):1571–1578. doi:10.2174/1381612023394232
11. Chaudhury S, Gray JJ (2008) Conformer selection and induced fit in flexible backbone protein-protein docking using computational and NMR ensembles. J Mol Biol 381(4):1068–1087. doi:10.1016/j.jmb.2008.05.042
12. Henzler-Wildman KA, Lei M, Thai V, Kerns SJ, Karplus M, Kern D (2007) A hierarchy of timescales in protein dynamics is linked to enzyme catalysis. Nature 450(7171):913–U27. doi:10.1038/nature06407

13. Lindorff-Larsen K, Piana S, Palmo K, Maragakis P, Klepeis JL, Dror RO, Shaw DE (2010) Improved side-chain torsion potentials for the Amber ff99SB protein force field. Proteins: Str, Funct, Bioinform 78(8):1950–1958. doi:10.1002/prot.22711
14. Klepeis JL, Lindorff-Larsen K, Dror RO, Shaw DE (2009) Long-timescale molecular dynamics simulations of protein structure and function. Curr Opin Struct Biol 19(2):120–127. doi:10.1016/j.sbi.2009.03.004
15. Goetz AW, Williamson MJ, Xu D, Poole D, Le Grand S, Walker RC (2012) Routine microsecond molecular dynamics simulations with AMBER on GPUs. 1. generalized born. J Chem Theory Comput 8(5):1542–1555. doi:10.1021/ct200909j
16. Luque I, Freire E (2000) Structural stability of binding sites: Consequences for binding affinity and allosteric effects. Proteins: Str Funct Genet 41(S4):63–71. doi:10.1002/1097-0134(2000) 41:4+<63::AID-PROT60>3.0.CO;2-6
17. Kroemer RT (2007) Structure-based drug design: docking and scoring. Curr Protein Pept Sci 8(4):312–328
18. B-Rao C, Subramanian J, Sharma SD (2009) Managing protein flexibility in docking and its applications. Drug Discov Today 14(7–8):394–400. doi:10.1016/j.drudis.2009.01.003
19. Jiang F, Kim SH (1991) Soft docking-matching of molecular—surface cubes. J Mol Biol 219(1):79–102. doi:10.1016/0022-2836(91)90859-5
20. Claussen H, Buning C, Rarey M, Lengauer T (2001) FlexE: efficient molecular docking considering protein structure variations. J Mol Biol 308(2):377–395. doi:10.1006/jmbi.2001.4551
21. Flick J, Tristram F, Wenzel W (2012) Modeling loop backbone flexibility in receptor-ligand docking simulations. J Comput Chem 33(31):2504–2515. doi:10.1002/jcc.23087
22. Cozzini P, Kellogg GE, Spyrakis F, Abraham DJ, Costantino G, Emerson A, Fanelli F, Gohlke H, Kuhn LA, Morris GM, Orozco M, Pertinhez TA, Rizzi M, Sotriffer CA (2008) Target flexibility: an emerging consideration in drug discovery and design. J Med Chem 51(20):6237–6255. doi:10.1021/jm800562d
23. Gamblin SJ, Skehel JJ (2010) influenza hemagglutinin and neuraminidase membrane glycoproteins. J Biol Chem 285(37):28403–28409. doi:10.1074/jbc.R110.129809
24. Lill MA (2011) Efficient incorporation of protein flexibility and dynamics into molecular docking simulations. Biochemistry 50(28):6157–6169. doi:10.1021/bi2004558
25. Russell RJ, Haire LF, Stevens DJ, Collins PJ, Lin YP, Blackburn GM, Hay AJ, Gamblin SJ, Skehel JJ (2006) The structure of H5N1 avian influenza neuraminidase suggests new opportunities for drug design. Nature 443(7107):45–49. doi:10.1038/nature05114
26. Cheng LS, Amaro RE, Xu D, Li WW, Arzberger PW, McCammon JA (2008) Ensemble-based virtual screening reveals potential novel antiviral compounds for avian influenza neuraminidase. J Med Chem 51(13):3878–3894. doi:10.1021/jm8001197
27. Grienke U, Schmidtke M, Kirchmair J, Pfarr K, Wutzler P, Durrwald R, Wolber G, Liedl KR, Stuppner H, Rollinger JM (2010) Antiviral Potential and Molecular Insight into Neuraminidase Inhibiting Diarylheptanoids from Alpinia katsumadai. J Med Chem 53(2):778–786. doi:10.1021/jm901440f
28. Vavricka CJ, Li Q, Wu Y, Qi JX, Wang MY, Liu Y, Gao F, Liu J, Feng EG, He JH, Wang JF, Liu H, Jiang HL, Gao GF (2011) Structural and functional analysis of laninamivir and its octanoate prodrug reveals group specific mechanisms for influenza NA inhibition. PloS Pathog 7(10):e1002249. doi:10.1371/journal.ppat.1002249
29. Li Q, Qi JX, Zhang W, Vavricka CJ, Shi Y, Wei JH, Feng EG, Shen JS, Chen JL, Liu D, He JH, Yan JH, Liu H, Jiang HL, Teng MK, Li XB, Gao GF (2010) The 2009 pandemic H1N1 neuraminidase N1 lacks the 150-cavity in its active site. Nat Struct Mol Biol 17(10):1266–1268. doi:10.1038/nsmb.1909
30. van der Vries E, Collins PJ, Vachieri SG, Xiong XL, Liu JF, Walker PA, Haire LF, Hay AJ, Schutten M, Osterhaus A, Martin SR, Boucher CAB, Skehel JJ, Gamblin SJ (2012) H1N1 2009 Pandemic influenza virus: Resistance of the I223R neuraminidase mutant explained by kinetic and structural analysis. PloS Pathog 8(9):e1002914. doi:10.1371/journal.ppat.1002914

31. Rudrawar S, Dyason JC, Rameix-Welti MA, Rose FJ, Kerry PS, Russell RJ, van der Werf S, Thomson RJ, Naffakh N, von Itzstein M (2010) Novel sialic acid derivatives lock open the 150-loop of an influenza A virus group-1 sialidase. Nat Comm 1 113. doi:10.1038/ncomms1114
32. Amaro RE, Swift RV, Votapka L, Li WW, Walker RC, Bush RM (2011) Mechanism of 150-cavity formation in influenza neuraminidase. Nat Comm 2388. doi:10.1038/ncomms1390
33. Wallnoefer HG, Lingott T, Gutierrez JM, Merfort I, Liedl KR (2010) Backbone flexibility controls the activity and specificity of a protein-protein interface: specificity in snake venom metalloproteases. J Am Chem Soc 132(30):10330–10337. doi:10.1021/ja909908y
34. Case DA, Darden TA, Cheatham III TE, Simmerling CL, Wang J, Duke RE, Luo R, Walker RC, Zhang W, Merz KM, Roberts B, Hayik S, Roitberg A, Seabra G, Swails J, Goetz AW, Kolossváry I, Wong KF, Paesani F, Vanicek J, Wolf RM, Wu X, Brozell SR, Steinbrecher T, Gohlke H, Cai Q, Ye X, Wang J, Hsieh M-J, Cui G, Roe DR, Mathews DH, Seetin MG, Salomon-Ferrer R, Sagui C, Babin V, Luchko T, Gusarov S, Kovalenko A, Kollman PA (2012) AMBER12. University of California, San Francisco
35. Roe DR, Cheatham TE (2013) PTRAJ and CPPTRAJ: software for processing and analysis of molecular dynamics trajectory data. J Chem Theory Comput 9(7):3084–3095. doi:10.1021/ct400341p
36. von Grafenstein S, Wallnoefer HG, Kirchmair J, Fuchs JE, Huber RG, Spitzer GM, Schmidtke M, Sauerbrei A, Rollinger JM, Liedl KR (2013) Interface dynamics explain assembly dependency of influenza neuraminidase catalytic activity. J Biomol Struct Dyn. Published online doi:10.1080/07391102.2013.855142
37. McCammon JA, Gelin BR, Karplus M (1977) Dynamics of folded proteins. Nature 267(5612):585–590. doi:10.1038/267585a0
38. Xu M, Lill MA (2011) Significant enhancement of docking sensitivity using implicit ligand sampling. J Chem Inf Model 51(3):693–706. doi:10.1021/ci100457t
39. Wang JM, Wolf RM, Caldwell JW, Kollman PA, Case DA (2004) Development and testing of a general amber force field. J Comput Chem 25(9):1157–1174
40. Fuchs JE, Huber RG, Von Grafenstein S, Wallnoefer HG, Spitzer GM, Fuchs D, Liedl KR (2012) Dynamic regulation of phenylalanine hydroxylase by simulated redox manipulation. PloS One 7(12):e53005. doi:10.1371/journal.pone.0053005
41. Lin J-H, Perryman AL, Schames JR, McCammon JA (2002) Computational drug design accommodating receptor flexibility: the relaxed complex scheme. J Am Chem Soc 124(20):5632–5633. doi:10.1021/ja0260162
42. Seeliger D, de Groot BL (2010) Conformational transitions upon ligand binding: holo-structure prediction from apo conformations. PloS Comp Biol 6(1):e1000634. doi:10.1371/journal.pcbi.1000634
43. Nichols SE, Baron R, Ivetac A, McCammon JA (2011) Predictive power of molecular dynamics receptor structures in virtual screening. J Chem Inf Model 51(6):1439–1446. doi:10.1021/ci200117n
44. McGovern SL, Shoichet BK (2003) Information decay in molecular docking screens against holo, apo, and modeled conformations of enzymes. J Med Chem 46(14):2895–2907. doi:10.1021/jm0300330
45. Mackey MD, Melville JL (2009) Better than random? The chemotype enrichment problem. J Chem Inf Model 49(5):1154–1162. doi:10.1021/ci8003978
46. Bolstad ESD, Anderson AC (2009) In pursuit of virtual lead optimization: pruning ensembles of receptor structures for increased efficiency and accuracy during docking. Proteins: Str, Funct, Bioinform 75(1):62–74. doi:10.1002/prot.22214
47. Armen RS, Chen J, Brooks CL III (2009) An evaluation of explicit receptor flexibility in molecular docking using molecular dynamics and torsion angle molecular dynamics. J Chem Theory Comput 5(10):2909–2923. doi:10.1021/ct900262t
48. Marelius J, Ljungberg KB, Aqvist J (2001) Sensitivity of an empirical affinity scoring function to changes in receptor-ligand complex conformations. Eur J Pharm Sci 14(1):87–95. doi:10.1016/s0928-0987(01)00162-2

49. Kirchmair J, Spitzer GM, Liedl KR (2011) Consideration of water and solvation effects in virtual screening. In: Sotriffer C (ed) Virtual screening: principals, challenges, and practical guidelines, vol 82. Wiley, Weinheim, pp 263–289
50. Cole JC, Korb O, Olsson TSG, Liebeschuetz J (2011) The basis for target-based virtual screening: protein structures. In: Sotriffer C (ed) Virtual screening: principals, challenges, and practical guidelines, vol 82. Wiley, Weinheim, pp 87–114
51. Borjesson U, Hunenberger PH (2001) Explicit-solvent molecular dynamics simulation at constant pH: methodology and application to small amines. J Chem Phys 114(22):9706–9719
52. Merski M, Shoichet BK (2013) The impact of introducing a histidine into an apolar cavity site on docking and ligand recognition. J Med Chem 56(7):2874–2884. doi:10.1021/jm301823g
53. Kim MO, Nichols SE, Wang Y, McCammon JA (2013) Effects of histidine protonation and rotameric states on virtual screening of M-tuberculosis RmlC. J Comput-Aided Mol Des 27(3):235–246. doi:10.1007/s10822-013-9643-9
54. Labute P (2009) Protonate3D: assignment of ionization states and hydrogen coordinates to macromolecular structures. Proteins: Str, Funct, Bioinform 75(1):187–205. doi:10.1002/prot.22234
55. Halperin I, Ma BY, Wolfson H, Nussinov R (2002) Principles of docking: an overview of search algorithms and a guide to scoring functions. Proteins: Str, Funct, Genet 47(4):409–443. doi:10.1002/prot.10115
56. Sinko W, Lindert S, McCammon JA (2013) Accounting for receptor flexibility and enhanced sampling methods in computer-aided drug design. Chem Biol Drug Des 81(1):41–49. doi:10.1111/cbdd.12051
57. Osguthorpe DJ, Sherman W, Hagler AT (2012) Exploring protein flexibility: incorporating structural ensembles from crystal structures and simulation into virtual screening protocols. J Phys Chem B 116(23):6952–6959. doi:10.1021/jp3003992
58. Fulle S, Gohlke H (2010) Molecular recognition of RNA: challenges for modelling interactions and plasticity. J Mol Recognit 23(2):220–231. doi:10.1002/jmr.1000
59. Bruno A, Amori L, Costantino G (2011) Addressing the conformational flexibility of serine racemase by combining targeted molecular dynamics, conformational sampling and docking studies. Mol Inform 30(4):317–328. doi:10.1002/minf.201000162
60. Abagyan R, Totrov M, Kuznetsov D (1994) ICM—a new method for protein modeling and design—applications to docking and structure prediction from the distorted native conformation. J Comput Chem 15(5):488–506. doi:10.1002/jcc.540150503
61. Sugita Y, Okamoto Y (1999) Replica-exchange molecular dynamics method for protein folding. Chem Phys Lett 314(1–2):141–151. doi:10.1016/s0009-2614(99)01123-9
62. Metropolis N, Rosenbluth AW, Rosenbluth MN, Teller AH, Teller E (1953) Equation of state calculations by fast computing machines. J Chem Phys 21(6):1087–1092. doi:10.1063/1.1699114
63. Gervasio FL, Laio A, Parrinello M (2005) Flexible docking in solution using metadynamics. J Am Chem Soc 127(8):2600–2607. doi:10.1021/ja0445950
64. Zacharias M (2004) Rapid protein-ligand docking using soft modes from molecular dynamics simulations to account for protein deformability: binding of FK506 to FKBP. Proteins: Str, Funct, Bioinform 54(4):759–767. doi:10.1002/prot.10637
65. Hinsen K (1998) Analysis of domain motions by approximate normal mode calculations. Proteins: Str, Funct, Genet 33(3):417–429. doi:10.1002/(sici)1097-0134(19981115)33:3<417::aid-prot10>3.0.co;2-8
66. May A, Zacharias M (2008) Protein-ligand docking accounting for receptor side chain and global flexibility in normal modes: evaluation on kinase inhibitor cross docking. J Med Chem 51(12):3499–3506. doi:10.1021/jm800071v
67. Leis S, Zacharias M (2011) Efficient inclusion of receptor flexibility in grid-based protein-ligand docking. J Comput Chem 32(16):3433–3439. doi:10.1002/jcc.21923
68. Cavasotto CN, Kovacs JA, Abagyan RA (2005) Representing receptor flexibility in ligand docking through relevant normal modes. J Am Chem Soc 127(26):9632–9640. doi:10.1021/ja042260c

69. Rueda M, Bottegoni G, Abagyan R (2009) Consistent improvement of cross-docking results using binding site ensembles generated with elastic network normal modes. J Chem Inf Model 49(3):716–725. doi:10.1021/ci8003732
70. Tran HT, Zhang S (2011) Accurate prediction of the bound form of the akt pleckstrin homology domain using normal mode analysis to explore structural flexibility. J Chem Inf Model 51(9):2352–2360. doi:10.1021/ci2001742
71. Cavasotto CN, Abagyan RA (2004) Protein flexibility in ligand docking and virtual screening to protein kinases. J Mol Biol 337(1):209–225. doi:10.1016/j.jmb.2004.01.003
72. Davis IW, Baker D (2009) ROSETTALIGAND docking with full ligand and receptor flexibility. J Mol Biol 385(2):381–392. doi:10.1016/j.jmb.2008.11.010
73. Meiler J, Baker D (2006) ROSETTALIGAND: protein-small molecule docking with full side-chain flexibility. Proteins: Str, Funct, Bioinform 65(3):538–548. doi:10.1002/prot.21086
74. Alonso H, Bliznyuk AA, Gready JE (2006) Combining docking and molecular dynamic simulations in drug design. Med Res Rev 26(5):531–568. doi:10.1002/med.20067
75. Knegtel RMA, Kuntz ID, Oshiro CM (1997) Molecular docking to ensembles of protein structures. J Mol Biol 266(2):424–440. doi:10.1006/jmbi.1996.0776
76. Polgar T, Keseru GM (2006) Ensemble docking into flexible active sites. Critical evaluation of flexE against JNK-3 and beta-secretase. J Chem Inf Model 46(4):1795–1805. doi:10.1021/ci050412x
77. Osterberg F, Morris GM, Sanner MF, Olson AJ, Goodsell DS (2002) Automated docking to multiple target structures: incorporation of protein mobility and structural water heterogeneity in AutoDock. Proteins: Str, Funct, Genet 46(1):34–40. doi:10.1002/prot.10028
78. Broughton HB (2000) A method for including protein flexibility in protein-ligand docking: improving tools for database mining and virtual screening. J Mol Graphics Modell 18(3):247–257. doi:10.1016/s1093-3263(00)00036-x
79. Cosconati S, Marinelli L, Di Leva FS, La Pietra V, De Simone A, Mancini F, Andrisano V, Novellino E, Goodsell DS, Olson AJ (2012) Protein flexibility in virtual screening: the BACE-1 case study. J Chem Inf Model 52(10):2697–2704. doi:10.1021/ci300390h
80. Lin JH, Perryman AL, Schames JR, McCammon JA (2003) The relaxed complex method: accommodating receptor flexibility for drug design with an improved scoring scheme. Biopolymers 68(1):47–62. doi:10.1002/bip.10218
81. Schames JR, Henchman RH, Siegel JS, Sotriffer CA, Ni HH, McCammon JA (2004) Discovery of a novel binding trench in HIV integrase. J Med Chem 47(8):1879–1881. doi:10.1021/jm0341913
82. Perryman AL, Lin J-H, McCammon JA (2006) Optimization and computational evaluation of a series of potential active site inhibitors of the V82F/I84V drug-resistant mutant of HIV-1 protease: an application of the relaxed complex method of structure-based drug design. Chem Biol Drug Des 67(5):336–345. doi:10.1111/j.1747-0285.2006.00382.x
83. Trott O, Olson AJ (2010) Software news and update autodock vina: improving the speed and accuracy of docking with a new scoring function, efficient optimization, and multithreading. J Comput Chem 31(2):455–461. doi:10.1002/jcc.21334
84. Amaro RE, Baron R, McCammon JA (2008) An improved relaxed complex scheme for receptor flexibility in computer-aided drug design. J Comput-Aided Mol Des 22(9):693–705. doi:10.1007/s10822-007-9159-2
85. Sivanesan D, Rajnarayanan RV, Doherty J, Pattabiraman N (2005) In-silico screening using flexible ligand binding pockets: a molecular dynamics-based approach. J Comput-Aided Mol Des 19(4):213–228. doi:10.1007/s10822-005-4788-9
86. Paulsen JL, Anderson AC (2009) Scoring ensembles of docked protein: ligand interactions for virtual lead optimization. J Chem Inf Model 49(12):2813–2819. doi:10.1021/ci9003078
87. Korb O, Olsson TSG, Bowden SJ, Hall RJ, Verdonk ML, Liebeschuetz JW, Cole JC (2012) Potential and limitations of ensemble docking. J Chem Inf Model 52(5):1262–1274. doi:10.1021/ci2005934
88. Nichols SE, Swift RV, Amaro RE (2012) Rational prediction with molecular dynamics for hit identification. Curr Top Med Chem 12(18):2002–2012

89. Shao JY, Tanner SW, Thompson N, Cheatham TE (2007) Clustering molecular dynamics trajectories: 1. characterizing the performance of different clustering algorithms. J Chem Theory Comput 3(6):2312–2334. doi:10.1021/ct700119m
90. Daura X, van Gunsteren WF, Mark AE (1999) Folding-unfolding thermodynamics of a beta-heptapeptide from equilibrium simulations. Proteins: Str, Funct, Genet 34 (3):269–280. doi:10.1002/(sici)1097-0134(19990215)34:3<269::aid-prot1>3.0.co;2-3
91. Durrant JD, Hall L, Swift RV, Landon M, Schnaufer A, Amaro RE (2010) Novel naphthalene-based inhibitors of trypanosoma brucei RNA editing ligase 1. PloS Neglect Trop D 4(8):e803 doi:10.1371/journal.pntd.0000803
92. Damm-Ganamet KL, Smith RD, Dunbar JB, Stuckey JA, Carlson HA (2013) CSAR benchmark exercise 2011–2012: evaluation of results from docking and relative ranking of blinded congeneric series. J Chem Inf Model. doi:10.1021/ci400025f
93. Ben Nasr N, Guillemain H, Lagarde N, Zagury J-F, Montes M (2013) Multiple structures for virtual ligand screening: defining binding site properties-based criteria to optimize the selection of the query. J Chem Inf Model 53(2):293–311. doi:10.1021/ci3004557
94. Sotriffer CA, Dramburg I (2005) "In situ cross-docking" to simultaneously address multiple targets. J Med Chem 48(9):3122–3125. doi:10.1021/jm050075j
95. Huang S-Y, Zou X (2007) Ensemble docking of multiple protein structures: considering protein structural variations in molecular docking. Proteins: Str, Funct, Bioinform 66(2):399–421. doi:10.1002/prot.21214
96. Bottegoni G, Kufareva I, Totrov M, Abagyan R (2009) Four-dimensional docking: a fast and accurate account of discrete receptor flexibility in ligand docking. J Med Chem 52(2):397–406. doi:10.1021/jm8009958
97. Kuntz ID, Blaney JM, Oatley SJ, Langridge R, Ferrin TE (1982) A geometric approach to macromolecule-ligand interactions. J Mol Biol 161(2):269–288. doi:10.1016/0022-2836(82)90153-x
98. Kitchen DB, Decornez H, Furr JR, Bajorath J (2004) Docking and scoring in virtual screening for drug discovery: methods and applications. Nature Rev Drug Discov 3(11):935–949. doi:10.1038/nrd1549
99. Jorgensen WL (2004) The many roles of computation in drug discovery. Science 303(5665):1813–1818. doi:10.1126/science.1096361
100. Rognan D (2011) Docking methods for virtual screening: principles and recent advances. In: Sotriffer C (ed) Virtual screening: principals, challenges, and practical guidelines, vol 82. Wiley, Weinheim, pp 153–176
101. Henzler AM, Rarey M (2011) Protein flexibility in structure-based virtual screening: from models to algorithms. In: Sotriffer C (ed) Virtual screening: principals, challenges, and practical guidelines, vol 82. Wiley, Weinheim, pp 223–244
102. Carlson HA, Dunbar JB Jr (2011) A call to arms: what you can do for computational drug discovery. J Chem Inf Model 51(9):2025–2026. doi:10.1021/ci200398g
103. Warren GL, Andrews CW, Capelli A-M, Clarke B, LaLonde J, Lambert MH, Lindvall M, Nevins N, Semus SF, Senger S, Tedesco G, Wall ID, Woolven JM, Peishoff CE, Head MS (2006) A critical assessment of docking programs and scoring functions. J Med Chem 49(20):5912–5931. doi:10.1021/jm050362n
104. Perola E, Charifson PS (2004) Conformational analysis of drug-like molecules bound to proteins: an extensive study of ligand reorganization upon binding. J Med Chem 47(10):2499–2510. doi:10.1021/jm030563w
105. Gohlke H, Hendlich M, Klebe G (2000) Predicting binding modes, binding affinities and 'hot spots' for protein-ligand complexes using a knowledge-based scoring function. Perspec Drug Discov Des 20(1):115–144. doi:10.1023/a:1008781006867
106. Velec HFG, Gohlke H, Klebe G (2005) DrugScore(CSD)-knowledge-based scoring function derived from small molecule crystal data with superior recognition rate of near-native ligand poses and better affinity prediction. J Med Chem 48(20):6296–6303. doi:10.1021/jm050436v

107. Kuhn B, Fuchs JE, Reutlinger M, Stahl M, Taylor NR (2011) Rationalizing tight ligand binding through cooperative interaction networks. J Chem Inf Model 51(12):3180–3198. doi:10.1021/ci200319e
108. Repasky MP, Murphy RB, Banks JL, Greenwood JR, Tubert-Brohman I, Bhat S, Friesner RA (2012) Docking performance of the glide program as evaluated on the Astex and DUD datasets: a complete set of glide SP results and selected results for a new scoring function integrating WaterMap and glide. J Comput-Aided Mol Des 26(6):787–799. doi:10.1007/s10822-012-9575-9
109. Abel R, Young T, Farid R, Berne BJ, Friesner RA (2008) Role of the active-site solvent in the thermodynamics of factor Xa ligand binding. J Am Chem Soc 130(9):2817–2831. doi:10.1021/ja0771033
110. Cummings MD, Arnoult É, Buyck C, Tresadern G, Vos AM, Wegner JK (2011) Preparing and filtering compound databases for virtual and experimental screening. In: Sotriffer C (ed) Virtual screening: principals, challenges, and practical guidelines, vol 82. Wiley, Weinheim, pp 35–59
111. Mason JS, Bortolato A, Congreve M, Marshall FH (2012) New insights from structural biology into the druggability of G protein-coupled receptors. Trends Pharmacol Sci 33(5):249–260. doi:10.1016/j.tips.2012.02.005
112. Smith RD, Dunbar JB Jr, Ung PM-U, Esposito EX, Yang C-Y, Wang S, Carlson HA (2011) CSAR benchmark exercise of 2010: combined evaluation across all submitted scoring functions. J Chem Inf Model 51(9):2115–2131. doi:10.1021/ci200269q
113. Eldridge MD, Murray CW, Auton TR, Paolini GV, Mee RP (1997) Empirical scoring functions: 1. The development of a fast empirical scoring function to estimate the binding affinity of ligands in receptor complexes. J Comput-Aided Mol Des 11(5):425–445. doi:10.1023/a:1007996124545
114. Bottegoni G, Rocchia W, Rueda M, Abagyan R, Cavalli A (2011) Systematic exploitation of multiple receptor conformations for virtual ligand screening. PloS One 6(5):e18845. doi:10.1371/journal.pone.0018845
115. Xu M, Lill MA (2012) Utilizing experimental data for reducing ensemble size in flexible-protein docking. J Chem Inf Model 52(1):187–198. doi:10.1021/ci200428t
116. Wei BQ, Weaver LH, Ferrari AM, Matthews BW, Shoichet BK (2004) Testing a flexible-receptor docking algorithm in a model binding site. J Mol Biol 337(5):1161–1182. doi:10.1016/j.jmb.2004.02.015
117. Eriksson AE, Baase WA, Zhang XJ, Heinz DW, Blaber M, Baldwin EP, Matthews BW (1992) Response of a protein-structure to cavity-creating mutations and its relation to the hydrophobic effect. Science 255(5041):178–183. doi:10.1126/science.1553543
118. Barril X, Morley SD (2005) Unveiling the full potential of flexible receptor docking using multiple crystallographic structures. J Med Chem 48(13):4432–4443. doi:10.1021/jm048972v
119. Minh DDL (2012) Implicit ligand theory: rigorous binding free energies and thermodynamic expectations from molecular docking. J Chem Phys 137(10). doi:10.1063/1.4751284
120. Feng JA, Marshall GR (2010) SKATE: a docking program that decouples systematic sampling from scoring. J Comput Chem 31(14):2540–2554. doi:10.1002/jcc.21545
121. Korb O, McCabe P, Cole J (2011) The ensemble performance index: an improved measure for assessing ensemble pose prediction performance. J Chem Inf Model 51(11):2915–2919. doi:10.1021/ci2002796
122. Nicholls A (2008) What do we know and when do we know it?. J Comput-Aided Mol Des 22(3–4):239–255. doi:10.1007/s10822-008-9170-2
123. Peterson W, Birdsall T, Fox W (1954) The theory of signal detectability. Trans IRE Prof Group Inform Theory 4(4):171–212
124. Okimoto N, Futatsugi N, Fuji H, Suenaga A, Morimoto G, Yanai R, Ohno Y, Narumi T, Taiji M (2009) High-performance drug discovery: computational screening by combining docking and molecular dynamics simulations. PloS Comp Biol 5(10):e1000528. doi:10.1371/journal.pcbi.1000528

125. Amaro R, Cheng L, McCammon JA, Li WW, Arzberber PW (2009) Ensemble-based virtual screening reveals novel antiviral compounds for avian influenza neuraminidase. US Patent WO2009128964 22 Jan 2009
126. Proctor EA, Yin S, Tropsha A, Dokholyan NV (2012) Discrete molecular dynamics distinguishes native like binding poses from decoys in difficult targets. Biophys J 102(1):144–151. doi:10.1016/j.bpj.2011.11.4008
127. Wallnoefer H, Fox T, Liedl K (2010) Challenges for computer simulations in drug design. In: Paneth P, Dybala-Defratyka A (eds) Kinetics and dynamics, challenges and advances in computational chemistry and physics. Springer Netherlands, Dordrecht, pp 431–463
128. Honig B, Nicholls A (1995) Classical electrostatics in biology and chemistry. Science 268(5214):1144–1149. doi:10.1126/science.7761829
129. Still WC, Tempczyk A, Hawley RC, Hendrickson T (1990) Semianalytical treatment of solvation for molecular mechanics and dynamics. J Am Chem Soc 112(16):6127–6129. doi:10.1021/ja00172a038
130. Negri M, Recanatini M, Hartmann RW (2011) Computational investigation of the binding mode of bis(hydroxylphenyl)arenes in 17 beta-HSD1: molecular dynamics simulations, MM-PBSA free energy calculations, and molecular electrostatic potential maps. J Comput-Aided Mol Des 25(9):795–811. doi:10.1007/s10822-011-9464-7
131. Aqvist J, Medina C, Samuelsson JE (1994) New method for predicting binding-affinity in computer-aided drug design. Protein Eng 7(3):385–391. doi:10.1093/protein/7.3.385
132. Stjernschantz E, Oostenbrink C (2010) Improved ligand-protein binding affinity predictions using multiple binding modes. Biophys J 98(11):2682–2691. doi:10.1016/j.bpj.2010.02.034

Chapter 16
Cheminformatics: At the Crossroad of Eras

Denis Fourches

Abstract In this chapter, we discuss how the profusion of experimental chemogenomics data available in public repositories is transforming the field of cheminformatics. In particular, we describe (i) both theoretical and technical challenges related to the management, analysis, and visualization of large and diverse chemical datasets, (ii) the unique opportunities offered by *Big Chemical Data* for designing molecules with the desired properties and expanding the use of cheminformatics in novel areas of research, and (iii) some innovative approaches that are likely to shape the future of cheminformatics.

The growing compendium of chemogenomics datasets available in publicly-accessible repositories is moving the field of Cheminformatics from the era of data scarcity into the era of "*Big Data*" [1, 2]. Only ten years ago, a set of chemicals with their associated activities was considered "large" when including *ca.* 100 compounds. Due to the lack of available experimental data, Quantitative Structure-Activity Relationships (QSAR) models [3] were built using 20–50 molecules [4], sometimes even less [5]. Meanwhile, combinatorial and high–throughput screening (HTS) technologies have been skyrocketing in both academia and industry [6]. Although pharmaceutical companies still run the biggest HTS platforms, incorporating libraries of several millions of compounds, there are more and more academic centers that not only conduct HTS but integrate their platform within academic drug discovery centers [7]. The following US institutions are notable for this: UCLA's Molecular Screening Shared Resources (MSSR), Stanford's HTBC, Northwestern University's High Throughput Analysis Laboratory, Molecular Libraries Screening Center Network (MLSCN), as well as UNC's NIMH Psychoactive Drug Screening Program (PDSP) and UNC's Center for Integrative Chemical Biology and Drug Discovery (CICBDD). Certain of these centers have the potential to screen *ca.*

D. Fourches (✉)
Laboratory for Molecular Modeling, University of North Carolina,
Chapel Hill, NC 27599, USA
e-mail: fourches@email.unc.edu

L. Gorb et al. (eds.), *Application of Computational Techniques in Pharmacy and Medicine*,
Challenges and Advances in Computational Chemistry and Physics 17,
DOI 10.1007/978-94-017-9257-8_16

250,000 compounds against several panels of targets (*e.g.,* GPCRs, kinases). The development of these facilities is leading to the public release of massive amounts of chemical and biological data in the form of published materials and/or deposited entries in online databases, which often mix raw and unstructured data with more standardized and processed ones. After being curated and integrated, these very large datasets can be considered for in-depth cheminformatics studies, including selective hit fishing, SAR analysis, QSAR modeling, and study-dependent combinations of similarity searches, molecular docking, and pharmacophore analysis for virtual screening and/or molecular design. One can also note that phenotypic screens (both *in vitro* and *in vivo*) can be analyzed using cheminformatics techniques.

Screening is thus one of the most important driving forces for the field of chemogenomics, as it can ultimately allow the actual exploration of all possible small molecule-target interactions [8]. For drug discovery purposes and side effect assessment and tuning, filling the gaps in this overall molecule-target interaction matrix enables the analysis of compounds' polypharmacology. The latter can be at the origin of drug's high efficacy and/or drug's undesired effects caused by off-target binding [9]. Although chemogenomics is developing and expanding based on such revolutionary experimental platforms, assays, and protocols, it is clear that computer-aided analysis and modeling will have an increasingly important role to (i) curate, integrate, and exploit these huge amounts of data; (ii) develop predictive approaches for prognosticating the complex profile of multi-level activities for novel molecules; and (iii) screen extremely large libraries of virtual compounds using both ligand-based QSARs and structure-based scoring functions for better prioritizing promising molecules to be synthesized and tested first.

This novel era of data profusion is drastically modifying the field of cheminformatics, from its most practical and technical aspects to its ultimate role of chemical exploration and discovery. Utilizing chemical biological databases freely-available on the internet is becoming a common, everyday practice allowing researchers to obtain very diverse information at a glance. This diverse information includes:

1. Three-dimensional structures of proteins using the Protein Data Bank (PDB, http://www.rcsb.org/pdb/), which is very close to incorporating a stunning 100,000 entries,
2. Experimental activity measurements (mainly in the form of Ki and IC_{50} values) using the ChEMBL database, which contains more than 1.5M compound records: 9,350 biological targets and 734,000 assays (version 17, August 2013). More than 31M compound entries are also available in PubChem [10], the largest repository of molecules and their associated activities against biological endpoints, including HTS data deposited by pharmaceutical companies (*e.g.,* antimalaria screen of ~13,000 compounds by GSK),
3. Expert-verified structures (26M unique compounds available as of September 2013) and associated physico-chemical data using the Chemspider webportal [11]; a list of 2.7M purchasable compounds can be found in the ZINC database [12],

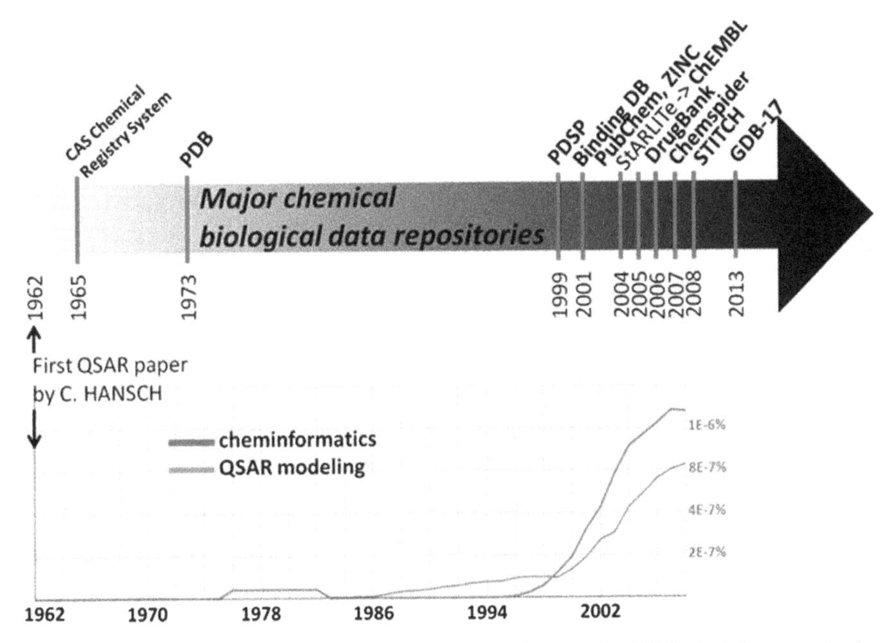

Fig. 16.1 Concurrent evolution of the availability of major chemical biological data repositories with the amount of books (or book chapters—data extracted from Google Scholar nGram Viewer) related to cheminformatics and QSAR modeling

4. Toxicity-relevant bioassays for thousands of chemicals from Toxcast [13] and Tox21 [14, 15] projects, which can be complemented by *in vivo* toxicity data from the ToxRefDB [16] database,
5. Chemical-protein interactions structured as systems chemical biology networks for different organisms using STITCH [17] and KEGG [18] databases.

All these different resources define a huge chemogenomics data matrix of several tens of millions of unique compounds (*row*) and a few thousands targets and other biological endpoints (*columns*). This sparse matrix is ideal for virtual screening but is still far too incomplete to represent the overall diversity of the chemical universe of drug-like molecules (recently estimated to be close to 10^{33} compounds) [19]. Recently, Reymond's research group attempted to enumerate all possible chemical structures solely incorporating C, O, N, S and halogen atoms. Their most recent efforts led to the generation of the GDB-17 database [20] containing more than 166 billion compounds, each including a maximum of 17 heavy atoms. Of course, only a fraction of GDB-17 compounds corresponds to drug-like molecules, whereas representative subsets of this database are ideal libraries for fragment-based molecular docking.

As illustrated in Fig. 16.1, the availability of chemical and biological data in the public domain played a critical role in enabling the development of cheminformatics approaches and QSAR modeling studies. This revolution of data availability has ended the days when multi-linear regression models were built on small congeneric sets of chemicals with no rigorous validation protocol, no applicability domain, and no experimental confirmation. Based on recent requirements set by the editorial board of the Journal of Chemical Information and Modeling [21], a significant portion of QSAR papers published over the past twenty years would not be accepted for publication today. This is one of the many consequences of today's profusion of experimental data: because it is now easier and inexpensive to measure experimental activities and properties for many chemicals, there is no need for non-predictive, non- or moderately reliable models built with few compounds that are not transparent and have not been proven to be efficient for virtual screening.

The vast majority of Cheminformatics approaches and associated computational tools are not ready yet for analyzing and modeling these new generations of very large, diverse, and complex data streams. Beyond the technical challenges due to the extreme size levels of the chemical databases requiring new analysis approaches, algorithms, and architectures (such as GPU computing), there are two major related issues that need to be addressed with priority. When it comes to data accuracy, cheminformaticians are indeed at the mercy of data providers who may inadvertently publish (partially) erroneous data. Several reports on the presence of errors and inaccuracies in public depositories [22, 23] were published recently, especially the study from [24] at Bayer describing that only 20–25 % of published data were in agreement with their in-house results when attempting to reproduce the exact same experiments. This lack of experimental reproducibility motivated Nature's editorial board to reinforce the journal's acceptance criteria in terms of statistical validation for reported results in submitted manuscripts as well as abolishing size constraints for method descriptions. Our group at UNC recently published a list of guidelines for chemical data curation prior to cheminformatics analysis [25]: we described several simple but important steps for cleaning chemical records in a database, including the removal of a fraction of the data that cannot be appropriately handled by conventional cheminformatics techniques. Chemical and biological curation is critical for enabling the accurate integration of data across different databases and assuring the correctness of models and hypotheses built on top of these data.

The profusion of chemical and biological data not only enables large scale cheminformatics studies but also novel areas of applications. Certain of these applications are just emerging and likely to significantly expand in the next few years. For example, one notable new approach involves the use of text mining of large collections of articles to extract information relevant to a specific compound, target, disease, or any other user's defined term. Recently, Chemotext [26] has been developed on top of Mesh annotations to derive compound-target-disease assertions. Our group at UNC showed that it was even feasible to derive predictive QSAR models from curated literature assertions extracted using text-mining protocols [27]. Another emerging trend is the development of hybrid QSAR models that involves a combination of chemical and biological descriptors. For instance, toxicogenomics profiles determined for 127 drugs were recently used together with chemical descriptors to generate hepatotoxicity QSAR models [28]. This strategy recently culminated in the development of the Chemical-Biological Read-Across (CBRA) approach [29] to infer a compound's toxicity from its closest chemical and biological analogues.

With several thousand nanotechnology-based consumer products available on the market, nanotechnology is drawing worldwide attention for its applications in various industrial areas, such as material science, medical research, and cosmetics [30–32]. Importantly, a significant portion of these efforts is directed towards the development of "green" products intended to achieve efficient and less polluting energy sources [33, 34]. In this context, cheminformatics has a role to play by: (i) facilitating the access, storage, search, and integration of all experimental results currently distributed in literature, databases, and other sources; (ii) achieving externally predictive QSAR models to compute nanomaterials' properties based on their structural characteristics; and (iii) boosting the development and testing processes by identifying the most promising nanomaterials that require focused experimental investigations. The latter point is especially of importance due to the concerns about the safety of certain nanomaterials and the development of nanomedicine [33, 35, 36]. In a recent proof-of-concept study, our research group introduced the terminology of Quantitative Nanostructure-Activity Relationships (QNARs) [37] that employs classical machine-learning methods for establishing links between chemical descriptors and various measured activities of nanomaterials. Published studies and developing trends in computational modeling of nanomaterials also have been summarized in a recent review [38].

With the forecasted, concomitant increase of both computing power and availability of additional chemical biological data, many new types of applications will appear and expand in the coming years. Integrated meta-databases are likely to emerge, combining structure-based, ligand-based, experimental and predicted properties, integrating the concepts and levels of systems chemical biology. Novel types of chemical descriptors (especially at 2.5D, 3.5D and 4D levels) will be developed to better take into account target–ligand interactions, notably molecular flexibility. One can also speculate on the development of system fingerprints to characterize, analyze, and compare complex chemical biological systems.

On a technical point of view, the performances of computer CPUs have improved as prognosticated by Moore's law. However, the most time-consuming calculations, such as molecular dynamics, quantum-chemical computation, and molecular docking, must still be performed on supercomputers and large clusters including thousands of CPUs in parallel. Interestingly, parallel calculations can also be conducted on modern video cards, also called graphics processing units (GPU) computing [39]. The most powerful individual workstations equipped with several high-end GPUs can incorporate up to 10,000 CUDA cores (*e.g.*, 4-way SLI configuration with Nvidia's Titan cards), allowing massively-parallel calculations as long as the software code incorporates specific CUDA instructions. However, to date, very few prototypes of cheminformatics software tools have been adapted to take advantage of GPU-computing capabilities. GPU-computing could represent an interesting solution to rapidly process, explore, and screen extremely large datasets of 10^9 compounds and more.

Finally, more and more cheminformatics software tools will become fully available and functional on tablets, smartphones and other mobile devices such as augmented reality glasses (*e.g.,* Google Glass Project). This new step in technology portability could realize one of the initial promises of cheminformatics, which was (and still is) to assist experimentalists for chemical-relevant decision making as close as possible to the lab bench. For instance, the following tablet applications are notable: *PyMol* (edited by Schrodinger; itunes.apple.com/us/app/pymol/) allows users to browse and visualize protein structures and protein-ligand complexes; *Chemspider* (edited by Molecular Materials Informatics; itunes.apple.com/us/app/chemspider/) enables structural and text queries on the Chemspider database; *Chemical Engineering AppSuite* (edited by John McLemore; itunes.apple.com/us/app/chemical-engineering-appsuite/) integrates a collection of chemistry related tools and databases; and *Elemental* (edited by Dotmatics Limited; itunes.apple.com/us/app/elemental/) and *Molprime* (edited by Molecular Materials Informatics; itunes.apple.com/us/app/molprime/) allows users to draw and export compound structures. These applications are the first representatives of an entire new generation of "cheminformatics apps" with user-friendly, tablet-ready graphical interfaces that will offer a direct and intuitive access to diverse chemogenomics data. It is very likely that the features offered by these applications will progressively integrate complex QSAR-based predictors coupled with sophisticated modules that will be capable of rapidly accessing and cross-searching chemical biological databases, visualizing [40] and analyzing HTS results, launching modeling and screening computations on remotely controlled workstations, and sharing chemical information in the cloud.

In summary, chemical biological data streams produced by modern drug discovery platforms are reaching unprecedented levels of size, diversity, and complexity that require the development of novel cheminformatics methods and tools to process and analyze them. This situation offers some unique perspectives for inventing a new generation of expert systems and decision making technologies. As the profusion of experimental data expands in the coming years to fully reach the level and requirements of so-called *Big Data*, joint efforts and collaborations between experimentalists and modelers will not only continue to strengthen but will become the key for successful drug discovery. Modelers will have to guarantee fast and accurate accesses to the correct and integrated chemical information and generate reliable experimentally-testable hypotheses (*e.g,* compound X is active toward both targets Y and Z to induce phenotype P). In return, experimentalists will have to keep on guiding modelers for better understanding their practical needs so that the computational technologies can be adapted accordingly. This transitioning period is extremely important for the whole cheminformatics research community: from the students starting to learn the basics and recognizing the need to know all molecular modeling approaches [41], to well-established researchers already facing the new scales of data complexity and associated opportunities. The era of *Big Chemical Data* is finally here and therefore it seems that modelers have never been so close to achieving the rational, computer-aided design of novel molecules with controlled polypharmacology and safety profiles.

Acknowledgments The author sincerely thanks Profs. Alexandre Varnek (University of Strasbourg, France) and Alexander Tropsha (University of North Carolina at Chapel Hill, USA) for fruitful discussions, training, support and trust. This chapter has been proofread by Dr. Laura Widman (University of North Carolina at Chapel Hill, USA). Financial support from NSF ABI 1147145, EPA RD832720, SRC/Sematech, and UNC Junior Faculty Award is also gratefully acknowledged.

References

1. Chute CG, Ullman-Cullere M, Wood GM, Lin SM, He M, Pathak J (2013) Some experiences and opportunities for big data in translational research. Genet Med 15:802–809
2. Moore KD, Eyestone K, Coddington DC (2013) The big deal about big data. Healthc Financ Manage 67:60–68
3. Tropsha A (2010) Best practices for QSAR model development, validation, and exploitation. Mol Inform 29:476–488
4. Varnek A, Fourches D, Sieffert N, Solov'ev VP, Hill C, Lecomte M (2007) QSPR Modeling of the Am III/Eu III separation factor: how far can we predict? J Solv Extr Ion Exch 25:1–26
5. Varnek A, Fourches D, Solov'ev V, Klimchuk O, Ouadi A, Billard I (2007) Successful "In Silico" design of new efficient uranyl binders. J Solv Extr Ion Exch 25:433–462
6. Carnero A (2006) High throughput screening in drug discovery. Clin Transl Oncol 8:482–490
7. Kozikowski AP, Roth B, Tropsha A (2006) Why academic drug discovery makes sense. Science 313:1235–1236
8. Bajorath J (2013) A perspective on computational chemogenomics. Mol Inform 32:1025–1028
9. Keiser MJ, Setola V, Irwin JJ et al (2009) Predicting new molecular targets for known drugs. Nature 462:175–181
10. Xie X-Q (2010) Exploiting PubChem for virtual screening. Expert Opin Drug Discov 5:1205–1220
11. Williams AJ (2014) Introduction to Chemspider. http://www.chemspider.com/help-what-can-i-do-with-chemspider.aspx. Accessed 8 Nov 2014.
12. Irwin JJ, Shoichet BK (2005) ZINC—a free database of commercially available compounds for virtual screening. J Chem Inf Model 45:177–182
13. Sipes NS, Martin MT, Kothiya P, Reif DM, Judson RS, Richard AM, Houck KA, Dix DJ, Kavlock RJ, Knudsen TB (2013) Profiling 976 ToxCast chemicals across 331 enzymatic and receptor signaling assays. Chem Res Toxicol 26:878–895
14. Huang R, Sakamuru S, Martin MT et al (2014) Profiling of the Tox21 10K compound library for agonists and antagonists of the estrogen receptor alpha signaling pathway. Sci Rep 4:5664–5673
15. Tice RR, Austin CP, Kavlock RJ, Bucher JR (2013) Improving the human hazard characterization of chemicals: a Tox21 update. Environ Health Perspect 121:756–765
16. Knudsen TB, Martin MT, Kavlock RJ, Judson RS, Dix DJ, Singh AV (2009) Profiling the activity of environmental chemicals in prenatal developmental toxicity studies using the U.S. EPA's ToxRefDB. Reprod Toxicol 28:209–219
17. Kuhn M, Szklarczyk D, Franceschini A, von Mering C, Jensen LJ, Bork P (2012) STITCH 3: zooming in on protein-chemical interactions. Nucleic Acids Res 40:D876–D880
18. Kanehisa M, Goto S, Sato Y, Furumichi M, Tanabe M (2012) KEGG for integration and interpretation of large-scale molecular data sets. Nucleic Acids Res 40:D109–D114
19. Polishchuk PG, Madzhidov TI, Varnek A (2013) Estimation of the size of drug-like chemical space based on GDB-17 data. J Comput Aided Mol Des 27:675–679
20. Ruddigkeit L, Blum LC, Reymond J-L (2013) Visualization and virtual screening of the chemical universe database GDB-17. J Chem Inf Model 53:56–65
21. Jorgensen WL (2006) QSAR/QSPR and proprietary data. J Chem Inf Model 46:937–937

22. Young D, Martin D, Venkatapathy R, Harten P (2008) Are the chemical structures in your QSAR correct? QSAR Comb Sci 27:1337–1345
23. Ekins S, Olechno J, Williams AJ (2013) Dispensing processes impact apparent biological activity as determined by computational and statistical analyses. PLoS One 8:e62325
24. Prinz F, Schlange T, Asadullah K (2011) Believe it or not: how much can we rely on published data on potential drug targets? Nat Rev Drug Discov 10:712
25. Fourches D, Muratov E, Tropsha A (2010) Trust, but verify: on the importance of chemical structure curation in cheminformatics and QSAR modeling research. J Chem Inf Model 50:1189–1204
26. Baker NC, Hemminger BM (2010) Mining connections between chemicals, proteins, and diseases extracted from Medline annotations. J Biomed Inform 43:510–519
27. Fourches D, Barnes JC, Day NC, Bradley P, Reed JZ, Tropsha A (2010) Cheminformatics analysis of assertions mined from literature that describe drug-induced liver injury in different species. Chem Res Toxicol 23:171–183
28. Low Y, Uehara T, Minowa Y et al (2011) Predicting drug-induced hepatotoxicity using QSAR and toxicogenomics approaches. Chem Res Toxicol 24:1251–1262
29. Low Y, Sedykh AY, Fourches D, Golbraikh A, Whelan M, Rusyn I, Tropsha A (2013) Integrative chemical-biological read-across approach for chemical hazard classification. Chem Res Toxicol 26:1199–1208
30. Donaldson K, Poland CA (2009) Nanotoxicology: new insights into nanotubes. Nat Nanotechnol 4:708–710
31. Balbus JM, Florini K, Denison RA, Walsh SA (2006) Getting it right the first time: developing nanotechnology while protecting workers, public health, and the environment. Ann N Y Acad Sci 1076:331–342
32. Hart P (2009) Nanotechnology, synthetic biology, & public opinion. The Woodrow Wilson International Center For Scholars
33. Bystrzejewska-Piotrowska G, Golimowski J, Urban PL (2009) Nanoparticles: their potential toxicity, waste and environmental management. Waste Manag 29:2587–2595
34. Jones R (2009) Nanotechnology, energy and markets. Nat Nanotechnol 4:75
35. Lockman PR, Mumper RJ, Khan MA, Allen DD (2002) Nanoparticle technology for drug delivery across the blood-brain barrier. Drug Dev Ind Pharm 28:1–13
36. Linkov I, Satterstrom FK, Corey LM (2008) Nanotoxicology and nanomedicine: making hard decisions. Nanomedicine 4:167–171
37. Fourches D, Pu D, Tassa C, Weissleder R, Shaw SY, Mumper RJ, Tropsha A (2010) Quantitative nanostructure-activity relationship modeling. ACS Nano 4:5703–5712
38. Fourches D, Pu D, Tropsha A (2011) Exploring quantitative nanostructure-activity relationships (QNAR) modeling as a tool for predicting biological effects of manufactured nanoparticles. Comb Chem High Throughput Screen 14:217–225
39. Heinzerling L, Klein R, Rarey M (2012) Fast force field-based optimization of protein-ligand complexes with graphics processor. J Comput Chem 33:2554–2565
40. Fourches D, Tropsha A (2013) Using graph indices for the analysis and comparison of chemical datasets. Mol Inform 32:827–842
41. Rognan D (2013) Towards the next generation of computational chemogenomics tools. Mol Inform 32:1029–1034

Index

L. Gorb et al. (eds.), *Application of Computational Techniques in Pharmacy and Medicine*, Challenges and Advances in Computational Chemistry and Physics 17,
DOI 10.1007/978-94-017-9257-8

CPSIA information can be obtained
at www.ICGtesting.com
Printed in the USA
LVHW02s1344110318
569451LV00005B/59/P

9 789402 406962